MEMBRANE
FABRICATION

MEMBRANE FABRICATION

Edited by
Nidal Hilal
Ahmad Fauzi Ismail
Chris J. Wright

CRC Press
Taylor & Francis Group
Boca Raton London New York

CRC Press is an imprint of the
Taylor & Francis Group, an **informa** business

CRC Press
Taylor & Francis Group
6000 Broken Sound Parkway NW, Suite 300
Boca Raton, FL 33487-2742

First issued in paperback 2018

ISBN 13: 978-1-138-89409-9 (pbk)
ISBN 13: 978-1-4822-1045-3 (hbk)

Library of Congress Cataloging-in-Publication Data

Membrane fabrication / editors, Nidal Hilal, Ahmad Fauzi Ismail, Chris Wright.
  pages cm
Includes bibliographical references and index.
ISBN 978-1-4822-1045-3 (hardcover : alk. paper) 1. Membranes (Technology)--Design and construction. 2. Membrane filters--Materials. I. Hilal, Nidal. II. Ismail, Ahmad Fauzi. III. Wright, Chris (Chemist)

TP159.M4M4425 2015
660'.28424--dc23   2015011616

Visit the Taylor & Francis Web site at
http://www.taylorandfrancis.com

and the CRC Press Web site at
http://www.crcpress.com

Dedication

---

*For Christine, Nadine, and Hani—you have
always been the fuel of my ambition.*

Nidal Hilal

*For Fadilah Abdullah, Faris Izzat, Faiz Azizi, and
Fatin Nabilah for their continuous support and love.*

Ahmad Fauzi Ismail

For Becky and Ophelia—you are always in my thoughts.

Chris J. Wright

Contents

Preface...xi
Editors...xix
List of Contributors...xxi

SECTION I Fabrication Processes for Polymeric Membrane

Chapter 1   Polymeric Membranes..3

Alberto Figoli, Silvia Simone, and Enrico Drioli

Chapter 2   Electrospinning: A Practical Approach for Membrane Fabrication......45

Luke Burke, Amir Keshvari, Nidal Hilal, and Chris J. Wright

Chapter 3   Control of Crystallization of Poly(Lactic Acid) Membranes............. 75

Shuichi Sato, Ryohei Shindo, Shinji Kanehashi, and Kazukiyo Nagai

Chapter 4   Innovative Methods to Improve Nanofiltration Performance through Membrane Fabrication and Surface Modification Using Various Types of Polyelectrolytes 103

Law Yong Ng and Abdul Wahab Mohammad

Chapter 5   Polysaccharides: A Membrane Material 161

Seema Shrikant Shenvi, Arun Mohan Isloor, and Ahmad Fauzi Ismail

Chapter 6   Cellulose and Its Derivatives for Membrane Separation Processes..... 193

Boor Singh Lalia, Farah E. Ahmed, Shaheen Fatima Anis, and Raed Hashaikeh

Chapter 7   PVDF Hollow-Fiber Membrane Formation and Production............ 215

Panu Sukitpaneenit, Yee Kang Ong, and Tai-Shung Chung

Chapter 8 PVDF Membranes for Membrane Distillation: Controlling Pore
Structure, Porosity, Hydrophobicity, and Mechanical Strength.......249

Rinku Thomas, Muhammad Ro'il Bilad, and Hassan Ali Arafat

Chapter 9 Membrane Contactor for Carbon Dioxide Absorption
and Stripping 285

*Rosmawati Naim, Masoud Rahbari-Sisakht,
and Ahmad Fauzi Ismail*

SECTION II   Fabrication Processes for Inorganic Membrane

Chapter 10 Microstructured Ceramic Hollow-Fiber Membranes:
Development and Application 317

Zhentao Wu, Benjamin F.K. Kingsbury, and Kang Li

Chapter 11 Ceramic Hollow-Fiber Support through a Phase Inversion-
Based Extrusion/Sintering Technique for High-Temperature
Energy Conversion Systems 347

*Mohd Hafiz Dzarfan Othman, Mukhlis A. Rahman, Kang Li,
Juhana Jaafar, Hasrinah Hasbullah, and Ahmad Fauzi Ismail*

Chapter 12 Development of Large-Scale Industrial Applications of
Novel Membrane Materials: Carbon Nanotubes, Aquaporins,
Nanofibers, Graphene, and Metal-Organic Frameworks 383

*Kailash Chandra Khulbe, Chaoyang Feng, Ahmad Fauzi
Ismail, and Takeshi Matsuura*

Chapter 13 Pd-Based Membranes and Membrane Reactors for Hydrogen
Production 437

Silvano Tosti

SECTION III Fabrication Processes for Composite Membrane

Chapter 14 Current Progress of Nanomaterial/Polymer Mixed-Matrix Membrane for Desalination 489

Goh Pei Sean, Ng Be Cheer, and Ahmad Fauzi Ismail

Chapter 15 Fabrication of Polymeric and Composite Membranes 511

Chun Heng Loh, Yuan Liao, Laurentia Setiawan, and Rong Wang

Chapter 16 Strategies to Use Nanoparticles in Polymeric Membranes 569

Bart Van der Bruggen, Ruixin Zhang, and Jeonghwan Kim

Chapter 17 Surface Modification of Inorganic Materials for Membrane Preparation 589

Dipak Rana, Takeshi Matsuura, and Ahmad Fauzi Ismail

Chapter 18 Fabrication of Low-Fouling Composite Membranes for Water Treatment 615

Victor Kochkodan and Nidal Hilal

Chapter 19 Fabrication of Polymer Nanocomposite Membrane by Intercalating Nanoparticles for Direct Methanol Fuel Cell 655

Juhana Jaafar, Ahmad Fauzi Ismail, Mohd Hafiz Dzarfan Othman, and Mukhlis A. Rahman

Chapter 20 Effects of Solvent and Blending on the Physical Properties of Sulfonated Poly(Ether Ether Ketone): A Promising Membrane Material for PEMFC 681

Amir-Al-Ahmed, Abdullah S. Sultan, and S.M. Javaid Zaidi

Index 711

Preface

In the next decade, the application of membranes used in separation technology is set to expand dramatically. The world's population tripled in the twentieth century, and is expected to increase by another 40%–50% in the next 50 years. Thus, technologies such as membrane separation, which improve the efficiency of resource management, must keep pace in order to meet the demands of the growing global society. Membranes play a crucial role in ensuring the optimum use and recovery of materials in industry. If we look at the case of clean water, there is a finite amount of water on the planet, and water as a resource cannot be replaced in the way that alternative fuel sources can replace petroleum. Global governance must manage water efficiently. Provision and efficient use of water, therefore, represents one of the most pressing challenges of the twenty-first century. The membrane separation industry is preparing for this challenge with the global desalination market expected to reach US $52.4 billion by 2020, up over 320% from US $12.5 billion in 2010, according to a recent report from SBI *Energy*. The same report further predicts that the reverse osmosis (RO) segment of membrane technology will see the largest growth, reaching US $40 billion by 2020.

This snapshot of the clean water market segment of membrane industries reflects the growth and development occurring in the other market segments such as pharmaceuticals, commodity chemicals, and process industries. The underlying theme of this expansion in application is increase in efficiency. This is in terms of improvement of the membranes properties, the membrane process scale-up coupled with the economic large-scale manufacture of the membrane. Central to this improvement in efficiency is membrane fabrication. Thus, this book is extremely timely. Membrane technology must optimize the fabrication of membranes in order to meet the challenges and opportunities presented to the associated industry by an increasingly resource-limited society. In a previous volume entitled *Membrane Modification—Technology and Applications* (ISBN-13: 978-1439866351), we brought together expert authors to examine the optimization of membrane technology through modification processes. In this book, we significantly extend this discussion and have asked prominent members of the membrane research and development community to appraise current membrane fabrication methods and how these methods are used in the optimization of membrane applications. To that end, we have split the book into three sections, namely, fabrication processes for (I) polymeric, (II) inorganic, and (III) composite membranes. This book covers the fabrication of all membrane types and will provide readers with a wealth of information, which can be used to optimize the fabrication of a membrane for application by membrane technologists.

The principle of membrane separation is based on the differential passage of solutes or suspensions in a solvent through a membrane achieved by the application of a driving force such as pressure. The material is either retained on the feed side and rejected from passage across the membrane or passed through the membrane. The movement across the membrane depends on the size, charge, activity, and partial pressure of the material and its interaction with the solvent and membrane surface. Differences in the material properties are used to separate mixtures and are the target of procedures to optimize membrane fabrication by controlling the physical and chemical attributes of the membrane. By changing the fabrication process, membranes can be enhanced not only for resistance of extremes in process conditions but also for improved resolution of mixtures during separation. In all types of membrane processes, it is membrane surface and pores that is the location for the separation and the mechanisms of control. Improvements to the membrane and its application through changes in the fabrication processes must consider how such changes impact on the membrane surface. A holistic approach must be adopted, linking membrane fabrication to membrane structure and efficiency within the final process application as characterized by performance metrics such as flux, material rejection, mechanical, and chemical stability, within the constraints of economic feasibility.

Adoption of membranes in separation processes continues to increase significantly due to the advantages they offer over conventional processes such as distillation solvent extraction and chromatography. These advantages include simplicity, versatility, high selectivity, operation at relatively lower temperatures, and pressures with commensurate reduction in energy costs and application to chemically/physically labile systems such as biological macromolecules and cells. Large numbers of materials that can be chosen for membrane fabrication also install the advantages of control over selectivity with the tailoring of membrane performance. In addition, membranes are not only better for the environment in that they allow more efficient use of resources and energy, but they are also generally fabricated from relatively simple and safe materials.

Membranes, however, do have disadvantages that arguably can be minimized at the membrane fabrication stage when such a holistic approach, as mentioned above, is adopted. Membranes can be susceptible to damage from process streams and environments during operation and cleaning regimes. Operation of membranes with typical chemical industry feed streams, if the membrane material is not chosen well, can cause the membrane to weaken and reduce the lifetime of the membrane to economically unacceptable levels. This is particularly true for polymeric membranes that are currently the dominant material used in membrane fabrication. Physical integrity of many membranes is also compromised by high temperature and pressures, which is the favored environment for many industrial processes, thus limiting the application of membrane systems. Encouragingly, the development of inorganic and composite membranes with their improved resilience, as compared to polymeric membranes, is overcoming these last two disadvantages of membrane systems. Fouling of membrane systems by the deposition of materials during normal process operation is a major problem in all membrane separation processes. If the fouling layer is not removed by process mechanisms such as hydrodynamic cross flow or cleaning, then the separation process efficiency will be reduced significantly. Membrane fouling can be reversible or irreversible depending on the interaction of the foulants with the membrane and compromises the separation process by reducing the size of the pores and altering the hydrophilicity, surface roughness, and charge of the membrane surface. Fouling of membranes is also currently limiting the adoption of membrane technology. A final issue that needs to be addressed to widen the use of membranes is scale-up, in terms of both the process application and the fabrication of the membrane.

Membrane fabrication is the initial focus for the reduction of the disadvantages associated with the application of membrane separation processes. Choice of membrane material, polymeric, inorganic, or composite, to fabricate a membrane is reliant on the process operating conditions and the nature of the feed stream. Polymeric membranes, the focus of the first section of this book, are relatively inexpensive, malleable for module construction, and there is a wide range of materials to choose from, including cellulose acetate, polysulfone, polyester, and polyamide. However, as mentioned above, these materials are susceptible to chemical and physical damage. On the other hand, inorganic membranes, as discussed in the second section of this book, are more robust and can operate in process extremes of temperature and pressure with greater resilience to chemical attack from liquids such as organic solvents or cleaning agents than polymeric membranes. However, inorganic membranes are expensive, and the membrane community has less experience with their operation and fabrication. Inorganic membranes are fabricated from ceramics and zeolites, including alumina, titanium, and zirconium. An interesting set of membranes, which feature in the final section of this book, are those fabricated as composite membranes, which bring together some of the advantages of the other two membrane groups, in that the blend of inorganic and organic materials provides the fabricated membrane with improved mechanical properties so that they are more robust to withstand pressure or more malleable to enable improved module construction. Blending of the materials may also improve functionality or process chemical compatibility. Examples of the materials used to produce the composites within a polymer matrix include carbon nanotubes and titanium oxide particles.

This book brings together experts from a number of disciplines working within membrane technology all with the same goal of the development and fabrication of new membranes for increased and efficient application of membrane separation processes. The authors of each chapter share their experience and insights in a specific membrane fabrication area. In many cases, this information extends into application of the membrane system as the research uses membrane performance data to improve subsequent membrane fabrication methods.

The first section of the book examines the fabrication methods employed to harness the properties of polymers in membrane separation. Chapter 1 describes the most common polymeric membrane preparation methods with a focus on phase inversion (PI) or phase separation. Thermodynamic principles of PI are discussed, as are the main factors that influence membrane morphology and properties. Details of hollow-fiber membrane fabrication through PI are also examined. The chapter covers key examples of membrane preparation for selected applications including RO, nanofiltration (NF), ultrafiltration (UF), and emerging processes such as membrane contactors (MCs). The authors argue that despite the impressive advancements in membrane preparation techniques for different applications, there is still a need for the research community to obtain a greater understanding of the fundamental mechanisms at the basis of membrane preparation. This emerges as one of the underlying themes of this book and is regarded as the only way to achieve perfect control over the membrane fabrication process in order to tailor membrane properties for separation process optimization.

The book then introduces the technology of electrospinning in Chapter 2, which takes a practical approach to describe the different techniques available for the membrane technologists. The chapter describes the basic techniques followed by an examination of more elaborate methods such as coaxial and bowl electrospinning. These emerging techniques have led to a renewed interest in electrospinning, as they open the door for a number of desirable membrane process improvements. These include the use of a broader range of materials with commensurate improvements in process versatility and fiber membrane functionality, coupled with enhanced membrane process performance and the possibility of scale-up for economically viable membrane fabrication. Chapter 2 is a companion with later chapters (Chapters 12 and 15) that discuss electrospinning from different perspectives in the context of composite membranes and membrane industry application.

Chapter 3 examines the use of poly(lactic acid) (PLA) in the fabrication of membranes and how crystallization can be controlled to improve the structure of the next generation of PLA membrane. PLA is an environment-friendly, biodegradable polymeric substance with a low melting point, high malleability, and high transparency. Crystallization of PLA can be controlled by cooling and heating conditions, various organic solvents, and vacuum ultraviolet irradiation. Chapter 3 also focuses on various controlled crystalline structures fabricated by thermal, solvent, and photo-induced crystallizations.

Chapter 4 introduces innovative methods to improve NF performance and starts with an excellent introduction to membrane materials in general, which we draw the readers' attention to. The discussion is then extended to argue the benefits of using polyelectrolytes to overcome some of the problems associated with membrane separation as discussed above. The chapter covers the different methods to functionalize the membrane surface with polyelectrolytes: UV-grafting, chemical cross-linking static self-adsorption, and dynamic self-adsorption. Polyelectrolytes offer a new solution to improve NF membrane performance through simple procedures with minimal resource requirements when compared to other modification or fabrication methods.

Drive for improved environmental credentials for membrane separations has meant that biopolymers have recently gained attention in fabrication research. These materials are biodegradable, biocompatible, nontoxic, and commonly available often as product from a biorefinery stratagem. Polysaccharides are an example of one such group of biopolymers. Chapter 5 gives a detailed account of the most commonly occurring polysaccharides that are used in membrane-based separations in the field of wastewater treatment, metal-ion removal, dye removal, and pervaporation. The authors discuss the properties of polysaccharides as membrane materials and identify how these can be improved by methods such as the addition of fillers, physical and chemical modification, and grafting. Chapter 6 continues the discussion on biopolymers and examines the use of cellulose and its derivatives for membrane separation processes. Cellulose is the most abundant organic raw material in nature and has been one of the primary choices for membrane fabrication material. There has been a recent resurgence in the use of cellulose as a membrane fabrication material—a consequence of new developments in cellulose extraction and modifica-tion. In addition, the development of nanocellulose and its interesting properties have also renewed interest in cellulose for membrane fabrication.

Polyvinylidene fluoride (PVDF) has received considerable attention as a prom-ising polymeric membrane material. This is due to its outstanding chemical and physical properties, which include its highly hydrophobic nature, good mechanical strength, and excellent chemical resistance. With these superior properties, PVDF hollow-fiber membranes have been widely applied in processes such as microfiltra-tion (MF), UF, NF, membrane distillation (MD), pervaporation, and MCs. Chapter 7 reviews the state-of-the-art technology for the production of PVDF hollow-fiber membranes, examining the basic principles of their fabrication through a nonsolvent-induced phase separation (NIPS) process. The chapter discusses the key spinning parameters of polymer concentration, nonsolvent additives, coagulant chemistry, air-gap distance, take-up speed, and dope rheology, and their influences on the for-mation of PVDF hollow fibers. Recent advances in PVDF hollow-fiber membrane fabrication are also assessed, and include dual-layer and new configuration spinneret configuration design, mixed-matrix membranes, and thermally induced phase sepa-ration. Chapter 8 takes a closer look at the use of PVDF membranes in MD starting with a discussion on the application of MD. Different polymers such as polytetra-fluoroethylene, polypropylene, polysulfone, and PVDF have been used to make MD membranes. The authors argue that PVDF has the potential to meet the requirements that exist for an appropriate MD membrane, which is currently commercially elu-sive. These include high hydrophobicity, chemical inertness, low thermal conductiv-ity, high porosity, narrow pore size distribution, specific average pore size range (0.2–0.5 μm), optimal thickness, and mechanical strength. The chapter reviews the recent attempts to optimize polymeric PVDF membranes for MD applications, using PI, in particular through immersion precipitation processes.

Chapter 9 deals with developments in the application of MCs and the associ-ated improvements in membrane fabrication to overcome current limitations due to membrane wetting problems, which lower the efficiency of mass transfer and reduce membrane life. There are numerous advantages of MCs over conventional methods for carbon dioxide absorption and stripping technology such as their compactness and high effective surface area per unit volume. MCs have also been developed for treatment of aqueous systems and they offer a combined membrane separation and absorption process in one physical setting.

The book then continues with a review of the current methods and development of fabrication processes for inorganic membrane, and Section II starts with a review of ceramic hollow-fiber membrane fabrication and application. Ceramic hollow-fiber membranes, sometimes called *capillary or microtubular membranes*, are relatively new types of ceramic membranes compared with planar and tubular counterparts. They have a high surface-area-to-volume ratio, with unique microchannels that have led to innovative applications in separation and catalytic reaction technologies. Chapter 10 introduces the fundamental principles of fabricating microstructured ceramic hollow-fiber membranes and discusses the processing parameters and their influence on the microstructures of the fabricated membranes. Chapter 11 deals with the fabrication of ceramic hollow fibers used as a support for energy conversion devices. The chapter discusses the fabrication of ceramic hollow fibers through a PI-based extrusion/sintering technique, which shows better control over the internal macrostructure of the fabricated hollow fiber when compared with other conventional extrusion techniques such as plastic mass ram extrusion. Versatility of the PI-based extrusion is also demonstrated by the preparation of dual-layer hollow fiber supports in a single step by using co-extrusion/co-sintering. Use of this technique dramatically simplifies the fabrication process and is applicable to the development of other inorganic membranes with advanced structures.

We continue the examination of inorganic membranes by looking at the use of novel membrane materials in large-scale industrial processes. Chapter 12 takes a pragmatic look at the development of membranes using novel materials that have received a lot of acclaim and attention from the membrane fabrication community. The chapter comprehensively reviews the use of carbon nanotubes, aquaporins, nanofibers, graphene, and metal organic framework, outlining the main features of each material and the current status of the corresponding membrane development and application. The latter focuses on the fabrication of relatively large membrane areas for module construction. However, the authors conclude that unfortunately the production of sizable membrane area at low cost with the capability to be installed in module geometries are still major technical challenges that have not yet been resolved for any of these novel membranes.

Chapter 13 discusses the properties, design, and fabrication of palladium (Pd)-based membranes and membrane reactors for hydrogen production and for conducting isotopic separation processes in fusion nuclear facilities. Membranes made of Pd-alloys are used for separating hydrogen from gaseous mixtures, whereas Pd-membrane reactors have been developed for producing hydrogen through dehydrogenation reactions. A Pd–Ag alloy with 20%–25% silver has been widely used and is commercially available for preparing membranes, and the chapter describes in detail how fabrication parameters influence the performance of such membranes. Design and manufacture of Pd-based membranes and membrane reactors have to take into account the properties of the Pd-alloys such as hydrogen uploading, strain, and mechanical strength. These properties vary significantly under hydrogenation, which can affect the stability and performance of the membrane modules.

The focus of this book then changes in Section III to look at fabrication processes for composite membranes and starts with an assessment in Chapter 14 of the current use in the desalination of nanomaterial composites in the form of mixed-matrix membranes. There is a tremendous potential for the fabrication of mixed-matrix membrane using various kinds of nanostructured materials, such as zeolite, silica, carbon nanotubes, and metal oxide nanoparticles, which will significantly impact desalination and water technologies. Incorporation of these advanced nanomaterials in the polymer matrix promises to resolve problems, such as fouling and poor process environment compatibility, encountered when using conventional polymer membranes. Mixed-matrix membrane has demonstrated outstanding desalination properties in terms of excellent water flux, high salt rejection, and low biofouling potential.

We begin Chapter 15 with an assessment of the methods used in the preparation of polymeric membranes with a focus on NIPS and electrospinning. The authors then discuss the fabrication of composite membranes identifying the advantage of using different materials for the selective layer and the support layer in a membrane as it is possible to tailor each layer individually to optimize membrane performance. The composite membranes can be fabricated in a single-step method or in two-step approaches using PI and electrospun membranes as the substrate, followed by applying a top selective layer on the substrate. The chapter also includes a discussion on membrane characterization techniques that are essential to assess the impact of changes in the fabrication process on the final membrane and its process operation.

The properties of nanoscale structures, based on their high surface-area-to-volume ratio, offer extensive opportunities to membrane manufacture and functionality. The effect on the membrane performance depends not only on the type and functionalities of the nanoparticles but also on the integration and interaction between nanoparticles and the membrane polymer. In Chapter 16, we explore the strategies that can be adopted to integrate nanoparticles in polymeric membranes. These include self-assembly, bulk addition, anchoring in or on the membrane surface, and layer-by-layer addition of nanoparticles. The choice of the type of nanoparticles depends on the application. After a comprehensive discussion of the topic, the chapter identifies that research challenges remain that are mainly related to the stability of the modification, and to the difficulty of defining an optimized synthesis procedure, which actually yields an improved membrane performance.

Chapter 17 examines two surface-modification strategies for the improved fabrication of membranes. First is the modification of the inorganic membrane surface to improve membrane performance by fine-tuning the surface chemistry and pore size of inorganic membranes to control the interaction of the feed components with the membrane. The second strategy is the surface modification of inorganic fillers to improve the compatibility between filler particles and polymeric membrane matrixes in the preparation of mixed-matrix membranes. This prevents the formation of voids between the inorganic filler particle and the surrounding polymer matrix by improving the compatibility between the two phases. This approach improves membrane selectivity by preventing the leakage of feed-fluid components through membrane voids.

Chapter 18 presents an overview of recent developments in the fabrication of low-fouling composite membranes for water treatment. The chapter begins with a discussion on membrane surface properties such as hydrophilicity, charge and surface roughness, which affect membrane biofouling and organic fouling. The discussion then focuses on recent studies on preparation of low-fouling composite membranes using interfacial polymerization, surface grafting, coating/adsorption of a protective layer on the membrane surface, and surface modification of polymer membranes with nanoparticles. Numerous studies have shown that increasing the membrane hydrophilicity, reducing the roughness or imposing charged groups and bactericidal agents on the membrane surface may reduce membrane fouling with organic compounds, colloids, and microorganisms. The authors conclude that despite the extensive knowledge that exists on how to fabricate a membrane for a particular application, there is still a challenge to produce reliable composite membranes with antifouling properties, high mechanical strength, and high flux.

The book finishes with two chapters examining the fabrication of composite membranes for the application in fuel cells. Chapter 19 describes the fabrication of homogeneous dispersions of nanometer-sized clay particles on polymer matrices by intercalating the clay in the presence of a compatibilizer for the use in direct methanol fuel cells. This fabrication method is employed due to its simplicity and reliability. The chapter discusses the technique of intercalation integrated with a compatibilizer, 2,4,6-triaminopyrimidine, and its advantages in improving the characteristics of polymer electrolyte membrane. Proton exchange membrane fuel cells (PEMFC) are considered to be one of the most promising technologies for clean and efficient power-generation systems. Proton exchange membranes (PEMs) are key components in this fuel cell system. Chapter 20 reviews how researchers have focused on the development of a PEM with high proton conductivity, low electronic conductivity, low permeability to fuel, low electro-osmotic drag coefficient, good chemical and thermal stability, good mechanical properties, and low cost. According to the membrane materials used in fabrication, PEMs can be classified in three categories: per-fluorinates, partially fluorinated, and nonfluorinated. The chapter discusses the blending and use of different solvents with sulfonated poly(ether ether ketone) (SPEEK) to create nonfluorinated for PEMFC. The authors argue that SPEEK 1.6 is a promising membrane material for PEMFC because of its moderate proton conductivity and process stability.

We thank all the authors who have contributed to this book. We are very grateful for their hard work and willingness to share their experience and knowledge. We are convinced that the book is a significant and important source of information that is invaluable to anyone working in the field of membrane separation processes, whether that is in an established industry or an emerging area of membrane application.

Nidal Hilal
Swansea University

Ahmad Fauzi Ismail
Universiti Teknologi Malaysia

Chris J. Wright
Swansea University

Editors

Professor Nidal Hilal holds a chair in nanomem-
branology and water technologies and the founding
director of Centre for Water Advanced Technologies
and Environmental Research at Swansea University
in the United Kingdom. He is also the editor-in-chief
for the international journal *Desalination*, on the editorial boards of a number of international journals,
and a member of the advisory boards of several multinational organizations. Professor Hilal obtained his
PhD in chemical engineering in 1988 from Swansea
University and, in recognition of his outstanding
research contribution in the field of nanotechnology
and membrane separation, he was awarded a doctorate of science degree (DSc) from
the University of Wales in 2005. He was also awarded the prestigious Kuwait Prize
of Applied Science for 2005. Professor Hilal is internationally recognized as a world
leader in developing and applying the force measurement capability of atomic force
microscopy (AFM) to the study of membrane separation and engineering processes
at the nanoscale level. He has published around 300 articles in the refereed scientific
literature, including 27 chapters and 6 handbooks. He is a chartered engineer in the
UK (CEng), a Fellow of the Institution of Chemical Engineers (FIChemE), and a
fellow of the Learned Society of Wales (FLSW). He has served and has carried out
extensive consultancy for the industry, government departments, research councils,
and universities on an international basis.

Professor Ahmad Fauzi Ismail is the founding direc-
tor of Advanced Membrane Technology Research
Center and also the dean of Research for Materials
and Manufacturing Research Alliance of Universiti
Teknologi Malaysia (UTM). Professor Fauzi obtained
his PhD in chemical engineering in 1997 from the
University of Strathclyde, Glasgow, Scotland, and
MSc and BSc degrees from the Universiti Teknologi
Malaysia in 1992 and 1989, respectively. He is the
author and co-author of over 290 refereed journals.
He has also authored 2 books, 25 book chapters, and
2 edited books, has had 3 patents granted, with 17 patents pending. He has won more
than 90 awards and among the outstanding awards are the Malaysia Young Scientist
Award in 2000; ASEAN Young Scientist Award in 2001; two times winner of the
National Intellectual Property Award (patent category), 2009 and 2013 (product category); two times winner of National Innovation Award (waste to wealth category),
2009 and 2011 (product category). Recently, he won the National Academic Award

(innovation and product commercialization category) in August 2013; and Malaysian Toray Science and Technology Foundation Award on November 28, 2013. He is a Fellow of the Academy of Sciences Malaysia, chartered engineer in the United Kingdom (CEng), and a fellow of the Institution of Chemical Engineers (FIChemE). At present, he is the editor of *Desalination*. Professor Fauzi's research focuses on the development of polymeric, inorganic, and novel mixed-matrix membranes for water desalination, wastewater treatment, gas separation processes, membranes for palm oil refining, photocatalytic membranes for removal of emerging contaminants, and polymer electrolyte membrane for fuel cell applications. He has been involved extensively in R&D&C for multinational companies related to membrane-based processes for industrial application.

Chris J. Wright is a reader in bionanotechnology and membrane separation within the Multidisciplinary Nanotechnology Centre (MNC) at Swansea University, Swansea, Wales. At Swansea, he is an executive member of the Centre for NanoHealth and associate director of the Centre for Complex Fluids Processing. He graduated from the University of Wales in 1996 with a PhD in biochemical engineering. In 2001, he was awarded a prestigious advanced research fellowship from the Engineering and Physical Research Council (EPSRC), United Kingdom, in recognition of his  innovative research applying atomic force microscopy (AFM) to the characterization of membrane and biological surfaces. This five-year award allowed him to establish an internationally recognized research group exploiting the capabilities of AFM. His innovative research developing AFM measurement capabilities to study biological interfaces has been adopted by many other researchers and industry. In 2006, he was appointed portfolio director for process engineering at Swansea University within the college of engineering and is now director of PhD studies in the MNC.

His research interests include the control of polymer surfaces for improved membrane separation and tissue engineering, the control of biofilms, and the combination of AFM with advanced light microscopy methods. An underlying theme of this research is the application of nanotechnology to health care. His research has been sustained through major grants from government, charity, and the industry. He is on the editorial board of the *Journal of Nanoengineering and Nanosystems* and is a member of the EPSRC College for assessment of research grants. He has over 80 peer-reviewed international publications, with 15 invited book chapters and review articles.

List of Contributors

Farah E. Ahmed
Institute Center for Water and
  Environment
Department of Chemical and
  Environmental Engineering
Masdar Institute of Science and
  Technology
Abu Dhabi, United Arab Emirates

Amir-Al-Ahmed
Center of Excellence in Renewable
  Energy
King Fahd University of Petroleum &
  Minerals
Dhahran, Saudi Arabia

Shaheen Fatima Anis
Institute Center for Water and
  Environment
Department of Chemical and
  Environmental Engineering
Masdar Institute of Science and
  Technology
Abu Dhabi, United Arab Emirates

Hassan Ali Arafat
Institute Center for Water and
  Environment
Department of Chemical and
  Environmental Engineering
Masdar Institute of Science and
  Technology
Abu Dhabi, United Arab Emirates

Muhammad Ro'il Bilad
Centre for Surface Chemistry and
  Catalysis
Faculty of Bioscience Engineering
Leuven, Belgium

Luke Burke
Multidisciplinary Nanotechnology
  Centre
Systems and Process Engineering
  Centre
College of Engineering
Swansea University
Singleton Park, Swansea, Wales

Ng Be Cheer
Advanced Membrane Technology
  Research Center
Universiti Teknologi Malaysia
UTM Skudai, Johor, Malaysia

Tai-Shung Chung
Department of Chemical &
  Biomolecular Engineering
National University of Singapore
Singapore

Enrico Drioli
Institute on Membrane Technology
National Research Council
Rende (CS), Italy

Chaoyang Feng
Department of Chemical Engineering &
  Applied Chemistry
University of Toronto
Toronto, Ontario, Canada

and

Chemical and Biological Engineering
  Department
University of Ottawa
Ottawa, Ontario, Canada

Alberto Figoli
Institute on Membrane Technology
National Research Council
Rende (CS), Italy

Hasrinah Hasbullah
Advanced Membrane Technology
    Research Center
Universiti Teknologi Malaysia
UTM Skudai, Johor, Malaysia

Raed Hashaikeh
Institute Center for Water and
    Environment
Department of Chemical and
    Environmental Engineering
Masdar Institute of Science and
    Technology
Abu Dhabi, United Arab Emirates

Nidal Hilal
Centre for Water Advanced Technologies
    and Environmental Research
College of Engineering
Swansea University
Singleton Park, Swansea, Wales

Arun Mohan Isloor
Membrane Technology Laboratory
Chemistry Department
National Institute of Technology
    Karnataka
Surathkal, Mangalore, India

Ahmad Fauzi Ismail
Advanced Membrane Technology
    Research Center
Universiti Teknologi Malaysia
UTM Skudai, Johor, Malaysia

Juhana Jaafar
Advanced Membrane Technology
    Research Center
Universiti Teknologi Malaysia
UTM Skudai, Johor, Malaysia

Shinji Kanehashi
Department of Applied Chemistry
Meiji University
Kawasaki, Japan

Amir Keshvari
Multidisciplinary Nanotechnology Centre
Systems and Process Engineering Centre
College of Engineering
Swansea University
Singleton Park, Swansea, Wales

Kailash Chandra Khulbe
Chemical and Biological Engineering
    Department
University of Ottawa
Ottawa, Ontario, Canada

Jeonghwan Kim
Department of Environmental
    Engineering
INHA University
Incheon, Republic of Korea

Benjamin F.K. Kingsbury
Department of Chemical Engineering
South Kensington Campus
Imperial College London
London, United Kingdom

Victor Kochkodan
Centre for Water Advanced Technologies
    and Environmental Research
College of Engineering
Swansea University
Singleton Park, Swansea, Wales

Boor Singh Lalia
Institute Center for Water and
    Environment
Department of Chemical and
    Environmental Engineering
Masdar Institute of Science and
    Technology
Abu Dhabi, United Arab Emirates

Kang Li
Department of Chemical Engineering
South Kensington Campus
Imperial College London
London, United Kingdom

Yuan Liao
Singapore Membrane Technology Centre
Nanyang Environment and Water
    Research Institute
Nanyang Technological University
Singapore

Chun Heng Loh
Singapore Membrane Technology
    Centre
Nanyang Environment and Water
    Research Institute
Nanyang Technological University
Singapore

Takeshi Matsuura
Chemical and Biological Engineering
    Department
University of Ottawa
Ottawa, Ontario, Canada

Abdul Wahab Mohammad
Centre for Sustainable Process
    Technology
Faculty of Engineering and Built
    Environment
Universiti Kebangsaan Malaysia
UKM Bangi Selangor, Malaysia

Kazukiyo Nagai
Department of Applied Chemistry
Meiji University
Kawasaki, Japan

Rosmawati Naim
Advanced Membrane Technology
    Research Center
Universiti Teknologi Malaysia
UTM Skudai, Johor, Malaysia

and

Faculty of Chemical and Natural
    Resources Engineering
Universiti Malaysia Pahang
Pahang, Malaysia

Law Yong Ng
Department of Chemical and Process
    Engineering
Faculty of Engineering and Built
    Environment
Universiti Kebangsaan Malaysia
UKM Bangi Selangor, Malaysia

Yee Kang Ong
Department of Chemical &
    Biomolecular Engineering
National University of Singapore
Singapore

Mohd Hafiz Dzarfan Othman
Advanced Membrane Technology
    Research Center
Universiti Teknologi Malaysia
UTM Skudai, Johor, Malaysia

Masoud Rahbari-Sisakht
Advanced Membrane Technology
    Research Center
Universiti Teknologi Malaysia
UTM Skudai, Johor, Malaysia

Mukhlis A. Rahman
Advanced Membrane Technology
    Research Center
Universiti Teknologi Malaysia
UTM Skudai, Johor, Malaysia

Dipak Rana
Chemical and Biological Engineering
    Department
University of Ottawa
Ottawa, Ontario, Canada

Shuichi Sato
Department of Applied Chemistry
Meiji University
Kawasaki, Japan

Goh Pei Sean
Advanced Membrane Technology
    Research Center
Universiti Teknologi Malaysia
UTM Skudai, Johor, Malaysia

Laurentia Setiawan
Singapore Membrane Technology Centre
Nanyang Environment and Water
    Research Institute
Nanyang Technological University
Singapore

Seema Shrikant Shenvi
Membrane Technology Laboratory
Chemistry Department
National Institute of Technology
    Karnataka
Surathkal, Mangalore, India

Ryohei Shindo
Department of Applied Chemistry
Meiji University
Kawasaki, Japan

Silvia Simone
Institute on Membrane Technology
National Research Council
Rende (CS), Italy

Panu Sukitpaneenit
Department of Chemical &
    Biomolecular Engineering
National University of Singapore
Singapore

Abdullah S. Sultan
Department of Petroleum
    Engineering
King Fahd University of Petroleum &
    Minerals
Dhahran, Saudi Arabia

Rinku Thomas
Institute Center for Water and
    Environment
Department of Chemical and
    Environmental Engineering
Masdar Institute of Science and
    Technology
Abu Dhabi, United Arab Emirates

Silvano Tosti
ENEA
Unita Tecnica Fusione
C.R. ENEA Frascati
Frascati, Italy

Bart Van der Bruggen
Department of Chemical Engineering
ProcESS—Process Engineering for
    Sustainable Systems
KU Leuven
Leuven, Belgium

Rong Wang
Singapore Membrane Technology
    Centre
Nanyang Environment and Water
    Research Institute
and
School of Civil and Environmental
    Engineering
Nanyang Technological University
Singapore

Chris J. Wright
Multidisciplinary Nanotechnology
    Centre
Systems and Process Engineering
    Centre
College of Engineering
Swansea University
Singleton Park, Swansea, Wales

Zhentao Wu
Department of Chemical Engineering
South Kensington Campus
Imperial College London
London, United Kingdom

S.M. Javaid Zaidi
School of Chemical Engineering
Faculty of Engineering
Architecture and Information
    Technology
The University of Queensland
Brisbane, Queensland, Australia

Ruixin Zhang
Department of Chemical Engineering
ProcESS—Process Engineering for
    Sustainable Systems
KU Leuven
Leuven, Belgium

Section I

Fabrication Processes for Polymeric Membrane

1 Polymeric Membranes

Alberto Figoli, Silvia Simone, and Enrico Drioli

CONTENTS

1.1   Introduction 3
1.2   Membrane Preparation Techniques 4
    1.2.1   Sintering, Stretching, and Track-Etching 4
1.3   Membrane Preparation by PI..7
    1.3.1   Membrane Preparation Techniques via PI...7
1.4   Thermodynamic Principles of PI..7
    1.4.1   Phase Diagrams for TIPS and DIPS..7
    1.4.2   Solubility Parameters..12
    1.4.3   Trade-Off between Thermodynamic, Kinetic,
           and Membrane Morphology 13
1.5   Peculiarities of Hollow-Fiber Membrane Preparation through PI................... 15
1.6   Examples of Membrane Preparation for Pressure-Driven Separation
    Processes.. 19
    1.6.1   Microfiltration...20
    1.6.2   Ultrafiltration 21
    1.6.3   NF and Solvent-Resistant NF 22
    1.6.4   Reverse Osmosis..24
    1.6.5   Gas Separation 26
    1.6.6   Pervaporation...27
1.7   Examples of Membrane Preparation for MC..28
    1.7.1   Gas/Liquid Contactors 29
    1.7.2   Liquid/Liquid Contactors..30
    1.7.3   Membrane Distillation 30
    1.7.4   OD and Membrane Crystallizers... 31
1.8   Conclusions—Outlook 35
References...35

1.1   INTRODUCTION

A membrane can be defined as an "interphase between two adjacent phases acting as a selective barrier, regulating the transport of substances between the two compartments" ([1], p. 2217). Membrane technologies are widely recognized as advanced separation/concentration processes, which are ideally placed to aid process intensification [2], thanks to the possibility of exploiting the synergy between different membrane operations in an integrated system [3]. Membrane processes are now widespread at the industrial level, a result of the advances in membrane performance connected to higher productivity, enhanced selectivity, and improved stability. Nowadays, membranes are prepared using a wide variety of techniques, mainly depending on the membrane material but also on the application. In this chapter, the most common membrane preparation methods are described, with peculiar focus on phase inversion (PI) or phase separation (PS), which is the foremost technique for preparing polymeric membranes. The thermodynamic principles of PI, the main factors affecting membrane morphology and properties as well as the peculiarities of hollow-fiber membranes preparation through PI, are examined. Furthermore, significant examples of membrane preparation for selected applications, spanning from *classical* pressure-driven processes, such as reverse osmosis (RO), nanofiltration (NF), ultrafiltration (UF), and microfiltration (MF), to more recent emerging processes, such as membrane contactors (MCs), are reported.

1.2 MEMBRANE PREPARATION TECHNIQUES

1.2.1 SINTERING, STRETCHING, AND TRACK-ETCHING

There are several techniques for preparing membranes; the selection of the appropriate method depends on the material and the final membrane application. Membrane properties can be modulated, to a certain extent, by properly choosing the preparation technique and acting on the key process conditions. In Table 1.1, the main membrane materials, preparation techniques, and applications are summarized. PI, the most used membrane preparation technique, is discussed in detail in Section 1.3. The other most used techniques, usually employed in membrane preparation, are introduced in this section.

TABLE 1.1
Main Membrane Materials, Preparation Techniques, and Applications

Materials	Techniques	Applications
Organic polymers	Sintering	MF
	Stretching	MF/MC
	Track-etching	MF
	Phase inversion	MF/UF/NF/RO
Inorganic materials	Sintering	MF/UF
α-Al_2O_3		NF
α-Al_2O_3/γ-Al_2O_3		UF
γ-Al_2O_3/TiO_2/ZrO_2		MF
Stainless steel		GS
Palladium		
Glass (polycarbonate)	Track-etching	MF

Source: A. Bottino et al., *C. R. Chimie*, 12, 882–888, 2009.

TABLE 1.2

Principle of Sintering Method, Membrane Materials and Properties

Schematic of the Process	Materials Used	
	Powders of polymers	Polyehtylene
		Polytetrafluoroethylene
		Polypropylene
Membrane pore-size distribution 0.1–10 μm	Powders of metals	Stainless steel
		Tungsten
Porosity	Powders of ceramics	Aluminum or
10%–20% with polymers		zirconium oxide
80% with metals	Powders of graphite	Carbon
	Powders of glass	Silicalite

The sintering technique allows the preparation of symmetric membranes and is generally used to prepare ceramic or metallic membranes for application in UF and MF (Table 1.1). A powder consisting of particles (of the material) of a certain size is pressed and heated, at or just below the melting temperature [4].

The principle of the sintering method is shown in Table 1.2 (from Ref. 5). The pore size and porosity of the membranes obtained are generally affected by two main factors, namely, particle size and sintering profile, but also by temperature, heating/cooling rates, and dwelling time [5,6]. Membranes prepared by sintering can be produced as disks, cartridges, or fine-bore tubes.

Hollow-fiber ceramic membranes can be prepared using a three-step process based on a combination of PI and sintering methods involving the (1) preparation of a spinning suspension, (2) spinning of ceramic hollow-fiber precursors, and (3) final sintering [7].

The stretching technique is also used for producing MF polymeric membranes. A homogeneous polymer of partial crystallinity, in the shape of a film or hollow fiber, is stretched perpendicularly to the axis of crystallite orientation [4]. Relatively uniform pores, with diameters of 0.2–20 μm, are formed as a result of a partial fracture of the film (Figure 1.1) [5]. Polytetrafluoroethylene (PTFE), polypropylene (PP), and polyethylene membranes can be prepared by this technique. For instance, Celgard PP membranes are obtained by monodirectional stretching, whereas Gore-Tex membranes are produced by bidirectional stretching [4]. The membranes produced generally show high permeability to vapor and gases, although, due to the intrinsic material hydrophobicity, they are quite impenetrable to aqueous streams. Therefore, they are interesting for application as water-repellent textiles and contactors [5].

Track-etching allows the preparation of membranes having uniform cylindrical pores. A thin dense polymer film is exposed to high-energy particle radiation, which damages the polymer matrix. The damaged polymeric material is then etched away

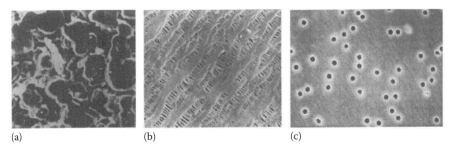

(a)   (b)   (c)

FIGURE 1.1 Membranes produced by (a) sintering, (b) stretching, and (c) track-etching. (Reprinted from H. Strathmann et al., *Basic Aspects in Polymeric Membrane Preparation*, In: E. Drioli and L. Giorno, eds., *Comprehensive Membrane Science and Engineering*, vol. 1: Basic Aspects of Membrane Science and Engineering. Elsevier, Amsterdam, the Netherlands, 2010, pp. 91–112, Copyright 2010, with permission from Elsevier.)

in an acid (or alkaline) bath [4,5]. Membrane porosity is generally around 10%, and is affected by residence time in the irradiator [5]. Pore dimensions are usually within the range 0.2–10 μm.

MF membranes, in silicon nitride, showing high porosity and narrow pore size distribution, coupled with very low flow resistance and minimal fouling tendency, were produced by laser interference lithography and silicon micromachining technology (Figure 1.2) [8]. Such membranes are often referred to as *microsieves*.

The template leaching technique is suitable for preparing porous membranes from polymers, which do not dissolve in common organic solvents [9], or from glass, metal alloys, and ceramics [5].

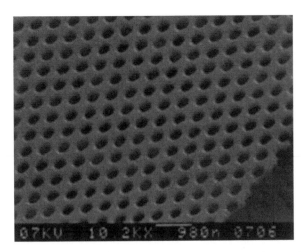

FIGURE 1.2 Field emission scanning electron microscopy image of a membrane produced by laser interference lithography. (Data from S. Kuiper et al., *Journal of Membrane Science*, 150, 1–8, 1998.)

1.3 MEMBRANE PREPARATION BY PI

1.3.1 MEMBRANE PREPARATION TECHNIQUES VIA PI

PI or PS is indeed the most common method for preparing polymeric membranes. It is based on the separation of an initially homogeneous system into two distinct phases, consisting of a polymer, a solvent, and, eventually, other additives. The solid phase, or polymer-rich phase, will give rise to the membrane matrix, whereas the solvent-rich liquid phase, or polymer-lean phase, will originate from the membrane pores. There are four techniques distinguished on the basis of the mechanism exploited to induce such separation, often called *demixing* or *precipitation*. These four techniques are evaporation-induced PS (EIPS), vapor-induced PS (VIPS), temperature-induced PS (TIPS), and nonsolvent-induced (or diffusion-induced) PS (NIPS or DIPS) [1,5]. In TIPS, precipitation is induced by lowering temperature. In DIPS, precipitation of the casting solution is obtained by immersion into a non-solvent bath. In VIPS, the nonsolvent is adsorbed from a vapor phase, which can also contain other gases such as air or nitrogen. In EIPS, precipitation is induced by the evaporation of a volatile solvent from the casting solution. According to some authors, there are only two types of PI, namely, temperature-induced and diffusion-induced, whereas immersion precipitation, vapor adsorption, and solvent evaporation are considered as three types of DIPS [10]. Phase inversion is extremely versatile and allows the preparation of membranes from several different polymers, as long as the polymer is soluble in a solvent, and the system shows a miscibility gap over a defined concentration and temperature range. Membranes, which have morphology and properties suitable for an impressive variety of processes, can be obtained.

1.4 THERMODYNAMIC PRINCIPLES OF PI

All the recipes reported in the literature for membrane preparation are based on the same principles, that is, thermodynamic and kinetic, such as the relationship between the chemical potentials and diffusivities of the individual components and Gibb's free energy of mixing of the entire system. Their interplay during membrane formation produces the final membrane structure; therefore, a better understanding of all these parameters is the optimum way to achieve a deeper knowledge of the membrane formation mechanisms, and how to tailor and optimize membrane morphology and properties. From a thermodynamic point of view, the two main mechanisms of PS, thermally induced and nonsolvent induced, are described with the aid of binary and ternary phase diagrams (Figure 1.3). Phase diagrams represent a useful instrument to better understand the mechanism of membrane formation. This is often called phenomenological description of the phase separation process [5].

1.4.1 PHASE DIAGRAMS FOR TIPS AND DIPS

TIPS is based on a *latent solvent* or diluent that behaves as a good solvent at temperatures close to the melting of the polymer, but that works as a nonsolvent at lower

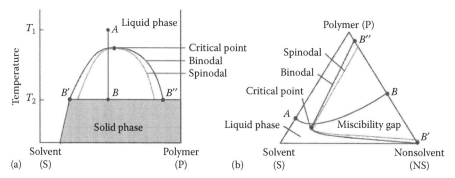

FIGURE 1.3   (a) Binary and (b) ternary phase diagrams describing TIPS and DIPS processes. (Reprinted from H. Strathmann et al., *Basic Aspects in Polymeric Membrane Preparation*, In: E. Drioli and L. Giorno, eds., *Comprehensive Membrane Science and Engineering*, vol. 1: Basic Aspects of Membrane Science and Engineering, Elsevier, Amsterdam, the Netherlands, 2010, pp. 91–112, Copyright 2010, with permission from Elsevier.)

temperatures. TIPS is often used for polymers that are not soluble at room temperature, such as polyolefins [1].

The TIPS technique consists of the following:

- Dissolving the polymer in the latent solvent
- Casting the solution in the desired shape
- Phase separating as a result of solution cooling
- Extracting the latent solvent by means of a more volatile substance
- Final drying of the membrane

The system must show a miscibility gap over a certain range of temperature and composition. In the binary phase diagram, the miscibility gap is surrounded by the spinodal curve, although the region in between the spinodal and the binodal curves is metastable. In Figure 1.3, it is seen that the temperature T_1 of an initially homogeneous system, located at the point A, decreases to reach T_2. The corresponding point B is located inside the miscibility gap and as a consequence the system will demix in two phases, which is indicated by B' and B''. The first one represents the polymer-rich phase and forms the solid membrane structure. The other phase forms liquid-filled membrane pores. Depending on the polymer type, dope composition (polymer concentration and solvent type), and the cooling rate, PI can proceed both through solid–liquid (S–L) and liquid–liquid (L–L) demixing, giving rise to different membrane structures and properties. Other phenomena, such as gelation and vitrification, can also take place.

L–L PS takes places when the temperature reaches the binodal curve. Two mechanisms of membrane formation may occur in this case: spinodal decomposition (SD) and nucleation and growth (NG). Although the latter occurs only in the metastable region comprised between the binodal and spinodal lines, SD takes place in the unstable region under the spinodal line. S–L PS takes place only if, during solution cooling, the crystallization temperature of the polymer is reached. If the polymer is amorphous, gelation through L–L separation takes place, which is arrested by the vitrification of the polymer-rich phase at the glass transition temperature [10].

As reported in the literature [10], in the binary phase diagram of an amorphous polymer, the intersection between the binodal curve and the glass transition boundary is defined as *Berghmans point*. The L–L demixing is interrupted by gelation, which leads to vitrification of the polymer-rich phase. The time between the beginning of the PS and the final vitrification is referred to as the gelation time. If the cooling rate is not infinitely slow, the final structure will be a porous glass.

When the system PS starts within the metastable region, which is found between the spinodal and binodal curves, it is commonly referred to as nucleation and growth, or NG. In the metastable region, indicated as II in the binary phase diagram (Figure 1.4), it can be noticed that the polymer-rich phase gives rise to a continuous matrix, whereas the polymer-lean phase produces isolated pores, that is, NG of the solvent phase in the polymer-rich phase. On the contrary, a suspension of polymer-rich phase in a continuous polymer-lean phase, NG of the polymer-rich phase in the solvent phase, can be obtained from region III of the phase diagram. When demixing starts in the unstable region, the mechanism is called SD, which is defined as a spontaneous process that does not need a nucleus [10].

As reported in the literature, the PS structures obtained by NG and SD will grow and coarsen during the gelation time. If this is infinite, two fully separate phases could be obtained. If the gelation time is short enough, the SD will give rise to a morphology with high interconnectivity, since the coalescence process is quickly

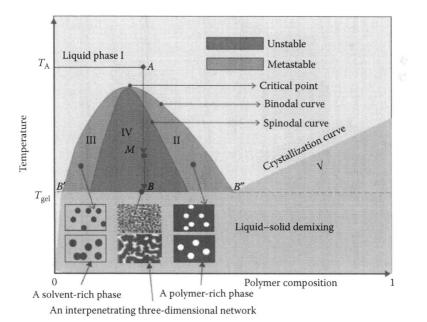

FIGURE 1.4 Schematic phase diagram of TIPS. (Data from A.G. Fane et al., Membrane technology for water: Microfiltration, ultrafiltration, nanofiltration, and reverse osmosis. In: P. Wilderer, ed., *Treatise on Water Science*, Academic Press, Oxford, 2011, pp. 301–335.)

stopped by vitrification. As the gelation time increases, the coalescence phenomena may result in a closed cell structure.

A good knowledge of the properties of the polymer/solvent system could help to adjust the gelation time, thus allowing control of the interconnectivity between pores and the final membrane morphology. Bicontinuous structures can be obtained through SD. However, the spinodal area can be directly reached during cooling only at the critical point, the point where the binodal and spinodal curves coincide. When composition is different, during cooling, the system must cross the metastable area first. In this case, in order to prevent demixing and improve pore interconnectivity, it can be useful to employ fast cooling [12]. Semicrystalline and crystalline polymers can also crystallize, giving rise to chain-folded lamellae and supramolecular architectures as axialites and spherulites [12].

NIPS consists in the preparation of a homogeneous polymer dope by dissolving the polymer in a suitable solvent. After casting in the desired shape, polymer precipitation is induced by immersion in a coagulation bath containing a nonsolvent. NIPS can be described by ternary phase diagrams. The system must exhibit a miscibility gap, for a defined range of polymer/solvent/nonsolvent (P/S/NS) compositions. Ternary phase diagrams always refer to a certain temperature. Similarly to what is described for TIPS phase diagrams, the metastable region is between the binodal and spinodal curves and the unstable region is delimited by the spinodal curve. Consider a point A as the initial system composition; this point is located in the stable region since only polymer and solvent are present. By adding a nonsolvent, the system composition will change, and point A will move toward point B. Going from A to B, the system composition changes due to S–NS exchange. Once the miscibility gap is reached, PS takes place. The upper boundary of the miscibility gap, B'', is the polymer-rich phase, and the lower boundary, B', is the polymer-lean phase.

In Figure 1.5, four regions can be recognized: region I, one solution phase; region II, two liquid–liquid phases; region III, two liquid–solid phases; region IV, one solid phase. Starting from a generic point A in region I, the system can follow four different paths. If the system follows path 1, it reaches region IV of the phase diagram after a glass transition; a homogeneous glassy film is obtained (vitrification).

When the system reaches the point S_1, L–L PS takes place, resulting in phase S_2 (polymer-rich phase) and S_3 (polymer-lean phase). When S_1 is located in the metastable region at high polymer concentration (path 2-1), similar to what is described for TIPS, membrane formation proceeds through NG of the polymer-lean phase, resulting in noninterconnected pores. Bicontinuous structure can be obtained following path 2-2, which enters directly in the unstable region (SD). When following path 2-3, the system enters in the metastable region at low polymer concentration, giving rise to low-integrity powdery agglomerates [11]. The point where the binodal and spinodal curves coincide is also here and is called the *critical point*. It represents the maximum solvent concentration in the coagulation bath that still allows the fabrication of a solidified membrane [10].

Two further important definitions are the *delay time* and *gelation time*. Delay time is defined as the time interval between the immersion in the coagulation bath and the beginning of the liquid–liquid demixing [10]. Looking at the phase diagram, it is easy to understand that, depending on the delay time, the system crosses

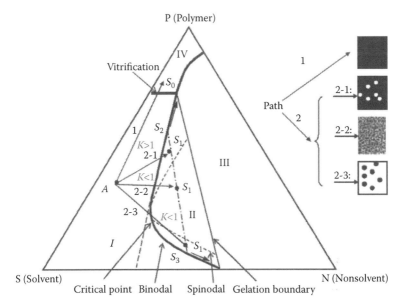

FIGURE 1.5 Ternary phase diagram describing membrane formation through immersion precipitation. (Data from A.G. Fane et al., Membrane technology for water: Microfiltration, ultrafiltration, nanofiltration, and reverse osmosis. In: P. Wilderer, ed., *Treatise on Water Science*, Academic Press, Oxford, 2011, pp. 301–335.)

different regions; this affects membrane morphology. For instantaneous demixing, the system immediately reaches the unstable region; when the delay time is longer, the system passes through the metastable region.

The gelation time is defined as the time interval between the onset of the demixing and the solidification of the polymer solution. This applies when the system enters region IV of the phase diagram, when the polymer-rich phase vitrifies after reaching the Berghmans point. This also influences membrane morphology since the system composition with solvent/nonsolvent exchange can proceed only before the vitrification.

The formation of the final membrane structure is indeed a complicated process, in which all the described mechanisms are involved. NG, SD, and gelation exert a significant influence on the final membrane. When PS proceeds through NG, the increase of the gelation time will promote the formation of interconnected pores.

The final membrane morphology also depends on the polymer type. For glassy polymers, such as cellulose acetate (CA), polyamide (PA), and polyimide, PI is mostly controlled by liquid/liquid demixing, whereas for semicrystalline polymers, such as polyvinylidene fluoride (PVDF), solid–liquid demixing and polymer crystallization can also take place during PI. Usually, L–L demixing gives rise to membranes having cellular morphology and/or finger-like macrovoids, with pores generated by the polymer-lean phase, surrounded by a matrix created by the polymer-rich phase. S–L demixing often results in particulate structure, which is made up of interlinked semicrystalline spherulites.

VIPS and EIPS are not treated in detail in this section. However, as reported in the literature, most of the concepts illustrated regarding TIPS and NIPS also apply to these two techniques.

1.4.2 Solubility Parameters

Solubility parameters are indexes usually employed to evaluate the interactions between the polymer, solvent, and nonsolvent, such as the solvent's ability to dissolve a given polymer, the miscibility between solvent and nonsolvent, and the coagulation power of a nonsolvent toward a polymer of interest. These interactions will strongly affect the path followed by the P/S/NS system during PI, and hence the final membrane morphology.

In general, the closer the solubility parameters of two chemical species, the more compatible they are.

The Hildebrand solubility parameter for a pure liquid substance is defined as the square root of the cohesive energy density:

$$\delta = \left(\frac{\Delta H_v - RT}{V_m} \right)^{1/2}$$

where:

ΔH_v is the heat of vaporization

V_m the molar volume

According to Hansen's theory [13], δ, the Hildebrand parameter, can be calculated using the three components: δ_d, which represents the energy from dispersion bonds; δ_h, which represents the energy from hydrogen bonds between molecules; and δ_p, which represents the energy from dipolar intermolecular forces:

$$\sqrt{\left(\delta_d^2 + \delta_p^2 + \delta_h^2 \right)}$$

The interactions between polymer (P), solvent (S), and nonsolvent (NS) can be evaluated by calculating the difference between their solubility parameters using the following equations [14,15]:

$$\text{P–S: } \delta_{P,S} = \sqrt{\left(\delta_{d,P} - \delta_{d,S} \right)^2 + \left(\delta_{p,P} - \delta_{p,S} \right)^2 + \left(\delta_{h,P} - \delta_{h,S} \right)^2}$$

$$\text{P–NS: } \Delta\delta_{P-NS} = \sqrt{\left(\delta_{d,P} - \delta_{d,NS} \right)^2 + \left(\delta_{p,P} - \delta_{p,NS} \right)^2 + \left(\delta_{h,P} - \delta_{h,NS} \right)^2}$$

$$\text{S–NS: } \Delta\delta_{S-NS} = \sqrt{\left(\delta_{d,S} - \delta_{d,NS} \right)^2 + \left(\delta_{p,S} - \delta_{p,NS} \right)^2 + \left(\delta_{h,S} - \delta_{h,NS} \right)^2}$$

The mutual interaction between P, S, and NS strongly affects the mechanism of membrane formation. For instance, when the difference between the solubility parameters of P and S is small, S has a strong dissolving capacity. As a consequence, the path followed during PI to reach the miscibility gap and finally, the membrane morphology, is affected, as discussed in Section 1.4.1. If the difference between the P and NS parameters is large, NS will have a strong coagulant power. As a consequence, fast L–L demixing could take place. Finally, the difference between the NS and S parameters influences the S–NS exchange during coagulation.

For example, Wang et al. [14] prepared PVDF MF membranes for wastewater treatment and studied the effects of different solvent compositions, in particular N,N dimethylformamide (DMF), N,N dimethylacetylamide (DMAC), triethyl phosphate (TEP), dimethyl sulfoxide (DMSO), and their mixtures (50/50) on the produced membranes features. These four solvents have affinity for PVDF that decreases in the following order: TEP > DMAC > DMF > DMSO. The properties of the produced membranes varied with the solvent type and, hence, with the difference of solubility parameters between P, S, and NS, which was water in all cases. When changing solvent type, S–NS diffusivities reduce as follows: DMF > DMAC > TEP > DMSO. The morphology of the top-layer was found to be more dependent on the affinity between P and NS, whereas the S–NS exchange rate was found to affect the pore structure in the sublayer. A larger difference between P and S parameter (PVDF–DMSO couple) will induce sudden PS after immersion in the coagulation bath. However, DMSO had lower diffusion rate in the coagulation bath, which delayed solidification and caused the development of a sublayer with macrovoids. This also increased membrane thickness. In contrast, when using pure DMF, DMAC, and TEP, the prepared membranes showed lower surface porosity and shorter finger-like pores in the sublayer. The mixture of DMF and TEP, which has the highest dissolving capacity for PVDF, delayed PS a lot, resulting in a nonporous top layer. The use of solvent mixtures could delay the S–NS exchange and, in general, promote the growth of macrovoids in the sublayer. However, for the DMF–TEP mixture, the growth of macrovoids is limited, due to the development of a skin layer that prevents the S–NS exchange.

1.4.3 Trade-Off between Thermodynamic, Kinetic, and Membrane Morphology

The morphology of membranes produced through immersion precipitation and, in particular, the dichotomy sponge versus finger-like structure is a clear example of interplay or, as also reported in the literature, trade-off between thermodynamic and kinetic factors. Finger-like macrovoids, generally formed during membrane preparation, represents unwanted morphology, being connected to low mechanical strength. Several studies proposed different mechanisms to explain and/or avoid macrovoids formation. Early studies suggested interfacial hydrodynamic instabilities, caused by surface tension gradients, as a possible origin of macrovoid initiation [16–18]. According to Ray et al. [19], macrovoids formation is connected to concentration gradients at the interface between the polymer solution and the nonsolvent bath. Smolders et al. [20] connected macrovoids formation to the type of demixing i.e., delayed or instantaneous. According to the mechanism proposed in their work, macrovoids are produced under the skin layer from newly formed nuclei of the diluted phase if the solvent concentration exceeds a certain threshold value and if the composition in front of the nuclei remains stable for a suitable period.

Macrovoids formation can be avoided by delayed demixing, and increasing polymer and/or nonsolvent concentration in the polymeric dope. Other studies suggested to introduce solvent into the coagulation bath [21–23], to increase solvent evaporation time [24], to work with a S–NS pair with low miscibility [25] or to use organic additives such as polyvinylpyrrolidone (PVP) [26,27].

Regarding the preparation of hollow fiber, Simone et al. [28] reported that the effect of PVP on macrovoids formation depends on concentration. At low concentration, the presence of PVP increases the dope instability, thus promoting faster demixing and enhancing macrovoids development (thermodynamic effect). At high concentration, PVP increases the dope viscosity, thus delaying demixing and avoiding macrovoids formation (kinetic effect). These findings are in agreement to what is described in the early work by Lee et al. [29], who investigated the trade-off between thermodynamic enhancement and kinetic hindrance during PI. They prepared polysulfone (PSU) membranes and analyzed the system PSU/DMF/PVP. The increase of PVP concentration reduces the thermodynamic stability of dope solutions and should induce faster L–L demixing. However, the increase of PVP concentration was able to induce the formation of macrovoids and also increase membrane permeability until a certain threshold value (7.5%). Further increase of PVP caused an increase of dope solution viscosity. This caused a rheological hindrance of the demixing. The overall diffusion between components was delayed due to kinetic factors. This study is a clear example of how, during membrane formation, the same factor (PVP concentration) could influence both the thermodynamic and kinetic properties of the system.

Sadrzadeh and Bhattacharjee [30] discussed complex systems composed of polyethersulfone/1-methyl-2-pyrrolidone/additive; the additive was either polyethylene glycol (PEG) or PVP with different molecular weights. They demonstrated that two dimensionless parameters can be calculated, for each system, to quantify the thermodynamic enhancement and the kinetic hindrance to PI due to additives; these parameters could be used to predict membrane morphology. A simple model [31] was used to calculate diffusion rates of the solvent and nonsolvent in the coagulation bath.

Another recent and interesting study [32] showed, by direct microscopic observation, the influence of solvent and nonsolvent type during PI. The polymer polysulfone (PSU), the solvents NMP and N,N-DMF, and the nonsolvents water and glycerol were studied. Although PSU/DMF/water system resulted in sponge-like morphology, the finger-like macrovoids developed when using PSU/NMP/water system. In both systems, at the polymer–coagulation bath interface, there was fast S–NS exchange, which caused the formation of a skin layer. However, the morphology of the membrane produced from PSU/DMF/water was mainly sponge-like, due to the slow nonsolvent influx, which was hindered by the formation of a skin layer. According to Hansen's solubility parameters, NMP is a better solvent for PSU than DMF, hence, the formation of the skin was slow. Void lengths were found to decrease exponentially with increasing polymer concentration. The thickness of the skin layer was reported to increase with PSU percentage. Macrovoids formation in the PSU/NMP/water membrane was avoided by inducing the formation of a viscous gel layer, which caused similar effects to polymer precipitation by VIPS, which was also found to inhibit macrovoids formation in the PSU/NMP/water system [33]. Authors proposed that void growth takes place by convective nonsolvent flow, through the polymer solution, driven by gradients in interfacial energy. An increase in viscosity might avoid voids formation by hindering the supply of nonsolvent. This confirmed the kinetic effect of viscosity on macrovoids formation.

Macrovoids formation was, finally, hindered by using a poor nonsolvent, a mixture of water and glycerol or water and NMP. The addition of solvent in the coagulation media delays the S–NS exchange and hence the nonsolvent influx, which is responsible for voids growth. The effect of glycerol is connected both to its lower nonsolvent power and to its viscosity, which further delayed the nonsolvent influx.

1.5 PECULIARITIES OF HOLLOW-FIBER MEMBRANE PREPARATION THROUGH PI

Depending on their dimensions, it is possible to distinguish hollow-fiber membranes (diameter < 0.5 mm), capillary membranes (0.5 mm < diameter < 5 mm), and tubular membranes (diameter > 5 mm) [34]. The preparation of hollow-fiber membranes through PI is more complex, with respect to flat sheet, due to the higher number of parameters involved. However, hollow-fiber modules are usually preferred, because they ensure space savings, more productivity, and reduction of costs, which is also connected to maintenance, as these modules can be backflushed [35].

Hollow-fiber preparation requires a polymeric dope of suitable viscosity, usually a few thousand of centipoises. As a consequence, the polymer and additive concentrations in the dope are usually higher, thus affecting porosity and pore size. Furthermore, this could result in non-Newtonian rheological behavior [36]. There are three main methods for preparing hollow fibers and capillaries, namely, wet spinning, melt spinning, and dry spinning [37].

There are several parameters that are known to affect fiber morphology, properties, and performance. These parameters can be divided into the following four categories:

- Parameters connected to the dope composition include polymer and additives type, concentration, viscosity, and temperature.
- Parameters connected to the spinning experiment include temperature, dope extrusion rate, spinneret type (double/triple), geometry and dimensions, air-gap length, and atmosphere (moisture).
- Parameters connected to the coagulation include temperature, composition and injection rate of the bore fluid (BF), and temperature and composition of the coagulation bath.
- Parameters connected to the eventual post-treatments include take-up speed (stretching), chemical or thermal posttreatments (additive leaching, thermal annealing), and drying techniques (hexane, glycerol, etc.).

The type of polymer determines some key membrane features, such as its hydrophilicity/hydrophobicity and fouling tendency, structure, mechanical properties, and chemical resistance [36]. Polymer concentration is a key parameter for the thermodynamics and kinetics of PI. Higher polymer concentration will reduce the solvent volume fraction, and as a result, less nonsolvent is required to achieve PS. Moreover, due to its effect on viscosity, it will affect the kinetics of S–NS exchange. Higher polymer concentration could result in the formation of a thicker skin, which will delay coagulation of the inner layers. Tasselli et al. [38] observed that polymer concentration affected the thermodynamics of PI in the preparation of modified poly(ether ether ketone) (PEEK-WC) hollow fibers; the binodal curve was found to shift toward the P–S axis, indicating less nonsolvent tolerance, as the polymer concentration increased. Polymer concentration affected the kinetics as well, due to its effect on dope viscosity. Fiber morphology was found to be affected by increasing polymer concentration; finger-like voids at the outer surface were reported to reduce significantly. The effect of nonsolvent was more pronounced at higher polymer concentration, as expected from the shift of the binodal curve toward the P–S axis. Polymer concentration was found to also affect fiber performance, with typical trade-off between flux and rejection. Sukitpaneenit and Chung [39] observed an increase of dope viscoelastics properties with polymer concentration in the preparation of PVDF hollow fibers, which influenced fiber morphology and, in particular, macrovoids formation. This was attributed to the increased shear and elongation viscosities, due to greater degree of chain entanglement, which reduced nonsolvent penetration during coagulation.

Pore-forming additives are known to affect the delicate balance between kinetics and thermodynamics. Additives with high molecular weight (Mw) are usually retained in the fiber structure, thus modifying hydrophilicity/hydrophobicity [30]. For instance, PVP is known to affect the thermodynamic and kinetics of the PI process due to its hydrophilicity (thermodynamic enhancement) and its effect on the dope viscosity (kinetic hindrance). Tasselli et al. [40] observed that macrovoids growth was suppressed in PEEK-WC hollow fibers, until complete sponge-like structure was obtained by increasing the PVP concentration. The increase of PVP concentration from 0 to 20 wt% reduced porosity from 84% to 74%. Water permeability was found to decrease, whereas dextran rejection and fiber mechanical strength were found to increase.

Temperature is a key parameter referring to both the dope and the coagulants [36]. Indeed, temperature affects dope viscosity. Peng et al. [41] observed that more macrovoids can be observed in the cross-sectional morphology of Torlon® polyamideimide fibers when increasing the spinneret temperature due to the reduction of dope viscosity. The temperature of both inner and outer coagulant will affect the interdiffusion between solvent and nonsolvent at the fiber walls, thus affecting the kinetics of PI. Chung and Kafchinski [42] observed that a more porous structure was formed in 6FDA/6FDAM polyimide fibers by increasing the external coagulant temperature due to delayed demixing (connected to increased solubility).

Fiber coagulation is much more complicated, with respect to flat-sheet membrane, since it involves two surfaces. The thermodynamics and kinetics of PI are affected by both the coagulants. As discussed in Section 1.4, the main factors are the nonsolvent power, its mutual affinity with the dope solvent, the solubility parameter differences between P–NS and S–NS, the solvent and nonsolvent diffusivities connected to their molecular size, and, obviously, temperature. Tasselli and Drioli [43] showed that hollow-fiber morphology, transport, and mechanical properties can be tailored by varying the composition of the BF. In particular, the effect of different R—OH BFs, with R=H; CH_3; C_2H_5; n—C_3H_7 or n—C_4H_9, on the properties of PEEK-WC hollow-fiber membranes was examined. Going from water to alcohols with progressively longer aliphatic chain, the binodal curve moved toward higher nonsolvent concentrations, indicating less nonsolvent power. Phase diagrams were found to be similar by adding low molecular weight PVP (Luviskol K-17, Mw 12 kDa) to the dope solution, showing that the additive influenced mostly the kinetics, rather than the thermodynamics of PI. The different composition of BFs also influenced the kinetics of PI, mainly due to the increasing nonsolvent molecular dimensions, which further reduced diffusivity.

Tasselli et al. [38] observed that increasing solvent (DMAc) percentages in the BF affected fiber dimensions, thus increasing the diameter and reducing the thickness. This was attributed to the effect of solvent percentage on the degree of fiber inflation, normally caused by BF injection. Higher solvent percentage induced a delayed onset of demixing, resulting in a softer skin at the inner surface, which was easier to inflate when compared to a rigid skin produced by sudden coagulation induced by water. Increase of solvent concentration in the BF was found to decrease rejection without affecting permeability due to a more open skin layer.

Fiber properties can be modulated by acting on the atmosphere of the air gap. Tasselli and Drioli [40] found that the relative humidity percentage in the air-gap atmosphere strongly affected the morphology of the outer layer of PEEK-WC hollow fibers. Although all membranes prepared under unsaturated conditions showed similar morphology and water permeability, the presence of supersaturated water vapor and microdroplets in the air gap induced the formation of a macroporous skin at the outer surface, which induced local PS at the outer surface of the fibers.

The rheology of spinning experiments is complicated and involves both the shear stresses experienced by the dope within the spinneret and the elongational stresses in the air-gap region connected to gravitational force or additional stretching during the take-up of fibers. Tasselli et al. [38] observed that an increase of the air gap-induced higher stretching of the nascent fiber due to the gravitational force. This resulted in elongation, higher spinning rate, and hence reduction of fiber dimensions. Peng et al. [36] pointed out that the supplementary elongational stress may be caused by the take-up device and influence the properties of hollow fibers by inducing extra phase instability, thereby enhancing PS and promoting orientation and packing. The air gap could modulate the effect of the rheological die-swell and affect the morphology of fibers. Peng et al. [44] found that there was a critical air-gap distance, which varied with the dope composition, above which the effects of die swell and chain relaxation were suppressed, and as a consequence nonsolvent intrusion and macrovoids formation were hindered.

The rheology, molecular orientation, and finally the fiber morphology are strongly affected by the spinneret architecture. The structure of a common spinneret, and the phenomena that take place during extrusion are depicted in Figure 1.6.

Peng et al. [36] pointed out that if fibers are extruded using a small annular gap, the shear rates will be higher, thus inducing molecular orientation, chain packing, and confounding macrovoids formation. According to Widjojo and Chung [45], shear stress within the spinneret could also help to eliminate the irregularities of the outer surface besides macrovoids formation.

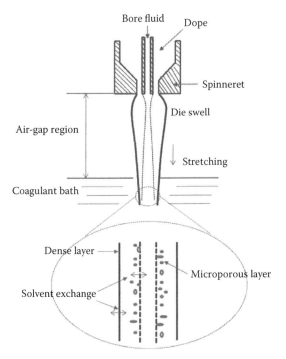

FIGURE 1.6 Schematic diagram of area near the spinneret and the formation of nascent hollow fiber during PI. (Data from N. Peng et al., *Progress in Polymer Science*, 37, 1401–1424, 2012.)

In the literature, different modifications of the spinneret design were proposed as follows: (1) spinnerets with different flow angles, to tailor pore size distribution and control pure water permeability [46] and (2) spinnerets with microstructured annulus or needle, to increase the active surface area [47,48].

It is possible to produce dual-layer hollow fibers, for several applications, using a triple orifice spinnerets. The production of dual-layer fibers is more complex due to the presence of two dope solutions. One of the most critical issues is delamination of the outer layer. A modified spinneret, with an indented middle tube, has been proposed to induce interdiffusion between the two fiber layers and prevent delamination [49].

Fiber posttreatments and drying procedures represent further important steps. Thermal treatments and cross-linking could be required to improve mechanical and chemical stabilities and/or selectivity [50–53]. After spinning, fibers are usually washed with deionized or milli-Q water, to exchange the residual solvent. Then a suitable drying procedure should be applied. Some authors suggest to gradually exchange water with low-surface tension liquids, such as methanol, ethanol, or hexane, in a multistep process, involving fibers soaked in different mixtures or pure liquids. See, for instance, Abed et al. [54] (ethanol and hexane), Xu et al. [55] (50% ethanol, pure ethanol, ethanol–hexane 1:1), Mansourizadeh et al. [56], and Wang et al. [57] (methanol and hexane). Other studies suggest the use of water/glycerol mixtures [28,38,40,43].

1.6 EXAMPLES OF MEMBRANE PREPARATION FOR PRESSURE-DRIVEN SEPARATION PROCESSES

Membrane pressure-driven processes, namely, MF, UF, NF, and RO are normally carried out in the liquid phase. Although water permeates through the membrane, other species are partially or completely rejected. According to Fane et al., The MF–UF range can be considered as a continuum [11]; both processes involve porous membranes. MF is carried out using symmetric membranes, with pore size ranging from 0.05 to 10 μm. UF, instead, requires asymmetric membranes, with pore size from 1 to 100 nm. The NF and RO spectrum is also considered as a continuum [11]. NF/RO membranes are usually thin-film composite (TFC) structures with nonporous skin. The most important features of pressure-driven membrane processes are resumed in Table 1.3.

MF is used to remove suspended solids, algae, and bacteria. Membranes can be prepared from either polymeric or inorganic materials. The structure is usually symmetric. UF also removes viruses, colloids, and macromolecules. The separation ability of a UF membrane is usually expressed in terms of molecular weight cut off (MWCO). This is defined as the molecular weight of the solute that is 90% retained by the membrane. Typical MWCO values for UF membranes ranges from 1 to 300 kDa [11]. Although for MF membranes permeability is usually affected by the entire membrane thickness, performance of the UF membranes mostly depends on the skin-layer properties.

RO membranes can remove even monovalent ions, as Na^+ and Cl^-; the typical application is seawater desalination. The literature defines these membranes as dense

TABLE 1.3
Typical Properties of Pressure-Driven Membrane Processes

	Microfiltration	Ultrafiltration	Nanofiltration	Reverse Osmosis
Pore size (nm)	50–10,000	1–100	~2	<2
Water permeability $(lm^{-2}h^{-1}bar^{-1})$	>500	20–500	5–50	0.5–10
Operating pressure (bar)	0.1–2.0	1.0–5.0	2.0–10	10–100
MWCO (Da)	Not applicable	1000–300,000	>100	>10
Targeted contaminants in water	Bacteria and algae, suspended solids	Bacteria, virus, colloids, and macromolecules	Di- and multivalent ions, natural organic matter, and small organic molecules	Dissolved ions and small molecules
Membrane materials	Polymeric, inorganic	Polymeric, some inorganic	Thin-film composite polyamide, cellulose acetate, and other materials	Thin-film composite polyamide and cellulose acetate

or with subnanometer pores. Depending on their performances, expressed in terms of permeability and rejection to NaCl, these membranes are usually divided into seawater RO (SWRO) membranes and brackish water RO (BWRO) membranes. NF can remove bi- and multivalent ions and small organic molecules, being located between the UF and RO range. Besides water softening, another relevant application of NF membranes is solvent separation, which requires materials with high chemical stability or solvent-resistance nanofiltration (SRNF) membranes. RO and NF membranes can be prepared by PI, to produce an integrally asymmetric structure, made up of one polymeric material, or by interfacial polymerization (IP), to obtain a TFC structure, with a skin of cross-linked aromatic PA over a support layer such as PSU on a reinforcing fabric.

1.6.1 MICROFILTRATION

MF represents a viable alternative to conventional processes for the treatment and recycling of water and wastewater [58,59]. Its increasing application is mainly connected to the progressively more severe regulation both regarding environmental safety and drinking water quality [60]. MF is also widely applied in the agro-industrial sector [61,62] and in biotechnology [63]. One of the main obstacles to the full application of MF is membrane fouling, which induces a rapid decline in productivity. Furthermore, the physical and chemical cleaning procedures, backflushing, and treatment with various cleaning agents are time-consuming and may reduce membrane lifetime, resulting in reduction of plant capacity and efficiency [64]. Apart from the adjustment of hydrodynamic conditions, most work has been devoted to the preparation of MF membrane with high productivity, desired selectivity, and low fouling, coupled with the enhancement of mechanical, chemical, and thermal stabilities.

Among the different polymeric materials, the most widely used for producing MF membranes through the PI process are PVDF, PSU, and PES, together with polycarbonate, CA, diacetate, triacetate, and their blends [11]. Hydrophilic surfaces are usually preferred for fouling reduction. However, although CA is naturally hydrophilic, PVDF, PSU, and PES can be modified using different additives. Susanto et al. [65] prepared MF membrane in PES, using triethylene glycol (TEG) and Pluronic®, a poly(ethylene oxide)-b-poly(propylene oxide)-b-poly(ethylene oxide) triblock copolymer, as nonsolvent and wettability modification agent, respectively. Membranes were prepared by combining the immersion precipitation technique with exposure to humid air before coagulation (combination of VIPS and NIPS). The best performances were observed for membranes prepared with the dope having the following composition: PES/NMP/TEG/Plu = 10/30/55/5 (wt%) with 3 min exposure in humid air (50%–60% RH) before coagulation (Figure 1.7).

Incorporation of nanoparticles (NPs) is reported as an effective method to improve the hydrophilicity, thus reducing the susceptibility to fouling of polymeric membranes. For instance, Hong and He [66] prepared PVDF MF membranes using PEG 600 as additive and ZnO particles as nanofiller. Similarly, Dong et al. [67] prepared PVDF membranes using PEG 600 as an additive and $Mg(OH)_2$ NPs as filler. PVDF MF membranes were prepared through TIPS using nano-TiO_2 particles as filler and dimethyl phthalate (DMP) as diluent by Shi et al. [68]. The same authors obtained interesting

FIGURE 1.7 Scanning electron microscopy images of the surface morphology (left) and cross-sectional morphology (right) of a PES membrane prepared from dope PES/NMP/Plu/ TEG = 10/30/5/55 wt% (Pluronic® PE6400) with 3 min of exposure to humid air. (Data from H. Susanto et al., *Journal of Membrane Science*, 342, 153–164, 2009.)

results by modifying the nano-TiO$_2$ particles using carboxyl-functional ionic liquid ([CH2COOHmim]Cl) [69]. Another strategy for improving the hydrophilicity of PVDF membranes is surface treatment. PVDF MF membranes with negative charge were produced by direct sulfonation with chlorosulfonic acid [70]. UV-photo-grafting modification on PVDF MF membranes was carried out using 2-dimethylaminoethyl methacrylate (qDMAEM) and 2-acrylamido-2-methyl-1-propanesulfonic acid (AMPS) and benzophenone as the photo-initiator by Hilal et al. [71].

1.6.2 ULTRAFILTRATION

Membranes for UF can be prepared through the PI process using polymers such as PVDF, PSU, PES, CA, but also polyacrylonitrile (PAN) and modified poly(ether ether ketone) [11,38]. The main difference with MF is the need for asymmetric membranes, with thin selective skin and support layer with reduced resistance. The main targets for preparing optimized membranes for UF are obtaining a sharp cut-off, improved mechanical and chemical resistances, and reduced susceptibility to fouling.

For instance, research on PVDF membrane production focuses mainly on the reduction of its intrinsic hydrophobicity to reduce fouling, by methods such as the introduction of nanosized alumina [72]. Al$_2$O$_3$–PVDF nanocomposite membranes have been successfully tested for the treatment of oily wastewater, showing better resistance to fouling [73]. Organic–inorganic PVDF–silica (SiO$_2$) composite hollow fibers showed improved hydrophilicity and permeability and enhanced antifouling properties in UF experiments [74]. PVDF membranes were modified by PVP and were used to purify flavonoids from crude *Ginkgo biloba* extraction products [75]. Coating with hydrophilic polymers, such as chitosan or poly(vinyl alcohol), has also been performed for fouling reduction [76,77]. In addition, other techniques, such as blending and grafting, have been tested to modify the surface properties of PVDF membranes [78–81].

Recently, the use of TiO_2 NPs for preparing mixed matrix membranes (MMMs) was reported by several authors, due to TiO_2 NPs commercial availability, good stability, and excellent photocatalytic, antibacterial, antifouling, and UV-cleaning properties. Song et al. [82] studied the preparation of photo-catalytically active PVDF/TiO_2 hybrid membranes, which were tested for natural organic matter removal in both dead-end and cross-flow UF experiments. PES is also widely used for the preparation of UF membranes [83,84]. Razmjou et al. [85] prepared PES UF membranes with TiO_2 NPs modified mechanically or both mechanically and chemically. PES MMMs were also prepared by coating TiO_2 NPs onto the membrane surface [86].

1.6.3 NF AND SOLVENT-RESISTANT NF

The most commonly used polymeric materials for preparing NF membranes are CA, polyimide and PA; other materials are polyvinyl alcohol (PVA), sulfonated PSU, and inorganic materials, such as some metal oxides.

NF membranes can be integrally asymmetric or TFC. A typical TFC membrane is made up of a dense selective skin (PA) on the top of a microporous PSU or PES layers; a nonwoven fabric acts as mechanical support (Figure 1.8).

The main advantages of TFC are connected to the possibility of controlling and optimizing the properties of each layer, in order to attain the desired selectivity and permeability coupled with excellent mechanical resistance. There are different techniques for preparing composite membranes, such as lamination, coating, and plasma polymerization. However, the IP technique is the most applied technique for preparing TFC membranes [5]. It is based on the reaction of monomers on the surface of the porous support film at the interface between two immiscible media. IP was first developed by Cadotte et al. [87,88], who found that composite membranes with high flux, high rejection to aqueous sulfate ions, coupled with low selectivity toward chloride ions, can be produced by interfacial cross-linking of piperazine (PIP) with trimesoyl chloride (TMC)/isophthaloyl chloride mixture. The basic mechanism for polymerization of PA layers, based on PIP and TMC, is shown in Figure 1.9.

Fully aromatic PA can be obtained from TMC and m-phenylenediamine (MPD), whereas semiaromatic PA can be obtained from TMC and PIP (Figure 1.10). The research on TFC membranes aims at enhancing membrane productivity and selectivity, improving resistance against chemicals as chlorine and solvents, and reducing

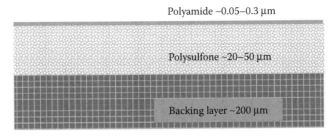

Polyamide ~0.05–0.3 µm

Polysulfone ~20–50 µm

Backing layer ~200 µm

FIGURE 1.8 Typical structure of a TFC membrane. (Data from A.G. Fane et al., Membrane technology for water: Microfiltration, ultrafiltration, nanofiltration, and reverse osmosis. In: P. Wilderer, ed., *Treatise on Water Science*, Academic Press, Oxford, 2011, pp. 301–335.)

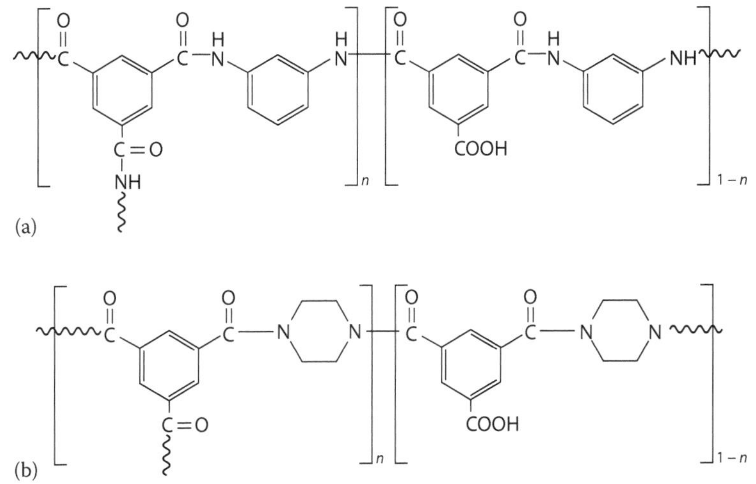

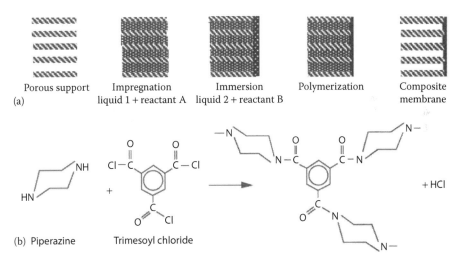

FIGURE 1.9 (a) Fully aromatic polyamide based on TMC and MPD. (b) Semiaromatic polyamide based on TMC and PIP. (Data from A.G. Fane et al., Membrane technology for water: Microfiltration, ultrafiltration, nanofiltration, and reverse osmosis. In: P. Wilderer, ed., *Treatise on Water Science*, Academic Press, Oxford, 2011, pp. 301–335.)

Porous support	Impregnation	Immersion	Polymerization	Composite
(a)	liquid 1 + reactant A	liquid 2 + reactant B		membrane

(b) Piperazine   Trimesoyl chloride   + HCl

FIGURE 1.10 (a) Basic mechanism for the preparation of TFC membranes by IP. (b) Reaction between piperazine (PIP) and trimesoyl chloride (TMC) for preparing semiaromatic polyamide layer. (Data from H. Strathmann et al., *Basic Aspects in Polymeric Membrane Preparation*, In: E. Drioli and L. Giorno, eds., *Comprehensive Membrane Science and Engineering*, vol. 1: Basic Aspects of Membrane Science and Engineering. Elsevier, Amsterdam, the Netherlands, 2010, pp. 91–112.)

the susceptibility to fouling. A recent review by Lau et al. summarized the development of TFC membrane technology over the last decade [89].

Other techniques for NF membrane preparation are photo- or thermal-grafting, dip-coating, electron beam irradiations, and plasma-initiated polymerization. Other strategies for improving the performance of NF membranes include blending and incorporation of nanofillers. Mansourpanah et al. [90] prepared NF membranes from a blend of PES/polyimide and modified them using ethylenediamine as a cross-linker and newly synthesized modifiers (PEG-triazine). Redox-initiated graft polymerization and sulfonation were compared as modifying methods for NF membranes by Van der Bruggen [91]. TiO_2 NPs were assembled on the surfaces of PES/polyimide NF membrane by Mansourpanah et al. [92]. PES/polyimide blend membranes were also –OH functionalized using different concentrations of diethanolamine.

Solvent-resistant NF (SRNF) represents a fairly new and interesting application of NF in different industrial fields (e.g., food, chemical, and pharmaceutical) for purification, recovery, or recycling of oligomers, catalysts, and solvents. This process requires membranes to be able to withstand aggressive environments, with high chemical resistance, coupled with desired permeability and selectivity. Not only TFC but also integrally skinned asymmetric membranes can be used for SRNF. The most widely used polymers for the preparation of SRNF membranes are polyimide, PAN, polyelectrolyte complex membranes (PECMs), and polydimethylsiloxane (PDMS).

Polyelectrolyte complexes (PECs) are a variegate group of multicomponent polymeric materials that can be used for preparing membranes for different applications. The main aspects regarding physical background and the preparation and application of PECMs were recently reviewed by Zhao et al. [93]. PECs are normally insoluble in common organic solvents; therefore, they represent promising materials for producing SRNF membranes [93]. Vankelecom et al. [94] prepared multilayered PEC SRNF membranes for filtration of polar aprotic solvents (DMF and tetrahydrofuran [THF]).

Recently, polymers belonging to the sulfone family, such as PSU, have also been reported for the preparation of SRNF membranes. Holda et al. [95] prepared SRNF membranes from PSU using a mixture of NMP and THF (70/30) as solvent.

Polyphenylsulfone (PPSU) (Figure 1.11) shows great potential for the preparation of SRNF membranes, since it has high chemical and mechanical resistance and capacity to operate at high temperatures. Darvishmanesh et al. [97] prepared PPSU hollow-fiber NF membranes and studied their isopropanol (IPA) permeability and rejection to Rose Bengal and Bromothymol blue, as well as their resistance to several solvents. Although the produced fibers were visually stable in most of the solvents except MEK, permeability tests showed that the membranes were not stable in acetone and toluene.

1.6.4   REVERSE OSMOSIS

RO is the most relevant membrane-based technique for seawater desalination [98]. Similar to NF, RO is carried out using asymmetric membranes with a nonporous skin layer. Membranes can be integrally skinned or TFC. The most important technique for the preparation of such membranes is IP, which has been already described in Section 1.6.3 devoted to NF membranes. As reported by Lee et al. [99], the studies about the preparation of polymeric membranes for RO application, from 1950 to 1980, focused on the search for optimum membrane materials. Subsequently, the performance of RO membranes was improved by controlling membrane formation reactions and using catalysts and additives.

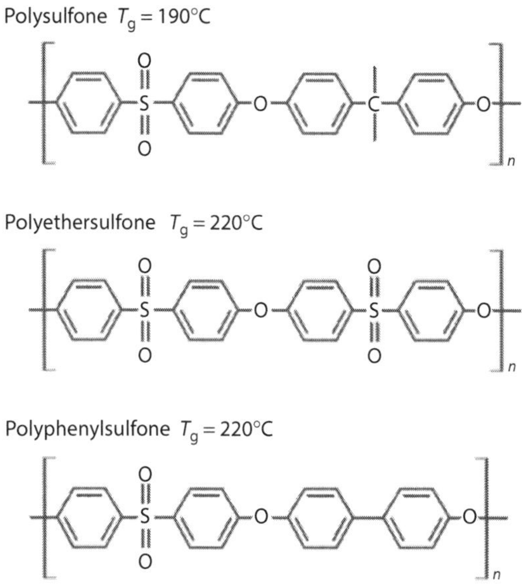

Polysulfone $T_g = 190°C$

Polyethersulfone $T_g = 220°C$

Polyphenylsulfone $T_g = 220°C$

FIGURE 1.11 Chemical structure of the sulfone family. (Data from S. Darvishmanesh et al., *Journal of Membrane Science*, 384, 89–96, 2011.)

As reported by Misdan et al. [100], the three main challenges of RO TFC membranes in the desalination industry are fouling propensity, boron rejection, and chlorination. The main strategies for the development of RO membranes with reduced tendency to fouling were reviewed by Kang and Cao [101]. These include (1) the development of new RO materials or improvement of IP process; (2) the surface modification of existing RO membranes by physical methods, adsorption and coating, or chemical methods, including grafting and plasma polymerization; and (3) the preparation of hybrid polymeric/inorganic membranes. For instance, Li et al. [102] reported on the preparation of TFC membranes using two novel synthesized tri- and tetra-functional biphenyl acid chloride—3,4′,5-biphenyl triacyl chloride 3,3′,5,5′-biphenyl tetraacyl chloride—and MPD. Lee et al. [99] proposed that zeolite membranes, thin-film nanocomposite membranes, carbon nanotube (CNT) membranes, and biomimetic membranes could offer an attractive alternative for improving the performance of RO membranes.

TiO_2 NPs are an interesting candidate to solve the problem of fouling and biofouling. Kwak et al. [103] prepared a polymeric/hybrid RO membrane composed of aromatic PA thin films underneath TiO_2 NPs. Kim et al. [104] prepared TFC aromatic PA membrane, with TiO_2 self-assembled on the surface. Moreover, Madaeni and Ghaemi [105] prepared PVA RO membranes with a coating of TiO_2 NPs. Buonomenna [106] highlighted that nanotechnology is one of the most promising strategies for producing the so-called nanoenhanced membranes (NEMs), which are RO membranes with enhanced properties. NEMs can be produced by exploiting inorganic materials, such as zeolites or TiO_2, as an alternative. Bioinspired NEMs can be produced using CNTs or aquaporins.

1.6.5   GAS SEPARATION

The application of polymeric membranes for selective separation of gas mixtures is becoming a viable alternative to traditional gas separation (GS) technologies, such as pressure swing adsorption and cryogenic separation. Membrane GS results in several advantages, as it does not require any phase transition and moving parts, and can, therefore, be also used in remote locations [107]. The development of new materials and membranes for GS is exhorted by the growing necessity for low emission plants and clean industrial processes.

As pointed out by Nunes and Peinemann [108], inorganic membranes are usually preferred because many processes at the industrial level are carried out at high temperature. However, polymeric membranes can be used for H_2/hydrocarbon separation in the platformer off gases from refineries and for CO_2 separation in coal plants. Polymeric membranes for GS can be symmetric or asymmetric, but should have a dense selective layer. Three types of membrane structures can be employed: (1) homogeneous dense membranes (symmetric); (2) integrally skinned asymmetric membranes; and (3) composite membranes.

The choice of the membrane material is of crucial importance, since the transport mechanism is based on the affinity between the membrane matrix and the permeating species. Besides investigating the most suitable material available, studies on GS membranes aim at finding the most appropriate techniques for producing defect-free membranes, showing high performance in terms of flux and selectivity (limited by typical trade-off), coupled with good mechanical/thermal/chemical resistance.

Brunetti et al. [107] produced integrally skinned asymmetric membranes from PEEK-WC and studied the influence of different preparation parameters, such as the composition and temperature of the coagulation bath and casting knife gap set, on the membrane morphology and transport properties, permeance, and selectivity. Membranes were prepared by immersion precipitation, using THF as solvent. Three different coagulation baths were tested: a mixture of methanol/water 70/30, pure methanol (MeOH), or pure IPA.

Interesting advances in the field of GS membrane materials are polymers with high-free volumes, such as poly(1-trimethylsilyl-1-propyne), poly(4-methyl-2-pentyne), and polymers of intrinsic microporosity [108]. Regarding the emerging application of CO_2 capture, the copolymer class Pebax® (Arkema) showed promising results. For instance, Bondar et al. [109] reported interesting values of CO_2/N_2 and CO_2/H_2 selectivity for different grades of Pebax membranes.

The preparation of MMMs using, for instance, zeolite fillers, seems another promising strategy for improving GS membrane performance and overcoming the typical trade-off between productivity and selectivity. Although the preparation of defect-free zeolite membranes is very difficult, MMMs offer the possibility of combining their superior transport properties to the simplicity of processing polymer membranes [110].

Besides zeolites, different types of carbon-based fillers, such as carbon molecular sieves (CMSs), fullerenes, and CNTs, have been recently reported as promising materials for preparing high-performance MMMs for GS. Vu et al. [111]

reported the preparation of polyimide membranes filled with CMS particles, which showed considerable improvement in permselectivity for CO_2/CH_4 and O_2/N_2 gas couples, and higher permeability to CO_2 and O_2, with respect to the neat membrane.

1.6.6 PERVAPORATION

Pervaporation (PV) is a membrane-based process used to separate the components of a liquid mixture. It requires dense membranes. The liquid feed is heated up and placed in contact with the active layer, whereas a vacuum or a sweep gas is applied downstream. The driving force is a chemical potential gradient through the membrane cross section. The separation phenomenon is explained according to the solution–diffusion model. The selective separation depends on the different dissolution of feed molecules into the membrane matrix and their diffusivity.

Based on the feed composition and the target of the separation, PV is generally classified into the following three categories: hydrophilic, hydrophobic, and organophilic. The first type is carried out to dehydrate organic compounds or to extract water from a mixture using hydrophilic membranes [112,113]. The second type is carried out using hydrophobic membranes and can be applied for recovering organic solvents, removing alcohol from alcoholic beverages, and recovering aroma compounds from fruit juices [114,115]. Finally, the third type is useful to separate organic/organic mixtures [116,117]. PV is now established as a good alternative to traditional separation processes, for instance, extraction or distillation, thanks to its ease of operation, lower costs, and reduced requirement of chemicals. An interesting application of PV is the separation of organic/organic mixtures made up of close-boiling points liquids or forming an azeotrope, because distillation is less-efficient and more expensive particularly in such cases.

Zereshki et al. [118] prepared poly(lactic acid) (PLA) homogenous dense membranes, which were used for selective separation of methanol/methyl *tert*-butyl ether (MeOH/MTBE). This separation is of great interest since the mixture of MeOH and MTBE forms a minimum boiling azeotrope. PLA/PVP membranes were applied for PV separation of ethanol/cyclohexane azeotropic mixture [119]. Ethanol/cyclohexane mixture separation through PV was also carried out using membranes produced from a blend between PEEK-WC and PVP [120]. Similar to what was observed for PLA/PVP membranes, although all the prepared membranes were ethanol selective, the addition of PVP improved membrane performance. Furthermore, PVP enhanced the selective adsorption of ethanol from the membrane, thus enhancing selectivity. PEEK-WC dense membranes were also used to carry out the separation of MeOH/MTBE azeotropic mixture through PV [121].

The search for new membrane materials is of peculiar interest to carry out organic/organic separations. When the feed contains particularly aggressive solvents, it is necessary to find membrane materials that are able to withstand such harsh environments. Simone et al. [122] prepared asymmetric dense membranes using a copolymer of ethylene chlorotrifluoroethylene (ECTFE), commercialized by Solvay Advanced Polymers as Halar®. ECTFE represents an exceptionally promising material for preparing membranes for separations involving organic solvents, due to its exceptional resistance to a broad group of aggressive and corrosive chemicals, coupled to high temperature resistance and durability.

The incorporation of nanofillers, for instance, zeolites, represents a promising strategy for improving membrane performance in PV. Dobrak et al. [123] prepared PDMS composite membranes and investigated the effect of two types of fillers, namely, commercial zeolite silicalite (CBV 3002) and laboratory-made colloidal silicalite-1, on membrane performance in the removal of ethanol from ethanol/water mixtures through PV. Filler incorporation increased membrane stability by cross-linking. Furthermore, the PDMS membrane filled with commercial zeolites showed a significant increase of selectivity. Incorporation of CBV 3002 fillers into a PDMS composite membrane was also found to enhance the performance in PV tests of toluene removal from water [124].

1.7 EXAMPLES OF MEMBRANE PREPARATION FOR MC

Among the new membrane operations, MCs have shown great potential for the development of environmentally sustainable technologies. An MC generally uses a polymer matrix to create an interface for the mass transfer and/or the reaction between two phases. MCs and related operations were recently described in detail by Drioli and Criscuoli [125–127]. As further references, the books by R. Baker [128] and N. Li [129] can be consulted.

MCs are an extension of the traditional concept of membrane operations. MCs can work with either hydrophobic or hydrophilic membranes, depending on the process and nature of the feed.

Depending on the physical state of the two phases, different types of MCs can be built up (Figure 1.12). Typical examples of MC include gas/liquid, liquid/liquid with immiscible phases, and liquid/liquid with miscible phases, including membrane distillation (MD), osmotic distillation (OD), and membrane crystallizers. There are

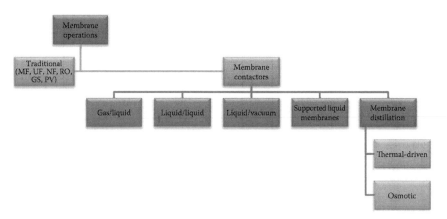

FIGURE 1.12 Membrane contactor categories.

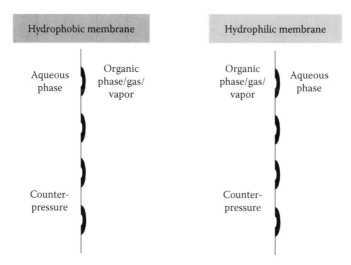

FIGURE 1.13 Hydrophobic and hydrophilic MC.

other possibilities, such as supercritical fluid/liquid systems, solid/fluid systems, and MCs with two immobilized phase interfaces. These systems are explained in the provided references [125–129].

If the membrane is hydrophilic, the aqueous/polar phase fills the membrane pores, whereas the gas/apolar phase is excluded. On the other hand, if the membrane is hydrophobic, the aqueous phase cannot penetrate the membrane pores. The phase that fills the membrane pores creates a meniscus inside each pore, due to capillary forces (Figure 1.13).

1.7.1 Gas/Liquid Contactors

In gas/liquid systems, a gas and a liquid stream are put in contact by means of a porous membrane. Usually the process employs a porous hydrophobic membrane and a nonwetting liquid phase. The main applications are oxygenations of aqueous solutions, degassing of water/aqueous solutions for ultrapure water production, carbonation of beverages and humidification/dehumidification of gas streams. Among the various hydrophobic polymers, PP and PTFE are the most popular membrane materials. PTFE and PP membranes are usually produced by stretching or thermal methods. On the contrary, PVDF flat-sheet and hollow-fiber membranes can be prepared through either TIPS or NIPS. PVDF is gaining considerable attention as a membrane material for application in different MC processes. Xu et al. [55] prepared PVDF hollow-fiber membranes using DMA as solvent and LiCl and PVP as additives. Asymmetric hollow-fiber membranes with inner skinless structures are favorable for carbon dioxide (CO_2) separation and absorption in gas–liquid MCs.

PVDF hollow-fiber membranes were prepared through TIPS from dopes containing GTA as diluent and glycerol or PEG as nonsolvent additives [130,131]. The prepared PVDF hollow-fiber membranes were tested as gas–liquid MCs for CO_2

absorption. It was shown that the performances of the prepared fibers are comparable to that of commercial PTFE fibers. Taking into account that PTFE membrane preparation is more difficult and expensive, PVDF membranes prepared by TIPS are promising candidates for use in MCs. In a recent work, hollow-fiber membranes prepared from a PVDF/GTA/glycerol ternary system by TIPS have also been successfully tested for propylene absorption [132].

1.7.2   LIQUID/LIQUID CONTACTORS

In liquid/liquid systems, one of the two phases is able to wet the membrane and fills the membrane pores. The other liquid is located at the other side of the membrane and an interface between two immiscible liquids is created at the entrance of each pore. Bey et al. [133] prepared PVDF hollow fibers using DMF as solvent and a mixture of water and PVP (K-17) as pore-forming additives, which were applied as contactors for the removal of As(V). The effect of BF composition and flowrate on fiber morphology and properties was examined. The MC performance was enhanced by working with thinner membranes, which was obtained by increasing the BF flowrate. A removal of about 70% in 6 h of operation was obtained working with a solution with arsenic 100 ppm as feed.

Hydrophilic PEEK-WC hollow fibers were applied as contactors for the removal of Cr(VI) [134]. Fibers were prepared by the dry/wet spinning technique using, for the first time, gamma-butyrolactone (GBL) as solvent. PVP (Luviskol K17) was used as additive. The prepared hydrophilic fibers were found to be very effective for Cr(VI) removal, as an extraction value up to 99% was achieved.

1.7.3   MEMBRANE DISTILLATION

MD is usually referred to as a specific type of liquid/liquid contactor working with miscible phases. MD can also work in other configurations, such as liquid-sweep gas, liquid-vacuum or liquid-air gap. In all cases, the membrane is not wetted by the liquid phases.

In addition to optimization of module design, which aims at improving flow geometry, and suppressing temperature and concentration polarization phenomena, the properties of the membrane material and membrane morphology are crucial for the process performance. Membranes for MD should have the following characteristics [135–137]:

- High hydrophobicity;
- Adequate pore size;
- Low thermal conductivity;
- High porosity;
- Suitable thickness;
- Good thermal stability and good chemical resistance to feed solution.

Membrane pore wetting is a fundamental problem in MD [139]. PTFE and PP membranes mostly fulfil the requisite of high hydrophobicity, but their preparation is complicated and expensive. PVDF meets the requirements of good hydrophobicity coupled with ease of manufacturing. Different studies reported several strategies for improving the hydrophobicity and properties of PVDF flat-sheet and hollow-fiber membranes for MD process.

For instance, Simone et al. [28] prepared PVDF hollow fibers using PVP and water as additives. The effect of PVP concentration on membrane morphology, in particular on macrovoids formation, was examined. Typical trade-off between thermodynamic enhancement and kinetic hindrance was observed; PVP concentration modulated macrovoids formation. The morphology of PVDF fibers can be tailored by varying the polymer molecular weight and concentration, the type and concentration of additives (water, PVP, maleic anhydride) in the dope and the composition of both coagulation bath and BF [139]. Wang et al. [57] prepared PVDF hollow fibers using NMP as solvent and ethylene glycol (EG) as additive. Hou et al. [140] prepared PVDF hollow fibers using DMA as solvent and exploiting the synergistic effect of LiCl and PEG 1500 as pore-forming additives.

The production of double-layer fibers, extruded using a triple spinneret, was also reported as a viable strategy for improving the performance of PVDF membranes in MD, to ensure mechanical support coupled with reduced resistance to transport, and to overcome the trade-off between fast diffusion and thermal insulation. The properties of different nanofillers, such as PTFE or hydrophobic/hydrophilic cloisite clay particles, can be used to better regulate the properties of both layers. Bonyadi and Chung [141] prepared double-layer fibers using hydrophobic cloisite 15A for the outer layer and hydrophilic cloisite NA+ for the inner layer. The produced fibers showed a direct contact MD (DCMD) flux as high as 55.2 kg/m^{2}h (feeding a solution of NaCl 3.5 wt%). Teoh et al. [142] produced composite double-layer fibers using PVDF and PTFE fillers for the outer layer. PTFE and cloisite fillers were used by Wang et al. [143] to tailor the properties of the outer and inner layer of PVDF hollow fibers, respectively.

Other strategies for producing hydrophobic membranes for MD are the modification of hydrophilic polymers or ceramic materials. Qtaishat et al. [144] produced two different types of hydrophobic surface-modifying macromolecules (SMMs) and prepared hydrophobic/hydrophilic polyetherimide composite membranes. The SMMs blended PEI membranes achieved better DCMD fluxes than those of a commercial PTFE membrane tested under the same conditions. Similarly fluorinated SMMs were used to modify hydrophilic poly(sulfone) [145]. Krajewski et al. [146] used 1H,1H,2H,2H-perfluorodecyltriethoxysilane to create a hydrophobic active layer on commercial tubular zirconia membranes supported on alumina. The produced membranes were tested in air-gap MD (AGMD). Hendren et al. [147] used 1H,1H,2H,2H-perfluorodecyltriethoxysilane, trichloromethylsilane, and trimethylchlorosilane to modify, by surface grafting, two types of alumina Anodisc™ ceramic membranes. The authors demonstrated that this surface treatment was effective and tested the produced membranes in DCMD.

1.7.4 OD and Membrane Crystallizers

OD is a specific type of MC working with miscible liquids; therefore, it requires hydrophobic membranes. OD has great potential for application in the food industry, because the absence of any thermal gradient mean that properties such as color, taste, flavor, and antioxidant activity are better preserved.

Commercial fibers or modules have been effectively used to carry out OD. For instance, Cassano et al. [148] used a Liqui-Cel® extra-flow MC equipped with microporous PP hollow fibers to concentrate UF-clarified cactus pear fruit juice. The same commercial HF-contactor was used to concentrate pomegranate juice (*Punica granatum* L.) [149]. PP membranes can be prepared through TIPS using suitable diluents as *N,N*-bis-(2-hydroxyethyl)tallow amine [5] and diphenylether [150].

Membrane crystallization (MCr) is an innovative technique based on contactors. It can be coupled with MD in desalination plants, allowing the recovery of the salts naturally present in sea- and brackish water, as high-quality crystals, leading to a more sustainable desalination process [3,99,151]. The correct choice of an NF/RO step to be integrated with an MD/MCr plant permits the lowering of the brine disposal cost [152]. Macedonio et al. [153] demonstrated that RO can be coupled with wind-aided intensified evaporation and membrane crystallizer technologies, allowing the reduction of brine discharge. Commercial HF modules equipped with hydrophobic PP hollow fibers, such as the membrane module MD020CP-2N, supplied by Microdyn, have been successfully used [3,151,153].

MCr can also be used on small molecules and proteins, resulting in high-quality crystals, which can be used for structural analysis or as pharmaceutical products [154]. The possibility of obtaining protein crystals with a sharp size distribution and high diffraction quality has been demonstrated [155,156]. The microporous membrane promotes heterogeneous nucleation, allowing the recovery of polymorphs that cannot be normally crystallized under standard conditions [157]. In other studies, commercial PP membranes were used as support for crystallization [155–157]. However, the preparation of PVDF membranes was reported as a suitable alternative to PP. For instance, Gugliuzza et al. [158] reported the preparation of functional perfluorinated membranes, based on PVDF and sulfonamide derivatives, by combining NIPS and wet chemistry techniques. The prepared membranes were tested for Hen Egg White Lysozyme (HEWL) crystallization. Curcio et al. [159] studied the influence of the structural properties of PVDF membranes on the heterogeneous nucleation rate of HEWL crystals. It was found that membrane properties influenced protein crystal growth rate.

Recently, the effects of material chemistry, surface nanostructure, and wetting regime effects on the heterogeneous nucleation of organic molecules were investigated by Di Profio et al. [160]. Different materials were tested, some membranes were lab-prepared through phase inversion techniques (Table 1.4); other membranes were commercial. An overview of the polymeric materials used in this investigation is given in Figure 1.14.

The ability of a membrane surface to promote, or hinder, heterogeneous nucleation during crystallization depends on an interplay between surface roughness and wettability. As shown by Curcio et al., hydrophobic surfaces are less effective in promoting heterogeneous nucleation [159]. Roughness can either enhance or prevent heterogeneous nucleation, depending on the contact angle. Hydrophilic surfaces have more tendency to adsorb molecules from the solution, in this case, increase of roughness enhances nucleation. On the other hand, when using hydrophobic supports, an increase of roughness reduces heterogeneous nucleation.

TABLE 1.4

Experimental Conditions Used by Di Profio et al. to Prepare Polymeric Membranes

Surface no.	Polymers (wt%)	Solvents (wt%)	Additives (wt%)	Techniques	Conditions
1	co-PVDF (14%)	DMA (86%)		VIPS	$23 \pm 3°C$; $50 \pm 5\%$ RH
2	co-PVDF (13.7%)	DMA (84.3%)	TBABr (2%)	VIPS	$23 \pm 3°C$; $50 \pm 5\%$ RH
3	co-PVDF (13.7%)	DMA (84.3%)	H_2O (2%)	VIPS	$23 \pm 3°C$; saturated water vapor atmosphere
4	co-PVDF (13.7%)	DMA (84.3%)	TBABr (2%)	VIPS	$23 \pm 3°C$; saturated water vapor atmosphere
5	co-PVDF (14%)	DMA (86%)		VIPS	$23 \pm 3°C$; saturated water vapor atmosphere
6	co-PVDF (14%)	DMA (86%)		SE + NIPS	SE: $23 \pm 3°C$; $15 \pm 1\%$ RH; NIPS: liquid water at $23 \pm 3°C$
9	PDMS (20%)	DCM (80%)		SE	$23 \pm 3°C$; $15 \pm 1\%$ RH
11	SPEEK-WC SD 1.0 (15%)	DMF (85%)		SE	$23 \pm 3°C$; $15 \pm 1\%$ RH
12	SPEEK-WC SD 0.8 (15%)	DMF (85%)		SE	$23 \pm 3°C$; $15 \pm 1\%$ RH
13	SPEEK-WC SD 0.4 (15%)	DMF (85%)		SE	$23 \pm 3°C$; $15 \pm 1\%$ RH
14	SPEEK-WC SD 0.2 (15%)	DMF (85%)		SE	$23 \pm 3°C$; $15 \pm 1\%$ RH
15	SPEEK-WC SD 0.1 (15%)	DMF (85%)		VIPS	$23 \pm 3°C$; $50 \pm 5\%$ RH
16	SPEEK-WC SD 0.1 (15%)	DMF (85%)		SE	$23 \pm 3°C$; $15 \pm 1\%$ RH
17	SPEES-WC SD 0.5	DMF (85%)		SE	$23 \pm 3°C$; $15 \pm 1\%$ RH
20	PI (12%)	NMP (88%)		NIPS	Liquid water at $23 \pm 3°C$
21	PI (18%)	NMP (82%)		NIPS	Liquid water at $23 \pm 3°C$
22	PI (15%)	NMP (85%)		NIPS	Liquid water at $23 \pm 3°C$
23	PI (18%)	DMA (82%)		NIPS	Liquid water at $23 \pm 3°C$
24	PI (21%)	NMP (79%)		NIPS	Liquid water at $23 \pm 3°C$
25	PI (23%)	NMP/dioxane (38.5%/38.5%)		SE + NIPS	SE: $23 \pm 3°C$; $15 \pm 1\%$ RH; NIPS: liquid water at $23 \pm 3°C$

Source: G. Di Profio et al., *Crystal Growth & Design*, 12, 3749–3757, 2012.

DCM, dichloromethane; DMA, *N,N* dimethylacetamide; DMF, *N,N* dimethylformamide; NMP, 1-methyl-2-pyrrolidone; PDMS, polydimethylsiloxane; PI, phase inversion; PVDF, polyvinylidene fluoride; RH, relative humidity; SD, spinodal decomposition; SE, solvent evaporation; SPEES-WC, sulfonated polyetherethersulfone with cardo group; SPEEK, Sulfonated polyether ether ketone; TBABr, tetrabutilamonium bromide; VIPS, vapor-induced PS.

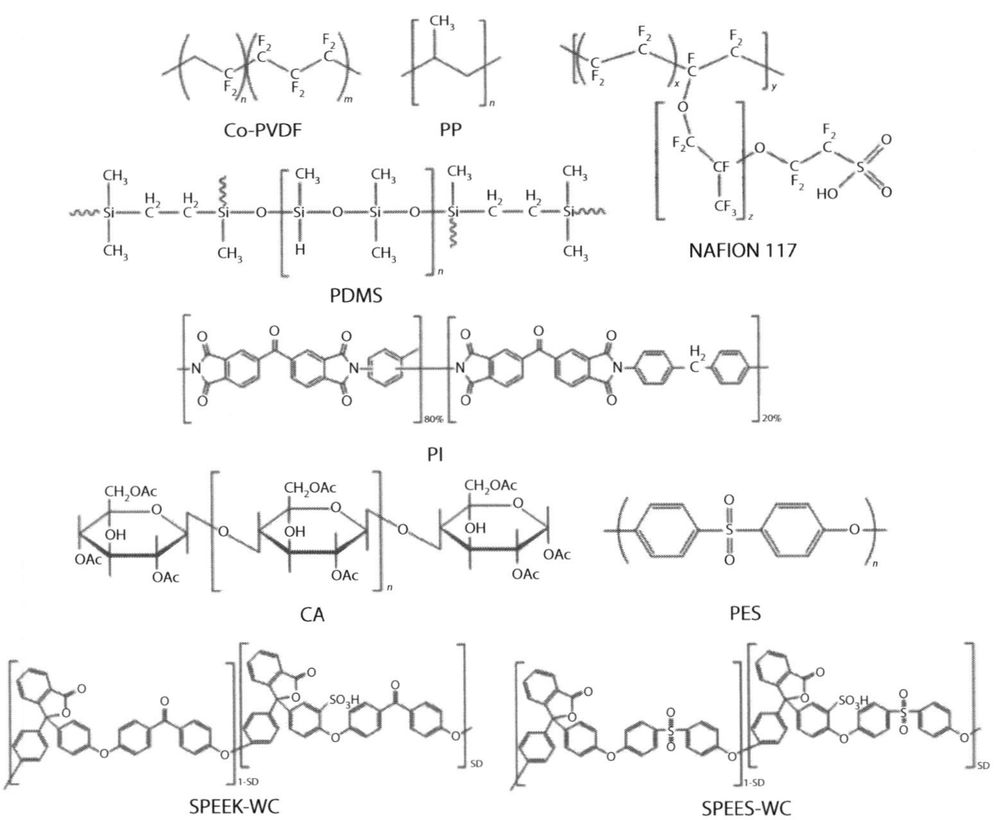

FIGURE 1.14 Polymeric membrane materials used by Di Profio et al. to study the effect of material chemistry on heterogeneous nucleation of acetaminophen, acetylsalicylic acid, and glycine. (Data from G. Di Profio et al., *Crystal Growth & Design*, 12, 3749–3757, 2012.)

1.8 CONCLUSIONS—OUTLOOK

In recent years, the improvement of membrane technology and its increasing cost effectiveness have resulted in a remarkable increase in the application of membrane technology at the industrial scale. This is mainly due to the development of selective and low-fouling membranes with higher efficiency. Numerous large-scale plants have been built to solve some of the most critical issues of our era, such as water shortage. Despite the impressive advancements in membrane preparation techniques for different applications, membrane scientists are expected to carry out important work in the coming years, especially regarding two aspects. The first one is obtaining a greater understanding of the fundamental mechanisms of membrane preparation. As shown in this chapter, this is the only way to achieve perfect control over the membrane-formation process and to tailor membrane features to obtain the best performance for each application. The second one is connected with concerns on the potential environmental impacts of large-scale membrane production plants. As widely accepted, membrane processes are more sustainable and are characterized by lower environmental impact. However, their preparation often requires toxic solvents that are expected to be banned in the next few years. Furthermore, most of the polymers are oil derived. Therefore, in the opinion of the authors, the only way to obtain green and sustainable alternatives to traditional processes by membrane technology is to look for low environmental-impact solvents as well as push research toward the use of biopolymers.

REFERENCES

1. M. Ulbricht, Advanced functional polymer membranes, *Polymer* 47 (2006) 2217–2262.
2. E. Drioli, G. Di Profio, and E. Fontananova, Membrane separations for process intensification and sustainable growth, *Fluid/Particle Separation Journal* 16 (2004) 1–18.
3. E. Drioli, E. Curcio, G. Di Profio, F. Macedonio, and A. Criscuoli, Integrating membrane contactors technology and pressure-driven membrane operations for seawater desalination energy, exergy and costs analysis, *Chemical Engineering Research and Design* 84 A3 (2006) 209–220.
4. A. Bottino, G. Capannelli, A. Comite, F. Ferrari, R. Firpo, and S. Venzano, Membrane technologies for water treatment and agroindustrial sectors, *C. R. Chimie* 12 (2009) 882–888.
5. H. Strathmann, L. Giorno, and E. Drioli, *Basic Aspects in Polymeric Membrane Preparation*, In: E. Drioli and L. Giorno (eds.), *Comprehensive Membrane Science and Engineering*, vol. 1: Basic Aspects of Membrane Science and Engineering. Amsterdam, the Netherlands: Elsevier, 2010; pp. 91–112.
6. Z. Wu, R. Faiz, T. Li, B.F.K. Kingsbury, and K. Li, A controlled sintering process for more permeable ceramic hollow fibre membranes, *Journal of Membrane Science*, http://dx.doi.org/10.1016/j.memsci.2013.05.040.
7. K. Li, *Ceramic Hollow Fiber Membranes and Their Applications*, In: E. Drioli and L. Giorno (eds.), *Comprehensive Membrane Science and Engineering*, vol. 1: Basic Aspects of Membrane Science and Engineering. Amsterdam, the Netherlands: Elsevier, 2010; pp. 253–273.
8. S. Kuiper, C.J.M. van Rijn, W. Nijdam, and M.C. Elwenspoek, Development and applications of very high flux microfiltration membranes, *Journal of Membrane Science* 150 (1998) 1–8.

9. J. Sa-nguanruksa, R. Rujiravanit, P. Supaphol, and S. Tokura, Porous polyethylene membranes by template-leaching technique: Preparation and characterization, *Polymer Testing* 23 (2004) 91–99.

10. J. Ren and R. Wang, Preparation of polymeric membranes, In: L.K. Wang, J.P. Chen, Y.T. Hung, and N.K. Shammas (eds.), *Handbook of Environmental Engineering*, vol. 13: Membrane and Desalination Technologies. Totowa, NJ: Humana Press, 2011.

11. A.G. Fane, C.Y. Tang, and R. Wang, Membrane technology for water: Microfiltration, ultrafiltration, nanofiltration, and reverse osmosis. In: P. Wilderer (ed.), *Treatise on Water Science*. Oxford: Academic Press, 2011; pp. 301–335.

12. P. van de Witte, P.J. Dijkstra, J.W.A. van den Berg, and J. Feijen, Phase separation processes in polymer solutions in relation to membrane formation, *Journal of Membrane Science* 117 (1996) 1–31.

13. C.M. Hansen, The three dimensional solubility-key to paint component affinities. 1. Solvent, plasticizers, polymers and resins, *Journal of Paint Technology* 39 (1967) 104.

14. Q. Wang, Z. Wang, and Z. Wu, Effects of solvent compositions on physicochemical properties and anti-fouling ability of PVDF microfiltration membranes for wastewater treatment, *Desalination* 297 (2012) 79–86.

15. Q.F. Alsalhy, K.T. Rashid, W.A. Noori, S. Simone, A. Figoli, and E. Drioli, Poly(vinyl chloride) hollow-fiber membranes for ultrafiltration applications: Effects of the internal coagulant composition, *Journal of Applied Polymer Science* 124 (2012) 2087–2099.

16. M.A. Frommer, R.M. Messalem, Mechanism of membrane formation. 6. Convective flows and large void formation during membrane precipitation, *Industrial and Engineering Chemistry Product Research and Development* 12 (1973) 328–333.

17. R. Matz, Hydronautics, inc., quarterly progress report 7005-4 to the office of saline water, April 1970.

18. R. Matz, The structure of cellulose-acetate membranes. 1. Development of porous structures in anisotropic membranes, *Desalination* 10 (1972) 1–15.

19. R.J. Ray, W.B. Krantz, and R.L. Sani, Linear-stability theory model for finger formation in asymmetric membranes, *Journal of Membrane Science* 23 (1985) 155–182.

20. C.A. Smolders, A.J. Reuvers, R.M. Boom, and I.M. Wienk, Microstructures in phase-inversion membranes. 1. Formation of macrovoids, *Journal of Membrane Science* 73 (1992) 259–275.

21. H. Strathmann, K. Kock, P. Amar, and R.W. Baker, The formation mechanism of asymmetric membranes, *Desalination* 16 (1975) 179–203.

22. J.M. Cheng, D.M. Wang, F.C. Lin, and J.Y. Lai, Formation and gas flux of asymmetric PMMA membranes, *Journal of Membrane Science* 109 (1996) 93–107.

23. H. Strathmann and K. Kock, Formation mechanism of phase inversion membranes, *Desalination* 21 (1977) 241–255.

24. F.G. Paulsen, S.S. Shojaie, and W.B. Krantz, Effect of evaporation step on macro- void formation in wet-cast polymeric membranes, *Journal of Membrane Science* 91 (1994) 265–282.

25. P. Neogi, Mechanism of pore formation in reverse-osmosis membranes during the casting process, *AIChE Journal* 29 (1983) 402–410.

26. R.M. Boom, I.M. Wienk, T. Vandenboomgaard, and C.A. Smolders, Microstructures in phase inversion membranes. 2. The role of a polymeric additive, *Journal of Membrane Science* 73 (1992) 277–292.

27. R.M. Boom, T. Vandenboomgaard, and C.A. Smolders, Mass-transfer and thermodynamics during immersion precipitation for a 2-polymer system—Evaluation with the system PES–PVP–NMP–water, *Journal of Membrane Science* 90 (1994) 231–249.

28. S. Simone, A. Figoli, A. Criscuoli, M.C. Carnevale, A. Rosselli, and E. Drioli, Preparation of hollow fibre membranes from PVDF/PVP blends and their application in VMD, *Journal of Membrane Science* 364(2010) 219–232, doi:10.1016/j.memsci.2010.08.013.

29. K.-W. Lee, B.-K. Se, S.-T. Nam, and M.-J. Han, Trade-off between thermodynamic enhancement and kinetic hindrance during phase inversion in the preparation of polysulfone membranes, *Desalination* 159 (2003) 289–296.

30. M. Sadrzadeh and S. Bhattacharjee, Rational design of phase inversion membranes by tailoring thermodynamics and kinetics of casting solution using polymer additives, *Journal of Membrane Science* 441 (2013) 31–44.

31. A.V. Patsis and E.H. Henriques, Interdiffusion in complex polymer systems used in the formation of microporous coatings, *Journal of Polymer Science Part B: Polymer Physics* 28 (1990) 2681–2689.

32. G.R. Guillen, G.Z. Ramon, H.P. Kavehpour, R.B. Kaner, and E.M.V. Hoek, Direct microscopic observation of membrane formation by nonsolvent induced phase separation, *Journal of Membrane Science* 431 (2013) 212–220.

33. H.C. Park, Y.P. Kim, H.Y. Kim, and Y.S. Kang, Membrane formation by water vapor induced phase inversion, *Journal of Membrane Science* 156 (1999) 169–178.

34. M. Mulder, *Basic Principles of Membrane Technology*. Dordrecht, the Netherlands: Kluwer Academic Publishers, 2003.

35. T.S. Chung, J.J. Qin, and J. Gu, Effect of shear rate within the spinneret on morphology, separation performance and mechanical properties of ultrafiltration polyethersulfone hollow fiber membranes, *Chemical Engineering Science* 55 (2000) 1077.

36. N. Peng, N. Widjojo, P. Sukitpaneenit, M.M. Teoh, G.G. Lipscomb, T.S. Chung, and J.Y. Lai, Evolution of polymeric hollow fibers as sustainable technologies: Past, present, and future, *Progress in Polymer Science* 37 (2012) 1401–1424.

37. G.R. Guillen, Y. Pan, M. Li, and E.M.V. Hoek, Preparation and characterization of membranes formed by nonsolvent induced phase separation: A review, *Industrial & Engineering Chemistry Research* 50 (2011) 3798–3817.

38. F. Tasselli, J.C. Jansen, and E. Drioli, PEEKWC Ultrafiltration hollow-fiber membranes: Preparation, morphology, and transport properties, *Journal of Applied Polymer Science*, 91 (2004) 841–853.

39. P. Sukitpaneenit and T.S. Chung, Molecular elucidation of morphology and mechanical properties of PVDF hollow fiber membranes from aspects of phase inversion, crystallization and rheology, *Journal of Membrane Science* 340 (2009) 192–205.

40. F. Tasselli, J.C. Jansen, F. Sidari, and E. Drioli, Morphology and transport property control of modified poly(ether ether ketone) (PEEKWC) hollow fiber membranes prepared from PEEKWC/PVP blends: Influence of the relative humidity in the air gap, *Journal of Membrane Science* 255 (2005) 13–22.

41. N. Peng, T.S. Chung, and J.Y. Lai, The rheology of Torlon® solutions and its role in the formation of ultra-thin defect-free Torlon® hollow fiber membranes for gas separation, *Journal of Membrane Science* 326 (2009) 608–617.

42. T.S. Chung and E.R. Kafchinski, The effects of spinning conditions on asymmetric 6FDA/6FDAM polyimide hollow fibers for air separation, *Journal of Applied Polymer Science* 65 (1997) 1555–1569.

43. F. Tasselli and E. Drioli, Tuning of hollow fiber membrane properties using different bore fluids, *Journal of Membrane Science* 301 (2007) 11–18.

44. N. Peng, T.S. Chung, and K.Y. Wang, Macrovoid evolution and critical factors to form macrovoid-free hollow fiber membranes, *Journal of Membrane Science* 318 (2008) 363–372.

45. N. Widjojo and T.S. Chung, Thickness and air-gap dependence of macrovoid evolution in phase-inversion asymmetric hollow fiber membranes, *Industrial & Engineering Chemistry Research* 45 (2006) 7618–7626.

46. K.Y. Wang, T. Matsuura, T.S. Chung, and W.F. Guo, The effects of flow angle and shear rate within the spinneret on the separation performance of poly(ethersulfone) (PES) ultrafiltration hollow fiber membranes, *Journal of Membrane Science* 240 (2004) 67–79.

47. W. Nijdam, J. de Jong, C.J.M. van Rijn, T. Visser, L. Versteeg, G. Kapantaidakis, G.H. Koops, and M. Wessling, High performance microengineered hollow fiber membranes by smart spinneret design, *Journal of Membrane Science* 256 (2005) 209–215.
48. P.Z. Culfaz, M. Wessling, and R.G.H. Lammertink, Hollow fiber ultrafiltration membranes with microstructured inner skin, *Journal of Membrane Science* 369 (2011) 221–227.
49. N. Widjojo, T.S. Chung, and W.B. Krantz, A morphological and structural study of Ultem/P84 copolyimide dual-layer hollow fiber membranes with delamination-free morphology, *Journal of Membrane Science* 294 (2007) 132–146.
50. K.K. Kopec, S.M. Dutczak, M. Wessling, and D.F. Stamatialis, Chemistry in a spinneret—On the interplay of crosslinking and phase inversion during spinning of novel hollow fiber membranes, *Journal of Membrane Science* 369 (2011) 308–318.
51. C. Cao, T.S. Chung, Y. Liu, R. Wang, and K.P. Pramoda, Chemical cross-linking modification of 6FDA-2,6-DAT hollow fiber membranes for natural gas separation, *Journal of Membrane Science* 216 (2003) 257–268.
52. L. Liu, R. Wang, and T.S. Chung, Chemically cross-linking modification of polyimide membranes for gas separation, *Journal of Membrane Science* 189 (2001) 231–239.
53. J. Peter and K.V. Peinemann, Multilayer composite membranes for gas separation based on crosslinked PTMSP gutter layer and partially crosslinked Matrimid® 5218 selective layer, *Journal of Membrane Science* 340 (2009) 62–72.
54. M.R. Moghareh Abed, S.C. Kumbharkar, Andrew M. Groth, and K. Li, Ultrafiltration PVDF hollow fibre membranes with interconnected bicontinuous structures produced via a single-step phase inversion technique, *Journal of Membrane Science* 407–408 (2012) 145–154.
55. A. Xu, A. Yang, S. Young, D. de Montigny, and P. Tontiwachwuthikul, Effect of internal coagulant on effectiveness of polyvinylidene fluoride membrane for carbon dioxide separation and absorption, *Journal of Membrane Science* 311 (2008) 153–158.
56. A. Mansourizadeh, A.F. Ismail, and T. Matsuura, Effect of operating conditions on the physical and chemical CO_2 absorption through the PVDF hollow fiber membrane contactor, *Journal of Membrane Science* 353 (2010) 192–200.
57. K.Y. Wang, T.-S. Chung, and M. Gryta, Hydrophobic PVDF hollow fiber membranes with narrow pore size distribution and ultra-thin skin for the freshwater production through membrane distillation, *Chemical Engineering Science* 63 (2008) 2587–2594.
58. W. Yuan and A.L. Zydney, Humic acid fouling during microfiltration, *Journal of Membrane Science* 157(1) (1999) 1–12.
59. H. Zhu, X. Wen, and X. Huang, Characterization of membrane fouling in a microfiltration ceramic membrane system treating secondary effluent, *Desalination* 284 (2012) 324–331.
60. N. Ma, Y. Zhang, X. Quan, X. Fan, and H. Zhao, Performing a microfiltration integrated with photocatalysis using an $Ag-TiO_2/HAP/Al_2O_3$ composite membrane for water treatment: Evaluating effectiveness for humic acid removal and anti-fouling properties, *Water Research* 44 (2010) 6104–6114.
61. B. Girard and L.R. Fukumoto, Membrane processing of fruit juices and beverages: A review, *Critical Reviews in Biotechnology* 20 (2000) 109–179.
62. Y. El Rayessa, C. Albasi, P. Bacchin, P. Taillandier, J. Raynal, M. Mietton-Peuchot, and A. Devatine, Cross-flow microfiltration applied to oenology: A review, *Journal of Membrane Science* 382 (2011) 1–19.
63. R. van Reis and A. Zydney, Bioprocess membrane technology, *Journal of Membrane Science* 297 (2007) 16–50.
64. M. Ulbricht, W. Ansorge, I. Danielzik, M. König, and O. Schuster, Fouling in microfiltration of wine: The influence of the membrane polymer on adsorption of polyphenols and polysaccharides, *Separation and Purification Technology* 68 (2009) 335–342.

65. H. Susanto, N. Stahra, and M. Ulbricht, High performance polyethersulfone microfiltration membranes having high flux and stable hydrophilic property, *Journal of Membrane Science* 342 (2009) 153–164.

66. J. Hong and Y. He, Effects of nano sized zinc oxide on the performance of PVDF microfiltration membranes, *Desalination* 302 (2012) 71–79.

67. C. Dong, G. He, H. Li, R. Zhao, Y. Han, and Y. Deng, Antifouling enhancement of poly(vinylidene fluoride) microfiltration membrane by adding $Mg(OH)_2$ nanoparticles, *Journal of Membrane Science* 387–388 (2012) 40–47.

68. F. Shi, Y. Ma, J. Ma, P. Wang, and W. Sun, Preparation and characterization of PVDF/TiO_2 hybrid membranes with different dosage of nano-TiO_2, *Journal of Membrane Science* 389 (2012) 522–531.

69. F. Shi, Y. Ma, J. Ma, P. Wang, and W. Sun, Preparation and characterization of PVDF/TiO_2 hybrid membranes with ionic liquid modified nano-TiO_2 particles, *Journal of Membrane Science* 427 (2013) 259–269.

70. G.N.B. Barona, B.J. Cha, and B. Jung, Negatively charged poly(vinylidene fluoride) microfiltration membranes by sulfonation, *Journal of Membrane Science* 290 (1/2) (2007) 46–54.

71. N. Hilal, V. Kochkodan, L. Al-Khatib, and T. Levadna, Surface modified polymeric membranes to reduce (bio)fouling: A microbiological study using *E. coli*, *Desalination* 167 (2004) 293–300.

72. Y. Ji-xiang, S. Wen-xin, Y. Shui-li, and L. Yan, Influence of DOC on fouling of a PVDF ultrafiltration membrane modified by nano-sized alumina, *Desalination* 239 (2009) 29–37.

73. L. Yan, S. Hong, M.L. Li, and Y.S. Li, Application of the Al_2O_3–PVDF nanocomposite tubular ultrafiltration (UF) membrane for oily wastewater treatment and its antifouling research, *Separation and Purification Technology* 66 (2009) 347–352.

74. L. Yu, Z. Xu, H. Shen, and H. Yang, Preparation and characterization of PVDF–SiO_2 composite hollow fiber UF membrane by sol–gel method, *Journal of Membrane Science* 337 (2009) 257–265.

75. Z. Xu, L. Li, F. Wu, S. Tan, and Z. Zhang, The application of the modified PVDF ultrafiltration membranes in further purification of Ginkgo biloba extraction, *Journal of Membrane Science* 255 (1/2) (2005) 125–131.

76. S. Boributh, A. Chanachai, and R. Jiraratananon, Modification of PVDF membrane by chitosan solution for reducing protein fouling, *Journal of Membrane Science* 342 (2009) 97–104.

77. J.R. Du, S. Pledszus, P.M. Huck, and X. Feng, Modification of poly/vinylidene fluoride) ultrafiltration membranes with poly(vinyl alchol) for fouling control in drinking water treatment, *Water Research* 43 (2009) 4559–4568.

78. G. Yuan, Z. Xua, and Y. Wei, Characterization of PVDF–PFSA hollow fiber UF blend membrane with low-molecular weight cut-off, *Separation and Purification Technology* 69 (2009) 141–148.

79. N. Li, C. Xiao, S. An, and X. Hu, Preparation and properties of PVDF/PVA hollow fiber membranes, *Desalination* 250 (2010) 530–537.

80. S.P. Nunes and K.V. Peinemann, Ultrafiltration membranes from PVDF/PMMA blends, *Journal of Membrane Science* 73(1) (1992) 25–35.

81. Y.C. Chiang, Y. Chang, A. Higuchi, W.Y. Chen, and R.C. Ruaan, Sulfobetaine grafted poly(vinylidene fluoride) ultrafiltration membranes exhibit excellent antifouling property, *Journal of Membrane Science* 339(1/2) (2009) 151–159.

82. H. Song, J. Shao, Y. He, B. Liu, and X. Zhong, Natural organic matter removal and flux decline with PEG–TiO_2-doped PVDF membranes by integration of ultrafiltration with photocatalysis, *Journal of Membrane Science* 405–406 (2012) 48–56.

83. H. Susanto and M. Ulbricht, Characteristics, performance and stability of polyether-sulfone ultrafiltration membranes prepared by phase separation method using different macromolecular additives, *Journal of Membrane Science* 327 (2009) 125–135.

84. A. Idris, N.M. Zain, and M.Y. Noordin, Synthesis, characterization and performance of asymmetric polyetehrsulfone (PES) ultrafiltration membranes with polyethylene glycol of different molecular weights as additives, *Desalination* 207 (2007) 324.

85. A. Razmjou, J. Mansouri, and V. Chen, The effects of mechanical and chemical modification of TiO$_2$ nanoparticles on the surface chemistry, structure and fouling performance of PES ultrafiltration membranes, *Journal of Membrane Science* 378 (2011) 73–84.

86. A. Razmjou, J. Mansouri, V. Chen, M. Lim, and R. Amal, Titania nanocomposite poly-ethersulfone ultrafiltration membranes fabricated using a low temperature hydrothermal coating process, *Journal of Membrane Science* 380 (2011) 98–113.

87. J.E. Cadotte, K.E. Cobian, R.H. Forester, and R.J. Petersen, Continued evaluation of insitu-formed condensation polymers for reverse osmosis membranes, Final report, San Francisco, CA: Office of Water Research and Technology, Contract No. 14-30-3298, PB-253193, April 1976.

88. J.E. Cadotte, M.J. Steuck, and R.J. Petersen, Research on in-situ-formed condensation polymers for reverse osmosis membranes, MRI Report No. PB-288387, pp. 10–17, March 1978.

89. W.J. Lau, A.F. Ismail, N. Misdan, and M.A. Kassim, A recent progress in thin film composite membrane: A review, *Desalination* 287 (2012) 190–199.

90. Y. Mansourpanah, S.S. Madaeni, A. Rahimpour, Z. Kheirollahi, and M. Adeli, Changing the performance and morphology of polyethersulfone/polyimide blend nanofiltration membranes using trimethylamine, *Desalination* 256 (2010) 101–107.

91. B. Van der Bruggen, Comparison of redox initiated graft polymerisation and sulfonation for hydrophilisation of polyethersulfone nanofiltration membranes, *European Polymer Journal* 45 (2009) 1873–1882.

92. Y. Mansourpanah, S.S. Madaeni, A. Rahimpour, A. Farhadian, and A.H. Taherin Formation of appropriate sites on nanofiltration membrane surface for binding TiO$_2$ photo-catalyst: Performance, characterization and fouling-resistant capability. *Journal of Membrane Science* 330 (2009) 297–306.

93. Q. Zhao, Q.F. Ana, Y. Ji, J. Qian, and C. Gao, Polyelectrolyte complex membranes for pervaporation, nanofiltration and fuel cell applications, *Journal of Membrane Science* 379 (2011) 19–45.

94. X.F. Li, S.D. Feyter, D.J. Chen, S. Aldea, P. Vandezande, F.D. Prez, and I.F.J. Vankelecom, Solvent-resistant nanofiltration membranes based on multilayered polyelectrolyte complexes, *Chemistry of Materials* 20 (2008) 3876–3883.

95. A.K. Hołda, B. Aernouts, W. Saeys, and I.F.J. Vankelecom, Study of polymer concentration and evaporation time as phase inversion parameters for polysulfone-based SRNF membranes, *Journal of Membrane Science* 442 (2013) 196–205.

96. Anonymous, Processing Guide for Polymer Membranes, Solvay Specialty Polymers, www.solvay.com.

97. S. Darvishmanesh, F. Tasselli, J.C. Jansen, E. Tocci, F. Bazzarelli, P. Bernardo, P. Luis, J. Degrève, E. Drioli, and B. Van der Bruggen, Preparation of solvent stable polyphe-nylsulfone hollow fiber nanofiltration membranes, *Journal of Membrane Science* 384 (2011) 89–96.

98. F. Macedonio, E. Drioli, A.A. Gusev, A. Bardow, R. Semiat, and M. Kurihara, Efficient technologies for worldwide clean water supply, *Chemical Engineering and Processing* 51 (2012) 2–17.

99. K.P. Lee, T.C. Arnot, and D. Mattia, A review of reverse osmosis membrane materials for desalination—Development to date and future potential, *Journal of Membrane Science* 370 (2011) 1–22.

100. N. Misdan, W.J. Lau, and A.F. Ismail, Seawater Reverse Osmosis (SWRO) desalination by thin-film composite membrane—Current development, challenges and future prospects, *Desalination* 287 (2012) 228–237.

101. G. Kang and Y. Cao, Development of antifouling reverse osmosis membranes for water treatment: A review, *Water Research* 46 (2012) 584–600.

102. L. Li, S.B. Zhang, X.S. Zhang, and G.D. Zheng, Polyamide thin film composite membranes prepared from 3,4′,5-biphenyl triacyl chloride, 3,3′,5,5′-biphenyl tetraacyl chloride and m-phenylenediamine, *Journal of Membrane Science* 289 (2007) 258–267.

103. S.Y. Kwak, S.H. Kim, and S.S. Kim, Hybrid organic/inorganic reverse osmosis (RO) membrane for bactericidal anti-fouling. 1. Preparation and characterization of TiO_2 nanoparticle self-assembled aromatic polyamide thin-film-composite (TFC) membrane, *Environmental Science & Technology* 35 (2001) 2388–2394.

104. S.H. Kim, S.Y. Kwak, B.H. Sohn, and T.H. Park, Design of TiO_2 nanoparticle self assembled aromatic polyamide thin-film composite (TFC) membrane as an approach to solve biofouling problem, *Journal of Membrane Science* 211 (2003) 157–165.

105. S.S. Madaeni and N. Ghaemi, Characterization of self-cleaning RO membranes coated with TiO_2 particles under UV irradiation, *Journal of Membrane Science* 303 (2007) 221–233.

106. M.G. Buonomenna, Nano-enhanced reverse osmosis membranes, *Desalination* 314 (2013) 73–88.

107. A. Brunetti, S. Simone, F. Scura, G. Barbieri, A. Figoli, and E. Drioli, Hydrogen mixture separation with PEEK-WC asymmetric membranes, *Separation and Purification Technology* 69 (2009) 195–204.

108. S.P. Nunes and K.V. Peinemann, *Advanced Polymeric and Organic–Inorganic Membranes for Pressure-Driven Processes*, In: E. Drioli and L. Giorno (eds.), *Comprehensive Membrane Science and Engineering*, vol. 1: Basic Aspects of Membrane Science and Engineering, Amsterdam, the Netherlands: Elsevier, 2010; pp. 113–129.

109. V. Bondar, B.D. Freeman, and I. Pinnau, Gas sorption and characterization of poly(ether-b-amide) segmented block copolymers, *Journal of Polymer Science Part B: Polymer Physics* 37 (1999) 2463–2475.

110. G. Clarizia, C. Algieri, and E. Drioli, Filler-polymer combination: A route to modify gas transport properties of a polymeric membrane, *Polymer* 45 (2004) 5671–5681.

111. D.Q. Vu, W.J. Koros, and S.J. Miller, Effect of condensable impurity in CO_2/CH_4 gas feeds on performance of mixed matrix membranes using carbon molecular sieves, *Journal of Membrane Science* 221 (2003) 233–239.

112. P. Gómez, R. Ibáñez, I. Ortiz, and I. Grossmann, Optimum design of PV processes for dehydration of organic mixtures, *Desalination* 193 (2006)152–159.

113. W. Kujawski, Pervaporative removal of organics from water using hydrophobic membranes. Binary mixtures, *Separation Science and Technology* 35 (2000) 89–108.

114. M. Catarino and A. Mendes, Dealcoholizing wine by membrane separation processes, *Innovative Food Science and Emerging Technologies* 12 (2011) 330–337.

115. M. Peng, L.M. Vane, and S.X. Liu, Recent advances in VOCs removal from water by pervaporation, *Journal of Hazardous Materials* B98 (2003) 69–90.

116. M. Niang and G. Luo, A triacetate cellulose membrane for the separation of methyl tert-butyl ether/methanol mixtures by pervaporation, *Separation and Purification Technology* 24 (2001) 427–435.

117. B. Smitha, D. Suhanya, S. Sridhar, and M. Ramakrishna, Separation of organic–organic mixtures by pervaporation—A review, *Journal of Membrane Science* 241 (2004) 1–21.

118. S. Zereshki, A. Figoli, S.S. Madaeni, S. Simone, and E. Drioli, Pervaporation separation of methanol/methyl tert-butyl ether with poly(lactic acid) membranes, *Journal of Applied Polymer Science* 118 (2010) 1364–1371.

119. S. Zereshki, A. Figoli, S.S. Madaeni, S. Simone, J.C. Jansen, M. Esmailinezhad, and E. Drioli, Poly(lactic acid)/poly(vinyl pyrrolidone) blend membranes: Effect of membrane composition on pervaporation separation of ethanol/cyclohexane mixture, *Journal of Membrane Science* 362 (2010) 105–112.

120. S. Zereshki, A. Figoli, S.S. Madaeni, S. Simone, M. Esmailinezhad, and E. Drioli, Effect of polymer composition in PEEKWC/PVP blends on pervaporation separation of ethanol/cyclohexane mixture, *Separation and Purification Technology* 75 (2010) 257–265.

121. S. Zereshki, A. Figoli, S.S. Madaeni, S. Simone, M. Esmailinezhad, and E. Drioli, Pervaporation separation of MeOH/MTBE mixtures with modified PEEK membrane: Effect of operating conditions, *Journal of Membrane Science* 371 (2011) 1–9.

122. S. Simone, A. Figoli, S. Santoro, F. Galiano, S. Alfadul, O.A. Al-Harbi, and E. Drioli, Preparation and characterization of ECTFE solvent resistant membranes and their application in pervaporation of water/toluene mixtures, *Separation and Purification Technology* 90 (2012) 147–161.

123. A. Dobrak, A. Figoli, S. Chovau, F. Galiano, S. Simone, I.F.J. Vankelecom, E. Drioli, and B. Van der Bruggen, Performance of PDMS membranes in pervaporation: Effect of silicalite fillers and comparison with SBS membranes, *Journal of Colloid and Interface Science* 346 (2010) 254–264.

124. S. Chovau, A. Dobrak, A. Figoli, F. Galiano, S. Simone, E. Drioli, S.K. Sikdar, and B. Van der Bruggen, Pervaporation performance of unfilled and filled PDMS membranes and novel SBS membranes for the removal of toluene from diluted aqueous solutions, *Chemical Engineering Journal* 159 (2010) 37–46.

125. A. Criscuoli, E. Drioli, and U. Moretti, Membrane contactors in the beverage industry for controlling the water gas composition, *Annals of the New York Academy of Sciences* 984 (2003) 1–16.

126. E. Drioli, A. Criscuoli, and E. Curcio, Membrane contactors and catalytic membrane reactors in process intensification, *Chemical Engineering & Technology* 26 (2003) 9.

127. E. Drioli, A. Criscuoli, and E. Curcio, Membrane contactors: Fundamentals, applications and potentialities, *Membrane Science and Technology Series 11*, Amsterdam, the Netherlands; Boston, MA: Elsevier, 2006.

128. R.W. Baker, *Membrane Technology and Applications*, 2nd Edition, chap. 13, New York: Wiley, 2004; p. 500.

129. N.N. Li, A.G. Fane, W.S.W. Ho, and T. Matsuura, *Advanced Membrane Technology and Applications*, chap. 26, Hoboken, NJ: Wiley, 2008; p. 687.

130. S. Rajabzadeh, T. Maruyama, T. Sotani, and H. Matsuyama, Preparation of PVDF hollow fibre membrane from a ternary polymer/solvent/nonsolvent system via thermally induced phase separation (TIPS) method, *Separation and Purification Technology* 63 (2008) 415–423.

131. S. Rajabzadeh, S. Yoshimoto, M. Teramoto, M. Al-Marzouqi, and H. Matsuyama, CO2 absorption by using PVDF hollow fibre membrane contactors with various membrane structures, *Separation and Purification Technology* 69 (2009) 210–220.

132. S. Rajabzadeh, M. Teramoto, M.H. Al-Marzouqi, E. Kamio, Y. Ohmukai, T. Maruyama, and H. Matsuyama, Experimental and theoretical study on propylene absorption by using PVDF hollow fiber membrane contactors with various membrane structures, *Journal of Membrane Science* 346 (2010) 86–97.

133. S. Bey, A. Criscuoli, A. Figoli, A. Leopold, S. Simone, M. Benamor, and E. Drioli, Removal of As(V) by PVDF hollow fibers membrane contactors using Aliquat-336 as extractant, *Desalination* 264 (2010) 193–200.

134. S. Bey, A. Criscuoli, S. Simone, A. Figoli, M. Benamor, and E. Drioli, Hydrophilic PEEK-WC hollow fibre membrane contactors for chromium (VI) removal, *Desalination* 283 (2011) 16–24.

135. M.S. El-Bourawi, Z. Ding, R. Ma, and M. Khayet, A framework for better understanding membrane distillation separation process, *Journal of Membrane Science* 285 (2006) 4.

136. S. Al-Obaidani, E. Curcio, F. Macedonio, G. Di Profio, H. Al-Hinai, and E. Drioli, Potential of membrane distillation in seawater desalination: Thermal efficiency, sensitivity study and cost estimation, *Journal of Membrane Science* 323 (2008) 85–98.

137. J. Zhang, N. Dow, M. Duke, E. Ostarcevic, J. Li, and S. Gray, Identification of material and physical features of membrane distillation membranes for high performance desalination, *Journal of Membrane Science* 349 (2010) 295–303.

138. M. Gryta and M. Barancewicz, Influence of morphology of PVDF capillary membranes on the performance of direct contact membrane distillation, *Journal of Membrane Science* 358 (2010) 158–167.

139. E. Drioli, A. Ali, S. Simone, F. Macedonio, S.A. AL-Jlil, F.S. Al Shabonah, H.S. Al-Romaih, O. Al-Harbi, A. Figoli, and A. Criscuoli, Novel PVDF hollow fiber membranes for vacuum and direct contact membrane distillation applications, *Separation and Purification Technology* 115 (2013) 27–38.

140. D. Hou, J. Wang, D. Qu, Z. Luan, and X. Ren, Fabrication and characterization of hydrophobic PVDF hollow fibre membranes for desalination through direct contact membrane distillation, *Separation and Purification Technology* 69 (2009) 78–86.

141. S. Bonyadi and T.S. Chung, Flux enhancement in membrane distillation by fabrication of dual layer hydrophilic–hydrophobic hollow fibre membranes, *Journal of Membrane Science* 306 (2007) 134–146.

142. M.M. Teoh, T.S. Chung, and Y.S. Yeo, Dual-layer PVDF/PTFE composite hollow fibers with a thin macrovoid-free selective layer for water production via membrane distillation, *Chemical Engineering Journal* 171 (2011) 684–691.

143. P. Wang, M.M. Teoh, and T.S. Chung, Morphological architecture of dual-layer hollow fiber for membrane distillation with higher desalination performance, *Water Research* 4 (2011) 5489–5500.

144. M. Qtaishat, D. Rana, M. Khayet, and T. Matsuura, Preparation and characterization of novel hydrophobic/hydrophilic polyetherimide composite membranes for desalination by direct contact membrane distillation, *Journal of Membrane Science* 327 (2009) 264–273.

145. M. Qtaishat, M. Khayet, and T. Matsuura, Novel porous composite hydrophobic/hydrophilic polysulfone membranes for desalination by direct contact membrane distillation, *Journal of Membrane Science* 341 (2009) 139–148.

146. S.R. Krajewski, W. Kujawski, M. Bukowska, C. Picard, and A. Larbot, Application of fluoroalkylsilanes (FAS) grafted ceramic membranes in membrane distillation process of NaCl solutions, *Journal of Membrane Science* 281 (2006) 253–259.

147. Z.D. Hendren, J. Brant, and M.R. Wiesner, Surface modification of nanostructured ceramic membranes for direct contact membrane distillation, *Journal of Membrane Science* 331 (2009) 1–10.

148. A. Cassano, C. Conidi, R. Timpone, M. D'Avella, and E. Drioli, A membrane-based process for the clarification and the concentration of the cactus pear juice, *Journal of Food Engineering* 80 (2007) 914–921.

149. A. Cassano, C. Conidi, and E. Drioli, Clarification and concentration of pomegranate juice (*Punica granatum* L.) using membrane processes, *Journal of Food Engineering* 107 (2011) 366–373.

150. W. Yave and R. Quijada, Preparation and characterization of porous microfiltration membranes by using tailor-made propylene/1-octadecene copolymers, *Desalination* 228 (2008) 150–158.

151. F. Macedonio, E. Curcio, and E. Drioli, Integrated membrane systems for seawater desalination: Energetic and exergetic analysis, economic evaluation experimental study, *Desalination* 203 (2007) 260–276.

152. F. Macedonio and E. Drioli, Hydrophobic membranes for salts recovery from desalination plants, *Desalination and Water Treatment*, 18 (2010) 224–234.
153. F. Macedonio, L. Katzirc, N. Geismac, S. Simone, E. Drioli, and J. Gilron, Wind-Aided Intensified eVaporation (WAIV) and Membrane Crystallizer (MCr) integrated brackish water desalination process: Advantages and drawbacks, *Desalination* 273 (2011) 127–135.
154. F. Rosenberg, Protein crystallization, *Journal of Crystal Growth* 166 (1996) 40.
155. E. Curcio, G. Di Profio, and E. Drioli, A new membrane-based crystallization technique: Tests on lysozyme, *Journal of Crystal Growth* 247 (2003) 166–176.
156. E. Curcio, S. Simone, G. Di Profio, E. Drioli, A. Cassetta, and D. Lamba, Membrane crystallization of lysozyme under forced solution flow, *Journal of Membrane Science* 257(1/2) (2005) 134–143.
157. S. Simone, E. Curcio, G. Di Profio, M. Ferraroni, and E. Drioli, Polymeric hydrophobic membranes as a tool to control polymorphism and protein–ligand interactions. *Journal of Membrane Science* 283(1/2) (2006) 123–132.
158. A. Gugliuzza, M. Aceto, S. Simone, E. Curcio, R. Madonna, G. Di Profio, and E. Drioli, Novel functional per-fluorinated membranes: Suitable nucleating systems for protein crystallization, *Desalination* 199 (2006) 200–203.
159. E. Curcio, E. Fontananova, G. Di Profio, and E. Drioli, Influence of the structural properties of poly(vinylidene fluoride) membranes on the heterogeneous nucleation rate of protein crystals, *Journal of Physical Chemistry B* 110 (2006) 12438–12445.
160. G. Di Profio, E. Fontananova, E. Curcio, and E. Drioli, From tailored supports to controlled nucleation: Exploring material chemistry, surface nanostructure, and wetting regime effects in heterogeneous nucleation of organic molecules, *Crystal Growth & Design* 12 (2012) 3749–3757.

2 Electrospinning
A Practical Approach for Membrane Fabrication

Luke Burke, Amir Keshvari,
Nidal Hilal, and Chris J. Wright

CONTENTS

2.1 Introduction 45
2.2 The Electrospinning Process 47
2.3 Materials 49
2.4 Electrospinning Parameters 50
    2.4.1 Applied Voltage 50
    2.4.2 Electrode Distance 52
    2.4.3 Solution Concentration and Conductivity 52
    2.4.4 Solution Mass Flow Rate 53
2.5 Modification of the Electrospinning System 54
    2.5.1 Emitter Modifications 54
        2.5.1.1 Coaxial Electrospinning 55
        2.5.1.2 Needleless Electrospinning 57
        2.5.1.3 Multiple Spinnerets 59
    2.5.2 Collector Modifications 60
        2.5.2.1 Controlled Deposition 60
        2.5.2.2 Electrostatic Deflection 62
    2.5.3 Electrospinning Polymer Modifications 63
2.6 Industrial and Commercial Applications 63
    2.6.1 Commercially Available Electrospinning Systems 65
2.7 Conclusions 66
References 67

2.1 INTRODUCTION

Electrospinning is a polymer fabrication technology that is enjoying a recent renaissance, and membrane technologists are at the forefront in applying this exciting and versatile technique. Electrospinning is capable of producing both polymer and composite membranes with many desirable characteristics for optimum performance in membrane separations of a wide range of industries. This chapter provides the reader with a practical background in the current *state of the art* of the technology.

The different electrospinning configurations are discussed highlighting their advantages and disadvantages, and the chapter concludes with a consideration of the future prospects of the technology and its application.

Nanofibers are generally defined as one-dimensional, flexible solid-state nanomaterials that are characterized by a diameter equal to or less than 100 nm and an aspect ratio greater than 100:1 (Zhou and Gong 2008). However, for many researchers, this definition is often extended to all submicron diameter fibers, as it is at these length scales that certain properties characteristic of nanostructured materials begin to arise (Nakajima et al. 1994).

The applications of submicron and nanofibers are wide ranging; not only are they a versatile group of materials for filtration (Barhate and Ramakrishna 2007), but they are also finding uses in regenerative medicine (Ramakrishna et al. 2010), tissue engineering (Pham et al. 2006), drug delivery (Sill and Von Recum 2008), wound dressing (Chen et al. 2008), advanced textiles (Lee and Obendorf 2007), contraception (Ball et al. 2012), and sensors (Choi et al. 2009a). One of the key advantages of nanofibers over microfibers is their increased surface area, which can be as much as 1000 times greater for nanofibers when compared to microfibers (Huang et al. 2003). Furthermore, fiber porosity, morphology, surface chemistry, mechanical performance, solvent resistance as well as the overall porosity of the nanofiber membrane can affect the performance of a nanofiber mat. Production of nanofibers has been achieved through a variety of techniques, including drawing (Xing et al. 2008b), template synthesis (Tao and Desai 2007; Grimm et al. 2008), liquid–liquid phase separation (Ma and Zhang 1999), self-assembly (Hartgerink et al. 2001; Zhang 2003), vapor-phase polymerization (Rollings et al. 2007; Rollings and Veinot 2008), doctor blading (Liewhiran et al. 2008), and electrospinning (Fong and Reneker 2001; Li and Xia 2004b). It should be noted that self-assembly allows the formation of nanofibers with a diameter of approximately 5–8 nm, which is significantly smaller than electrospun fibers. However, the complexity of the technique, coupled with low productivity and lack of structural and mechanical strength of the produced matrices, limits its application (Ma et al. 2005; Beachley and Wen 2010). In addition, drawing requires the polymers to be highly resistive against deformation, while producing only single fibers (Huang et al. 2003). Unlike self-assembly, phase separation is a relatively facile technique; however, it is effective only with a limited number of polymers and is not well suited to scale-up (Beachley and Wen 2010).

By contrast, with these methods, the electrospinning process allows control over fiber porosity, morphology, size, and overall nanofiber mat density; can be applied to a wide variety of materials; and is well suited to scale-up, all with a relatively inexpensive and straightforward setup. These excellent properties have encouraged researchers to date to apply electrospun fibers to an assortment of research areas. The overall advantages and disadvantages of each method are summarized in Table 2.1 (Ramakrishna et al. 2005).

Foundations of electrospinning can be traced back as far as the seventeenth century to William Gilbert's work "On the Magnet and Magnetic Bodies and on That Great Magnet the Earth," wherein it was reported that the application of an appropriately charged piece of amber to a fluid caused a cone to form on the surface and small droplets to be ejected from the cone's tip. The phenomenon was described

TABLE 2.1

A Comparison of the Most Popular Nanofiber Fabrication Methods

Process	Advances	Scalability	Repeatability	Convenience	Process Control
Drawing	Laboratory	No	Yes	Yes	No
Template synthesis	Laboratory	No	Yes	Yes	Yes
Phase separation	Laboratory	No	Yes	Yes	No
Self-assembly	Laboratory	No	Yes	No	No
Electrospinning	Laboratory, commercial	Yes	Yes	Yes	Yes

Source: Ramakrishna, S. et al., *An Introduction to Electrospinning and Nanofibres*, World Scientific Publishing, Singapore, 2005.

as *electrospraying*, and was later developed by C.V. Boys (1887) in the nineteenth century as *electrical spinning*. However, in 1934, Anton Formhals was the first to report the successful electrospinning of an artificial cellulose acetate thread from acetone solution (Formhals 1934). Further research into electrical spinning apparatus and the theoretical underpinnings continued over the following years until in 1964 when Sir Geoffrey Taylor (1964, 1966, 1969) published a series of papers, which provided a robust mathematical model of the process; a conical formation was observed when a fluid was subjected to sufficient electrical charge; this was named the *Taylor cone*. Finally during the beginning of the twentieth century, several research groups, most notably the Reneker group at the University of Akron, demonstrated the formation of fibers of a wide variety of organic polymers and popularized the term *electrospinning*. Since then publications on electrospinning have been increasing exponentially year on year (Li and Xia 2004b).

2.2 THE ELECTROSPINNING PROCESS

Electrospinning is a technique used to extrude micro to nanoscale fibers from solutions or melts by utilizing the electrostatic repulsion forces generated by applying high-intensity electric fields to the fluid in the vicinity of a grounded collection plate. A typical electrospinning setup is shown in Figure 2.1.

Briefly, a solution is fed at a controlled rate to maintain a droplet at the end of a conductive capillary, such as a blunt-end metallic needle. This capillary is directly connected to a high-voltage power supply in the range of 1–50 kV and acts as an electrode to charge the solution, although in certain situations a separate internal electrode is used to charge the solution instead. A grounded *collector*, usually a solid metallic target or grounded foil, is maintained between 1 and 30 cm away, exposing the highly charged droplet to a powerful electrostatic force. As the voltage is increased, like charges on the surface of the droplet repel each other, directly opposing surface tension and causing the pendant-like droplet to distend into a cone,

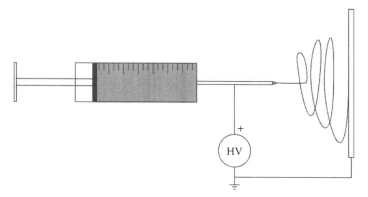

FIGURE 2.1   Single-emitter electrospinning apparatus.

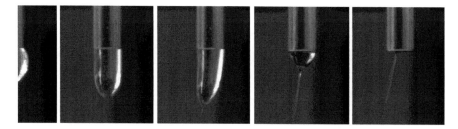

FIGURE 2.2   The formation of a Taylor cone from poly(ethylene oxide) (PEO) and repulsion of material at the tip of the needle of an electrospinner as applied voltage increases (profactor).

known as the *Taylor cone*, which is directed at the grounded collector (Taylor 1969) (Figure 2.2). At a critical voltage, the electrostatic force acting on the surface of the droplet overcomes surface tension, causing an expulsion of material from the cone's tip. Provided that the electrospinning solution is sufficiently viscous so as to prevent breakup into droplets due to Rayleigh instabilities, a consistent jet of the solution is formed (Yarin et al. 2001). This jet then travels through the air gap between the emitter and collector while undergoing a whipping instability, which allows the jet to thin and dry before depositing randomly orientated fibers comprised of the dried solute material on the grounded collector (Deitzel et al. 2001).

Figure 2.1 presents the most basic electrospinning system, that of single-needle electrospinning. The most commonly cited comprehensive review of this standard electrospinning process is that of Doshi and Reneker (1995), which describes the electrospinning process as follows: An electric field is used to create a charged jet of polymer solution. As this jet travels in air, the solvent evaporates leaving behind a charged fiber that can be electrically deflected or collected on a metal screen. Fibers with a variety of cross-sectional shapes and sizes were produced from different polymers. The diameter of these fibers was in the range from 0.05 to 5 microns. This relatively cheap and simple system is capable of producing a nonwoven fibrous mat from a wide variety of polymers within a vast range of diameters, between 10 nm and 5 μm (Liao et al. 2006) (Figure 2.3). Even with this simple electrospinning setup, the user

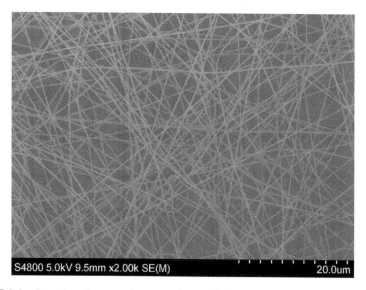

FIGURE 2.3 Scanning electron microscopy image of electrospun polymer, poly(acrylonitrile) nonwoven nanofiber mat produced by electrospinning.

is able to closely control fiber properties by altering variables such as acceleration, voltage, humidity, temperature, solution concentration, and composition, among others (Deitzel et al. 2001). However, this setup has some inherent drawbacks, for example, the deposition of the nanofibers is always random, the fiber material is either a single material or a homogenous mix and the process is rate limited to the flow rate of the solution through the syringe or capillary, which can be limiting for fast-drying or highly viscous polymers (Leach et al. 2011).

2.3 MATERIALS

The primary strength of electrospinning is its ability to form nanofibers from a wide variety of materials, provided the viscosity is sufficient. To date polymers (Reneker and Chun 1996), metals (Hui et al. 2010), and ceramics (Sigmund et al. 2006) have all been successfully electrospun in nano or microfibers. By far the most common category of materials that have been electrospun are medium- to high-molecular weight polymers, both synthetic and natural (Sell et al. 2010), and includes melts (Lyons and Ko 2005), blends (Bognitzki et al. 2001), and nanoparticle (Son et al. 2006) or drug-loaded polymers (Katti et al. 2004). A wide range of polymers are commonly used in electrospinning; however, certain polymer/solvent systems are more common, usually due to their appropriate molecular weight (and therefore viscosity), conductivity, and volatility of the solvent. These include polyurethanes (Schreuder-Gibson et al. 2002), polyamides (Tsai et al. 2002), polyester (Reneker and Chun 1996), poly(ethylene oxide) (Son et al. 2004), polystyrene (Megelski et al. 2002), poly(vinyl pyrrolidone) (Yu et al. 2009), poly(methylmethacrylate) (Gupta et al. 2005), poly(vinyl alcohol) (PVA) (Lee et al. 2004), poly(lactic-*co*-glycolic acid)

(Shin et al. 2006), polyacrylonitrile (Gupta et al. 2003), and poly(caprolactone) (Yoshimoto et al. 2003), as well as biological polymers, including chitosan (Bhattarai et al. 2005), silk fibroin (Min et al. 2004a), collagen (Matthews et al. 2002), and gelatin (Huang et al. 2004). In addition to these polymer-based systems, it is possible to electrospin precursor materials with or without polymer carriers, for example, Yang et al. (2004) electrospun zinc acetate/polymer poly(vinyl alcohol) (PVA)-blend fibers and used heat treatment to form zinc oxide (ZnO) nanowires; this technique is far faster than the standard hydrothermal growth used to produce ZnO nanowires, although the average wire diameter is greater and wires cannot be as accurately deposited.

2.4 ELECTROSPINNING PARAMETERS

Electrospinning is a versatile process, with many environmental and input variables contributing to the properties of the produced fibers (Figure 2.4). These parameters, when properly controlled, allow even a simple electrospinning setup to be used to produce a wide variety of fiber compositions, architectures, and densities. In general, the processes affecting fiber formation can be divided into three groups: (1) system parameters, comprising solution properties such as concentration of the solution, electrical conductivity, molecular weight, viscosity and surface tension, and also any equipment modifications such as diameter of the spinneret outlet, single or multiple spinnerets, and spinneret morphology (discussed later in Section 2.5); (2) process parameters such as applied voltage, solution flow rate, emitter–collector separation distance, temperature, humidity, and air velocity in the chamber; and (3) ambient parameters, which are less controllable, such as temperature, humidity, and pressure. Due to the emergence of nanotechnology and the popularity and versatility of the electrospinning process, over 200 research papers were published in the last decade on the fundamental underpinnings of electrospinning (Hohman et al. 2001a, 2001b; Yarin et al. 2001; Rutledge and Fridrikh 2007; Jaworek and Sobczyk 2008), as well as the effect of altering these conditions on the nanofiber diameter and morphology (Tan et al. 2005; Deitzel et al. 2001; Zhang et al. 2005; Fridrikh et al. 2003; Haghi and Akbari 2007). In particular the parameters used to control, and usually to reduce, the fiber diameter are of great interest (Theron et al. 2004; Ramakrishna et al. 2005; Tan et al. 2005; Baji et al. 2010; Bhardwaj and Kundu 2010).

2.4.1 APPLIED VOLTAGE

The driving force of fiber formation in electrospinning is the electrostatic force generated between the spinning material and collector. The role of voltage is linked to fiber formation process through the properties of the electrospun material, for example, solution conductivity, as well as the experimental setup, such as the emitter–collector separation distance. As far back as 1971, it was observed (Baumgarten 1971) that, in general, as the potential difference between the spinning solution and the collector is increased, the fiber diameter initially decreases and then increases as the voltage is raised further. This phenomenon is likely due to the increased electrostatic stress on the electrospinning solution, leading initially to an increased thinning of the polymer jet and resultant nanofibers, while also drawing an increased volume of material into

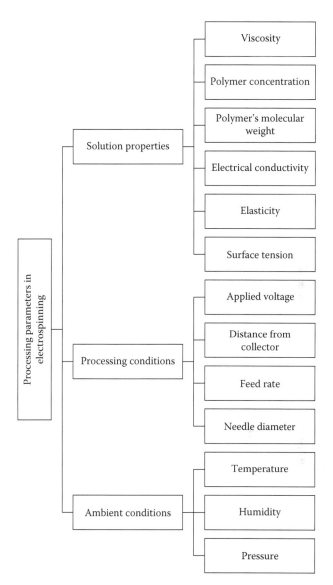

FIGURE 2.4 The electrospinning control parameters split into the cohorts of solution, processing, and ambient conditions. (Data from Haghi, A.K., ed., *Electrospinning of Nanofibres in Textiles*, Apple Academic Press, Toronto, Ontario, Canada, 2011.)

the electrospun fibers. This has been confirmed by further research groups (Yördem et al. 2008; Zhang et al. 2005; Buchko et al. 1999). Deitzel et al. (2001) electrospun a 7wt% poly(ethylene oxide) (PEO)/water solution and observed that at lower voltages, the initiation of the jet happened from a droplet formed on the syringe tip but at higher voltages, this phenomenon happened directly from the tip.

In addition to altering the magnitude of the supplied voltage, the effect of the electrical polarity was investigated in relation to the properties of electrospun fibers.

Kessick et al. (2004) found that using alternating current (AC) instead of the more common direct current (DC) produced changes in the surface charge of the electrospun jet, leading to a reduction in the whipping instability portion of fiber flight. More importantly, in DC electrospinning, the collector device must be conductive or semiconductive in order to dissipate the charge carried by the electrospun material, otherwise the collector will charge preventing further fiber deposition. However, using an AC charge buildup on the collector can be avoided, allowing nanofibers to be continuously deposited on even highly nonconductive materials (Kessick et al. 2004). Droplet formation and jet breakup can also be reduced by surface charge density (Fong et al. 1999), which can be increased by introducing electrolytes into the solution to increase conductivity (Zong et al. 2002).

Furthermore, Suppahol et al. (2005) investigated the effect of charging the electrospinning solution with a negative polarity relative to the collector plate. They found that when using a negatively charged emitter with a grounded collector, the fibers thus produced displayed a ribbon-like morphology with flat cross section. The suggested cause of this ribbon-like architecture of the fibers was due to the polymer drying as an outer layer of the jet and forming a skin, leaving the remaining solvent within the jet to evaporate, and thus allowing the hardened skin to be flattened by atmospheric pressure into the ribbon shape (Koombhongse et al. 2001). Although not stated by the researchers, it is worth noting that the formation of a flat ribbon-like fiber would increase the surface-to-volume ratio, potentially improving the fiber's properties in certain applications.

Finally, Killic et al. (2008) investigated the possibility of leaving the solution grounded, and instead charging the collector plate to produce the electrostatic attraction required for electrospinning. The group observed that the production efficiency, nanofiber diameters, and pore sizes in the electrospun nonwoven mat were all larger and more heterogeneously distributed than in the standard electrospinning setup.

2.4.2 ELECTRODE DISTANCE

The separation of the charged electrode and collector plate affect both the shape and intensity of the electric field and the total flight time available to the electrospinning jet. Reduction of the electrode distance can result in suppression of the later stages of fiber elongation and solvent evaporation. Generally this leads to increased fiber diameter and adhesion both to other fibers and the collector, due to incomplete solvent evaporation. This in turn leads to the formation of a denser, more interconnected fiber mesh (Buchko et al. 1999). In addition, by greatly reducing the distance, it is possible to remove the whipping instability stage of the electrospinning jet, allowing for more precise control over the position of deposited fibers at the expense of increased fiber diameter (Sun et al. 2006).

2.4.3 SOLUTION CONCENTRATION AND CONDUCTIVITY

The selection of solvent strongly affects the conductivity of a polymer solution as well as the morphology of fibers and electrospun membranes. A relatively simple method for reducing fiber diameter is to decrease the concentration of the electrospinning solution. However, in order to maintain electrospinning of fibers with a uniform diameter, a high enough concentration must be maintained in order to prevent jet deformations due to Rayleigh instability (Thompson et al. 2007). In lower concentrations, electrospinning will convert to electrospraying, causing the formation of microparticles due to lower viscosity and higher surface tensions in the polymer solution, and in some cases formation of beaded nanofibers (Yener and Jirsak 2012). The relationship between concentration and viscosity, which consequently affects the quality of fibers, can be estimated using Mark–Houwink equation

$$[\eta] = K M_w^{\alpha}$$

where:

η is the intrinsic viscosity

M_w is the molecular weight

K and α are constants that are a function of polymer and solvent viscosity and temperature and are empirically determined by measuring the molecular weight and viscosity over a wide range of molecular weights and fitting the best line in Mark–Houwink equation

Generally, polymers used for electrospinning have α value between 0.5 and 0.8 (Kwaambwa et al. 2007).

For instance, Tacx et al. (2000) have reported the following relationship for PVA in water.

$$[\eta] = 6.51 \times 10^{-4} M_w^{0.628}$$

In higher concentrations, formation of a mixture of beads and fibers frequently occurs and in very high concentrations helix-shaped microribbons starts to appear (Li and Wang 2013). In some cases, fibers with a so-called *bead-on-a-string* morphology, or even *electrosprayed* nanospheres are desired, which can be achieved by intentionally reducing the solution concentration (Jaworek and Sobczyk 2008).

2.4.4 Solution Mass Flow Rate

There are few systematic studies on the relationship between flow rate of the electrospinning solution to the capillary tip and the effect on fiber diameter or morphology. However, Megelski et al. (2002) varied the solution flow rate of a polystyrene/tetrahydrofuran polymer system and found that higher flow rates lead to increased fiber diameter and fiber porosity up to a critical value as a result of which large beads began to appear along the fibers. It should be noted that the correlation of increasing pore size with increasing flow rate may not be a generally applicable rule, as the experiment was specifically designed to result in porous fibers. More recent research by Theron et al. (2004) into the effect of solution flow rate yielded similar results; however, their theoretical analysis illustrates that the cause of the increase in fiber diameter is due to the decrease in charge density of the fiber jet, as flow rate is increased and follows a power-law relationship. The authors note that their current measurements were performed as averages and that by measuring the tip-collector current and voltage in real time, more accurate modeling may be possible. In addition, Haghi et al. (2011) conclude that when the flow rate exceeds a critical value, the rate of solution removal from the capillary tip limits the rate of solution delivery to the tip, which leads to the formation of unstable jets, and thus largely beaded nanofibers.

2.5 MODIFICATION OF THE ELECTROSPINNING SYSTEM

The electrospinning system discussed thus far is comprised of three components, namely, a capillary emitter, an electrically grounded collector, and a high-voltage source (Doshi and Reneker 1995). Although this basic setup is capable of producing an impressive range of nanostructured materials without modification, it is the modular and modifiable nature of the electrospinning apparatus that is arguably its greatest strength.

As described in Section 2.2, the basic electrospinning system produces a nonwoven, randomly deposited mat comprised of nanofibers of the electrospun material. However, certain applications in a variety of fields often require further adaptations such as ordered arrays of nanofibers (Li et al. 2004); twin-component fibers comprised of multiple, discrete materials (McCann et al. 2005a); or controlled pore size of the resulting nanofiber mat (McCann et al. 2005b). Furthermore the electrospinning process, though much faster than other nanofiber assembly methods, has a much lower throughput than is required at a commercial or industrial level, and the capillary mechanism is prone to blocking, requiring frequent maintenance (Teo and Ramakrishna 2006). Each of these situations can be addressed by making specific modifications to the standard electrospinning setup, allowing the researcher or industrialist to further tailor the already-excellent results of electrospinning to suit their desired outcome. These modifications can generally be divided into two sections, namely, emitter modifications, the capillary through which the electrospun material is charged and ejected (Section 2.5.1), and collector modifications, where the material is deposited and electrical charge is released. Section 2.5.1 explores some well-established as well as experimental adaptations that have been applied to the electrospinning system.

2.5.1 EMITTER MODIFICATIONS

Emitter, also known as the tip or spinneret of the electrospinning system, is the point from which the electrospun material is ejected. In the basic electrospinning system, the emitter is usually a blunt-end needle that is directly connected to a high-voltage source. The emitter is required to expose the electrospinning material to a sufficiently powerful electric field to cause the Taylor cone formation and to create an electrically forced jet of material to be ejected in the direction of the collector, as well as to maintain a constant supply of material to the Taylor cone to allow fiber formation to continue for as long a time period as possible. A frequent problem with emitters, particularly capillaries or needles, is blockage caused by the hardening of electrospinning material before being ejected. It is also usually the *bottleneck* of an electrospinning system, limiting the maximum speed of fiber formation for the whole system. By modifying the emitter, however, it is possible to overcome these issues as well as produce more advanced fiber compositions and architectures.

2.5.1.1 Coaxial Electrospinning

Coaxial electrospinning is the technique of forming a nanofiber comprised of two or more distinct layers, which are arranged around a common central longitudinal axis, similar to the insulation around an electrical cable. An example of such a fiber can be seen in Figure 2.5.

There are two methods of achieving this fiber architecture; the first is to use phase-separating or immiscible materials, such as polymer blends (Bazilevsky et al. 2007). The architecture of these fibers then depends on the blend ratio of the solutes, with higher percentage content of the sheath material leading to a thicker sheath layer; similarly for the core component, there is, however, a critical upper value for each, with levels exceeding this value leading to a co-continuous fiber of a single material containing localized pockets of the less-concentrated solute (Wei et al. 2006). Coaxial electrospinning requires careful selection of electrospun materials and blend ratios, and allows limited control over the eventual architecture of the fiber. For example, the material with lower viscosity invariably migrates to the outer (sheath) layer of the nanofiber, with the more viscous material forming the core. In addition, as two materials are dissolved in a single solvent, it is necessary for both the materials to be soluble in the same solvent, with the varying degrees of solubility dictating whether a material is located at the core or sheath portion of the fiber. Nonetheless, this technique is powerful as it allows higher levels of fiber architecture to be achieved without modifying the electrospinning apparatus whatsoever and has been used to improve several aspects of

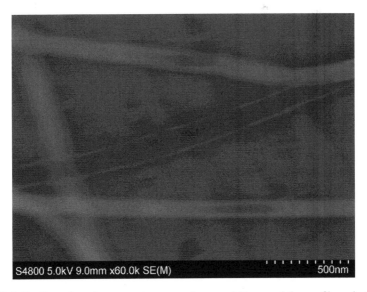

S4800 5.0kV 9.0mm x60.0k SE(M)   500nm

FIGURE 2.5 Scanning electron microscopy image of the coaxial nanofiber electrospun from PEO/chitosan—the contrast between the fiber's edges and center reveal the coaxial structure.

electrospun fibers, including biocompatibility and hemostatic behavior of PEO fibers by coating in the biopolymer chitosan (Zhang et al. 2009).

Alternatively, the use of a compound spinneret comprising of inner and outer nonmixing lumens to electrospin compound core–shell fibers was first described by Sun et al. (2003). The overall system is nearly identical to the standard electrospinning setup, with the only change being applied to the capillary or needle through which the electrospinning solution is delivered. A schematic diagram of the coaxial electrospinning system setup is shown in Figure 2.6.

By delivering two distinct materials to the inner and outer portions of the capillary separately, a compound Taylor cone forms from the two materials. From this point, the electrospinning process is remarkably similar; the compound Taylor cone erupts at a critical voltage to form a compound jet, which undergoes a whipping instability, removing the solvent and producing nanofibers with a core–sheath architecture of the two materials (McCann et al. 2005a). Although the coaxial electrospinning technique is not severely restrictive in terms of compatible materials, some considerations of which material/solvent combinations are compatible is necessary. Naturally, immiscibility is an important factor to ensure a sharp boundary between the two materials in the produced nanofiber (Xin et al. 2008); however, polymer blends can also lead to further complexities in the whipping instability of the jet caused by electrostatic repulsion forces between the two materials (Yu et al. 2011), which can result in tightly coiled or otherwise deformed nanofibers.

Coaxial nanofibers have several advantages over conventional, single-component fibers. Perhaps the most notable advantage is the ability of coaxial spinnerets to electrospin a usually unsuitable material using an easily spinnable but sacrificial material as a sheath (Zhang et al. 2006; Greiner et al. 2006; Kowalczyk et al. 2008). Similarly,

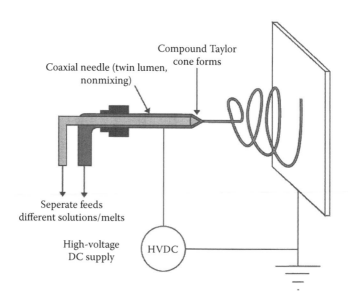

FIGURE 2.6 Schematic representation of the coaxial electrospinning system for the formation of core/sheath nanofibers.

using a sacrificial core material, hollow nanotubes may be formed of inorganic and ceramic materials (Li and Xia 2004a), including the much-publicized carbon nanotubes (Yeo and Friend 2006). It is notable that such nanotubes or nanotube templates interact similarly to standard electrospun fibers and as such can also be subjected to collector modifications, for example, to collect uniaxially aligned arrays or bundles of nanotubes. These modifications are discussed in detail in Section 5.2. Such nanotubes have potential uses in nanofluidics, catalysis, encapsulation, drug delivery, bioseparation (Mitchell et al. 2002), and filtration (Li et al. 2005a) applications. Furthermore, materials with desirable surface chemical properties, such as biocompatibility or antibacterial attributes, are often not as mechanically strong as other polymers; therefore, using a structurally strong synthetic polymer as a core material, coupled with a sheath of active material, it is possible to harness the strengths of both materials in a single product (Zhang et al. 2004). This is also advantageous for materials, in which the active material is highly expensive, or in limited supply as all the material is in contact with the fiber's surroundings, leaving very little active material to be wasted on the internal fiber structure. Finally, by utilizing a sheath with controlled thickness produced from a material with a known degradation rate, controlled release of an active component from the core layer can be achieved (Jiang et al. 2006).

Another technique, which is similar to coaxial electrospinning, is that of aero-electrospinning, as described by He et al. (2007). This technique, which is primarily used in solvent-less melt electrospinning, uses a sheath of hot air passed through the outer portion of the coaxial needle. This hot air not only assists in preventing the melted material from cooling and hardening in the needle but also serves to extend the linear portion of the electrospinning jet once it has exited the needle. The authors claim that the use of aero-electrospinning yielded reduced fiber diameters and increased control over the instability and eventual collector contact point.

A significant drawback of the coaxial electrospinning method is the increased complexity of the coaxial capillary or needle compared to the standard single-lumen electrospinning setup. This can result in more complex cleaning procedures and an increased tendency for blockage (Liu and Hsieh 2002).

2.5.1.2 Needleless Electrospinning

A major concern with conventional electrospinning setups is the tendency of the electrospun solution to dry within the needle or aperture, thereby causing blockage (Kim et al. 2007; Liu and Hsieh 2002). The problem is so pervasive in the electrospinning technique that certain solute/solvent combinations are termed *un-spinnable* and ruled out entirely. In addition, the single-jet electrospinning methodology is very limited in terms of throughput, usually operating in 0.1–1 g/hr in terms of collected fiber weight (Zhou et al. 2009). To counter this, it is possible to design an electrospinning system to spin directly from a reservoir without using an aperture. This is known as the *needleless electrospinning* method (Zhou et al. 2010).

The general principle of needleless electrospinning is to initiate the formation of a Taylor cone on the surface of the electrospinning solution by increasing the electric field strength at a localized point. This can be achieved either by reducing the distance between the charged polymer solution and the grounded collection point or by inducing curvatures in the fluid surface. The electric field density (D) of a sphere of radius (R) and charge (Q) is calculated as follows:

$$D = \frac{Q}{4\pi R^2}$$

Therefore, areas with higher degrees of curvature (equivalent to lower radius) experience a higher electric field density, causing Taylor cones to form preferentially at these points (Niu and Lin 2012). These areas of higher electric field density are termed *spinning sites* (Thoppey et al. 2010); generally speaking, the more spinning sites formed by a technique, the greater the production rate of fibers (Figure 2.7).

There are two primary variations on the needleless electrospinning method, the first uses a structure such as an inclined surface or plate (Thoppey et al. 2011), cone, rotating cylinder, or other geometric object to carry the electrospinning solution into the proximity of the grounded collector and provides an area of high curvature. Schematics of both plate-edge and drum electrospinning are illustrated in Figure 2.8. These represent just two possible electrode configurations of needleless electrospinning; other popular variations include a sharp-edged rotating disc rather than a drum to further increase curvature at the disc edge, thereby increasing the electric field density and increasing spinning sites (Yee et al. 2008), saw-toothed discs (Formhals 1934) to further increase areas of high curvature, rotating cones (Lu et al. 2010), spheres (Niu et al. 2012), wire-coils (Lin et al. 2010), and overflowing bowls, where the spinning sites are formed as the solution flows over the sharp-edge bowl (Thoppey et al. 2011). In some experimental work, multiple spinnerets of these types have been employed, for example, a *waterfall* system of multiple plates stacked underneath each other or several parallel discs rotating within a single reservoir, to further increase fiber production rate (Thoppey et al. 2010).

FIGURE 2.7 Multiple spinning sites or Taylor cones being formed by the spinning mandrel of a bowl electrospinner (El Marco NanoSpider™ Lab 200 system)—electrospinning fibers upward onto a collector above the bath.

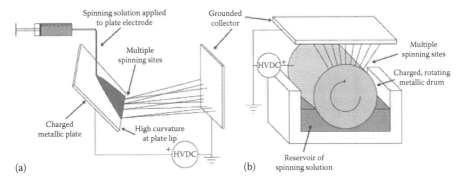

FIGURE 2.8 Schematic representation of (a) plate and (b) rotating-drum variations of the needleless electrospinning system.

The second method of needleless electrospinning is to induce perturbations in the surface of the charged fluid to momentarily create points of high curvature as a result of which a Taylor cone is formed. This has been achieved using bubbles (Yang et al. 2009), which is shown to form the smallest electrospun fibers to date with a 20 nm diameter (Yang et al. 2011). Another method is to use surface vibrations (Wan et al. 2006), which has the added advantage of effectively lowering the surface tension of the fluid, allowing the formation of Taylor cones more easily. He et al. (2010) state that the threshold voltage ($E_{\text{threshold}}$) displays a power-law relationship with surface tension (γ) of the form

$$E_{\text{threshold}} \propto \gamma^{1/2}$$

Each of these methods overcome the difficulties of aperture-based systems and produce electrospun fibers at a much greater rate due to the formation of multiple Taylor cones simultaneously. Thus, such systems are more amenable to scale-up for membrane fabrication. However, the lack of an aperture reduces control over the process, for example, twin-lumen or coaxial needles that can produce core–sheath fibers cannot be implemented in a bowl electrospinning setup, and the flow rate of electrospinning solution can be harder to accurately control, making the formation of more advanced fiber architectures such as beaded structures or electrosprayed micro or nanospheres difficult.

2.5.1.3 Multiple Spinnerets

Another technique used to overcome the insufficient throughput of the standard electrospinning technique is to use multiple spinnerets. The use of multiple spinnerets can increase throughput as well as allow a nonwoven nanofiber mesh with fibers of different materials (Theron et al. 2005; Min et al. 2004b). However, the use of multiple spinnerets produces multiple charged jets of electrospun material in close proximity, and as such can induce complex electrostatic forces between the jets in flight. This was demonstrated by Varesano et al. (2009), who altered both the fiber structure and deposition location, both of which are strongly dependent on the orientation of spinnerets. The effect of electric repulsion between jets has been mitigated using auxiliary electrodes, positioned along the flight path of the jet, using basic electrostatic principles to focus and maintain jet stability, even when in proximity to the other charged jets (Kim et al. 2006). Furthermore, it is possible to combine electrospinning and electrospraying processes in order to embed microspheres of a secondary material. To date, this has been primarily used to produce microtissues in tissue engineering by electrospraying live-cell suspensions into a polymer matrix, as it is electrospun, ensuring full-depth penetration of the cells into the matrix (Stankus et al. 2006; Ekaputra et al. 2008). Alternatively, as different materials with dissimilar conductivities tend to form fibers of different sizes, combinations can lead to thicker microfibers for structural support with an interdispersed nanofiber matrix more suited to filtration (Tuzlakoglu et al. 2005).

2.5.2 COLLECTOR MODIFICATIONS

It has long been observed that the nature of the grounded collector can greatly influence the characteristics of electrospun fiber (Zhang and Chang 2007; Kim et al. 2005). The scope for collector types is almost unlimited, with materials as diverse as paper, metallic foils, and liquids being previously employed (Liu and Hsieh 2002), among many others.

In all the cases discussed so far, the collector considered has been a simple flat plate of a conductive material, resulting in a tightly packed, nonwoven mat of randomly oriented nanofibers. In many instances, this is sufficient; however, for certain applications, it is desirable to control the deposition of the electrospun fibers more tightly, for example, to produce well-ordered grids of fibers with highly consistent pore sizes; this is often termed *writing* (Pham et al. 2006). To achieve greater levels of control, several adaptations to the standard electrospinning apparatus are possible; however, by far the most popular is the adaptation of the collector. By altering the properties of the collector, it is possible to tailor the fiber's eventual orientation either mechanically or through the control of the electric field acting upon the jet and, therefore, the overall properties of the nanofiber mat (Li et al. 2005b). Furthermore, the porosity of the produced fibers can be influenced by the chemical or thermal properties of the collector surface, for example, increased micropore density on electrospun fibers has been observed when electrospinning is performed directly onto the surface of a cryogenic fluid (liquid nitrogen) due to rapid phase separation of the solvent and solute (McCann et al. 2005b).

2.5.2.1 Controlled Deposition

2.5.2.1.1 Rotating Drum

The rotating drum collector is one of the earliest adaptations to the basic electrospinning setup (Formhals 1934). The technique employs a grounded drum powered by an electric motor such that it rotates as fibers are deposited. At high-enough speeds, usually around a few thousand rpm, the drum rotation causes fibers to be deposited aligned along the direction of the drum rotation, with the degree of alignment being determined by the rotation speed of the drum (Kim et al. 2004). Furthermore, by increasing the drum speed, it is possible to stretch fibers as they are deposited, thereby further reducing fiber diameter (Theron et al. 2001). Often the drum is mounted on a screw-thread axel, causing it to translate horizontally past the emitter, thereby preventing new fibers from being deposited over existing ones (Bhattarai et al. 2004).

2.5.2.1.2   Rotating Disk

A common variation on the rotating drum collector is the rotating disk. The use of a disk as opposed to a drum causes the electrospun jet to proceed in a conical path followed by an inverse conical path. Initially, the loops of the whipping jet grow in diameter, enveloping a steadily expanding cone; however, at a point above the collector, this process reverses and the jet begins to converge on the sharp-edged disk (Baji et al. 2010). This is due to the increased electric field at the sharp edge of the disk, as discussed in Section 5.1.2, which in turn leads to reduced fiber diameters. A drawback of the rotating disk collector is the severely reduced area onto which fibers can be deposited. This has been partially overcome by winding a conductive copper wire around an insulating drum, resulting in the increased electric field properties characteristic of the disk collector, while maintaining the larger collection area of the drum model (Zussman et al. 2003).

2.5.2.1.3   Near-Field Electrospinning

A more recent development in controlled deposition of nanofibers is the near-field electrospinning method. For greater control over nanofiber positioning, the separation distance between the emitter and collector can be drastically reduced in order to precisely deposit fibers at a given location, as described by Sun et al. (2006). The authors used a separation distance between the emitter and collector of 500 μm, and a mechanical probe to draw an initial fiber from a polymer droplet charged to slightly subelectrospinning voltages. After the initial mechanical drawing of fiber, the electrospinning process continues at much lower throughput rates than standard electrospinning; this allows the fiber to dry in the 500 μm before hitting the collector. The collector is then controlled through a programmable x–y stage to produce highly controlled fiber structures. Although this technique overcomes some of the weakness of random deposition electrospinning, it also has a number of drawbacks. Due to the decreased distance between the emitter and collector, the electrospun material has less flight time to allow solvent evaporation and jet thinning. This leads to a restricted choice of materials capable of drying sufficiently quickly as well as increasing the fibers diameter dramatically. The technique of mechanically initiating the electrospinning process alleviates this, with fiber diameters of 38 ± 4 nm being reported (Sun et al. 2006); however, the added complication of this process and the overall reduced throughput limits its application on an industrial scale. The technique was recently developed by Bisht et al. (2011) using super-elastic polymer ink, allowing continuous electrospinning at only 200 V, three times less than Sun's group, and allowed three-dimensional patterning of electrospun structures, with fiber diameters as little as 16.2 nm. However, fiber continuity was impeded using this approach with fibers breaking up into continuous lines of nanodroplets at the lowest fiber diameters. A further consideration of near-field electrospinning is that some physical effects observed in the standard electrospinning differ from the near-field setup, for example, a drop in the electric field strength results in decreased fiber diameter. Chang et al. (2008) describe some of these observed differences; however, further research is required to characterize the process completely.

2.5.2.2 Electrostatic Deflection

Regardless of any further modifications to the electrospinning system, jet formation is, by definition, caused by electrostatic forces. This results in the jet of electrospun material traveling to the collector always possessing an intense electric charge. By harnessing electric field properties, it is possible to deflect the jet in a known direction and magnitude, allowing the path of the jet to be altered and increasing control over fiber deposition.

In their seminal review article, Doshi and Reneker (1995) first observed electrostatic deflection of electrospun nanofibers. Here, two metal plates were used to produce a transverse electric field across the fiber's flight path. This transverse field was able to deflect fibers up to 15° onto a collecting screen. More recently, Bellan and Craighead (2006) employed a series of electrical lenses and deflectors to obtain single nanofibers accurately deposited for applications to functional electronic devices. A variation of this technique is to use multiple collectors, each of which can be charged or grounded by an automated system, as described by Ishii et al. (2008). This system allowed the authors to not only control the orientation of electrospun fibers but also to stop the electrospinning process after a very well-defined time interval, meaning the number of produced fibers could be controlled precisely.

The potential for the control of nanofibers deposition through electrostatic deflection is promising, as the technique does not compromise on electrospinning strengths such as fiber diameter and material choices, while theoretically allowing for greater control of fiber deposition. Figure 2.9 presents the electrostatic deflection system for the controlling position of electrospun fibers.

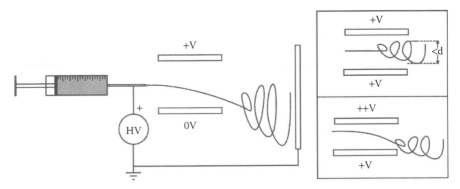

FIGURE 2.9   Electrostatic deflection system for the controlling position of electrospun fibers.

2.5.3 ELECTROSPINNING POLYMER MODIFICATIONS

Although the choice of electrospinnable materials is very wide, it is occasionally desirable to produce nanofibers from materials that are not inherently compatible with the process. To this end, post-electrospinning procedures can be used to treat electrospun precursor fibers to obtain the desired material or material properties (Schiffman & Schauer 2007). A common example is using coaxial electrospinning to produce nanotubes. In this technique, a coaxial fiber comprised of the desired material as the sheath layer and a sacrificial core is electrospun using a coaxial setup, as described in Section 5.1.1. The core material is then removed with a material-specific solvent, leaving a nanotube of the desired material (Li et al. 2005a). This technique can also be reversed using the desired material as the core and a sacrificial sheath, thereby assisting in electrospinning materials poorly suited to being electrospun alone (Loscertales et al. 2006). In addition, it is common to electrospin precursor fibers, for example, Yang et al. (2004) used a blend of zinc acetate and PVA to produce a solution of the required viscosity to electrospin fibers that were then calcined at 700°C to produce ZnO nanofibers. The heat treatment removed the PVA entirely, as demonstrated by X-ray diffraction studies, and also reduced the fiber diameter by approximately five times (based on published scanning electron microscopy micrographs, exact measurements unpublished). Finally Choi et al. (2009b) sputter coated ZnO onto polyvinyl-acetate fibers and calcined to produce ZnO nanotubes. Although the authors focused entirely on ZnO, both of these methods have potential applications in forming nanofibers from any materials with thermal stability greater than electrospinnable polymers.

The morphology and mechanical properties of an electrospun nanofiber are very much dependent on the solution, process, and ambient conditions. It was identified that membrane selectivity, porosity, and mechanical strength were modifiable either through pre-electrospinning modification of polymer solutions or through post-electrospinning treatments of nanofibers (Nasreen et al. 2013). Such modifications of electrospun fiber for membrane fabrication are covered comprehensively in Chapters 12 and 15.

2.6 INDUSTRIAL AND COMMERCIAL APPLICATIONS

With sufficient control of input variables and considered design of collectors and emitters, it is possible to produce electrospun nanofibers from a variety of materials. These fibers are deposited in a given area as a nonwoven mat, with the potential for pore sizes in the micro-, nano-, and ultrafiltration domains. Generally, using nanofibers in filtration applications has shown a great promise and demonstrated performance increases (Kosmider and Scott 2002). The advantage of an electrospinning system is that due to the number of influential variables, pore sizes can be highly controlled with basic alterations such as conductivity or shape of the collector (Li et al. 2005b). Additionally, as a bottom-up system, material waste is minimal, assisting in cost-effectiveness. In their review of electrospinning for filtration applications, Bjorge et al. (2009) estimated a 2.5× decrease in cost/m^2 for electrospun verses other commercial membranes (€50 for commercial vs. €20 for electrospun membranes). Furthermore, they speculated lowered running costs of a nanofiber filtration system due to a lower transmembrane pressure required in electrospun compared to existing commercial membranes.

The versatility of the electrospinning technique, as applied to filtration, are seen in the publications of Chu et al. (2009, 2006, 2012). In these articles, the researchers produced high-flux, potentially micro-, ultra-, and nanofiltration polymeric membranes, respectively, from a remarkably similar electrospinning setup by simply altering the concentration and chamber humidity of the polymer solution. These simple changes yielded membranes with pore sizes ranging from 5 nm to 100 µm.

At present, few commercially available membranes are fabricated via the electrospinning technique, however a number of research groups have attempted to apply the electrospinning technique in order to achieve commercially viable results. For instance, a new type of high-flux membrane with a thin coating layer of water-resistant and hydrophilic chitosan on a support layer of electrospun asymmetric polyacrylonitrile (PAN) scaffolds consisting of two layers of 4 wt% and 10 wt% PAN and a nonwoven polyethylene terephthalate (PET) substrate was introduced by Yoon et al. (2006) for ultra-/nanofiltration. The performance of the electrospun membrane was improved as compared with a commercial nanofiber membrane (NF270). The chitosan/PAN/PET membrane had a higher membrane flux, but a similar rejection performance (>99.9%) for an oily wastewater feed (Yoon et al. 2006).

Zhao et al. (2012) carried out a similar set of experiments on a modified chitosan-coated layer made by casting a solution of chitosan, cross-linked with glutaraldehyde (GA) and terephthaloyl chloride (TPC), on electrospun polyvinylidene difluoride (PVDF) scaffolds. They compared their membrane performance with Sepro's ultrafiltration (UF) membrane (PES10), where they reported a higher flux and higher rejection for bovine serum albumin (BSA) at 0.2 MPa, thus demonstrating the commercial potential of the membrane (Zhao et al. 2012).

More recently, a new class of high-performance-thin-film nanocomposite (TFNC) ultrafiltration membrane was introduced by You et al. (2013). In their research, electrospun PAN nanofibrous substrates, coupled with a thin hydrophilic PVA barrier layer incorporating surface-oxidized multiwalled nanotubes (MWNTs) cross-linked by GA, were used to fabricate a composite membrane. As a result of incorporating MWNTs into the PVA barrier layer, there was a twofold improvement in the permeate flux, with maintenance of a high rejection rate (99.5%) at low pressure (0.1 MPa) during oil/water emulsion separation (You et al. 2013).

Bjorge et al. (2009) in their critical review highlighted weaknesses of electrospun membranes compared to current commercial options, primarily the structural weakness of the electrospun membrane and the rapidity and irreversibility of fouling when used for bacterial filtration, although the work does note the increased effectiveness of fibers containing silver nanoparticles. Such fibers are relatively simple to produce by including silver nitrate or presynthesized silver nanoparticles into the electrospinning feedstock in the presence of appropriate solvents. An example of such fibers is seen in Figure 2.10. It should be noted that these drawbacks were less in the study by Chu et al. (2012), whereby a supporting structure formed from a tougher polymer, poly(ethylene terephthalate), was used as a supporting structure for the poly(ether sulfone) nanofibers filtration membrane. In another study, PAN and polysulfone (PSu) electrospun nanofibers were modified by polydopamine,

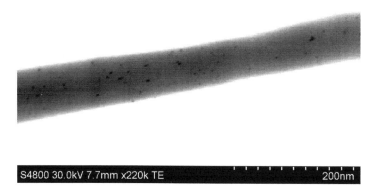

FIGURE 2.10 Transmission electron microscopy image of electrospun poly(vinyl pyrrolidone) nanofiber containing silver nanoparticles for antibacterial control.

where the resultant membrane exhibited 100% improvement of tensile strength; the Young's modulus increased by 80% and 210% for PAN and PSu, respectively (Huang et al. 2014). The development of membranes using electrospun nanofibers for water treatment has recently had an extensive focus on developing solutions for fouling and biofouling control. For instance, Goethals et al. (2014) successfully improved pathogen removal efficiency of polyamide membranes functionalized with a photosensitizer compound (phthalocyanines), which has biocide capability through the production of singlet oxygen.

2.6.1 Commercially Available Electrospinning Systems

Despite the long history of electrospinning, commercially available electrospinning systems have only recently begun to emerge. Indeed, within the authors' laboratory, as with many research groups, an in-house system has been used for research purposes. This arguably demonstrates the straightforwardness of the electrospinning process where the most important factor is operator experience. There are only a few off-the-shelf electrospinners. This is in part due to the difficulty in scale-up and frequency of blockages in needle-type systems. However, the advances in needle-free electrospinning have alleviated both of these issues, as such several commercial spinning systems have become available in the past five years. The first use of needleless electrospinning systems was in 2004 when a magnetic-field-assisted electrospinning was reported. This progress led to invention of a rotating roller-fiber spinneret. The El Marco company has developed a range of pilot- to industrial-scale devices based on needleless electrospinning, which is known as the NanoSpider™ range. Figure 2.11 presents a scanning electron microscopy image of Ultramid B24 N03 polymer solution electrospun with NanoSpider NS-200.

Despite the drawbacks, there are also a number of commercially available needle-based electrospinning systems currently on the market, including the modular Spraybase® system, marketed at lab-scale production levels, with a visualization system for observing the formation of the Taylor cone and jet in real time. Further

S4800 10.0kV 10.5mm x2.20k SE(M)   20.0um

FIGURE 2.11   Scanning electron microscopy image of the electrospun Ultramid B24 N03 (Nylon-6) nanofiber fabricated using needleless electrospinning setup (El Marco NanoSpider).

needle-based systems are also available from Linari Biomedical and E-Spin Nanotech; the advantages of these systems are their modular nature, which allows simple swapping of spinnerets, collectors, and voltage sources, depending on the requirements of the research or production project.

2.7 CONCLUSIONS

To date, there has been an exponential increase in the use of electrospinning for a wide variety of applications. To be concise, this chapter has dealt only with the general aspects of the electrospinning process and its modifications, with a view of producing nanoporous membranes. Many further reviews of electrospinning may be found in a variety of journals spanning many disciplines, and new modifications, applications, and theoretical underpinnings are published frequently on the subject. Electrospinning allows the researcher or industrialist with a simple and low-cost route to the fabrication of highly porous nanostructures, which can be tightly controlled in terms of their material composition, internal structure, and general topography. Two primary challenges remain to further expose the power of the electrospinning technique, these are scale-up and control of fiber deposition, as discussed above in Sections 2.5.1 and 2.5.2 respectively. Both of these issues have seen impressive breakthroughs in the past five years; however, the combination of several advanced electrospinning techniques into a single unit remains elusive. Once this is overcome, the potential for industrial-scale production of multilayered nanofibers, which are printed at nanoscale resolution, will make electrospinning the preferred production technique for many membrane applications. With the number of researchers, both academic and private, currently working in the field, it appears to be only a matter of time until such a breakthrough is realized.

REFERENCES

Baji, A. et al. 2010. Electrospinning of polymer nanofibres: Effects on oriented morphology, structures and tensile properties. *Composites Science and Technology*, 70, pp. 703–18.

Ball, C., Krogstad, E., Chaowanachan, T., and Woodrow, K.A., 2012. Drug-eluting fibres for HIV-1 inhibition and contraception. *PLoS One*, 7(11), e49792.

Barhate, R.S. and Ramakrishna, S., 2007. Nanofibrous filtering media: Filtration problems and solutions from tiny materials. *Journal of Membrane Science*, 296, pp. 1–8.

Baumgarten, P.K., 1971. Electrostatic spinning of acrylic microfibres. *Journal of Colloid and Interface Science*, 36(1), pp. 71–79.

Bazilevsky, A.V., Yarin, A.L., and Megaridis, C.M., 2007. Co-electrospinning of core-shell fibres using a single-nozzle technique. *Langmuir*, 23(5), pp. 2311–14.

Beachley, V. and Wen, X., 2010. Polymer nanofibrous structures: Fabrication, biofunctionalization, and cell interactions. *Progress in Polymer Science*, 35(7), pp. 868–92.

Bellan, L.M. and Craighead, H.G., 2006. Control of an electrospinning jet using electric focusing and jet-steering fields. *Journal of Vacuum Science & Technology B: Microelectronics and Nanometer Structures*, 24(6), pp. 3179–83.

Bhardwaj, N. and Kundu, S.C., 2010. Electrospinning: A fascinating fibre fabrication technique. *Biotechnology Advances*, 28, pp. 325–47.

Bhattarai, N. et al., 2005. Electrospun chitosan-based nanofibres and their cellular compatibility. *Biomaterials*, 26(31), pp. 6176–84.

Bhattarai, S.R. et al., 2004. Novel biodegradable electrospun membrane: Scaffold for tissue engineering. *Biomaterials*, 25(13), pp. 2595–602.

Bisht, G.S. et al., 2011. Controlled continuous patterning of polymeric nanofibres on three-dimensional substrates using low-voltage near-field electrospinning. *Nano Letters*, 11(4), pp. 1831–37.

Bjorge, D. et al., 2009. Performance assessment of electrospun nanofibres for filtration applications. *Desalination*, 249(3), pp. 942–48.

Bognitzki, M. et al., 2001. Preparation of fibres with nanoscaled morphologies: Electrospinning of polymer blends. *Polymer Engineering & Science*, 41(6), pp. 982–89.

Boys, C.V., 1887. On the production, properties, and some suggested uses of the finest threads. *Proceedings of the Physical Society of London*, 9(1), pp. 8–19.

Buchko, C.J., Chen, L.C., Shen, Y., and Martin, D.C., 1999. Processing and microstructural characterization of porous biocompatible protein polymer thin films. *Polymer*, 40, pp. 7397–407.

Chang, C., Limkrailassiri, K., and Lin, L., 2008. Continuous near-field electrospinning for large area deposition of orderly nanofibre patterns. *Applied Physics Letters*, 93(12), pp. 123111–13.

Chen, J.P., Chang, G.Y., and Chen, J.K., 2008. Electrospun collagen/chitosan nanofibrous membrane as wound dressing. *Colloids and Surfaces: A Physicochemical and Engineering Aspects*, 313/314, pp. 183–88.

Choi, S.H. et al., 2009a. Hollow ZnO nanofibres fabricated using electrospun polymer templates and their electronic transport properties. *American Chemical Society Nano*, 3(9), pp. 2623–31.

Choi, S.W., Park, J.Y., and Kim, S.S., 2009b. Synthesis of SnO_2–ZnO core–shell nanofibres via a novel two-step process and their gas sensing properties. *Nanotechnology*, 20(46), pp. 2623–2631.

Chu, B. et al., 2006. High flux ultrafiltration membranes based on electrospun nanofibrous PAN scaffolds and chitosan coating. *Polymer*, 47(7), pp. 2434–41.

Chu, B. et al., 2009. Design and fabrication of electrospun polyethersulfone nanofibrous scaffold for high-flux nanofiltration membranes. *Journal of Polymer Science Part B: Polymer Physics*, 47(22), pp. 2288–300.

Chu, B. et al., 2012. Electrospun nanofibrous membranes for high flux microfiltation. *Journal of Membrane Science*, 392/393, pp. 167–74.

Deitzel, J.M., Kleinmeyer, J., Harris, D., and Beck Tan, N.C., 2001. The effect of processing variables on the morphology of electrospun nanofibres and textiles. *Polymer*, 42(1), pp. 261–72.

Doshi, J. and Reneker, D.H., 1995. Electrospinning process and applications of electrospun fibres. *Journal of Electrostatics*, 35(2/3), pp. 151–60.

Ekaputra, A.K., Prestwich, G.D., Cool, S.M., and Hutmacher, D.W., 2008. Combining electrospun scaffolds with electrosprayed hydrogels leads to three-dimensional cellularization of hybrid constructs. *Biomacromolecules*, 9(8), pp. 2097–103.

Fong, H., Chun, I., and Reneker, D.H., 1999. Beaded nanofibres formed during electrospinning. *Polymer*, 40(16), pp. 4585–92.

Fong, H. and Reneker, D.H., 2001. Electrospinning and the formation of nanofibres. In D.R. Salem (ed.), *Structure Formation in Polymeric Fibres*. Cincinnati, OH: Hanser Gardner Publications, pp. 225–46.

Formhals, A., 1934. Process and apparatus for preparing artificial threads. US Patent 1975504.

Fridrikh, S.V., Yu, J.H., Brenner, M.P., and Rutledge, G.C., 2003. Controlling the fibre diameter during electrospinning. *Physical Review Letters*, 90, pp. 144502–504.

Goethals, A. et al., 2014. Polyamide nanofibre membranes functionalized with zinc phthalocyanines. *Journal of Applied Polymer Science*, 131(1), pp. 1–7, doi:10.1002/app.40486.

Greiner, A., Wendorff, J.H., Yarin, A.L., and Zussman, E., 2006. Biohybrid nanosystems with polymer nanofibres and nanotubes. *Applied Microbiology and Biotechnology*, 71, pp. 387–93.

Grimm, S. et al., 2008. Nondestructive replication of self-ordered nanoporous alumina membranes via cross-linked polyacrylate nanofibre arrays. *NanoLetters*, 8(7), pp. 1954–59.

Gupta, P., Elkins, C., Long, T.E., and Wilkes, G.L., 2005. Electrospinning of linear homopolymers of poly (methyl methacrylate): Exploring relationships between fibre formation, viscosity, molecular weight and concentration in a good solvent. *Polymer*, 46(13), pp. 4799–810.

Gupta, P. and Wilkes, G.L., 2003. Some investigations on the fibre formation by utilizing a side-by-side bicomponent electrospinning approach, *Polymer*, 44 (20), pp. 6353–59.

Haghi, A.K. (ed.), 2011. *Electrospinning of Nanofibres in Textiles*. Toronto, Ontario, Canada: Apple Academic Press.

Haghi, A.K. and Akbari, M., 2007. Trends in electrospinning of natural nanofibers. *Physica Status Solidi*, 204, pp. 1830–44.

Hartgerink, J.D., Beniash, E., and Stupp, S.I., 2001. Self-assembly and mineralization of peptide-amphiphile nanofibres. *Science*, 294(5547), pp. 1684–88.

He, J.H., Liu, Y., and Xu, L., 2010. Apparatus for preparing electrospun nanofibres: A comparative review. *Materials Science and Technology*, 26(11), pp. 1275–87.

He, J.H., Wan, Y.Q., and Xu, L., 2007. Nano-effects, quantum-like properties in electrospun nanofibres. *Chaos, Solitons & Fractals*, 33(1), pp. 26–37.

Hohman, M.M., Shin, M., Rutledge, G., and Brenner, M., 2001a. Electrospinning and electrically forced jets. I. Stability Theory. *Physics of Fluids*, 13, pp. 2201–20.

Hohman, M.M., Shin, M., Rutledge, G., and Brenner, M., 2001b. Electrospinning and electrically forced jets. II. Applications. *Physics of Fluids*, 13, pp. 2221–36.

Huang, Z.M., Zhang, Y.Z., Kotaki, M., and Ramakrishna, S., 2003. A review on polymer nanofibres by electrospinning and their applications in nanocomposites. *Composites Science and Technology*, 63(15), pp. 2223–53.

Huang, Z.M., Zhang, Y.Z., Ramakrishna, S., and Lim, C.T., 2004. Electrospinning and mechanical characterization of gelatin nanofibres. *Polymer*, 45(15), pp. 5361–68.

Huang, L. et al., 2014. Improved mechanical properties and hydrophilicity of electrospun nanofibre membranes for filtration applications by dopamine modification. *Journal of Membrane Science*, 460(1), pp. 241–49.

Hui, W. et al., 2010. Electrospun metal nanofibre webs as high-performance transparent electrode. *Nano Letters*, 10(10), pp. 4242–48.

Ishii, Y., Sakai, H., and Murata, H., 2008. A new electrospinning method to control the number and a diameter of uniaxially aligned polymer fibres. *Materials Letters*, 62(19), pp. 3370–72.

Jaworek, A. and Sobczyk, A.T., 2008. Electrospraying route to nanotechnology: An overview. *Jorunal of Electrostatics*, 66(3/4), pp. 197–219.

Jiang, H. et al., 2006. Modulation of protein release from biodegradable core–shell structured fibres prepared by coaxial electrospinning. *Journal of Biomedical Materials Research*, 79B(1), pp. 50–57.

Katti, D.S., Robinson, K.W., Ko, F.K., and Laurencin, C.T., 2004. Bioresorbable nanofibre-based systems for wound healing and drug delivery: Optimization of fabrication parameters. *Journal of Biomedical Materials Research*, 70B(2), pp. 286–96.

Kessick, R., Fenn, J., and Tepper, G., 2004. The use of AC potentials in electrospraying and electrospinning processes. *Polymer*, 45(9), pp. 2981–84.

Killic, A., Oruc, F., and Demir, A., 2008. Effects of polarity on electrospinning process. *Textile Research Journal*, 78(6), pp. 532–39.

Kim, G.H., Cho, Y.S., and Kim, W.D., 2006. Stability analysis for multi-jets electrospinning process modified with a cylindrical electrode. *European Polymer Journal*, 42, pp. 2031–38.

Kim, H.S., Kim, K., Jin, H.J., and Chin, I.J., 2005. Morphological characterization of electrospun nano-fibrous membranes of biodegradable poly(L-lactide) and poly(lactide-co-glycolide). *Macromolecular Symposia*, 224(1), pp. 145–54.

Kim, S.J. et al., 2007. Preparation and characterization of antimicrobial polycarbonate nanofibrous membrane. *European Polymer Journal*, 43(8), pp. 3146–52.

Kim, W.K. et al., 2004. The effect of molecular weight and the linear velocity of drum surface on the properties of electrospun poly(ethylene terephthalate) nonwovens. *Fibres and Polymers*, 5(2), pp. 122–27.

Koombhongse, S., Liu, W., and Reneker, D.H., 2001. Flat polymer ribbons and other shapes by electrospinning. *Journal of Polymer Science Part B: Polymer Physics*, 39(21), pp. 2598–606.

Kosmider, K. and Scott, J., 2002. Polymeric nanofibres exhibit an enhanced air filtration performance. *Filtration & Separation*, 39(6), pp. 20–22.

Kowalczyk, T., Nowicka, A., Elbaum, D., and Kowalewski, T.A., 2008. Electrospinning of bovine serum albumin. Optimization and the use for production of biosensors. *Biomacromolecules*, 9(7), pp. 2087–90.

Kwaambwa, H.M. et al., 2007. Viscosity, molecular weight and concentration relationships at 298K of low molecular weight cis-polyisoprene in a good solvent. *Colloids and Surfaces A: Physicochemical and Engineering Aspects*, 294(1), pp. 14–19.

Leach, M.K., Feng, Z.Q., Tuck, S.J., and Corey, J.M., 2011. Electrospinning fundamentals: Optimizing solution and apparatus parameters. *Journal of Visualized Experiments*, 47, pp. 1–4, doi:10.3791/2494.

Lee, J.S. et al., 2004. Role of molecular weight of atactic poly(vinyl alcohol) (PVA) in the structure and properties of PVA nanofabric prepared by electrospinning. *Journal of Applied Polymer Science*, 93(4), pp. 1638–46.

Lee, S. and Obendorf, S.K., 2007. Use of electrospun nanofibre web for protective textile materials as barriers to liquid penetration. *Textile Research Journal*, 77, pp. 696–702.

Li, D., McCann, J.T., and Xia, Y., 2005a. Use of electrospinning to directly fabricate hollow nanofibres with functionalized inner and outer surfaces. *Small*, 1(1), pp. 83–86.

Li, D., Ouyang, G., McCann, J.T., and Xia, Y., 2005b. Collecting electrospun nanofibres with patterned electrodes. *Nano Letters*, 5(5), pp. 913–16.

Li, D., Wang, Y., and Xia, Y., 2004. Electrospinning nanofibres as uniaxially aligned arrays and layer-by-layer stacked films. *Advanced Materials*, 16(4), pp. 361–66.

Li, D. and Xia, Y., 2004a. Direct fabrication of composite and ceramic hollow nanofibres by electrospinning. *Nano Letters*, 4(5), pp. 933–38.

Li, D. and Xia, Y., 2004b. Electrospinning of nanofibres: Reinventing the wheel? *Advanced Materials*, 16(14), pp. 1151–70.

Li, Z. and Wang, C., 2013. *One-Dimensional nanostructures: Electrospinning Technique and Unique Nanofibres*, Berlin, Germany: Springer.

Liao, S. et al., 2006. Biomimetic electrospun nanofibres for tissue regeneration. *Biomedical Materials*, 1(3), R45–R53.

Lin, T., Wang, X., Wang, X., and Niu, H., 2010. Electrostatic spinning assembly. Patent no: WO2010043002 (A1).

Liu, H. and Hsieh, Y.L., 2002. Ultrafine fibrous cellulose membranes from electrospinning of cellulose acetate. *Journal of Polymer Science Part B: Polymer Physics*, 40(18), pp. 2119–29.

Liewhiran, C. and Phanichphant, S., 2008. Doctor-bladed thick films of flame-made Pd/ZnO nanoparticles for ethanol sensing. *Current Applied Physics*, 8(3/4), pp. 336–339.

Loscertales, I.G. et al., 2006. Coaxial electrospinning for nanostructured advanced materials. *MRS Proceedings*, 948, 0948-B06-01 doi:10.1557/PROC-0948-B06-01.

Lu, B. et al., 2010. Superhigh-throughput needleless electrospinning using a rotary cone as spinneret. *Small*, 6(15), pp. 1612–16.

Lyons, J. and Ko, F., 2005. Melt electrospinning of polymers: A review. *Polymer News*, 30(6), pp. 170–78.

Ma, P.X. and Zhang, R., 1999. Synthetic nano-scale fibrous extracellular matrix. *Journal of Biomedical Materials Research*, 46(1), pp. 60–72.

Ma, Z., Kotaki, M., Inai, R., and Ramakrishna, S., 2005. Potential of nanofibre matrix as tissue-engineering scaffolds. *Tissue Engineering*, 11(1/2), pp. 101–9.

Matthews, J.A., Wnek, G.E., Simpson, D.G., and Bowling, G.L., 2002. Electrospinning of collagen nanofibres. *Biomacromolecules*, 3(2), pp. 232–38.

McCann, J.T., Li, D., and Xia, Y., 2005a. Electrospinning of nanofibres with core-sheath, hollow, or porous structures. *Journal of Materials Chemistry*, 15, pp. 735–38.

McCann, J.T., Marquez, M., and Xia, Y., 2005b. Highly porous fibres by electrospinning into a cryogenic liquid. *Journal of the American Chemical Society*, 128(5), pp. 1436–37.

Megelski, S., Stephens, J.S., Chase, D.B., and Rabolt, J.F., 2002. Micro- and nanostructured surface morphology on electrospun polymer fibres. *Macromolecules*, 35(22), pp. 8456–66.

Min, B.M. et al., 2004a. Electrospinning of silk fibroin nanofibres and its effect on the adhesion and spreading of normal human keratinocytes and fibroblasts in vitro. *Biomaterials*, 25(7/8), pp. 1289–97.

Min, B.M. et al., 2004b. Formation of nanostructured poly(lactic-co-glycolic acid)/chitin matrix and its cellular response to normal human keratinocytes and fibroblasts. *Carbohydrate Polymers*, 57(3), pp. 285–92.

Mitchell, D.T. et al., 2002. Smart nanotubes for bioseparations and biocatalysis. *Journal of the American Chemical Society*, 124(40), pp. 11864–65.

Nakajima, T., Kajiwara, K., and McIntyre, J.E., 1994. *Advanced Fibre Spinning Technology*. Cambridge: Woodhead Publishing, p. 187.

Nasreen, S.A.A.N. et al., 2013. Advancement in electrospun nanofibrous membranes modification and their application in water treatment. *Membranes*, 3(1), pp. 266–284.

Niu, H. and Lin, T., 2012. Fibre generators in needleless electrospinning. *Journal of Nanomaterials*, pp. 1–13, doi:10.1155/2012/725950.

Niu, H., Wang, X., and Lin, T., 2012. Needleless electrospinning: Influences of fibre generator geometry. *Journal of the Textile Institute*, 103(7), pp. 787–94.

Pham, Q., Sharma, U., and Mikos, A., 2006. Electrospinning of polymeric nanofibres for tissue engineering applications: A review. *Tissue Engineering*, 12, pp. 1197–211.

Ramakrishna, S. et al., 2005. *An Introduction to Electrospinning and Nanofibres*. Singapore: World Scientific Publishing.

Ramakrishna, S. et al., 2010. Science and engineering of electrospun nanofibres for advances in clean energy, water filtration, and regenerative medicine. *Journal of Materials Science*, 45, pp. 6283–312.

Reneker, D.H. and Chun, I., 1996. Nanometre diameter fibres of polymer, produced by electrospinning. *Nanotechnology*, 7(3), pp. 216–223.

Rollings, D.E., Tsoi, S., Sit, J.C., and Veinot, J.G.C., 2007. Formation and aqueous surface wettability of polysiloxane nanofibres prepared via surface initiated, vapor-phase polymerization of organotrichlorosilanes. *Langmuir*, 23(10), pp. 5275–78.

Rollings, D.E. and Veinot, J.G.C., 2008. Polysiloxane nanofibres via surface initiated polymerization of vapor phase reagents: A mechanism of formation and variable wettability of fibre-bearing substrates. *Langmuir*, 24(23), pp. 13653–62.

Rutledge, G.C. and Fridrikh, S.V., 2007. Formation of fibres by electrospinning. *Advanced Drug Delivery Review*, 59, pp. 1384–91.

Schiffman, J.D. and Schauer, C.L., 2007. Cross-linking chitosan nanofibres. *Biomacromolecules*, 8(1), pp. 594–601.

Schreuder-Gibson, H. et al., 2002. Protective textile materials based on electrospun nanofibres. *Journal of Advanced Materials*, 34(3), pp. 44–55.

Sell, S.A. et al., 2010. The use of natural polymers in tissue engineering: A focus on electrospun extracellular matrix analogues. *Polymers*, 24(4), pp. 522–53.

Shin, J.H. et al., 2006. Electrospun PLGA nanofibre scaffolds for articular cartilage reconstruction: Mechanical stability, degradation and cellular responses under mechanical stimulation *in vitro*. *Journal of Biomaterials Science, Polymer Edition*, 17(1/2), pp. 103–119.

Sigmund, W. et al., 2006. Processing and structure relationships in electrospinning of ceramic fibre systems. *Journal of the American Chemical Society*, 89(2), pp. 395–407.

Sill, T.J. and Von Recum, H.A., 2008. Electrospinning: Applications in drug delivery and tissue engineering. *Biomaterials*, 29, pp. 1989–2006.

Son, W.K., Youk, J.H., Lee, T.S., and Park, W.H., 2004. The effects of solution properties and polyelectrolyte on electrospinning of ultrafine poly (ethylene oxide) fibres. *Polymer*, 45(9), pp. 2959–66.

Son, W.K., Youk, J.H., and Park, W.H., 2006. Antimicrobial cellulose acetate nanofibres containing silver nanoparticles. *Carbohydrate Polymers*, 65(4), pp. 430–34.

Stankus, J.J., Guan, J., Fujimoto, K., and Wagner, W.R., 2006. Microintegrating smooth muscle cells into a biodegradable, elastomeric fibre matrix. *Biomaterials*, 27(5), pp. 735–44.

Sun, D., Chang, C., Li, S., and Lin, L., 2006. Near-field electrospinning. *Nano Letters*, 6(4), pp. 839–42.

Sun, Z. et al., 2003. Compound core-shell polymer nanofibres by co-electrospinning. *Advanced Materials*, 15(22), pp. 1929–32.

Suppahol, P., Mit-uppatham, C., and Nithitanakul, M., 2005. Ultrafine electrospun polyamide-6 fibres: Effects of solvent system and emitting electrode polarity on morphology and average fibre diameter. *Macromolecular Materials and Engineering*, 290(9), pp. 933–42.

Tacx, J.C.J.F., Schoffeleers, H.M., Brands, A.G.M., and Teuwen, L., 2000. Dissolution behavior and solution properties of polyvinylalcohol as determined by viscometry and light scattering in DMSO, ethyleneglycol and water. *Polymer*, 41(3), pp. 947–957.

Tan, S.H., Inai, R., Kotaki, M., and Ramakrishna, S., 2005. Systematic parameter study for ultra-fine fibre fabrication via electrospinning process. *Polymer*, 46, pp. 6128–34.

Tao, S.L. and Desai, T.A., 2007. Aligned arrays of biodegradable poly (ε-caprolactone) nanowires and nanofibres by template synthesis. *Nano Letters*, 7(6), pp. 1463–68.

Taylor, G., 1964. Disintegration of water drops in an electric field. *Proceedings of the Royal Society A: Mathematical, Physical and Engineering Sciences*, 280(1382), pp. 383–97.

Taylor, G., 1966. The force exerted by an electric field on a long cylindrical conductor. *Proceedings of the Royal Society A: Mathematical, Physical and Engineering Sciences*, 291(1425), pp. 145–58.

Taylor, G., 1969. Electrically driven jets. *Proceedings of the Royal Society A: Mathematical, Physical and Engineering Sciences*, 313(1515), pp. 453–75.

Teo, W.E. and Ramakrishna, S., 2006. A review on electrospinning design and nanofibre assemblies. *Nanotechnology*, 17(14), R89–R106.

Theron, A., Zussman, E., and Yarin, A.L., 2001. Electrostatic field-assisted alignment of electrospun nanofibres. *Nanotechnology*, 12, pp. 384–90.

Theron, S.A., Yarin, A.L., Zussman, E., and Kroll, E., 2005. Multiple jets in electrospinning: Experiment and modeling. *Polymer*, 46(9), pp. 2889–99.

Theron, S.A., Zussman, E., and Yarin, A.L., 2004. Experimental investigation of the governing parameters in the electrospinning of polymer solutions. *Polymer*, 45, pp. 2017–30.

Thompson, C.J., Chase, G.G., Yarin, A.L., and Reneker, D.H., 2007. Effects of parameters on nanofibre diameter determined from electrospinning model. *Polymer*, 48(23), pp. 6913–22.

Thoppey, N.M., Bochinski, J.R., Clarke, L.I., and Gorga, R.E., 2010. Unconfined fluid electrospun into high quality nanofibres from a plate edge. *Polymer*, 51(21), pp. 4928–36.

Thoppey, N.M., Bochinski, J.R., Clarke, L.I., and Gorga, R.E., 2011. Edge electrospinning for high throughput production of quality nanofibres. *Nanotechnology*, 22(34), doi:10.1088/0957-4484/22/34/345301.

Tsai, P.P., Schreuder-Gibson, H., and Gibson, P., 2002. Different electrostatic methods for making electret filters. *Journal of Electrostatics*, 54, pp. 333–341.

Tuzlakoglu, K. et al., 2005. Nano- and micro-fibre combined scaffolds: A new architecture for bone tissue engineering. *Journal of Materials Science*, 16, pp. 1099–104.

Varesano, A., Carletto, R.A., and Mazzuchetti, G., 2009. Experimental investigations on the multi-jet electrospinning process. *Journal of Materials Processing Technology*, 209, pp. 5178–85.

Wan, Y.Q., He, J.H., and Wu, Y.Y.J.Y., 2006. Vibrorheological effect on electrospun polyacrylonitrile (PAN) nanofibres. *Materials Letters*, 60(27), pp. 3296–300.

Wei, M., Kang, B., Sung, C., and Mead, J., 2006. Core-sheath structure in electrospun nanofibres from polymer blends. *Macromolecular Materials and Engineering*, 291(11), pp. 1307–14.

Xin, Y. et al., 2008. Core–sheath functional polymer nanofibres prepared by co-electrospinning. *European Polymer Journal*, 44(4), pp. 1040–45.

Xing, X., Wang, Y., and Li, B., 2008b. Nanofibre drawing and nanodevice assembly in poly(trimethylene terephthalate). *Optics Express*, 16(14), pp. 10815–22.

Yang, R., He, J., Xu, L., and Yu, J., 2009. Bubble-electrospinning for fabricating nanofibres. *Polymer*, 50(24), pp. 5846–50.

Yang, R.R., He, J.H., Yu, J.Y., and Xu, L., 2011. Bubble-electrospinning for Fabrication of Nanofibres with Diameter of about 20 nm. *International Journal of Nonlinear Sciences and Numerical Simulation*, 11, pp. 163–64.

Yang, X. et al., 2004. Preparation and characterization of ZnO nanofibres by using electrospun PVA/zinc acetate composite fibre as precursor. *Inorganic Chemistry Communications*, 7(2), pp. 176–78.

Yarin, A.L., Koombhongse, S., and Reneker, D.H., 2001. Taylor cone and jetting from liquid droplets in electrospinning of nanofibres. *Journal of Applied Physics*, 90(9), pp. 4836–46.

Yee, W.A. et al., 2008. Stress-induced structural changes in electrospun polyvinylidene difluoride nanofibres collected using a modified rotating disk. *Polymer*, 49(19), pp. 4196–203.

Yener, F. and Jirsak, O., 2012. Comparison between the needle and roller electrospinning of polyvinylbutyral. *Journal of Nanomaterials*, 1(2012), pp. 1–6, doi:10.1155/2012/839317.

Yeo, L.Y. and Friend, J.R., 2006. Electrospinning carbon nanotube polymer composite nanofibres. *Journal of Experimental Nanoscience*, 1(2), pp. 177–209.

Yoon, K. et al., 2006. High flux ultrafiltration membranes based on electrospun nanofibrous PAN scaffolds and chitosan coating. *Polymer*, 47(1), pp. 2434–41.

Yoon, K. et al., 2009. Formation of functional polyethersulfone electrospun membrane for water purification by mixed solvent and oxidation processes. *Polymer*, 50(1), pp. 2893–99.

Yördem, O.S., Papila, M., and Menceloğlu, Y.Z., 2008. Effect of electrospinning parameters on polyacrylonitrile nanofibre diameter: An investigation by response surface methodology. *Materials & Design*, 29, pp. 34–44.

Yoshimoto, H., Shin, Y.M., Terai, H., and Vacanti, J.P., 2003. A biodegradable nanofibre scaffold by electrospinning and its potential for bone tissue engineering. *Biomaterials*, 24(12), pp. 2077–82.

Yu, D.G. et al., 2009. Ultrafine ibuprofen-loaded polyvinylpyrrolidone fibre mats using electrospinning. *Polymer International*, 58(9), pp. 1010–13.

Yu, D.G. et al., 2011. Polyacrylonitrile nanofibres prepared using coaxial electrospinning with LiCl solution as sheath fluid. *Nanotechnology*, 22(43), pp. 1–7, doi:10.1088/0957-4484/22/43/435301.

You, H. et al., 2013. High flux low pressure thin film nanocomposite ultrafiltration membranes based on nanofibrous substrates. *Separation and Purification Technology*, 108(1), pp. 143–151.

Zhang, C. et al., 2005. Study on morphology of electrospun poly(vinyl alcohol) mats. *European Polymer Journal*, 41, pp. 423–32.

Zhang, D. and Chang, J., 2007. Patterning of electrospun fibres using electroconductive templates†. *Advanced Materials*, 19(21), pp. 3664–67.

Zhang, J.F., Yang, D.Z., Xu, F., Zhang, Z.P., Yin, R.X., and Nie, J., 2009. Electrospun core–shell structure nanofibres from homogeneous solution of poly (ethylene oxide)/chitosan. *Macromolecules*, 42(14), pp. 5278–84.

Zhang, S., 2003. Fabrication of novel biomaterials through molecular self-assembly. *Nature Biotechnology*, 21, pp. 1171–78.

Zhang, Y. et al., 2004. Preparation of core-shell structured PCL-r-gelatin bi-component nanofibres by coaxial electrospinning. *Chemistry of Materials*, 16(18), pp. 3406–409.

Zhang, Y.Z. et al., 2006. Coaxial electrospinning of (fluorescein isothiocyanate-conjugated bovine serum albumin)-encapsulated poly (ε-caprolactone) nanofibres for sustained release. *Biomacromolecules*, 7(4), pp. 1049–57.

Zhao, Z. et al., 2012. High performance ultrafiltration membrane based on modified chitosan coating and electrospun nanofibrous PVDF scaffolds. *Journal of Membrane Science*, 394/395(1), pp. 209–217.

Zhou, F.L. and Gong, R.H., 2008. Manufacturing technologies of polymeric nanofibres and nanofibre yarns. *Polymer International*, 57, pp. 837–45.

Zhou, F.L., Gong, R.H., and Porat, I., 2009. Mass production of nanofibre assemblies by electrostatic spinning. *Polymer International*, 58(4), pp. 331–42.

Zhou, F.L., Gong, R.H., and Porat, I., 2010. Mass production of nanofibre assemblies by electrostatic spinning. *Journal of Applied Polymer Science*, 115(5), pp. 2591–98.

Zong, X. et al., 2002. Structure and process relationship of electrospun bioabsorbable nanofibre membranes. *Polymer*, 43(16), pp. 4403–12.

Zussman, E., Theron, A., and Yarin, A.L., 2003. Formation of nanofibre crossbars in electrospinning. *Applied Physics Letters*, 82(6), pp. 973–75.

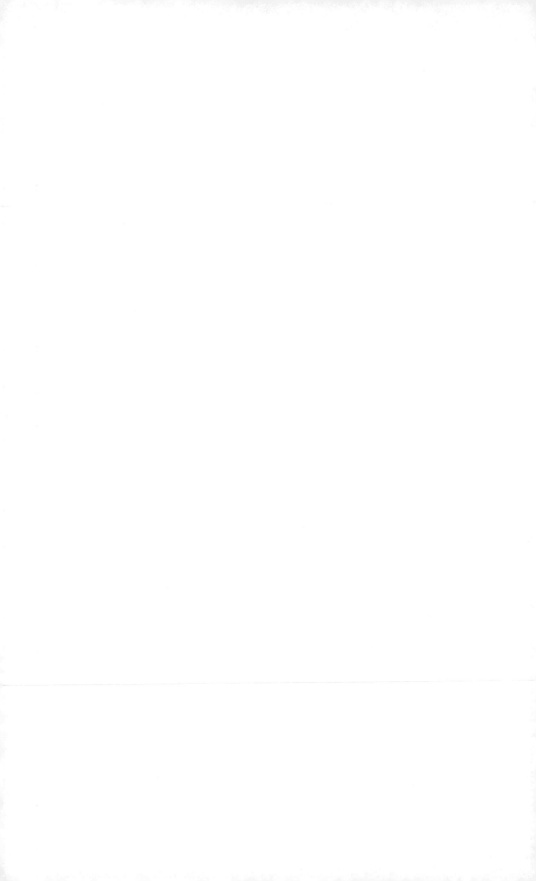

3 Control of Crystallization of Poly(Lactic Acid) Membranes

Shuichi Sato, Ryohei Shindo,
Shinji Kanehashi, and Kazukiyo Nagai

CONTENTS

3.1 Introduction 75
3.2 Crystallization of the PLA Membrane 76
    3.2.1 Thermally Induced Crystallization.. 76
    3.2.2 Solvent-Induced Crystallization 81
        3.2.2.1 Solubility Tests of the PLA Membrane............................. 81
        3.2.2.2 Physical Properties of Solvent-Induced Crystallized
                 PLA Membrane.. 83
    3.2.3 Photo-Induced Crystallization... 93
3.3 Conclusions... 99
References... 100

3.1 INTRODUCTION

Poly(lactic acid) (PLA, Figure 3.1) is an environment-friendly, biodegradable polymeric substance with a low melting point, high moldability, and high transparency. PLA is often used in packages, automobiles, and electronics. There are material and chemical recycling methods available for petroleum-based conventional plastics such as poly-ethylene, polystyrene, and poly(ethylene terephthalate). For such applications, polymer materials are exposed to various conditions during use, such as varying temperatures with cooling and heating conditions, organic solvents, and ultraviolet (UV) irradiation.

Figure 3.2 presents a bar graph comparing the number of references having *poly(lactic acid)* as the keyword for each year. This figure shows the result of SciFinder®. The number of references increased from 2000. PLA with biorecyclable behavior has drawn interest because of the "Act Concerning the Promotion of the Utilization of Recyclable Food Waste" from 2000. Furthermore, PLA production and distribution by NatureWorks LLC started in 2002. About 1500 references were reported on PLA over the last year, arguably because of its advantages, including its enhanced biological degradation behavior for a biorecyclable material. PLA is also expected to replace petroleum-based plastics.

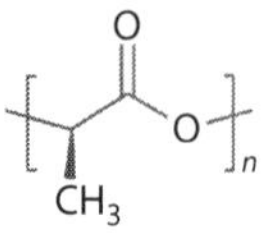

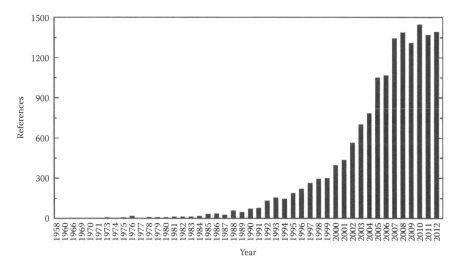

FIGURE 3.1   Chemical structure of PLA.

FIGURE 3.2   Number of references whose keyword for each year is *poly(lactic acid)*.

The physical structure of PLA has been systematically investigated. The structure of crystallized PLA is especially important for next-generation advanced PLA membrane. PLA crystallization can be controlled by cooling and heating conditions, various organic solvents, and vacuum UV (VUV) irradiation [1–3]. This chapter focuses on various controlled crystalline structures fabricated by thermal, solvent, and photo-induced crystallizations.

3.2   CRYSTALLIZATION OF THE PLA MEMBRANE

3.2.1   THERMALLY INDUCED CRYSTALLIZATION

PLA membranes with different crystalline structures were prepared by regulating the heating and cooling conditions during membrane preparation [1]. Thermally induced crystallization of PLA membranes was conducted by casting under vacuum for 48 hours at 70°C, 80°C, 90°C, and 150°C. The thermally treated PLA membranes were then cooled to room temperature under atmospheric pressure. The properties of the PLA membranes prepared in the study of Sawada et al. are summarized in Table 3.1 [1]. The membrane density increased from 1.257 to 1.273 g cm^{-3} as the treatment temperature increased from 70°C to 150°C.

The wide-angle X-ray diffraction (WAXD) patterns in the PLA membranes are shown in Figure 3.3. A sharp halo near the diffraction angles 15°, 16°, and 19° appeared

TABLE 3.1
Characterization of Thermally Treated PLA Membranes

Temperature (°C)	70	80	90	150
Density (g cm⁻³)	1.257 ± 0.001	1.264 ± 0.001	1.267 ± 0.001	1.273 ± 0.001
X_{C-WAXD} (%)	0.0	19.2 ± 0.3	33.0 ± 0.2	47.9 ± 0.4
T_g (°C)	60.3 ± 2.1	56.3 ± 0.3	57.1 ± 1.1	59.2 ± 0.3
T_c (°C)	–	124.2 ± 0.1	129.5 ± 0.1	124.2 ± 0.1
T_m (°C)	–	154.3 ± 0.1	154.1 ± 0.1	158.1 ± 0.1
ΔH_c (J/g)	–	-10.4 ± 0.3	-2.0 ± 0.1	2.4 ± 0.5
ΔH_m (J/g)	–	18.6 ± 0.5	21.6 ± 0.2	34.6 ± 0.3
X_{C-DSC} (%)	0.0	8.8 ± 0.4	21.1 ± 0.2	39.9 ± 0.2
T_d (%)	94.0 ± 0.1	93.9 ± 0.1	93.3 ± 0.1	93.3 ± 0.1
T_t (%)	0.3 ± 0.1	4.1 ± 0.1	29.7 ± 0.1	48.7 ± 0.1
Haze (%)	0.3 ± 0.1	4.4 ± 0.3	31.8 ± 0.7	52.2 ± 0.4

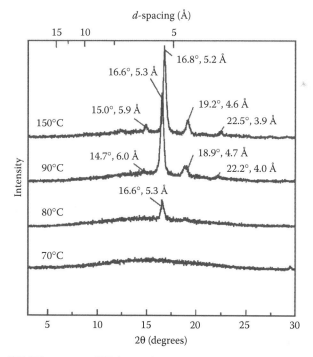

FIGURE 3.3 WAXD patterns of PLA membranes treated at 70°C, 80°C, 90°C, and 150°C. (Reprinted from *Transactions of the Materials Research Society of Japan*, 35: 241–46, H. Sawada et al., Gas transport properties and crystalline structures of poly[lactic acid] membranes. Copyright 2010, with permission from Materials Research Society of Japan.)

for the PLA membrane treated at temperatures higher than 80°C, whereas only a broad halo was observed for the PLA membranes that were treated at 70°C. The PLA membranes that were treated at 70°C exhibited an amorphous structure, whereas those treated at temperatures higher than 80°C showed crystalline structures. As treatment temperature increased, crystallinity (X_{C-WAXD}) increased from 19.2% to 47.9% (Table 3.1). However, the crystalline structures of the crystalline PLA membranes varied because the position of the sharp halo slightly shifted with heating.

The differential scanning calorimetry (DSC) thermograms of the PLA membranes during the first heating scan are shown in Figure 3.4. The sample preparation conditions of the PLA products significantly influenced their thermal properties [4]. The literature indicates that the glass transition temperature (T_g) of PLA membranes ranges from 55°C to 69°C [4–11]. The T_g values of the membranes were 60.3°C, 56.3°C, 57.1°C, and 59.2°C for heat treatments at 70°C, 80°C, 90°C, and 150°C, respectively, which were within the range mentioned in the literature. The crystallization temperature (T_c) for PLA in the literature varied from 79°C to 118°C [4–11]. The T_c values of the membranes were 124.2°C, 129.5°C, and 124.2°C for heat treatments at 80°C, 90°C, and 150°C, respectively, which were slightly higher than the values stated in the literature. The melting temperature (T_m) for PLA in the literature varied from 149°C to 192°C [4–11]. The T_m values of the membranes were 154.3°C, 154.1°C, and 158.1°C for heat treatments at 80°C, 90°C, and 150°C, respectively, which were within the range mentioned in the literature. Thermal analysis showed that the PLA membrane treated at 70°C exhibited an amorphous structure, whereas

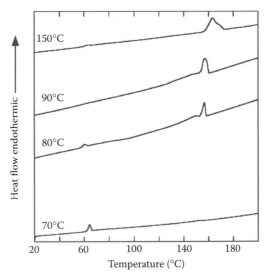

FIGURE 3.4 DSC thermograms of the first heating scan of PLA membranes treated at 70°C, 80°C, 90°C, and 150°C. (Reprinted from *Transactions of the Materials Research Society of Japan*, 35: 241–46, H. Sawada et al., Gas transport properties and crystalline structures of poly[lactic acid] membranes. Copyright 2010, with permission from Materials Research Society of Japan.)

the membranes treated at higher than 80°C showed a crystalline structure. As thermal treatment temperature increased, crystallinity (X_{C-DSC}) increased from 8.8% to 39.9% (Table 3.1). X_{C-DSC} was estimated by

$$X_{C\text{-}DSC} = \frac{\Delta H_m + \Delta H_c}{\Delta H_m^0} \times 100 \qquad (3.1)$$

where:

ΔH_m and ΔH_c represent the enthalpies of melting and crystallization of a polymer (J/g), respectively

ΔH_m^0 denotes the enthalpy of the PLA (L-donor 100%) crystal with an infinite crystal thickness and value of 93 J g^{-1} [12].

The X_{C-DSC} values of the PLA membrane were 8.8% at 80°C, 21.1% at 90°C, and 39.9% at 150°C. As mentioned previously, the X_{C-WAXD} value of the PLA membrane was 19.2% at 80°C, 33.0% at 90°C, and 47.9% at 150°C. The X_{C-DSC} value differed from the X_{C-WAXD} value. The X_{C-DSC} value was directly determined with the enthalpy changes, whereas the X_{C-WAXD} value was estimated with halo pattern division. Given this consideration, the X_{C-DSC} value had higher accuracy compared with the X_{C-WAXD} value. However, both WAXD and DSC analyses showed that the crystallinity of the PLA membrane increased as treatment temperature increased.

Figure 3.5 shows that the PLA membranes gradually become cloudy with the increase in treatment temperatures. Regardless of crystallinity, the total light transmission (T_t) of all PLA membranes was almost constant at about 93%. However, as treatment temperature increased, the diffused light transmission (T_d) increased from 0.3% to 48.7%. Thus, as crystallinity increased, the haze value increased from 0.3% to 52.2% (Table 3.1).

Polarization microscopy (POM) images, which are the standard for characterization of crystal domain dispersion, are presented in Figure 3.5. The PLA membrane treated at 70°C exhibited an amorphous structure and no crystal domains were observed. By contrast, the PLA membranes treated at higher than 80°C exhibited domains with dispersed color variation. As treatment temperature increased, the size of one unit of the color variation domains increased by about 3, 10, and 15 μm at 80°C, 90°C, and 150°C, respectively.

Scanning electron microscopy (SEM) images of all PLA membranes are presented in Figure 3.5. A smooth surface was observed in the amorphous PLA membrane treated at 70°C. In the PLA membranes treated at temperatures higher than 80°C, as crystallinity increased, the PLA crystal growth gradually branched out radially. The PLA membranes treated at 80°C and 90°C had spherical discontinuous units, whereas the PLA membrane treated at 150°C had continuous crystal branches. In Figure 3.5, the size of one unit of the color variation domains in the POM images expresses the uneven surface of the branched crystalline structures from the SEM images.

To conclude this section, the preparation of crystalline PLA membranes by regulating the heating and cooling conditions has been described. The crystalline structure of the PLA membranes has been systematically investigated. The PLA

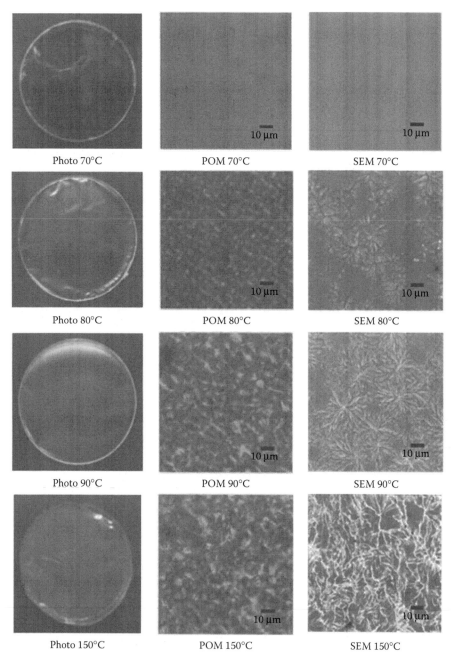

FIGURE 3.5 Photographs, POM, and SEM images of PLA membranes treated at 70°C, 80°C, 90°C, and 150°C. (Reprinted from *Transactions of the Materials Research Society of Japan*, 35: 241–46, H. Sawada et al., Gas transport properties and crystalline structures of poly[lactic acid] membranes. Copyright 2010, with permission from Materials Research Society of Japan.)

membrane treated at 70°C exhibited amorphous structures, whereas the membranes treated at temperature higher than 80°C exhibited crystalline structures. As treatment temperatures increases, the crystallinity also increased.

3.2.2 SOLVENT-INDUCED CRYSTALLIZATION

3.2.2.1 Solubility Tests of the PLA Membrane

Organic solvents significantly affect PLA properties; however, these effects are rarely reported. In other study, benzene-, toluene-, and xylene-induced crystallization in PLA membranes, and the sorption properties of ethyl lactate and aromatic hydrocarbon solvents were reported [13–15]. However, these reports only focused on a single group of solvents. Systematic investigations on other groups of solvents have not been conducted. The effects of 60 liquid organic solvents on PLA were systematically investigated using the Hansen solubility parameter (HSP), one of the digitizing methods for analyzing the interaction between polymer materials and organic solvents [2]. In HSP analysis, all solvents have three parameters, namely, energy from dispersion bonds between the molecules (δ_d), dipolar intermolecular force between the molecules (δ_p), and hydrogen bonds between molecules (δ_h). All solvents were characterized by a point in a three-dimensional structure, in which δ_d, δ_p, and δ_h are plotted on three mutually perpendicular axes. If the HSP values of the various organic solvents are near that of the given polymer, the solvent is considered compatible with the polymer material. Based on the HSP, the effects of organic solvents on the physical properties of PLA membrane, such as the degree of swelling, film density, crystallinity, and crystal structure were systematically investigated [2].

The solubility test results are listed in Table 3.2. These results are classified into three types: soluble (O), strongly swollen (ρ), and insoluble (⊆). The PLA membranes are insoluble in 11 of 16 polar solvents (protic acids and alcohols) and 9 of 11 nonpolar solvents (aromatic hydrocarbons and paraffins), but soluble in 28 of 32 aprotic solvents (amines and esters). Therefore, aprotic polar solvents can be used for PLA membranes.

Generally, polymer solubility can be estimated using the solubility parameters calculated by the group contribution method. However, solubility parameters do not accurately reflect polymer solubility. The solubility of PLA membranes was analyzed using the HSP. The solubility of PLA in 60 organic solvents using δ_d, δ_p, and δ_h is shown in Figure 3.6. The soluble solvent area ranged from 17.0 to 20.0 MPa$^{1/2}$ in δ_d, 7.0 to 11.0 MPa$^{1/2}$ in δ_p, and 5.0 to 9.0 MPa$^{1/2}$ in δ_h. On the basis of these results, R could be determined using the following equation:

$$R = \left[4\left(\delta_{dS} - \delta_{dP}\right)^2 + \left(\delta_{pS} - \delta_{pP}\right)^2 + \left(\delta_{hS} - \delta_{hP}\right)^2 \right]^{1/2} \tag{3.2}$$

where:

δ_{dS}, δ_{pS}, and δ_{hS} represent the solubility parameter of the solvent
δ_{dP}, δ_{pP}, and δ_{hP} represent the solubility parameter of the polymer

TABLE 3.2
HSP of Various Organic Solvents

Group	Solvent	Solvent Type	δ_d	δ_p	δ_h	δ_t	Result
Acid	Formic acid	Polar protic	14.3	11.9	16.6	25	ρ
	Acetic acid	Polar protic	14.5	8	13.5	21.3	ρ
Alcohol	Methanol	Polar protic	15.1	12.3	22.3	29.7	⊆
	Ethanol	Polar protic	15.8	8.8	19.4	26.6	⊆
	1-Propanol	Polar rotic	16	6.8	17.4	24.6	⊆
	2-Propanol	Polar protic	15.8	6.1	16.4	23.5	⊆
	1-Butanol	Polar protic	16	5.7	15.8	23.1	⊆
	3-Methyl-1-butanol	Polar protic	15.33	4.59	13.55	20.97	⊆
	Cyclohexanol	Polar protic	17.4	4.1	13.5	22.5	⊆
	Benzyl alcohol	Polar protic	18.4	6.3	13.7	23.7	ρ
	m-Cresol	Polar protic	18	5.1	12.9	22.7	O
Amine	Pyridine	Polar aprotic	19	8.8	5.9	21.7	O
	Aniline	Polar protic	19.4	5.1	10.2	22.5	ρ
	N-methylpyrrolidone	Polar aprotic	18	12.3	7.2	22.9	O
Anhydride	Acetic anhydride	Polar aprotic	18.4	16.4	10.2	26.6	ρ
Aromatic	Benzene	Nonpolar	18.4	0	2	18.6	O
hydrocarbon	Toluene	Nonpolar	18	1.4	2	18.2	ρ
	o-Xylene	Nonpolar	18.0	1.4	2.9	18.3	⊆
	m-Xylene	Nonpolar	17.8	0.8	2.7	18.0	⊆
	p-Xylene	Nonpolar	17.8	0.0	2.7	18.0	⊆
	Ethylbenzene	Nonpolar	17.8	0.6	1.4	17.8	⊆
Ester	Methyl acetate	Polar aprotic	15.5	7.2	7.6	18.8	ρ
	Ethyl formate	Polar aprotic	15.5	8.4	8.4	19.6	ρ
	γ-Butyrolactone	Polar aprotic	19.0	16.6	7.4	26.2	O
	Ethylacetate	Polar aprotic	15.8	5.3	7.2	18.2	O
	Methyl methacrylate	Polar aprotic	16.07	4.63	8.13	18.60	ρ
	Propylene-1,2-carbonate	Polar aprotic	20.1	18	4.1	27.2	O
	n-Butyl acetate	Polar aprotic	15.8	3.7	6.3	17.4	⊆
	Ethyl lactate	Polar aprotic	16	7.6	12.5	21.7	⊆
	Diethyl phthalate	Polar aprotic	17.6	9.6	4.5	20.5	ρ
	Di-n-butyl phthalate	Polar aprotic	17.8	8.6	4.1	20.3	⊆
Ether	Tetrahydrofuran	Polar aprotic	16.8	5.7	8	19.4	O
	1,3-Dioxolane	Polar aprotic	17.26	8.14	9.29	21.22	O
	1,4-Dioxane	Polar aprotic	19	1.8	7.4	20.5	O
	Isopropyl ether	Nonpolar	13.75	2.84	4.61	14.78	⊆
Organic chloride	Acetyl chloride	Polar aprotic	15.80	10.60	3.90	19.40	ρ
Chlorinated	Dichloromethane	Polar aprotic	18	12.3	7.2	22.9	O
solvent	Chloroform	Polar aprotic	17.8	3.1	5.7	19	O
	o-Dichlorobenzene	Polar aprotic	19.2	6.3	3.3	20.5	ρ

TABLE 3.2 (*Continued*)
HSP of Various Organic Solvents

Group	Solvent	Solvent Type	δ_d	δ_p	δ_h	δ_t	Result
Ketone	Acetone	Polar aprotic	15.5	10.4	7	20.1	○
	2-Butanone	Polar aprotic	16	9	5.1	19	ρ
	Cyclohexanone	Polar aprotic	17.8	6.3	5.1	19.6	ρ
	Methyl isobutyl ketone	Polar aprotic	15.3	6.1	4.1	17	⊆
	Acetophenone	Polar aprotic	19.6	8.6	3.7	21.7	ρ
Nitrogen-	Nitrobenzene	Polar aprotic	20.1	8.6	4.1	22.1	○
containing	Acetonitrile	Polar aprotic	15.3	18	6.1	24.6	○
	Formamide	Polar protic	17.2	26.2	19	36.6	⊆
	Dimethylformamide	Polar aprotic	17.4	13.7	11.3	24.8	ρ
	Dimethylacetamide	Polar aprotic	16.8	11.5	10.2	22.7	○
Organophosphate	Trimethyl phosphate	Polar aprotic	17.09	16.15	10.54	25.77	ρ
	Triethyl phosphate	Polar aprotic	16.54	11.12	8.75	21.76	ρ
Paraffinic	Hexane	Nonpolar	14.9	0	0	14.9	⊆
hydrocarbon	Heptane	Nonpolar	15.3	0	0	15.3	⊆
	Octane	Nonpolar	15.6	0	0	15.6	⊆
	2,2,4-Trimethylpentane	Nonpolar	14.3	0	0	14.3	⊆
Polyhydric	Ethylene glycol	Polar protic	17	11	26	32.9	⊆
alcohol	Glycerol	Polar protic	17.4	12.1	29.3	36.2	⊆
Sulfur-containing	Dimethyl sulfoxide	Polar aprotic	18.2	6.3	6.1	20.3	ρ
Water	Water	Polar protic	15.5	16	42.4	47.9	⊆

○; Soluble, ρ; Strongly swollen, ⊆; Insoluble (obtained membrane).

Equation 3.2 calculates the distance between the solvent and the polymer. When total R of these solvents was minimal, the solubility parameters of PLA were determined as follows: $\delta_d = 17.5$ MPa$^{1/2}$, $\delta_p = 9.5$ MPa$^{1/2}$, and $\delta_h = 7.3$ MPa$^{1/2}$. The total HSP (δ_{tP}) was 21.2 MPa$^{1/2}$ and was obtained using the following equation:

$$\delta_{tP}^2 = \delta_{dP}^2 + \delta_{pP}^2 + \delta_{hP}^2 \tag{3.3}$$

3.2.2.2 Physical Properties of Solvent-Induced Crystallized PLA Membrane

The results of the swelling test are listed in Table 3.3. The relationship between swelling and δ_{tP} is shown in Figure 3.7. The solvents that dissolved and strongly swelled PLA in the solubility tests are plotted above 300 wt%. As shown in Figure 3.7, solvents with solubility parameters near 21.2 exhibited greater swelling. The relationship between swelling and the three HSP parameters are shown in Figure 3.7. The solvents with δ_d less than 16 MPa$^{1/2}$ and δ_p less than 9 MPa$^{1/2}$ exhibited less swelling (Figures 3.7b and c), whereas the solvents with δ_h near 7.3 MPa$^{1/2}$ exhibited more swelling (Figure 3.7d). As shown in Figure 3.7 the hydrogen bonding solubility parameter more effectively reflects the solubility of the PLA based on the three

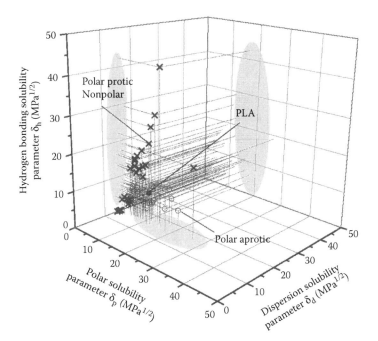

FIGURE 3.6 Solubility region of PLA (■) in various organic solvents in the Hansen space. Solvent type: polar aprotic (○), polar protic, and nonpolar (×). (Reprinted from S. Sato et al., Effects of various liquid organic solvents on solvent-induced crystallization of amorphous poly[lactic acid] film, *Journal of Applied Polymer Science*, 2013, 129, 1607–17. Copyright Wiley-VCH Verlag GmbH & Co. KGaA.)

cohesive parameters. However, the hydrogen bonding solubility parameter depends not only on hydrogen bonding but also on dispersion and polar parameters. These results are consistent with the solubility tests.

The chemical structure of the insoluble PLA membranes immersed in the solvents at 35°C for 24 h was confirmed by nuclear magnetic resonance (NMR) and Fourier transform infrared (FTIR) analyses. These insoluble membranes immersed in the solvents were dried under a vacuum for 48 h at 70°C. And then, NMR and FTIR analyses confirmed the chemical structure and removal of the residual solvent—PLA: ^{1}H-NMR (500 MHz, CDCl$_3$-d, δ); 1.57–1.59 (3H, H^1); and 5.10–5.19 (H, H^2). IR, 2960 and 2870 cm^{-1} (C–H stretching); 1750 cm^{-1} (C=O stretching); 1470 cm^{-1} (C–H stretching and C–H bending); and 1180 and 1080 cm^{-1} (C–O–C bending). No difference in peak position was observed between the immersed PLA membranes and the nonimmersed PLA membranes in NMR and FTIR spectra. This result indicates that no changes in chemical structure occurred under solvent immersion.

The photographs of the PLA membranes in this study are shown in Figure 3.8. The PLA membranes immersed in 24 separate solvents were prepared (i.e., insoluble) for solubility testing. The degree of cloudiness can be classified into four types: clear transparency membranes (Type I), slightly cloudy membranes (Type II), cloudy membranes (Type III), and creamy white membranes (Type IV). Type I membranes are clear, similar to nonimmersed membranes, and swell by less than 2 wt%. Type II membranes are

TABLE 3.3

Density and Degree of Swelling of PLA Membranes Immersed in Various Organic Solvents

Group	Solvents	Swelling (wt%)	Density (g cm^{-3})
	Non	–	1.257 ± 0.001
Water	Water	0.7 ± 0.1	1.256 ± 0.002
Alcohol	Methanol	14.0 ± 1.4	1.241 ± 0.002
	Ethanol	8.7 ± 0.5	1.252 ± 0.001
	1-Propanol	7.5 ± 0.1	1.252 ± 0.001
	2-Propanol	8.3 ± 0.3	1.254 ± 0.002
	1-Butanol	20.4 ± 0.1	1.255 ± 0.003
	3-Methyl-1-butanol	7.0 ± 1.8	1.250 ± 0.001
	Cyclohexanol	8.4 ± 0.8	1.251 ± 0.001
Nitrogen-containing	Formamide	2.5 ± 0.1	1.251 ± 0.001
Aromatic hydrocarbon	o-Xylene	274.0 ± 37.3	0.488 ± 0.050
	m-Xylene	159.3 ± 5.8	0.851 ± 0.025
	p-Xylene	147.4 ± 2.2	0.833 ± 0.014
	Ethylbenzene	204.3 ± 35.2	0.532 ± 0.044
Ester	n-Butyl acetate	161.9 ± 1.9	0.835 ± 0.033
	Ethyl lactate	192.4 ± 25.2	0.801 ± 0.010
	Di-n-butyl phthalate	47.4 ± 2.8	1.251 ± 0.002
Ether	Isopropyl ether	7.5 ± 1.1	1.250 ± 0.001
Ketone	Methyl isobutyl ketone	149.1 ± 33.4	0.708 ± 0.009
Polyhydric alcohol	Glycerol	4.2 ± 2.2	1.250 ± 0.002
	Ethylene glycol	2.1 ± 0.6	1.256 ± 0.003
Paraffin	Hexane	0.39 ± 0.31	1.257 ± 0.001
	Heptane	0.27 ± 0.11	1.261 ± 0.001
	Octane	0.29 ± 0.21	1.260 ± 0.002
	2,2,4-trimethylpentane	0.23 ± 0.11	1.256 ± 0.002

slightly cloudy and swell by 2–5 wt%. Type III membranes are cloudier than Type II and swell by 7–20 wt%. Type IV membranes are creamy white and swell by more than 140 wt%. These cloudy membranes depend on crystallization for light diffusion.

The membrane density of these PLA membranes is summarized in Table 3.3. The sample preparation conditions of the PLA products significantly influenced their membrane densities. The literature indicates that PLA membranes have density values ranging from 1.248 to 1.27 g cm^{-3} [9,16–18]. The density of the nonimmersed amorphous PLA membrane in this study was 1.257 g cm^{-3}, which falls within the range reported in the literature [2]. Type I and Type II membranes had densities of approximately 1.257 g cm^{-3}. However, the densities of Type III membranes decreased, ranging from 1.241 to 1.252 g cm^{-3}. The membranes that swelled more had lower densities. The densities of Type IV membranes sharply decreased, ranging from 0.488 to 0.851 g cm^{-3}.

The POM and SEM images of these four PLA film types are shown in Figure 3.8. Type I membranes have an amorphous structure and no crystalline

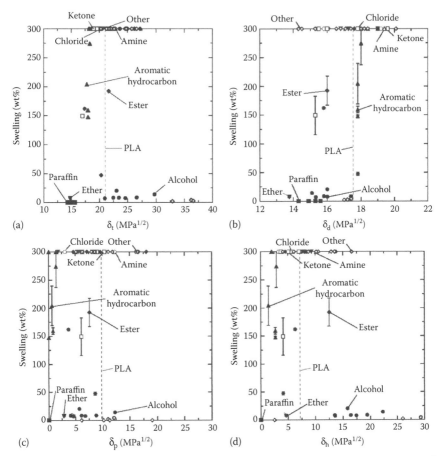

FIGURE 3.7 Relationship between swelling and (a) total Hansen solubility parameter, δ_t, (b) δ_d, (c) δ_p, (d) δ_h. Organic solvent: alcohol (●), paraffin (■), aromatic hydrocarbon (▲), ester (◆), ether (▼), chloride (○), ketone (□), amine (△), and others (◇). (Reprinted from S. Sato et al., Effects of various liquid organic solvents on solvent-induced crystallization of amorphous poly[lactic acid] film, *Journal of Applied Polymer Science*, 2013, 129, 1607–17. Copyright Wiley-VCH Verlag GmbH & Co. KGaA.)

domains, whereas Type II and Type III membranes have dispersed crystal domains. The maximum size of one unit of the color variation domains of Type II and Type III membranes approximately ranged from 1 to 5 μm. Type IV membranes have unobservable crystal morphologies because light does not pass through the membrane.

Smooth surfaces were observed in Type I membranes, whereas changes in surface structure were observed in Type II to Type IV membranes immersed in organic solvents. Microasperities were observed in Type II membranes. As crystallinity increased, crystal growth gradually branched out radially in Type III membranes. In Type IV membranes, numerous microscale pores were observed. High-swelling PLA membranes exhibit a porous structure. These results indicate that the reduction

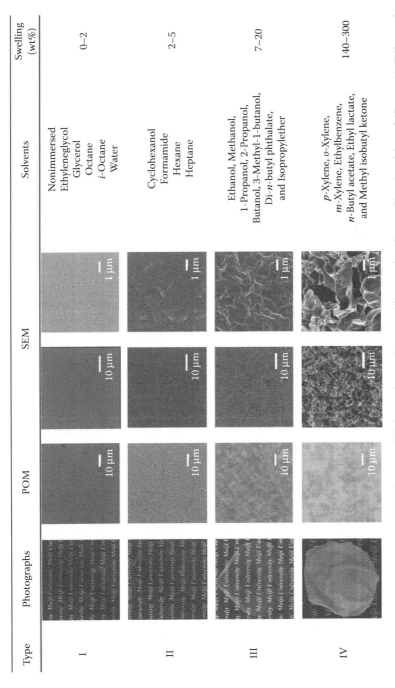

Type	Photographs	POM	SEM	SEM	Solvents	Swelling (wt%)
I			10 μm	1 μm	Nonimmersed Ethyleneglycol Glycerol Octane *i*-Octane Water	0–2
II		10 μm	10 μm	1 μm	Cyclohexanol Formamide Hexane Heptane	2–5
III		10 μm	10 μm	1 μm	Ethanol, Methanol, 1-Propanol, 2-Propanol, Butanol, 3-Methyl-1-butanol, Di-*n*-butyl phthalate, and Isopropylether	7–20
IV		10 μm	10 μm	1 μm	*p*-Xylene, *o*-Xylene, *m*-Xylene, Ethylbenzene, *n*-Butyl acetate, Ethyl lactate, and Methyl isobutyl ketone	140–300

FIGURE 3.8 Photographs, POM, and SEM images of PLA membranes immersed in organic solvents. (Reprinted from S. Sato et al., Effects of various liquid organic solvents on solvent-induced crystallization of amorphous poly[lactic acid] film, *Journal of Applied Polymer Science*, 2013, 129, 1607–17. Copyright Wiley-VCH Verlag GmbH & Co. KGaA.)

in membrane density depends on organic solvent permeation through the PLA membranes because of high solubility.

The DSC thermograms of the PLA membranes immersed into four types of solvents at 35°C for 24 h are shown in Figure 3.9. The T_g, T_c, T_m, ΔH_c, and ΔH_m values determined using the DSC thermograms of the PLA membranes are summarized in Table 3.4. T_c and T_m peaks were not observed in Type I membranes. The sample preparation conditions of the PLA products significantly influenced their thermal properties [4–11].

The T_g values of Type II to Type IV PLA membranes ranged from 56.1°C to 63.1°C, which is within the range stated in the literature [4–11]. The T_c values of Type II to Type IV PLA membranes varied from 107.9°C to 115.7°C, which is within the range mentioned in the literature [4–11]. The T_m values for Type II to Type IV PLA membranes ranged from 146.3°C to 150.3°C; this range is lower than that mentioned in the literature [4–11]. The T_m values of Type II and Type IV PLA membranes peaked at approximately 150°C, whereas those of Type III PLA membranes peaked at approximately 144°C and 150°C. Double melting endothermic peaks generally indicate recrystallization behavior or differences in crystalline structure, depending on the melting behavior.

Crystallinity as a function of density is shown in Figure 3.10. The density of Type I and Type II films was 1.257 g cm^{-3}, which is similar to that of nonimmersed amorphous PLA membranes. Crystallinity does not depend on membrane density at low-range crystallinity (0%–5%). Meanwhile, the crystallinity of Type III PLA membranes was approximately 25% at membrane densities, ranging from 1.242 to 1.257 g cm^{-3}. Density slightly decreased with increasing crystallinity. The crystallinity of

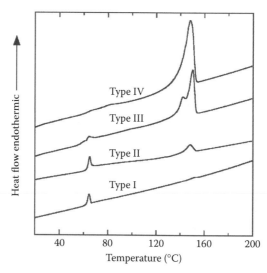

FIGURE 3.9 DSC curves of PLA membranes immersed in various organic solvents. (Reprinted from S. Sato et al., Effects of various liquid organic solvents on solvent-induced crystallization of amorphous poly[lactic acid] film, *Journal of Applied Polymer Science*, 2013, 129, 1607–17. Copyright Wiley-VCH Verlag GmbH & Co. KGaA.)

TABLE 3.4
Thermal Properties of Solvent-Induced Crystallized PLA Membranes

Group	Organic Solvent	T_g (°C)	T_c (°C)	T_m (°C)	ΔH_c (J/g)	ΔH_m (J/g)	X_{c-DSC} (%)
	Non	60.3 ± 2.1	—	—	—	—	0.0
Alcohol	Methanol	62.6 ± 0.4	110.8 ± 0.3	145.4 ± 0.3, 150.1 ± 0.2	-0.3 ± 0.1	26.2 ± 2.0	27.9 ± 2.1
	Ethanol	62.6 ± 0.3	109.0 ± 1.2	143.4 ± 0.6, 150.3 ± 0.3	-0.5 ± 0.1	24.7 ± 1.3	25.0 ± 1.5
	1-Propanol	62.1 ± 0.3	108.7 ± 0.3	142.1 ± 0.2, 150.0 ± 0.1	-0.7 ± 0.1	24.8 ± 1.9	26.0 ± 2.0
	2-Propanol	60.2 ± 0.5	107.9 ± 0.2	142.3 ± 0.1, 149.5 ± 0.3	-0.5 ± 0.1	24.4 ± 1.4	25.7 ± 1.5
	1-Butanol	57.5 ± 1.8	109.0 ± 0.3	142.1 ± 1.4, 150.0 ± 0.4	-0.3 ± 0.1	24.1 ± 1.6	25.6 ± 1.7
	3-Methyl-1-butanol	60.6 ± 0.2	110.0 ± 0.5	148.6 ± 0.2	-1.0 ± 0.2	21.0 ± 0.6	21.6 ± 0.4
	Cyclohexanol	62.4 ± 0.1	115.7 ± 0.8	148.2 ± 0.1	-0.1 ± 0.0	2.0 ± 0.2	2.2 ± 0.2
Nitrogen-containing	Formamide	62.4 ± 0.2	114.8 ± 0.5	147.8 ± 0.0	-0.4 ± 0.0	4.4 ± 0.3	4.3 ± 0.3
Aromatic hydrocarbon	o-Xylene	61.8 ± 0.1	109.7 ± 0.2	147.5 ± 0.3	-0.5 ± 0.1	38.9 ± 1.0	41.3 ± 1.2
	m-Xylene	62.1 ± 1.3	110.3 ± 0.8	148.6 ± 0.8	-0.4 ± 0.1	36.5 ± 2.0	38.8 ± 2.0
	p-Xylene	59.5 ± 0.2	109.1 ± 1.3	148.2 ± 0.2	-0.3 ± 0.2	36.2 ± 1.4	38.6 ± 1.4
	Ethylbenzene	60.6 ± 0.8	108.7 ± 3.1	148.7 ± 0.5	-0.5 ± 0.1	40.3 ± 2.6	42.8 ± 2.7
	Non	60.3 ± 2.1	—	—	—	—	0.0
Ester	n-Butyl acetate	59.4 ± 0.2	110.3 ± 0.8	147.8 ± 0.4	-0.4 ± 0.1	39.2 ± 3.5	41.7 ± 3.7
	Ethyl lactate	60.6 ± 2.1	109.8 ± 0.5	148.3 ± 0.7	-0.4 ± 0.1	40.1 ± 2.7	42.6 ± 2.9
	Di-n-butyl phthalate	45.1 ± 0.3	110.2 ± 0.4	146.3 ± 0.4	-0.3 ± 0.2	25.8 ± 2.2	27.4 ± 2.2
Ether	Isopropyl ether	56.1 ± 0.3	109.4 ± 0.1	141.3 ± 0.2, 149.8 ± 0.2	-0.5 ± 0.0	23.9 ± 2.0	25.1 ± 2.1
Ketone	Methyl isobutyl ketone	62.1 ± 1.0	110.0 ± 0.5	148.9 ± 0.3	-0.5 ± 0.1	38.0 ± 3.0	40.4 ± 3.2
Poly hydricalcohol	Glycerol	61.5 ± 0.1	111.1 ± 0.1	148.3 ± 0.1	-0.3 ± 0.1	2.6 ± 0.1	2.5 ± 0.1
	Ethylene glycol	62.5 ± 1.0	115.7 ± 0.8	148.8 ± 0.5	-0.1 ± 0.0	5.5 ± 0.1	5.8 ± 0.1
Paraffin	Hexane	62.3 ± 0.0	110.2 ± 0.3	148.2 ± 0.1	-0.2 ± 0.0	5.8 ± 0.1	6.0 ± 0.0
	Heptane	62.6 ± 0.4	111.0 ± 0.4	148.4 ± 0.1	-0.3 ± 0.1	6.6 ± 0.8	6.8 ± 0.9
	Octane	63.1 ± 0.1	111.8 ± 4.8	148.6 ± 0.4	-0.1 ± 0.0	2.9 ± 0.5	3.0 ± 0.5
	i-Octane	62.8 ± 0.1	113.0 ± 3.5	148.3 ± 0.1	-0.1 ± 0.0	3.6 ± 0.5	3.8 ± 0.5
Water	Water	61.6 ± 0.1	120.8 ± 0.1	149.1 ± 0.1	-0.1 ± 0.1	1.4 ± 0.1	1.4 ± 0.1

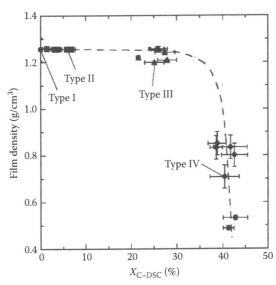

FIGURE 3.10 Membrane density as a function of crystallinity under DSC. (Reprinted from S. Sato et al., Effects of various liquid organic solvents on solvent-induced crystallization of amorphous poly[lactic acid] film, *Journal of Applied Polymer Science*, 2013, 129, 1607–17. Copyright Wiley-VCH Verlag GmbH & Co. KGaA.)

the PLA membranes was approximately 40% at membrane densities, ranging from 0.5 to 0.8 g cm⁻³. This significant density reduction is caused by the porous structures observed in Type IV PLA membranes. Crystalline polymer membranes are denser than amorphous membrane; however, PLA membranes crystallized using organic solvents are sparser than amorphous membranes. No liner relationship was observed between crystallinity and membrane density, as indicated in Figure 3.10. An inflection point was observed in Type III area. The crystalline structure depends on the relationship between crystallinity and membrane density. Consequently, crystalline structures were investigated using X-ray analysis as now discussed (WAXD).

The WAXD patterns of the four types of PLA membranes are shown in Figure 3.11. The WAXD patterns of Type I membranes show only broad peaks. By contrast, several sharp peaks near the diffraction angles, 16° and 19°, were observed in Type II to Type IV membranes. This result indicates that the nonimmersed PLA membranes are amorphous, whereas the solvent-immersed PLA membranes exhibit a crystalline structure. By contrast, Type I PLA membranes showed broad peaks. Despite the appearance of crystalline domains, small crystalline peaks may be masked inside the broad amorphous peaks.

The conditions for the sample preparation of the PLA products significantly influenced their crystalline structure (α-, β-, and γ-form) [19]. The Miller index of the α-form crystal diffraction peaks was observed at 16.7° and 19.2°, which were consistent with (200), (110) and (100), (203), respectively [20–22]. The unit cell of the α-crystal structure has a space group of $P2_12_12_1$ and the following dimensions: $a = 10.68$Å, $b = 6.17$ Å, $c = 28.86$ Å, and $\alpha = \beta = \gamma = 90°$ [23]. Moreover, the peaks of the β-crystal structure

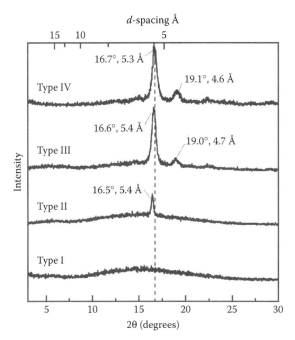

FIGURE 3.11 WAXD patterns of PLA membranes immersed in various organic solvents. (Reprinted from S. Sato et al., Effects of various liquid organic solvents on solvent-induced crystallization of amorphous poly[lactic acid] film, *Journal of Applied Polymer Science*, 2013, 129, 1607–17. Copyright Wiley-VCH Verlag GmbH & Co. KGaA.)

were observed at 16.5°, 18.8°, and 28.8°. The β-crystal structure is a trigonal unit cell with a space group of $P3_2$ and the following dimensions: a = b = 10.52Å, c = 8.80Å, and α = β = 90°, γ = 120° [24,25]. The β-crystal structure is less stable than the α-crystal structure. In addition, the β-crystal structure reportedly exhibits reduced cohesiveness. Type IV membranes formed the α-crystal structures because their WAXD patterns show sharp peaks at 16.7° and 19.1°. This result indicates that Type II and Type III PLA membranes formed β-crystal structures because the observed peak at approximately 16.7° shifted to 16.5°. However, the crystalline structure of Type IV is almost an α-form structure because no shift in the observed peak occurred.

The X_{C-WAXD} values that were determined using the WAXD patterns of the PLA membranes are summarized in Tables 3.5. The crystallinity of Type I membranes could not be determined, whereas that of Type II membranes was approximately 7%. The crystallinity of Type III membranes significantly increased from 32% to 37%. Subsequently, crystallinity gradually increased above 200 wt% swelling. Finally, crystallinity stabilized at 42%–55%. The X_{C-DSC} value differed from the X_{C-WAXD} value. The X_{C-WAXD} value was calculated by the peak fitting method, whereas the X_{C-DSC} value was directly determined by enthalpy change. However, a similar tendency was observed. Crystallinity increased as the membrane density decreased.

The solvent-induced crystallization of the PLA membrane formed a mixture of α- and β-crystals. The diffraction angles 16.5° and 16.7° are considered the maximum

TABLE 3.5
X-Ray Analysis of Solvent-Induced Crystallized PLA Membranes

Group	Organic Solvent	$X_{C\text{-WAXD}}$ (%)	$X_{C\alpha\text{-WAXD}}$ (%)	$X_{C\beta\text{-WAXD}}$ (%)	$X_{C\alpha\text{-WAXD}}/(X_{C\alpha\text{-WAXD}} + X_{C\beta\text{-WAXD}})$ (%)
Alcohol	Non	0	0	0	0
	Methanol	35.5 ± 3.5	35.5 ± 3.5	0	100
	Ethanol	37.1 ± 3.1	19.3 ± 18.5	17.8 ± 17.8	51.0 ± 49.0
	1-Propanol	32.5 ± 6.5	19.7 ± 7.8	12.8 ± 7.7	61.0 ± 23.0
	2-Propanol	35.8 ± 2.4	23.9 ± 5.8	11.9 ± 7.4	67.0 ± 19.0
	1-Butanol	32.8 ± 5.8	23.5 ± 10.3	9.3 ± 4.5	70.0 ± 21.0
	3-Methyl-1-butanol	6.5 ± 3.5	1.3 ± 2.2	5.3 ± 3.6	21.0 ± 29.0
	Cyclohexanol	0	0	0	0
Nitrogen-containing	Formamide	0	0	0	0
aromatic hydrocarbon	o-Xylene	55.3 ± 3.6	52.9 ± 2.5	2.4 ± 4.8	96.0 ± 8.0
	m-Xylene	47.2 ± 6.4	38.4 ± 6.9	6.9 ± 8.7	82.0 ± 18.0
	p-Xylene	47.5 ± 2.2	45.6 ± 2.3	2.0 ± 3.9	96.0 ± 8.0
	Ethylbenzene	49.1 ± 1.3	44.4 ± 8.3	4.7 ± 9.5	91.0 ± 19.0
Ester	Non	0	0	0	0
	n-Butyl acetate	48.1 ± 1.5	48.1 ± 1.5	0	100
	Ethyl lactate	46.2 ± 1.6	34.1 ± 9.2	12.1 ± 10.4	74.0 ± 22.0
	Di-n-butyl phthalate	42.2 ± 4.5	24.8 ± 12.2	17.3 ± 7.7	58.0 ± 22.0
Ether	Isopropyl ether	32.4 ± 2.8	20.9 ± 3.7	11.5 ± 4.8	65.0 ± 14.0
Ketone	Methyl isobutyl ketone	52.0 ± 5.4	50.7 ± 6.8	1.4 ± 2.7	97.0 ± 05.0
Polyhydric hydrocarbon	Glycerol	0	0	0	0
	Ethylene glycol	2.4 ± 0.2	0.2 ± 0.4	2.2 ± 0.4	8.0 ± 16.0
Paraffin	Hexane	4.0 ± 0.9	4.0 ± 0.9	0	100
	Heptane	6.9 ± 2.0	2.7 ± 0.2	4.2 ± 2.2	44.0 ± 16.0
	Octane	0	0	0	0
	i-Octane	0	0	0	0
Water	Water	0	0	0	0

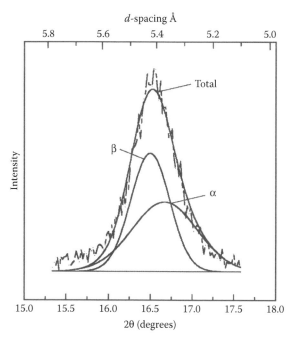

FIGURE 3.12 Waveform separation of Gaussian functions using WAXD patterns. (Reprinted from S. Sato et al., Effects of various liquid organic solvents on solvent-induced crystallization of amorphous poly[lactic acid] film, *Journal of Applied Polymer Science*, 2013, 129, 1607–17. Copyright Wiley-VCH Verlag GmbH & Co. KGaA.)

peak and the performed waveform separation, respectively (Figure 3.12). The $X_{C\alpha-WAXD}$, $X_{C\beta-WAXD}$, and X_{C-WAXD} values evaluated using the WAXD patterns are summarized in Table 3.5. Type I membranes show only broad peaks, whereas X_{C-WAXD} cannot be determined. Type II membranes were mostly β-form structures, however the data have a large margin of error. Type III membranes formed a crystallized mixture of α- and β-form in each half ration. Type IV membranes formed almost α-form crystalline structures. Membranes with higher crystallinity mainly have α-form crystals.

This section discusses the systematic investigation of the effects of 60 organic solvents on PLA by the HSP. The hydrogen bonding solubility parameter accurately reflects the solubility of the PLA membranes using HSP but depends on hydrogen bonding as well as dispersion and polar parameters. The PLA membranes immersed in organic solvent became cloudy and indicated no changes in chemical structure. However, solvent-induced crystallization of the PLA membranes was observed. The structures depend not on the organic solvent but on the degree of swelling. The organic solvent-induced crystallization formed a crystallized mixture of α- and β-forms. The density of the crystalline PLA membranes was lower than that of the amorphous PLA membranes.

3.2.3 PHOTO-INDUCED CRYSTALLIZATION

Surface modification by UV irradiation has been applied to various materials such as polymers, glasses, ceramics, and metals [26]. For such a surface modification technology, VUV by excimer lamp irradiation enables irradiation under atmospheric pressure and lowers temperature, with no damage to the materials. The VUV modification process by an excimer lamp is also faster than that of a low-pressure mercury lamp because of its high generative capacity of oxygen radicals [27]. Surface modification was used to develop characteristics such as adhesion of materials and to remove surface organic compounds [28–30]. We developed the surface photo-oxidation, scission reaction, and crystallization of PLA membrane by a 172-nm VUV excimer lamp irradiation from 3 to 100 min. The physical properties of the VUV-irradiated PLA membrane were also systematically investigated [3]. This is the so-called photo-induced crystallization of PLA.

The properties of the VUV-irradiated membrane were systematically studied for 0–100 min. The UV–visible (UV–vis) spectra of these PLA membranes are shown in Figure 3.13. The transmittance of the PLA membranes showed no changes in the 400–700 nm range, and the values exceeded 90%. However, almost no transmittance was found at values below 200 nm. This result indicated that the photons at 172 nm were completely absorbed at the PLA membrane surface, and VUV irradiation did not depend on the color center and the bulk of the PLA membrane because transmission would not be possible inside the PLA membrane. Meanwhile, photon absorbance from 240 to 310 nm increased as irradiation time increased. In an organic compound, this result indicates that the C=C double bond with high conductivity depends on this absorbance. Ikada reported that structural modifications were observed on PLA membranes when submitted to UV irradiation [31–34]. Figure 3.14 is a schematic representation of structural changes such as chain cleavage and formation of C=C double bonds

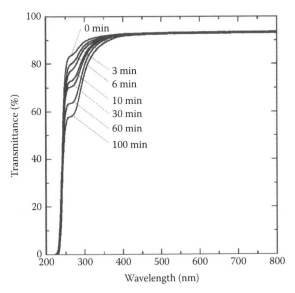

FIGURE 3.13   UV–vis spectra of VUV-irradiated PLA membranes for 3, 6, 10, 30, 60, and 100 min. (Reprinted from *Desalination*, 287, S. Sato et al., Effects of irradiation with vacuum ultraviolet xenon excimer lamp at 172 nm on water vapor transport through poly[lactic acid] membranes, 290–300. Copyright 2012, with permission from Elsevier B.V.)

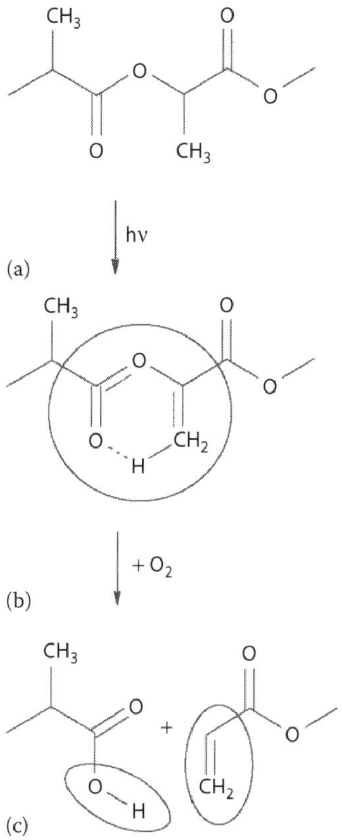

FIGURE 3.14 Norrish II mechanism for photo-oxidation of PLA: (a) the chain of PLA under VUV irradiation; (b) photophysical excitation; and (c) oxidation and scission reactions in PLA chains. (Reprinted from *Desalination*, 287, S. Sato et al., Effects of irradiation with vacuum ultraviolet xenon excimer lamp at 172 nm on water vapor transport through poly[lactic acid] membranes, 290–300. Copyright 2012, with permission from Elsevier B.V.)

and hydroperoxide O–H at newly formed chain terminals by a Norrish II mechanism for photon-oxidation of PLA [35]. A particular bond or group in a macromolecule can be activated by the absorption of a photon of UV irradiation. The macromolecule is generally formed by two kinds of bonds. The primary bonds represent the covalent intramolecular bonds, and the weaker bonds are the so-called secondary bonds, hydrogen, dipole, and van der Waals, with dissociation energy. The primary bonds of the molecule remain intact at low doses of UV irradiation; however, secondary bonds are easily altered by molecular motions during chain scission. VUV irradiation also affects the weakest bonds of the polymer structure as well as UV irradiation.

The difference attenuated total reflection-FTIR (ATR-FTIR) spectra of the VUV-irradiated PLA for 3, 6, 10, 30, 60, and 100 min based on the unirradiated PLA are shown in Figure 3.15. The important peaks are PLA: IR: 2960 and 2870 cm^{-1} (C–H stretching), 1750 cm^{-1} (C=O stretching), 1470 cm^{-1} (C–H stretching, C–H bending), and 1180 cm^{-1} and 1080 cm^{-1} (C–O–C stretching). Each peak intensity of these

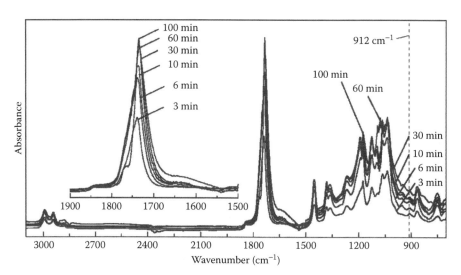

FIGURE 3.15 Difference ATR-IR spectra of VUV-irradiated PLA membranes for 3, 6, 10, 30, 60, and 100 min are based on unirradiated membrane. (Reprinted from *Desalination*, 287, S. Sato et al., Effects of irradiation with vacuum ultraviolet xenon excimer lamp at 172 nm on water vapor transport through poly[lactic acid] membranes, 290–300. Copyright 2012, with permission from Elsevier B.V.)

spectra increased as irradiation time increased because of the degree of adhesion to ZeSe for internal reflective element. In addition, the C=O bond peak near 1730 cm⁻¹ slightly shifted to a lower wavenumber side as irradiation time increased. The absorbance peak for ketone carbonyl typically occurs at 1715 cm⁻¹ and that for ester carbonyl occurs at 1735 cm⁻¹ [36]. This result indicates that PLA chemical structure changes with chain cleavage and formation of C=C double bonds and hydrophilicity hydroperoxide O–H at newly formed chain terminals by a Norrish II mechanism (Figure 3.14) using photo-oxidation and scission reaction of VUV irradiation. This tendency was in good agreement with that of the UV–vis spectra.

The WAXD patterns of the VUV-irradiated and unirradiated PLA membranes are shown in Figure 3.16. All PLA membranes showed one broad halo, indicating that all membranes were amorphous structures. This result indicated that VUV irradiation did not depend on the bulk structure of the PLA membranes, and VUV with 172 nm was absorbed on the surface of the PLA membrane.

The DSC thermograms of the VUV-irradiated and unirradiated PLA membranes are shown in Figure 3.17. The T_g, T_c, T_m, ΔH_c, and ΔH_m values determined by the DSC thermograms of the VUV-irradiated and unirradiated PLA membranes are summarized in Table 3.6. Endothermic peaks were observed near the 50°C–60°C range in all PLA membranes. These peaks corresponded to the glass transition temperature. In the VUV-irradiated PLA membranes, peaks for crystallization and melting temperature were observed. The sample preparation conditions of the PLA products significantly influenced their thermal properties. The literature data for PLA membranes showed that T_g values range from 55°C to 69°C [4–11]. The T_g values of the VUV-irradiated and unirradiated PLA membranes

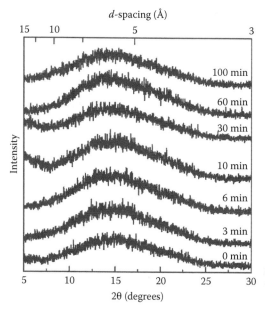

FIGURE 3.16 WAXD spectra of VUV-irradiated PLA membranes for 3, 6, 10, 30, 60, and 100 min. (Reprinted from *Desalination*, 287, S. Sato et al., Effects of irradiation with vacuum ultraviolet xenon excimer lamp at 172 nm on water vapor transport through poly[lactic acid] membranes, 290–300. Copyright 2012, with permission from Elsevier B.V.)

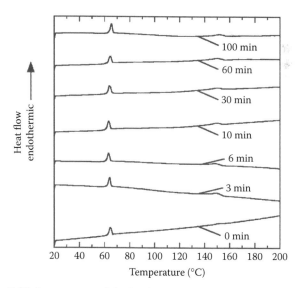

FIGURE 3.17 DSC thermograms of the first heating scan of the VUV-irradiated for 3, 6, 10, 30, 60, and 100 min and unirradiated PLA membrane. (Reprinted from *Desalination*, 287, S. Sato et al., Effects of irradiation with vacuum ultraviolet xenon excimer lamp at 172 nm on water vapor transport through poly[lactic acid] membranes, 290–300. Copyright 2012, with permission from Elsevier B.V.)

TABLE 3.6
Thermal Properties of Photo-Induced Crystallized PLA Membranes

VUV Irradiation Time (minutes)	T_g (°C)	T_c (°C)	T_m (°C)	ΔH_c (J/g)	ΔH_m (J/g)	X_{C-DSC} (%)
0	60.3 ± 2.1	–	–	–	–	0.0
3	60.4 ± 2.4	109.0 ± 9.7	148.9 ± 0.8	−1.8 ± 1.4	4.7 ± 2.1	3.1 ± 0.4
6	58.6 ± 1.4	109.6 ± 2.0	148.3 ± 0.4	−1.1 ± 0.8	6.5 ± 1.1	5.8 ± 1.1
10	60.2 ± 0.4	114.5 ± 1.4	148.9 ± 0.6	−1.0 ± 0.3	5.0 ± 1.1	4.3 ± 0.4
30	59.6 ± 0.5	112.9 ± 2.0	149.2 ± 1.2	−0.9 ± 0.5	5.1 ± 1.1	4.5 ± 1.2
60	60.3 ± 0.4	115.0 ± 0.7	147.8 ± 3.3	−1.4 ± 1.3	6.2 ± 1.1	5.2 ± 1.0
100	61.2 ± 1.2	127.9 ± 0.1	149.9 ± 1.8	−1.0 ± 0.5	5.0 ± 1.2	4.3 ± 1.0

ranged from 58.6°C to 61.2°C. The data obtained from the current study were within the range reported in the literature [4–11].

Slight exothermic peaks were observed near the 109.0°C–127.9°C range. The data of the VUV-irradiated PLA for 100 min were higher than that mentioned in the literature [4–11]. However, other PLA membranes had data within the range of the literature values. By contrast, the T_m values of the VUV-irradiated and unirradiated PLA membranes varied from 147.8°C to 149.9°C, which were within the range reported in the literature [4–11]. Thermal analysis indicated that the unirradiated PLA membrane exhibited an amorphous structure, whereas the VUV-irradiated PLA membrane showed a surface crystalline structure.

The X_{C-DSC} values of the VUV-irradiated and unirradiated PLA membranes evaluated by Equation 3.1 are summarized in Table 3.6. The X_{C-DSC} values clearly increased in the range of irradiation time from 0 to 6 min and settled to 4.3% as irradiation time increased from 6 to 100 min. Figures 3.17 and 3.18 show that X_{C-DSC} result differs from the WAXD pattern. In the WAXD pattern, crystallinity can be estimated by dividing the halo pattern, whereas crystallinity estimated by DSC value is directly determined by enthalpy change. Given this factor, the X_{C-DSC} values obtained higher accuracy when compared with the values obtained by WAXD analysis. A sharp halo of crystalline PLA was buried in the broad halo of amorphous PLA.

The SEM images of all PLA membranes in the current study are presented in Figure 3.18. A smooth surface was observed in the unirradiated amorphous PLA membrane. The unevenness of the PLA crystal growth gradually branched radially, and some cracks for photolysis occurred on the surface of the VUV-irradiated PLA membrane. The size of unevenness increased as VUV irradiation time increased, which was 2–5 μm in the time range of 3–30 min, and 10 μm in the time range of 60–100 min. Figures 3.1 and 3.2 indicate that the roughness of these PLA membranes is estimated to be less than 300 nm because no change was observed in the photograph images and the UV–vis spectra in the range of 300–800 nm. Surface photo-oxidation, scission reaction, and crystallization of the PLA membrane by VUV irradiation as well as UV irradiation occurred.

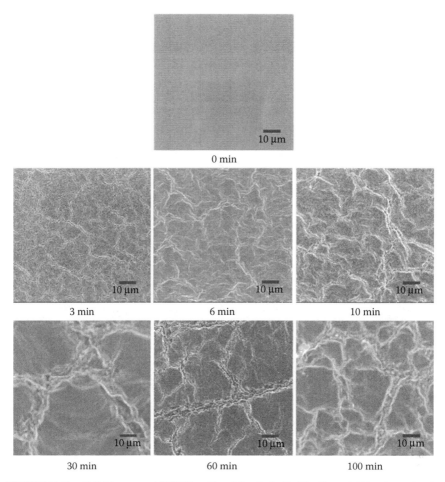

FIGURE 3.18   SEM images of VUV-irradiated for 3, 6, 10, 30, 60, and 100 min and unir-radiated PLA membranes. (Reprinted from *Desalination*, 287, S. Sato et al., Effects of irradiation with vacuum ultraviolet xenon excimer lamp at 172 nm on water vapor transport through poly[lactic acid] membranes, 290–300. Copyright 2012, with permission from Elsevier B.V.)

3.3   CONCLUSIONS

The crystal structure affects the performance of polymer membranes. PLA is an environmental-friendly, biodegradable polymer substance with a low melting point and good moldability. It is often used in packages, automobiles, and electronics. We can even expect that PLA would replace petroleum-based plastic. In this chapter, we have summarized the fabrication of the PLA membrane, focusing on the control of crystallization based on thermally induced, solvent-induced, and photo-induced crystallization methods. Interestingly, PLA crystallization can be controlled by cooling and heating organic solvents and VUV irradiation. The PLA crystalline structure depended on their sample preparation conditions. New methods controlling PLA crystalline structure are important in next-generation advanced PLA membranes in such applications.

REFERENCES

1. Sawada H. et al. 2010. Gas transport properties and crystalline structures of poly(lactic acid) membranes. *Transactions of the Materials Research Society of Japan* 35: 241–46.
2. Sato S. et al. Effects of various liquid organic solvents on solvent-induced crystallization of amorphous poly(lactic acid) film. *Journal of Applied Polymer Science* 129: 1607–17.
3. Sato S. et al. 2012. Effects of irradiation with vacuum ultraviolet xenon excimer lamp at 172 nm on water vapor transport through poly(lactic acid) membranes. *Desalination* 287: 290–300.
4. Weir N.A. et al. 2004. Processing, annealing and sterilisation of poly-L-lactide. *Biomaterials* 25: 3939–49.
5. Chen C.C. et al. 2003. Preparation and characterization of biodegradable PLA polymeric blends. *Biomaterials* 24: 1167–73.
6. Duek E.A.R., C.A.C. Zavaglia, and W.D. Belangero. 1999. In vitro study of poly(lactic acid) pin degradation. *Polymer* 40: 6465–73.
7. Komatsuka T., A. Kusakabe, and K. Nagai. 2008. Characterization and gas transport properties of poly(lactic acid) blend membranes. *Desalination* 234: 212–20.
8. Lee J.H. et al. 2003. Thermal and mechanical characteristics of poly(L-lactic acid) nanocomposite scaffold. *Biomaterials* 24: 2773–78.
9. Sarazin P., X. Roy, and B.D. Favis. 2004. Controlled preparation and properties of porous poly(L-lactide) obtained from a co-continuous blend of two biodegradable polymers. *Biomaterials* 25: 5965–78.
10. Tsuji H. and K. Suzuyoshi. 2002. Environmental degradation of biodegradable polyesters 1. Poly(epsilon-caprolactone), poly(R)-3-hydroxybutyrate, and poly (L-lactide) films in controlled static seawater. *Polymer Degradation and Stability* 75: 347–55.
11. Yao F.L. et al. 2004. Synthesis and characterization of functional L-lactic acid/citric acid oligomer. *European Polymer Journal* 40: 1895–901.
12. Fischer E.W., H.J. Sterzel, and G. Wegner. 1973. Investigation of structure of solution grown crystals of lactide copolymers by means of chemical-reactions. *Kolloid-Zeitschrift and Zeitschrift Fur Polymere* 251: 980–90.
13. Iwata T. and Y. Doi. 1998. Morphology and enzymatic degradation of poly(l-lactic acid) single crystals. *Macromolecules* 31: 2461–67.
14. Auras R., B. Harte, and S. Selke. 2006. Sorption of ethyl acetate and d-limonene in poly(lactide) polymers. *Journal of the Science of Food and Agriculture* 86: 648–56.
15. Colomines G. et al. 2010. Barrier properties of poly(lactic acid) and its morphological changes induced by aroma compound sorption. *Polymer International* 59: 818–26.
16. Mikos A.G. et al. 1994. Preparation and characterization of poly(l-lactic acid) foams. *Polymer* 35: 1068–77.
17. Pillin I., N. Montrelay, and Y. Grohens. 2006. Thermo-mechanical characterization of plasticized PLA: Is the miscibility the only significant factor. *Polymer* 47: 4676–82.
18. Suryanegara L., A.N. Nakagaito, and H. Yano. 2009. The effect of crystallization of PLA on the thermal and mechanical properties of microfibrillated cellulose-reinforced PLA composites. *Composites Science and Technology* 69: 1187–92.
19. Yasuniwa M. et al. 2006. Crystallization behavior of poly(L-lactic acid). *Polymer* 47: 7554–63.
20. Wang Y.M., S.S. Funari, and J.F. Mano. 2006. Influence of semicrystalline morphology on the glass transition of poly(L-lactic acid). *Macromolecular Chemistry and Physics* 207: 1262–71.
21. Zhang J.M. et al. 2005. Crystal modifications and thermal behavior of poly(L-lactic acid) revealed by infrared spectroscopy. *Macromolecules* 38: 8012–21.

22. Pan P. et al. 2008. Effect of crystallization temperature on crystal modifications and crystallization kinetics of poly(L-lactide). *Journal of Applied Polymer Science* 107: 54–62.

23. Sasaki S. and T. Asakura. 2003. Helix distortion and crystal structure of the alpha-form of poly(L-lactide). *Macromolecules* 36: 8385–90.

24. Puiggali J. et al. 2000. The frustrated structure of poly(L-lactide). *Polymer* 41: 8921–30.

25. Sawai D. et al. 2003. Preparation of oriented beta-form poly(L-lactic acid) by solid-state coextrusion: Effect of extrusion variables. *Macromolecules* 36: 3601–605.

26. Ozdemir M., C.U. Yurteri, and H. Sadikoglu. 1999. Physical polymer surface modification methods and applications in food packaging polymers. *Critical Reviews in Food Science and Nutrition* 39: 457–77.

27. Mathieson I. and R.H. Bradley. 1996. Improved adhesion to polymers by UV/ozone surface oxidation. *International Journal of Adhesion and Adhesives* 16: 29–31.

28. Jang J. and Y. Jeong. 2006. Nano roughening of PET and PTT fabrics via continuous UV/O$_3$ irradiation. *Dyes and Pigments* 69: 137–43.

29. Ko Y.G. et al. 2001. Immobilization of poly(ethylene glycol) or its sulfonate onto polymer surfaces by ozone oxidation. *Biomaterials* 22: 2115–23.

30. Vig J.R. 1985. UV/ozone cleaning of surfaces. *Journal of Vacuum Science and Technology A* 3: 1027–34.

31. Ikada E. and M. Ashida. 1991. Promotion of photodegradation of polymers for plastic waste treatment. *Journal of Photopolymer Science and Technology* 4: 247–54.

32. Ikada E. 1993. Role of the molecular structure in the photodecomposition of polymers. *Journal of Photopolymer Science and Technology* 6: 115–22.

33. Ikada E. 1997. Photo- and biodegradable polyesters. Photodegradation behaviors of aliphatic polyesters. *Journal of Photopolymer Science and Technology* 10: 265–69.

34. Ikada E. 1999. Relationship between photodegradability and biodegradability of some aliphatic polyesters. *Journal of Photopolymer Science and Technology* 12: 251–56.

35. Belbachir S. et al. 2010. Modelling of photodegradation effect on elastic-viscoplastic behaviour of amorphous polylactic acid films. *Journal of the Mechanics and Physics of Solids* 58: 241–55.

36. Ammala A. et al. 2011. An overview of degradable and biodegradable polyolefins. *Progress in Polymer Science* 36: 1015–49.

4 Innovative Methods to Improve Nanofiltration Performance through Membrane Fabrication and Surface Modification Using Various Types of Polyelectrolytes

Law Yong Ng and Abdul Wahab Mohammad

CONTENTS

4.1 Introduction to Membrane Materials... 104
    4.1.1 Introduction to NF Application 105
    4.1.2 Drawbacks in NF Application 107
        4.1.2.1 Low Solute Rejection and Selectivity in NF...................... 107
        4.1.2.2 NF Membrane Fouling 108
        4.1.2.3 Membrane Life-Time Limitation 108
    4.1.3 Modification or Fabrication Methods Previously Applied to Produce NF Membranes 109
    4.1.4 Polyelectrolytes as the New Alternative Materials for NF Membrane Modification or Fabrication 112
    4.1.5 Polyelectrolyte Solution Parameters for NF Membrane Modification or Fabrication 114
4.2 Methods Used during the NF Membrane Surface Modification Using Polyelectrolytes... 115
    4.2.1 UV-Grafting of Polyelectrolytes onto the Membrane Surface 117
        4.2.1.1 Acrylic Acid Grafted on PES NF Membranes 117
        4.2.1.2 Sodium *p*-Styrenesulfonate Grafted on PSF UF Membranes... 119
    4.2.2 Chemical Cross-Linking of PECs with the Membrane Surface....... 119
    4.2.3 Static Self-Adsorption of Polyelectrolytes onto the Membrane Surface 119

    4.2.4   Dynamic Self-Adsorption of Polyelectrolytes onto the
            Membrane Surface... 122
4.3   Alteration on NF Membrane Characteristics and Performances Using
      Polyelectrolytes.. 125
    4.3.1   Membrane Surface Hydrophilicity 125
    4.3.2   Membrane Surface and Cross-Sectional Morphology 127
    4.3.3   Membrane Surface Charge Property 129
    4.3.4   Membrane Pure Water Permeability or Flux.................................. 130
    4.3.5   Solute Rejection and Selectivity Capability
            of Polyelectrolyte-Modified Membrane... 134
    4.3.6   Changes in Membrane Separation Layer Thickness 143
    4.3.7   Performance Stability of the Polyelectrolyte-Modified
            Membranes 145
4.4   Drawbacks on the Applications of Polyelectrolytes in NF Membrane
      Modification or Fabrication 147
4.5   Proposed Future Studies on the NF Membrane Modification or
      Fabrication Using Polyelectrolytes 150
4.6   Summary 151
References.. 152

4.1   INTRODUCTION TO MEMBRANE MATERIALS

Membrane technology has been widely integrated into separation systems that normally involve mixtures of substances in liquid or gaseous states (Semsarzadeh and Ghalei 2013). Membranes have become increasingly important because their employment in the water treatment processes (Peiris et al. 2010), such as wastewater reclamation, drinking water processes, brackish water treatment, water softening, desalination, and so forth. Membranes have been used to remove the water contaminants such as natural organic matter (Zhang et al. 2011), bacteria, viruses (Lv et al. 2006), particulates (Zhang et al. 2011), and inorganic ions. Membrane technology has become an important part of the separation technology over the past decades, as it can work in the absence of chemical addition unlike conventional purification systems.

Commonly used membrane materials can be divided into two general categories, namely, polymeric and inorganic (or ceramic) materials. Inorganic materials that are commonly employed in membrane fabrication include titanium dioxide, zirconium dioxide, and alumina (Guerra and Pellegrino 2013). Inorganic-based membranes have excellent resistance toward chemical attack, high mechanical strength, and a high tolerance to extremes of pH and temperature. These properties are especially useful during the membrane cleaning process, especially for membrane applications, which deal with feed streams that have high propensity for membrane fouling. High durability of the inorganic membranes makes them an attractive choice of membrane materials, as the membrane replacement and maintenance costs can be greatly minimized. However, inorganic membranes are not broadly employed in the industries due to their high production cost and high weight. Moreover, inorganic membranes have high rigidity, which restricts the ease of membrane integration into any membrane filtration system. This limitation makes the inorganic membranes process difficult to be scaled-up in comparison to polymeric membrane processes. As inorganic membranes have good resistance toward chemicals, the modification of inorganic membranes is not as easy as polymeric membranes. This has further reduced the suitability and employment of inorganic membranes for various applications, where different properties of the membrane surfaces are desired to improve the separation processes. Although inorganic membranes have high temperature tolerances, they are not widely employed in water purification systems, as most of the polymeric membranes can withstand the operating temperature of the water purification processes and the cost for polymeric membranes used is much lower when compared to the possible inorganic membranes.

On the other hand, polymeric membranes have gained more popularity in recent years due to ample materials for selection to meet the requirements of applications. Polymeric membranes can be produced in extremely thin films or in any other shapes depending on the requirements of the processes that make them less space-consuming. Polymeric membranes also have better pore size distribution control and can be designed to suit the desired separation performances. Polymeric materials used in membrane fabrication include polysulfone (PSF) (Ng and Mohammad 2010), polyurethane (PU) (Chen et al. 2007), polyvinylidene fluoride (PVDF) (Li et al. 2008b), polyethersulfone (PES) (Wang et al. 2011a; Zhao et al. 2011; Ahmad et al. 2013), and cellulose (Ruan et al. 2004; Jie et al. 2005). The majority of the commercially available membranes for separation processes are made of polymeric materials.

In recent years, alternative membrane materials that combine both organic–organic, inorganic–inorganic, and organic–inorganic compounds have contributed to the fabrication of newly designed composite membranes (Ng et al. 2013). Composite membranes that combine the advantages from their respective materials of construction can improve the membrane performances as reported. For instance, PU–silica/polyvinyl alcohol (PVA) composite membranes produced were reported to have better selectivity in gas separation (Semsarzadeh and Ghalei 2013).

Throughout the discussion in this chapter, the employment of membranes (mainly polymeric) for nanofiltration (NF) applications is elaborated. The drawbacks during the employment of polymeric membranes and several inorganic membranes in NF processes are also discussed, as well as the ways to overcome these drawbacks or improve the membrane using polyelectrolytes. For a better understanding on the subject, previous studies and research works will be included and referred throughout the discussion.

4.1.1 INTRODUCTION TO NF APPLICATION

NF membranes have properties between ultrafiltration (UF) and reverse osmosis (RO) membranes. In comparison to UF membrane, NF membrane has better inorganic ion-retention properties. UF membranes mainly separate the solute components based on the size or molecular weight differences, which is known as the *sieving effect*. UF is best employed in fractionation processes, which normally involve high molecular weight solutes (Arunkumar and Etzel 2013).

In comparison to RO membranes, NF membranes offer more advantages such as higher fluxes, higher rejection of multivalent salt ions, lower operating costs, and lower maintenance costs. NF membranes differ from the RO membranes mainly due to the fact that the former are specifically designed for the retention of multivalent ions or organic contaminants, although allowing the permeation of others (Košutić and Kunst 2002; Liikanen et al. 2003). Lower operating pressure of NF membranes is another advantage when compared to RO membranes, due to the fact that the compounds that pass through NF membranes will not elevate the osmotic pressure in the system (Košutić and Kunst 2002).

Depending on the type of NF membrane and solute involved, several mechanisms are used to explain the solute-rejection performances observed. The mechanisms employed in explaining NF membrane performances include physical sieving, electrostatic exclusion or adsorption, and diffusion limitation. When neutrally charged solutes are present in the feed solution of a filtration system, the main mechanism contributing to the rejection performance will be based on the steric hindrance. According to a research conducted by Košutić et al. (2000), membrane porosity was confirmed to be the key factor for the rejection of neutrally charged solutes, where the solute molecular weights and the membrane pore size distributions can be well correlated to each other. Some researchers employed electrostatic repulsion in explaining the rejection mechanism for negatively charged solutes in the presence of negatively charged membrane surfaces (Kiso et al. 2001).

Classification of NF membrane performance still involves some complexity, as the NF membrane rejection properties are not solely reliant on one specific parameter. One of the important parameters considered is the molecular weight cut-off of the NF membrane. However, this parameter is quite disputable, as various types of solutes can be selected and the rejection percentages are not standardized; these are mostly dependent on the membrane manufacturers and can range from 60% to 90%. Determination of molecular weight cut-off values is also highly dependent on the solute concentration and the solvent used. Some previous work also verified that it is not practical to correlate NF membrane performances using molecular weight cut-off value alone (Van der Bruggen et al. 1998, 1999). Thus, most researchers are now using the salt-rejection characterization as a basis for comparison between different membrane performances (Bowen et al. 2000; Ballet et al. 2004; Krieg et al. 2005; Murthy and Chaudhari 2009; Abuhabib et al. 2012). However, in most cases, the operating pressure; solution pH; filtration unit, either cross-flow or dead-end stirred cell; and filtration operation temperature are still not well standardized for the salt-rejection characterization evaluation.

NF membranes are widely applied in groundwater treatment to reduce water hardness (Schaep et al. 1998), using either self-fabricated NF membranes (Rahimpour et al. 2010) or commercialized membranes. When NF membranes were evaluated using a cross-flow unit, they recorded high rejection toward magnesium sulfate (85%–90%), but low rejection toward sodium chloride (64%–67%). NF membranes were also employed in the study of deacidifying and demineralizing acid cheese whey as a function of temperature, pressure, and pH (Alkhatim et al. 1998). Other possible applications of NF include the color removal (Koyuncu 2002) and reduction of biological oxygen and chemical oxygen demands (Das et al. 2006). However, the application of NF membranes is not limited only to these. NF applications are getting more attention and wider employment in several industries. However, NF has not reached the matured technology stage yet and research is required to improve some of its drawbacks.

4.1.2 DRAWBACKS IN NF APPLICATION

The use of NF membranes has several advantages that overcome the limitations of UF and RO processes. However, NF membranes have several impediments for application, which are worth a short discussion that further justifies the requirement for innovative NF membrane fabrication and modification methods.

4.1.2.1 Low Solute Rejection and Selectivity in NF

NF membrane has been widely applied in water-related applications. However, low solute-removal rate for certain applications is still not effectively solved. For instance, monovalent nitrate removal in water treatment is considered as nonideal, as the nitrate rejection rate using commercial NF membrane is still considered low (Bohdziewicz et al. 1999; Van der Bruggen et al. 2001; Garcia et al. 2006). Removal of nitrate ions from the water stream is important as synergistic toxicity effects potentially occur in combination with some other compounds such as pesticides (L'Haridon et al. 1993). Low rejection of NF membrane toward other elements or compounds based on boron was also reported by Geffen et al. (2006) and Dydo et al. (2005).

In addition to the solute rejection capability, solute selectivity in NF processes plays an important role in determining the NF efficiency. The selectivity of solute 1 over solute 2, termed as *separation factor* by certain researchers, was calculated using Equation (4.1) (Ahmadiannamini et al. 2010; Umpuch et al. 2010):

$$S_{1/2} = \left(\frac{C_{p,1}}{C_{f,1}}\right)\left(\frac{C_{f,2}}{C_{p,2}}\right) = \left(\frac{100\% - R_1}{100\% - R_2}\right) \tag{4.1}$$

where:
$C_{p,1}$ represents permeate concentration for component 1
$C_{f,2}$ represents feed concentration for component 2
$C_{f,1}$ represents feed concentration for component 1
$C_{p,2}$ represents permeate concentration for component 2
R_1 represents the rejection of component 1
R_2 represents the rejection of component 2

This selectivity equation is popularly used for both charged and noncharged solutes (Umpuch et al. 2010).

During the salt fractionation or selective separation (Oumar et al. 2001; Hilal et al. 2007), the removal of a certain solute over another is one of the most important applications of NF membranes. Unlike RO processes where almost none of the solute is permeable through the membrane, NF processes can partially allow the permeation of monovalent ions, but restrict the permeation of multivalent ions through the NF membrane. This feature makes NF highly preferable in comparison to RO processes, as lower NF membrane fouling rate can contribute to higher solution flux with lower energy consumption and longer membrane life. NF membrane can be used to enhance the fractionation process in food industry when applied after a UF membrane (Mohammad et al. 2012).

4.1.2.2   NF Membrane Fouling

In any membrane separation process, the accumulation of undesired foulants onto the membrane surface will normally contribute to a phenomenon known as *membrane fouling*. However, the occurrence of NF membrane fouling at the nanoscale involves complexities that are still far from being fully understood (Agenson and Urase 2007; Her et al. 2007; Jarusutthirak et al. 2007). NF membrane fouling can be caused by biological organisms, colloids, and organic and inorganic solutes (Xu et al. 2010).

Fouling, mainly caused by biological organisms, is more specifically termed as *membrane biofouling*. The biological organisms, fungi and bacteria, generally contribute to biofouling (Ivnitsky et al. 2010). However, colloidal fouling is normally caused by the accumulation of colloidal particles that form a cake-like layer on the membrane surface or clog membrane pores, which mainly depends on the relative size between membrane pores and colloidal particles (Tarabara et al. 2004; Warczok et al. 2004; Lee et al. 2005). Colloidal fouling is generally promoted or affected by the colloidal types, concentrations, pH, ionic strengths, and sizes as reported by Tarabara et al. (2004); Chun et al. (2002); Singh and Song (2005); Zhang and Song (2000); Li and Elimelech (2006); Boussu et al. (2007); and Wu and Sun (2005). Inorganic solute fouling can also commonly occur by the precipitation of the salts on the membrane surface known as *scaling*. This phenomenon can be observed when the retention of ions during the NF process reaches the solubility limit.

Membrane fouling is definitely not desired during filtration processes, as it lowers membrane performance recoveries, raises energy consumptions, shortens membrane lifetime with higher operating or maintenance costs, requires laborious membrane cleaning methods, and installs pretreatment processes, with higher chemical consumption during the membrane cleaning. In the last few years, several works have evaluated the fouling tendency of NF processes using mathematical approaches (Khirani et al. 2006; Listiarini et al. 2009a, 2009b). However, mathematical predictions of fouling are tools for process optimization and do not contribute significantly to lowering or avoiding the overall NF membrane fouling. Membrane modification is still the best solution at the moment to generate NF membranes that are capable of reducing the membrane fouling tendency (Al-Amoudi and Lovitt 2007).

4.1.2.3   Membrane Life-Time Limitation

The overall lifespan of a membrane is governed by the membrane chemical resistance and its antifouling property. Common membrane-cleaning processes employ acidic and alkaline solutions for removing salt precipitates and organic foulants, respectively, to recover the membrane permeability and other performances, which are important to minimize the energy consumption and operating costs (Li and

TABLE 4.1

Drawbacks of NF Membranes and Methods Employed to Improve Their Performances

Drawbacks of NF Membrane	Improvement Methods	References
1. Colloidal fouling	Develop membranes with lower surface roughness to reduce the accumulation of foulants onto the membrane surface	Al-Amoudi and Lovitt (2007)
2. Hydrophobic membrane surface that encourages organic fouling	Hydrophilic NF membranes can be developed through grafting process using hydrophilic materials	Akbari et al. (2006); Qiu et al. (2006); Morão et al. (2005)
3. Polymeric membrane biofouling	Biofilm formation or biofouling can be avoided through membrane modification using nanoparticles	Ng et al. (2013); Zodrow et al. (2009); Zhang et al. (2012); Liu et al. (2013a, 2013b)

Elimelech 2004; Religa et al. 2013; Simon et al. 2013). However, the deterioration of membrane structures due to the acidic or alkaline solution can greatly shorten the membrane lifespan. The ability of the polymeric membrane to withstand an extreme pH condition will definitely determine the life expectancy of the membrane in certain applications. In addition, fouling control of polymeric membranes, through either operating condition adjustment or membrane material selection, is important in order for the wider employment of NF processes to become feasible.

In summary, researchers have employed several methods to solve the problems and reduce the drawbacks that restrict the further development of NF membranes. Some of the improvement methods that were applied are listed in Table 4.1.

4.1.3 MODIFICATION OR FABRICATION METHODS PREVIOUSLY APPLIED TO PRODUCE NF MEMBRANES

As discussed, NF process simulation offers some possible process optimization by predicting the membrane-fouling mechanisms and sources of performance deterioration. However, membrane modification and fabrication still remains as one of the most attractive research strategies, offering a complete solution to the drawbacks associated with NF membrane applications. Development of new membrane materials with special features such as antifouling, biocompatibility, and functionality has become the current trend in NF research.

There are several methods that have been previously employed to produce NF membranes with targeted function. One of the most used methods to produce NF membranes is interfacial polymerization (Gohil and Ray 2009; Wang et al. 2011b), which is the main method used commercially to produce thin-film composite membrane (Rahimpour et al. 2010). In this method, a porous substrate or membrane support is normally required before the formation of a dense layer. However, there are several complexities for the interfacial polymerization technique, such as the right partition coefficient of the reactants in the two-phase solution; a suitable diffusion rate of the reactants is required to achieve the ideal degree of densification of the thin separation layer (Rahimpour et al. 2010). In addition, some other critical parameters need to be optimized to control the performance of the fabricated NF membrane, including coating duration, monomer composition, and membrane posttreatment temperature (Rahimpour et al. 2010). Interfacial polymerization of 1,3-phenylenediamine (PDA) with trimesoyl chloride (TMC) using PES membrane as substrate (Rahimpour et al. 2010) produces NF membrane with reasonable flux stability, as shown in Figure 4.1. Conversely, another research by Jahanshahi et al. (2010) revealed that after interfacial polymerization (i.e., polymerization of piperazine with TMC), PES membrane surface seemed to be rougher (Figure 4.2), which was not desired in NF applications.

Another NF membrane fabrication method is through the blending of polymeric host materials with other organic or inorganic materials. For instance, researchers embedded acid-oxidized multiwalled carbon nanotubes during the preparation of PES membrane through phase inversion method (Vatanpour et al. 2011). Embedding nanoparticles in polymeric membranes also become an attractive option recently, as the modified membranes exhibited significant variations in membrane performances from those of bare polymers (Ng et al. 2013). However, researchers discovered that the performance of the nanofillers-embedded polymeric membrane declined over the time (<200 minutes) when tested with various salt solutions at a pressure of 4 bars (as shown in Figure 4.3) (Vatanpour et al. 2011). Apart from carbon nanotubes, blending

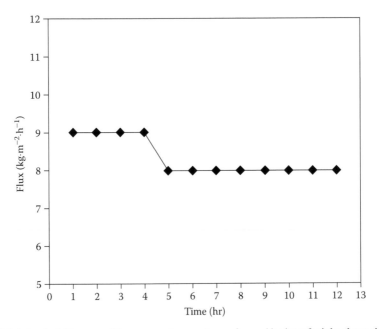

FIGURE 4.1 Stability test of the composite membrane formed by interfacial polymerization of 1,3-PDA with TMC using PES membrane as substrate. (Reprinted with permission from Rahimpour, A. et al. *Appl. Surf. Sci.*, 256, 1657–1663, 2010.)

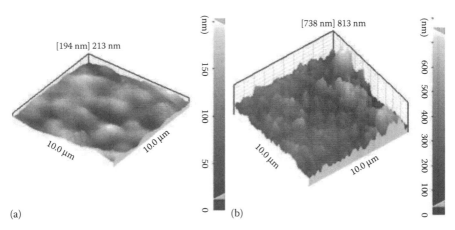

FIGURE 4.2 Membrane surface atomic force microscopy images (a) PES pure membrane and (b) poly(piperazine-amide) composite membrane. (Reprinted with permission from Jahanshahi, M. et al., *Desalination* 257, 129–136, 2010.)

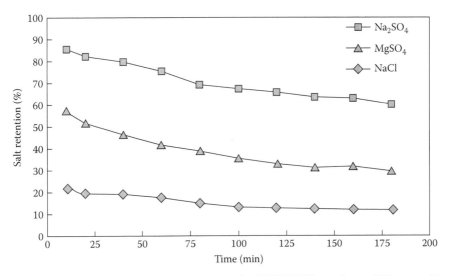

FIGURE 4.3 Salt rejection tests using 0.04 wt% MWCNT/PES membrane (200 ppm salt). (Reprinted with permission from Vatanpour, V. et al., *J. Memb. Sci.*, 375, 284–294, 2011.)

polymeric membrane with other nanoparticles (such as TiO_2) have also been well studied in Rajesh et al. (2013).

Plasma treatment or plasma-induced grafting (Buonomenna et al. 2007; Kim et al. 2011; Wang et al. 2012) also becomes one of the famously used surface modification methods to improve the NF membrane performances. However, researchers found that the employment of plasma treatment for NF membrane fabrication needs to be well controlled, as this method can induce pure water flux reduction when the higher plasma treatment power is used (Kim et al. 2011). Other methods to improve

NF membrane performances include UV-induced grafting (Qiu et al. 2005; Bilongo et al. 2010; Khayet et al. 2010), assembly of nanoparticles on membrane surfaces (Mansourpanah et al. 2009), and so forth.

Recently, another branch of simple and versatile method that involved mainly the employment of polyelectrolytes during the fabrication and modification of NF membranes has been introduced. The development involving the employment of polyelectrolytes in NF membrane researches is mainly fueled by the newly produced polyelectrolytes and deposition techniques, which will be discussed in Sections 4.1.4 and 4.2.

4.1.4 POLYELECTROLYTES AS THE NEW ALTERNATIVE MATERIALS FOR NF MEMBRANE MODIFICATION OR FABRICATION

Polyelectrolytes have been widely used in various applications, such as the preparation of microcapsules (Wang et al. 2008; Antipina and Sukhorukov 2011), biomedical application (Kwon and Gong 2006; Vergaro et al. 2011), purification of fat-containing wastewater (Shulevich et al. 2013), and so forth. However, researchers still have a poor understanding of this subject (Schmitz Kenneth 1993) and its interactions with the surrounding materials. Polyelectrolytes and polyelectrolyte complexes (PECs) can be differentiated in terms of the molecular structure, weak or strong electrolyte activity, molecular weight, and charge density. PECs are normally produced in aqueous solution through the electrostatic interactions between polycation and polyanion molecules. Under common circumstances, PECs formed are not water soluble, but they become soluble in certain organic solvent such as n-butanol solution (Guenet 2005). Dissolution of PECs in organic solvent was postulated to affect their structures (Guenet 2005), as shown in Figure 4.4.

Polyelectrolytes can be used to supply extremely thin but effective barriers in NF membrane applications. Development of NF membranes has been greatly benefited by the incorporation of polyelectrolytes onto the existing membrane substrate surface (either self-fabricated or commercially obtainable membranes). This is due to the fact that, in most cases, the charges imposed by the polyelectrolytes can be used effectively in rejecting the permeation of charged solutes through the polyelectrolyte-modified membranes. In comparison to the NF membranes with smaller pore sizes, polyelectrolyte-modified membranes can provide higher solution fluxes, as this type of membrane no longer relies on the size exclusion theory alone. The presence of polyelectrolytes is believed to improve the Donnan effect of NF membranes (Levenstein et al. 1996). Polyelectrolytes can be used in membrane surface modification with good control (in the nanometer range) over the distribution of ionized molecules within the thin film. The exceptionally thin layers created on NF membranes using polyelectrolytes is also important to reduce flux deterioration, which is a significant problem posed by other modification methods (Wei et al. 2006).

Blending a charged polymeric material into the casting solution can also be used to alter membrane surface charges. However, blending is not highly recommended as it is associated with many weaknesses. The main weakness of blending is the

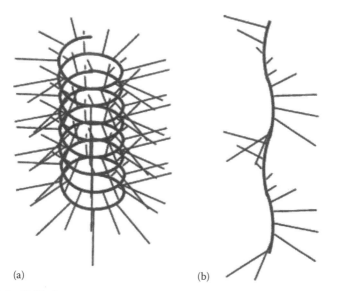

(a)   (b)

FIGURE 4.4 Helical structures of the polyelectrolyte moiety postulated through data fittings (a) helix 1 with cross-section radius of 2.7 nm and (b) helix 2 with cross-section radius of 1.75 nm. (Reprinted with permission from Guenet, J.-M., *J. Mol. Liq.*, 120, 3–6, 2005.)

complicated chemical reactivity. Membrane properties, such as the mechanical strength, contact angle, thermal resistance, and so forth, can also be changed easily. Thus, this method is seldom used when compared to the self-adsorption of polyelectrolytes onto the membrane surfaces.

Polyelectrolytes are usually used in membrane modification through the surface coating method. Among the two most important membrane surface modification methods, coating has more advantages over grafting for the following reasons:

1. Coating can be applied to existing polymeric membranes or existing installed membrane systems using static, dynamic, or dynamic-static adsorption methods.
2. Membranes modified by the coating process are not changed in terms of their structural properties.
3. Less (or no) expensive equipment is required for the coating process to take place.
4. This process does not require a prolonged chemical reaction time.

Due to the fact that polyelectrolytes offer a cheap, easy, and fast solution in the membrane modification process, the deposition of polyelectrolytes onto the membrane surface has gained popularity to improve NF membrane performance. Polyelectrolyte-modified membranes have also been reported to be able to provide high water permeability (Hong and Bruening 2006; Li et al. 2008a). Additionally, polyelectrolyte-modified membranes can be easily tailored to meet the requirements of the filtration process. There are numerous types of commercially available polyelectrolytes that can be selected to provide different membrane surface change density, hydrophilicity, morphology, and fouling resistivity. As the preparation of antifouling NF membranes through coating has rarely been reported, several research studies have been compiled here for the future development of NF membrane research.

4.1.5 POLYELECTROLYTE SOLUTION PARAMETERS FOR NF MEMBRANE MODIFICATION OR FABRICATION

There are several major parameters that need to be well considered before the polyelectrolyte solutions are employed for the fabrication or modification of NF membranes. These adjustments are especially important when the polyelectrolyte solutions need to be self-adsorbed onto the membrane surfaces. One of the parameters studied by researchers was the solution pH. A detailed study on the role of polyelectrolyte solution pH was carried out using weak polyelectrolytes in the construction of polyelectrolyte multilayers (PEMs) using poly(acrylic acid) and poly(allylamine hydrochloride) (PAH) (Shiratori and Rubner 2000). Researchers postulated that through the polyelectrolyte solution pH adjustment, the thickness of an adsorbed polyanion or polycation layer can be varied from 5Å to 80Å. Besides, researchers also found that a significant change in the polyelectrolyte adsorption behavior, and an alteration in the polyelectrolyte layer thickness can be detected over a very narrow pH range. This research work has provided some insightful ideas in the development of NF membranes with controllable adsorption behaviors and characteristics.

In addition, molecular weight or molar mass of the polyelectrolyte used is another influential parameter that needs to be considered, as previous work suggested that it affects the bilayer thickness significantly (Sui et al. 2003; Kujawa et al. 2005; Lingström and Wågberg 2008). Researchers used two polysaccharides (i.e., hyaluronan and chitosan [CHI]) during the growth of multilayers. It was shown that when high molecular weight polysaccharides (hyaluronan 360000, CHI 160000) were used, the thicknesses obtained were two times larger than that obtained using low molecular weight polysaccharides (hyaluronan 30000, CHI 31000), when the number of bilayers was fixed at 12. Moreover, they observed that the optical layer thickness increased exponentially with the number of deposited bilayers, regardless of the molecular weight of the polyelectrolytes used. However, the effects of molecular weight were found to be small compared with the effects of charge density during the sequential adsorption of oppositely charged polyelectrolytes (Kolarik et al. 1999).

Another important parameter is the presence and concentration of monovalent electrolyte that can affect the conformation of the adsorbed polyelectrolyte chains (Milkova and Radeva 2013). This can further induce a change in the thickness of polyelectrolyte bilayers formed and the multilayer growth (Sui et al. 2003). Moreover, researchers found that the integrity of film fabrication improved with increasing solution ionic strength, confirming the key role of counterions (Kolarik et al. 1999). All of these parameters are important for the construction of the PEMs as effective separation layers for NF membranes.

4.2 METHODS USED DURING THE NF MEMBRANE SURFACE MODIFICATION USING POLYELECTROLYTES

NF membranes usually require more porous and higher charge density materials for the best performance. NF membranes are generally categorized as charged or noncharged membranes, depending on their surface material charges and the operating conditions (such as solution pH). However, charged NF membranes are preferable for better membrane performance (Teixeira et al. 2005), as they can impose charges to repel solutes with the same charge. Additionally, higher membrane surface hydrophilicity is preferred for fouling reduction (Hilal et al. 2005), as it can increase the membrane wettability and thus provides a better cleaning effect. Thus, polyelectrolytes appear to be one of the best candidates to provide these characteristics to a membrane surface. However, the conditions of membrane surface modification using polyelectrolytes have a strong influence on the properties of the produced membrane.

Production of NF membranes has benefited from polyelectrolyte film coatings. It has been shown in the past research studies that the adsorption of positively and negatively charged polyelectrolytes onto a material can contribute to the formation of extremely thin films and, at the same time, greatly alter the properties of the material (Lvov et al. 1999; Klitzing and Tieke 2003). When polyelectrolytes with opposite charges are alternatively adsorbed onto the substrate through the layer-by-layer (LbL) method, the thin film formed is sometimes designated as *PEM film* (Decher et al. 1992). LbL adsorption of polyelectrolytes onto the membrane surfaces has been successfully performed by several researchers (Lenk and Meier-Haack 2002; Yılmaztürk et al. 2009; Li et al. 2010; Ouyang et al. 2010). This technique has been well accepted in several research studies (especially regarding membrane modification) due to its simplicity and low production cost (Farhat and Hammond 2005). It also provides the desired performances in designated applications. However, not all polyelectrolyte-modified membranes are constructed using the LbL method. In some applications, researchers have used, and needed, only a single layer of the polyelectrolyte in order to produce a membrane with the desired properties and performances.

Polyelectrolyte-modified membranes have been reported to resist the adsorption of various kinds of proteins (Meier-Haack et al. 2000; Müller et al. 2001), thus providing better membrane-fouling control. Moreover, polyelectrolyte films provide a good selective barrier compared to porous membranes when the modified membranes are used in NF applications (Harris et al. 2000; Lajimi et al. 2004). A wide range of commercially available polyelectrolytes are now easily obtainable, which vary mainly in terms of functional group, molecular weight, and charge density. These are usually low cost and water soluble. These properties have encouraged the development of simple dip-coating methods in the modification or fabrication of composite polymeric membranes with lower cost (Farhat and Hammond 2005). Table 4.2 shows a list of polyelectrolytes that are commercially available and have been used in NF membrane modification studies.

Polyelectrolyte-based or polyelectrolyte-modified membranes can be easily tailored to suit various applications, such as fuel cell applications (Farhat and Hammond

TABLE 4.2
Polyelectrolytes that are Commercially Available and Used in NF Membrane Modification and Fabrication

Polyelectrolytes Used	Concentrations Used	Applications	References
Polystyrene sulfonate and poly(allylamine hydrochloride)	Started with 0.02 M PSS	Na^+, Ca^{2+}, and Mg^{2+} rejection	Ouyang et al. (2008)
Poly(diallyldimethyl ammonium chloride), poly(allylamine hydrochloride), polyacrylic acid, and polystyrene sulfonate	0.02 M	$MgSO_4$ rejection test and SiO_2 colloidal fouling evaluation	Shan et al. (2010)
Quaternary ammonium cellulose ether and sodium carboxymethyl cellulose	0.01 M	Inorganic salts and dye molecule rejection	Ji et al. (2010)
Polystyrene sulfonate and poly(diallyldimethylammonium) chloride	0.02 M	NaCl, NaF, and Na_2SO_4 salt rejection	Malaisamy et al. (2011)

2005), gas separation (Quinn et al. 1997), NF (Ahmadiannamini et al. 2012), and pervaporation separation (Lenk and Meier-Haack 2002). Surface coating or adsorption of membranes using polyelectrolytes (regardless of whether the coating is single-layered or multilayered) has mostly been performed under ambient conditions. In other words, no specific temperature or pressure (except for the dynamic deposition method) is required in order to obtain the coating layers on top of a selected membrane as the support. This can be explained by the fact that the temperature or pressure applied during polyelectrolyte adsorption onto a membrane surface has no significant or direct effect on the electrostatic force produced. The electrical force, which holds atoms, molecules, and solids together (Saslow 2002), has been used to explain the adhesion of polyelectrolytes onto the membrane surface. When there are two different objects with charges q and Q (measured in Coulombs), their electric force (represented by Equation 4.2 in the units of Newton) changes with the inverse square of their distance (r in terms of meters) (Saslow 2002).

$$|F| = \frac{(k \mid qQ \mid)}{r^2} \tag{4.2}$$

where:

$$k = \frac{1}{4\pi\varepsilon_0}$$

In this well-known equation, k represents a constant that can also be related to the permittivity constant ($\varepsilon_0 = 8.85418781762 \times 10^{-12}$ $C^2 \cdot N^{-1} \cdot m^{-2}$) (Saslow 2002). Through Coulomb's equation, it is seen that the requirement for a particular temperature and pressure is not necessary and thus the adsorption of polyelectrolytes can be done in a much cheaper and faster way. These considerations have made polyelectrolyte-based surface modification through electrostatic forces commercially feasible. However, elevated temperatures during the deposition of polyelectrolytes onto the membrane surface are believed to increase the mobility of the polyelectrolyte molecules and thus can increase the performance of the modified membrane.

4.2.1 UV-Grafting of Polyelectrolytes onto the Membrane Surface

UV-initiated grafting is one of the most popular methods for polymeric membrane surface modification as it offers many advantages, such as lower cost, and can be applied on an existing membrane surface without much effort. However, this method also comes with some weaknesses, which need to be overcome first. For example, the simplicity of the UV-initiated membrane surface modification is only applicable when poly(aryl sulfone) membranes are used (Seman et al. 2010), as these membranes can easily produce active sites or free radicals without the help of a photo-initiator (the most common photo-initiator used is benzophenone). In other words, when polymeric membranes other than poly(aryl sulfone) are used, the addition of a photo-initiator is required and thus an additional stage (known as the *photo-activation stage*) is required before the grafting process can take place. As a result, this incurs additional cost and time to the polyelectrolyte grafting process if UV-initiated grafting is used. Regardless of the membrane materials used, UV-initiated grafting is still worth investigating, as it is useful for membrane surface modification in a cheap and easy way.

During the UV-grafting of polyelectrolytes onto the polymeric membrane surfaces, both porous UF and NF membranes can be served as membrane substrate for the grafting process to take place. The discussion here is based only on two different works, using PES NF and PSF UF membranes as polymeric membrane substrate during the grafting process, as both of them are good candidates for UV-grafting using polyelectrolytes.

4.2.1.1 Acrylic Acid Grafted on PES NF Membranes

UV-initiated grafting using polyelectrolytes can alter the separation layer performance of membranes, as reported previously by Seman et al. (2010). It was observed that the irreversible fouling (using humic acid molecules in a synthetic solution) can be reduced when acrylic acid molecules are grafted onto a PES NF membrane surface. A UV lamp with a wavelength of 365 nm was used throughout that research, as the researchers tried to avoid damages to the membrane support. Short wavelengths such as 254 nm may contribute to bond scission, leading to polymer degradation and lower reproducibility. The proposed mechanism of UV-initiated grafting on a PES NF membrane using acrylic acid is shown in Figure 4.5.

In this work, researchers employed attenuated total reflection-Fourier transform infrared spectroscopy (ATR-FTIR) (Figure 4.6) analysis to support their postulation on UV-initiated grafting. In their proposed mechanism, the grafting process of the PES membrane started with the absorption of UV-light by phenoxy phenyl sulfone chromophores in the backbone of the polymer chain. Two radical sites were produced at the end of each polymer chains as a result of hemolytic cleavage of carbon–sulfur bond (in sulfone linkage). The produced radicals (aryl and sulfonyl) thus induced the polymerization of acrylic acid at the reactive sites of the radicals. Sulfonyl radical

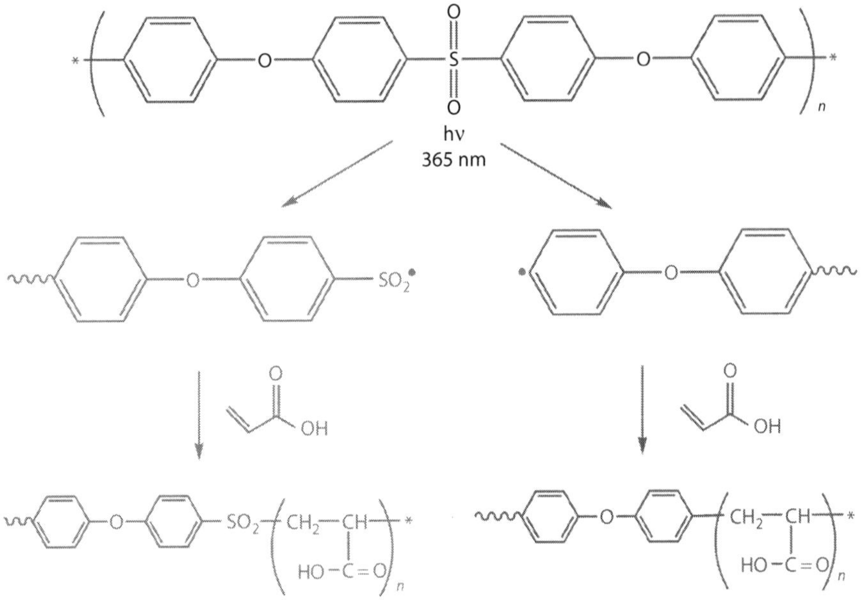

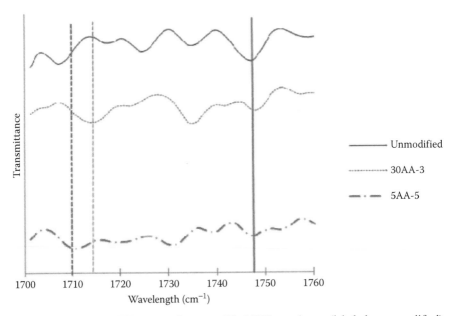

FIGURE 4.5 Postulated UV-initiated grafting on a PES NF membrane using acrylic acid. (Adapted with permission from Seman, M. N. et al., *J. Memb. Sci.*, 355, 133–141, 2010.)

FIGURE 4.6 ATR-FTIR spectra for unmodified PES membrane (labeled as unmodified), membrane modified using 5 g L^{-1} acrylic acid with UV-irradiation of 5 minutes (labeled as 5AA-5), and membrane modified using 30 g L^{-1} acrylic acid with UV-irradiation of 3 minutes (labeled as 30AA-3). (Reprinted with permission from Seman, M. N. et al., *J. Memb. Sci.*, 355, 133–141, 2010.)

may also generate an additional aryl radical by losing sulfur dioxide, which was proved by an additional peaks at 1710–1713 cm^{-1} (correspond to carboxylic acid group bands) (Seman et al. 2010).

4.2.1.2 Sodium p-Styrenesulfonate Grafted on PSF UF Membranes

It has been reported that the NF membrane developed by UV-photo-polymerization of sodium p-styrenesulfonate (SSS) monomers on a PSF UF membrane can be used in dye effluent treatment (Akbari et al. 2002). Researchers used phase inversion method to prepare the PSF UF substrate. The photo-reactor was a cylindrical chamber, in which the membrane was placed on the wall together with a suitable concentration of SSS after washing with distilled water. A UV lamp was installed at the center of the chamber. Irradiation of the membrane was conducted for a certain period of time before it was removed from the reactor and washed with distilled water. It was postulated that the sulfonate group found in the polyelectrolyte molecules can repel the sulfonate group in dye molecules, thus reducing membrane fouling.

4.2.2 Chemical Cross-Linking of PECs with the Membrane Surface

Although LbL adsorption of polyelectrolytes onto the membrane surface has been considered as one of the most common practices among membrane researchers, other methods have also been suggested, such as cross-linking through chemical reaction. In 2010 (Ji et al. 2010), a group of researchers tried to prepare chemically cross-linked PECs onto the surface of commercial PSF UF membranes. Initially, the PECs were prepared by mixing quaternary ammonium cellulose ether (QCMC) and sodium carboxymethyl cellulose (CMCNa). Hydrochloric acid (HCl) was used as the protonation reagent for both the QCMC and CMCNa solutions separately. Precipitated or solid PECs created during the mixing process was filtered out and washed with deionized water to remove free polyelectrolytes and ions. Then, the precipitates were dried in a vacuum at 60°C for 24 hours. Then, researchers dispersed a defined amount of PECs in NaOH for 48 hours at room temperature. In the next step, glutaraldehyde (as a cross-linker) and sulfuric acid (as a curing accelerator) were added to the PECs solution to form the casting solution. This casting solution was then cast onto the PSF UF membrane, which was then heat-cured for a certain period of time. The proposed mechanism is shown in Figure 4.7. The resulting NF composite membranes were then stored in deionized water before use.

4.2.3 Static Self-Adsorption of Polyelectrolytes onto the Membrane Surface

Static deposition is a process where the deposition of polyelectrolytes onto the membrane support is done without any external force to make the polyelectrolyte solution flow through or stay on the membrane support. This also implies that the static deposition method is more inexpensive and easier to be broadly applied in membrane surface modification because less energy is required to complete the modification or fabrication process. Most of the membrane fabrication or modification employing

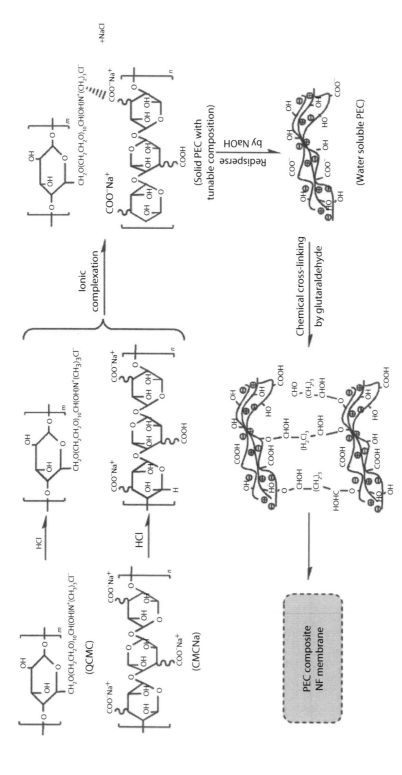

FIGURE 4.7 Proposed mechanism for the PECs composite NF membrane. (Reprinted with permission from Ji, Y. et al., *J. Memb. Sci.*, 357, 80–89, 2010.)

polyelectrolytes reported were based on static deposition method, as it is easier, less time consuming, and well-controlled. This method allows for good control of the membrane thickness and thus controlled solution flux by altering the number of layers to be deposited onto the membrane surface.

There are basically two methods used in static LbL deposition. First, the membrane support can be immersed into the polyelectrolyte solution, so that both sides of the membrane support are exposed and deposited with polyelectrolyte molecules (double-sided method). In the double-sided method, when a membrane is immersed into a polycation solution followed by a polyanion solution, four monolayers of polyelectrolytes will be formed (two monolayers on the active side of the membrane and two monolayers on the opposite side of the membrane). Another possible method is the deposition of polyelectrolytes onto the membrane's active side only (single-sided deposition). However, this method requires a specific membrane-holding cell that can prevent the polyelectrolyte solution from coming into direct contact with the support side of the membrane. In the single-sided deposition method, when a membrane is immersed into a polycation solution followed by a polyanion solution, there are only two monolayers of polyelectrolyte formed (two monolayers on the active side of the membrane and no layers on the opposite side of the membrane). Figure 4.8 shows a schematic diagram of the single-sided and double-sided methods to modify membrane surfaces using different polyelectrolyte solutions.

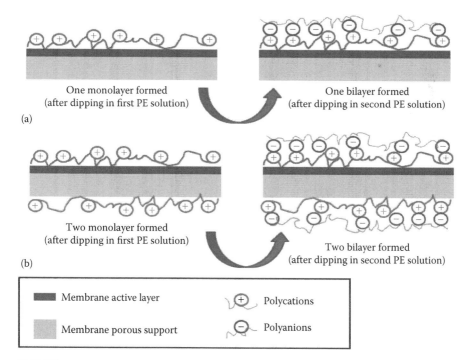

FIGURE 4.8 Dipping methods using two different polyelectrolyte (polycation and polyanion) solutions to produce (a) single-sided and (b) double-sided modified membranes.

Theoretically, the LbL method uses the differences in the molecular weights of the polyelectrolytes to form a porous thin layer, which at the same time possesses certain surface charges to increase the rejection of salt ions. The first layer of the polyelectrolyte that is adsorbed onto the support membrane is critical, depending on the application. In certain applications (especially pervaporation), the separation process requires a nonporous membrane, as a highly porous membrane can lead to a decline in separation performance (Meier-Haack et al. 2001). When this is the case, polyelectrolytes with a lower molecular weight in comparison to the molecular weight cut-off value of the support membrane are highly recommended. Low-molecular-weight polyelectrolyte molecules can fill up the exposed pores on the membrane support, thus reducing the pore size and the permeability of the final membrane. Conversely, polyelectrolytes with a comparatively higher molecular weight relative to the molecular weight cut-off value of the membrane support are strongly recommended in order to produce a highly porous final membrane. By following this procedure, the blockage of pores on the membrane support can be avoided and thus a porous final membrane can be easily obtained. This technique is especially useful in NF membrane modification where the main objective of the NF membrane is to reject salt ions effectively, but at the same time increase the permeability of the membrane.

Various studies using the static deposition method are well reported previously. Researchers are using various combinations of polycations, polyanions, and membrane substrates during the previous study. Here are some examples of the combinations employed in the past research works:

1. Polystyrene sulfonate (PSS)/poly(diallyldimethylammonium) chloride (PDADMAC) deposited on polyamide (PA) NF membranes (Malaisamy et al. 2011) and PSF UF membranes (Baowei et al. 2012).
2. Self-fabricated polyacrylonitrile (PAN) membranes were used and modified with PDADMAC/sulfonatedpoly(ether ether ketone) (SPEEK) polyelectrolyte-pairs to produce solvent-resistant NF membranes (Li et al. 2008a, 2010).
3. Self-fabricated PAN membranes were modified using PDADMAC/polyacrylic acid (PAA) (Ahmadiannamini et al. 2012). An automated dip-coater was employed to control the dipping duration, number of cycles, and oscillation speed for improved consistency of the produced membranes.
4. CHI/sodium alginate (SA) on NF cellulose acetate (CA) membranes (Lajimi et al. 2004).
5. PSS/PAH and PSS/PDADMAC were deposited onto the porous alumina supports (Miller and Bruening 2004; Hong and Bruening 2006; Cheng et al. 2013).

4.2.4 Dynamic Self-Adsorption of Polyelectrolytes onto the Membrane Surface

Dynamic deposition implies a process where the polyelectrolyte solution is forced to move on/through the support membrane (Ji et al. 2008). Dynamic deposition might offer a better choice in the deposition process as the distribution of the polyelectrolyte molecules on the support membrane would be better. Thus, the deposition of polyelectrolytes onto the support membrane is evenly distributed and is expected to provide better membrane separation performance. Dynamic deposition method normally involves the use of a cross-flow or vacuum filtration unit (Deng et al. 2008) that forces the polyelectrolyte solutions (in alternative manners if various layers of polyelectrolytes are desired) to pass through the membrane substrates (Baowei et al. 2012). In certain cases, only one polyelectrolyte layer is sufficient to be dynamically deposited to produce the desired performances (Ba et al. 2010). Here are some of the research works that employed dynamic deposition method:

1. PDADMAC/PSS pair using PSF UF membranes (Baowei et al. 2012).
2. Poly(4-styrenesulfonic acid-*co*-maleic acid) sodium salt (PSSMA), PAH, and PSS on PAN UF membranes (Deng et al. 2008). In this work, each polycation and polyanion solution was vacuum-filtered for 5 min and rinsed with deionized water before being placed in an oven at 50°C.
3. Single layer of PVA, PAA, and polyvinyl sulfate (PVS)–potassium salt on self-fabricated P84-PEI (polyethyleneimine) membranes (Ba et al. 2010).

Other than static and dynamic deposition methods, researchers also tried dynamic-static deposition method (Baowei et al. 2012). In this method, dynamic deposition occurred when a PDADMAC solution was used. However, the PSS solution flowed over the membrane surface at atmospheric pressure (no PSS solution permeated through the UF membrane). This method actually combined both the dynamic and static deposition methods during the deposition of each polyelectrolyte layer using a cross-flow unit.

The advantages of dynamically modified membranes over the statically modified membranes are as follows:

1. Better sealing effect to reduce the membrane pore diameter and thus improve the rejection capability
2. Higher charge distribution
3. Membrane coated with polyelectrolyte layers with better uniformity (well-distributed polyelectrolyte molecules on the membrane surface)
4. Fewer number of polyelectrolyte layers are required compared to the statically deposited membranes
5. Shorter preparation time is required

Although the advantages of the dynamically modified membrane prevail, this method also comes along with some weaknesses. According to this specific study (Deng et al. 2008), the rejection data obtained through the dynamic deposition method was not much improved compared to the static deposition method (Table 4.3). The differences between the dynamically and statically modified membranes were not more than 12% in terms of salt rejection. Thus, it can be concluded that the weaknesses of dynamically modified membranes are as follows:

TABLE 4.3
Membrane Performance Using the Dynamic and Static Modification Methods

Method	NaCl		Na$_2$SO$_4$		MgCl$_2$		MgSO$_4$	
	R (%)	F (L m^{-2}·h)	R (%)	F (L m^{-2}·h)	R (%)	F (L m^{-2}·h)	R (%)	F (L m^{-2}·h)
Dynamic	32.4 ± 1.3	47.0 ± 0.3	91.2 ± 0.7	42.0 ± 0.3	64.6 ± 1.0	46.8 ± 0.3	85.8 ± 0.5	45.8 ± 0.3
Static	31.0 ± 0.8	48.3 ± 0.4	83.1 ± 1.5	45.4 ± 0.3	58.7 ± 0.9	48.8 ± 0.4	76.3 ± 1.2	46.3 ± 0.3
			Analyzed data					
Actual difference[a]	1.4	1.3	8.1	3.4	5.9	2.0	9.5	0.5
Difference in %[b]	4.32	2.77	8.88	8.10	9.13	4.27	11.07	1.09

Source: Deng, H.-Y. et al., J. Memb. Sci., 323, 125–133, 2008.

a Actual difference = |Dynamic−Static|

b Difference in % = ((|Dynamic − Static|/|Dynamic|) × 100%

1. Higher energy consumption, as external force is required to dynamically modify the membrane surface
2. Higher production costs
3. Insignificant developments, as the results obtained were not much improved compared to statically modified membranes
4. Lower fluxes in comparison to statically modified membranes, as the polyelectrolyte molecules would tend to aggregate at membrane pores, thus reducing the membrane pore diameter

4.3 ALTERATION ON NF MEMBRANE CHARACTERISTICS AND PERFORMANCES USING POLYELECTROLYTES

As discussed, there are various methods available to modify the membrane surfaces for NF applications. Most of the modification or fabrication methods employed can contribute to some significant changes to the membrane performances. Some of the changes made to the membrane performances or characteristics is discussed in Section 4.3.

4.3.1 MEMBRANE SURFACE HYDROPHILICITY

Variations in the number of bilayers and types of polyelectrolytes (Ji et al. 2010) can contribute to changes in membrane surface hydrophilicity (Malaisamy et al. 2011). As a common practice in the past research work, membrane surface hydrophilicity values can be determined based on the contact angle measurement using goniometer. Figure 4.9 shows that the membrane hydrophilicity was only greatly improved when more than five bilayers were adsorbed onto the membrane surface. For every single layer of polyelectrolyte adsorbed, the membrane contact angle was not constant due to the fact that the PDADMAC surface is considered hydrophobic and the PSS surface is considered hydrophilic (Elzbieciak et al. 2008). Thus, it is highly suggested that the PSS (or any hydrophilic polyelectrolyte) should be used in the LbL method as the last surface coating to increase the overall membrane hydrophilicity and thus reduce membrane fouling based on the findings in this study. However, the observed hydrophilicity property of PSS, and whether this is also true for other types of polyanions (with the same or different molecular weight), is worth further investigation.

Another similar study (Baowei et al. 2012) was conducted to compare the surface hydrophilicity changes due to different deposition methods. However, the method used to deposit the polyelectrolyte layers did not significantly alter the membrane surface hydrophilicity, as can be seen in Figure 4.10. Dynamic deposition, used throughout this research, can be regarded as a new approach for membrane surface modification by employing polyelectrolyte. Another similar work that was conducted using dynamic deposition method verified that the membrane surface hydrophilicity can be improved using monolayer of polyelectrolytes (Ba et al. 2010). However, this method produces membranes with almost the same property as statically or dynamic-statically deposited membranes (Baowei et al. 2012).

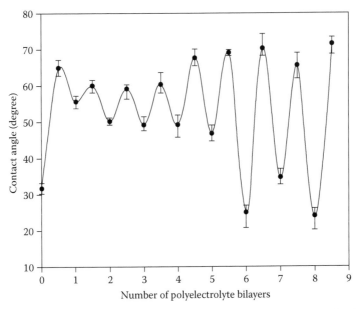

FIGURE 4.9 Membrane surface hydrophilicity study based on the number of polyelectrolyte bilayers. (Reprinted with permission from Malaisamy, R. et al., *Sep. Purif. Technol.*, 77, 367–374, 2011.)

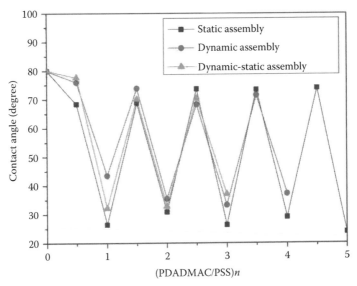

FIGURE 4.10 Variation in the contact angle with the deposition of PDADMAC/PSS on a PSF substrate prepared by different methods. (Reprinted with permission from Baowei, S. U. et al., *J. Memb. Sci.*, 2012.)

Dynamic deposition method is also energy and time consuming in comparison to other two methods. At the current stage, dynamic deposition method using polyelectrolytes is still not suitable to be regarded as the best technique to improve the membrane surface hydrophilicity. In brief, it can be postulated that the membrane final contact angle values are strongly dependent on the type of polyelectrolyte deposited.

4.3.2 MEMBRANE SURFACE AND CROSS-SECTIONAL MORPHOLOGY

In the previous work (Ji et al. 2010), researchers conducted the atomic force microscopy (AFM) to evaluate the root mean square (RMS) roughness values of the polyelectrolytes and PEC-modified membranes. The RMS values were found to be increasing with increasing ionic complexation degree (designated as ICD_{theory} by the researchers). ICD_{theory} is defined as the theoretical value of ionic cross-linking between oppositely charged polyelectrolytes within the PECs, which can be calculated using Equation 4.3 as follows:

$$ICD_{theory} = \frac{[PE^-] - [HCl]}{[PE^-]} \quad (4.3)$$

where:
[PE⁻] was the mole concentration of negatively charged polyelectrolyte
[HCl] was the concentration of the hydrochloric acid

According to this equation, higher concentration of the HCl could actually lead to lower ICD_{theory} value or lower ionic complexation. This further reduced the PEC particle sizes thus lowering RMS values. RMS values were calculated based on AFM result from Figure 4.11 from (a) to (f) was 3.527, 4.063, 4.519, 5.585, 8.088, and 8.436.

Researchers also claimed that PEC-modified membrane can have better antifouling behavior compared to polyelectrolyte-modified membrane (Ji et al. 2010). For clearer understanding on the antifouling behavior, the scanning electron microscopy images were adapted and shown in Figure 4.12. PEC-modified membrane showed less foulants on the membrane separation layer before and after cleaning with water. However, the complexity involved and the long duration of the membrane preparation steps used in this research work require further improvement.

In another work using static LbL deposition method (Li et al. 2008a), it was observed that membranes with rougher surfaces could be obtained by increasing the number of deposited layers (refer Figure 4.13). The rougher membrane surfaces found in this work were one of the weaknesses that had to be overcome to avoid the fouling tendency of the membrane.

According to another similar work (Figure 4.14), Li et al. (2010) found that different salt concentrations in the polyelectrolyte solutions can actually manipulate the membrane surface roughness produced. However, the ionic strength of polyelectrolyte solutions was not in a linear relationship with membrane surface roughness. The membrane

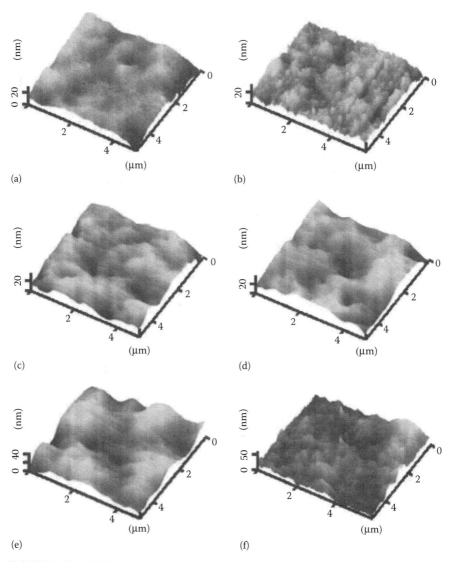

FIGURE 4.11   AFM morphological study of membrane with ICD_{theory} value (a) 0, (b) 0.05, (c) 0.1, (d) 0.5, and (e) 0.9. AFM morphology of blank polysulphone UF membrane is shown in (f). (Reprinted with permission from Ji, Y. et al., *J. Memb. Sci.*, 357, 80–89, 2010.)

surface morphology images were displayed in Figure 4.15. This posed another problem to be solved in the future, that is, to relate the ionic strength of the polyelectrolyte solution and membrane surface roughness with membrane rejection performance. However, this research work had successfully produced membranes with excellent solvent stability by adding supporting electrolytes in the polyelectrolyte solutions.

In another similar work, PSS/PAH multilayers were statically deposited onto a porous alumina support (Cheng et al. 2013). Researchers claimed that an increased

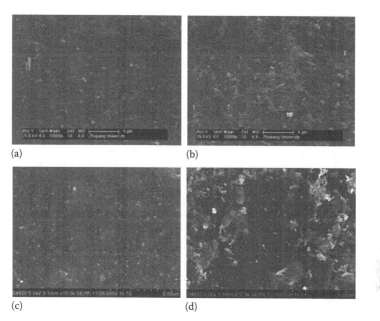

FIGURE 4.12 (a) and (b) are PEC-modified membrane and polyelectrolyte-modified membrane after fouled with solutes. (c) and (d) are the observation made on the same kind of fouled membrane after cleaned with water (for solutes removal study). (Reprinted with permission from Ji, Y. et al., *J. Memb. Sci.*, 357, 80–89, 2010.)

selectivity can only be observed after the membrane separation layer (not the membrane interior pores) is fully enclosed or sealed by the polyelectrolytes. Their postulation was supported by the field-emission scanning electron microscopy (FESEM) images obtained by several groups of researchers (refer Figures 4.16 through 4.18) (Harris et al. 2000; Hong et al. 2006b). Based on the images obtained, they found that the PSS/PAH multilayers can be formed on the membrane separation layers. However, the polyelectrolyte molecules would not fill up the membrane pores.

In another similar work (Miller and Bruening 2004) where PSS/PDADMAC or PSS/CHI are to be statically deposited onto the porous alumina support, researchers were successful to obtain membranes with high rejections toward neutrally charged molecules such as sucrose and raffinose. In certain cases, selectivities as high as 50–60 were obtained with only 3.5–5.5 bilayers. According to the SEM images obtained in that research work (refer Figure 4.19), five bilayers of PSS/PDADMAC with a thickness of approximately 30 nm could produce 95%–99% rejections toward sucrose and raffinose. Besides, only 4.5 bilayers of PSS/CHI were sufficient to cover the 20 nm pores on the alumina support (no visible pores in Figure 4.19[b]).

4.3.3 Membrane Surface Charge Property

In 2011, composite PA NF membranes were used in a study of polyelectrolyte adsorption using LbL manner (Malaisamy et al. 2011). The researchers employed measurements to assess the zeta potentials of the polyelectrolyte-modified and

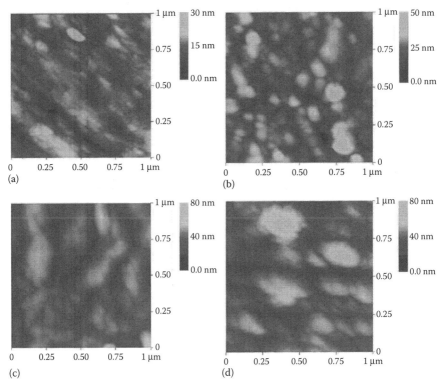

FIGURE 4.13 Membrane surface morphologies obtained using AFM after modified by (a) 5 bilayers, (b) 10 bilayers, (c) 15 bilayers, and (d) 20 bilayers. (Reprinted with permission from Li, X. et al., *Chem. Mater.*, 20, 3876–3883, 2008a. Copyright 2008 American Chemical Society.)

unmodified PA membranes. As the zeta-potential measurements are seldom conducted in LbL manner for polyelectrolyte-modified membranes, this research work provided extremely valuable information for the membrane surface charge analysis. The initial zeta-potential reading of the PA membrane was reported as −25 mV, and it varied throughout the analysis depending on the type of polyelectrolyte and number of bilayers. The result of zeta potentials with respect to the number of polyelectrolyte bilayers is depicted in Figure 4.20. In brief, the magnitude of the zeta potential increased with an increase in the number of bilayers. However, the final zeta-potential reading of membrane was also dependent on the last polyelectrolyte layer on top of the membrane surface. Thus, it can be postulated that by controlling the type of final polyelectrolyte layer and the number of bilayers, the membrane zeta potential, separation performance, and application suitability can be controlled.

4.3.4 Membrane Pure Water Permeability or Flux

UV-grafting by employing polyelectrolyte was conducted previously for humic acid rejection test (Seman et al. 2010). Researchers found that the membranes modified through UV-grafting by employing PAA can lead to some changes in membrane

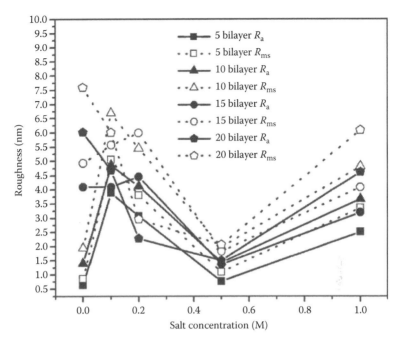

FIGURE 4.14 Membrane surface roughness study of multilayered PEC membranes prepared from PE solutions with different salt concentrations. (Reprinted with permission from Li, X. et al., *J. Memb. Sci.*, 358, 150–157, 2010.)

permeability. With a certain duration of UV exposure, high membrane permeability could be obtained (refer to Figure 4.21). However, this work did not include an optimization study in terms of the duration of irradiation with respect to the concentration of the acrylic acid used.

Based on another similar work using UV-irradiation, it can be seen that this method produced a membrane with lower solution permeability when the irradiation time increased or when a higher concentration of the grafting solution (SSS) was used (Table 4.4). The lower permeability recorded after exposure to longer irradiation time can be caused by the morphological changes within the PSF substrate itself. Besides, higher concentration of the grafting solution can contribute to thicker film formation thus higher resistance toward the solution permeation. Researchers have confirmed that the membrane fouling during the application was reduced after the grafting process. However, this study did not report the membrane characterization results in details, nor the chemistry of the grafting process. Moreover, the optimization of the parameters was not conducted for the highest membrane performance.

In another study where researchers used static deposition method (Malaisamy et al. 2011), a 50% loss in the pure water flux was recorded in comparison to the membrane before modification with PEMs. The authors claimed that the reduction in the pure water flux was due to the increased total membrane resistance as a result of polymer chains built on the membrane surface. In another work (Malaisamy

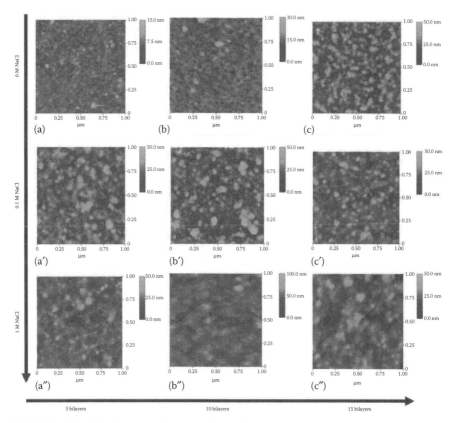

FIGURE 4.15 Membrane surface morphology images obtained for membranes modified by polyelectrolyte solutions of 0, 0.1, and 1.0 M NaCl. (Reprinted with permission from Li, X. et al., *J. Memb. Sci.*, 358, 150–157, 2010.)

and Bruening 2005) that employed PSS/PAH and PSS/PDADMAC multilayers, researchers suggested that, in certain cases, the flux reduction observed can be contributed by the polyelectrolyte adsorption within the pores of porous support as well. However, they were unable to provide solid evidence through the FESEM imaging.

Li et al. (2010) found that modified membranes with higher solution fluxes can be obtained when the polyelectrolyte solutions were added with higher salt concentrations (Figure 4.22) due to the formation of looser structures. However, when 20 bilayers of polyelectrolyte with various salt concentrations were deposited, all of them displayed similar rejection capability toward the Rose Bengal in isopropyl alcohol solutions. This showed that the membranes modified with higher salt concentrations can be potentially used to increase the membrane selectivity, as they produced better solution fluxes but similar rejection capability. It might be reasoned with the loopy structures that consisted of higher surface charge density. However, no specific reason was given for the observation made. Thus, it can be postulated that the loopy structures formed can be contributed by the changes in molecular structures by modifying the polyelectrolyte solution ionic strength, which will be discussed in Section 4.3.6.

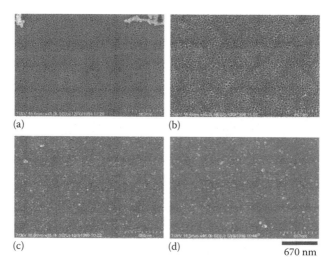

(a)   (b)

(c)   (d)   670 nm

FIGURE 4.16   Separation layer FESEM images of porous alumina substrate (a) before coated with polyelectrolytes, after coated with (b) two, (c) four, and (d) five PAH/PSS bilayers. (Reprinted with permission from Harris, J. J. et al., *Chem. Mater.* 12, 1941–1946, 2000. Copyright 2000 American Chemical Society.)

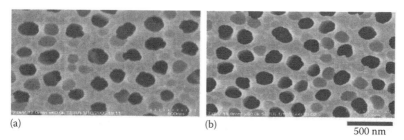

(a)   (b)   500 nm

FIGURE 4.17   FESEM images on the permeate side of a bare porous alumina substrate (a) before and (b) after adsorption of 10 PAH/PSS bilayers on the membrane separation layer. (Reprinted with permission from Harris, J. J. et al., *Chem. Mater.* 12, 1941–1946, 2000. Copyright 2000 American Chemical Society.)

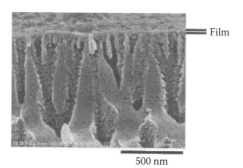

— Film

500 nm

FIGURE 4.18   FESEM of cross-sectional morphology of a porous alumina support coated with PSS/PAH multilayers. (Reprinted with permission from Hong, S. U., *Ind. Eng. Chem. Res.*, 45, 6284–6288, 2006b. Copyright 2006 American Chemical Society.)

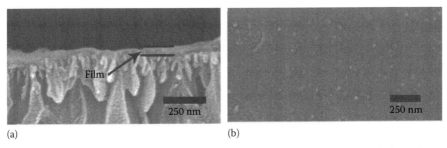

(a)   (b)

FIGURE 4.19 SEM images for alumina porous membrane modified using polyelectrolytes. (a) cross section of five bilayers of PSS/PDADMAC (~30 nm) (b) membrane separation surface using 4.5 bilayers of PSS/CHI. (Reprinted with permission from Miller, M. D. and M. L. Bruening, *Langmuir*, 20, 11545–11551, 2004. Copyright 2004 American Chemical Society.)

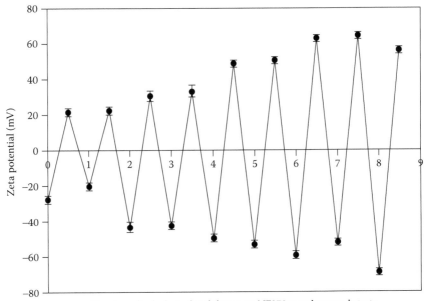

Number of polyelectrolyte bilayers on NF270 membrane substrate

FIGURE 4.20 Zeta potential of unmodified and polyelectrolyte-modified NF membranes with different numbers of bilayers. (Reprinted with permission from Malaisamy, R. et al., *Sep. Purif. Technol.*, 77, 367–374, 2011.)

4.3.5 SOLUTE REJECTION AND SELECTIVITY CAPABILITY OF POLYELECTROLYTE-MODIFIED MEMBRANE

Dynamically formed membranes were first investigated in 1997 using monolayer of PAA (a weak acid polyelectrolyte) that was deposited on a macroporoustitania layer sintered on the inside of a porous stainless steel tubular module (Xu and Spencer 1997). This work used the produced membrane to investigate the separation of Cibacron blue (a kind of dye) from $NaNO_3$. It was found that the separation factor

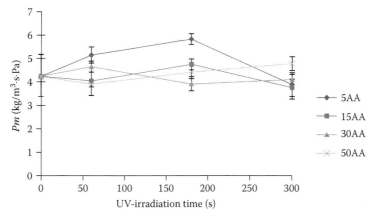

FIGURE 4.21 Membrane permeability produced using various acrylic acid concentrations and UV-irradiation time. (Reprinted with permission from Seman, M. N. A. et al., *J. Memb. Sci.*, 355, 133–141, 2010.)

TABLE 4.4
Water Permeability for Unmodified and Modified PSF Membranes

Membrane	M0	M1	M2	M3	M4	M5
Sodium *p*-styrenesulfonate (wt%)	–	1	3	3	4	4
Irradiation time (min)	–	10	10	15	15	20
Permeability ($L \cdot h^{-1} \cdot m^{-2} \cdot bar^{-1}$)	50	19	12	8	5	2

Source: Akbari, A. et al., *Desalination*, 149, 101–107, 2002.

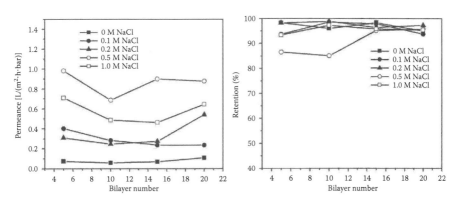

FIGURE 4.22 Filtration performances of the modified membranes for Rose Bengal (RB) in isopropyl alcohol (IPA) solutions. (Reprinted with permission from Li, X. et al., *J. Memb. Sci.*, 358, 150–157, 2010.)

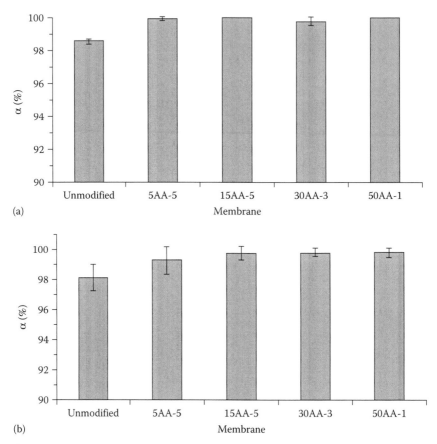

FIGURE 4.23 Humic acid rejections by nonmodified and UV-modified membranes when humic acid solutions at (a) pH 7 and (b) pH 3 were used. (Reprinted with permission from Seman, M. N. A. et al., *J. Memb. Sci.*, 355, 133–141, 2010.)

increased with decreasing flux and ionic strength, increasing pH, cross-flow velocity, and dye-to-salt concentration ratio.

In another work, the rejection toward humic acid was found to be improved after the membranes were UV-modified with acrylic acid (Figure 4.23). The improvement in membrane rejection was not sufficient to be considered good (i.e., not more than 10% overall) using the UV-irradiation method. However, this work provided some evidences to verify the possibility of employing polyelectrolyte through UV-grafting method to improve the membrane rejection property toward certain solutes.

For the study through static deposition of polyelectrolytes (Malaisamy et al. 2011), researchers recorded the rejection of chloride and fluoride ions (with different concentrations) in a mixed salt solution (composed of sulfate, fluoride, and chloride ions), which was then expressed in terms of selectivity for chloride over fluoride ions. Selectivity for solute 1 over solute 2 was calculated using the selectivity equation (Malaisamy et al. 2011) as explained previously and analyzed using ion chromatography. It was observed that the variations in the number of bilayer

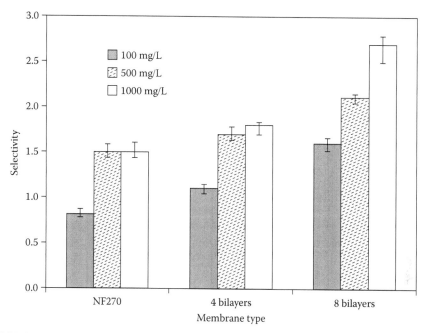

FIGURE 4.24 Modified and unmodified membrane selectivity for mixed solutions (at different concentrations). (Reprinted with permission from Malaisamy, R. et al., *Sep. Purif. Technol.*, 77, 367–374, 2011.)

contributed to the differences in ion selectivity, as shown in Figure 4.24. Better selectivity for chloride over fluoride ions was observed when the number of poly-electrolyte bilayers was higher. When the number of polyelectrolyte bilayer was increased, the excessive charges present in the membrane separation layers contributed to more effective repulsion toward the permeating ions (Krasemann and Tieke 1999), as described in Figure 4.25. It can be concluded that higher charge density

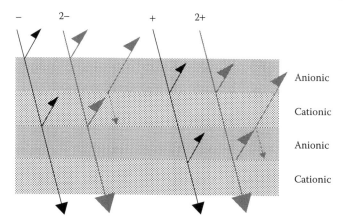

FIGURE 4.25 Proposed repulsion model for multibipolar membrane. (Adapted with permission from Urairi, M. et al., *J. Memb. Sci.*, 70, 153–162, 1992.)

in the membrane separation layer can lead to higher separation factor or selectivity. This type of selectivity would be highly useful in drinking water and water softening processing. Lower rejection toward chloride ions makes the NF membrane fouling tendency to be potentially reduced and thus prolonged the membrane life expectancy.

Polyelectrolyte-modified membrane in LbL manner allows the membrane surface charge to be easily fine-tuned, which helps to improve the membrane selectivity. For instance, when self-fabricated PAN membranes were used and modified with PDADMAC/SPEEK polyelectrolyte pairs to produce solvent-resistant NF membranes (Li et al. 2008a), researchers found that the modified membranes showed high selectivity for solutes that were of the same charge as the last deposited polyelectrolyte layer. The researchers found that the retention of negatively charged methyl orange (MO) was much higher compared to the positively charged crystal violet (CV), although the molar volume of MO is much smaller compared to CV. Clearly, the PEMs contributed to the Donnan exclusion effect between the sulfonic acid groups and the negatively charged solutes.

In another work using similar static deposition method (Ahmadiannamini et al. 2012), researchers tried to manipulate the membrane retention property by adjusting the salt concentrations and pH of the polyelectrolyte solutions. Higher salt concentration in the polyelectrolyte solution was found to produce membrane with lower thickness, lower retention capability, and lower permeance (Figure 4.26). The reason suggested for this finding was the instability of the multilayers built up in the presence of high salt concentrations in polyelectrolyte solutions. The instability of the multilayers can decompose to form some loosely attached structures or layers, which thus caused lower restrictions to the passage of the solutes. Decomposition of the layers also can lead to the formation of thinner separation layers.

Additionally, reduction of the membrane thickness when the pH changed from 2 to 4 (Ahmadiannamini et al. 2012) contributed to increased charge density of the PAA chains and thus required less material to compensate for the charges present due to the previously deposited layer. This caused lower restrictions to the passage of the solutes through the membrane separation layer, thus higher permeance but lower retention capability (Figure 4.27). This work presented solid evidences as to how the ionic strength and the pH of the polyelectrolyte solutions could affect membrane thickness. This could be highly useful for future membrane modification research, as solution fluxes could greatly benefit from this well-controlled mechanisms.

In 2006, the PSS/PAH pair was once again statically deposited onto the porous alumina support, which was then applied in the separation of neutral amino acids in aqueous solution (Hong and Bruening 2006). The amino acids studied included neutral amino acids such as L-glutamine, L-serine, L-alanine, L-glycine, and a basic amino acid (L-lysine). The authors claimed that NF of the mixture of the four neutral amino acids could be completed with high selectivity for L-glycine over L-glutamine (about 50-fold) with a solution flux of 1.3 m^3 (m^{-2}· day^{-1}). It was concluded that the LbL-modified membrane using the PSS/PAH pair could be used to separate neutrally charged amino acids through the size exclusion mechanism. The Donnan effect is not utilized in neutrally charged molecule separation, as described in this research. In the same work (Hong

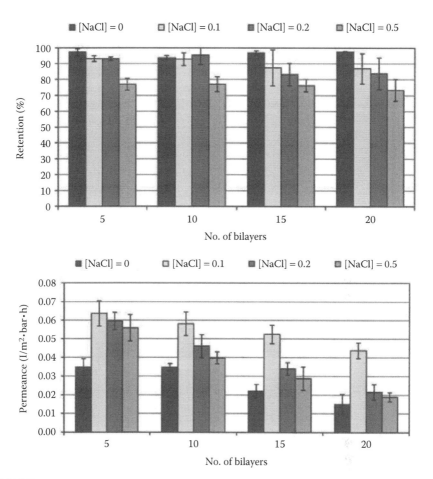

FIGURE 4.26 Retention and permeance performances for membranes modified using polyanion solutions with different salt concentration. (Reprinted with permission from Ahmadiannamini, P. et al., *J. Memb. Sci.*, 394–395, 98–106, 2012.)

and Bruening 2006), the researchers used only few bilayers (as shown in Table 4.5) to achieve the desired rejections toward glycine, lysine, and glutamine. Higher number of bilayers that adsorbed contributed to better size exclusion mechanism can be explained using the three-zone model (Ladam et al. 1999). In the proposed three-zone model (refer to Figure 4.28), Zones I and III preserved their specific characteristics. Only Zone III, the outermost layers, displays classic polyion-like behavior. As the number of bilayers increases, the only significant difference would be in Zone II. Because Zone II grows in thickness with increasing number of bilayers, it contributes to better pore-sealing effect and size-exclusion effect. These combined effects thus contributed to higher rejection toward the solutes and thus selectivity.

In 2013, porous alumina support was once again statically deposited with PSS/ PAH multilayers by Cheng et al. (2013). This work revealed that the surface charge is still not a dominant factor in controlling the solute transport. Besides, selectivity

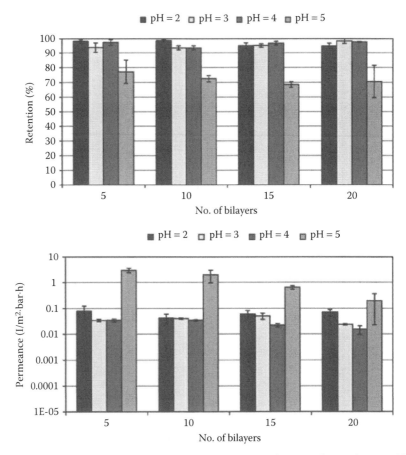

FIGURE 4.27 Effect of the polyanion pH toward the thickness and retention capability of the modified membranes. (Reprinted with permission from Ahmadiannamini, P. et al., *J. Memb. Sci.*, 394–395, 98–106, 2012.)

TABLE 4.5
Rejections and Selectivities from NF Experiments Using Membrane Modified by Polyelectrolytes (with NaCl as Supporting Electrolyte)

	Glycine Rejection (%)	Lysine Rejection (%)	Glutamine Rejection (%)	Glycine/ Glutamine Selectivity	Lysine/ Glutamine Selectivity	Glycine/ Lysine Selectivity
(PSS/PAH)₄PSS	10.4 ± 4.1	31.8 ± 4.6	44.2 ± 5.5	1.7 ± 0.0	1.3 ± 0.0	1.3 ± 1.0
(PSS/PAH)₆	10.4 ± 0.9	50.9 ± 6.1	85.2 ± 0.8	6.1 ± 0.4	3.3 ± 0.6	1.8 ± 0.2
(PSS/PAH)₆PSS	14.7 ± 2.4	39.5 ± 2.3	72.6 ± 1.6	3.1 ± 0.4	2.2 ± 0.2	1.3 ± 0.1
(PSS/PAH)₇	19.7 ± 6.7	64.6 ± 6.6	92.8 ± 0.9	11.1 ± 0.5	4.9 ± 0.8	2.3 ± 0.3
(PSS/PAH)₇	14.2 ± 2.9	65.5 ± 3.3	92.3 ± 0.6	11.1 ± 0.7	4.5 ± 0.4	2.5 ± 0.2

Source: Hong, S. U. and M. L. Bruening, *J. Memb. Sci.*, 280, 1–5, 2006.

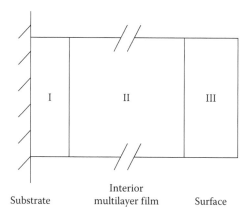

Substrate · Interior multilayer film · Surface

FIGURE 4.28 Proposed three-zone model in explaining the selective separation. (Reprinted with permission from Ladam, G. et al., *Langmuir*, 16, 1249–1255, 1999. Copyright 1999 American Chemical Society.)

TABLE 4.6
Selectivities and Ion Fluxes in the Diffusion Dialysis of KCl and MgCl$_2$ through Bare Alumina and Polyelectrolytes-Modified Membranes

Membrane	Ion	Single Salt Ion Flux (nmol cm^{-2} s^{-1})	Selectivity (K$^+$/Mg^{2+})
Bare alumina	K$^+$	6.43 ± 0.65	1.47 ± 0.15
	Mg^{2+}	4.36 ± 0.04	
(PSS/PAH)$_4$-coated	K$^+$	2.39 ± 0.50	>350
alumina	Mg^{2+}	<0.007	
(PSS/PAH)$_4$PSS-coated	K$^+$	3.09 ± 0.18	276 ± 93
alumina	Mg^{2+}	0.011 ± 0.004	

Source: Cheng, C. et al., *Langmuir*, 29, 1885–1892, 2013. Copyright 2013 American Chemical Society.

between potassium and magnesium ions (Table 4.6) was most probably contributed by the differences in hydrated ion sizes or salvation energies. According to another work done on the transport of neutral molecules (using sucrose, glucose, and glycerol) (Liu and Bruening 2003) in measuring the effective pore diameter, the effective pore diameter in (PSS/PAH)$_7$ films is around 0.8–1.0 nm. This makes the dense PEMs impermeable to the magnesium ions (hydrated diameter of 8Å), but more permeable to potassium ions (hydrated diameter of 3Å).

Besides, another group of researchers found some interesting results using PSS/PAH multilayers with porous alumina as their support (Stanton et al. 2003). They postulated that the increased supporting electrolyte (using MnCl$_2$) concentration during the deposition of the outer polyelectrolyte layer can contribute to higher charge density thus better rejection. They successfully proved this postulation based on the NF experiments that were carried out at 4.8 bar using salt solution of 50 ppm in the ion of interest.

TABLE 4.7
Ion Rejections (%) and Permeate Fluxes (in Parentheses in m^3 m^{-2} day^{-1}) for PSS- and PAH (Added with Different Amounts of Supporting Electrolyte)-Modified Membranes

Membrane	Na^+ (NaCl)	SO_4^{2-} (Na_2SO_4)	Ca^{2+} ($CaCl_2$)	Mg^{2+} ($MgSO_4$)
(PSS/	60 ± 7	44 ± 7	94 ± 1	96 ± 1
PAH)$_4$PSS[a]	(2.0 ± 0.5)	(2.1 ± 0.3)	(2.3 ± 0.3)	(1.6 ± 0.1)
(PSS/PAH)$_5$	81 ± 3	42 ± 5	96.6 ± 0.7	97.6 ± 0.5
	(1.1 ± 0.2)	(1.3 ± 0.0)	(1.4 ± 0.1)	(1.7 ± 0.1)
(PSS/	51 ± 6	48 ± 3	94 ± 1	94 ± 2
PAH)$_4$PSS[b]	(2.2 ± 0.2)	(1.9 ± 0.2)	(2.0 ± 0.1)	(1.8 ± 0.4)
(PSS/	31.0 ± 5	91 ± 4	64 ± 7	88 ± 5
PAH)$_4$PSS[c]	(2.1 ± 0.2)	(1.7 ± 0.1)	(1.7 ± 0.1)	(1.5 ± 0.3)

Source: Stanton, B. W. et al., *Langmuir*, 19, 7038–7042, 2003. Copyright 2003 American Chemical Society.

[a] 0.5 M $MnCl_2$ in final PSS layer deposition solution

[b] 0 M $MnCl_2$ in final PSS layer deposition solution

[c] 2.5 M $MnCl_2$ in final PSS layer deposition solution

Na_2SO_4 rejections increased from about 50% to 90% when the $MnCl_2$ in final PSS layer deposition solutions were increased from 0 to 2.5 M (refer to Table 4.7), but having the same permeate fluxes. The high permeate flux and rejection of negatively charged anions suggested that more Cl^- ions were adsorbed in the membrane than Mn^{2+} ions.

Porous alumina support was also used for static LbL polyelectrolyte modification (Hong et al. 2009) using PSS/PDADMAC pair to construct NF membranes, which were then applied to recover phosphate from a chloride-containing solution. As the structure of phosphate can be changed according to the feed solution pH, higher phosphate rejection was observed when a higher feed solution pH was used. This was explained by the fact that the chloride rejection by the polyelectrolyte-modified membrane was not affected by the feed solution pH.

In another similar work (Miller and Bruening 2004) using PSS/PDADMAC or PSS/CHI with porous alumina support, researchers successfully obtained membranes with high rejections toward neutrally charged molecules such as sucrose and raffinose. Researchers obtained selectivities of as high as 50 to 60, with only 3.5 to 5.5 bilayers. According to the three-zone model explained in this section, few bilayers were sufficient to produce membrane with the desired Zone II for size exclusion mechanism to take place (Donnan effect is not utilized). In another similar work (Hong et al. 2007), researchers coincidentally have similar findings when they tested PSS/PDADMAC-modified porous alumina supports. When monovalent ions were used in the separation process, the obtained selectivity results suggested that the discrimination among monovalent ions is based on the Stokes radius or sizes.

As depicted in Figure 4.29, it was claimed that the dynamically or dynamic-statically deposited polyelectrolyte layers could contribute to better Na_2SO_4 rejection

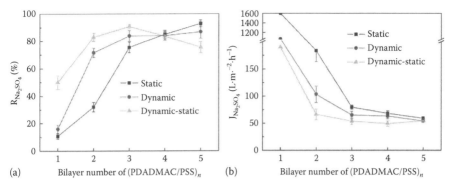

FIGURE 4.29 Different preparation methods with manipulated number of bilayers: (a) variation in Na_2SO_4 rejection and (b) flux of $(PDADMAC/PSS)_n$ modified membranes. (Reprinted with permission from Baowei, S. U. et al., *J. Memb. Sci.*, 2012.)

(Baowei et al. 2012). Additionally, greater differences in salt rejection were observed when two bilayers of PDADMAC/PSS were deposited using three different methods. Nevertheless, the differences in the salt rejection capability were not significant when the number of bilayers was increased over three, regardless of the method used. Thus, it can be postulated that the method of deposition is less significant in improving the membrane rejection capability when higher salt rejection is desired. As a result, the cheapest and easiest way should be preferred to make it commercially feasible. Besides, when five bilayers of polyelectrolyte were deposited, the fluxes and rejections generated using these three different methods were roughly the same. This further verified that the method of depositing the polyelectrolyte layers was less significant in improving the membrane salt rejection capability.

4.3.6 CHANGES IN MEMBRANE SEPARATION LAYER THICKNESS

In several studies conducted previously, researchers analyzed the thicknesses of the multilayers formed using polyelectrolytes. Thicknesses of the multilayers formed can be related to the other membrane performances, whether in a direct or indirect manner. An understanding of the changes in polyelectrolyte layer thicknesses in several aspects is considered as an important factor and worth a short discussion here.

In 2010 (Li et al. 2010), a work was conducted in order to investigate the influence of the polyelectrolyte preparation conditions on membrane performance. This research used a similar membrane support and materials, as reported in Li et al. (2008a). The researchers included some supporting electrolyte in the polyelectrolyte solution in the hope of altering the polyelectrolyte charge density and its configuration on the PAN membrane surface. In this specific study, PDADMAC and SPEEK were alternatively deposited onto a glass slide using polyelectrolyte solutions of different concentrations (NaCl solutions from 0 to 1 M). This was conducted in order to confirm the structure of the polyelectrolyte molecules when the ionic strength of the solution was changed. However, it could only be concluded that the thicknesses of the polyelectrolyte layers detected by the UV–vis device increased with increasing ionic strength.

The increased salt concentrations that contributed to the increased polyelectrolyte thin-film thicknesses was previously explained in another multilayer deposition study (Ladam et al. 1999). When salt concentration in the polyelectrolyte solution is higher, the Debye length of an electrical potential is lower due to the screening effect by the salt ions. The correlation between the Debye length and the salt concentration was well investigated by Tadmor et al. (2002). According to the work, Debye length is vital in describing the solution containing charged polyelectrolytes. By referring to the Debye length (κ^{-1}) equation (given by Equation 4.4), higher salt or ion concentration (n_i) would contribute to lower Debye length value. This screening effect contributed to lower self-repulsion and thus forming dense globule. The more coiled form of the polyelectrolyte chains would eventually lead to thicker thin film (in multilayered systems) on the membrane substrate (as displayed in Figure 4.30).

$$\kappa^{-1} = \left(\frac{\varepsilon\, \varepsilon_o k_B T}{\sum_i n_i z_i^{\,2} e^2} \right)^{1/2} \tag{4.4}$$

where:
  ε is the dielectric constant of water
  ε_o is the permittivity of free space
  k_B is the Boltzmann constant
  T is the temperature
  z_i is the valency of ion species
  e is for the electronic charge

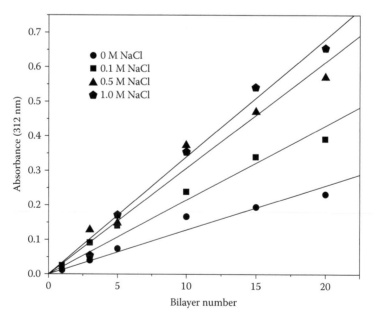

FIGURE 4.30 Effect of salt addition toward polyelectrolyte chains and thin-film thicknesses. (Reprinted with permission from Li, X. et al., *J. Memb. Sci.*, 358, 150–157, 2010.)

Besides, when the polyelectrolytes in the coiled form (polyelectrolyte solution in the presence of salt ions) adsorbed onto the membrane substrate, higher amount of the deposited polyelectrolytes in the next layer is expected in order to compensate the excessive build-up of the surface charge. This encouraged the formation of thicker PEMs, which can be represented by

$$\delta Q = a \cdot C^{\alpha} + b \tag{4.5}$$

where:

δQ is the amount of polyelectrolyte

C is the salt concentration

a, b, and α are the fitting parameters for the fitting function (Ladam et al. 1999)

In another work where self-fabricated PAN membranes were once again used (Ahmadiannamini et al. 2012), an automated dip coater was employed to control the dipping duration of PAN membranes in PDADMAC/PAA solution, number of cycles, and oscillation speed for improved consistency of the produced membranes. The researchers also postulated that the increase in membrane thickness at higher salt concentration or ionic strength was due to the formation of more coiled polyelectrolytes, which thus occupied a smaller area (Ahmadiannamini et al. 2012). However, this research work found some contradictory results with the previous work (Li et al. 2010) in terms of reduction in membrane thickness when the salt concentration was further increased (Figure 4.31). The differences might be caused by different material selection (type of polyelectrolyte) and the employment of the dip-coater in the research work (Ahmadiannamini et al. 2012). Besides, the deterioration of the polyelectrolyte structures at high salt concentration can be the reason for the reduction in membrane thicknesses. Self-fabricated membranes used in this work also can be the reason for the changes in membrane thicknesses measured. However, these factors are not clearly investigated and explained previously. In the same work, Ahmadiannamini et al. (2012) also provided some evidences on the possibility of manipulating membrane thicknesses through the control of polyelectrolyte solution pH. In brief, solution pH needs to be maintained in certain range in order to obtain thinner PEMs, or vice versa.

4.3.7 Performance Stability of the Polyelectrolyte-Modified Membranes

Stability of the multilayer formation on the membrane substrates is seldom studied in many of the previous works. The reason for this observation can be due to the fact that there is no any well-organized or standardized method to evaluate the multilayer formation stability. However, the multilayer formation stability is important to evaluate the membrane performances in long-term applications. As a result, several studies that evaluated the multilayer stability from different perspectives are discussed here.

Selection of polyelectrolyte materials plays a key role in producing membranes with good performance stability. For instance, CA NF membranes were used as the support membranes during the static LbL deposition method in 2004 (Lajimi et al. 2004). The

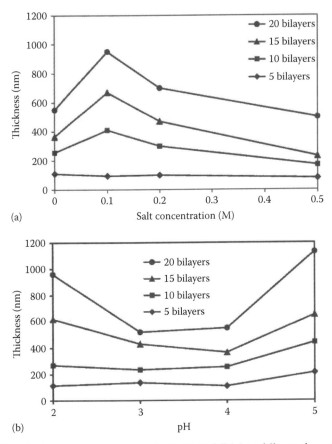

FIGURE 4.31 Thickness measurements of PDADMAC/PAA multilayered membranes prepared using different (a) salt concentrations and (b) pH levels. (Reprinted with permission from Ahmadiannamini, P. et al., *J. Memb. Sci.*, 394–395, 98–106, 2012.)

investigators applied CHI and SA polyelectrolyte layers with high molecular weights during the membrane modification. It was discovered that the permeation rates of salt solutions were higher than that of pure water after the membranes were modified with those weakly charged polyelectrolytes. Thus, they postulated that the observed increase in the solution flux could have been due to the detachment of the polyelectrolyte layers from the surface, as the charge density for both the polyelectrolytes and CA membranes were low. This would suggest that the polyelectrolyte molecules or the membrane surface must possess a higher charge density for greater coating stability. In another work, Li et al. (2010) claimed that the salt ions could induce thicker multilayers and fewer interpenetrating layers. As a result, every single polyelectrolyte molecule possessed higher mobility, which lowered the stability of the multilayers formed.

Polymeric membranes modified through the dynamic single layer of polyelectrolyte adsorption method, as reported by Ba et al. (2010), provided some useful information about the membrane performance stability. In this study, a cross-flow system was used for dynamic coating and for fouling tests (using synthetic solutions of bovine serum albumin [BSA], SA, and humic acid sodium salt [HA]). Researchers used 6 L of PVA, PAA, and PVS–potassium salt solutions (50 mg L^{-1}) as feeds in the cross-flow system and were circulated for 8–12 hours until the steady state was achieved. By doing so, it was assumed that all of the polyelectrolyte adsorption sites on the polymeric membrane were saturated. The membranes were then rinsed with deionized water. The ability of membrane to remove the coating at low pressure (4.8 bars) with an HCl solution at pH 2 for 10–12 hours was also tested. Ba et al. (2010) studied that PVA was adsorbed onto the membrane surface, thereby forming a protective layer that could be removed easily together with the foulants that accumulated on the PVA coating layer by acid cleaning. Instability of the thin-film formation using neutrally charged PVA has donated to another possible application for NF. The removal of the PVA thin film would thus produce a fresh membrane surface for the next application. However, researchers found that PVS formed a more stable coating layer (one not easily removed by acid cleaning), as the PVS charges are opposite to those of the polymeric membrane used. Thus, ionic attraction between PVS and the polymeric membrane surface can be used to explain the phenomenon observed. For the PAA coating, unstable layer formation was observed during the desalination process as the water flux increased over the time. The PAA layer was also not completely removed using acid cleaning. The pH of the PAA solution or the filtration environment was reported to be significant toward the formation or desorption of the coating layer. At pH 4.7, a thick coating layer of PAA was formed as PAA molecules were partially charged and could be easily adsorbed onto the oppositely charged membrane. However, when desalination was carried out at pH 7, PAA was almost completely ionized, thus producing repulsion forces among the PAA chains. This caused desorption of the PAA or thinning of the PAA coating layer, thus increased fluxes were observed. This suggests that polyelectrolyte-modified membranes should be used in applications where the pH of the solution does not ionize the polyelectrolyte layer completely, if the stability of the coating layer is very much desired during the NF processes. The PVA coating showed better fouling resistance toward BSA, but not HA or SA, which are negatively charged. The PVS-modified membrane restored almost 100% of the solution flux after it was tested with BSA and SA, as this membrane can remain almost neutral under acidic conditions. Lastly, it was observed that the PAA-modified membrane would become positively charged after being treated with acid. This resulted in poor desorption of negatively charged foulants and the membrane became severely fouled.

4.4 DRAWBACKS ON THE APPLICATIONS OF POLYELECTROLYTES IN NF MEMBRANE MODIFICATION OR FABRICATION

One of the minor drawbacks by employing polyelectrolytes in NF membrane modification or fabrication is the need for a pretreatment stage. Pretreatment methods have been used before the adsorption of polyelectrolytes onto membrane surfaces for several reasons. A pretreatment process can be used to clean the membrane surface as commercialized polymeric membranes contain various kinds of

TABLE 4.8
Membrane Pretreatment Methods Suggested Before Polyelectrolyte Adsorption Processes

Pretreatment Methods	Other Conditions Used	Reasons for Pretreatment	References
Membrane immersion in 0.1 M NaOH for 2 h	Followed by immersion in clean water for 8 h	For membrane cleaning purpose	Egueh et al. (2010)
Membrane immersion in 0.1 M NaOH for 3 h	Followed by immersion in deionized water for 24 h at 4°C with the water exchanged after the first 12 h of storage	–	Shan et al. (2010)
NF membranes soaked in water for 3 h	Water was replaced every hour, then rinsed with water before use	Cleaning purposes	Malaisamy et al. (2011)
UV/O$_3$-cleaned porous alumina support	–	Cleaning purposes	Hong and Bruening (2006); Hong et al. (2009)
PAN UF membrane modified with NaOH 1.5 M for 1.5 h at 45°C	Intensive washing with distilled water until pH 7	To hydrolyze the membrane to make it negatively charged and to reduce the pore size for better deposition and adhesion	Deng et al. (2008)
UF PSF membrane immersed in deionized water	–	To remove impurities	Baowei et al. (2012)

preservatives. Pretreatment process can also be used to improve the membrane surface charges thus the electrostatic interactions with the accompanying poly-electrolyte layers. Based on Table 4.8, most of the pretreatment processes suggest soaking the membrane in an NaOH solution at a certain concentration for a defined period of time. All membranes are rinsed in clean water before they are used for polyelectrolyte adsorption.

Another drawback of applying polyelectrolytes in NF membrane modification or fabrication is the need to control the polyelectrolyte solution ionic strength and pH. As discussed, ionic strength and the pH of the polyelectrolyte solution can affect the molecular structure of the polyelectrolyte molecules. Alteration of the poly-electrolyte molecules can result in variable charge densities, which in turn affects the performance of the NF membrane. Table 4.9 shows some of the polyelectrolyte solution preparation steps that have been used in the adsorption process. However, it was also reported that the solution ionic strength for a specific polyelectrolyte

TABLE 4.9
Preparation of Polyelectrolyte Solutions Before the Adsorption Process

Polyelectrolyte Solution Preparations	Other Descriptions	Reasons	Reference
PAH and PSS were prepared in 0.1 M NaCl	Various pH values were obtained with different concentration of the polyelectrolytes	–	Eguech et al. (2010)
PDADMAC, PAH, PSS, and PAA were prepared in 1–2.5 M NaCl	The pH of the polyelectrolyte solutions was adjusted using 1 M HCl and 1 M NaOH	For solution pH and ionic strength adjustment	Shan et al. (2010)
PVA, PVS, and PAA in 50 mg/L were prepared in DI water at 80°C for overnight	No pH adjustment	–	Ba et al. (2010)
0.5 M of $CaCl_2$ was added into PSS and PSSMA solution	Solution was adjusted to pH 2.5 with HCl	For solution pH and ionic strength adjustment	Deng et al. (2008)
0.02 M PSS in 0.5 M $MnCl_2$ or NaCl, 0.02 M PAH in 0.5 M NaBr or NaCl	Solutions pH were adjusted using NaOH and HCl	For solution pH and ionic strength adjustment	Hong and Bruening (2006)
0, 0.1, 0.2, and 0.5 M NaCl were separately added into PDADMAC and PAA solution of 0.2 wt%	Polyelectrolyte solutions with pH 2, 3, and 4 and the original pH of the solution were obtained using NaOH and HCl if necessary	For solution pH and ionic strength adjustment	Ahmadiannamini et al. (2012)
0.5–2.5 M of $MnCl_2$ was included into the PSS and PAH solution	–	To increase the sulfate rejection	Hong et al. (2006a)

(e.g., PDADMAC) should be maintained at a low level, as PDADMAC films can be dissociated at high ionic strength (Dubas and Schlenoff 2001). As a result, a PDADMAC solution with low ionic strength was prepared for polyelectrolyte adsorption onto the PA membrane NF270 (commercial membrane produced by FilmTec, Dow Chemical Company, Midland, Michigan) (Shan et al. 2010).

In most membrane modification processes, posttreatment processes remain one of the important steps for ensuring better, or even outstanding, membrane performance. However, a posttreatment process is not a compulsory step in the polyelectrolyte adsorption process (Shan et al. 2010). This can be an important advantage to compensate the drawbacks involved during the adsorption of the polyelectrolytes onto the polymeric membranes. In most cases, the polyelectrolyte-modified membranes can be used directly for membrane characterizations and applications.

4.5   PROPOSED FUTURE STUDIES ON THE NF MEMBRANE MODIFICATION OR FABRICATION USING POLYELECTROLYTES

Although various polyelectrolytes have been successfully incorporated in the surface modification of different types of membranes, as discussed, the stability of polyelectrolytes still exerts a significant effect on long-term usage. Polyelectrolytes have been classified as highly water-soluble compounds, although water might not be the best solvent for some polyelectrolytes. However, the solubility of polyelectrolytes in water might contribute to the contamination of the feed solution in some cases. Further exploration of the effect of polyelectrolytes toward the feed solution is highly suggested for future studies.

The LbL deposition method used in most polyelectrolyte-modified membranes requires too much repetition. These processes require excessive energy and are time consuming for industrial purposes. Suggested methods to overcome this issue include changing the concentration of the polyelectrolyte solution, the ionic strength of the polyelectrolyte solution, and deposition temperatures. However, investigations are still in progress with the hope that a better solution will be introduced soon to reduce the time required and to make this process commercially feasible. One of the significant progresses at the current research trend is the use of single polyelectrolyte layer through dynamic deposition method.

Besides, material matching between various polycations and polyanions is important to produce an optimized membrane. However, it has been little discussed in past research as to how to determine the best polyelectrolyte pairs and how to evaluate their actual performance. Thus, this could be a study that should be carried out in the future for better membrane improvement using polyelectrolyte-based membrane modification.

Because polyelectrolytes have been widely used as flocculants or coagulants (especially anionic polyelectrolytes) for the treatment of industrial effluents or potable water sources, they are rarely considered to be prospective toxicants. However, several studies revealed that the polyelectrolytes are toxic substances to certain aquatic organisms, especially cationic polyelectrolytes (Goodrich et al. 1991; Liber et al. 2005; Harford et al. 2011). Besides, another group of researchers found that the cytotoxic effect is the strongest for PAH, followed by PDADMAC and PSS (Chanana et al. 2005). The health hazards posed by using polyelectrolytes in the membrane surface modifications are not well addressed and investigated. Thus, toxicity data should be established based on the type and concentration of the polyelectrolyte before polyelectrolyte-based membranes are commercialized for drinking water purification.

Reproducibility of the deposition of polyelectrolytes onto membranes should be improved to reduce the deviations in the data. In most cases, the performance reproducibility of the double-sided polyelectrolyte deposition method is a bit low and should be improved in comparison to the single-sided method to make it commercially feasible. This has been explained by the morphology of the membrane support materials, poorly controlled thickness of the support, unknown factors affecting the membrane support during the hydrolyzation process, and so forth.

Besides, production using lab-scale surface modification of the membrane is simple. However, in terms of larger scaling and to make the processes commercially feasible, special methods or processes using polyelectrolytes during the membrane modification need to be reconsidered, as this could result in lower production costs over the long term.

Simplicity of surface modification of polymeric membranes through polyelectrolyte adsorption has been restricted by complex membrane characterization methods. In the polyelectrolyte adsorption process, polymeric membrane properties in the wet state are important for several reasons. Polymeric membranes possess different or inconsistent pore sizes according to the degree of swelling when the membranes are immersed into water or under certain operating conditions. Moreover, the surface charges or zeta potentials and surface morphologies might vary for a membrane in the wet or dry state. Thus, wet state analyses, for example, AFM in tapping mode (Percec et al. 1998), optical fixed-angle reflectometry (Buron et al. 2009), and neutron reflectivity measurement (David et al. 2004), or dry state analyses, for example, ellipsometry (Mendelsohn et al. 2000) and X-ray reflectivity (Steitz et al. 2000) should be undertaken accordingly to improve membrane characterization for better separation performance.

4.6 SUMMARY

Following the discussions throughout the chapter, the authors have made some conclusions based on some of the findings presented here.

In the LbL adsorption method using charged polyelectrolytes (whether static, dynamic, or combination of both methods), the produced membranes would possess a charged surface that could help in enhancing the Donnan effect, thus providing better separation performance and selectivity. Additionally, the membrane pore size could be well controlled through variations in the number of polyelectrolyte deposition layers, polyelectrolyte solution pH, polyelectrolyte solution ionic strength, type of polyelectrolytes employed, and so forth. These are significant considerations during the production of multilayer polyelectrolyte-modified membranes for certain applications. For instance, when membrane with better size exclusion characteristic is desired (sieving effect), higher number of polyelectrolyte layers is preferred in order to increase the thickness of polyelectrolyte (Zone II) layer. However, controlling the growth of Zone II layer is important to prevent the deterioration of membrane permeability. Besides, the type of polyelectrolyte used for the construction of membrane effective separation layer is important in determining the membrane surface hydrophilicity, zeta potentials, and the type of solute to be rejected. Membrane thickness also can be well controlled by manipulating the polyelectrolyte solution pH or the concentration of supporting electrolyte due to the stretching behavior of the polyelectrolyte molecules.

During the polyelectrolyte adsorption process onto a membrane surface, the selection of the membrane molecular weight cut-off is quite important. A large membrane molecular weight cut-off value can help to improve membrane permeability after the polyelectrolyte layers are deposited. However, if the molecular weight cut-off value of the support membrane is too large, polyelectrolyte molecules might completely permeate through the membrane support and thus fewer molecules can be adsorbed onto the membrane surface. This contributes to a lower sealing effect and thus lower solute rejection capability. In contrast, when the molecular weight cut-off value of the membrane is too small, more polyelectrolyte molecules can be easily adsorbed onto the membrane surface. This encourages a better sealing effect, thus better rejection capability. However, lower membrane permeability can be observed too, as the considerable buildup of the coating layer of the polyelectrolyte can restrict the permeation of other molecules.

Supporting membrane materials are important to determine the effectiveness of the polyelectrolyte coating. The PAN support membrane has been widely reported to be used in the polyelectrolyte coating, as this kind of membrane possesses a high charge density when hydrolyzed in NaOH solution. The negatively charged surface of the PAN membrane could promote stronger ionic strength with polycations and thus better coating stability. However, other types of membrane materials have been studied in less detail. Thus, the suitability of the other membrane materials to be used in the polyelectrolyte deposition method still remains unclear.

Undeniably, polyelectrolytes offer a new solution to improve NF membrane performance. In comparison to other modification or fabrication methods, this has remained an attractive choice, as it requires simple steps, cheaper devices, as well as less energy and time consumption. However, the reactions involved and factors affecting the effectiveness of the selective barriers formed on the supporting membrane are still far from being fully understood. Thus, further research is required in order to obtain a good grasp of the mechanisms involved in polyelectrolyte-modified NF membranes.

REFERENCES

Abuhabib, A. A., A. W. Mohammad, N. Hilal, R. A. Rahman, and A. H. Shafie. 2012. Nanofiltration membrane modification by UV grafting for salt rejection and fouling resistance improvement for brackish water desalination. *Desalination* 295:16–25.

Seman, M. N. A., M. Khayet, Z. I. B. Ali, and N. Hilal. 2010. Reduction of nanofiltration membrane fouling by UV-initiated graft polymerization technique. *J. Memb. Sci.* 355:133–141.

Agenson, K. O. and T. Urase. 2007. Change in membrane performance due to organic fouling in nanofiltration (NF)/reverse osmosis (RO) applications. *Sep. Purif. Technol.* 55:147–156.

Ahmad, A. L., A. A. Abdulkarim, B. S. Ooi, and S. Ismail. 2013. Recent development in additives modifications of polyethersulfone membrane for flux enhancement. *Chem. Eng. J.* 223:246–267.

Ahmadiannamini, P., X. Li, W. Goyens, B. Meesschaert, and I. F. J. Vankelecom. 2010. Multilayered PEC nanofiltration membranes based on SPEEK/PDDA for anion separation. *J. Memb. Sci.* 360:250–258.

Ahmadiannamini, P., X. Li, W. Goyens, N. Joseph, B. Meesschaert, and I. F. J. Vankelecom. 2012. Multilayered polyelectrolyte complex based solvent resistant nanofiltration membranes prepared from weak polyacids. *J. Memb. Sci.* 394–395:98–106.

Akbari, A., S. Desclaux, J. C. Remigy, and P. Aptel. 2002. Treatment of textile dye effluents using a new photografted nanofiltration membrane. *Desalination* 149:101–107.

Akbari, A., S. Desclaux, J. C. Rouch, P. Aptel, and J. C. Remigy. 2006. New UV-photografted nanofiltration membranes for the treatment of colored textile dye effluents. *J. Memb. Sci.* 286:342–350.

Al-Amoudi, A. and R. W. Lovitt. 2007. Fouling strategies and the cleaning system of NF membranes and factors affecting cleaning efficiency. *J. Memb. Sci.* 303:4–28.

Alkhatim, H. S., M. I. Alcaina, E. Soriano, M. I. Iborra, J. Lora, and J. Arnal. 1998. Treatment of whey effluents from dairy industries by nanofiltration membranes. *Desalination* 119:177–183.

Antipina, M. N. and G. B. Sukhorukov. 2011. Remote control over guidance and release properties of composite polyelectrolyte based capsules. *Adv. Drug Deliv. Rev.* 63:716–729.

Arunkumar, A. and M. R. Etzel. 2013. Fractionation of α-lactalbumin from β-lactoglobulin using positively charged tangential flow ultrafiltration membranes. *Sep. Purif. Technol.* 105:121–128.

Ba, C., D. A. Ladner, and J. Economy. 2010. Using polyelectrolyte coatings to improve fouling resistance of a positively charged nanofiltration membrane. *J. Memb. Sci.* 347:250–259.

Ballet, G. T., L. Gzara, A. Hafiane, and M. Dhahbi. 2004. Transport coefficients and cadmium salt rejection in nanofiltration membrane. *Desalination* 167:369–376.

Baowei, S. U., T. Wang, Z. Wang, X. Gao, and C. Gao. 2012. Preparation and performance of dynamic layer-by-layer PDADMAC/PSS nanofiltration membrane. *J. Memb. Sci.* 423–424(0): 324–331.

Bilongo, T. G., J. C. Remigy, and M. J. Clifton. 2010. Modification of hollow fibers by UV surface grafting. *J. Memb. Sci.* 364:304–308.

Bohdziewicz, J., M. Bodzek, and E. Wąsik. 1999. The application of reverse osmosis and nanofiltration to the removal of nitrates from groundwater. *Desalination* 121:139–147.

Boussu, K., A. Belpaire, A. Volodin, C. Van Haesendonck, P. Van der Meeren, C. Vandecasteele, and B. Van der Bruggen. 2007. Influence of membrane and colloid characteristics on fouling of nanofiltration membranes. *J. Memb. Sci.* 289:220–230.

Bowen, W.R., M. G. Jones, J. S. Welfoot, and H. N. S. Yousef. 2000. Predicting salt rejections at nanofiltration membranes using artificial neural networks. *Desalination* 129: 147–162.

Buonomenna, M. G., L. C. Lopez, P. Favia, R. d'Agostino, A. Gordano, and E. Drioli. 2007. New PVDF membranes: The effect of plasma surface modification on retention in nanofiltration of aqueous solution containing organic compounds. *Water Res.* 41:4309–4316.

Buron, C. C., C. Filiâtre, F. Membrey, C. Bainier, L. Buisson, D. Charraut, and A. Foissy. 2009. Surface morphology and thickness of a multilayer film composed of strong and weak polyelectrolytes: Effect of the number of adsorbed layers, concentration and type of salts. *Thin Solid Films* 517:2611–2617.

Chanana, M., A. Gliozzi, A. Diaspro, I. Chodnevskaja, S. Huewel, V. Moskalenko, K. Ulrichs, H.-J. Galla, and S. Krol. 2005. Interaction of polyelectrolytes and their composites with living cells. *Nano Lett.* 5:2605–2612.

Chen, Y., Y. Liu, H. Fan, H. Li, B. Shi, H. Zhou, and B. Peng. 2007. The polyurethane membranes with temperature sensitivity for water vapor permeation. *J. Memb. Sci.* 287:192–197.

Cheng, C., A. Yaroshchuk, and M. L. Bruening. 2013. Fundamentals of selective ion transport through multilayer polyelectrolyte membranes. *Langmuir* 29:1885–1892.

Chun, M.-S., H. I. Cho, and I. K. Song. 2002. Electrokinetic behavior of membrane zeta potential during the filtration of colloidal suspensions. *Desalination* 148:363–368.

Das, C., P. Patel, S. De, and S. DasGupta. 2006. Treatment of tanning effluent using nanofiltration followed by reverse osmosis. *Sep. Purif. Technol.* 50:291–299.

David, C., R. Krastev, and M. Schönhoff. 2004. Oscillations in solvent fraction of polyelectrolyte multilayers driven by the charge of the terminating layer. *Langmuir* 20:11465–11472.

Decher, G., J. D. Hong, and J. Schmitt. 1992. Buildup of ultrathin multilayer films by a self-assembly process. III. Consecutively alternating adsorption of anionic and cationic polyelectrolytes on charged surfaces. *Thin Solid Films* 210–211, Part 2:831–835.

Deng, H.-Y., Y.-Y. Xu, B.-K. Zhu, X.-Z. Wei, F. Liu, and Z.-Y. Cui. 2008. Polyelectrolyte membranes prepared by dynamic self-assembly of poly (4-styrenesulfonic acid-co-maleic acid) sodium salt (PSSMA) for nanofiltration (I). *J. Memb. Sci.* 323:125–133.

Dubas, S. T. and J. B. Schlenoff. 2001. Polyelectrolyte multilayers containing a weak poly-acid: Construction and deconstruction. *Macromolecules* 34:3736–3740.

Dydo, P., M. Turek, J. Ciba, J. Trojanowska, and J. Kluczka. 2005. Boron removal from land-fill leachate by means of nanofiltration and reverse osmosis. *Desalination* 185:131–137.

Egueh, A.-N. D., B. Lakard, P. Fievet, S. Lakard, and C. Buron. 2010. Charge properties of membranes modified by multilayer polyelectrolyte adsorption. *J. Colloid Interface Sci.* 344:221–227.

Elzbieciak, M., M. Kolasinska, and P. Warszynski. 2008. Characteristics of polyelectrolyte multilayers: The effect of polyion charge on thickness and wetting properties. *Colloids Surf. A Physicochem. Eng. Asp.* 321:258–261.

Farhat, T. R., and P. T. Hammond. 2005. Designing a new generation of proton-exchange membranes using layer-by-layer deposition of polyelectrolytes. *Adv. Funct. Mater.* 15:945–954.

Garcia, F., D. Ciceron, A. Saboni, and S. Alexandrova. 2006. Nitrate ions elimination from drinking water by nanofiltration: Membrane choice. *Sep. Purif. Technol.* 52:196–200.

Geffen, N., R. Semiat, M. S. Eisen, Y. Balazs, I. Katz, and C. G. Dosoretz. 2006. Boron removal from water by complexation to polyol compounds. *J. Memb. Sci.* 286:45–51.

Gohil, J. M., and P. Ray. 2009. Polyvinyl alcohol as the barrier layer in thin film composite nanofiltration membranes: Preparation, characterization, and performance evaluation. *J. Colloid Interface Sci.* 338:121–127.

Goodrich, M. S., L. H. Dulak, M. A. Friedman, and J. J. Lech. 1991. Acute and long-term toxicity of water-soluble cationic polymers to rainbow trout (*Oncorhynchus mykiss*) and the modification of toxicity by humic acid. *Environ. Toxicol. Chem.* 10:509–515.

Guenet, J.-M. 2005. Molecular structure of polyelectrolyte/surfactant complexes vs. polymer tacticity. *J. Mol. Liq.* 120:3–6.

Guerra, K. and J. Pellegrino. 2013. Development of a techno-economic model to compare ceramic and polymeric membranes. *Sep. Sci. Technol.* 48:51–65.

Harford, A. J., A. C. Hogan, D. R. Jones, and R. A. van Dam. 2011. Ecotoxicological assess-ment of a polyelectrolyte flocculant. *Water Res.* 45:6393–6402.

Harris, J. J., J. L. Stair, and M. L. Bruening. 2000. Layered polyelectrolyte films as selective, ultrathin barriers for anion transport. *Chem. Mater.* 12:1941–1946.

Her, N., G. Amy, A. Plottu-Pecheux, and Y. Yoon. 2007. Identification of nanofiltration mem-brane foulants. *Water Res.* 41:3936–3947.

Hilal, N., H. Al-Zoubi, N. A. Darwish, and A. W. Mohammad. 2007. Performance of nano-filtration membranes in the treatment of synthetic and real seawater. *Sep. Sci. Technol.* 42:493–515.

Hilal, N., O. O. Ogunbiyi, N. J. Miles, and R. Nigmatullin. 2005. Methods employed for con-trol of fouling in MF and UF membranes: A comprehensive review. *Sep. Sci. Technol.* 40:1957–2005.

Hong, S. U. and M. L. Bruening. 2006. Separation of amino acid mixtures using multilayer polyelectrolyte nanofiltration membranes. *J. Memb. Sci.* 280:1–5.

Hong, S. U., R. Malaisamy, and M. L. Bruening. 2006a. Optimization of flux and selectivity in Cl–/SO42– separations with multilayer polyelectrolyte membranes. *J. Memb. Sci.* 283:366–372.

Hong, S. U. R. Malaisamy, and M. L. Bruening. 2007. Separation of fluoride from other mon-ovalent anions using multilayer polyelectrolyte nanofiltration membranes. *Langmuir* 23:1716–1722.

Hong, S. U., M. D. Miller, and M. L. Bruening. 2006b. Removal of dyes, sugars, and amino acids from NaCl solutions using multilayer polyelectrolyte nanofiltration membranes. *Ind. Eng. Chem. Res.* 45:6284–6288.

Hong, S. U., L. Ouyang, and M. L. Bruening. 2009. Recovery of phosphate using multilayer polyelectrolyte nanofiltration membranes. *J. Memb. Sci.* 327:2–5.

Ivnitsky, H., D. Minz, L. Kautsky, A. Preis, A. Ostfeld, R. Semiat, and C. G. Dosoretz. 2010. Biofouling formation and modeling in nanofiltration membranes applied to wastewater treatment. *J. Memb. Sci.* 360:165–173.

Jahanshahi, M., A. Rahimpour, and M. Peyravi. 2010. Developing thin film composite poly(piperazine-amide) and poly(vinyl-alcohol) nanofiltration membranes. *Desalination* 257:129–136.

Jarusutthirak, C., S. Mattaraj, and R. Jiraratananon. 2007. Factors affecting nanofiltration performances in natural organic matter rejection and flux decline. *Sep. Purif. Technol.* 58:68–75.

Ji, S., G. Zhang, Z. Liu, Y. Peng, and Z. Wang. 2008. Evaluations of polyelectrolyte multilayer membranes assembled by a dynamic layer-by-layer technique. *Desalination* 234:300–306.

Ji, Y., Q. An, Q. Zhao, H. Chen, J. Qian, and C. Gao. 2010. Fabrication and performance of a new type of charged nanofiltration membrane based on polyelectrolyte complex. *J. Memb. Sci.* 357:80–89.

Jie, X., Y. Cao, J.-J. Qin, J. Liu, and Q. Yuan. 2005. Influence of drying method on morphology and properties of asymmetric cellulose hollow fiber membrane. *J. Memb. Sci.* 246:157–165.

Khayet, M., M. N. A. Seman, and N. Hilal. 2010. Response surface modeling and optimization of composite nanofiltration modified membranes. *J. Memb. Sci.* 349:113–122.

Khirani, S., R. B. Aim, and M.-H. Manero. 2006. Improving the measurement of the Modified Fouling Index using nanofiltration membranes (NF–MFI). *Desalination* 191:1–7.

Kim, E.-S., Q. Yu, and B. Deng. 2011. Plasma surface modification of nanofiltration (NF) thin-film composite (TFC) membranes to improve anti organic fouling. *Appl. Surf. Sci.* 257:9863–9871.

Kiso, Y., T. Kon, T. Kitao, and K. Nishimura. 2001. Rejection properties of alkyl phthalates with nanofiltration membranes. *J. Memb. Sci.* 182:205–214.

Klitzing, R. V. and B. Tieke. 2003. Polyelectrolyte membranes. In: *Filler-Reinforced Elastomers/ Scanning Force Microscopy*. Berlin, Germany: Springer.

Kolarik, L., D. N. Furlong, H. Joy, C. Struijk, and R. Rowe. 1999. Building assemblies from high molecular weight polyelectrolytes. *Langmuir* 15:8265–8275.

Košutić, K., L. Kaštelan-Kunst, and B. Kunst. 2000. Porosity of some commercial reverse osmosis and nanofiltration polyamide thin-film composite membranes. *J. Memb. Sci.* 168:101–108.

Košutić, K. and B. Kunst. 2002. Removal of organics from aqueous solutions by commercial RO and NF membranes of characterized porosities. *Desalination* 142:47–56.

Koyuncu, I. 2002. Reactive dye removal in dye/salt mixtures by nanofiltration membranes containing vinylsulphone dyes: Effects of feed concentration and cross flow velocity. *Desalination* 143:243–253.

Krasemann, L. and B. Tieke. 1999. Selective ion transport across self-assembled alternating multilayers of cationic and anionic polyelectrolytes. *Langmuir* 16:287–290.

Krieg, H. M., S. J. Modise, K. Keizer, and H. W. J. P. Neomagus. 2005. Salt rejection in nanofiltration for single and binary salt mixtures in view of sulphate removal. *Desalination* 171:205–215.

Kujawa, P., P. Moraille, J. Sanchez, A. Badia, and F. M. Winnik. 2005. Effect of molecular weight on the exponential growth and morphology of hyaluronan/chitosan multilayers: A surface plasmon resonance spectroscopy and atomic force microscopy investigation. *J. Am. Chem. Soc.* 127:9224–9234.

Kwon, H. J., and J. P. Gong. 2006. Negatively charged polyelectrolyte gels as bio-tissue model system and for biomedical application. *Curr. Opin. Colloid Interface Sci.* 11:345–350.

Ladam, G., P. Schaad, J. C. Voegel, P. Schaaf, G. Decher, and F. Cuisinier. 1999. In situ determination of the structural properties of initially deposited polyelectrolyte multilayers. *Langmuir* 16:1249–1255.

Lajimi, R. H., A. B. Abdallah, E. Ferjani, M. S. Roudesli, and A. Deratani. 2004. Change of the performance properties of nanofiltration cellulose acetate membranes by surface adsorption of polyelectrolyte multilayers. *Desalination* 163:193–202.

Lee, S., J. Cho, and M. Elimelech. 2005. Combined influence of natural organic matter (NOM) and colloidal particles on nanofiltration membrane fouling. *J. Memb. Sci.* 262:27–41.

L'Haridon, J., M. Fernandez, V. Ferrier, and J. Bellan. 1993. Evaluation of the genotoxicity of *n*-nitrosoatrazine, *n*-nitrosodiethanolamine and their precursors *in vivo* using the newt micronucleus test. *Water Res.* 27:855–862.

Lenk, W. and J. Meier-Haack. 2002. Polyelectrolyte multilayer membranes for pervaporation separation of aqueous-organic mixtures. *Desalination* 148:11–16.

Levenstein, R., D. Hasson, and R. Semiat. 1996. Utilization of the Donnan effect for improving electrolyte separation with nanofiltration membranes. *J. Memb. Sci.* 116:77–92.

Li, Q. and M. Elimelech. 2004. Organic fouling and chemical cleaning of nanofiltration membranes: Measurements and mechanisms. *Environmental Science & Technology* 38:4683–4693.

Li, Q. and M. Elimelech. 2006. Synergistic effects in combined fouling of a loose nanofiltration membrane by colloidal materials and natural organic matter. *J. Memb. Sci.* 278:72–82.

Li, X., S. De Feyter, D. Chen, S. Aldea, P. Vandezande, F. Du Prez, and I. F. J. Vankelecom. 2008a. Solvent-resistant nanofiltration membranes based on multilayered polyelectrolyte complexes. *Chem. Mater.* 20:3876–3883.

Li, X., W. Goyens, P. Ahmadiannamini, W. Vanderlinden, S. De Feyter, and I. Vankelecom. 2010. Morphology and performance of solvent-resistant nanofiltration membranes based on multilayered polyelectrolytes: Study of preparation conditions. *J. Memb. Sci.* 358:150–157.

Li, X., Y. Wang, X. Lu, and C. Xiao. 2008b. Morphology changes of polyvinylidene fluoride membrane under different phase separation mechanisms. *J. Memb. Sci.* 320:477–482.

Liber, K., L. Weber, and C. Lévesque. 2005. Sublethal toxicity of two wastewater treatment polymers to lake trout fry (*Salvelinus namaycush*). *Chemosphere* 61:1123–1133.

Liikanen, R., I. Miettinen, and R. Laukkanen. 2003. Selection of NF membrane to improve quality of chemically treated surface water. *Water Res.* 37:864–872.

Lingström, R. and L. Wågberg. 2008. Polyelectrolyte multilayers on wood fibers: Influence of molecular weight on layer properties and mechanical properties of papers from treated fibers. *J. Colloid Interface Sci.* 328:233–242.

Listiarini, K., W. Chun, D. D. Sun, and J. O. Leckie. 2009a. Fouling mechanism and resistance analyses of systems containing sodium alginate, calcium, alum and their combination in dead-end fouling of nanofiltration membranes. *J. Memb. Sci.* 344:244–251.

Listiarini, K., D. D. Sun, and J. O. Leckie. 2009b. Organic fouling of nanofiltration membranes: Evaluating the effects of humic acid, calcium, alum coagulant and their combinations on the specific cake resistance. *J. Memb. Sci.* 332:56–62.

Liu, X. and M. L. Bruening. 2003. Size-selective transport of uncharged solutes through multilayer polyelectrolyte membranes. *Chem. Mater.* 16:351–357.

Liu, X., S. Qi, Y. Li, L. Yang, B. Cao, and C. Y. Tang. 2013. Synthesis and characterization of novel antibacterial silver nanocomposite nanofiltration and forward osmosis membranes based on layer-by-layer assembly. *Water Res.* 47:3081–3092.

Liu, Y., E. Rosenfield, M. Hu, and B. Mi. 2013. Direct observation of bacterial deposition on and detachment from nanocomposite membranes embedded with silver nanoparticles. *Water Res.* 47:2949–2958.

Lv, W., X. Zheng, M. Yang, Y. Zhang, Y. Liu, and J. Liu. 2006. Virus removal performance and mechanism of a submerged membrane bioreactor. *Process Biochem.* 41:299–304.

Lvov, Y., K. Ariga, M. Onda, I. Ichinose, and T. Kunitake. 1999. A careful examination of the adsorption step in the alternate layer-by-layer assembly of linear polyanion and polycation. *Colloids Surf. A Physicochem. Eng. Asp.* 146:337–346.

Malaisamy, R. and M. L. Bruening. 2005. High-flux nanofiltration membranes prepared by adsorption of multilayer polyelectrolyte membranes on polymeric aupports. *Langmuir* 21:10587–10592.

Malaisamy, R., A. Talla-Nwafo, and K. L. Jones. 2011. Polyelectrolyte modification of nano-filtration membrane for selective removal of monovalent anions. *Sep. Purif. Technol.* 77:367–374.

Mansourpanah, Y., S. S. Madaeni, A. Rahimpour, A. Farhadian, and A. H. Taheri. 2009. Formation of appropriate sites on nanofiltration membrane surface for binding TiO_2 photo-catalyst: Performance, characterization and fouling-resistant capability. *J. Memb. Sci.* 330:297–306.

Meier-Haack, J., W. Lenk, D. Lehmann, and K. Lunkwitz. 2001. Pervaporation separation of water/alcohol mixtures using composite membranes based on polyelectrolyte multilayer assemblies. *J. Memb. Sci.* 184:233–243.

Meier-Haack, J., T. Rieser, W. Lenk, D. Lehmann, S. Berwald, and S. Schwarz. 2000. Effect of polyelectrolyte complex layers on the separation properties and the fouling behavior of surface and bulk modified membranes. *Chem. Eng. Technol.* 23:114–118.

Mendelsohn, J. D., C. J. Barrett, V. V. Chan, A. J. Pal, A. M. Mayes, and M. F. Rubner. 2000. Fabrication of microporous thin films from polyelectrolyte multilayers. *Langmuir* 16:5017–5023.

Milkova, V. and T. Radeva. 2013. Effect of ionic strength and molecular weight on electrical properties and thickness of polyelectrolyte bi-layers. *Colloids Surf. A Physicochem. Eng. Asp.* 424:52–58.

Miller, M. D. and M. L. Bruening. 2004. Controlling the nanofiltration properties of multilayer polyelectrolyte membranes through variation of film composition. *Langmuir* 20:11545–11551.

Mohammad, A. W., C. Y. Ng, Y. P. Lim, and G. H. Ng. 2012. Ultrafiltration in food processing industry: Review on application, membrane fouling, and fouling control. *Food Bioprocess. Tech.* 5:1143–1156.

Morão, A., I. C. Escobar, M. T. P. De Amorim, A. Lopes, and I. C. Gonçalves. 2005. Postsynthesis modification of a cellulose acetate ultrafiltration membrane for applications in water and wastewater treatment. *Environ. Prog.* 24:367–382.

Müller, M., T. Rieser, P. L. Dubin, and K. Lunkwitz. 2001. Selective interaction between proteins and the outermost surface of polyelectrolyte multilayers: Influence of the polyanion type, pH and salt. *Macromol. Rapid Commun.* 22:390–395.

Murthy, Z. V. P. and L. B. Chaudhari. 2009. Rejection behavior of nickel ions from synthetic wastewater containing Na_2SO_4, $NiSO_4$, $MgCl_2$ and $CaCl_2$ salts by nanofiltration and characterization of the membrane. *Desalination* 247:610–622.

Ng, L. Y. and A. W. Mohammad. 2010. Optimization of polysulfone/poly(vinyl alcohol) membranes incorporated with silicon dioxide nanoparticles. In *World Engineering Congress.* Kuching, Sarawak, Malaysia.

Ng, L. Y., A. W. Mohammad, C. P. Leo, and N. Hilal. 2013. Polymeric membranes incorporated with metal/metal oxide nanoparticles: A comprehensive review. *Desalination* 308:15–33.

Oumar, A., C. D. Trébouet, P. Jaouen, and F. Quéméneur. 2001. Nanofiltration of seawater: Fractionation of mono- and multi-valent cations. *Desalination* 140:67–77.

Ouyang, L., D. M. Dotzauer, S. R. Hogg, J. Macanás, J.-F. Lahitte, and M. L. Bruening. 2010. Catalytic hollow fiber membranes prepared using layer-by-layer adsorption of polyelectrolytes and metal nanoparticles. *Catal. Today* 156:100–106.

Ouyang, L., R. Malaisamy, and M. L. Bruening. 2008. Multilayer polyelectrolyte films as nanofiltration membranes for separating monovalent and divalent cations. *J. Memb. Sci.* 310:76–84.

Peiris, R. H., C. Hallé, H. Budman, C. Moresoli, S. Peldszus, P. M. Huck, and R. L. Legge. 2010. Identifying fouling events in a membrane-based drinking water treatment process using principal component analysis of fluorescence excitation-emission matrices. *Water Res.* 44:185–194.

Percec, V., C. H. Ahn, G. Ungar, D. J. P. Yeardley, M. Moller, and S. S. Sheiko. 1998. Controlling polymer shape through the self-assembly of dendritic side-groups. *Nature* 391:161–164.

Qiu, C., F. Xu, Q. T. Nguyen, and Z. Ping. 2005. Nanofiltration membrane prepared from cardo polyetherketone ultrafiltration membrane by UV-induced grafting method. *J. Memb. Sci.* 255:107–115.

Qiu, C., Q. T. N. L. Zhang, and Z. Ping. 2006. Nanofiltration membrane preparation by photomodification of cardo polyetherketone ultrafiltration membrane. *Sep. Purif. Technol.* 51:325–331.

Quinn, R., D. V. Laciak, and G. P. Pez. 1997. Polyelectrolyte-salt blend membranes for acid gas separations. *J. Memb. Sci.* 131:61–69.

Rahimpour, A., M. Jahanshahi, N. Mortazavian, S. S. Madaeni, and Y. Mansourpanah. 2010. Preparation and characterization of asymmetric polyethersulfone and thin-film composite polyamide nanofiltration membranes for water softening. *Appl. Surf. Sci.* 256: 1657–1663.

Rajesh, S., S. Senthilkumar, A. Jayalakshmi, M. T. Nirmala, A. F. Ismail, and D. Mohan. 2013. Preparation and performance evaluation of poly (amide–imide) and TiO_2 nanoparticles impregnated polysulfone nanofiltration membranes in the removal of humic substances. *Colloids Surf. A Physicochem. Eng. Asp.* 418:92–104.

Religa, P., A. Kowalik-Klimczak, and P. Gierycz. 2013. Study on the behavior of nanofiltration membranes using for chromium(III) recovery from salt mixture solution. *Desalination* 315:115–123.

Ruan, D., L. Zhang, Y. Mao, M. Zeng, and X. Li. 2004. Microporous membranes prepared from cellulose in NaOH/thiourea aqueous solution. *J. Memb. Sci.* 241:265–274.

Saslow, W. M. 2002. Chapter 2—Coulomb's law for static electricity, principle of superposition. In A. McDonald (ed.), *Electricity, Magnetism, and Light*. San Diego, CA: Academic Press.

Schaep, J., B. Van der Bruggen, S. Uytterhoeven, R. Croux, C. Vandecasteele, D. Wilms, E. Van Houtte, and F. Vanlerberghe. 1998. Removal of hardness from groundwater by nanofiltration. *Desalination* 119:295–301.

Schmitz, K. S. (ed.). 1993. An overview of polyelectrolytes. In *Macro-Ion Characterization*. Washington, DC: American Chemical Society.

Semsarzadeh, M. A. and B. Ghalei. 2013. Preparation, characterization and gas permeation properties of polyurethane–silica/polyvinyl alcohol mixed matrix membranes. *J. Memb. Sci.* 432:115–125.

Shan, W., P. Bacchin, P. Aimar, M. L. Bruening, and V. V. Tarabara. 2010. Polyelectrolyte multilayer films as backflushable nanofiltration membranes with tunable hydrophilicity and surface charge. *J. Memb. Sci.* 349:268–278.

Shiratori, S. S. and M. F. Rubner. 2000. pH-dependent thickness behavior of sequentially adsorbed layers of weak polyelectrolytes. *Macromolecules* 33:4213–4219.

Shulevich, Y. V., T. H. Nguyen, D. S. Tutaev, A. V. Navrotskii, and I. A. Novakov. 2013. Purification of fat-containing wastewater using polyelectrolyte–surfactant complexes. *Sep. Purif. Technol.* 113:18–23.

Simon, A., W. E. Price, and L. D. Nghiem. 2013. Impact of chemical cleaning on the nanofiltration of pharmaceutically active compounds (PhACs): The role of cleaning temperature. *J. Taiwan Inst. Chem. Eng.* 44(5): 713–723.

Singh, G. and L. Song. 2005. Quantifying the effect of ionic strength on colloidal fouling potential in membrane filtration. *J. Colloid Interface Sci.* 284:630–638.

Stanton, B. W., J. J. Harris, M. D. Miller, and M. L. Bruening. 2003. Ultrathin, multilayered polyelectrolyte films as nanofiltration membranes. *Langmuir* 19:7038–7042.

Steitz, R., V. Leiner, R. Siebrecht, and R. V. Klitzing. 2000. Influence of the ionic strength on the structure of polyelectrolyte films at the solid/liquid interface. *Colloids Surf. A Physicochem. Eng. Asp.* 163:63–70.

Sui, Z., D. Salloum, and J. B. Schlenoff. 2003. Effect of molecular weight on the construction of polyelectrolyte multilayers: Stripping versus sticking. *Langmuir* 19:2491–2495.

Tadmor, R., E. Hernández-Zapata, N. Chen, P. Pincus, and J. N. Israelachvili. 2002. Debye length and double-layer forces in polyelectrolyte solutions. *Macromolecules* 35:2380–2388.

Tarabara, V. V., I. Koyuncu, and M. R. Wiesner. 2004. Effect of hydrodynamics and solution ionic strength on permeate flux in cross-flow filtration: Direct experimental observation of filter cake cross-sections. *J. Memb. Sci.* 241:65–78.

Teixeira, M. R., M. J. Rosa, and M. Nyström. 2005. The role of membrane charge on nanofiltration performance. *J. Memb. Sci.* 265:160–166.

Umpuch, C., S. Galier, S. Kanchanatawee, and H. Roux-de Balmann. 2010. Nanofiltration as a purification step in production process of organic acids: Selectivity improvement by addition of an inorganic salt. *Process Biochem.* 45:1763–1768.

Urairi, M., T. Tsuru, S.-I. Nakao, and S. Kimura. 1992. Bipolar reverse osmosis membrane for separating mono-and divalent ions. *J. Memb. Sci.* 70:153–162.

Van der Bruggen, B., K. Everaert, D. Wilms, and C. Vandecasteele. 2001. Application of nanofiltration for removal of pesticides, nitrate and hardness from ground water: Rejection properties and economic evaluation. *J. Memb. Sci.* 193:239–248.

Van der Bruggen, B., J. Schaep, W. Maes, D. Wilms, and C. Vandecasteele. 1998. Nanofiltration as a treatment method for the removal of pesticides from ground waters. *Desalination* 117:139–147.

Van der Bruggen, B., J. Schaep, D. Wilms, and C. Vandecasteele. 1999. Influence of molecular size, polarity and charge on the retention of organic molecules by nanofiltration. *J. Memb. Sci.* 156:29–41.

Vatanpour, V., S. S. Madaeni, R. Moradian, S. Zinadini, and B. Astinchap. 2011. Fabrication and characterization of novel antifouling nanofiltration membrane prepared from oxidized multiwalled carbon nanotube/polyethersulfone nanocomposite. *J. Memb. Sci.* 375:284–294.

Vergaro, V., F. Scarlino, C. Bellomo, R. Rinaldi, D. Vergara, M. Maffia, F. Baldassarre et al. 2011. Drug-loaded polyelectrolyte microcapsules for sustained targeting of cancer cells. *Adv. Drug Deliv. Rev.* 63:847–864.

Wang, D., W. Zou, L. Li, Q. Wei, S. Sun, and C. Zhao. 2011a. Preparation and characterization of functional carboxylic polyethersulfone membrane. *J. Memb. Sci.* 374:93–101.

Wang, L., D. Li, L. Cheng, L. Zhang, and H. Chen. 2011b. Preparation of thin film composite nanofiltration membrane by interfacial polymerization with 3,5-diaminobenzoylpiperazine and trimesoyl chloride. *Chin. J. Chem. Eng.* 19:262–266.

Wang, X.-L., J.-F. Wei, Z. Dai, K.-Y. Zhao, and H. Zhang. 2012. Preparation and characterization of negatively charged hollow fiber nanofiltration membrane by plasma-induced graft polymerization. *Desalination* 286:138–144.

Wang, Z., H. Zhu, D. Li, and X. Yang. 2008. Preparation and application of single polyelectrolyte microcapsules possessing tunable autofluorescent properties. *Colloids Surf. A Physicochem. Eng. Asp.* 329:58–66.

Warczok, J., M. Ferrando, F. López, and C. Güell. 2004. Concentration of apple and pear juices by nanofiltration at low pressures. *J. Food Eng.* 63:63–70.

Wei, X., R. Wang, Z. Li, and A. G. Fane. 2006. Development of a novel electrophoresis-UV grafting technique to modify PES UF membranes used for NOM removal. *J. Memb. Sci.* 273:47–57.

Wu, M. and D. D. Sun. 2005. Characterization and reduction of membrane fouling during nanofiltration of semiconductor indium phosphide (InP) wastewater. *J. Memb. Sci.* 259:135–144.

Xu, P., C. Bellona, and J. E. Drewes. 2010. Fouling of nanofiltration and reverse osmosis membranes during municipal wastewater reclamation: Membrane autopsy results from pilot-scale investigations. *J. Memb. Sci.* 353:111–121.

Xu, X., and H. G. Spencer. 1997. Dye-salt separations by nanofiltration using weak acid polyelectrolyte membranes. *Desalination* 114:129–137.

Yılmaztürk, S., H. Deligöz, M. Yılmazoğlu, H. Damyan, F. Öksüzömer, S. N. Koç, A. Durmuş, and M. A. Gürkaynak. 2009. A novel approach for highly proton conductive electrolyte membranes with improved methanol barrier properties: Layer-by-layer assembly of salt containing polyelectrolytes. *J. Memb. Sci.* 343:137–146.

Zhang, M. and L. Song. 2000. Mechanisms and parameters affecting flux decline in cross-flow microfiltration and ultrafiltration of colloids. *Environ. Sci. Technol.* 34:3767–3773.

Zhang, M., K. Zhang, B. De Gusseme, and W. Verstraete. 2012. Biogenic silver nanoparticles (bio-Ag0) decrease biofouling of bio-Ag0/PES nanocomposite membranes. *Water Res.* 46:2077–2087.

Zhao, W., J. Huang, B. Fang, S. Nie, N. Yi, B. Su, H. Li, and C. Zhao. 2011. Modification of polyethersulfone membrane by blending semi-interpenetrating network polymeric nanoparticles. *J. Memb. Sci.* 369:258–266.

Zhang, W., X. Zhang, Y. Li, J. Wang, and C. Chen. 2011. Membrane flux dynamics in the submerged ultrafiltration hybrid treatment process during particle and natural organic matter removal. *J. Environ. Sci.* 23:1970–1976.

Zodrow, K., L. Brunet, S. Mahendra, D. Li, A. Zhang, Q. Li, and P. J. J. Alvarez. 2009. Polysulfone ultrafiltration membranes impregnated with silver nanoparticles show improved biofouling resistance and virus removal. *Water Res.* 43:715–723.

5 Polysaccharides
A Membrane Material

*Seema Shrikant Shenvi, Arun Mohan Isloor,
and Ahmad Fauzi Ismail*

CONTENTS

5.1   Introduction 161
5.2   Alginic Acid.. 162
    5.2.1   Applications of Alginic Acid in Membrane Technology................. 162
        5.2.1.1   Pervaporation... 162
        5.2.1.2   Removal of Metal Ions: By Membrane-Based
                  Separation or by Adsorption... 164
5.3   Chitosan 165
    5.3.1   Application of Chitosan in Membrane Technology......................... 168
        5.3.1.1   Pervaporation... 168
        5.3.1.2   Ultrafiltration and NF Membranes 170
        5.3.1.3   Removal of Metal Ion and Dyes 174
5.4   Cellulose 176
    5.4.1   Applications of Cellulose in Membrane Technology 177
        5.4.1.1   UF and NF Membrane 177
        5.4.1.2   Pervaporation... 179
        5.4.1.3   Removal of Metal Ions and Dye from Aqueous
                  Solutions 180
5.5   Cyclodextrins 181
5.6   Lignins 182
5.7   Conclusions 183
References... 184

5.1   INTRODUCTION

Natural polysaccharides form a major class of the most extensively used biopolymers for different applications, one of which includes membrane-based separation. The major advantages offered by these materials for membrane-based separation include the following:

- Hydrophilicity of the polymer owing to the presence of hydroxyl groups present in the glucose units that comprise them
- Presence of large numbers of other functional groups such as carboxyl, amine, and hydroxyl groups

- Flexible structure of the polymer chains and their chemical reactivity
- Their abundant occurrence in nature and relatively low cost
- Biocompatibility and biodegradability

Over the decades, these environmental-benign materials have been used for the removal of heavy metal ions from industrial wastes, toxic dyes from textile industries, pervaporation, and protein separations. This chapter gives a detailed account of the commonly used biopolymers in membrane technology, including chitosan (CH), alginic acid, and cellulose.

5.2 ALGINIC ACID

For many years after the discovery of alginic acid in 1884, the only thing known was that this polysaccharide was an acid whose soluble salts such as calcium and sodium formed viscous solutions; from these solutions, the acid form could be precipitated with the help of mineral acids (Tallis 1950). Today, the potency of this polysaccharide has been explored to a great extent, making it one of the most widely used water-soluble polysaccharides after CH, owing to its good film-forming property. Alginate is a water-soluble, linear, unbranched polysaccharide that occurs mainly in the cell walls of brown algae. The structure of alginic acid consists of a linear chain made up of (1–4) linked β-D-mannuronic acid (M) and α-L-guluronic acid (G) arranged in a blockwise fashion (Figure 5.1). These blocks can be either homolymeric or heteropolymeric arranged sequentially (Fischer and Dorfel 1955; Haung et al. 1966; Bhat and Aminabhavi 2007).

5.2.1 Applications of Alginic Acid in Membrane Technology

5.2.1.1 Pervaporation

Alginic acid, being a highly hydrophilic polysaccharide, is very useful to develop membranes primarily for pervaporation applications. Pervaporation is a membrane-based separation technique for selective removal of solvents from aqueous media. It has also been used for organic–organic mixture separations (Adoor et al. 2008). However, the major concern in its use is its high swelling nature in the presence of hydrophilic solvents such as water (Li et al. 2011). Under these circumstances, the sodium alginate form of alginic acid serves as a potential candidate for pervaporation (PV) applications, having reasonably good mechanical strength and hydrophilicity (Uragami and Saito 1989; Shi et al. 1998). Cross-linking of sodium alginate becomes an essential part of membrane preparation because it is water soluble;

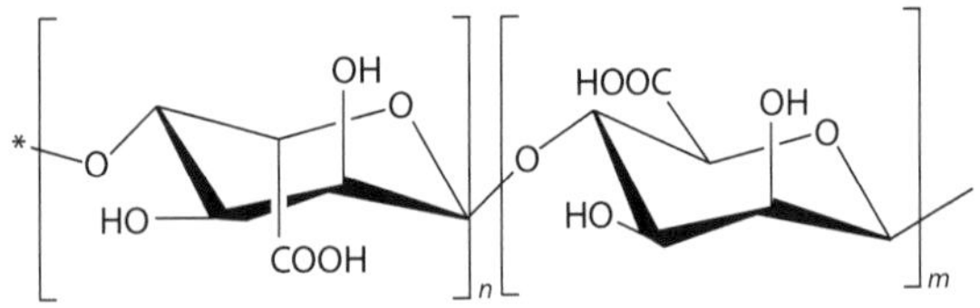

FIGURE 5.1 Structure of alginic acid. The symbol "*" indicates the repeating units of the polymer chain.

moreover, the membranes prepared of pristine sodium alginate are not mechanically strong enough. A number of cross-linking agents have been used for this purpose, including glutaraldehyde, hexane diamine, and urea–formaldehyde, in the presence of sulfuric acid, as catalyst, Ca^{2+}, and polyvinyl alcohol (PVA). These cross-linkers strengthen the resultant membranes with a simultaneous improvement in separation performance. With the aim of enhancing the flux and selectivity, sodium alginate membranes are generally modified by using cross-linkers, by blending it with some other polymers, or by incorporating some particulate matrix (Bhat and Aminabhavi 2007).

Performance of sodium alginate membrane and its blend with relatively hydrophilic cellulose derivatives for the separation of water–dioxane mixtures and water–tetrahydrofuran (THF) mixtures have been studied (Naidu et al. 2005) (Figure 5.2). It has been observed that there was a remarkable improvement in pervaporation

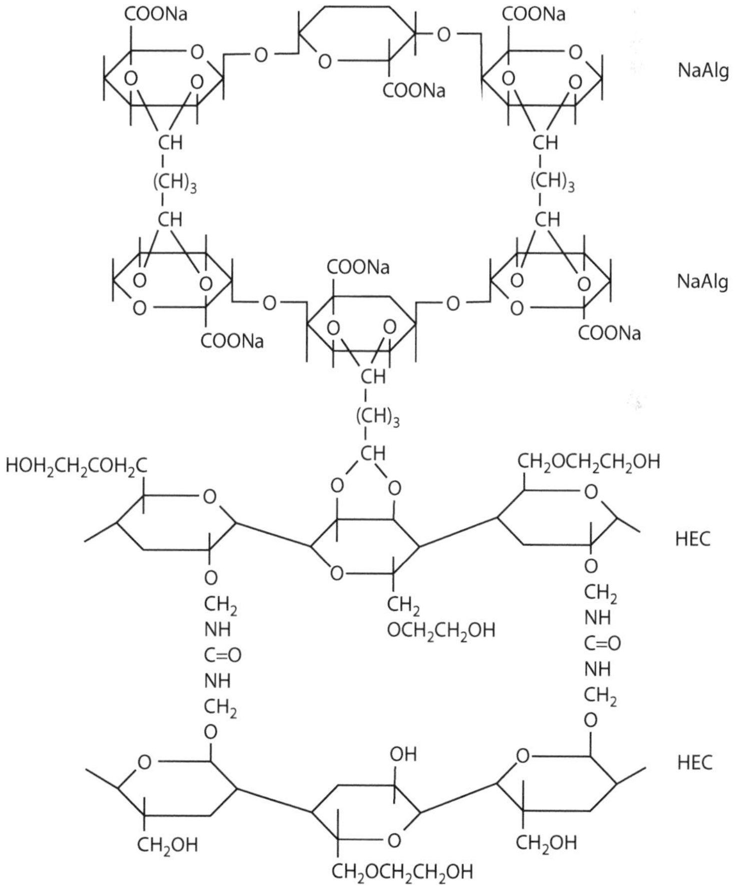

FIGURE 5.2 Structure of sodium alginate and HEC cross-linked membrane used for pervaporation of water-dioxane and water-THF mixture. (Data from Naidu, V. B. et al., *J. Memb. Sci.*, 260, 131–141, 2005.)

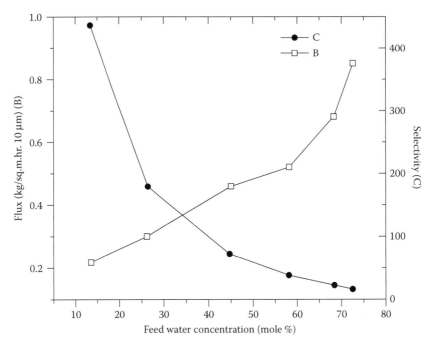

FIGURE 5.3 Performance of sodium alginate–chitosan blend membrane on dehydration of ethanol, effect of feed composition on the polyion complex membrane. (Data from Kanti, P. et al., *Sep. Purif. Technol.*, 40, 259–266, 2004.)

performance for water–THF mixtures for the blend membranes compared to the neat sodium alginate membrane.

Use of additives such as carbon nanotubes (CNTs) with the objective of enhancing the flux without compromising the selectivity has also been suggested. Introduction of porous fillers such as CNTs in the sodium alginate matrix are known to improve the separation performance of the membranes due to the combined effects of molecular sieving action, selective adsorption, and difference in diffusion rates (Peng et al. 2005; Sajjan et al. 2013). Other hydrophilic polymers such as PVA, CH, cellulose, and guar gum have also been used to improve the performance of alginate membranes (Yeom and Lee 1998; Moon et al. 1999; Wang 2000; Kanti et al. 2004) (Figure 5.3). Zeolites have also proved to be potential fillers in alginate membranes, which are known to augment the pervaporation performance (Patil et al. 2009). Thus, alginate membranes have proved to be commercially important for pervaporation separation.

5.2.1.2 Removal of Metal Ions: By Membrane-Based Separation or by Adsorption

Presence of abundant carboxyl groups in the backbone of sodium alginate makes its interaction possible with low molecular weight species, especially metal ions through complexation. This helps in the efficient removal of heavy and toxic metal ions from aqueous media, thus ensuring clean and safe drinking water. For adsorption applications, beads are commonly preferred over flat-sheet membranes, as they offer relatively more surface area for adsorption, thereby guaranteeing the efficient removal of metal ions; there is sufficient literature explaining the use of sodium alginate as porous adsorbent for the removal of these ions. As previously stated, sodium alginate, being water soluble, does not show very good adsorption performance toward metal ions. Hence there is lot of scope to enhance its adsorption capacity by suitable modification.

The multilayer silica–zirconia (ZrO_2) and alumina tubular nanofiltration (NF) membranes modified with alginic acid for the removal of cadmium have also been suggested (Athanasekou et al. 2009). Addition of nanoparticles to the alginate matrix for dye removal has recently been studied. The nanoparticles used in this work consisted of magnetite–ferrite nanoparticles. The study concluded that the nanoparticle–alginate composite served as an eco-friendly adsorbent for dye removal from colored wastewater (Mahmoodi 2013). Moreover, in recent years, the use of sodium alginate to prepare composite NF membranes for the separation of divalent ions from low molecular weight organics is gaining grounds. Chen et al. (2010b) reported the preparation of sodium alginate/polysulfone (PSf) composite NF membrane that was crosslinked by glutaraldehyde. It was observed that the membranes showed rejection up to 87% for inorganic electrolytes such as Na_2SO_4. Thus, the potency of sodium alginate is being explored, and it would soon be competing with the existing membrane materials for the preparation of not just pervaporation membranes but also NF membranes.

5.3 CHITOSAN

CH is the second-most abundant natural polymer on earth after cellulose, which is obtained by deacetylation of chitin (Figure 5.4). Chitin occurs in the shells of crustaceans such as shrimps, crabs, and prawns. CH exhibits excellent resistance to most organic solvents, hence it has tremendous potential in separation processes (Table 5.1).

CH offers excellent advantages as a membrane material because it has reactive amine and hydroxyl groups in its backbone, which provide lots of scope for chemical modification. The amine groups are responsible for the hydrophilicity and complexation ability of this material. Chemical modification is essential, as it enables the introduction of different functionalities, which may be useful in expanding the commercial utilization of this material. Advantages of CH membranes can be extended from separation membranes to biomedical membranes, as it possesses biological properties such as antibacterial (Tanigawa et al. 1992) and wound-healing activity (Kweon et al. 2003). Indeed, CH is probably

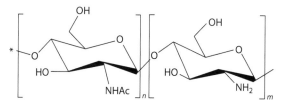

FIGURE 5.4 Structure of CH. The symbol "*" indicates the repeating units of the polymer chain.

TABLE 5.1

List of the Chitosan Derivatives Known So Far

Chitosan Derivatives	References	Structure of Chitosan Derivatives
N-Phthaloyl chitosan	Kurita 2001; Padaki et al. 2011	
N-Carboxymethyl chitosan	Muzzarelli et al. 1982	
N,O-Carboxymethyl chitosan	Chen et al. 2004; Miao et al 2006	
O-Sulfate N-Sulfate	Miao et al. 2005 Holme and Perkin 1997	
Phosphonic chitosan	Heras et al. 2001	
Quaternized chitosan	Domard et al. 1986	
O-Succinyl chitosan	Zhang et al. 2003	
Cinnamate chitosan	Wu et al. 2007	

TABLE 5.1 (*Continued*)
List of the Chitosan Derivatives Known So Far

Chitosan Derivatives	References	Structure of Chitosan Derivatives
Crown ether-bound chitosan derivative	Tang et al. 2002	
Cyclodextrin derivative	Tojima et al. 1999	
O-Pegylated chitosan	Gorochovceva and Makuska 2004	
N-propylphosphonyl chitosan	Kumar et al. 2013b	
Silylated chitosan	Kurita et al. 2004	

the most profoundly and widely used polysaccharides, displaying a vast array of applications, ranging from drug delivery, tissue engineering, controlled release of fertilizers, metal and dye adsorption, pervaporation, fuel cell application, protein separation, photography, chromatographic separations, and so on (Dutta et al. 2004). A brief account of the application of CH in membrane technology is discussed in Section 5.3.1.

5.3.1   APPLICATION OF CHITOSAN IN MEMBRANE TECHNOLOGY

5.3.1.1   Pervaporation

CH is capable of forming strong hydrogen bonds with water molecules owing to the close proximity of its solubility parameter value to that of water (Ravindra et al. 1998). Due to this affinity toward water molecules, it is primarily used for the dehydration of alcohol solutions. Also, it exhibits high hydrophilicity and good film-forming capabilities. These properties make CH a promising material for pervaporation applications. Kanti et al. (2004) performed experiments on the dehydration of ethanol through CH–alginate blends. The studies demonstrated that this polyionic complex of CH and alginate could yield sufficiently good flux, better separation factors, thermal, and mechanical stability. The sorption studies confirmed greater affinity of the membrane toward water than ethanol. With the increase in feed water concentration, the separation factor decreased and flux increased because of the increase in swelling due to the hydrophilic nature of both the polymers. A number of studies on the dehydration of alcohol and other organic solvents using CH have been reported in the past few decades (Uragami et al. 1994; Lee et al. 1997; Moon et al. 1999; Nawawi and Huang 1997; Svang-Ariyaskul et al. 2006). CH/PSf composite membrane was prepared with PSf serving as a porous support and the CH surface was cross-linked by glutaraldehyde to improve the permselectivity and permeation rate for the dehydration of alcohol mixtures (Huang et al. 1999). Another CH/PSf composite hollow-fiber membrane was reported by Liu et al. (2008). It was concluded from the study that the membranes were stable and useful for long-term operations. In this work, the presence of poly (styrene sulfonic acid) on the surface of CH/PSf hollow fibers formed chemical bonds with CH chains that ensured good adhesion between CH and PSf, which was ultimately responsible for its stability. Zhang et al. (2007) worked on the surface modification of cross-linked CH membranes to study the pervaporation of various organic/water mixtures (Table 5.2).

Adding inorganic fillers or reinforcements into CH membranes is one of the ways of improving their mechanical stability because they are prone to swelling. γ-(glycidyloxypropyl)trimethoxysilane was used as a cross-linker in the preparation of CH membranes for pervaporation dehydration of isopropanol–water mixtures. High flux and permselectivity accompanied by long-term stability in these membranes suggested their potential in practical applications (Liu et al. 2005b).

Use of nanosized silica (SiO_2) particles carrying sulfonic acid groups to cross-link CH to form $CH–SiO_2$ complex membranes have recently been reported by

TABLE 5.2

Performance of Chitosan–Glutaraldehyde Cross-Linked Membranes Sutrface Modified by Maleic Anhydride toward 90 wt% Aqueous Solution of Various Organic Solvents

Organic Solvent	t (°C)	Separation Factor (α)		Permeation Flux (g/m$_2$h)	
		CS-GA	CS-GA-MA	CS-GA	CS-GA-MA
Methanol	0	46	74	100	186
	40	34	97	152	253
	50	23	68	230	301
	60	19	31	288	356
Ethanol	50	127	991	201	238
	60	105	634	250	300
Isopropanol	30	1050	1376	102	142
	40	1116	1491	139	167
	50	337	491	169	178
	60	196	366	197	211
1,4-Dioxane	30	809	1277	86	126
	40	634	891	132	189
	50	491	591	198	323
	60	400	520	247	420
Acetone	40	1276	1791	84	117

Source: Zhang, W. et al., *J. Membr. Sci.*, 295, 130–138, 2007.

Liu et al. (2005a). The functionalized SiO$_2$ nanoparticles served as spacers for CH polymeric chains, thereby providing extra space for water to permeate, and hence resulting in high permeation rates. Nanosized titania (TiO$_2$) was dispersed homogeneously into a CH matrix by an *in situ* sol-gel process (Yang et al. 2009). No aggregation was observed even at high concentrations of TiO$_2$. The study suggested that the CH/TiO$_2$ nanocomposite membrane prepared by this method showed better performance than CH/TiO$_2$ membrane prepared by blending during pervaporation dehydration of ethanol (Figure 5.5).

Novel zeolite was another inorganic filler that was used in a CH matrix to separate ethanol–water mixture (Chen et al. 2001). It was reported that the separation factor of ethanol–water mixture was improved by the addition of zeolite filler.

Recently, a novel quaternized CH composite membrane using sodium montmorillonite (Na$^+$-MMT) as filler for the separation of water–isopropanol mixtures was reported (Choudhari and Kariduraganavar 2009). Addition of MMT not only improved the thermal stability of the composite membrane but also improved flux and separation efficiency in comparison with pure quaternized membrane. Increase in MMT in the membrane led to a decrease in flux with an increase in separation factor as anticipated, which was due to decrease in the free volume and the establishment of a tortuous path. Very recent work on CH wrapped

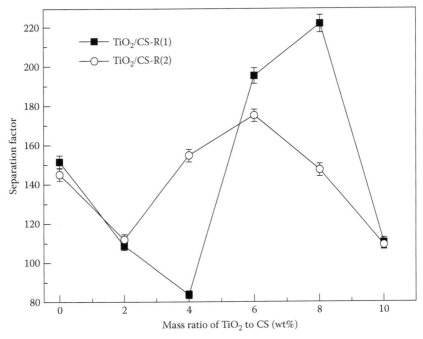

FIGURE 5.5 Effect of TiO$_2$ concentration on the separation performance of 90 wt% aqueous ethanol solution in the CH hybrid membrane: R(1) *in situ* generation of TiO$_2$ nanoparticles, and R(2) blending of TiO$_2$ nanopaticles. (Data from Yang, D. et al., *Chem. Eng. Sci.*, 64, 3130–3137, 2009.)

multiwalled CNT (MWCNT) incorporated into sodium alginate membrane was reported for the separation of water–isopropanol mixtures. Many such pervaporation membranes with CH forming blends with PVA, sodium carboxymethyl cellulose (CMCNa), hydroxyethyl cellulose (HEC), and cellulose acetate (CA) are available in the literature (Chanachai et al. 2000; Svang-Ariyaskul et al. 2006; Zhao et al. 2009a).

5.3.1.2 Ultrafiltration and NF Membranes

NF membranes are classified as those membranes that are capable of rejecting divalent ions up to 98% and monovalent ions ranging from 30% to 85% depending on the charge carried by the ions and the membrane surface. CH being a polycationic material has proved to be a very promising material in the synthesis of NF membranes. Apart from using neat CH, a number of modifications are suggested in the literature to introduce a specific functionality on the CH structure to improve the performance of the prepared membrane. As mentioned earlier, cross-linking of CH and its derivatives becomes essential in order to ensure chemical stability in acidic medium or in case the derivative is water soluble. These CH films are mainly used in the synthesis of thin-film composite membranes (TFC). A novel amphoteric composite NF membrane was synthesized from sulfated chitosan (SCS) using glutaraldehyde as the cross-linking agent and polyacrylonitrile (PAN) as the support membrane

(Miao et al. 2005). The salt rejection of these membranes toward different monovalent and divalent inorganic electrolytes was tested. The studies suggested that the membranes acquired a partial negative charge due to the adsorption of anions from the electrolyte solution, and it was this charge that governed the membrane performance, showing rejection in the order $K_2SO_4 > Na_2SO_4 > KCl > NaCl > MgSO_4 > MgCl_2$. In another similar work with SCS, epichlorohydrin (ECH) was used as the cross-linking agent for which the results were similar (Miao et al. 2013) (Figure 5.6).

Quaternized CH was used to synthesize positively charged NF membrane cross-linked by toluene diisocyanate (TDI) (Huang et al. 2008) (Figure 5.7). The results in this work were in accordance with the Donnan exclusion mechanism, showing rejection of the membrane in the order $MgCl_2 > NaCl > MgSO_4 > Na_2SO_4$. These results confirmed the positive surface of the membrane.

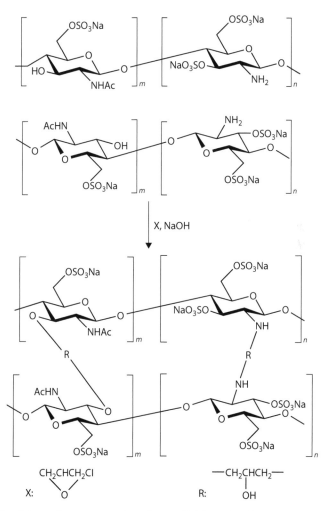

FIGURE 5.6 Schematic representation of cross-linking of sulfated CH with epichlorohydrin. (Data from Miao, J. et al., *Chem. Eng. Sci.*, 87, 152–159, 2013.)

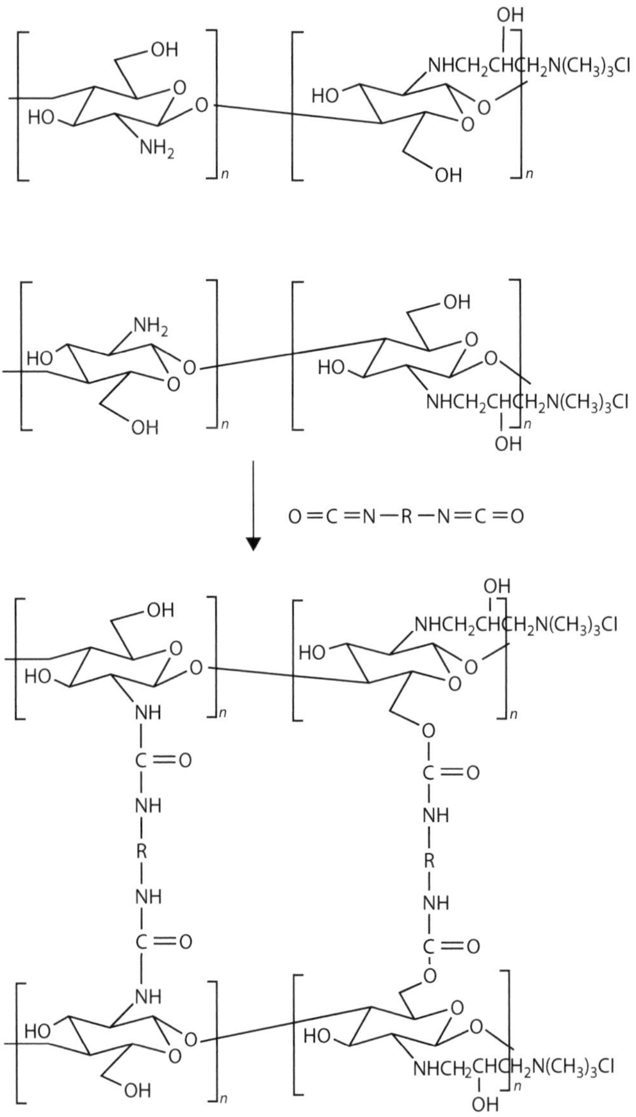

FIGURE 5.7 Schematic representation of the cross-linking of quaternized CH by TDI. (Data from Huang, R. et al., *Sep. Purif. Technol.*, 61, 424–429, 2008.)

Other modifications of CH leading to positive and negative membrane surface have been suggested such as graft copolymer of trimethylallyl ammonium chloride onto CH and *N,O*-carboxymethyl CH (Huang et al. 2006; Miao et al. 2008). Synthesis of xanthated CH, succinated CH, *N*-propylphosphonic CH has been reported but its performance in the field of NF membrane still remains to be a matter of research (Zhang et al. 2003; Sankararamakrishnan and Sanghi 2006; Kumar et al. 2013b).

UF membranes form another class of membranes that cover fractionation and concentration in pure water production and water treatment (Baker 2004). Boributh et al. (2009) worked on poly (vinylidene fluoroide) (PVDF) ultrafiltration (UF)-modified using CH by three different methods. The results demonstrated that the modified membranes exhibited better antifouling properties. The performance of the PSf UF membranes was improved by the incorporation of N-succinyl CH, N-propylphosphonic CH, and CH as an additive with an aim to control fouling (Kumar et al. 2013a, 2013b) (Figure 5.8). The membranes showed enhanced flux-recovery ratio in comparison with neat PSf membrane. PAN/CH UF membranes were prepared by filtering CH solution through a porous PAN support. The water permeation rate, molecular weight cut off, and the pore size distribution of the membrane indicated reduction in pore size

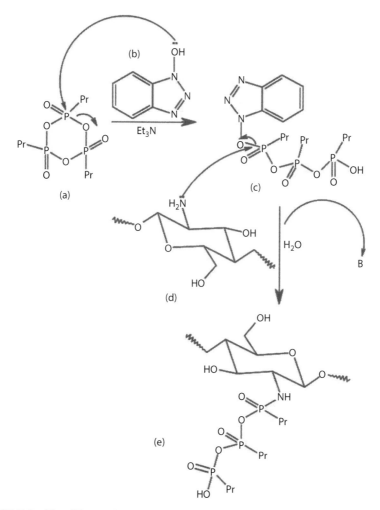

FIGURE 5.8 Plausible mechanism for the formation of N-propylphosphonic CH. (Data from Kumar et al. 2013b.)

of the membrane due to the formation of a CH layer on the inner surface of the PAN membrane (Musale et al. 1999).

In a recent work, high flux UF membrane was prepared using electrospun nano-fibers of PAN, which was coated with CH solution. The nanofibrous PAN scaffold increased the surface porosity of the membrane, resulting in higher flux with the hydrophilic CH layer (Yoon et al. 2006). In the future, the use of nanofibers made up of CH may help in improving the performance of the subsequently fabricated UF and NF membranes.

5.3.1.3   Removal of Metal Ion and Dyes

CH exhibits cationicity in acidic media due to the presence of amino groups in its structure. This property of CH renders it advantageous in terms of metal cations, especially transition metal chelation in near neutral solution. Protonation of amino groups may also be helpful in electrostatic interaction with anionic dyes, and thus help in their removal from an aqueous stream (Figure 5.9).

Another way of removing metal ions from water streams using CH is through polymer-enhanced or polymer-assisted UF (PEUF). In this method, a metal-bound chitosan complex is retained by a UF membrane, based on the principle that the size of the metal-CH complex is large enough to be retained on the surface of the UF membrane thereby helping in its removal (Guibal et al. 2005). A number of modifications are suggested in the literature to improve the metal sorption capacity of CH.

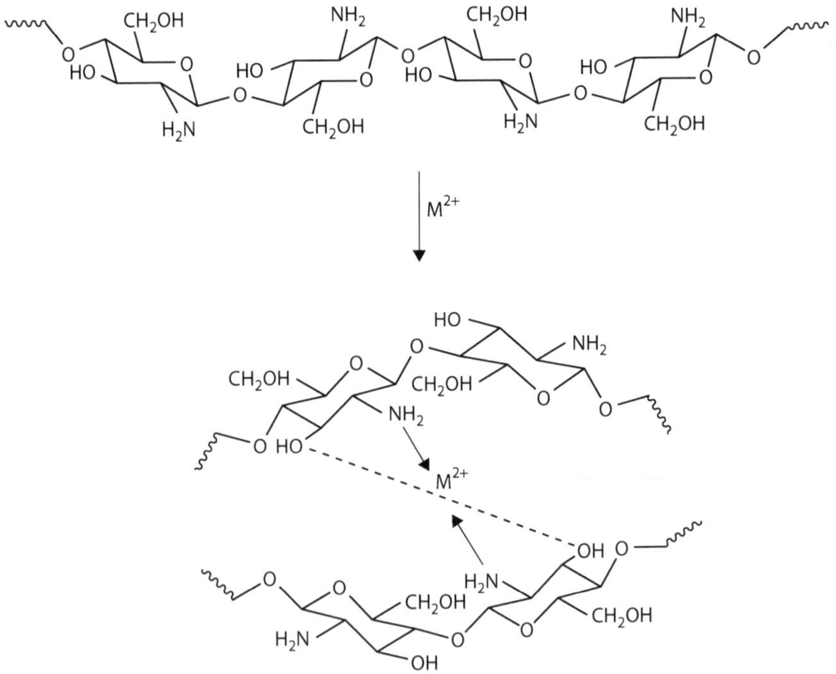

FIGURE 5.9   Possible interaction of CH molecule with metal ions. (Data from Wang, X. et al., *Polym. Bull.*, 55, 105–113, 2005.)

These modifications include reactions of CH with carboxylic anhydrides, crown ethers, Schiff base reactions, ethylenediaminetetraacetic acid, diethylenetriamine-pentaacetic acid, phosphorous derivatives, and sulfur derivatives (Muzzarelli et al. 1984; Nishi et al. 1987; Peng et al. 1998; Nagib et al. 1999; Shigemasa et al. 1999; Cárdenas et al. 2001). The affinity of CH toward metals approximately follows the following series Pd > Au > Hg > Pt > Cu > Ni > Zn > Mn > Pb > Co > Cr > Cd > Ag (Krajewska 2001, 2005). A detailed review on the modification and mechanism of interaction of CH with metal ions is given by Guibal (2004). Multilayer composite membranes using electrostatic self-assembly was prepared from CH/polyacrylic acid, which showed enhanced Cu^{2+} metal ion removal. This technique is proposed to be a very promising tool to prepare functional membranes that have high metal-binding capacities (Fu and Kobayashi 2010). Blend membranes comprising of CH and acrylonitrile butadiene styrene were prepared to study the separation of mercury (Hg) (Boricha and Murthy 2009). The study suggested that the prepared membranes were capable of rejecting Hg ions up to 96%. A novel method for the synthesis of CH/CA hollow-fiber membrane was reported by Han et al. (2007). Rather than using acidic solution to dissolve CH, the use of nonacidic dope solvent was suggested in this work. CH nanoparticles were synthesized and their dope solution with CA was used to make hollow-fiber membrane with an aim to augment Cu^{2+} ion adsorption. The improved accessibility of CH in the form of nanoparticles present in the blend enhanced Cu^{2+} adsorption. Chemical modification of CH with amine has been used for the removal of Hg^{2+} ions (Jeon and Holl 2003). In another study, sorption of palladium on CH through grafting of sulfur compounds such as urea was investigated (Guibal et al. 2002). Application of CH and acetylated CH membranes for the removal of a number of metal ions such as Cr(VI), Mn(II), Fe(III), Ni(II), Cu(II), Zn(II), and Cd(II) was proposed by Kaminski and Modrzejewska (1997). Currently, CH beads, more than flat-sheet membranes, are the prevalent method of removal of metal ions, as they offer a relatively greater surface area (Ngah et al. 2005; Ngah and Fatinathan 2008, 2010; Kumar et al. 2009).

Modification of CH with MWCNTs has recently been suggested for the removal of uranium (Chen et al. 2013). Magnetic CH/graphene oxide (GO) has also recently been suggested for the removal of Pb (Fan et al. 2013). Such modifications in the future, with the addition of inorganic fillers such as zeolites and clay, may enhance the metal-binding capacity of CH. Transition metal ions were also retained by PEUF using CH as the complexing agent in a near neutral solution (Juang and Chiou 2000a; Kuncoro et al. 2005). CH-enhanced membrane filtration was reported for the removal of metal ions, which included Cu(II), Co(II), Ni(II), and Zn(II). It was concluded that the removal of metal ions was enhanced 6–10 times by the addition of CH at acidic pH. Also, the removal of Cu(II) was greater in comparison with other metals (Juang and Chiou 2000a,b). Recovery of Hg ions by the addition of CH enhanced UF up to 95% under optimum experimental conditions (Kuncoro et al. 2003). The effect of pH, metal ion concentration, and polymer concentration was studied on the permeation flux and retention was studied in this work. The increase in pH of the solution increased the metal-binding capacity of CH.

In the case of textile dyes, reports are available on the use of CH beads, where the pH at which maximum removal of dye occurs remains the critical parameter, because at acidic pH, the removal of anionic dyes will be maximum (Chiou and Li 2002,

2003; Chiou et al. 2004). However, the removal of basic dyes, that is, cationic dyes, is not very significant (Crini 2005). CH-based nanocomposites with MMT, polyurethane, and bentonite have been reportedly used for the removal of dyes from aqueous streams (Ngah et al. 2011). Removal of cationic and anionic dye using CH/nanoclay-mixed matrix on PVDF membrane was discussed by Daraei et al. (2013). The study concluded that the dye removal of the prepared TFC membranes was predominantly due to adsorption phenomenon, in addition to other mechanisms such as sieving effects, concentration polarization, and inertia effects that accompany a TFC membrane filtration process. Similarly, CH/MMT adsorbent membrane was also prepared for the efficient removal of bezactiv orange V-3R dye (Nesic et al. 2012). Thus, CH and its derivatives have been competently used in the removal of reactive dyes (Guibal et al. 2003; Largura et al. 2010).

5.4 CELLULOSE

Cellulose is the most abundant naturally occurring material on earth. It is a polysaccharide found in cell walls of green plants, wood, hemp, cotton, and other plant-based materials. Structurally, it consists of a linear chain of several hundreds of anhydroglucose units joined by β-1,4 linkages (Figure 5.10).

Cellulose-based materials are the most primitive and traditional membrane materials that still continue to enjoy the popularity of being one of the most widely used membrane materials owing to its abundance and reusability. Other properties include its hydrophilicity, exhibiting a contact angle of 20°, chirality, low cost, and biodegradability (Kamel et al. 2006). However, unlike most of its hydrophilic counterparts, cellulose is insoluble in water or most organic solvents due to the extensive intra-molecular hydrogen bonding and intra-strand hydrogen bonding, which is attributed to the presence of a large number of hydroxyl groups in its backbone (Figure 5.11). Thus, processing of this material for membrane application becomes tedious.

This emphasizes the need for suitable modification or derivatization of cellulose for its versatile and widespread use as a membrane material. Chemical modification alters cellulose properties with respect to elasticity, hydrophilicity, ion exchange or adsorption capacity, water uptake capacity, thermal, and microbial resistance. The reactive primary and secondary hydroxyl groups in its backbone structure serve as potential sites for the introduction of different functional groups. The possible modifications include etherification, esterification (Marchetti et al. 2000), oxidation,

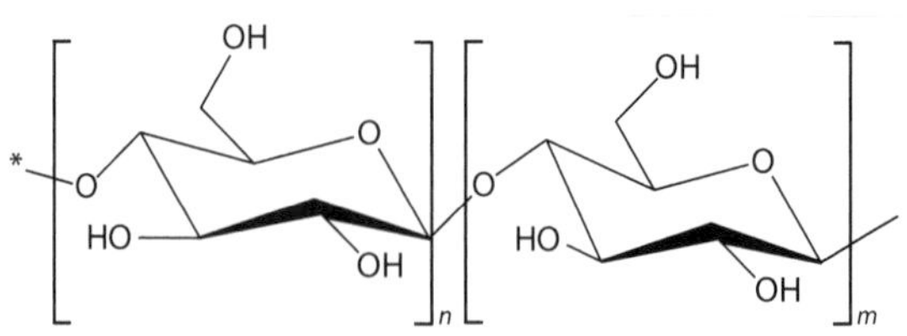

FIGURE 5.10 Structure of cellulose molecules. The symbol "*" indicates the repeating units of the polymer chain.

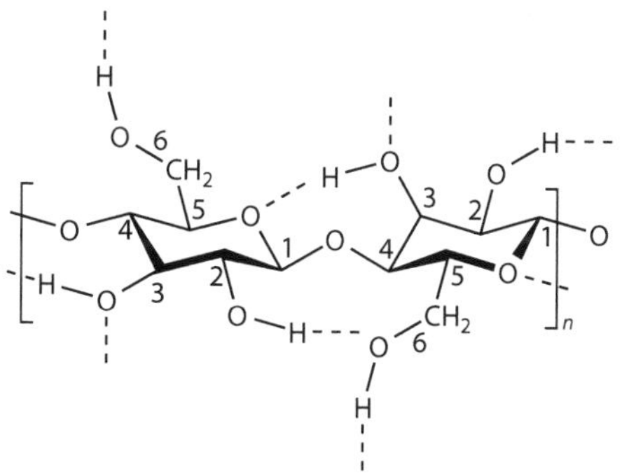

FIGURE 5.11 Hydrogen bonding existing in cellulose.

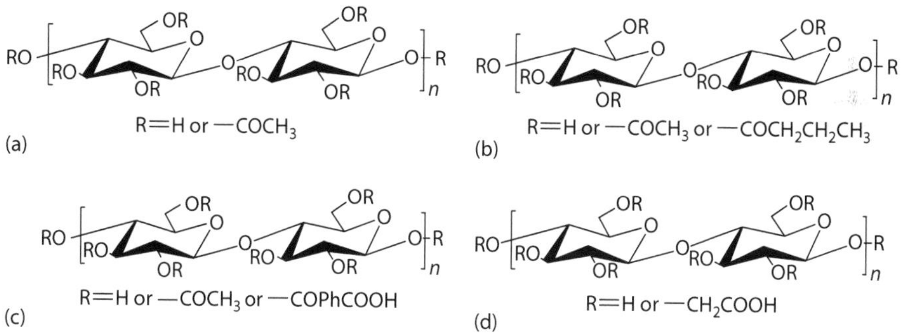

R=H or —COCH₃

(a)

R=H or —COCH₃ or —COCH₂CH₂CH₃

(b)

R=H or —COCH₃ or —COPhCOOH

(c)

R=H or —CH₂COOH

(d)

FIGURE 5.12 Structure of (a) cellulose acetate, (b) cellulose acetate butyrate, (c) cellulose acetate phthalate, (d) carboxymethyl cellulose.

grafting of different polymers on cellulose, and halogenation (Tashiro and Shimura 1982; O'Connell et al. 2008). The commercially available cellulose derivatives include CA, cellulose triacetate, cellulose butyrate, water-soluble HEC, methyl cellulose, hydroxypropyl cellulose (HPC), and CMC (Figure 5.12). Also, cationic cellulose derivatives containing nitrogen have been synthesized in recent years and whose applications in membrane field are yet to be explored (Heinze and Liebert 2001; Liesiene 2010).

5.4.1 Applications of Cellulose in Membrane Technology

5.4.1.1 UF and NF Membrane

The first critical breakthrough in the field of membrane preparation for water purification was achieved in 1959 when Sourirajan and Loeb synthesized CA membranes through a phase inversion process. Emergence of CA reverse osmosis membranes paved the way for other membrane-based separations, including microfiltration, UF, NF, and pervaporation. Traditionally, CA was used as a prime cellulose derivative for membrane preparation owing to its reasonably good toughness, biocompatibility, potential flux, low cost, easy solubility in most of the commonly used organic solvents, and hydrophilicity, which helps in reducing membrane fouling (Dunweg et al. 1995; Qin et al. 2003; Idris and Yet 2006; Saljoughi et al. 2009). Extensive work has been carried out studying the performance of this widely used cellulose derivative with respect to morphology and permeation properties by variation in concentration, coagulation bath, coagulation bath temperature, blending counterparts, and water-soluble and insoluble additives (Mahendran et al. 2004; Sivakumar et al. 2006; Saljoughi et al. 2009; Chen et al. 2010a). Studies have been carried out on the addition of ceramic fillers such as bentonite, kalonite, silica, and alumina in cellulose membranes, which has resulted in an improvement in the thermal properties, permeate flux, salt rejection, and mechanical strength of the resulting membranes (Goossens and Van Haute 1976; Finken 1983; Doyen et al. 1990). Addition of such fillers particularly affected the macrovoid formation in the membrane during the phase inversion step, which is governed by the kinetic and thermodynamic behavior of the polymer system, in addition to the stability of the fillers (Wara et al. 1995). In recent years, some researchers have incorporated nanoparticles in CA membranes with the aim to improve their overall performance. Although switching from microscale to nanoscale, there is an increase in the ratio of surface area to volume, and the particle size is affected by quantum effects. Arthanareeswaran and Thanikaivelan (2010) studied the effect of the incorporation of ZrO_2 nanoparticles in a CA matrix. The studies concluded that such organomineral membranes improved the pure water flux, mechanical stability, membrane resistance, and antifouling property of the resultant membranes. Similar results were reported by other research groups who incorporated TiO_2, SiO_2, and silver (Ag) (Chou et al. 2005; Chen et al. 2010a; Abedini et al. 2011).

Similar to CA, other water insoluble cellulose derivatives such as CA butyrate and CA phthalate have also been used recently to prepare UF membrane; however, they lack the popularity as enjoyed by CA due to their relatively poorer mechanical stability. These derivatives will also be used on a wide scale in membrane preparation in the near future (Sabde et al. 1997; Rahimpour and Madaeni 2007; Hashino et al. 2011). Water-soluble derivatives of cellulose, mainly CMCNa, are used in the preparation of NF membranes. Water solubility makes cross-linking of these membranes extremely essential. Yu et al. (2012b) reported a simple modification of polypropylene hollow-fiber microfiltration membrane to NF by dip-coating with CMCNa followed by cross-linking with $AlCl_3$ (Figure 5.13). These membranes had reported rejection of inorganic salts in the order $MgCl_2 < CaCl_2 < MgSO_4 < NaCl < KCl < Na_2SO_4$, which is a characteristic of negatively charged membrane and the rejection property being governed by the Donnan exclusion mechanism.

Use of CMCNa cross-linked with $FeCl_3$ and ECH has also been recently suggested for the separation of divalent ions from low molecular weight organics and for dye rejection (Miao et al. 2007; Yu et al. 2012a). Apart from these commonly reported modification techniques, new ways of modifying cellulose and its derivatives are gaining momentum. For example, the modification of CNTs with CA, and synthesis of CA nanoparticles or nanocrystalline cellulose have already been investigated; however, their utilization in wastewater treatment has not been widely explored yet (Ke 2009; Kulterer et al. 2011; Zhou et al. 2012).

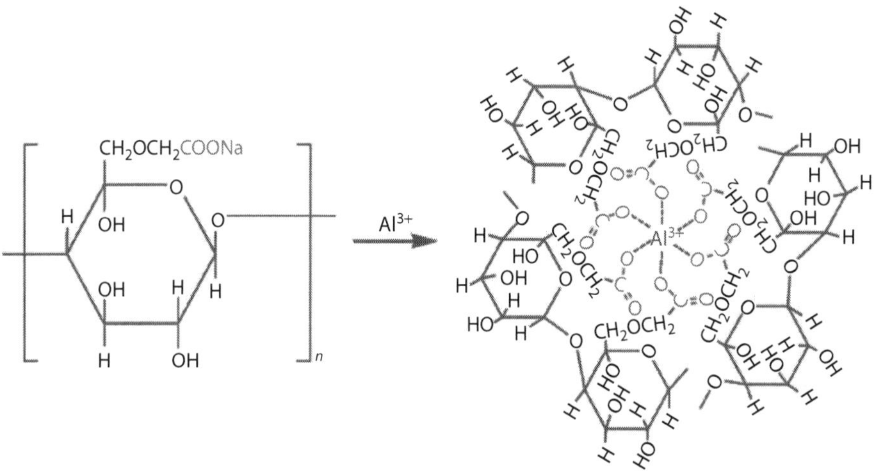

FIGURE 5.13 Schematic representation of the surface of CMCNa-dip-coated polypropylene hollow fiber cross-linked with $AlCl_3$. (Data from Yu, S. et al. 2012b.)

5.4.1.2 Pervaporation

It is well known that a pervaporation membrane should have good hydrophilicity, mechanical strength, and thermal resistance. However, to ensure its selectivity in separation, it is absolutely essential that the degree of swelling is not excessive. In order to achieve this, cross-linking of cellulose and its derivatives is adopted. Cellulose and its derivatives form an important class of pervaporation membranes due to the presence of abundant hydroxyl groups and other functional groups on its backbone. The most commonly used cross-linking agents for this purpose include glutaraldehyde and mixtures of urea + formaldehyde + sulfuric acid. Most of these membranes are particularly used for dehydration of alcohols (Chanachai et al. 2000; Jiraratananon et al. 2002). Cellulosic compounds having hydroxyl groups tend to preferentially remove water due to their interaction with water molecules, resulting in high flux and selectivity (Sridhar et al. 2004; Veerapur et al. 2007). Water-soluble cellulose ethers such as HEC and HPC are capable of forming compatible blends with a vast number of other water-soluble polymers (Chanachai et al. 2000). HEC was used for pervaporation, and particularly for the desulfurization, of gasoline components, which were the major contributors of SO_x air pollution (Qu et al. 2010). In this work, 1,6-hexanediol diacrylate was used as the cross-linker with benzoyl peroxide as the initiator. The cross-linking density greatly influenced the sorption and transport of small molecules through the membrane (Wang et al. 2000; Lin et al. 2008). HEC was also used in the pervaporation of 1,4 dioxane + water mixtures and THF + water mixtures. THF and dioxane are water soluble at all proportions and form azeotropic mixtures with water. Their separation is essential because they form an important class of organic solvents and find application in chemical and pharmaceutical industries (Naidu et al. 2005). In another similar case, HEC was grafted with acrylamide and then blended with sodium alginate to study dehydration of acetic acid by pervaporation (Rao et al. 2006). These membranes, being hydrophilic in nature, successfully recovered 89% of water from acetic acid solutions. HPC blended with CH was prepared for pervaporation dehydration of isopropanol. Addition of hydrophilic HPC, to the already hydrophilic CH, converted the blend CH–HPC to even higher hydrophilicity, enabling the achievement of higher degrees of dehydration of the isopropanol (Veerapur et al. 2007). Some literature has also reported on the use of CMCNa polyelectrolyte complexes for pervaporation dehydration of alcohols such as ethanol and isopropanol (Zhao et al. 2009a, 2009b; Jin et al. 2010). There is still scope for the use of other cellulose derivatives in the ever-expanding field of pervaporation, especially for dehydration application.

5.4.1.3 Removal of Metal Ions and Dye from Aqueous Solutions

Adsorption membranes are also another class of membranes, in which cellulose and its derivatives are used for the removal of dyes and heavy metal ions from aqueous solutions. Presence of different functional groups helps in the adsorption of metal ions and organic dyes, depending on the type of cellulose derivatives; CMC, in particular, proves as an efficient biosorbent (Dewangan et al. 2011). CMC was used in PEUF of copper ions from wastewater solutions. In PEUF, the metal ions are first ionically bound to the polymeric chain; these large metal–polymer complexes are then filtered through UF membranes. In this particular work by Mundkur et al. (1993), the concentration of copper in simulated wastewater reduced from 100 ppm to less than 1 ppm by PEUF, in which CMC was used as complexing agent. The complexation ability of this biopolymer proved to be more proficient than sodium polystyrene sulfonate in this study. The removal of anionic dyes such as Congo red and Methyl blue by polypropylene hollow fiber coated with CMCNa, with long-term performance stability and antifouling property, was documented by Yu et al. (2012b). A great amount of literature is available on the removal by cellulose additives of heavy metals such as copper (Cu[II]), lead (Pb[II]), nickel (Ni[II]), cadmium (Cd[II]), and mercury (Hg[II]), with an aim to improve the removal capacity of cellulose (Maekawa and Koshijima 1990; Aoki et al. 1999; Marchetti et al. 2000; Dewangan et al. 2010). These modification techniques included halogenation, oxidation, etherification, esterification, photo-grafting, high-energy radiation grafting, and chemical grafting (O'Connell et al. 2008). A recent article by Ting et al. (2013) investigated the preparation and characterization of cellulose-based electrospun nanofiber membranes and studied their adsorptive properties using Cu(II), Cd(II), and Pb(II) as model heavy metal foulants. These nanofiber membranes were successfully modified with thiol groups using thioglycolic acid. The results demonstrated that the thiol group played an important role during the adsorption process and the type of adsorption was chemisorption. Another such modification of cellulose was suggested by Musyoka et al. (2011), in which cellulose was modified by ethylenediamne for the removal of Pb and Cd from aqueous solutions. It is well known that amine groups serve as efficient anchoring sites, leading to the complexation of metal ions. Cost-effective functionalization of cellulose using succinic anhydride and maleic anhydride, which possess carboxyl functional groups, has also been reported for the removal of Cd from contaminated waters (Belhalfaoui et al. 2009; de Melo et al. 2009). Recently, the eco-friendly synthesis of a new nanocomposite film comprising of CMC and GO was reported and its properties were investigated (Yadav et al. 2013). Although only structural, mechanical, and thermal properties of these films were investigated in this work, there is future scope to study adsorption properties of these films, considering the presence of abundant carboxyl and hydroxyl functional groups in GO, in addition to the already-existing functional groups present in CMC. Such modification and the use of electrospun membranes or other additives such as zeolite and GO to improve the surface area can be undertaken in the future to enhance the adsorption capacity of cellulose membranes.

Even though there is exhaustive literature available on popular polysaccharides such as CH, cellulose, and sodium alginate, the focus of related membrane research is steadily shifting to less popular polysaccharides such as starch, cyclodextrins (CDs), and lignin.

5.5 CYCLODEXTRINS

CDs are torus-shaped cyclic oligosaccgarides generally containing 6–12 glucopyranose units linked by α-(1,4) bonds. CDs are formed by the degradation of starch by the enzyme cyclodextrin glucanotransferase (CGTase) (Szetjli 1998). The three smallest CDs include α-, β-, and γ-CDs, which contain six, seven, and eight glucose units, respectively. The cyclic structure of this molecule is positioned in such a way that it results in a molecular structure having a hydrophilic exterior surface with an apolar or hydrophobic interior cavity (Figure 5.14). This apolar cavity enables CDs to form inclusion complexes with various molecules especially aromatic compounds. The factors that govern the formation of this *guest–host* complex are van der Waal interaction, hydrophobic interaction, hydrogen bonding, and size (Norkus 2009). Formation of inclusion complexes greatly alters the physical and chemical properties of the incoming guest molecule (Szejtli 1998; Schneiderman and Stalcup 2000; Crini and Morcellet 2002).

Being water soluble in their native form, CDs are usually modified to form insoluble derivatives. It is known that CH and CDs have a common set of applications; hence, efforts are usually directed to couple their properties without any compromise on their individual adsorption property. Many reports are available which sheds

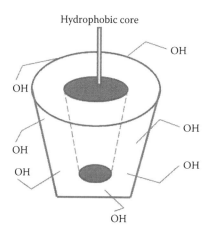

FIGURE 5.14 Graphical representation of the toroid structure of CD.

considerable light upon coupling of CDs and CH (Tojima et al. 1999; Martel et al. 2001; Aoki et al. 2003). Although the literature available on the use of CD as membrane material is not as vast as other polysaccharides, few noteworthy mentions of this material are available. A comprehensive review of the interaction of CDs with pentavalent, trivalent, and divalent ions is available (Martin Del Valle 2004; Norkus 2009). Permeation of xylene isomers through liquid membranes containing CD was reported by Lue et al. (2002). The study concluded that the addition of CD not only increased the selectivity toward *p*-xylene but also resulted in the enhancement of mass transfer flux. Separation of isomers by UF using modified CDs has also been reported. In this work, a detailed study on the separation of structural and optical isomers by suitably modifying the CD was carried out. Acrylated CD has been reported to be used in aromatic/aliphatic separation (Rölling et al. 2010). In the last decade, the concept of CD nanosponges gained popularity in scientific circles. The idea behind this concept is that, native CDs are suitably cross-linked to form nanostructured material containing hyper-cross-linked CD. These nanosponges have tremendous capacity to encapsulate a variety of materials, which makes it a potent candidate for the treatment of contaminated water (Ma and Li 1999; Trotta and Tumiatti 2003; Trotta and Cavalli 2009). The potency of this polysaccharide in the removal of dyes and other organics is still to be explored in future.

5.6 LIGNINS

Lignin is the second-most abundant polymer after cellulose occurring in the cell wall of plants and some algae. Lignins are amorphous, chemical resistant, water insoluble, aromatic biopolymers that are relatively hydrophobic in nature (Lalvani et al. 2000). It is an unusual biopolymer mainly because unlike other polymers, it does not have a well-defined primary structure. The basic unit contains a number of functional groups such as primary hydroxyl groups, secondary hydroxyl groups, phenolic hydroxyl groups, carbonyl groups, and labile methoxy groups (Figure 5.15). These functional groups are known to serve as binding sites as well as cation exchange sites for heavy metals (Pahlman and Khalafalla 1988; Verma et al. 1990; Srivastava et al. 1994; Lalvani et al. 1997).

Removal of Cd, Zn, Pb, and Cu through adsorption by modified lignin was reported by Celik and Demirbaş (2005). The detailed study concluded that the adsorption by lignin was higher at higher pH, which indicated that an ion-exchange mechanism prevailed between metal ions and lignin. Platt and Clysdale (1985) reported on the adsorption of iron in the presence of other metal ions. Composite membrane prepared from CA by the incorporation of lignin for the improved removal of copper was recently investigated by Nakanishi et al. (2011). The CA used in this work was synthesized from cellulose present in sugarcane bagasse, from which dense composite membrane was obtained after the addition of lignin by an evaporation–precipitation method. To date, lignin has been used for the adsorption of metal ions and its application as a membrane material has not been explored much, a consequence of its limited solubility. But with the presence of phenolic hydroxyl and primary hydroxyl groups in its backbone, it is bound to be another eco-friendly membrane material in the making.

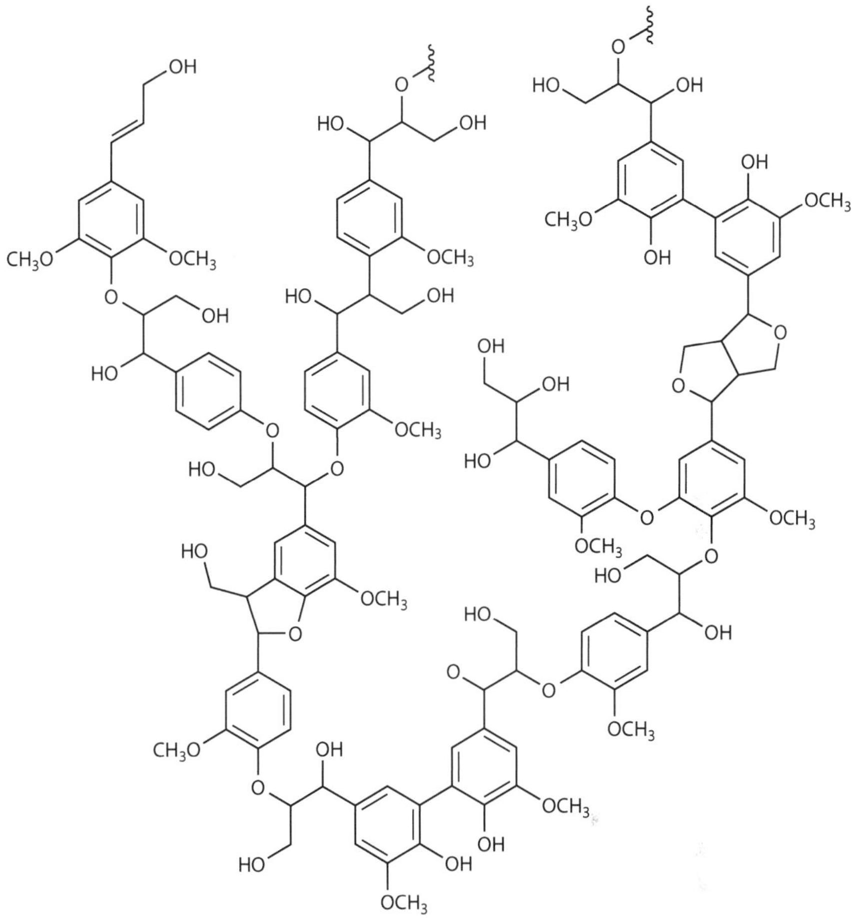

FIGURE 5.15 Possible structure of lignin.

Other than the above-mentioned biopolymers, research continues to examine the usefulness of other polysaccharides as well. The use of hyaluronic acid and chondroitin sulfate as membranes has so far been limited to tissue engineering and biomedical applications. However, these materials will also be seen in the mainstream membrane market for water purification in future.

5.7 CONCLUSIONS

Polysaccharides form an important class of membrane materials in the field of wastewater treatment, metal ion removal, dye removal, and pervaporation. It is still an ever-expanding field where there is always scope to improve the performance in comparison with its predecessors by methods such as the addition of some fillers, physical and chemical modification, and grafting. Research is being actively conducted on the lesser known polysaccharides where currently the main obstacle in their wide scale use is the raw material extraction cost. Once this obstacle is overcome, there will be a bounty of useful, environmental-benign, and highly separation-efficient membrane materials, which will surpass the existing conventional techniques that are currently prevalent in separation technology.

REFERENCES

Abedini, R., Mousavi, S. M., and Aminzadeh, R. 2011. A novel cellulose acetate (CA) membrane using TiO_2 nanoparticles: Preparation, characterization and permeation study. *Desalination* 277: 40–45.

Adoor, S. G., Manjeshwar, L. S., Bhat, S. D., and Aminabhavi, T. M. 2008. Aluminum-rich zeolite beta incorporated sodium alginate mixed matrix membranes for pervaporation dehydration and esterification of ethanol and acetic acid. *J. Memb. Sci.* 318: 233–246.

Aoki, N., Fukushima, K., Kurakata, H., Sakamoto, M., and Furuhata, K. 1999. 6-Deoxy-6-mercaptocellulose and its S-substituted derivatives as sorbents for metal ions. *React. Funct. Polym.* 42: 223–233.

Aoki, N., Nishikawa, M., and Hattori, K. 2003. Synthesis of chitosan derivatives bearing cyclo-dextrin and adsorption of p-nonylphenol and bisphenol A. *Carbohydr. Polym.* 52: 219–223.

Arthanareeswaran, G. and Thanikaivelan, P. 2010. Fabrication of cellulose acetate–zirconia hybrid membranes for ultrafiltration applications: Performance, structure and fouling analysis. *Sep. Purif. Technol.* 74: 230–235.

Athanasekou, C. P., Papageorgiou, S. K., Kaselouri, V., Katsaros, F. K., Kakizis, N. K., Sapalidis, A. A., and Kanellopoulos, N. K., 2009. Development of hybrid alginate/ceramic membranes for Cd^{2+} removal. *Microporous Mesoporous Mater.* 120: 154–164.

Baker, R. W. 2004. *Membrane Technology and Applications*, 2nd edn., Wiley, Chichester.

Belhalfaoui, B., Aziz, A., Elandaloussi, E. H., Ouali, M. S., and L. C. Me'norval. 2009. Succinate-bonded cellulose: A regenerable and powerful sorbent for cadmium-removal from spiked high-hardness groundwater. *J. Hazard. Mater.* 169: 831–837.

Bhat, S. D. and Aminabhavi, T. M. 2007. Pervaporation separation using sodium alginate and its modified membranes—A review. *Sep. Purif. Rev.* 36: 203–229.

Boributh, S., Chanachai, A., and Jiraratananon, R. 2009. Modification of PVDF membrane by chitosan solution for reducing protein fouling. *J. Memb. Sci.* 342: 97–104.

Boricha, A. G. and Murthy, Z. V. P. 2009. Acrylonitrile butadiene styrene/chitosan blend membranes: Preparation, characterization and performance for the separation of heavy metals. *J. Memb. Sci.* 339: 239–249.

Cárdenas, G., Orlando, P., and Edelio, T. 2001. Synthesis and applications of chitosan mercaptanes as heavy metal retention agent. *Int. J. Biol. Macromol.* 28: 167–174.

Celik, A. and Demirbaş, A. 2005. Removal of heavy metal ions from aqueous solutions via adsorption onto modified lignin from pulping wastes. *Energy Sources* 27: 1167–1177.

Chanachai, A., Jiraratinanon, R., Uttapap, D., Moon, G. Y., Anderson, W. A., and Huang, R. Y. M. 2000. Pervaporation with chitosan/hydroxyethyl cellulose (CS/HEC) blended membranes. *J. Memb. Sci.* 166: 271–280.

Chen, J.-H., Lu, D.-Q., Chen, B., and OuYang, P.-K. 2013. Removal of U(VI) from aqueous solutions by using MWCNTs and chitosan modified MWCNTs. *J. Radioanal. Nucl. Chem.* 295: 2233–2241.

Chen, S.-C., Wu, Y.-C., Mi, F.-L., Lin, Y.-H., Yu, L.-C., and Sung, H.-W. 2004. A novel pH-sensitive hydrogel composed of N,O-carboxymethyl chitosan and alginate cross-linked by genipin for drug delivery. *J. Controlled Release* 96: 285–300.

Chen, W., Su, Y., Zhang, L., Shi, Q., Peng, J., and Jiang, Z. 2010a. In situ generated silica nanoparticles as pore-forming agent for enhanced permeability of cellulose acetate membranes. *J. Memb. Sci.* 348: 75–83.

Chen, X., Gao, X., Wang, W., Wang, D., and Gao, C. 2010b. Study of sodium alginate/polysulfone composite nanofiltration membrane. *Desalination Water Treat.* 18: 198–205.

Chen, X., Yang, H., Gu, Z., and Shao, Z. 2001. Preparation and characterization of HY zeolite-filled chitosan membranes for pervaporation separation. *J. Appl. Polym. Sci.* 79: 1144–1149.

Chiou, M. S., Ho, P. Y., and Li, H. Y. 2004. Adsorption of anionic dyes in acid solutions using chemically cross-linked chitosan beads. *Dyes Pigm.* 60: 69–84.

Chiou, M. S. and Li, H. Y. 2002. Equilibrium and kinetic modeling of adsorption of reactive dye on cross-linked chitosan beads. *J. Hazard. Mat.* B93: 233–248.

Chiou, M. S. and Li, H. Y. 2003. Adsorption behavior of reactive dye in aqueous solution on chemical cross-linked chitosan beads. *Chemosphere*, 50: 1095–1105.

Chou, W. L., Yu, D. G., and Yang, M. C. 2005. The preparation and characterization of silver-loading cellulose acetate hollow fiber membrane for water treatment. *Polym. Adv. Technol.* 16: 600–607.

Choudhari, S. K. and Kariduraganavar, M. Y. 2009. Development of novel composite membranes using quaternized chitosan and Na⁺-MMT clay for the pervaporation dehydration of isopropanol. *J. Colloid Interface Sci.* 338: 111–120.

Crini, G. 2005. Recent developments in polysaccharide-based materials used as adsorbents in wastewater treatment. *Prog. Polym. Sci.* 30: 38–70.

Crini, G. and Morcellet, M. 2002. Synthesis and applications of adsorbents containing cyclodextrins. *J. Sep. Sci.* 25: 789–813.

Daraei, P., Madaeni, S. S., Salehi, E., Ghaemi, N., Ghari, H. S., Khadivi, M. A., and Rostami, E. 2013. Novel thin film composite membrane fabricated by mixed matrix nanoclay/chitosan on PVDF microfiltration support: Preparation, characterization and performance in dye removal. *J. Memb. Sci.* 436: 97–108.

Del Valle, E. M. M. 2004. Cyclodextrins and their uses: A review. *Process Biochem.* 39: 1033–1046.

de Melo, J. C. P., da Silva Filho, E. C., Santana, S. A. A., and Airoldi, C. 2009. Maleic anhydride incorporated onto cellulose and thermodynamics of cation-exchange process at the solid/liquid interface. *Colloids Surf. A* 346: 138–145.

Dewangan, T., Tiwari, A., and Bajpai, A. K. 2010. Adsorption of Hg(II) Ions onto binary biopolymeric beads of carboxymethyl cellulose and alginate. *J. Dispersion Sci. Technol.* 31: 844–851.

Dewangan, T., Tiwari, A., and Bajpai, A. K. 2011. Removal of chromium(VI) ions by adsorption onto binary biopolymeric beads of sodium alginate and carboxymethyl cellulose. *J. Dispersion Sci. Technol.* 32: 1075–1082.

Domard, A., Rinaudo, M., and Terrassin, C. 1986. New method for the quaternization of chitosan. *Int. J. Macromol.* 8: 105–107.

Doyen, W., Leysen, R., Mortar, J., and Waes, G. 1990. New composite tubular membranes for ultrafiltration. *Desalination* 79: 163–179.

Dunweg, G., Lother, S., and Wolfgang, A. 1995. Dialysis membrane made of cellulose acetate. US Patent 5,403,485.

Dutta, P. K., Dutta, J., and Tripathi, V. S. 2004. Chitin and chitosan: Chemistry, properties and applications. *J. Sci. Ind. Res.* 63: 20–31.

Fan, L., Luo, C., Sun, M., Li, X., and Qiu, H. 2013. Highly selective adsorption of lead ions by water-dispersible magnetic chitosan/graphene oxide composites. *Colloids Surf. B* 103: 523–529.

Finken, H. 1983. Bentonite-stabilized CDA/CTA membranes: I. Improved long-term transport properties. *Desalination* 48: 207–221.

Fischer, F. G. and Dorfel, H. 1955. The polyuranic acids of brown algae. *Hoppe Seylers Z. Physiol. Chem.* 302: 186–203.

Fu, H. and Kobayashi, T. 2010. Self-assembly functionalized membranes with chitosan microsphere/polyacrylic acid layers and its application for metal ion removal. *J. Mater. Sci.* 45: 6694–6700.

Goossens, I. and Van Haute, A. 1976. The influence of mineral fillers on the membrane properties of high flux asymmetric cellulose acetate reverse osmosis membranes. *Desalination*, 18: 203–214.

Gorochovceva, N. and Makuska, R. 2004. Synthesis and study of water- soluble chitosan-O-poly(ethylene glycol) graft copolymers. *Eur. Polym. J.* 40: 685–691.

Guibal, E. 2004. Interactions of metal ions with chitosan-based sorbents: A review. *Sep. Purif. Technol.* 38: 43–74.

Guibal, E., McCarrick, P., and Tobin, J. M. 2003. Comparison of the sorption of anionic dyes on activated carbon and chitosan derivatives from dilute solutions. *Sep. Sci. Technol.* 38: 3049–3073.

Guibal, E., Sweeney, N. V. O., Vincent, T., and Tobin, J. M. 2002. Sulfur derivatives of chitosan for palladium sorption. *React. Funct. Polym.* 50: 149–163.

Guibal, E., Touraud, E., and Roussy, J. 2005. Chitosan interactions with metal ions and dyes: Dissolved-state vs. solid-state application. *World J. Microbiol. Biotechnol.* 21: 913–920.

Han, W., Liu, C., and Bai, R. 2007. A novel method to prepare high chitosan content blend hollow fiber membranes using a non-acidic dope solvent for highly enhanced adsorptive performance. *J. Membr. Sci.* 302: 150–159.

Hashino, M., Hirami, K., Katagiri, T., Kubota, N., Ohmukai, Y., Ishigami, T., Maruyama, T., and Matsuyama, H. 2011. Effects of three natural organic matter types on cellulose acetate butyrate microfiltration membrane fouling. *J. Memb. Sci.* 379: 233–238.

Haung, A., Larean, B., and Smolder, O. 1966. A study of construction of alginic acid by partial acid hydrolysis. *Acta Chem. Scand.* 20: 183–190.

Heinze, T. and Liebert, T. 2001. Unconventional methods in cellulose functionalization. *Prog. Polym. Sci.* 26: 1689–1762.

Heras, A., Rodríguez, N., Ramos, V., and Agulló, E. 2001. N-Methylene phosphonic chitosan: A novel soluble derivative. *Carbohydr. Polym.* 44: 1–8.

Holme, K. R. and Perkin, A. S. 1997. Chitosan N-sulfate: A water-soluble polyelectrolyte. *Carbohydr. Res.* 302: 7–12.

Huang, R., Chen, G., Sun, M., and Gao, C. 2006. A novel composite nanofiltration (NF) membrane prepared from graft copolymer of trimethylallyl ammonium chloride onto chitosan (GCTACC)/poly(acrylonitrile) (PAN) by epichlorohydrin cross-linking. *Carbohydr. Res.* 341: 2777–2784.

Huang, R., Chen, G., Yang, B., and Gao, C. 2008. Positively charged composite nanofiltration membrane from quaternized chitosan by toluene diisocyanate cross-linking. *Sep. Purif. Technol.* 61: 424–429.

Huang, R. Y. M., Pal, R., and Moon, G. Y. 1999. Crosslinked chitosan composite membrane for the pervaporation dehydration of alcohol mixtures and enhancement of structural stability of chitosan/polysulfone composite membranes. *J. Memb. Sci.* 160: 17–30.

Idris, A. and Yet, L. K. 2006. The effect of different molecular weight PEG additives on cellulose acetate asymmetric dialysis membrane performance. *J. Memb. Sci.* 280: 920–927.

Jeon, C. and Holl, W. H. 2003. Chemical modification of chitosan and equilibrium study for mercury ion removal. *Water Res.* 37: 4770–4780.

Jin, H., An, Q., Zhao, Q., Qian, J., and Zhu, M. 2010. Pervaporation dehydration of ethanol by using polyelectrolyte complex membranes based on poly (N-ethyl-4-vinylpyridinium bromide) and sodium carboxymethyl cellulose. *J. Memb. Sci.* 347: 183–192.

Jiraratananon, R., Chanachai, A., Huang, R. Y. M., and Uttapap, D. 2002. Pervaporation dehydration of ethanol–water mixtures with chitosan/hydroxyethylcellulose (CS/HEC) composite membranes I. Effect of operating conditions. *J. Memb. Sci.* 195: 143–151.

Juang, R.-S. and Chiou, C.-H. 2000a. Metal removal from aqueous solutions using chitosan-enhanced membrane filtration. *J. Memb. Sci.* 165: 159–167.

Juang, R.-S. and Chiou, C.-H. 2000b. Ultrafiltration rejection of dissolved ions using various weakly basic water-soluble polymers. *J. Memb. Sci.* 177: 207–214.

Kamel, S., Hassan, E. M., and El-Sakhawy, M. 2006. Preparation and application of acrylonitrile-grafted cyanoethyl cellulose for the removal of copper (II) ions. *J. Appl. Polym. Sci.* 100: 329–334.

Kaminski, W. and Modrzejewska, Z. 1997. Application of chitosan membranes in separation of heavy metal ions. *Sep. Sci. Technol.* 32: 2659–2668.

Kanti, P., Srigowri, K., Madhuri, J., Smitha, B., and Sridhar, S. 2004. Dehydration of ethanol through blend membranes of chitosan and sodium alginate by pervaporation. *Sep. Purif. Technol.* 40: 259–266.

Ke, G. 2009. Homogeneous modification of carbon nanotubes with cellulose acetate. *Chin. Chem. Lett.* 20: 1376–1380.

Krajewska, B. 2001. React. Diffusion of metal ions through gel chitosan membrane. *React. Funct. Polym.* 47: 37–47.

Krajewska, B. 2005. Membrane-based processes performed with use of chitin/chitosan materials. *Sep. Purif. Technol.* 41: 305–312.

Kulterer, M. R., Reischl, M., Reichel, V. E., Hribernik, S., Wu, M., Köstler, S., Kargl, R., and Ribitsch, V. 2011. Nanoprecipitation of cellulose acetate using solvent/nonsolvent mixtures as dispersive media. *Colloids Surf. A* 375: 23–29.

Kumar, M., Tripathi, B. P., and Shahi, V. K. 2009. Crosslinked chitosan/polyvinyl alcohol blend beads for removal and recovery of Cd(II) from wastewater. *J. Hazard. Mater.* 172: 1041–1048.

Kumar, R., Isloor, A. M., Ismail, A. F., and Matsuura, T. 2013a. Performance improvement of polysulfone ultrafiltration membrane using N-succinyl chitosan as additive. *Desalination* 318: 1–8.

Kumar, R., Isloor, A. M., Ismail, A. F., and Matsuura, T. 2013b. Synthesis and characterization of novel water soluble derivative of Chitosan as an additive for polysulfone ultrafiltration membrane. *J. Memb Sci.* 440: 140–147.

Kuncoro, E. K., Lehtonen, T., Roussy, J., and Guibal, E. 2003. Mercury removal by polymer-enhanced ultrafiltration using chitosan as the macroligand, *Proceedings of the International Biohydrometallurgy Symposium*, Tsezos, M., Hatzikioseyian, A., and Remoudaki, E. (Eds.), National Technical University of Athens Part I, Athens, Greece, pp. 621–630.

Kuncoro, E. P., Roussy, J., and Guibal, E. 2005. Mercury recovery by polymer-enhanced ultrafiltration: Comparison of chitosan and Poly(ethylenimine) used as macroligand. *Sep. Sci. Technol.* 40: 659–684.

Kurita, K. 2001. Controlled functionalization of the polysaccharide chitin. *Prog. Polym. Sci.* 26: 1921–1971.

Kurita, K., Hirakawa, M., Kikuchi, S., Yamanaka, H., and Yang, J. 2004. Trimethylsilylation of chitosan and some properties of the product. *Carbohydr. Polym.* 56: 333–337.

Kweon, D. K., Song, S. B., and Park, Y. Y. 2003. Preparation of water-soluble chitosan/heparin complex and its application as wound healing accelerator. *Biomaterials* 24: 1595–1601.

Lalvani, S. B., Hübner, A., and Wiltowski, T. S. 2000. Chromium adsorption by lignin. *Energy Sources* 22: 45–56.

Lalvani, S. B., Wiltowski, T. S., Murphy, D., and Lalvani. L. S. 1997. Metal removal from process water by lignin. *Environ. Technol.* 18: 1163–1168.

Largura, M. C. T., Debrassi, A., Santos, H. H., Marques, A. T., and Rodrigues, C. A. 2010. Adsorption of rhodamine B onto O-carboxymethylchitosan-N-lauryl. *Sep. Sci. Technol.* 45: 1490–1498.

Lee, Y. M., Nam, S. Y., and Woo, D. J. 1997. Pervaporation of ionically surface crosslinked chitosan composite membranes for water-alcohol mixtures. *J. Memb. Sci.* 133: 103–110.

Li, Y. F., Jia, H. P., Cheng, Q. L., Pan, F. S., and Jiang, Z. Y. 2011. Sodium alginate–gelatin polyelectrolyte complex membranes with both high water vapor permeation and high permselectivity. *J. Memb. Sci.* 375: 304–312.

Liesiene, J. 2010. Synthesis of water-soluble cationic cellulose derivatives with tertiary amino groups. *Cellulose* 17: 167–172.

Lin, L., Kong, Y., and Zhang, Y. 2008. Sorption and transport behavior of gasoline components in polyethylene glycol membranes. *J. Memb. Sci.* 325: 438–445.

Liu, Y. L., Hsu, C. H., Su, Y. H., and Lai, J. Y. 2005a. Chitosan-silica complex membranes from sulfonic acid functionalized silica nanoparticles for pervaporation dehydration of ethanol-water solutions. *Biomacromolecules*, 6: 368–373.

Liu, Y. L., Su, Y. H., Lee, K. R., and Lai, J. Y. 2005b. Crosslinked organic–inorganic hybrid chitosan membranes for pervaporation dehydration of isopropanol–water mixtures with a long-term stability. *J. Membr. Sci.* 251: 233–238.

Liu, Y. L., Yu, C. H., Ma, L. C., Lin, G. C., Tsai, H. A., and Lai, J. Y. 2008. The effects of surface modifications on preparation and pervaporation dehydration performance of chitosan/polysulfone composite hollow-fiber membranes. *J. Memb. Sci.* 311: 243–250.

Lue, S. J., Juang, H. J., and Hou, S. Y. 2002. Permeation of xylene isomers through supported liquid membranes containing cyclodextrins. *Sep. Sci. Technol.* 37: 463–480.

Ma, M. and Li, D. 1999. New organic nanoporous polymers and their inclusion complexes. *Chem. Mater.* 11: 872–876.

Maekawa, E. and Koshijima, T., 1990. Preparation and characterisation of hydroxamic acid derivatives and its metal complexes derived from cellulose. *J. Appl. Polym. Sci.* 40: 1601–1613.

Mahendran, R., Malaisamy, R., and Mohan, D. 2004. Preparation, characterization and effect of annealing on performance of cellulose acetate/sulfonated polysulfone and cellulose acetate/epoxy resin blend ultrafiltration membranes. *Eur. Polym. J.* 40: 623–633.

Mahmoodi, N. M. 2013. Magnetic ferrite nanoparticle–alginate composite: Synthesis, characterization and binary system dye removal. *J. Taiwan Inst. Chem. Eng.* 44: 322–330.

Marchetti, M., Clement, A., Loubinoux, B., and Gerardin, P., 2000. Decontamination of synthetic solutions containing heavy metals using chemically modified sawdusts bearing polyacrylic acid chains. *J. Wood Sci.* 46: 331–333.

Martel, B., Devassine, M., Crini, G., Weltrowski, M., Bourdonneau, M., and Morcellet, M. 2001. Preparation and sorption properties of a b-cyclodextrin-linked chitosan derivative. *J. Appl. Polym. Sci.* 39: 169–176.

Miao, J., Chen, G., and Gao, C. 2005. A novel kind of amphoteric composite nanofiltration membrane prepared from sulfated chitosan (SCS). *Desalination* 181: 173–183.

Miao, J., Chen, G., Gao, C., and Dong, S. 2008. Preparation and characterization of N,O-carboxymethyl chitosan/polysulfone composite nanofiltration membrane crosslinked with epichlorohydrin. *Desalination* 233: 147–156.

Miao, J., Chen, G., Gao, C., Lin, C., Wang, D., and Sun, M. 2006. Preparation and characterization of N,O-carboxymethyl chitosan (NOCC)/polysulfone (PS) composite nanofiltration membranes. *J. Membr. Sci.* 280: 478–484.

Miao, J., Chen, G., Li, L., and Dong, S. 2007. Formation and characterization of carboxymethyl cellulose sodium (CMC-Na)/poly (vinylidene fluoride) (PVDF) composite nanofiltration membranes. *Sep. Sci. Technol.* 42: 3085–3099.

Miao, J., Zhang, L., and Lin, H. 2013. A novel kind of thin film composite nanofiltration membrane with sulfated chitosan as the active layer material. *Chem. Eng. Sci.* 87: 152–159.

Moon, G. Y., Pal, R., and Huang, R. Y. M. 1999. Novel two ply composite membranes of chitosan and sodium alginate for the pervaporation dehydration of isopropanol and ethanol. *J. Memb. Sci.* 156: 17–27.

Mundkur, S. D. and Watters, J. C. 1993. Polyelectrolyte-enhanced ultrafiltration of copper from a waste stream. *Sep. Sci. Technol.* 28: 1157–1168.

Musale, D. A., Kumar, A., and Pleizier, G. 1999. Formation and characterization of poly(acrylonitrile)/chitosan composite ultrafiltration membranes. *J. Memb. Sci.* 154: 163–173.

Musyoka, S. M., Ngila, J. C., Moodley, B., Petrik, L., and Kindness, A. 2011. Synthesis, characterization, and adsorption kinetic studies of ethylenediamine modified cellulose for removal of Cd and Pb. *Anal. Lett.* 44: 1925–1936.

Muzzarelli, R. A. A., Tanfani, F., and Emanuelli, M. 1984. Chelating derivatives of chitosan obtained by reaction with ascorbic acid. *Carbohydr. Polym.* 4: 137–151.

Muzzarelli, R. A. A., Tanfani, F., Emanuelli, M., and Mariotti, S. 1982. N-(Carboxymethylidene) chitosans and N-(carboxymethyl)chitosans: Novel chelating polyampholytes obtained from chitosan glyoxylate. *Carbohydr. Res.* 107: 199–214.

Nagib, S., Inoue, K., Yamaguchi, T., and Tamaru, T. 1999. Recovery of Ni from a large excess of Al generated from spent hydrodesulfurization catalyst using picolylamine type chelating resin and complexane types of chemically modified chitosan. *Hydrometallurgy* 51: 73–85.

Naidu, V. B., Rao, K. S. V. K., and Aminabhavi, T. M. 2005. Pervaporation separation of water + 1,4-dioxane and water + tetrahydrofuran mixtures using sodium alginate and its blend membranes with hydroxyethylcellulose—A comparative study. *J. Memb. Sci.* 260: 131–141.

Nakanishi, S. C., Gonçalvesa, A. R., Rocha, G., Ballinas, M. L., and Gonzalez, G. 2011. Obtaining polymeric composite membranes from lignocellulosic components of sugarcane bagasse for use in wastewater treatment. *Desalination Water Treat.* 27: 66–71.

Nawawi, M. G. M. and Huang, R. Y. M. 1997. Pervaporation dehydration of isopropanol with chitosan membranes. *J. Memb. Sci.* 124: 53–62.

Nesic, A. R., Velickovic, S. J., and Antonovic, D. G. 2012. Characterization of chitosan/montmorillonite membranes as adsorbents for bezactiv orange V-3R dye. *J. Hazard. Mat.* 209–210: 256–263.

Ngah, W. S. and Fatinathan, S. 2008. Adsorption of Cu(II) ions in aqueous solution using chitosan beads, chitosan–GLA beads and chitosan–alginate beads. *Chem. Eng. J.* 143: 62–72.

Ngah, W. S., and Fatinathan, S. 2010. Adsorption characterization of Pb(II) and Cu(II) ions onto chitosan-tripolyphosphate beads: Kinetic, equilibrium and thermodynamic studies. *J. Environ. Manage.* 91: 958–969.

Ngah, W. S., Ghani, S. A., and Kamari, A. 2005. Adsorption behaviour of Fe(II) and Fe(III) ions in aqueous solution on chitosan and cross-linked chitosan beads. *Bioresour. Technol.* 96: 443–450.

Ngah, W. S., Teong, L. C., and Hanafiah, M. A. K. M. 2011. Adsorption of dyes and heavy metal ions by chitosan composites: A review. *Carbohydr. Polym.* 83: 1446–1456.

Nishi, N., Maekita, Y., Nishimura, S., Hasegawa, O., and Tokura, S. 1987. Highly phosphorylated derivatives of chitin, partially deacetylated chitin and chitosan as new functional polymers: Metal binding property of the insolubilized materials. *Int. J. Biol. Macromol.* 9: 109–114.

Norkus, E. 2009. Metal ion complexes with native cyclodextrins. An overview. *J. Incl. Phenom. Macrocycl. Chem.* 65: 237–248.

O'Connell, D. W., Birkinshaw, C., and O'Dwyer, T. F. 2008. Heavy metal adsorbents prepared from the modification of cellulose: A review. *Bioresour. Technol.* 99: 6709–6724.

Padaki, M., Isloor, A. M., and Wanichapichart, P. 2011. Polysulfone/N-phthaloylchitosan novel composite membranes for salt rejection application. *Desalination* 279: 409–414.

Pahlman, J. E. and Khalafalla, S. E. 1988. Use of lignochemicals and humic acids to remove heavy metals from process waste streams. Bureau of Mines, U. S. Department of the Interior, Report RI 9200.

Patil, M. B., Veerapur, R. S., Bhat, S. D., Madhusoodana, C. D., and Aminabhavi, T. M. 2009. Hybrid composite membranes of sodium alginate for pervaporation dehydration of 1,4-dioxane and tetrahydrofuran. *Desalination Water Treat.* 3: 11–20.

Peng, C., Wang, Y., and Tang, Y. 1998. Synthesis of crosslinked chitosan-crown ethers and evaluation of these products as adsorbents for metal ions. *J. Appl. Polym. Sci.* 70: 501–506.

Peng, F. B., Lu, L. Y., and Hu, C. L. 2005. Significant increase of permeation flux and selectivity of poly(vinylalcohol) membranes by incorporation of crystalline flake graphite. *J. Memb. Sci.* 259: 65–73.

Platt, S. R. and Clysdale, F. M. 1985. Binding of iron by lignin in the presence of various concentrations of calcium, magnesium and zinc. *J. Food Sci.* 50: 1322–1326.

Qin, J. J., Li, Y., Lee, L. S., and Lee, H. 2003. Cellulose acetate hollow fiber ultrafiltration membranes made from CA/PVP 360K/NMP/water. *J. Memb. Sci.* 218: 173–183.

Qu, H., Kong, Y., Lv, H., Zhang, Y., Yang, J., and Shi, D. 2010. Effect of crosslinking on sorption, diffusion and pervaporation of gasoline components in hydroxyethyl cellulose membranes. *Chem. Eng. J.* 157: 60–66.

Rahimpour, A. and Madaeni, S. S. 2007. Polyethersulfone (PES)/cellulose acetate phthalate (CAP) blend ultrafiltration membranes: Preparation, morphology, performance and antifouling properties. *J. Memb. Sci.* 305: 299–312.

Rao, K. S. V. K., Naidu, B. V. K., Subha, M. C. S., Sairam, M., Mallikarjuna, N. N., and Aminabhavi, T. M. 2006. Novel carbohydrate polymeric blend membranes in pervaporation dehydration of acetic acid. *Carbohydr. Polym.* 66: 345–351.

Ravindra, R., Krovvidi, K. R., and Khan, A. A. 1998. Solubility parameter of chitin and chitosan. *Carbohydr. Polym.* 36: 121–127.

Rölling, P., Lamers, M., and Staudt, C. 2010. Cross-linked membranes based on acrylated cyclodextrins and polyethylene glycol dimethacrylates for aromatic/aliphatic separation. *J. Memb. Sci.* 362: 154–163.

Sabde, A. D., Trivedi, M. K., Ramachandran, V., Hanra, M. S., and Misra, B. M. 1997. Casting and characterization of cellulose acetate butyrate based UF membranes. *Desalination* 114: 223–232.

Sajjan, A. M., JeevanKumar, B. K., Kittur, A. A., and Kariduraganavar, M. Y. 2013. Novel approach for the development of pervaporation membranes using sodium alginate and chitosan-wrapped multiwalled carbon nanotubes for the dehydration of isopropanol. *J. Memb. Sci.* 425–426: 77–88.

Saljoughi, E., Sadrzadeh, M., and Mohammadi, T. 2009. Effect of preparation variables on morphology and pure water permeation flux through asymmetric cellulose acetate membranes. *J. Memb. Sci.* 326: 627–634.

Sankararamakrishnan, N. and Sanghi, R. 2006. Preparation and characterization of a novel xanthated chitosan. *Carbohydr. Polym.* 66: 160–167.

Schneiderman, E. and Stalcup, A. M. 2000. Cyclodextrins: A versatile tool in separation science. *J. Chromatogr. B*, 745: 83–102.

Shi, Y., Wang, X., Chen, G., Golemme, G., Zhang, S., and Drioli, E. 1998. Preparation and characterization of high-performance dehydrating pervaporation alginate membranes. *J. Appl. Polym. Sci.* 68: 959–968.

Shigemasa, Y., Usui, H., Morimoto, M., Saimoto, H., Okamoto, Y., Minami, S., and Sahiwa, H. 1999. Chemical modification of chitin and chitosan 1: Preparation of partially deacetylated chitin derivatives via a ring-opening reaction with cyclic acid anhydrides in lithium chloride/N,N-dimethylacetamide. *Carbohydr. Polym.* 39: 237–243.

Sivakumar, M., Mohan, D. R. and Rangarajan, R. 2006. Studies on cellulose acetate-polysulfone ultrafiltration membranes. II. Effect of additive concentration. *J. Membr. Sci.* 268: 208–219.

Sridhar, S., Smitha, B., Latha, U. S. M., and Ramakrishna, M. 2004. Pervaporation of 1-4 dioxane water mixtures using poly(vinylalcohol) membranes crosslinked with toluylene-2,4-diisocyanate. *J. Polym. Mater.* 21: 181–188.

Srivastava, S. K., Singh, A. K., and Sharma, A. 1994. Studies on the uptake of lead and zinc by lignin obtained from black liquor: A paper industry waste material. *Environ. Technol.* 15: 353–361.

Svang-Ariyaskul, A., Huang, R. Y. M., Douglas, P. L., Pal, R., Feng, X., Chen, P., and Liu, L. 2006. Blended chitosan and polyvinyl alcohol membranes for the pervaporation dehydration of isopropanol. *J. Memb. Sci.* 280: 815–823.

Szejtli, J. 1998. Introduction and general overview of cyclodextrin chemistry. *Chem. Rev.* 98: 1743–1753.

Tallis, E. E. 1950. The structure of alginate fibres. *J. Text. Institute Trans.* 41: T151–T158.

Tang, X. H., Tan, S. Y., and Wang, Y. T. 2002. Study of the synthesis of chitosan derivatives containing benzo-21-crown-7 and their adsorption properties for metal ions. *J Appl. Polym. Sci.* 83: 1886–91.

Tanigawa, T., Tanaka, Y., Sashiwa, H., Saimoto, H., and Shigemasa, Y. 1992. Various biological effects of chitin derivatives. In: Brine, C. J., Sandford, P. A., Zikakis, J. P. (Eds.), *Advances in Chitin and Chitosan.* New York: Elsevier Science Publishers. pp. 206–215.

Tashiro, T. and Shimura, Y. 1982. Removal of mercuric ions by systems based on cellulose derivatives. *J. Appl. Polym. Sci.* 27: 747–756.

Ting, X., ZuLei, Z., HaiQing, L., ZhengZhi, Y., Lei, L., and XiaoMing, L. 2013. Characterization of cellulose-based electrospun nanofiber membrane and its adsorptive behaviours using Cu(II), Cd(II), Pb(II) as models. *Sci. China. Chem.* 56: 567–575.

Tojima, T., Katasura, H., Nishiki, M., Nishi, N., Tokura, S., and Sakairi, N. 1999. Chitosan beads with α-cyclodextrin: Preparation and inclusion property to nitrophenolates. *Carbohydr. Polym.* 40: 17–22.

Trotta, F. and Cavalli, R. 2009. Characterization and applications of new hyper-cross-linked cyclodextrins. *Compos. Interfaces* 16: 39–48.

Trotta, F. and Tumiatti, W. 2003. Cross-linked polymers based on cyclodextrin for removing polluting agents. WO 03/085002.

Uragami, T., Masuda, T., and Miyata, T. 1994. Structure of chemically modified chitosan membranes and their characteristics of permeation and separation of aqueous ethanol solutions. *J. Memb. Sci.* 88: 243–251.

Uragami, T. and Saito, M. 1989. Studies on synthesis and permeabilities of special polymer membranes. 68. Analysis of permeation and separation characteristics and new technique for separation of aqueous alcoholic solutions through alginic acid membranes. *Sep. Sci. Technol.* 24: 541–554.

Veerapur, R. S., Gudasi, K. B., and Aminabhavi, T. M. 2007. Pervaporation dehydration of isopropanol using blend membranes of chitosan and hydroxyl propyl cellulose. *J. Memb. Sci.* 304: 102–111.

Verma, K. V. R., Swaminathan, T., and Subrahmanyam, P. V. R. 1990. Heavy metal removal with lignin. *J. Environ. Sci. Heal. A.* 25: 243–265.

Wang, H., Ugomori, T., and Wang, Y. 2000. Sorption and pervaporation properties of cross-linked membranes of poly(ethylene oxide imide) segmented copolymer to aromatic/nonaromatic hydrocarbon mixtures. *J. Polym. Sci. B: Polym. Phys.* 38: 1800–1811.

Wang, X., Du, Y., Fan, L., Liu, H., and Hu, Y. 2005. Chitosan-metal complexes as antimicrobial agent: Synthesis, characterization and Structure-activity study. *Polym. Bull.* 55: 105–113.

Wang, X. P. 2000. Modified alginate composite membranes for the dehydration of acetic acid. *J. Memb. Sci.,* 170: 71–79.

Wara, N. M., Francis, L. F., and Velamakanni, B. V. 1995. Addition of alumina to cellulose acetate membranes. *J. Memb. Sci.* 104: 43–49.

Wu, Y. S., Hisada, K., Maeda, S., Sasaki, T., and Sakurai, K. 2007. Fabrication and structural characterization of the Langmuir-Blodgett films from a new chitosan derivative containing cinnamate chromophores. *Carbohyd. Polym.* 68: 766–772.

Yadav, M., Rhee, K. Y., Jung, I. H., and Park, S. J. 2013. Eco-friendly synthesis, characterization and properties of a sodium carboxymethyl cellulose/graphene oxide nanocomposite film. *Cellulose* 20: 687–698.

Yang, D., Li, J., Jiang, Z., Lu, L., and Chen, X. 2009. Chitosan/TiO$_2$ nanocomposite pervaporation membranes for ethanol dehydration. *Chem. Eng. Sci.* 64: 3130–3137.

Yeom, C. K. and Lee, K. H. 1998. Characterization of sodium alginate and poly(vinyl alcohol) blend membranes in pervaporation separation. *J. Appl. Polym. Sci.* 67: 949–959.

Yoon, K., Kim, K., Wang, X., Fang, D., Hsiao, B. S., and Chu, B. 2006. High flux ultrafiltration membranes based on electrospun nanofibrous PAN scaffolds and chitosan coating. *Polymer* 47: 2434–2441.

Yu, S., Chen, Z., Cheng, Q., Lü, Z., Liu, M., and Gao, C. 2012a. Application of thin-film composite hollow fiber membrane to submerged nanofiltration of anionic dye aqueous solutions. *Sep. Purif. Technol.* 88: 121–129.

Yu, S., Zheng, Y., Zhou, Q., Shuai, S., Lü, Z., and Gao, C. 2012b. Facile modification of polypropylene hollow fiber microfiltration membranes for nanofiltration. *Desalination* 298: 49–58.

Zhang, C., Ping, Q., Zhang, H., and Shen, J. 2003. Synthesis and characterization of water-soluble O-succinyl-chitosan. *Eur. Polym. J.* 39: 1629–1634.

Zhang, W., Li, G., Fang, Y., and Wang, X. 2007. Maleic anhydride surface-modification of crosslinked chitosan membrane and its pervaporation performance. *J. Membr. Sci.* 295: 130–138.

Zhao, Q., Qian, J., An, Q., Gao, C., Gui, Z., and Jin, H. 2009a. Synthesis and characterization of soluble chitosan/sodium carboxymethyl cellulose polyelectrolyte complexes and the pervaporation dehydration of their homogeneous membranes. *J. Membr. Sci.* 333: 68–78.

Zhao, Q., Qian, J., An, Q., Gui, Z., Jin, H., and Yin, M. 2009b. Pervaporation dehydration of isopropanol using homogeneous polyelectrolyte complex membranes of poly(diallyldimethylammonium chloride)/sodium carboxymethyl cellulose. *J. Membr. Sci.* 329: 175–182.

Zhou, Y., Zhao, H., Bai, H., Zhang, L., and Tang, H. 2012. Papermaking effluent treatment: A new cellulose nanocrystalline/polysulfone composite membrane. *Procedia Environ. Sci.* 16: 145–151.

6 Cellulose and Its Derivatives for Membrane Separation Processes

Boor Singh Lalia, Farah E. Ahmed,
Shaheen Fatima Anis, and Raed Hashaikeh

CONTENTS

6.1 Introduction 193
6.2 Forms of Cellulose 195
    6.2.1 Nanocrystalline Cellulose 195
    6.2.2 Microfibrillated Cellulose 197
    6.2.3 Cellulose Derivatives 198
6.3 Preparation and Properties of Cellulose Membranes 199
    6.3.1 Cellulose Membranes 200
    6.3.2 Cellulose Fillers 203
6.4 Conclusions 210
References 210

6.1 INTRODUCTION

Membrane technology is widely used for producing drinking water from different sources, such as groundwater, brackish water, and seawater. This is of particular interest in arid regions where natural water resources are scarce. Membrane technology has attractive features such as continuous separation process, low energy consumption, no chemical uses, and tunable membrane properties. Due to these unique characteristics, membrane technology is also widely incorporated in other areas such as pharmaceuticals, biotechnology, food, and chemical industries. Membranes can be broadly classified into two categories: biological and synthetic membranes. Synthetic membranes can be subdivided into organic and inorganic membranes. The basic components of organic membranes are polymers and that of inorganic membranes are ceramics. This chapter focuses on the fabrication of polymeric membranes based on cellulose and its derivatives.

Polymers have dominated membrane fabrication because of their low cost, formability, flexibility, and chemical stability. There are different fabrication techniques

TABLE 6.1

Summary of the Commonly Used Polymers and Fabrication Techniques for the Preparation of Polymeric Membranes for Water-Treatment Processes

Water-Treatment Processes	Polymers Used for Membranes Fabrication	Fabrication Techniques	Average Pore Size of the Membrane
Reverse osmosis	Cellulose acetate/triacetate	Phase inversion	1Å–3Å
	Aromatic polyamide	Interfacial	
	Polypiperazine	polymerization	
	Polybenzimidazoline		
Nanofiltration	Polyamide	Interfacial	0.001–0.01 μm
	Polysulfone	polymerization	
	Polyol	Phase inversion	
	Polyphenol		
Ultrafiltration	Polyacrylonitrile	Phase inversion	0.001–0.1 μm
	Polyethersulfone		
	Polysulfone		
	Poly(phthalazineone ether sulfone ketone)		
	Poly(vinyl butyral)		
	Polyvinylidene fluoride		
Microfiltration	Polyvinylidene fluoride	Phase inversion	0.1–10 μm
	Poly(tetrafluorethylene)	Stretching	
	Polypropylene	Track-etching	
	Polyethylene		
	Polyethersulfone		
	Polyetheretherketone		
Membrane distillation	Poly(tetrafluorethylene)	Phase inversion	0.1–1 μm
	Polyvinylidene fluoride	Stretching	
		Electrospinning	

and polymers used for the preparation of polymeric membranes. Table 6.1 provides a list of polymeric materials used, corresponding fabrication techniques, and the intended membrane application. Details of fabrication techniques and structural characteristics of the materials are discussed elsewhere [1].

In addition to the intrinsic properties of the polymer, the film-forming capabilities are important. When selecting a membrane material, two main properties are targeted, namely, stability and formability. Stability reflects on the durability of the membrane and its ability to perform under harsh operation conditions. This includes flexibility, which is important for membrane installation and handling. Formability is related to the film-forming capabilities of the material with control over porosity, pore size, and pore size distribution. Material hydrophobicity is also an important factor depending on the intended application. Selection of membrane fabrication technique depends on the choice of the polymer and the desired structure of the membrane. The most commonly used techniques for the preparation of polymeric membranes include phase inversion, interfacial polymerization, stretching, track-etching, and electrospinning.

Cellulose is the basic building block of green plants as it provides structural integrity. It belongs to the polysaccharide family and consists of linear chains of D-glucose $(C_6H_{10}O_5)_n$ attached together with glycosidic bonding. These linear chains bundle together to form microfibrils. In each microfibril, the chains are strongly bonded together to form a crystalline structure; relatively weak bonding results in the formation of amorphous regions. The major polymorphs of cellulose are native or cellulose I and regenerated or cellulose II. The later polymorph could be obtained by the dissolution and regeneration of cellulose I and is the more stable form of cellulose.

Cellulose is an abundant and biodegradable natural raw material characterized by interesting properties such as amphiphilicity, chirality, broad-chemical modification capacity, and the formation of different polymorphs [2]. Even though cellulose exhibits such fascinating properties, the use of cellulose in membrane application is largely hindered by the inability of common solvents to dissolve cellulose, which is seen as a necessary step in membrane formation. This is mainly why cellulose derivatives such as cellulose acetate (CA) have been used for membrane fabrication. During the last few years, there has been a tremendous amount of research and development in further exploring the potential of cellulose. Nano forms of cellulose, particularly cellulose nanocrystals (CNs) and microfibrillated cellulose (MFC), have received considerable interest. These nano forms of cellulose have attractive mechanical properties such as high strength and stiffness as well as large surface area, which allow them to be used as fillers in membrane fabrication.

6.2 FORMS OF CELLULOSE

6.2.1 NANOCRYSTALLINE CELLULOSE

As a polymer, cellulose is a partially crystalline material. It consists of amorphous and crystalline regions. Unlike synthetic polymers, cellulose is built by nature in a way that the crystalline portion is in the form of whiskers. Battista proposed the use of hydrochloric acid to separate the amorphous and crystalline regions of cellulose in the late 1950s [3]. Subsequently, extensive research was carried out to extract the crystalline phase of cellulose from different sources using mineral acids. Acid hydrolysis is a facile route for the extraction of nanocrystalline cellulose (NCC). In this method, raw cellulose is hydrolyzed using a strong acid, such as sulfuric acid, under controlled conditions of temperature, agitation, and time. Compared to the crystalline regions, amorphous domains of cellulose are more vulnerable to strong acid attack and are hydrolyzed as a result. The crystalline regions have greater resistance to acid attack and remain intact or partially hydrolyzed (Figure 6.1).

The nature of the acid and the acid-to-cellulosic fibers ratio are also important parameters that affect the preparation of NCC [4]. The resulting suspension after the hydrolysis is diluted with water and washed with successive centrifugations. Dialysis against distilled water is then performed to remove any free acid molecules from the dispersion.

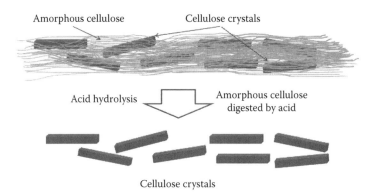

Amorphous cellulose   Cellulose crystals

Acid hydrolysis   Amorphous cellulose
digested by acid

Cellulose crystals

FIGURE 6.1   Schematic showing the extraction of NCC using acid hydrolysis.

Due to acid hydrolysis, a negative surface charge develops on NCC. This charge helps to form a stable NCC suspension in aqueous media and also assists its dispersion in polymer matrices. Surface charge was found to be high in NCC suspensions extracted using sulfuric acid as compared to other acids, namely, hydrochloric and hydrobromic acids. Araki et al. investigated a two-step method to introduce surface charge on cellulose crystals and studied its influence on the viscosity of cellulose crystal suspensions [5]. The conductometric studies revealed that surface charge on cellulose crystals obtained by hydrochloric acid hydrolysis (low surface charge acid) can be controlled by posttreatment with sulfuric acid.

Lima et al. extracted CNs or whiskers from different cellulose sources such as wood, cotton, and animal origins [6]. Nanowhiskers obtained from cotton are typically 8–10 nm long with diameters between 100 and 300 nm. On the other hand, tunicate nanowhiskers are 100 nm to a few microns in length and 10–20 nm in diameter. Above the critical concentration of nanowhiskers in water, ordered nanowhiskers showed liquid crystal properties.

Our research group extracted NCC from microcrystalline cellulose (MCC) and Kimwipe tissue paper and used it as a reinforcement for electrospun fibers [7,8]. Atomic force microscopy (AFM) images of the NCC crystals at different magnifications are shown in Figure 6.2.

Bondeson et al. optimized the extraction process of NCC from MCC by varying the concentration of MCC and sulfuric acid, hydrolysis time, temperature, and ultrasonic treatment time [9]. It was found that an acid concentration of 63.5% (w/w) sulfuric acid can be used to produce cellulose whiskers of 200–400 nm length and ~10 nm in diameter with 30% yield. Low yield of cellulose nanowhiskers has been attributed to the disintegration of amorphous cellulose and degradation of crystalline region during acid hydrolysis.

Bai et al. studied the effect of the centrifugation technique on the size distribution of NCC [10]. NCC suspension obtained from acid hydrolysis was separated into six fractions using different relative centrifugation forces for a fixed centrifugal time. It was observed that, by increasing the relative centrifugation force, the narrowest crystal size distribution can be separated from the suspension. This was concluded from the inverse relationship between the volume of cellulose whiskers and the

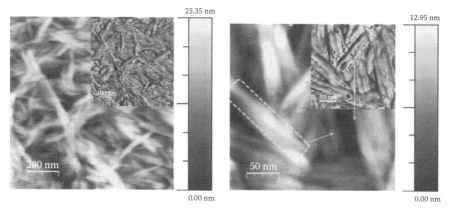

FIGURE 6.2 AFM images of NCC crystals of Kimwipe tissue paper extracted using sulfuric acid hydrolysis. (Data from Lalia, B.S. et al., *Journal of Applied Polymer Science*, **126**(S1), E442–E448, 2012.)

relative centrifugation force. Sadeghifar et al. used hydrobromic acid for the production of NCC using Whatman filter paper as starting material. They also studied the surface functionalization of NCC with a click chemistry reaction [11]. The optimized conditions used for the extraction of NCC were found to be 2.5 M HBr, with a hydrolysis time set to 3 h at 100°C. The diameter and length of NCC obtained in this work is in the range of 7–8 and 100–200 nm, respectively. NCC acts as a reinforcing filler in nanocomposites and polymers due to its high Young's modulus, which was found to be between 130 and 250 GPa [12,13]. These values of Young's modulus are close to the theoretical value 167.5 GPa calculated by Toshira et al. [14]. Sturcova et al. measured the elastic modulus of the tunicate cellulose whiskers using a Raman spectroscopic technique and found it to be 145 GPa [15]. In addition to its nanosize and chiral nematic behavior, NCC has high tensile strength and aspect ratio, which makes it an attractive material for polymeric membranes.

6.2.2 MICROFIBRILLATED CELLULOSE

Natural fibers consist of aggregated cellulose chains arranged in a hierarchical structure. These elementary fibrils are composed of cellulose chains called *cellulose macro fibrils* [16]. Figure 6.3 shows a transmission electron microscopy (TEM) image of cellulose microfibrils, or MFC [17].

MFC has both crystalline and amorphous regions and is different from NCC, which is typically a single crystal of cellulose. The extraction process of MFC involves the breaking down of the hierarchical macrostructure of cellulose either mechanically or chemically, namely, mechanical shearing or enzymatic hydrolysis [18–20]. MFC consists of crystalline and amorphous regions different from the crystal structure of native cellulose. In a review article by Siro and Plackett, different preparation and characterization techniques have been discussed in detail [21]. Mechanical disintegration for generating MFCs consists of energy-intensive processes such as grinding, mechanical homogenization at high pressures, mechanical

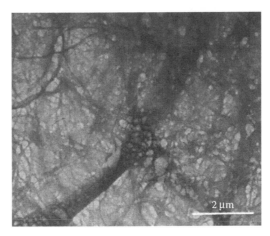

FIGURE 6.3 TEM image of cellulose microfibrils. (Data from Lu, J. et al., *Polymer*, **49**, 1285–1296, 2008.)

filtration, and cryo-crushing. Chemical processes involve the hydrolysis of natural cellulose in concentrated mineral acids under controlled conditions of temperature, agitation, and time followed by the regeneration of nano/microsized fibrils in water or any other antisolvent [22]. MFC-based materials have many potential applications in food, pharmaceuticals, cosmetics, paints, and coatings. One important characteristic of MFCs is their hydrophilic nature, which prevents their use in nonpolar media, but also makes them very promising for membrane technology. The major challenge with MFC is their tedious and costly extraction methods. The physical properties of MFCs depend on the method of formation and the source from which it is extracted. Taniguchi et al. prepared MFC from wood pulp, chitosan, cotton cellulose, and tunicin cellulose using mechanical processes, and developed translucent films using a solution-casting method [23]. The tensile strength of films obtained from tunicin cellulose showed the highest strength compared to others.

6.2.3 CELLULOSE DERIVATIVES

Cellulose forms are not soluble in most organic and aqueous solvents except in dimethyl acetamide (DMAc)/lithium chloride (LiCl) solution and a few ionic liquids. The insolubility of cellulose makes its processing challenging. On the other hand, cellulose derivatives such as CA, caboxyalkyl cellulose, and hydroxyalkyl cellulose can be dissolved in water or common organic solvents. The basic unit of cellulose is D-glucose, which contains three hydroxyl groups at C1, C3, and C4 positions and assists in the formation of cellulose derivatives under different preparation conditions. Native cellulose is indicated by its degree of polymerization (DP), that is, the average number of monomers. DP affects the mechanical, solution, and physiological properties of cellulose and helps in designing effective cellulose derivatives [24]. Abundant hydroxyl groups in cellulose lead to the formation of inter- and intra-hydrogen bonding between the linear cellulose chains; together with the amphiphilic nature of cellulose chain, aggregates are formed, which are responsible for cellulose

TABLE 6.2
Some Commercially Available Cellulose Derivatives

Cellulose Derivatives	DS Range	Solubility	Applications
Cellulose nitrate	1.5–3.0	Methanol, ethanol, and ether	Films, fibers, and explosives
Cellulose acetate	1.0–3.0	Acetone	Films, fibers, coatings, and heat- and rot-resistant fabrics
Methyl cellulose	1.5–2.4	Hot H_2O	Food additives, films, cosmetics, and greaseproof papers
Carboxymethyl cellulose	0.5–1.2	H_2O	Food additives, fibers, coatings, oil well drilling muds, paper size, paints, and detergents
Ethyl cellulose	2.3–2.6	Organic solvents	Plastics and lacquers
Hydroxyethyl cellulose	Low DS	H_2O	Films
Hydroxypropyl cellulose	1.5–2.0	H_2O	Paints
Hydroxypropylmethyl cellulose	1.5–2.0	H_2O	Paints
Cynoethyl cellulose	2.0	Organic solvents	Products with high dielectric constants, fabric with heat and rot resistance

Source: Allen, G. and J.C. Bevington, *Comprehensive Polymer Science*, Vol. 6., Pergamon, Oxford, 1986.

bundling fibers. Hydroxyl groups in the amorphous regions are readily available for chemical modification [25]. Cellulose nitrate was the first cellulose derivative discovered by Henri Braconnot in 1832. It was formed by treating cellulose with nitric acid, but it was found to be very unstable. In the late 1980s, photographic films were produced by plasticizing cellulose nitrate with camphor. The most extensively studied cellulose derivative, CA, was first discovered by Paul Schützenberger in 1865. The preparation process involved the acetylation of cellulose with acetic anhydride. The extent of acetylation and the degree of substitution (DS) determines the solubility, molecular weight, and melt properties of CA [26]. The most common applications of CA include textiles, plastics, photographic films, and surface coatings [27]. A summary of cellulose derivatives and applications are shown in Table 6.2.

Other cellulose derivatives are cellulose ethers that include methyl cellulose (MC), ethyl cellulose (EC) carboxymethyl cellulose (CMC), hydroxyethyl cellulose, and hydroxypropyl cellulose. Cellulose ethers are used in pharmaceutical, paper, adhesive, and personal-care products.

6.3 PREPARATION AND PROPERTIES OF CELLULOSE MEMBRANES

In this section, the usefulness of cellulose or cellulose derivatives as a membrane matrix and as reinforcement filler is discussed in detail.

6.3.1 CELLULOSE MEMBRANES

Cellulose and its derivatives have been widely used as a matrix for membrane applications since the early 1960s [29,30] because of the many attractive properties of cellulose that include optimal strength, flexibility, toughness, good desalting capability [31], high flux [32], and low cost [33]. Loeb and Sourirajan were the first to fabricate membranes from CA. The membrane fabrication process included casting of the solution and then subsequent heat treatment. These membranes showed promising properties as well as high water fluxes and separation efficiencies for reverse osmosis (RO) applications [34].

Since then, various materials were reported to be mixed with cellulose derivatives as matrix for diversified membrane processes [35]. Many cellulose derivatives have been exploited as membranes, but CA still remains the most extensively used. Among the various membrane separation processes, CA is widely used for pressure driven RO and ultrafiltration (UF) processes [36]. In 1966, Reid and Breton investigated the performance of RO membranes for long-term performance. They found a decline in the rejection of saltwater over a period of time with continuous feed water. The decline was attributed to the hydrolysis of the acetate membrane [37]. In 1969, Charles et al. patented CA membranes prepared in conjunction with critical acetyl content for RO applications. The casting solution was dispersed as a thin film on a casting surface, and the membrane was gelled by immersing it in cold water. This was followed by heat treatment to achieve compactness in the membrane to allow for efficient salt rejection [38]. Seawater desalination through cellulose triacetate (CTA) for RO membranes was investigated by Joshi et al. High salt-rejection rates were found when a high polymer concentration of about 13% was used [39].

In 1977, Ronald et al. patented a method of fabricating asymmetric CTA membranes for an RO application. It was found that the choice of solvent plays a critical role in the rejection capability of the membrane. CA was dissolved in dioxane and acetone and the membranes were solution cast. A high salt-rejection rate of 98.5%–99% was obtained through these membranes [40].

A phase inversion method has been used to prepare membranes from cellulose diacetate and CTA dissolved in dipropylene glycol [41]. Phase inversion has also been used to prepare asymmetric membranes from CA with polyvinylpyrrolidone (PVP) as an additive. The process includes casting a film out of the desired polymer. The polymer is usually mixed with a solvent and a nonsolvent. Low miscibility between the polymer and nonsolvent helps in initiating precipitation, whereas high miscibility between the solvent and nonsolvent is responsible for diffusional flow [36].

CA was reported to be blended with carboxymethyl CA (CMCA) for making UF membranes. These membranes were prepared by the phase inversion method. It was found that the antifouling property of the membrane was enhanced by the presence of CMCA in CA with the optimal ratio of the blend being 80/20 (CMCA/CA). Polyethylene glycol (PEG) when used as an additive decreased the contact angle and increased the water flux. In addition, a high protein-rejection rate of 86.3% was observed together with improved morphology [29]. CTA membranes with activated carbon (AC-CTA) were investigated by Rodríguez et al. for UF to remove uranium from water. The membranes for this purpose were fabricated by the

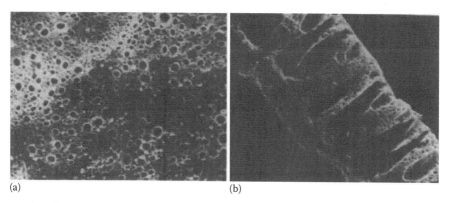

(a)   (b)

FIGURE 6.4 SEM images of membranes for 5 wt% PVP at (a) 500× (left) and (b) 750× (right). (Data from Sivakumar, M. et al., *Journal of Membrane Science*, **169**, 215–228, 2000.)

evaporation–precipitation method. Solutions of AC and CTA were prepared using different concentrations, and it was concluded that filtration was successful regardless of the different AC concentrations [42]. Membranes for UF have also been prepared from CA and polyurethane (PU) with PVP as an additive. The polymer blend of CA/PU (75/25) was cast using a casting blade. The CA/PU membranes exhibited increased flux with increasing PU. Scanning electron microscopy (SEM) images of CA/PU membranes shown in Figure 6.4 attest to the asymmetric nature of the membranes together with cavities in the structure. The membranes were effectively used for various treatments, including separation of proteins by UF [43].

Filtration of olefin/paraffin is an important process especially in the oil industry. $AgBF_4$–CA membranes were fabricated successfully through solution casting for separating mixtures of ethylene/ethane and propylene/propane [44]. It was reported that polymer solvent also has an effect on the performance of CA membranes for filtration. Acetone, dimethyl–formamide (DMF), and N-methylpyrrolidone (NMP) were investigated as solvents with CA. The membranes were solution cast and studied for the filtration of methanol with methyl tertiary butyl ether. The SEM images show their morphology after use (Figure 6.5). It was noted that for the same volume of the polymer solvent, the CA–acetone membranes were thicker. All membranes contained voids of different characteristics in as-cast condition, but the voids in the CA–acetone membranes disappeared after use. The differences in the three membranes may be attributed to the mechanism through which the membranes are formed during the casting process. This in turn relates to the physical properties of the solvents, as acetone evaporates much faster than the other two. The latter create a nonthermodynamic situation and produce an unstable membrane with voids and porosities. DMF evaporates much more slowly and produces a stable membrane. For NMP, the selectivity of the membrane is low due to plasticization. Overall, the volatility of the solvent plays a significant role in the morphology of a CA membrane [45].

Effect of the concentration of CA on the membranes cannot be ignored. It was shown that the selectivity of the CA–acetone membranes increased with increasing CA concentration. Also, the voids tend to disappear with increasing concentrations as the membranes become denser [45]. CA was also investigated for the treatment of palm oil.

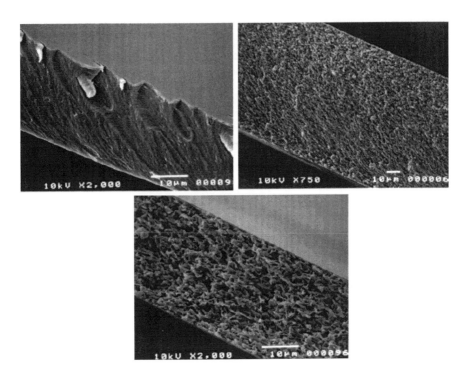

FIGURE 6.5 SEM images of CA membranes with acetone (top left), DMF (top right), and NMP (bottom) as a solvent—after use. (Data from Tabe-Mohammadi, A. et al., *Journal of Applied Polymer Science*, **82**, 2882–2895, 2001.)

The natural polymer was blended with polyethersulfone (PES) and the solution was prepared by a microwave heating technique. After microwaving the blend for 5 min, the membranes were further fabricated through the phase inversion method. The 19% CA-1 to 3% PES prepared membranes showed far more superiority than those prepared through the conventional heating technique [46]. CA has also been investigated with polysulfone (PSF) to form membranes for separating aqueous solutions of cadmium metal in humic acid. PVP was used to enhance pore formation. The thermal stability, flux, and solute rejection of the CA/PSF membranes improved with increasing the weight percent of PVP in the CA/Psf polymer blend. [47]. Regenerated cellulosic membranes have also been investigated for separating organic liquids. The cellulose solution was solution cast, coagulated in an alkaline solution, and then the cellulose was regenerated using an acidic solution. It was found that the membrane possessed higher selectivity and showed higher flux compared to that prepared using just aqueous solutions [48].

Cellulose derivatives are also used for preparing membranes for pervaporation. Membranes using homogeneous CTA membranes were reported to be prepared by the solution-casting method for separating isopropanol–aqueous mixtures. After casting, the surface of the membrane was modified with gaseous plasma [49]. Cellulose ester membranes were fabricated through solution casting for separating organic compounds such as isomeric xylenes [50]. Cellulose was incorporated with polyacrylamide for the pervaporation of water–ethanol mixtures. The membranes were fabricated by free-radical polymerization of the acrylamide. A novel interpenetrating polymer network was formed, and it was observed that the membranes were selective over various concentrations of ethanol [51].

CA has also been widely used for synthesizing dialysis membranes. PEG and CA membranes were prepared through solution casting [31] and a phase inversion method [52]. CA membranes have also been investigated for biomedical applications. Membranes were made from poly(vinyl) alcohol/polyethylene oxide and CMC for drug delivery-applications. The polymers were blended and fabricated through solvent casting and freeze drying [53]. CA membranes for hemofiltration were investigated using a novel blend of CA and 2-methacryloyloxyethyl phosphorylcholine. The polymer blend was specifically used to improve blood compatibility of CA membrane. Porous membranes were prepared by a phase inversion process. The resulting membranes showed good blood compatibility with excellent protein adsorption resistivity [54]. Bacterial cellulose membranes, which possess superior mechanical strength in comparison to native cellulose, have also been used [55]. Cellulose/chitin membranes fabricated through a solution pre-gelation method were reported by Junjie et al. The method was successful in producing denser membranes compared to that from casting. These membranes showed superior performance applicable to systems such as tissue engineering and bioseparation [56]. Table 6.3 highlights some of the membrane processes together with the cellulose derivatives used in fabrication.

6.3.2 Cellulose Fillers

As the need for low-cost and environmental-friendly technologies rises, composite membranes based on natural fibers, such as cellulose, have been the focus of many studies. Although cellulosic matrices have been widely incorporated in membrane technologies, cellulosic fillers are also emerging as promising reinforcing materials for polymer composite membranes. Cellulosic fillers are also a popular choice for modifying surface characteristics such as hydrophilicity. Composites with superior properties require good dispersion and compatibility of the filler material in the polymer matrix. Cellulosic fillers adhere well to hydrophilic matrices, but require additional treatment for compatibility in hydrophobic matrices. Cellulosic fillers can be in the form of MFC or CNs.

Recently, cellulosic fillers have been thoroughly investigated as reinforcement of polymer membranes used in barrier and separation applications. Many polymers with varying sources and properties have been used with cellulosic fillers to alter barrier characteristics of membranes. Garcia et al. [57] fabricated composite membranes by incorporating purified cellulose fibers into three different biodegradable thermoplastic polymers: poly(lactic acid) (PLA), polyhydroxybutyrate-*co*-valerate (PHBV), and polycaprolactones (PCL). Using chloroform as solvent, they added a dispersion of purified α-cellulose fibers at different concentrations to each polymer solution and applied the casting–evaporation method to prepare membranes. As each host polymer exhibited a different structure and thermal behavior, the cellulosic filler modified each material in a different way. Cellulose fibers acted as a nucleating agent to enhance crystallization in PLA, while showing the opposite behavior in PCL and

TABLE 6.3

Some Common Cellulosic Membranes Used in Different Applications

Application (Membrane Process)	Material + Cellulose Derivative	Fabrication Process
Dialysis membrane	PEG + cellulose acetate	Solution casting
		Phase inversion
Pervaporation	Cellulose triacetate	Solution casting-surface modification by gaseous plasma
	Cellulose ester	Solution casting
	Cellulose–polyacrylamide/ polyacrylic acid	Free-radical polymerization
Reverse osmosis	Dipropylene glycol + cellulose diacetate and cellulose triacetate	Phase inversion
	Cellulose acetate + acetyl	Solution casting
	Cellulose acetate + polyvinylpyrrolidone (additive)	Phase inversion
	Cellulose triacetate	Solution casting
Ultrafiltartion	Cellulose acetate + carboxymethyl cellulose acetate	Phase inversion
	Cellulose triacetate + activated carbon	Evaporation–precipitation method
	Cellulose acetate + polyurethane	Casting using a casting blade.
Filtration/treatment	Cellulose acetate + $AgBF_4$	Solution casting
	Cellulose acetate + (acetone, dimethylformamide and N-methylpyrrolidone as solvents	Solution casting
	Cellulose acetate–polyethersulfone	Microwave heating and phase inversion
	Regenerated cellulose	Solution casting
Biomedical membranes	carboxymethyl cellulose and poly(vinyl) alcohol/polyethylene oxide	Solvent casting Freeze drying
	Cellulose acetate + 2-methacryloyloxyethyl phosphorylcholine	Phase inversion
	Cellulose + chitin	Pre-gelation

PHBV. Strong fiber-matrix attachment and good dispersion of filler in the host matrix were observed in all three composites. However, at higher filler concentrations, self-association of fibers severely deteriorated membrane properties. Figure 6.6 shows the different morphologies of PHBV composites with different cellulose contents. This study demonstrated that cellulose fibers could be used to enhance barrier properties of certain biopolymers, as permeability in composite films was significantly less than that in neat polymers. Biodegradable polymer matrices combined with purified

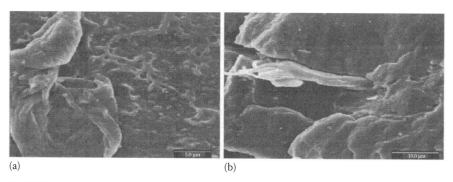

(a)   (b)

FIGURE 6.6   Cross-sectional SEM images of PHBV–cellulose composite films with (a) low filler concentration—good dispersion and phase continuity and (b) high filler concentration—phase discontinuity and reduced homogeneity. (Data from Sanchez-Garcia, M.D., *Carbohydrate Polymers*, **71**, 235–244, 2008.)

cellulose fibers offer superior barrier performance and can be used as sustainable, environmental-friendly materials for packaging and membrane applications.

Ibrahim et al. [58] investigated the effect of cellulose from rice straw or CMC as filler material on poly(vinyl alcohol) (PVA) membranes. They prepared membranes with a filler content of 12.5%. Cellulose was isolated from rice straw through alkaline pulping. The resulting cellulose fibers were washed and air dried. To obtain CMC, the cellulose powder was stirred in a sodium hydroxide solution and the resulting aqueous suspension was filtered and washed with ethanol. The filter residue was mixed with monochloroacetic acid in an aqueous NaOH bath, followed by washing and drying. PVA–cellulose composites were made by adding a cellulose solution in DMAc/LiCl solvent system to aqueous PVA solution. Composite membranes were obtained upon drying. For CMC, aqueous PVA solution was added to CMC suspension. The composite membranes had lower thermal stability than neat PVA membranes. PVA/cellulose membranes showed a rough surface, whereas PVA/CMC membranes showed a smooth surface, indicating that filler modification has an effect on membrane structure and morphology. Table 6.4 shows the effect of each filler on water sorption and salt-rejection properties of PVA membranes. When tested for desalination using a solution with 0.2% NaCl, PVA/cellulose membranes were more effective, reducing NaCl content by 25%, whereas PVA/CMC membranes decreased NaCl content by 15%. Pure PVA membranes had no effect on salt concentration. However, water flux through PVA membranes was significantly higher than both composite membranes.

CNs have also been incorporated into poly(ethylene-*co*-vinyl acetate) (EVA) to make EVA/CN composite membranes. Elanthikkal et al. [59] investigated the effect of CN on barrier properties of EVA membranes. Cellulose from banana waste was exposed to alkali treatment, bleaching, and acid hydrolysis to obtain CNs. The resulting CNs were approximately 300 nm long with an average fiber diameter of 30 nm. Tetrahydrofuran was used as solvent to dissolve EVA and to disperse aqueous suspension of CNs. Membranes were obtained by casting–evaporation method. Cellulose fibers were exposed to alkali treatment and bleaching followed by acid hydrolysis. An aqueous suspension obtained after dialysis and ultrasonication was

TABLE 6.4

Desalination Performance of PVA/Cellulose and PVA/CMC Membranes

Membrane Sample	Water Sorption (%)	Desalinity (%)
PVA	97.0	0
PVA/cellulose	82.8	25.0
PVA/CMC	61.6	15.0

Source: Ibrahim, M.M. et al., *Carbohydrate Polymers*, **95**(1), 414–420, 2013.

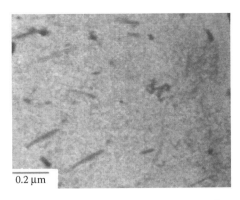

0.2 μm

FIGURE 6.7 TEM image of cellulose whiskers obtained from banana waste. (Data from Elanthikkal, S. et al., *Polymer Engineering & Science*, **52**, 2140–2146, 2012.)

freeze dried. The resulting CNs were approximately 300 nm long, with an average fiber diameter of 30 nm. Figure 6.7 shows a TEM image of the obtained CNs.

Barrier properties of the composite membranes were tested using carbon tetrachloride, chloroform, and dichloromethane as penetrants. Vapor transport through the membrane decreased when cellulose content was increased. This is explained by the greater presence of crystalline cellulose fibers that make the path more tortuous for vapors to pass through. CNs increase the permeation path of vapors and act as obstructions in the normally porous structure of pure EVA membranes, thus decreasing vapor permeability. A schematic of this behavior is shown in Figure 6.8. The effect of filler content on permeation coefficient for different vapors is shown in Figure 6.9. Barrier properties increase until 7.5% w/w cellulose loading, after which agglomeration of cellulose particles results in poor performance.

PVA is a biodegradable polymer widely studied as a membrane material for separation and packaging. However, barrier properties of PVA are limited, as it tends to swell in water. Paralikar et al. [60] used poly(acrylic acid) (PAA) as a cross-linking agent to impart moisture stability to PVA membranes. CNs were investigated as a filler for the membranes. CNs were obtained from cotton through partial hydrolysis with sulfuric acid solution. Solutions of PVA and PAA in distilled water were blended with a CN dispersion followed by casting and air drying to obtain membranes. Membranes were heat

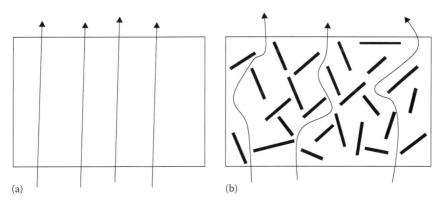

(a)   (b)

FIGURE 6.8 Illustration of permeation behavior through (a) EVA and (b) EVA–cellulose membranes. (Data from Elanthikkal, S. et al., *Polymer Engineering & Science*, **52**, 2140–2146, 2012.)

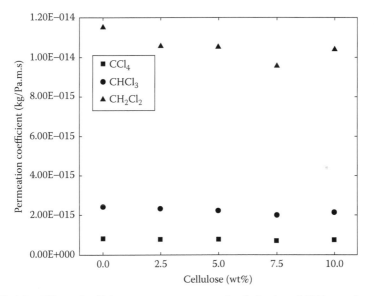

FIGURE 6.9 Effect of cellulose content on permeation behavior of EVA membranes for different permeates. (Data from Elanthikkal, S. et al., *Polymer Engineering & Science*, **52**, 2140–2146, 2012.)

treated for effective cross-linking. Optimum thermal and mechanical properties were obtained for membranes made with 80% PVA/10% PAA/10% CNs by weight. Tensile strength, modulus, and toughness in the composites showed extraordinary improvement as compared to neat PVA films, as is the case with nanoparticle-filled composites.

In the same study, CNs modified with carboxylation were also used, and although this had little effect on transport properties, better dispersion of filler in the polymer matrix was obtained. Interestingly, degradation for modified cellulose composites began at much higher temperatures than the unmodified composites, which the authors attributed to greater cross-linking in the modified composite. Transport properties through the membrane were studied for water as well as for toxic trichloroethylene (TCE). Although neat PVA membranes showed good resistance to TCE vapor transfer, CNs further reduced TCE permeation through the membrane. Hence, PVA/PAA/CN composite membranes showed improved barrier resistance to toxic chemicals.

The effect of adding CNs to biopolymers was also investigated by Saxena et al. [61]. They prepared composite films based on xylan–cellulose whiskers and xylan–softwood kraft fibers in order to evaluate the effect on water-transmission properties of CNs. Nanocrystals prepared from hydrolysis in sulfuric acid were less than 20 nm wide and had an average length of 150–200 nm. Composite films prepared with 10 wt% sulfonated cellulose whiskers demonstrated greatest reduction in water-transmission properties as compared to xylan-softwood kraft fibers and xylan alone. A strong hydrogen-bonding network of CNs caused mechanical properties to improve in the composite membrane. The CNs form an integrated matrix, improving barrier properties through the membrane.

Recently, high-flux UF membranes for dialysis were fabricated by reinforcing polysulfone with CN [62]. High-flux dialysis membranes differ from conventional membranes in that they have larger pore sizes and are more permeable to low as well as intermediate molecular weight proteins. Polysulfone is commonly used as a separation material due to its superior mechanical properties, durability as well as chemical and thermal stability. The pore size in polysulfone membranes can be adjusted to control water by adding polymers such as PEG. Polysulfone is increasingly being adapted as a medical membrane for blood purification and dialysis. Li et al. successfully fabricated polysulfone–CN composite membranes through a Loeb–Sourirajan (L–S) phase inversion method. The resulting membranes showed well-dispersed CNs in the polymer matrix. CN was obtained from wood pulp through acid hydrolysis, and the suspension was added to a polysulfone solution before casting–evaporation method to obtain the membranes. They found that CN altered the structure of the membrane, changing its properties. Addition of CN drastically improved the hydrophilicity of the membranes, allowing more efficient filtration. Optimum permeability and tensile properties were reached when filler content was 0.3 wt%, after which agglomeration caused membrane performance to decline. Previously, Noorani et al. [63] incorporated CN from cotton in a polysulfone matrix through a solvent exchange method followed by casting and soaking. Good dispersion of CN was observed for low filler loadings. Addition of CN greatly enhanced tensile strength and modulus, although mechanical properties began to decline for loadings above 7%. Water flux through the membranes also increases continuously up until 7%. Mechanical performance as well as vapor transmission was improved through this composite material based on polysulfone and CNs, confirming its potential for efficient filtration technologies.

UF composite membranes based on polyvinylidene fluoride (PVDF) and CNs were also reported by Bai et al. [64]. PVDF membranes are commonly used for UF, but their inherently hydrophobic nature promotes fouling, resulting in decreased permeability. Bai et al. used an L–S phase inversion method and investigated the effect of CN content on membrane performance. CN was obtained by acid hydrolysis and high-pressure homogenization of cellulose pulp. At a filler concentration of 0.1% w/w, water flux through the composite membranes was increased by over 47% and protein rejection was over 90%. Addition of CN resulted in higher crystallinity, significantly enhancing mechanical properties. The composite membranes showed an asymmetric morphology and had a more porous structure with larger pores than PVDF alone.

Although CNs have been incorporated into polymer matrices, yielding excellent properties due to their higher aspect ratios, the process of extracting single crystal is difficult and costly. On the other hand, MFC can be obtained through a simple process consisting of homogenizing paper pulp at high shear. MFC consists of a web-like structure with alternating crystalline and amorphous regions. Chiappone et al. [65] fabricated composite membranes by copolymerizing Bisphenol A ethoxylate dimethacrylate (BEMA) and poly(ethylene glycol) methyl ether methacrylate (PEGMA) using a fast and reliable UV-curing technique, with *in situ* addition of MFC. Resulting membranes were self-standing, flexible, and showed excellent mechanical and electrochemical stability. Ease of preparation and low fabrication costs made the resulting membranes promising to be used as gel polymer electrolytes for Li-ion batteries. Figure 6.10 shows the appearance of the composite polymer membrane with 3 wt% MFC.

Fibrous composite membranes have also been fabricated by electrospinning a mixture of CNs dispersed in a polymer solution. We have used NCC to reinforce electrospun poly(vinylidenefluoride-*co*-hexafluoropropylene) (PVDF-HFP) mats and investigated their use as separators for Li-ion batteries [7]. The effect of filler concentration on the mechanical properties was studied, and it was found that mats reinforced with 2 wt% NCC yielded optimum mechanical properties, with the tensile modulus being 75% greater than PVDF-HFP. Higher concentrations of NCC led to the deterioration of mechanical and thermal behaviors. The electrochemical performance of the nanocomposite mats was also studied, and it was found that reinforcement by NCC led to improved conductivity and storage modulus [8]. Ping et al. also fabricated fibrous membranes by electrospinning NCC-loaded PAA-ethanol mixtures [66]. CNs were obtained through acid hydrolysis of cotton. Fiber diameter decreased with rising cellulose content. CNs also contributed to fiber uniformity at low filler concentrations. It was found that the nanocrystals dispersed well in

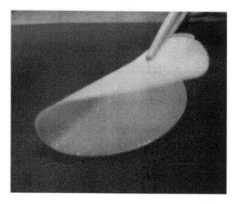

FIGURE 6.10 Appearance of BEMA/PEGMA composite polymer membrane with 3 wt% MFC. (Data from Chiappone, A. et al., *Journal of Power Sources*, **196**, 10280–10288, 2011.)

the fibers. Young's modulus and tensile strength of the composite membranes were increased by many times as compared to PAA alone. Heat treatment was also used for physical cross-linking at the matrix-filler interfaces, adding moisture stability and improved mechanical performance to the PAA-CN membrane. The electrospun PAA/cellulose nanocomposites have shown excellent mechanical properties and morphologies and can be used in advanced membrane

Ago et al. prepared bifibrous mats by electrospinning mixtures of biodegradable polymers lignin and PLA with CNs. Fiber morphology and thermal stability were studied as a function of filler concentration. Strong interaction between the lignin–PLA matrix and the CNs led to improved thermal stability of the membranes. Electrospun nanofibrous composites that use cellulosic fillers have tremendous potential for the development of novel multifunctional materials [67].

6.4   CONCLUSIONS

As the most abundant organic raw material in nature, cellulose and its unique properties have been increasingly explored for the development of sustainable systems. Cellulose has the capability to form different structures and morphologies, and it can also be modified into several derivatives. Cellulose has been the primary choice of material for membrane fabrication. Even though it lost some ground to synthetic polymers such as polyamide, recent developments in nanocellulose has opened new doors and renewed interest in cellulose for membrane fabrication. In addition, cellulosic fillers have been found to drastically improve membrane structure and performance and are, therefore, increasingly being used as reinforcements for polymeric membranes. Renewed interest in cellulose-based technologies has paved the way for multifunctional composite membranes for various applications.

REFERENCES

1. Lalia, B.S. et al., A review on membrane fabrication: Structure, properties and performance relationship. *Desalination*, 2013. **326**: 77–95.
2. Klemm, D. et al., Nanocelluloses as innovative polymers in research and application, in *Polysaccharides II*, D. Klemm, Ed., 2006, Springer, Berlin, Germany. pp. 49–96.
3. Battista, O.A., Hydrolysis and crystallization of cellulose. *Industrial & Engineering Chemistry*, 1950. **42**(3): 502–507.
4. Dong, X.M., J.-F. Revol, and D.G. Gray, Effect of microcrystallite preparation conditions on the formation of colloid crystals of cellulose. *Cellulose*, 1998. **5**(1): 19–32.
5. Araki, J. et al., Influence of surface charge on viscosity behavior of cellulose microcrystal suspension. *Journal of Wood Science*, 1999. **45**(3): 258–261.
6. de Souza Lima, M.M. and R. Borsali, Rodlike cellulose microcrystals: Structure, properties, and applications. *Macromolecular Rapid Communications*, 2004. **25**(7): 771–787.
7. Lalia, B.S., Y.A. Samad, and R. Hashaikeh, Nanocrystalline-cellulose-reinforced poly(vinylidenefluoride-co-hexafluoropropylene) nanocomposite films as a separator for lithium ion batteries. *Journal of Applied Polymer Science*, 2012. **126**(S1): E442–E448.
8. Lalia, B., Y. Samad, and R. Hashaikeh, Nanocrystalline cellulose-reinforced composite mats for lithium-ion batteries: Electrochemical and thermomechanical performance. *Journal of Solid State Electrochemistry*, 2013. **17**(3): 575–581.

9. Bondeson, D., A. Mathew, and K. Oksman, Optimization of the isolation of nanocrystals from microcrystalline cellulose by acid hydrolysis. *Cellulose*, 2006. **13**(2): 171–180.

10. Bai, W., J. Holbery, and K. Li, A technique for production of nanocrystalline cellulose with a narrow size distribution. *Cellulose*, 2009. **16**(3): 455–465.

11. Sadeghifar, H. et al., Production of cellulose nanocrystals using hydrobromic acid and click reactions on their surface. *Journal of Materials Science*, 2011. **46**(22): 7344–7355.

12. Sakurada, I., Y. Nukushina, and T. Ito, Experimental determination of the elastic modulus of crystalline regions in oriented polymers. *Journal of Polymer Science*, 1962. **57**(165): 651–660.

13. Zimmermann, T., E. Pöhler, and T. Geiger, Cellulose fibrils for polymer reinforcement. *Advanced Engineering Materials*, 2004. **6**(9): 754–761.

14. Tashiro, K. and M. Kobayashi, Theoretical evaluation of three-dimensional elastic constants of native and regenerated celluloses: Role of hydrogen bonds. *Polymer*, 1991. **32**(8): 1516–1526.

15. Šturcová, A., G.R. Davies, and S.J. Eichhorn, Elastic modulus and stress-transfer properties of tunicate cellulose whiskers. *Biomacromolecules*, 2005. **6**(2): 1055–1061.

16. Nguyen, X.T. and Z. Tan, Surface treatment with texturized microcrystalline cellulose microfibrils for improved paper and paper board. United States Patent 2006: US 7037405B2.

17. Lu, J., P. Askeland, and L.T. Drzal, Surface modification of microfibrillated cellulose for epoxy composite applications. *Polymer*, 2008. **49**(5): 1285–1296.

18. Henriksson, M. et al., An environmentally friendly method for enzyme-assisted preparation of microfibrillated cellulose (MFC) nanofibers. *European Polymer Journal*, 2007. **43**(8): 3434–3441.

19. Spence, K.L. et al., The effect of chemical composition on microfibrillar cellulose films from wood pulps: Mechanical processing and physical properties. *Bioresource Technology*, 2010. **101**(15): 5961–5968.

20. Jiang, F. and Y.-L. Hsieh, Chemically and mechanically isolated nanocellulose and their self-assembled structures. *Carbohydrate Polymers*, 2013. **95**(1): 32–40.

21. Siró, I. and D. Plackett, Microfibrillated cellulose and new nanocomposite materials: A review. *Cellulose*, 2010. **17**(3): 459–494.

22. Hashaikeh, R. and H. Abushammala, Acid mediated networked cellulose: Preparation and characterization. *Carbohydrate Polymers*, 2011. **83**(3): 1088–1094.

23. Taniguchi, T. and K. Okamura, New films produced from microfibrillated natural fibres. *Polymer International*, 1998. **47**(3): 291–294.

24. Hon, D.N.S., *Chemical Modification of Lignocellulosic Material*. 1996, Marcel Dekker, New York.

25. Coffey, D.G., D.A. Bell, and A. Henderson, *Food Polysaccharides and Their Application*. 1995, Marcel Dekker, New York.

26. Edgar, K., Organic cellulose esters, in *Encyclopedia of Polymer Science and Technology*, H.F. Mark, Ed., 2004, Wiley, New York. pp. 129–158.

27. Reverley, A., A review of cellulose derivatives and their industrial applications, in *Cellulose and Its Derivatives: Chemistry, Biochemistry and Applications* J.F. Kennedy, Ed., 1985, Ellis Horwood, Chichester. p. 211.

28. Allen, G. and J.C. Bevington, *Comprehensive Polymer Science*, Vol. 6. 1986, Pergamon, Oxford.

29. Han, B. et al., Preparation and characterization of cellulose acetate/carboxymethyl cellulose acetate blend ultrafiltration membranes. *Desalination*, 2013. **311**: 80–89.

30. Strathmann, H., P. Scheible, and R.W. Baker, A rationale for the preparation of Loeb-Sourirajan-type cellulose acetate membranes. *Journal of Applied Polymer Science*, 1971. **15**(4): 811–828.

31. Idris, A. and L.K. Yet, The effect of different molecular weight PEG additives on cellulose acetate asymmetric dialysis membrane performance. *Journal of Membrane Science*, 2006. **280**(1/2): 920–927.

32. Bokhorst, H., F. Altena, and C. Smolders, Formation of asymmetric cellulose acetate membranes. *Desalination*, 1981. **38**: 349–360.

33. Malaeb, L. and G.M. Ayoub, Reverse osmosis technology for water treatment: State of the art review. *Desalination*, 2011. **267**(1): 1–8.

34. Loeb, S. and S. Sourirajan, *Sea Water Demineralization by Means of an Osmotic Membrane.* 1962, ACS Publications, Washington, DC.

35. Kołtuniewicz, A., The history and state of arts in membrane technologies, 2006. Materials of VIII Spring Membrane School: Membrane, membrane processes and their applications. Opole-Turawa, Poland.

36. Saljoughi, E. and T. Mohammadi, Cellulose acetate (CA)/polyvinylpyrrolidone (PVP) blend asymmetric membranes: Preparation, morphology and performance. *Desalination*, 2009. **249**(2): 850–854.

37. Vos, K.D., A.P. Hatcher, and U. Merten, Lifetime of cellulose acetate reverse osmosis membranes. *I&EC Product Research and Development*, 1966. **5**(3): 211–218.

38. Cannon, C.R. Ed., Cellulose acetate membranes, U.S. Patent No. 3,460,683, August 12, 1969.

39. Joshi, S.V. and A.V. Rao, Cellulose triacetate membranes for seawater desalination. *Desalination*, 1984. **51**(3): 307–312.

40. Fox, R.L. and C.R. Mungle, Method of making asymmetric cellulose triacetate membranes, U.S. Patent No. 4,026,978. May 31, 1977.

41. Nolte, M.C.M. et al., Cellulose acetate reverse osmosis membranes made by phase inversion method: Effects of a shear treatment applied to the casting solution on the membrane structure and performance. *Separation Science and Technology*, 2011. **46**(3): 395–403.

42. Villalobos-Rodríguez, R. et al., Uranium removal from water using cellulose triacetate membranes added with activated carbon. *Applied Radiation and Isotopes*, 2012. **70**(5): 872–881.

43. Sivakumar, M. et al., Preparation and performance of cellulose acetate–polyurethane blend membranes and their applications—II. *Journal of Membrane Science*, 2000. **169**(2): 215–228.

44. Ryu, J.H., Lee, H., Kim, Y.J., Kang, Y.S., and Kim, H.S. (2001). Facilitated olefin transport by reversible olefin coordination to silver ions in a dry cellulose acetate membrane. *Chemistry–A European Journal*, 2001. **7**(7): 1525–1529.

45. Tabe-Mohammadi, A. et al., Effects of polymer solvents on the performance of cellulose acetate membranes in methanol/methyl tertiary butyl ether separation. *Journal of Applied Polymer Science*, 2001. **82**(12): 2882–2895.

46. Idris, A., I. Ahmed, and W.J. Ho, Performance of cellulose acetate-polyethersulfone blend membrane prepared using microwave heating for palm oil mill effluent treatment, *Water Science and Technology*, 2007. **56**(8): 169–177.

47. Kumari, A., G. Sarkhel, and A. Choudhury, Effect of polyvinylpyrrolidone on separation performance of cellulose acetate-polysulfone blend membranes. *Journal of Macromolecular Science, Part A*, 2013. **50**(7): 692–702.

48. Hafez, M.M. and H.W. Pauls, Method for preparing thin regenerated cellulose membranes of high flux and selectivity for organic liquids separations, 1985, Exxon Research & Engineering Co. U.S. Patent No. 4,496,456.

49. Bhat, N. and D. Wavhal, Preparation of cellulose triacetate pervaporation membrane by ammonia plasma treatment. *Journal of Applied Polymer Science*, 2000. **76**(2): 258–265.

50. Mulder, M.H.V., F. Kruitz, and C.A. Smolders, Separation of isomeric xylenes by pervaporation through cellulose ester membranes. *Journal of Membrane Science*, 1982. 349–363.

51. Buyanov, A.L. et al., Cellulose–poly(acrylamide or acrylic acid) interpenetrating polymer network membranes for the pervaporation of water–ethanol mixtures. *Journal of Applied Polymer Science*, 1998. **69**(4): 761–769.

52. Saljoughi, E., M. Amirilargani, and T. Mohammadi, Asymmetric cellulose acetate dialysis membranes: Synthesis, characterization, and performance. *Journal of Applied Polymer Science*, 2010. **116**(4): 2251–2259.

53. Agarwal, R., M.S. Alam, and B. Gupta, Polyvinyl alcohol-polyethylene oxide-carboxymethyl cellulose membranes for drug delivery. *Journal of Applied Polymer Science*, 2013. **129**(6): 3728–3736.

54. Ye, S.H. et al., Novel cellulose acetate membrane blended with phospholipid polymer for hemocompatible filtration system. *Journal of Membrane Science*, 2002. **210**(2): 411–421.

55. Sokolnicki, A.M. et al., Permeability of bacterial cellulose membranes. *Journal of Membrane Science*, 2006. **272**(1/2): 15–27.

56. Wu, J. et al., Structure and properties of cellulose/chitin blended hydrogel membranes fabricated via a solution pre-gelation technique. *Carbohydrate Polymers*, 2010. **79**(3): 677–684.

57. Sanchez-Garcia, M.D., E. Gimenez, and J.M. Lagaron, Morphology and barrier properties of solvent cast composites of thermoplastic biopolymers and purified cellulose fibers. *Carbohydrate Polymers*, 2008. **71**(2): 235–244.

58. Ibrahim, M.M. et al., Evaluation of cellulose and carboxymethyl cellulose/poly(vinyl alcohol) membranes. *Carbohydrate Polymers*, 2013. **95**(1): 414–420.

59. Elanthikkal, S. et al., Barrier properties of poly(ethylene-co-vinyl acetate)/cellulose composite membranes. *Polymer Engineering & Science*, 2012. **52**(10): 2140–2146.

60. Paralikar, S.A., J. Simonsen, and J. Lombardi, Poly(vinyl alcohol)/cellulose nanocrystal barrier membranes. *Journal of Membrane Science*, 2008. **320**(1/2): 248–258.

61. Saxena, A. and A.J. Ragauskas, Water transmission barrier properties of biodegradable films based on cellulosic whiskers and xylan. *Carbohydrate Polymers*, 2009. **78**(2): 357–360.

62. Li, S. et al., Preparation and characteristics of polysulfone dialysis composite membranes modified with nanocrystalline cellulose. *Bioresources*, 2011. **6**(2): 1670–1680.

63. Noorani, S., J. Simonsen, and S. Atre, Nano-enabled microtechnology: Polysulfone nanocomposites incorporating cellulose nanocrystals. *Cellulose*, 2007. **14**(6): 577–584.

64. Bai, H. et al., Preparation and characterization of poly(vinylidene fluoride) composite membranes blended with nano-crystalline cellulose. *Progress in Natural Science: Materials International*, 2012. **22**(3): 250–257.

65. Chiappone, A. et al., Microfibrillated cellulose as reinforcement for Li-ion battery polymer electrolytes with excellent mechanical stability. *Journal of Power Sources*, 2011. **196**(23): 10280–10288.

66. Ping, L. and H. You-Lo, Cellulose nanocrystal-filled poly(acrylic acid) nanocomposite fibrous membranes. *Nanotechnology*, 2009. **20**(41): 415–604.

67. Ago, M. et al., Lignin-based electrospun nanofibers reinforced with cellulose nanocrystals. *Biomacromolecules*, 2012. **13**(3): 918–926.

7 PVDF Hollow-Fiber Membrane Formation and Production

*Panu Sukitpaneenit, Yee Kang Ong,
and Tai-Shung Chung*

CONTENTS

7.1 Introduction 216
7.2 Fundamentals 218
    7.2.1 Basic Understanding of Hollow-Fiber Spinning 218
    7.2.2 Phase Inversion Mechanisms of PVDF 219
7.3 Key Spinning Parameters on PVDF Hollow-Fiber Formation and the
    Development of Macrovoid-Free Membrane Morphology 221
    7.3.1 Polymer Concentration 221
    7.3.2 Nonsolvent Additives in Spinning Dopes 223
    7.3.3 Internal (Bore Fluid) and External Coagulant Chemistry 224
    7.3.4 Air-Gap Distance and Take-Up Speed 228
    7.3.5 Dope Rheology 229
7.4 Emerging R&D on PVDF Hollow-Fiber Fabrication Technology 232
    7.4.1 Spinneret Design in Hollow-Fiber Fabrication 232
        7.4.1.1 Dual-Layer Spinneret 232
        7.4.1.2 Multichannel Rectangular Spinneret 233
        7.4.1.3 Multibore Spinneret 235
    7.4.2 Mixed-Matrix Hollow Fibers 236
        7.4.2.1 Single-Layer Mixed-Matrix Hollow Fibers 236
        7.4.2.2 Multilayer Mixed-Matrix Hollow Fibers 238
    7.4.3 TIPS in PVDF Hollow-Fiber Spinning 239
7.5 Conclusion and Perspectives 240
Acknowledgments 241
References 241

7.1   INTRODUCTION

The advances and breakthroughs in molecular design of membrane materials and membrane fabrication are of paramount importance to expand membrane technologies in modern separation processes. Because the first polymeric hollow-fiber membrane was patented as a separation device by Mahon in 1966 [1], research on hollow-fiber membranes has received worldwide attention from both academia and industry, and hollow-fiber membranes made from different polymeric materials have progressively penetrated into various separation processes and applications. Compared with conventional flat-sheet membranes, the hollow-fiber configuration offers several advantages due to its inherent characteristics and module design such as (1) a larger membrane area per unit volume of membrane modules, which results in a higher productivity; (2) good self-mechanical support to withstand backwashing for liquid separation; and (3) ease of handling during module fabrication and process operation [2–5]. Nowadays, hollow-fiber membranes are widely employed as the alternative to traditional separation techniques in a broad spectrum of applications related to energy, water production, environmental, and health sciences.

Poly(vinylidene fluoride) (PVDF) is one of the promising polymeric materials that has prominently emerged in membrane research and development (R&D) due to its excellent chemical and physical properties such as highly hydrophobic nature, robust mechanical strength, good thermal stability, and superior chemical resistance. To date, PVDF hollow-fiber membranes have dominated the production of modern microfiltration (MF); ultrafiltration (UF); membrane bioreactor (MBR) membranes for municipal water and wastewater treatment; and separation in food, beverage, dairy, and wine industries. In the last two decades, increasing effort has been made in the development of PVDF hollow fibers in other separation applications such as membrane contractors [6,7], membrane distillation (MD) [8–11], and pervaporation [12,13].

There are several methods for the fabrication of PVDF membranes [14,15], including sintering, stretching, track-etching, template leaching, and phase inversion. Among these, phase inversion using a nonsolvent-induced phase separation (NIPS) is the most commonly used process. Most of the commercially available membranes are prepared by NIPS, as it offers simplicity and flexibility to scale-up the membrane manufacture. In addition, hollow-fiber membranes with various morphological structures and physicochemical properties for diverse application can be attained by manipulating and optimizing the polymer dope chemistry and fabrication parameters.

Despite the fact that there has been extensive study on the preparation and characterizations of flat-sheet PVDF membranes [16–20] fabricated by NIPS, limited studies have been devoted to the fabrication and characterization of PVDF hollow-fiber membranes [21–24]. It is well accepted that the phase inversion process of hollow-fiber membranes is much more complex than that of flat-sheet membranes, and the controlling factors for the former during membrane formation are distinctly different from those for the latter. One essential difference is the dope formulation that affects the dope viscosity. Usually, a polymer dope possessing a viscosity of a few hundred centipoises ($\times 10^{-1}$ Pa·s) is sufficient for casting a flat-sheet membrane. However, the minimum dope viscosity required for spinning hollow fibers is an order of magnitude higher, that is, a few thousand centipoises ($\times 100$ Pa·s) [4,5]. Moreover, during membrane formation, the phase inversion mechanism of hollow-fiber membranes is considerably more complicated than that of flat-sheet membrane formation. Neglecting the moisture-induced phase inversion involved in the membrane fabrication, phase inversion starts from the top surface of a cast film after immersing in a coagulation bath. On the contrary, for hollow-fiber membranes produced through NIPS, phase inversion starts immediately at the lumen side of the nascent fiber upon extrusion from a spinneret by the internal coagulant (hereafter referred to as bore fluid). Subsequently, the nascent fibers contact with the external coagulant and induce phase separation at the outer surface of the hollow fibers.

Fabrication of PVDF hollow-fiber membranes with macrovoid-free structure and good separation performance is very challenging. Generally, the presence of macrovoids in hollow-fiber membranes is undesirable, because they are considered as weak mechanical regions, which possibly lead to membrane failure under high pressures or vibrational operation. It is worthy to note that, as compared to the conventional MF and UF membranes full of finger-like macrovoids two decades ago, most commercially available MF and UF membranes today have the preferred sponge-like structure without macrovoids. Therefore, significant efforts have been taken to advance the spinning sciences and technologies that aim at fabricating PVDF hollow fibers with a favorable macrovoid-free structure. Although the development of macrovoid-free hollow-fiber membranes made of various membrane materials has been reported and investigated, the majority of the aforementioned studies are focused on glassy polymers such as cellulose acetate, polyamide, and polyimide [25–27], and the concept of membrane formation may not be fully applicable to PVDF, which is a semicrystalline polymer. In fact, for glassy polymers, the precipitation mechanism during phase inversion is predominantly governed by liquid–liquid demixing, whereas both liquid–liquid demixing and solid–liquid demixing accompanying crystallization control the precipitation of semicrystalline polymers such as PVDF [28–30]. Typically, liquid–liquid demixing results in a cellular morphology with pores created from the polymer-lean phase, which are surrounded by the polymeric matrix formed by the polymer-rich phase. Conversely, solid–liquid demixing results in interlinked semicrystalline particles or a globular structure. Clearly, the formation mechanism of glassy polymer-based hollow-fiber membranes cannot simply be extended for PVDF hollow-fiber membranes.

The main objectives of this chapter are to review the science and engineering of membrane formation as well as fabrication technologies developed for PVDF hollow-fiber membranes with desirable membrane morphology and separation performance. The chapter will cover (1) fundamental and basic understanding of PVDF hollow-fiber membrane formation using NIPS; (2) key spinning parameters influencing PVDF hollow-fiber membrane formation and macrovoid suppression; and (3) recent progress on PVDF hollow-fiber membrane fabrication technology and emerging areas such as new spinneret design, mixed-matrix membranes, and thermally induced phase separation (TIPS).

7.2 FUNDAMENTALS

7.2.1 Basic Understanding of Hollow-Fiber Spinning

The hollow-fiber spinning process comprises of a large number of control parameters throughout the entire process chain of dope formulation, coagulation chemistry, and spinneret design, to the spinning parameters such as air-gap and take-up speed. Commercially available polymeric hollow-fiber membranes are usually spun from a hot spinneret with a certain air-gap distance, coupled with a moderate take-up speed [4,5]. Figure 7.1 portrays a typical hollow-fiber spinning line for the fabrication of polymeric hollow fibers through NIPS. Once the polymer dope solution is mixed homogeneously and then allowed to degas overnight, the hollow-fiber spinning process is usually conducted as follows: (1) metering the polymer solution and bore fluid by different precision pumps; (2) extruding the dope solution and bore fluid through a spinneret; (3) experiencing internal coagulation when the dope contacts with the bore fluid exiting from the spinneret; (4) encountering solvent evaporation/moisture-induced phase separation at the outer surface of nascent fibers in the air-gap region; (5) stretching of fibers by gravity and elongation tensions induced by the air-gap and take-up units; (6) completing phase inversion and the solidification of as-spun fibers in the external coagulation bath; (7) collecting as-spun fibers at the roller that controls the take-up speed of the spinning process; and (8) employing solvent exchange or additional posttreatments to remove residual solvents/additives and control membrane pore size. The enlarged region near the spinneret during the membrane formation is illustrated in Figure 7.2. The polymer dope solution experiences at least

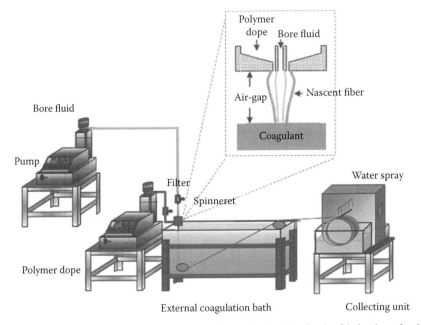

FIGURE 7.1 Schematic diagram of a hollow-fiber spinning line for the fabrication of polymeric hollow fibers through NIPS.

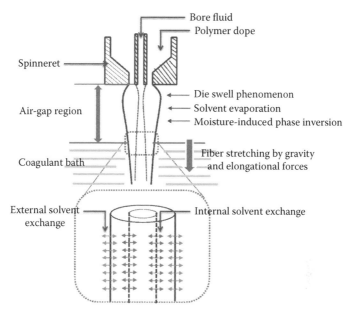

FIGURE 7.2 Schematic representation of the enlarged region near the spinneret and the formation of a nascent hollow fiber through NIPS.

three external stresses during the spinning process, which subsequently affect the morphology and performance of the as-spun fiber. These stresses are classified as follows: (1) shear and elongation stresses at the spinneret; (2) stresses induced by gravity at the air-gap; and (3) stresses induced by the take-up roller and coagulant bath.

7.2.2 PHASE INVERSION MECHANISMS OF PVDF

As a semicrystalline polymer, the phase inversion mechanism of PVDF is unique and much more complicated than that of glassy polymers such as cellulose acetate, polyamide, polysulfone, and polyimide. PVDF typically exhibits two types of demixing during the phase inversion, namely, liquid–liquid and solid–liquid accompanying crystallization [5,19,31,32], which subsequently controls the membrane morphological structure. In a rapid phase inversion process, liquid–liquid demixing primarily governs the precipitation path and results in membranes comprising of a relatively dense skin layer and a cellular microstructure with macrovoids. In contrast, solid–liquid demixing is known as crystallization dominant when a slow (delayed) phase inversion rate takes place. The resultant membranes possess an interlinked PVDF spherulitic crystal or globular structure, which may be macrovoid-free but weak in mechanical strength [33,34]. Figure 7.3 illustrates a hypothetical phase diagram of semicrystalline polymers proposed by Mulder [14]. Solid–liquid demixing usually occurs at high polymer concentrations, whereas liquid–liquid demixing preferably takes place at moderate polymer and nonsolvent concentrations. Both liquid–liquid and solid–liquid demixings coexist in the shaded area, as highlighted in Figure 7.3. Nevertheless, the actual phase diagram varies depending upon temperature, dope

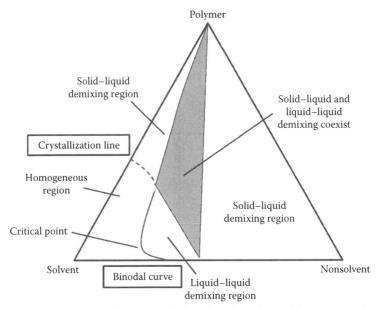

Critical point: The point at which the binodal and spinodal curves coincide

FIGURE 7.3 Hypothetic phase diagram of semicrystalline polymers. (Data from M. Mulder, *Basic Principles of Membrane Technology*, Kluwer Academic Publishers, Dordrecht, the Netherlands, 1996.)

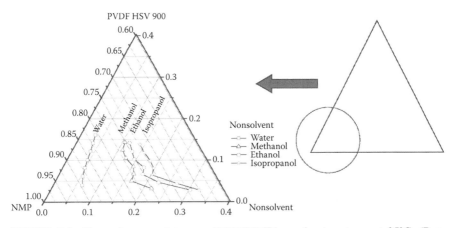

FIGURE 7.4 Phase diagram of ternary PVDF/NMP/nonsolvent systems at 25°C. (Data from P. Sukitpaneenit and T.S. Chung, *J. Memb. Sci.*, 340, 192–205, 2009.)

composition as well as solvent and nonsolvent types in the system [30,31]. Figure 7.4 depicts a typical phase diagram of PVDF/*N*-methyl-2-pyrrolidone (NMP)/nonsolvent systems at 25°C, which is experimentally constructed based on the cloud-point measurements [34]. The cloud points represent the gelation points, which are results of the synergistic effects of both liquid–liquid and solid–liquid demixings of the polymer solution [35]. As revealed in Figure 7.4, the gelation boundary of the PVDF/NMP/water system is closer to the polymer–solvent axis, which implies that only a small amount of water is required to induce the phase inversion, as compared with the other PVDF/NMP/nonsolvent alcohol systems. This result signifies that the thermodynamic stability of the PVDF/NMP/nonsolvent systems follows the consecutive order water < methanol < ethanol < isopropanol, whereby water is a strong nonsolvent and alcohols are weak nonsolvents for the PVDF/NMP system. Nevertheless, it is important to note that the phase diagram only describes the thermodynamic equilibrium between the interrelated polymer, solvent, and nonsolvent components. It does not cover the kinetics factors such as solvent and nonsolvent exchange rates and crystallization rate during phase inversion. These other factors also play vital roles during the actual demixing process [36]. In fact, it is plausible that the demixing process switches from being thermodynamically to kinetically dominant if the occurrence time is reasonably short. In summary, the final PVDF membrane structure is strongly dependent on the two governing paths of the phase inversion mechanisms, liquid–liquid and solid–liquid demixings, as well as crystallization rate and solvent exchange rate involved during the membrane formation.

7.3 KEY SPINNING PARAMETERS ON PVDF HOLLOW-FIBER FORMATION AND THE DEVELOPMENT OF MACROVOID-FREE MEMBRANE MORPHOLOGY

7.3.1 POLYMER CONCENTRATION

It is well known that polymer concentration plays an essential role in membrane formation through the phase inversion process. The effect of polymer concentration on the morphology of PVDF hollow-fiber membranes has been demonstrated in many studies [21,34,37–42]. Generally, a concept of critical polymer concentration is often employed as a guideline for a proper selection of polymer content in dope solutions [4,5]. The critical polymer concentration can be determined from the correlation of viscosity versus polymer concentration at a specific shear rate and temperature. An example of the critical polymer concentration of PVDF/NMP dope solutions is illustrated in Figure 7.5. Depending on the specific application, PVDF dopes with a polymer content below the critical concentration are usually adopted to fabricate microporous PVDF hollow fibers for water-related applications such as MF [43], UF [40], and MD [8,10]. On the other hand, dopes possessing a polymer concentration above the critical value are chosen to fabricate PVDF hollow fibers with a relatively dense selective skin for gas separation and pervaporation [12,13,39].

This differentiation of PVDF membrane application can be drawn due to the fact that polymer dopes with a higher polymer concentration generally have a higher viscosity and tend to induce polymer chain entanglement, which can effectively reduce the probability of defects in the dense selective layer [44]. Figure 7.6 illustrates the cross section and outer surface morphologies of PVDF hollow-fiber membranes spun from three different PVDF/NMP dope concentrations (15, 17, and 19 wt%) and coagulated in a water bath. At 15 wt% PVDF (Figure 7.6), the water intrusion is very severe, resulting in a cross-sectional morphology full of large finger-like macrovoids penetrating through the entire membrane wall. When the polymer content increases

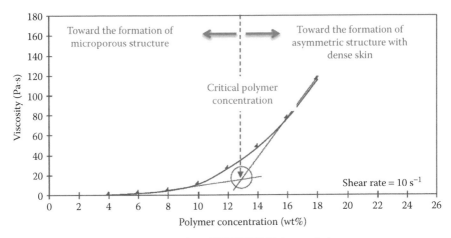

FIGURE 7.5 The critical concentration of PVDF/NMP dope solution.

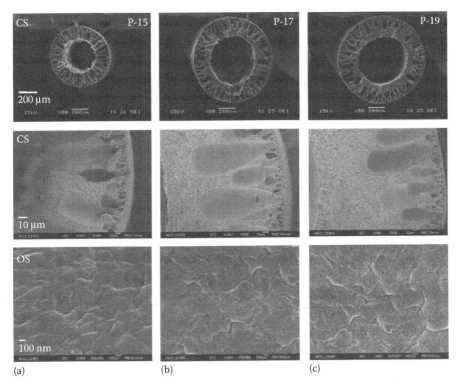

(a)   (b)   (c)

FIGURE 7.6 Cross section and outer surface morphologies of hollow-fiber membranes spun from PVDF/NMP with different polymer concentrations (wt%) (a) P-15, (b) P-17, and (c) P-19. CS, cross section; OS, outer surface. (Data from P. Sukitpaneenit and T.S. Chung, *J. Memb. Sci.*, 340, 192–205, 2009.)

from 15 to 19 wt%, the water intrusion is suppressed to a certain extent and the macrovoids near the lumen side of the fibers disappear and become a sponge-like structure. This phenomenon is mainly attributed to the greater viscoelasticity of a concentrated polymer solution that prohibits the immediate convective-type solvent exchange in an instantaneous liquid–liquid phase inversion process [45]. However, without optimizing other spinning parameters, by solely increasing the polymer concentration, this can only reduce, but cannot completely eliminate the macrovoids formation. Moreover, in practice, there are other constraints to increasing the polymer dope concentration because gelation may occur. In addition, it is difficult to degas as well as to pump the spinning dope.

7.3.2 Nonsolvent Additives in Spinning Dopes

The influence of adding a nonsolvent additive into spinning dopes on phase inversion, membrane morphology, separation performance, and macrovoid suppression have been extensively studied. The nonsolvent additives employed in the fabrication of PVDF hollow-fiber membranes can be classified into three different categories: (1) low or small molecular weight additives such as water [23,34], alcohols [23,34], ethylene glycol [11,40,46], glycerol [47–49], and inorganic salts such as lithium chloride (LiCl) [23,48–50] and lithium perchlorate [51]; (2) high molecular weight additives, often referred to as polymeric additives, such as polyvinylpyrrolidone (PVP) [43,52,53] and polyethylene glycol (PEG) [38,49,54,55]; and (3) other types of additives such as phosphorous acid [47,49] and 1,2-ethanediol [56]. In most cases, additives function as a pore former, enhance the dope viscosity, increase hydrophilicity or accelerate the phase inversion process, improve membrane morphology, flux, and selectivity as well as eliminate macrovoids.

Sukitpaneenit and Chung [34] investigated the impact of small molecular additives such as water, methanol, and ethanol in spinning dopes on membrane morphology. When water is used as the additive, the membrane structure comprises finger-like macrovoids with a dense skin surface. Compared with membranes spun without additives, the membranes spun with the incorporation of water as the additive showed a slight increase in the number of macrovoids. It is believed that the addition of water may suppress polymer interaction and favor liquid–liquid demixing, thus resulting in macrovoid formation. This finding is consistent with the results reported by Wang et al. for a PVDF/dimethylacetamide (DMAc) system [23]. When methanol and ethanol are used as the additives, the membrane structure is significantly different. In both cases, the membranes consist of packed spherulitic globule structure in the inner substrate and small voids underneath the outer surface of the fiber, resulting from the solid–liquid demixing or crystallization. The addition of inorganic LiCl salt can remarkably impact the morphology of PVDF membranes. It has been reported that LiCl additive may result in a porous structure with less or more macrovoids depending on LiCl concentration [23,57]. Fontananova et al. [57] observed that LiCl additive can induce delayed demixing (i.e., kinetic effect) and suppress macrovoids because the addition of highly concentrated LiCl can increase solution viscosity and delay the mutual diffusion between solvents and nonsolvents during the phase inversion. However, at relatively low concentrations, LiCl may behave inversely and facilitate a rapid phase inversion process due to the enhancement of thermodynamic immiscibility (i.e., thermodynamic effect), and thus result in finger-like macrovoids.

PVP and PEG are among the most widely used polymeric additives for fabricating PVDF hollow fibers with a highly porous structure. As they are highly hydrophilic in nature and well miscible with water, the existence of these polymers in polymer dopes usually enhances liquid–liquid demixing during the phase inversion, and thus enables the formation of large finger-like macrovoids. The impact of different molecular weights of PVP additives on the PVDF morphology and performance was studied by Wang et al. [21]. They reported that PVP with a low molecular weight of 10,000 or less tends to create small pores and easily leaches out from the membrane matrix, whereas most of the PVP with a high molecular weight of about 360,000 tends to remain in the membrane and may obstruct the path of the interconnected pores/voids. The fibers spun from a low molecular weight PVP have a much higher water flux and a better solute retention than those prepared from a high molecular weight PVP.

Naim et al. [49] systematically investigated and compared the influence of LiCl, glycerol, PEG-400, methanol, and phosphoric acid as additives in spinning dopes on microporous PVDF hollow-fiber membranes for CO_2 stripping through membrane contactors. They found that the hydrophobicity and gas permeability of the membranes reduced with the addition of a nonsolvent additive. However, the liquid entry pressure of the membranes was improved more due to the sponge-like structures developed near the inner layer of the fibers. Among the additives used, the introduction of PEG-400 into the dope resulted in the membranes with the highest stripping flux, which can be correlated to its high effective surface porosity and superior gas permeation. Hou et al. employed LiCl and PEG-1500 as the additives in PVDF/DMAc dope solutions to fabricate hollow-fiber membranes for MD applications [54], the coupling effects of both additives resulted in hollow-fiber membranes with a porous sponge-like structure, narrow pore size distribution, high porosity, and good hydrophobicity.

7.3.3 Internal (Bore Fluid) and External Coagulant Chemistry

During hollow-fiber spinning, there are two coagulations taking place at the internal and external surfaces of nascent fibers. One can manipulate the internal surface morphology by properly choosing the bore-fluid chemistry, flow rate, and controlling the internal coagulation process. Similarly, the outer surface morphology can be optimized by adjusting the outer coagulant chemistry and coagulation conditions. Typically, when a strong coagulant such as water is employed, a relatively dense and smooth surface is formed. On the contrary, when a weaker coagulant is used, for example, alcohols, mixtures of water with various alcohols or with various solvents, a relatively porous and rough surface is obtained. In addition, many studies have proven that both bore-fluid and external coagulants play a vital role in controlling the phase inversion mechanism, macrovoid formation, and dense layer thickness in the PVDF membranes [5,8,10,15,33,34].

Figure 7.7 depicts the effect of NMP/water bore-fluid composition on the PVDF fiber-membrane structure spun under an air-gap length of 1 cm, using water as the external coagulant and free fall condition (i.e., without take-up stretching) [13].

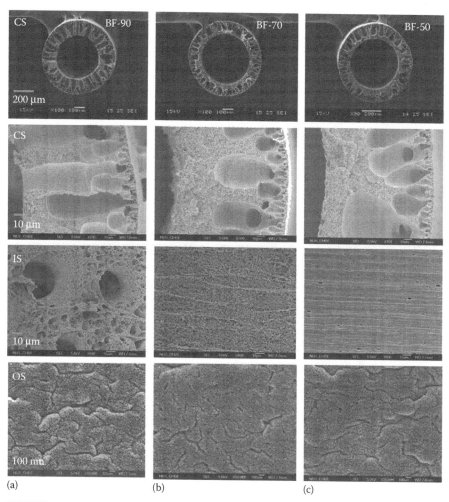

FIGURE 7.7 The membrane morphology of PVDF hollow-fiber membranes spun from different NMP/water bore-fluid compositions (a) BF-90 (90/10 wt% NMP/water), (b) BF-70 (70/30 wt% NMP/water), and (c) BF-50 (50/50 wt% NMP/water). CS, cross section; IS, inner surface; OS, outer surface. (Data from P. Sukitpaneenit and T.S. Chung, *J. Memb. Sci.*, 374, 67–82, 2011.)

Apparently, when a 90/10 wt% NMP/water mixture is used as the bore fluid, the as-spun membrane shows finger-like macrovoids across the entire cross section. Big holes are observed in the inner surface where the holes existing in the inner surface are directly connected through these finger-like macrovoids. This unique cross-sectional morphology may arise from the use of a relatively high NMP content in the bore fluid. Not only does it induce a delayed demixing that results in a very soft inner skin but also enables the rapid intrusion of bore fluid into the nascent fiber during the membrane formation. On the contrary, when 70/30 and 50/50 wt% NMP/water mixtures are employed as bore fluids, it is observed that the water intrusion is suppressed to a certain degree and the finger-like structure near the lumen side of the fiber transforms into a fully sponge-like structure. Moreover, no large holes in the inner surface are observed for both cases. Instead, a uniform and highly porous inner-skin structure is obtained for 70/30 wt% NMP/water, whereas a relatively denser structure is observed when 50/50 wt% NMP/water is utilized.

The effects of external coagulants on PVDF membrane morphology are represented in Figure 7.8 where the hollow fibers were spun from a 15 w/w% PVDF/NMP solution using 0, 10, 20, and 50 wt% ethanol/water mixtures as coagulants [34]. When water is employed as the external coagulant, the cross section of the resultant membrane comprises mainly large finger-like macrovoids with a small portion of cellular morphology. This morphological feature exemplifies that the precipitation mechanism is dominated by instantaneous liquid–liquid demixing. With the application of 10 and 20 wt% ethanol/water mixtures as the external coagulants, the size of finger-like macrovoids is reduced as compared to the spinning condition of using solely water as the external coagulant. When the ethanol content is further increased to 50 wt%, a macrovoid-free structure is attained. Additionally, it is apparent that the membrane structure is gradually transformed from an interconnected-cellular type (P-15, no ethanol content) to an interconnected-globule transition type (E-10, E-20) and finally to a globule-type structure (E-50). It has been reported that the globule-type structure consists of spherical globules made of PVDF [30,32,33]. The significant alteration in membrane morphology can be elucidated by considering the strength and amount of different coagulants that cause the delayed demixing accompanying the crystallization during the phase inversion. When pure water is utilized as the external coagulant, the demixing process occurs rapidly and there is inadequate time to induce PVDF crystallization. As a result, the liquid–liquid demixing dominates the phase separation. However, by introducing ethanol in the external coagulant, the liquid–liquid demixing process is delayed, and the crystallization process takes place, often referred to as solid–liquid demixing. Thus, through increasing ethanol content in the external coagulant, the phase separation process is eventually controlled by the solid–liquid demixing.

It is interesting to note that the addition of alcohol in the external coagulation bath intensifies the delayed demixing, which allows a greater extent of crystallization. The enhancement in delayed demixing by the nonsolvents follows the order of IPA > ethanol > methanol > water, which is consistent with the phase diagram detailed in Figure 7.4. Table 7.1 lists the solubility parameters for the PVDF, nonsolvents, and solvents [34,58,59]. The difference in solubility parameters between PVDF and the nonsolvents is in the order of (PVDF–water) > (PVDF–methanol) > (PVDF–ethanol) > (PVDF–IPA). A greater difference in solubility parameter typically implies a faster precipitation rate and a duration too short for crystallization to occur. From the kinetics viewpoint, a higher diffusion of a nonsolvent in a solvent results in a faster precipitation. The diffusion coefficients of nonsolvents, such as water and alcohols, with respect to the solvent (i.e., NMP) follow the order water > methanol > ethanol > IPA. Therefore, both thermodynamic and kinetic aspects conclude that water induces the most rapid precipitation among the nonsolvents employed for the fabrication of PVDF hollow fibers, whereas alcohols effectively delay the demixing process and finally result in the formation of spherulitic crystallites in the PVDF hollow fibers.

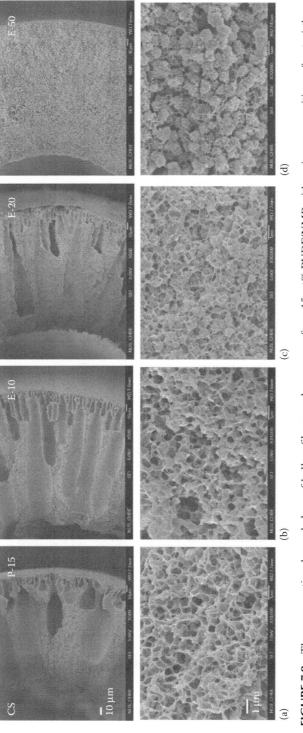

FIGURE 7.8 The cross-sectional morphology of hollow-fiber membranes spun from 15 wt% PVDF/NMP with various compositions of water/ethanol (w/w%) external coagulant (a) P-15 (water), (b) E-10 (water/ethanol 90/10), (c) E-20 (water/ethanol 80/20), and (d) E-50 (water/ethanol 50/50). (Data from P. Sukitpaneenit and T.S. Chung, *J. Memb. Sci.*, 340, 192–205, 2009.)

TABLE 7.1
Solubility Parameters of Polymers, Solvents, and Nonsolvents

Chemicals	Solubility parameters (MPa$^{1/2}$)			
	δ_d	δ_p	δ_h	δ_t
Water	15.60	16.00	42.30	47.80
Methanol	15.10	12.30	22.30	29.60
Ethanol	15.80	8.80	19.40	26.50
Isopropanol (IPA)	15.80	6.10	16.40	23.50
N-methyl-2-pyrrolidone	18.00	12.30	7.21	22.90
Poly(vinylidene fluoride)	–	–	–	14.15[a]
	–	–	–	17.75[b]
	17.20	12.50	9.20	23.20[c]

Source: C.M. Hansen. Hansen solubility parameter. In: *A User's Handbook*, CRC Press, New York, 1999.

δ_d, dispersive parameter; δ_p, polar parameter; δ_h, hydrogen bonding parameter; and δ_t, total solubility parameter.

[a] Simulated using material studio (synthia) based on the Van Krevelen equation.
[b] Simulated using material studio (synthia) based on the Fedors equation.
[c] From reference [59] based on solubility test.

7.3.4 Air-Gap Distance and Take-Up Speed

Generally, a reasonably high take-up speed is preferentially employed in the industrial-scale hollow-fiber fabrication in order to maximize the productivity and minimize the production cost. In general, the diameter of hollow fibers decreases when air-gap or take-up speed is increased. An increase in air-gap length or take-up speed usually results in a reduction in permeation flux because a high air-gap or take-up speed not only induces different precipitation paths but also facilitates chain orientation and packing owing to the gravity and elongation forces [4,5]. However, if a high air-gap or take-up speed is employed that is beyond the viscoelastic region of nascent fibers, it may also create defects and increase flux because of chain scission or chains being pulled apart.

The influences of air-gap distance and take-up speed on membrane morphology and separation performance of PVDF hollow fiber have been investigated by many researchers [8,13,21,24,42]. Wang et al. [21] prepared PVDF UF membranes spun with different air-gaps (5, 10, 15 cm). They reported that the water permeation flux decreased with increasing air-gap distance, whereas only a slight change on the rejection factor of dextran was observed. Khayet [24] conducted a systematic study on the effects of air-gap on the morphology and performance of PVDF hollow fibers spun from a dope solution containing PVDF/ethylene glycol/DMAc (23/4/73 w/w%) and a 50 v/v% aqueous ethanol solution was used as the internal and external coagulants. The air-gap length was varied from 1 to 80 cm. The researcher reported that an increase in air-gap length results in hollow-fiber membranes with a thinner and denser sponge-like structure due to a greater molecular orientation and chain packing [24]. Consistent with the general trends, their permeation flux decreased and the solute rejection increased with increasing air-gap distance. As demonstrated in the recent work by Sukitpaneenit and Chung [13], an increase in the air-gap distance or take-up speed not only helps in suppressing the formation of macrovoids in PVDF fibers but also aids in creating sponge-like structure. The reduction of macrovoids may be attributed to the rapid shrinkage process of the fiber diameter occurring during the elongation tension by its own gravity (for the case of air-gap) or the elongation stretch by the take-up unit. These stretching processes may induce a radial outflow, with negative normal stress, which hinders the capillary intrusion of coagulant diffusion, thus eliminating the chance of forming macrovoids [4,5,26,61]. Despite the fact that membranes with a macrovoid-free morphology produced in most studies typically consist of interlinked PVDF semicrystalline particles (or globular structure), which are usually the result of solid–liquid demixing, the PVDF hollow-fiber membranes developed, as described in the work of Sukitpaneenit and Chung [13], exhibit a nearly perfect macrovoid-free morphology composing of an interconnected-cellular structure. This morphology usually occurs when the phase invasion is controlled by a liquid–liquid demixing mechanism.

The trend of increasing air-gap distance or take-up speed toward promoting the liquid–liquid demixing is also supported by X-ray diffractometer characterization. The membranes spun with a higher air-gap distance or take-up speed exhibit a lower degree of crystallinity. It is interesting to note that hollow-fiber membrane that is macrovoid-free and with a cellular structure, produced from liquid–liquid demixing, is preferable, as it has a superior mechanical strength than those constructed with PVDF–globule networks [34,39]. Thus, introducing the external elongation stress through either increasing air-gap distance or take-up speed is one of the possible effective approaches to facilitate liquid–liquid demixing (suppress solid–liquid demixing) and produce a robust and mechanically stable membrane.

7.3.5 DOPE RHEOLOGY

During the formation of hollow fiber, polymer solutions experience (1) both shear and elongation stresses within the spinneret depending on its geometry and design, (2) die swell and elongation stresses at the exit of the spinneret, and (3) elongation stresses in the air-gap region. All these factors affect polymer orientation, chain packing, and structural morphology of the final hollow-fiber membrane, as demonstrated by Chung and coworkers [34,60–69]. The influences of PVDF dope rheology on membrane morphology was investigated by Ren et al. [70]. They found that increasing shear rate in the spinning process resulted in PVDF membranes with a larger mean pore size and a wider pore size distribution. In addition to the shear rate experienced within the spinneret, the elongation rate or stress in the air-gap induced by the gravity of the nascent fibers or external take-up speed is inevitable. In fact, the elongation viscosity of polymer solutions pertaining to the elongation rate must be considered. Interestingly, few studies have demonstrated that the elongation viscosity of polymer solutions affects the membrane morphology and mechanical properties to a greater extent as compared to shear viscosity. They found that polymer

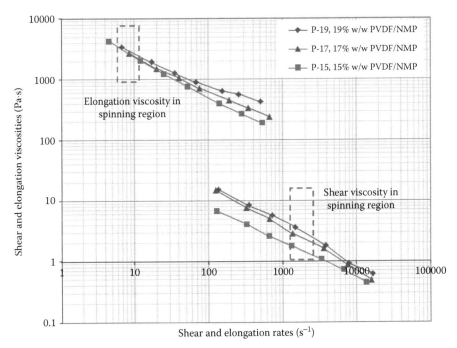

FIGURE 7.9 Shear and elongation viscosities as a function of shear and elongation rates for PVDF/NMP solutions with different polymer concentrations. (Data from P. Sukitpaneenit and T.S. Chung, *J. Memb. Sci.*, 340, 192–205, 2009.)

dopes with elongation hardening behavior may produce hollow fibers with a denser structure, lesser macrovoids, and better mechanical properties [71,72].

Figure 7.9 illustrates the shear and elongation viscosities as a function of shear and elongation rates for PVDF/NMP dope solutions with different polymer concentrations [34]. All dope solutions exhibit shear and strain–thinning behavior over the test ranges of shear and elongation rates. With increasing polymer concentration, the shear and elongation viscosities tend to increase. This is attributed to the enhanced viscoelastic properties of the dopes when the polymer concentration increases. At a higher polymer concentration, there is a greater degree of chain entanglement that creates more resistance to stretching and sliding of polymer chains, thereby increasing the shear and elongation viscosities. When the shear and elongation viscosities are high, it is more difficult for the coagulant to penetrate into the polymer matrix. Therefore, the macrovoids are suppressed, as illustrated in Figure 7.6.

Figure 7.10 illustrates the shear and elongation viscosities as a function of shear and elongation rates for PVDF/NMP dope solutions with different nonsolvent additives such as water, methanol, and ethanol [34]. The dotted rectangular areas represent the shear and elongation rates experienced during the spinning process. These results suggest that all dope solutions exhibit shear-thinning behavior regardless of the type of additives employed. The addition of nonsolvents enhances the shear viscosity of polymer dopes. This finding can be explained by considering the initial state of the polymer dope. The nonsolvent likely induces phase separation occurring

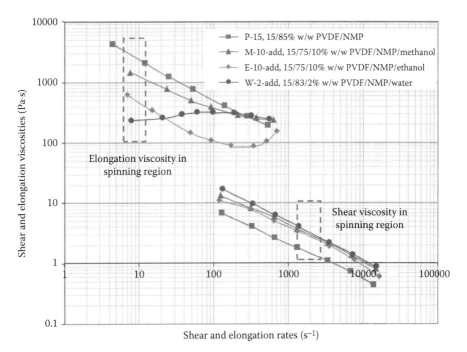

FIGURE 7.10 Shear and elongation viscosities as a function of shear and elongation rates for PVDF/NMP solutions with different nonsolvent additives. (Data from P. Sukitpaneenit and T.S. Chung, *J. Memb. Sci.*, 340, 192–205, 2009.)

within the polymer dope. It increases molecular friction among polymeric chains during shear flow and results in a high shear viscosity [35].

In contrast to the influence of nonsolvent additives on shear viscosity, the addition of nonsolvent additives deteriorates elongation viscosity, as shown in Figure 7.10. This phenomenon can be satisfactorily explained by considering the behavior of polymer chains in the mixed solvent/nonsolvent environment. Typically, polymer chains tend to be extended and entangled with neighboring chains when in contact with a good solvent. Conversely, the polymer chains tend to exist in a highly coiled conformation state with limited interchain entanglement in the presence of a poor solvent. Because the degree of chain entanglement is reduced when the polymer is in contact with a poor solvent, the elongation viscosity is lower. Figure 7.11 depicts snapshots from the molecular dynamics simulations displaying the PVDF chain conformation in respective mixed solvent/nonsolvent environments [34]. The radius of gyration (R_g) of the polymer chain and solubility parameter (δ) can be calculated. R_g represents the distance from the center of gravity of a polymer chain to the end of the chain. If the polymer chain is in a highly coiled state, the corresponding radius of gyration will be small. With reference to Figure 7.11, the radius of gyration for PVDF/NMP/methanol is greater than PVDF/NMP/ethanol and PVDF/NMP/water. This finding supports the postulation that a larger radius of gyration results in a higher elongation viscosity. The simulated solubility parameter (δ) of PVDF/mixed solvent systems is also in agreement with the aforementioned simulated results.

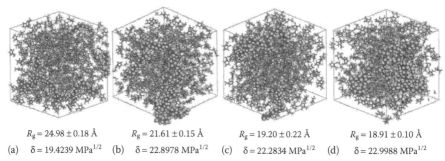

$R_g = 24.98 \pm 0.18$ Å	$R_g = 21.61 \pm 0.15$ Å	$R_g = 19.20 \pm 0.22$ Å	$R_g = 18.91 \pm 0.10$ Å
(a) $\delta = 19.4239$ MPa$^{1/2}$	(b) $\delta = 22.8978$ MPa$^{1/2}$	(c) $\delta = 22.2834$ MPa$^{1/2}$	(d) $\delta = 22.9988$ MPa$^{1/2}$

FIGURE 7.11 Snapshots from the molecular dynamic simulations showing the PVDF chain conformation in different mixed solvents environment (δ is the calculated solubility parameter in the whole system, including PVDF and solvents): (a) NMP, (b) methanol/NMP, (c) ethanol/NMP, and (d) water/NMP. (Data from P. Sukitpaneenit and T.S. Chung, *J. Memb. Sci.*, 340, 192–205, 2009.)

The solubility parameter difference is the highest for the PVDF/NMP/water system, which accounts for the highest shear viscosity. This indicates that the molecular simulation data is valid for the investigated PVDF/NMP/nonsolvent systems.

Another important observation is that all the above spinning dopes, except the one containing water as an additive, exhibit strain-thinning behavior. This may be attributed to weak molecular interactions in the conditions of chain stretching and sliding in these systems under elongation. However, for the PVDF/NMP/water system, the elongation viscosity increases with an increase in elongation rate. This may likely arise from hydrogen bonding or strong interactions between NMP and water that result in a greater resistance for chain stretching and sliding in the system.

7.4 EMERGING R&D ON PVDF HOLLOW-FIBER FABRICATION TECHNOLOGY

7.4.1 SPINNERET DESIGN IN HOLLOW-FIBER FABRICATION

7.4.1.1 Dual-Layer Spinneret

Spinneret design plays a vital role in determining the morphology and consequently the performance of the as-spun hollow fibers. The concepts on fabricating multilayer composite hollow fibers by means of single-step coextrusion of polymer solutions as well as bore fluid through a multiorifice spinneret have been extensively studied since the 1980s [61,73–78]. The dual-layer hollow fiber possesses the advantages of cost reduction as the relatively expensive functional polymer is extruded at the outer layer, whereas the inner-layer material is often chosen from inexpensive materials, which have good mechanical strength as the support. In addition, the simultaneous coextrusion approach of the aforementioned spinning process can simplify the fabrication process for composite hollow-fiber membranes by eliminating the secondary coating step during the fabrication process. This spinning technique may provide higher degrees of freedom for the design and customization of materials and

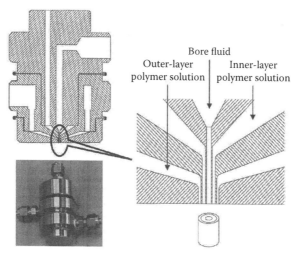

FIGURE 7.12   Schematic of the dual-layer spinneret and its flow channels for bore fluid as well as polymer solutions. (Data from D.F. Li, T.S. Chung, R. Wang, and Y. Liu, *J. Memb. Sci.*, 198, 211–223, 2002.)

morphology in both selective and supporting layers [79]. The schematic of the dual-layer spinneret is shown in Figure 7.12

Use of a dual-layer spinneret to create a highly porous outer-surface morphology on polymeric hollow fibers was reported by various groups [80,81]. In this process, the solvent was concurrently extruded at the outer-layer channel of the spinneret with the polymer solution flowing through the inner-layer channel. As a result, it creates a delay-demixing at their interface and reduces the nonsolvent intrusion into the nascent fiber during the phase inversion process. Bonyadi and Chung [81] employed the abovementioned technique to fabricate highly porous macrovoid-free PVDF hollow-fiber membranes for MD. The flux and energy efficiency of the hollow-fiber membranes were reported to enhance MD applications by two to three times by increasing the surface porosity of the membranes.

On the other hand, Ong and Chung [82] proposed an alternative single-step hollow-fiber fabrication technique; namely, immiscibility-induced phase separation process (I^2PS) to fabricate high performance dual-layer hollow fibers for the dehydration of ethanol through pervaporation. The as-spun dual-layer hollow fiber possesses an outer protective layer on top of the selective layer that shields the selective layer from the swelling phenomenon of the organic feed mixtures. PVDF was selected as one of the promising protective layer materials due to its outstanding chemical and physical properties.

7.4.1.2   Multichannel Rectangular Spinneret

The concept of multichannel rectangular membranes is a result of the aspiration to harvest and to combine the advantages of both flat-sheet and hollow-fiber membranes such as (1) enhanced mechanical strength, (2) higher membrane surface per volume ratio, and (3) ease to assemble into membrane modules. The fabrication of multichannel rectangular membranes through NIPS was reported by Chung and coworkers [83,84]. The multichannel rectangular spinneret used in the aforementioned studies consists of a rectangular slit for the polymer solution and seven injectors aligned in series for the bore fluid, as shown in Figure 7.13.

Teoh et al. [84] used the abovementioned spinneret to fabricate multichannel rectangular PVDF membranes for seawater desalination through an MD process, and the cross-sectional morphology of the membrane is portrayed in Figure 7.13. The multichannel rectangular membranes were observed to possess a grooved outer surface mainly due to the occurrence of hydrodynamic instability (Marangoni instability) and solidification-induced shrinkage coupled with bulking instability during the phase inversion process, as illustrated in Figure 7.14.

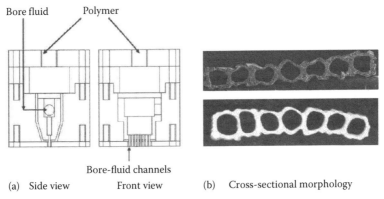

(a)   Side view   Front view   (b)   Cross-sectional morphology

FIGURE 7.13   Schematic of (a) a rectangular spinneret and (b) cross-sectional morphology of the rectangular membranes. (Data form M.M. Teoh et al., *Ind. Eng. Chem. Res.*, 50, 14046–14054, 2011.)

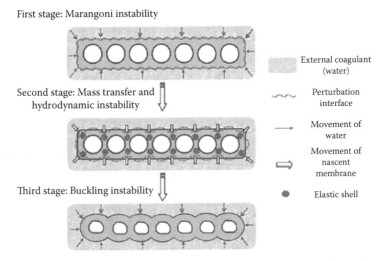

FIGURE 7.14   Proposed mechanism in forming the grooved outer surface. (Data from M.M. Teoh et al., *Ind. Eng. Chem. Res.*, 50, 14046–14054, 2011)

Compared to traditional cylindrical hollow-fiber membranes, the grooved pattern at the outer surface of rectangular membranes plays a vital role in enhancing distillate flux by promoting the formation of eddies and turbulent flows at the outer surface of the membrane. In addition, the grooved surface can function as the built-in spacers to prevent the membranes from attaching to each other. These newly spun multichannel rectangular membranes displayed a promising distilled flux of 54.7 kg m^{-2}s^{-1} [83], using a hot feed brine solution of 80°C through direct contact MD (DCMD).

7.4.1.3 Multibore Spinneret

As compared with single-bore hollow fibers, the multibore hollow fibers are capable of providing higher mechanical properties. This membrane configuration was inspired from the cross-sectional morphology of a lotus root that is porous and mechanically strong and was implemented in the fabrication of inorganic ceramic membranes [85,86] and polymeric membranes [87] for various applications. The number of bore channels in the multibore hollow fiber usually varies from 7 to 37. Recently, Wang and Chung [88] explored the fabrication of PVDF lotus-root-like multibore hollow fiber for MD. The multibore spinneret consists of seven bore needles and its schematic is portrayed in Figure 7.15. The flows of the polymer solution and bore fluid were concurrently from top and side of the spinneret to ensure the even distribution of polymer solution

It is worthwhile to note that a fiber-like network layer is observed at the inner surface of the as-spun fiber, and it is hypothesized to be caused by the high-pressure difference between the polymer solution and the bore fluid during the spinning process. In addition, the lotus-root-like structure of the multibore hollow fiber gradually transforms into a wheel structure when the fiber is subjected to take-up stretching

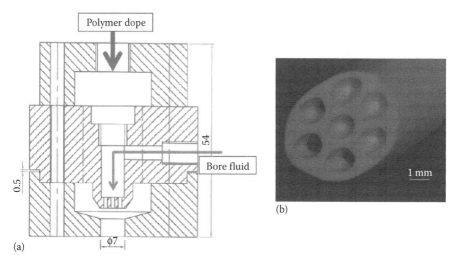

(a)

(b)

FIGURE 7.15 (a) Schematic of multibore spinneret and (b) cross-sectional morphology of the lotus-root-like multibore hollow fiber. (Data from P. Wang and T.S. Chung, *J. Memb. Sci.,* 421–422, 361–374, 2012.)

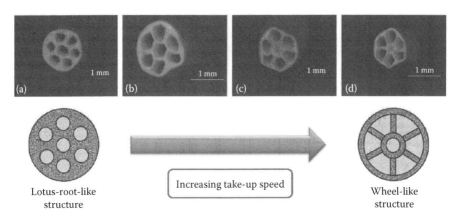

FIGURE 7.16 (a–d) Evolution of cross-sectional morphology from lotus-root-like structure to wheel-like structure as the function of take-up speed. (Data from P. Wang and T.S. Chung, *J. Memb. Sci.*, 421–422, 361–374, 2012.)

in the axial direction, as shown in Figure 7.16. The as-spun multibore hollow fibers exhibited superior mechanical strength with a slightly lower or comparable performance as compared with single-bore hollow fibers.

7.4.2 MIXED-MATRIX HOLLOW FIBERS

Researchers have been exploring the possibility to synergistically combine the advantages of organic membranes, such as low cost and ease of fabrication, and inorganic membranes, such as high performance, through the fabrication of mixed-matrix membranes. Incorporation of inorganic fillers such as alumina (Al_2O_3), calcium carbonate ($CaCO_3$), silica (SiO_2), silver nitrate ($AgNO_3$), titanium dioxide (TiO_2), and zirconium dioxide (ZrO_2) into the polymer matrix is believed to enhance the chemical and mechanical properties, as well as the performance of the composite membranes [89,90].

The distribution of fillers across the membrane is vital to determine the morphology and performance of the composite hollow fibers. Jiang et al. [91] reported that (1) the shear rate within the spinneret, (2) the die swell at the outlet of the spinneret, and (3) the elongation draw at the air-gap region were the factors affecting the filler distribution in the mixed-matrix composite hollow fibers, whereas Xiao et al. [92] observed that a high elongation draw ratio can radially displace the fillers toward the bore and shell side of the composite hollow fibers.

7.4.2.1 Single-Layer Mixed-Matrix Hollow Fibers

Among the abovementioned inorganic fillers, TiO_2 has been extensively studied due to its hydrophilic and antibacterial properties that are predominantly useful in the fabrication of antifouling membranes. Yu et al. [93] investigated the fabrication of PVDF–TiO_2 composite hollow fibers for UF through the physical blending and sol-gel methods. Incorporation of TiO_2 into the polymer dope solution has effectively suppressed macrovoids formation of the as-spun hollow fibers. In addition, the as-spun

PVDF–TiO_2 composite hollow fibers exhibit superior hydrophilicity and mechanical properties as compared with the pristine PVDF hollow fibers. The composite hollow fibers prepared through the TiO_2 sol-gel technique were observed to have better particle dispersion and interaction between the particle–polymer matrixes. Yuliwati et al. [94] observed that the addition of a small amount (1.95 wt%) of TiO_2 can greatly improve the performance of PVDF–TiO_2 composite hollow fibers. The aforementioned composite hollow fibers possess smaller pore sizes coupled with higher hydrophilicity and porosity.

On the other hand, Yu et al. [95] fabricated PVDF–SiO_2 composite hollow fibers through the sol-gel route. The β-phase crystal was found to be the prevailing crystal phase for composite hollow fibers, whereas the α-phase crystal dominated in the pristine PVDF hollow fibers. The authors attributed the abovementioned phenomena to the confinement of PVDF polymer chains in the SiO_2 inorganic network. Hashim et al. [96] investigated the posttreatment effects of PVDF–SiO_2 composite hollow fibers using both acid and alkali. Removal of SiO_2 by hydrofluoric acid produced PVDF hollow fibers with a relatively high water permeability without deteriorating their mechanical properties. Conversely, several studies have shown that the composite membranes with the correct combination of multinanoparticles can result in better performance than the composite membranes consisting of single nanoparticles [97,98]. Han et al. [99] reported that the PVDF/multinanoparticle composite hollow fibers comprising of a particle mixture of TiO_2–SiO_2–Al_2O_3 (1 wt%, respectively) and TiO_2–Al_2O_3 in a weight ratio of 2:1 exhibited the best performance in both pure water flux and bovine serum albumin rejection during UF experiments.

The application of PVDF–$CaCO_3$ composite hollow-fiber membranes in MD was explored by Hou et al. [100]. The $CaCO_3$ was treated with octadecyl dihydrogen phosphate to increase its hydrophobicity. The addition of $CaCO_3$ nanoparticles reduced the pore diameter and improved the porosity of the as-spun composite hollow fibers. Their study revealed that the surface roughness and hydrophobicity of the composite hollow fibers increase with the addition of $CaCO_3$ nanoparticles. Hydroxyapatite (HAP) [$Ca_{10}(PO_4)_6(OH)_2$] also known as a bone mineral, possesses similar chemical and structural properties as bone. Lang et al. [101] incorporated HAP into the PVDF polymer to fabricate PVDF composite hollow-fiber membranes reinforced with HAP nanocrystal whiskers. The resultant hollow-fiber membranes have shown significant improvements in Young's modulus, tensile strength, and elongation at break. In addition, the hydroxyl functional group on HAP is believed to contribute to the enhancement of permeation flux in UF experiments.

Chung and coworkers fabricated a wide array of PVDF composite hollow fibers predominantly for MD by incorporating Teflon (PTFE) [102–104] and clay [105] as the fillers. Teoh and Chung [102] reported that the incorporation of PTFE particles into the PVDF matrix can eliminate macrovoids formation in the composite hollow-fiber membranes, as represented in Figure 7.17. The presence of PTFE particles in the polymer matrix can also minimize the heat loss during the MD process. Wang et al. [105] observed that the addition of hydrophobic clay particles improves the mechanical stability of the composite hollow-fiber membranes in long-term tests.

Particles loading: 0 wt%   Particles loading: 30 wt%   Particles loading: 50 wt%

FIGURE 7.17 Cross-sectional morphology of PVDF–PTFE hollow fiber as the function of particles loading. (Data from M.M. Teoh and T.S. Chung, *Sep. Purif. Technol.*, 66, 229–236, 2009.)

7.4.2.2 Multilayer Mixed-Matrix Hollow Fibers

The dual-layer spinneret is used to fabricate multilayer mixed-matrix hollow fibers. The fabrication of multilayer mixed-matrix hollow fibers can significantly reduce the production cost of the composite membranes by only incorporating the fillers into the selective layer. Sukitpaneenit and Chung [106] fabricated PVDF/nanosilica dual-layer hollow-fiber membranes for the recovery of ethanol through pervaporation. The inorganic filler was loaded at the outer-layer polymer solution to enhance the hydrophobicity and performance of the composite hollow-fiber membranes. Figure 7.18 displays the distribution of nanosilica in the fiber cross section captured by scanning electron microscopy–X-ray energy dispersive spectrometry, which indicates a high concentration of nanosilica at the outer selective layer of the dual-layer hollow-fiber membranes.

On the other hand, the hydrophobicity or hydrophilicity of the membrane at different layers of multilayers mixed-matrix PVDF hollow fibers can be manipulated by the addition of various fillers at the respective layers. Chung and coworkers

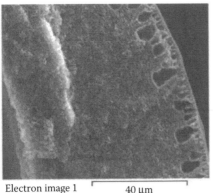 

Electron image 1 |——— 40 µm ———| Si Kα1

FIGURE 7.18 Distribution of nanosilica particles at the cross section of PVDF/nanosilica dual-layer hollow-fiber membranes. (Data from P. Sukitpaneenit and T.S. Chung, *Ind. Eng. Chem. Res.*, 51, 978–993, 2012.)

[10,107–109] reported that the incorporation of hydrophobic and hydrophilic fillers in the respective hydrophobic and hydrophilic layers of the dual-layer hollow fibers can enhance membrane performance as well as mechanical integrity for MD.

7.4.3 TIPS IN PVDF HOLLOW-FIBER SPINNING

As opposed to NIPS, the phase separation mechanism of TIPS is induced by the removal of thermal energy from high temperature homogeneous polymer solutions. TIPS is usually applied as an alternative membrane fabrication method for semicrystalline polymers or polymers that have poor solubility in common solvents [28,110]. The typical TIPS for hollow-fiber spinning consists of the following sequences [15,28,110]:

- Preparing the molten polymer solution by blending the polymer with a diluent at the temperature above the melting temperature of the polymer. The diluent must be a high boiling point and low molecular weight solid or liquid that is stable at the aforementioned temperature.
- Extruding the molten polymer solution through a spinneret.
- Removing the temperature from the extruded solution by either cooling the solution at a controlled rate or through a rapid cooling thermal quenching process to induce the phase separation process.
- Extracting the diluent through solvent extraction to produce microporous hollow fibers.

The phase inversion or demixing mechanism of TIPS can be further expanded into solid–liquid demixing and liquid–liquid demixing. The major factor controlling the demixing mechanism lies on the interaction of the polymer–diluent system [28]. A strong interaction between the polymer and the diluent tends to initiate the solid–liquid demixing through crystallization when the temperature is removed from the nascent fiber, whereas a weak interaction favors the liquid–liquid demixing process [28].

TIPS was reported to possess the ability to produce macrovoids-free membranes with superior mechanical properties and narrow pore size distribution [111]. In addition, the aforementioned process was able to produce a relatively thick symmetric microporous membrane for controlled release applications [28]. Although some studies claimed that TIPS has the advantage of fewer control parameters over NIPS, the high processing temperature and limited choice of high boiling point diluents are the main hurdles of TIPS [28,110].

Selection of diluents is crucial not only in controlling the phase separation mechanism but also in affecting the morphology and performance of the hollow-fiber membranes. Diluents with high boiling points such as triacetin, γ-butyrolactone, cyclohexanone, diethyl phthalate, dibutyl phthalate (DBP), and dioctyl phthalate are among the commonly used diluents in the fabrication of PVDF membranes through TIPS [28,112–114]. Ji et al. [115] reported the effects of mixed diluents to the morphology and performance of the PVDF hollow fibers. The increase of the DBP ratio in the mixed diluent of DBP and di(2-ethylhexyl) phthalate has altered the membrane morphology from an interconnected sponge-like structure to an asymmetric spherulitic bulk structure. The fibers consist of interconnected sponge-like structure that shows superior water permeability and elasticity as compared with the fibers that possess an asymmetric spherulitic bulk structure.

The effects of incorporating additives and nonsolvents in the molten PVDF solution were systematically investigated by Matsuyama and coworkers [114,116,117]. Rajabzadeh et al. [114] compared the effects of blending PVP and poly(methyl methacrylate) (PMMA) as the additives in PVDF solutions to fabricate hollow fibers through TIPS. The addition of PVP into the PVDF molten solution enhanced the mechanical strength and hydrophilicity of the hollow-fiber membranes as compared with those fabricated with PMMA as the additive. In addition, the incorporation of glycerol as the nonsolvent in the molten PVDF solution was reported to alter the phase inversion process from solid–liquid demixing to liquid–liquid demixing to produce a hollow fiber with a higher water permeability [114].

Ghasem and coworkers studied the effects of polymer concentration, extrusion temperature, and quenching temperature in TIPS on the performance of PVDF hollow fibers as contactors for the removal of carbon dioxide (CO_2) by adsorption [37,118,119]. They observed that the increase of PVDF concentration in the spinning dope reduces the mean pore size and the effective surface porosity [37]. In addition, fibers spun at a higher extrusion temperature displayed improvements in mechanical properties, porosity, and higher water contact angle, which enhanced the suitability of PVDF fibers for CO_2 removal [118]. Similar effects were also observed for fibers spun with a higher quench bath temperature. However, the quench bath temperature has insignificant effects on the water contact angle of the as-spun fibers [119]. Li et al. [120] reported that the mechanical properties of the as-spun fibers increase with an increment in take-up speed during the hollow-fiber spinning process.

7.5 CONCLUSION AND PERSPECTIVES

In this chapter, the basic principle of fabricating PVDF hollow-fiber membranes through NIPS was reviewed. Key spinning parameters such as polymer concentration, nonsolvent additives, coagulant chemistry, air-gap distance, take-up speed, dope rheology and their influences in the formation of PVDF hollow fibers, and the development of favorable macrovoid-free morphology were discussed. The recent advances and breakthroughs in PVDF hollow-fiber fabrication were also examined, which include dual-layer and new spinneret configuration design, mixed-matrix membranes, and TIPS. It is undeniable that PVDF will continue to serve as an indispensable material for hollow-fiber fabrication due to its excellent physicochemical properties and processability. Nevertheless, in-depth studies with a better understanding on the relationships between PVDF and nonsolvent additives, processing conditions, rheology, molecular simulation, and advanced spinning technologies are essential to enhance the quality and quantity of PVDF hollow fibers in the existing manufacture and to provide the basis for the development of the next-generation of PVDF hollow-fiber membranes.

ACKNOWLEDGMENTS

This research is supported by the National Research Foundation, Prime Minister's Office, Singapore, under its Competitive Research Program (CRP Award No. NRF-CRP5-2009-5) (NUS grant number R-279-000-311-281).

REFERENCES

1. H.I. Mahon. (1966). Permeability separatory apparatus and membrane element, method of making the same and process utilizing the same. US Patent 3,228,876.
2. R.E. Kesting. (1985). *Synthetic Polymeric Membranes: A Structural Perspective.* Wiley, New York.
3. T. Matsuura. (1994). *Synthetic Membranes and Membrane Separation Processing.* CRC Press, Boca Raton, FL.
4. T.S. Chung. (2008). Fabrication of hollow fiber membranes by phase inversion. In: *Advanced Membrane Technology and Applications*, N. Li et al. (Eds.), John Wiley & Sons, Inc., Hoboken, NJ, pp. 821–841.
5. N. Peng, N. Widjojo, P. Sukitpaneenit, M.M. Teoh, G.G. Lipscomb, T.S. Chung, and J.Y. Lai. (2012). Evolution of polymeric hollow fibers as sustainable technologies: Past, present, and future, *Prog. Polym. Sci.* 37: 1401–1424.
6. S. Atchariyawut, R. Jiraratananon, and R. Wang. (2007). Separation of CO_2 from CH_4 by using gas-liquid membrane contacting process, *J. Memb. Sci.* 304: 163–172.
7. S. Rajabzadeh, S. Yoshimoto, M. Teramoto, M. Al-Marzouqi, and H. Matsuyama. (2009). CO_2 absorption by using PVDF hollow fiber membrane contactors with various membrane structures, *Sep. Purif. Technol.* 69: 210–220.
8. M. Khayet and T. Matsuura. (2011). *Membrane Distillation: Principles and Applications.* Elsevier, New York.
9. K. Schneider and T.S. Van Gassel. (1984). Membrane distillation, *Chem. Eng. Technol.* 56: 514–521.
10. S. Bonyadi and T.S. Chung. (2007). Flux enhancement in membrane distillation by fabrication of dual layer hydrophilic-hydrophobic hollow fiber membranes, *J. Memb. Sci.* 306: 134–146.
11. K.Y. Wang, T.S. Chung, and M. Gryta. (2008). Hydrophobic PVDF hollow fiber membranes with narrow pore size distribution and ultra-thin skin for the fresh water production through membrane distillation, *Chem. Eng. Sci.* 63: 2587–2594.
12. P. Sukitpaneenit, T.S. Chung, and L.Y. Jiang. (2010). Modified pore-flow model for pervaporation mass transport in PVDF hollow fiber membranes for ethanol-water separation, *J. Memb. Sci.* 362: 393–406.
13. P. Sukitpaneenit and T.S. Chung. (2011). Molecular design of the morphology and pore size of PVDF hollow fiber membranes for ethanol-water separation employing the modified pore-flow concept, *J. Memb. Sci.* 374: 67–82.
14. M. Mulder. (1996). *Basic Principles of Membrane Technology*, Kluwer Academic Publishers, Dordrecht, the Netherlands.
15. F. Liu, N.A. Hashim, Y. Liu, M.R.M. Abed, and K. Li. (2011). Progress in the production and modification of PVDF membranes, *J. Memb. Sci.* 375: 1–27.
16. M. Sugihara, M. Fujimoto, and T. Uragami. (1979). Effect of casting solvent on permeation characteristics of polyvinylidene fluoride membranes, *Polym. Prepr., Am. Chem. Soc., Div. Polym. Chem.* 20: 999–1008.
17. T. Uragami, M. Fujimoto, and M. Sugihara. (1980). Studies on syntheses and permeabilites of special polymer membranes-24. Permeation characteristics of poly(vinylidene fluoride) membranes, *Polymer* 21: 1047–1051.

18. A. Bottino, G. Capannelli, and S. Munari. (1986). Factors affecting the structure and properties of asymmetric polymeric membranes. In: *Membrane and Membrane Processes*, E. Drioli and M. Nakagaki (Eds.), Plenum Press, New York.

19. H. Matsuyama, M. Teramoto, R. Nakatani, and T. Maki. (1999). Membrane formation via phase separation induced by penetration of nonsolvent from vapor phase. II. Membrane morphology, *J. Appl. Polym. Sci.* 74: 171–178.

20. M.L. Yeow, Y.T. Liu, and K. Li. (2004). Morphological study of poly(vinylidene fluoride) asymmetric membranes: Effects of the solvent, additive, and dope temperature, *J. Appl. Polym. Sci.* 92: 1782–1789.

21. D. Wang, K. Li, and W.K. Teo. (1999). Preparation and characterization of polyvinylidene fluoride (PVDF) hollow fiber membranes, *J. Memb. Sci.* 163: 211–220.

22. J. Kong and K. Li. (2001). Preparation of PVDF hollow-fiber membranes via immersion precipitation, *J. Appl. Polym. Sci.* 81: 1643–1653.

23. D. Wang, K. Li, and W.K. Teo. (2000). Porous PVDF asymmetric hollow fiber membranes prepared with the use of small molecular additives, *J. Memb. Sci.* 178: 13–23.

24. M. Khayet. (2003). The effects of air gap length on the internal and external morphology of hollow fiber membranes, *Chem. Eng. Sci.* 58: 3091–3104.

25. O.M. Ekiner and G. Vassilatos. (1990). Polyaramide hollow fibers for hydrogen/methane separation-spinning and properties, *J. Memb. Sci.* 53: 259–273.

26. N. Peng, T.S. Chung, and K.Y. Wang. (2008). Macrovoid evolution and critical factors to form macrovoid-free hollow fiber membranes, *J. Memb. Sci.* 318: 363–372.

27. D.T. Clausi and W.J. Koros. (2000). Formation of defect-free polyimide hollow fiber membranes for gas separations, *J. Memb. Sci.* 167: 79–89.

28. D.R. Lloyd, K.E. Kinzer, and H.S. Tseng. (1990). Microporous membrane formation via thermally induced phase separation. I. Solid–liquid phase separation, *J. Memb. Sci.* 52: 239–261.

29. T.H. Young, D.J. Lin, J.J. Gau, W.Y. Chuang, and L.P. Cheng. (1999). Morphology of crystalline Nylon-6,10 membranes prepared by the immersion-precipitation process: Competition between crystallization and liquid–liquid phase separation, *Polymer* 40: 5011–5021.

30. L.P. Cheng, T.H. Young, L. Fang, and J.J. Gau. (1999). Formation of particulate microporous poly(vinylidene fluoride) membranes by isothermal immersion precipitation from the 1-octanol/dimethylformamide/poly(vinylidene fluoride) system, *Polymer* 40: 2395–2403.

31. P. van de Witte, P.J. Dijkstra, J.W.A. van den Berg, and J. Feijen. (1996). Phase separation processes in polymer solutions in relation to membrane formation, *J. Memb. Sci.* 117: 1–31.

32. T.H. Young, L.P. Cheng, D.J. Lin, L. Fane, and W.Y. Chuang. (1999). Mechanisms of PVDF membrane formation by immersion-precipitation in soft (1-octanol) and harsh (water) nonsolvents, *Polymer* 40: 5315–5323.

33. M.G. Buonomenna, P. Macchi, M. Davoli, and E. Drioli. (2007). Poly(vinylidene fluoride) membranes by phase inversion: The role the casting and coagulation conditions play in their morphology, crystalline structure and properties, *Eur. Polym. J.* 43: 1557–1572.

34. P. Sukitpaneenit and T.S. Chung. (2009). Molecular elucidation of morphology and mechanical properties of PVDF hollow fiber membranes from aspects of phase inversion, crystallization and rheology, *J. Memb. Sci.* 340: 192–205.

35. L.P. Cheng, D.J. Lin, C.H. Shih, A.H. Dwan, and C.C. Gryte. (1999). PVDF membrane formation by diffusion-induced phase separation-morphology prediction based on phase behavior and mass transfer modeling, *J. Polym. Sci. B: Polym. Phys.* 37: 2079–2092.

36. A.M.W. Bulte, B. Folkers, M.H.V. Mulder, and C.A. Smolders. (1993). Membranes of semicrystalline aliphatic polyamide nylon 4,6: Formation by diffusion-induced phase separation, *J. Appl. Polym. Sci.* 50: 13–26.

37. N. Ghasem, M. Al-Marzouqi, and A. Duidar. (2012). Effect of PVDF concentration on the morphology and performance of hollow fiber membrane employed as gas–liquid membrane contactor for CO_2 absorption, *Sep. Purif. Technol.* 98: 174–185.

38. Y. Tang, N. Li, A. Liu, S. Ding, C. Yi, and H. Liu. (2012). Effect of spinning conditions on the structure and performance of hydrophobic PVDF hollow fiber membranes for membrane distillation, *Desalination* 287: 326–339.

39. S.H. Choi, F. Tasselli, J.C. Jansen, G. Barbieri, and E. Drioli. (2010). Effect of the preparation conditions on the formation of asymmetric poly(vinylidene fluoride) hollow fibre membranes with a dense skin, *Eur. Polym. J.* 46: 1713–1725.

40. M. Khayet, C.Y. Feng, K.C. Khulbe, and T. Matsuura. (2002). Preparation and characterization of polyvinylidene fluoride hollow fiber membranes for ultrafiltration, *Polymer* 43: 3879–3890.

41. M.C. Garcia-Payo, M. Essalhi, and M. Khayet. (2010). Effects of PVDF-HFP concentration on membrane distillation performance and structural morphology of hollow fiber membranes, *J. Memb. Sci.* 347: 209–219.

42. L. Shi, R. Wang, Y. Cao, C. Feng, D.T. Liang, and J.H. Tay. (2007). Fabrication of poly(vinylidene fluoride-co-hexafluropropylene) (PVDF-HFP) asymmetric microporous hollow fiber membranes, *J. Memb. Sci.* 305: 215–225.

43. B.J. Cha and J.M. Yang. (2006). Effect of high-temperature spinning and PVP additive on the properties of PVDF hollow fiber membranes for microfiltration, *Macromol. Res.* 14: 596–602.

44. T.S. Chung, S.K. Teoh, and X.D. Hu. (1997). Formation of ultrathin high-performance polyethersulfone hollow fiber membranes, *J. Memb. Sci.* 133: 161–175.

45. C.A. Smolders, A.J. Reuvers, R.M. Boom, and I.M. Wienk. (1992). Microstructures in phase inversion membranes. Part 1. Formation of macrovoids, *J. Memb. Sci.* 73: 259–275.

46. F. Edwie and T.S. Chung. (2012). Development of hollow fiber membranes for water and salt recovery from highly concentrated brine via direct contact membrane distillation and crystallization, *J. Memb. Sci.* 421–422: 111–123.

47. S. Atchariyawut, C. Feng, R. Wang, R. Jiraratananon, and D.T. Liang. (2006). Effect of membrane structure on mass-transfer in the membrane gas-liquid contacting process using microporous PVDF hollow fibers, *J. Memb. Sci.* 285: 272–281.

48. L. Shi, R. Wang, Y. Cao, D.T. Liang, and J.H. Tay. (2008). Effect of additives on the fabrication of poly(vinylidene fluoride-co-hexafluoropropylene) (PVDF-HFP) asymmetric microporous hollow fiber membranes, *J. Memb. Sci.* 315: 195–204.

49. R. Naim, A.F. Ismail, and A. Mansourizadeh. (2012). Effect of non-solvent additives on the structure and performance of PVDF hollow fiber membrane contactor for CO_2 stripping, *J. Memb. Sci.* 423–424: 503–513.

50. A. Mansourizadeh, A.F. Ismail, M.S. Abdullah, and B.C. Ng. (2010). Preparation of polyvinylidene fluoride hollow fiber membranes for CO_2 absorption using phase-inversion promoter additives, *J. Memb. Sci.* 355: 200–207.

51. M.L. Yeow, Y. Liu, and K. Li. (2005). Preparation of porous PVDF hollow fibre membrane via a phase inversion method using lithium perchlorate ($LiClO_4$) as an additive, *J. Memb. Sci.* 268: 16–22.

52. Z. Yuan and X. Dan-Li. (2008). Porous PVDF/TPU blends asymmetric hollow fiber membranes prepared with the use of hydrophilic additive PVP (K30), *Desalination* 223: 438–447.

53. S. Simone, A. Figoli, A. Criscuoli, M.C. Carnevale, A. Rosselli, and E. Drioli. (2010). Preparation of hollow fibre membranes from PVDF/PVP blends and their applications in VMD, *J. Memb. Sci.* 364: 219–232.

54. D. Hou, J. Wang, D. Qu, Z. Luan, and X. Ren. (2009). Fabrication and characterization of hydrophobic PVDF hollow fiber membranes for desalination through direct contact membrane distillation, *Sep. Purif. Technol.* 69: 78–86.

55. S. Wongchitphimon, R. Wang, R. Jiraratananon, L. Shi, and C.H. Loh. (2011). Effect of polyethylene glycol (PEG) as an additive on the fabrication of polyvinylidene fluoride-co-hexafluropropylene (PVDF-HFP) asymmetric microporous hollow fiber membranes, *J. Memb. Sci.* 369: 329–338.

56. M. Khayet, C.Y. Feng, K.C. Khulbe, and T. Matsuura. (2002). Study on the effect of a non-solvent additive on the morphology and performance of ultrafiltration hollow-fiber membranes, *Desalination* 148: 321–327.

57. E. Fontananova, J.C. Jansen, A. Cristiano, E. Curcio, and E. Drioli. (2006). Effect of additives in the casting solution on the formation of PVDF membranes, *Desalination* 192: 190–197.

58. C.M. Hansen. (1999). Hansen solubility parameter. In: *A User's Handbook*, CRC Press, New York.

59. A. Bottino, G. Capannelli, S. Munari, and A. Turturro. (1988). Solubility parameters of poly(vinylidene fluoride), *J. Polym. Sci. B: Polym. Phys.* 26: 785–794.

60. D. Li, T.S. Chung, and R. Wang. (2004). Morphological aspects and structure control of dual-layer asymmetric hollow fiber membranes formed by a simultaneous co-extrusion approach, *J. Memb. Sci.* 243: 155–175.

61. K.Y. Wang, D.F. Li, T.S. Chung, and S.B. Chen. (2004). The observation of elongation dependent macrovoid evolution in single and dual-layer asymmetric hollow fiber membranes, *Chem. Eng. Sci.* 59: 4657–4660.

62. T.S. Chung, W.H. Lin, and R.H. Vora. (2000). The effect of shear rates on gas separation performance of 6FDA-durene polyimide hollow fibers, *J. Memb. Sci.* 167: 55–66.

63. T.S. Chung, S.K. Teo, W.W.Y. Lau, and M.P. Srinivasan. (1998). Effect of shear stress within the spinneret on hollow fiber membrane morphology and separation performance, *Ind. Eng. Chem. Res.* 37: 3930–3938.

64. T.S. Chung, J.J. Qin, and J. Gu. (2000). Effect of shear rate within the spinneret on morphology, separation performance and mechanical properties of ultrafiltration poly-ethersulfone hollow fiber membranes, *Chem. Eng. Sci.* 55: 1077–1091.

65. K.Y. Wang, T. Matsuura, T.S. Chung, and W.F. Guo. (2004). The effects of flow angle and shear rate within the spinneret on the separation performance of poly(ethersulfone) (PES) ultrafiltration hollow fiber membranes, *J. Memb. Sci.* 240: 67–79.

66. N. Widjojo and T.S. Chung. (2006). Thickness and air gap dependence of macrovoid evolution in phase-inversion asymmetric hollow fiber membranes, *Ind. Eng. Chem. Res.* 45: 7618–7626.

67. N. Peng and T.S. Chung. (2008). The effects of spinneret dimension and hollow fiber dimension on gas separation performance of ultra-thin defect-free Torlon® hollow fiber membranes, *J. Memb. Sci.* 310: 455–465.

68. C. Cao, T.S. Chung, S.B. Chen, and Z. Dong. (2004). The study of elongation and shear rates in spinning process and its effect on gas separation performance of Poly(ether sulfone) (PES) hollow fiber membranes, *Chem. Eng. Sci.* 59: 1053–1062.

69. N. Widjojo, T.S. Chung, D.Y. Arifin, M. Weber, and V. Warzelhan. (2010). The elimination of die swell and spinning instability in the hyperbranched polyethersulfone (HPES) hollow fiber spinning process via novel spinneret designs and precise spinning conditions, *Chem. Eng. J.* 163: 143–153.

70. J. Ren, R. Wang, H.Y. Zhang, Z. Li, D.T. Liang, and J.H. Tay. (2006). Effect of PVDF dope rheology on the structure of hollow fiber membranes used for CO_2 capture, *J. Memb. Sci.* 281: 334–344.

71. O.M. Ekiner and G. Vassilatos. (2001). Polyaramide hollow fibers for H_2/CH_4 separation II. Spinning and properties, *J. Memb. Sci.* 186: 71–84.

72. N. Peng, T.S. Chung, and J.Y. Lai. (2009). The rheology of Torlon® solutions and its role in the formation of ultra-thin defect-free Torlon® hollow fiber membranes for gas separation, *J. Memb. Sci.* 326: 608–617.

73. E. Kuzumoto and K. Nitta. (1989). Production of permselective compound hollow yarn membranes. JP patent JP01015104.

74. T. He, M.H.V. Mulder, H. Strathmann, and M. Wessling. (2002). Preparation of composite hollow fiber membranes: Co-extrusion of hydrophilic coating onto porous hydrophobic support structures. *J. Memb. Sci.* 207: 143–56.

75. C.C. Pereira, R. Nobrega, K.V. Peinemann, and C.P. Borges. (2003). Hollow fiber membranes obtained by simultaneous spinning of two polymer solutions: A morphological study. *J. Memb. Sci.* 226: 35–50.

76. F.Y. Li, Y. Li, T.S. Chung, H. Chen, Y.C. Jean, and S. Kawi. (2011). Development and positron annihilation spectroscopy (PAS) characterization of polyamide imide (PAI) polyethersulfone (PES) based defect-free dual-layer hollow fiber membranes with an ultrathin dense-selective layer for gas separation, *J. Memb. Sci.* 378: 541–550.

77. Q. Yang, K.Y. Wang, and T.S. Chung. (2009). Dual-layer hollow fibers with enhanced flux as novel forward osmosis membranes for water production, *Environ. Sci. Technol.* 43: 2800–2805.

78. N. Widjojo and T.S. Chung. (2009). Pervaporation dehydration of C2–C4 alcohols by 6FDA-ODA-NDA/Ultem® dual-layer hollow fiber membranes with enhanced separation performance and swelling resistance, *Chem. Eng. J.* 155: 736–743.

79. D.F. Li, T.S. Chung, R. Wang, and Y. Liu. (2002). Fabrication of fluoropolyimide/polyethersulfone (PES) dual-layer asymmetric hollow fiber membranes for gas separation, *J. Memb. Sci.* 198: 211–223.

80. T. He, M.H.V. Mulder, and M. Wessling. (2003). Preparation of porous hollow fiber membranes with a triple-orifice spinneret, *J. Appl. Polym. Sci.* 87: 2151–2157.

81. S. Bonyadi and T.S. Chung. (2009). Highly porous and macrovoid-free PVDF hollow fiber membranes for membrane distillation by a solvent-dope solution co-extrusion approach, *J. Memb. Sci.* 331: 66–74.

82. Y.K. Ong and T.S. Chung. (2012). High performance dual-layer hollow fiber fabricated via novel immiscibility induced phase separation (I2PS) process for dehydration of ethanol, *J. Memb. Sci.* 421–422: 271–282.

83. N. Peng, M.M. Teoh, T.S. Chung, and L.L. Koo. (2011). Novel rectangular membranes with multiple hollow holes for ultrafiltration, *J. Memb. Sci.* 372: 20–28.

84. M.M. Teoh, N. Peng, T.S. Chung, and L.L. Koo. (2011). Development of novel multichannel rectangular membranes with grooved outer selective surface for membrane distillation, *Ind. Eng. Chem. Res.* 50: 14046–14054.

85. Y. Zhang, C. Qin, and J. Binner. (2006). Processing multi-channel alumina membranes by tape casting latex-based suspensions, *Ceram. Int.* 32: 811–818.

86. A.F. Ismail and K. Li. (2008). From polymeric precursors to hollow fiber carbon and ceramic membranes. In: *Inorganic Membranes: Synthesis, Characterization and Application*, R. Mallada and M. Menendez (Eds.), Elsevier, Amsterdam, the Netherlands, pp. 81–119.

87. K.A. Bu-Rashid and W. Czolkoss. (2007). Pilot tests of multibore UF membrane at addur SWRO desalination plant, Bahrain, *Desalination* 203: 229–242.

88. P. Wang and T.S. Chung. (2012). Design and fabrication of lotus-root-like multi-bore hollow fiber membrane for direct contact membrane distillation, *J. Memb. Sci.* 421–422: 361–374.

89. T.S. Chung, L.Y. Jiang, Y. Li, and S. Kulprathipanja. (2007). Mixed matrix membranes (MMMs) comprising organic polymers with dispersed inorganic fillers for gas separation, *Prog. Polym. Sci.* 32: 483–507.

90. L.Y. Ng, A.W. Mohammad, C.P. Leo, and N. Hilal. (2013). Polymeric membranes incorporated with metal/metal oxide nanoparticles: A comprehensive review, *Desalination* 308: 15–33.

91. L.Y. Jiang, T.S. Chung, C. Cao, Z. Huang, and S. Kulprathipanja. (2005). Fundamental understanding of nano-sized zeolite distribution in the formation of the

mixed matrix single- and dual-layer asymmetric hollow fiber membranes, *J. Memb. Sci.* 252: 89–100.

92. Y.C. Xiao, K.Y. Wang, T.S. Chung, and J. Tan. (2006). Evolution of nano-particle distribution during the fabrication of mixed matrix TiO_2-polyimide hollow fiber membranes, *Chem. Eng. Sci.* 61: 6228–6233.

93. L.Y. Yu, H.M. Shen, and Z.L. Xu. (2009). PVDF-TiO_2 composite hollow fiber ultrafiltration membranes prepared by TiO_2 sol-gel method and blending method, *J. App. Polym. Sci.* 113: 1763–1772.

94. E. Yuliwati, A.F. Ismail, T. Matsuura, M.A. Kassim, and M.S. Abdullah. (2011). Effect of modified PVDF hollow fiber submerged ultrafiltration membrane for refinery wastewater treatment, *Desalination* 283: 214–220.

95. L.Y. Yu, Z.L. Xu, H.M. Shen, and H. Yang. (2009). Preparing and characterization of PVDF–SiO_2 composite hollow fiber UF membrane by sol–gel method, *J. Memb. Sci.* 337: 257–265.

96. N.A. Hashim, Y. Liu, and K. Li. (2011). Preparation of PVDF hollow fiber membranes using SiO_2 particles: The effect of acid and alkali treatment on the membrane performances, *Ind. Eng. Chem. Res.* 50: 3035–3040.

97. S.M. Liu and K. Li. (2003). Preparation of TiO_2/Al_2O_3 composite hollow fibre membranes, *J. Memb. Sci.* 218: 269–277.

98. T. Van Gestel, C. Vandecasteele, A. Buekenhoudt, C. Dotremont, J. Luyten, R. Leysen, B. Van der Bruggen, and G. Maes. (2002). Alumina and titania multilayer membranes for nanofiltration: Preparation, characterization and chemical stability, *J. Memb. Sci.* 207: 73–89.

99. L.F. Han, Z.L. Xu, L.Y. Yu, Y.M. Wei, and Y. Cao. (2010). Performance of PVDF/multi-nanoparticles composite hollow fiber ultrafiltration membranes, *Iran. Polym. J.* 19: 553–565.

100. D. Hou, J. Wang, X. Sun, Z. Ji, and Z. Luan. (2012). Preparation and properties of PVDF composite hollow fiber membranes for desalination through direct contact membrane distillation, *J. Memb. Sci.* 405–406: 185–200.

101. W.Z. Lang, Q. Ji, J.P. Shen, Y.J. Guo, and L.F. Chu. (2013). Modified poly(vinylidene fluoride) hollow fiber composite membranes reinforced by hydroxyapatite nanocrystal whiskers, *J. App. Polym. Sci.* 127: 4564–4572.

102. M.M. Teoh and T.S. Chung. (2009). Membrane distillation with hydrophobic macro-void-free PVDF–PTFE hollow fiber membranes, *Sep. Purif. Technol.* 66: 229–236.

103. M.M. Teoh and T.S. Chung. (2009). Micelle-like macrovoids in mixed matrix PVDF-PTFE hollow fiber membranes, *J. Memb. Sci.* 338: 5–10.

104. P. Wang and T.S. Chung. (2012). A conceptual demonstration of freeze desalination-membrane distillation (FD-MD) hybrid desalination process utilizing liquefied natural gas (LNG) cold energy, *Water Res.* 46: 4037–4052.

105. K.Y. Wang, S.W. Foo, and T.S. Chung. (2009). Mixed matrix PVDF hollow fiber membranes with nanoscale pores for desalination through direct contact membrane distillation, *Ind. Eng. Chem. Res.* 48: 4474–4483.

106. P. Sukitpaneenit and T.S. Chung. (2012). PVDF/nanosilica dual-layer hollow fibers with enhanced selectivity and flux as novel membranes for ethanol recovery, *Ind. Eng. Chem. Res.* 51: 978–993.

107. M.M. Teoh, T.S. Chung, and Y.S. Yeo. (2011). Dual-layer PVDF/PTFE composite hollow fibers with a thin macrovoid-free selective layer for water production via membrane distillation, *Chem. Eng. J.* 171: 684–691.

108. P. Wang, M.M. Teoh, and T.S. Chung. (2011). Morphological architecture of dual-layer hollow fiber for membrane distillation with higher desalination performance, *Water Res.* 45: 5489–5500.

109. F. Edwie, M.M. Teoh, and T.S. Chung. (2012). Effects of additives on dual-layer hydrophobic-hydrophilic PVDF hollow fiber membranes for membrane distillation and continuous performance, *Chem. Eng. Sci.* 68: 567–578.

110. D.R. Lloyd, S.S. Kim, and K.E. Kinzer. (1991). Microporous membrane formation via thermally-induced phase separation. II. Liquid-liquid phase separation, *J. Memb. Sci.* 64: 1–11.

111. Z. Song, M. Xing, J. Zhang, B. Li, and S. Wang. (2012). Determination of phase diagram of a ternary PVDF/γ-BL/DOP system in TIPS process and its application in preparing hollow fiber membranes for membrane distillation, *Sep. Purif. Technol.* 90: 221–230.

112. B.J. Cha and J.M. Yang. (2007). Preparation of poly(vinylidene fluoride) hollow fiber membranes for microfiltration using modified TIPS process, *J. Memb. Sci.* 291: 191–198.

113. Y. Su, C. Chen, Y. Li, and J. Li. (2007). Preparation of PVDF membranes via TIPS method: The effect of mixed diluents on membrane structure and mechanical property, *J. Macromol. Sci. A* 44: 305–313.

114. S. Rajabzadeh, T. Maruyama, T. Sotani, and H. Matsuyama. (2008). Preparation of PVDF hollow fiber membrane from a ternary polymer/solvent/nonsolvent system via thermally induced phase separation (TIPS) method, *Sep. Purif. Technol.* 63: 415–423.

115. G.L. Ji, L.P. Zhu, B.K. Zhu, C.F. Zhang, and Y.Y. Xu. (2008). Structure formation and characterization of PVDF hollow fiber membrane prepared via TIPS with diluent mixture, *J. Memb. Sci.* 319: 264–270.

116. S. Rajabzadeh, C. Liang, Y. Ohmukai, T. Maruyama, and H. Matsuyama. (2012). Effect of additives on the morphology and properties of poly(vinylidene fluoride) blend hollow fiber membrane prepared by the thermally induced phase separation method, *J. Memb. Sci.* 423–424: 189–194.

117. S. Rajabzadeh, T. Maruyama, Y. Ohmukai, T. Sotani, and H. Matsuyama. (2009). Preparation of PVDF/PMMA blend hollow fiber membrane via thermally induced phase separation (TIPS) method, *Sep. Purif. Technol.* 66: 76–83.

118. N. Ghasem, M. Al-Marzouqi, and N. Abdul Rahim. (2012). Effect of polymer extrusion temperature on poly(vinylidene fluoride) hollow fiber membranes: Properties and performance used as gas-liquid membrane contactor for CO_2 absorption, *Sep. Purif. Technol.* 99: 91–103.

119. N. Ghasem, M. Al-Marzouqi, and A. Duaidar. (2011). Effect of quenching temperature on the performance of poly(vinylidene fluoride) microporous hollow fiber membranes fabricated via thermally induced phase separation technique on the removal of CO_2 from CO_2-gas mixture, *Int. J. Greenh. Gas Con.* 5: 1550–1558.

120. X. Li, H. Liu, C. Xiao, S. Ma, and X. Zhao. (2013). Effect of take-up speed on polyvinylidene fluoride hollow fiber membrane in a thermally induced phase separation process, *J. App. Polym. Sci.* 128: 1054–1060.

8 PVDF Membranes for Membrane Distillation

Controlling Pore Structure, Porosity, Hydrophobicity, and Mechanical Strength

Rinku Thomas, Muhammad Ro'il Bilad, and Hassan Ali Arafat

CONTENTS

8.1 Introduction 250
8.2 Membranes in MD—Ideal Structural Requirements...................................250
8.3 PVDF—An Extensively Researched Polymer for Various Applications 252
    8.3.1 PVDF Structure and Crystallinity...252
    8.3.2 Properties of PVDF 252
    8.3.3 PVDF as Membrane Materials 253
    8.3.4 Organic Solvents for PVDF...253
8.4 Membrane Preparation through Phase Inversion 254
8.5 Rationalization of Immersion Precipitation Process 255
    8.5.1 Thermodynamics Aspects 256
    8.5.2 Kinetic Aspects..260
8.6 Factors Affecting PVDF Membrane Fabrication 262
    8.6.1 Phase Inversion Parameters 262
        8.6.1.1 Effect of Exposure Time between Casting and
                Precipitation 262
        8.6.1.2 Effect of Solvent Choice 263
        8.6.1.3 Effect of PVDF Grade 267
        8.6.1.4 Effect of PVDF Concentration..268
        8.6.1.5 Effect of Coagulation Bath Composition and Temperature ...268
        8.6.1.6 Effect of PVDF Blending with Other Polymers 274
    8.6.2 Effects of Additives 274
    8.6.3 Optimization of Immersion Precipitation Parameters...................276
    8.6.4 Other Parameters 276
        8.6.4.1 Effect of PVDF Dissolving Temperature..........................276
        8.6.4.2 Rasonic Irradiation 277

  8.6.4.3 TiO$_2$ Coating 278
  8.6.4.4 Posttreatment 278
  8.6.4.5 Effect of Nonwoven Support... 278
8.7 Conclusions... 278
Bibliography 279

8.1 INTRODUCTION

Membrane distillation (MD) is a separation technique involving the nonisothermal transport of water vapor through a porous, hydrophobic membrane. It is a promising desalination technology that may help resolve the global freshwater shortage. To date, no large-scale MD plants have been implemented yet for desalination (Saffarini et al. 2012a, 2012b), as several scientific and technological challenges still hamper its industrial applications (Curcio and Drioli 2005; Khayet 2008). The major barriers include MD membrane and module design, membrane pore wetting, low permeate flow rate, flux decay, as well as uncertain energy and economic performance figures (Saffarini et al. 2012b). These challenges have attracted scientists and engineers striving to achieve the best MD membrane, performance and/or module and process design (Khayet and Matsuura 2001; Suk et al. 2002; Li and Sirkar 2004; Gilron et al. 2007). Among these research attempts, the development or selection of appropriate membrane materials was very important (Guillen-Burrieza et al. 2013; Saffarini et al. 2013).

A key factor for MD membranes is the fabrication procedures that influence the morphology and porosity of the membranes, thereby defining their performance efficiency. An exhaustive review on the progress in the production of poly(vinylidene fluoride) (PVDF) membranes was published by Liu et al. (2011). Since then, more than 100 publications have added more information and interpretation on the interplay of different parameters in the fabrication process of flat-sheet PVDF membranes.

8.2 MEMBRANES IN MD—IDEAL STRUCTURAL REQUIREMENTS

Souhaimi and Matsuura (2011) have laid down the main criteria that porous membranes should satisfy to be considered as good candidates for MD applications. These criteria were also reiterated by other researchers in the MD field (Curcio and Drioli 2005; Khayet 2011; Alkhudhiri et al. 2012):

1. The membrane may be comprised of one or more layers, but the top layer in contact with the liquid feed should be made of a hydrophobic material and be porous.
2. The pore size range may be from several nanometers to a few micrometers. The pore size distribution (PSD) should be as narrow as possible and the feed liquid should not penetrate into the pores. The liquid entry pressure (LEP), defined as the minimum transmembrane pressure that is required for the feed solution to enter into the pores, by overcoming the hydrophobic forces, should be as high as possible. A high LEP can be achieved by the material of low surface energy (i.e., high hydrophobicity) and small maximum pore size. On the other hand, a small maximum pore size parallels

a small mean pore size and, consequently, low membrane permeability. Therefore, a compromise between a high LEP and a high productivity should be made by choosing an appropriate pore size and PSD.

3. The tortuosity factor, the measure of the deviation of the structure from straight cylindrical pores normal to the surface, should be small as it is inversely proportional to the MD membrane permeability.

4. The porosity, the void volume fraction open to MD vapor flux, of the single-layer membrane or that of the hydrophobic layer in the case of the multilayered membrane should be as high as possible, as it is directly proportional to the MD membrane permeability.

5. The thickness of the single-layer membrane has an optimum value, as it is inversely proportional to the rate of both mass and heat (by conduction) transport through the membrane. Although a high mass transport is favored for the MD process, a high heat transport rate leads to undesired heat loss. Therefore, a compromise should be made between the mass and heat transfer by properly adjusting the membrane thickness. In the case of the multilayered membrane, the thickness of hydrophobic layer should be as thin as possible. One advantage of the multilayered membrane is that high mass transfer is enabled by making the hydrophobic layer as thin as possible, whereas heat transfer is suppressed by making the overall membrane thickness, hydrophobic and hydrophilic layers, as thick as possible.

6. The thermal conductivity of the membrane material should be as low as possible. As most hydrophobic polymers have similar conductivities, within the same order of magnitude, it is possible to diminish the conductive heat transfer of the membrane using membranes of high porosities. The conductive heat transfer coefficients of the air entrapped in the pores are an order of magnitude smaller than most of the used membrane materials.

7. The membrane should have good mechanical strength.

8. The membrane surface contacting the feed solution should be made of a material of high fouling and scaling resistance.

9. The membrane should exhibit good long-term thermal stability at temperatures as high as 100°C.

10. The membrane material should have excellent chemical resistance to various feed solutions. If the membrane has to be cleaned, resistance to cleaning chemicals is also necessary.

11. The membrane should be cost-effective and have a long life with stable MD performance, as demonstrated by permeability and selectivity under commercial application.

Though most polymer matrices satisfy key MD criteria, such as chemical resistance, process and thermal stability, and barrier properties, refinement in certain areas is still needed to enable such polymer matrices to become good MD membranes. Most of the membranes tested for suitability in MD have proven their worth in other widely commercialized separation techniques, particularly in microfiltration and ultrafiltration (Guillen-Burrieza et al. 2013).

8.3 PVDF—AN EXTENSIVELY RESEARCHED POLYMER FOR VARIOUS APPLICATIONS

8.3.1 PVDF Structure and Crystallinity

PVDF is a fluoropolymer with alternating $-CH_2$ and $-CF_2$ groups along the polymer chains that create a polarity allowing the polymers to be dissolved in certain solvents. Homopolymers of PVDF are semicrystalline, with typically 35%–70% crystallinity, and long chain macromolecules, which contain 59.4 wt% fluorine and 3 wt% hydrogen (Ameduri 2009). The crystalline phase of PVDF has three different molecular conformations and five distinct crystal polymorphs. These different crystalline phase structures and polymorphs were already reported by Dillon et al. (2006) and Hasegawa et al. (1972). Molecular weight, molecular weight distribution, extent of irregularities along the polymer chain, crystallinity, and the crystalline form are the major factors influencing the properties of PVDF. The high level of crystallinity yields PVDF stiffness, toughness, and creep-resistance (Cui et al. 2013). Chemical structures of PVDF homo and copolymers are shown in Figure 8.1.

8.3.2 Properties of PVDF

PVDF is inert but can dissolve in various solvents, oils, and acids. The glass transition (T_g) and melting temperatures (T_m) of the amorphous and crystalline PVDF regions are in the ranges of $-40°C$ to $-30°C$ and $155°C$ to $192°C$, respectively.

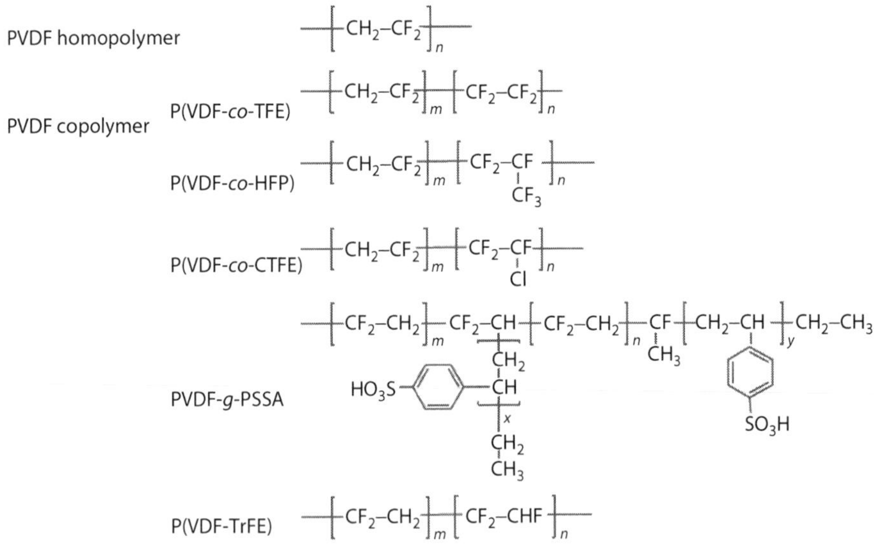

FIGURE 8.1 Chemical structures of PVDF homo and copolymers. (Data from Cui, Z. et al., *Progress in Polymer Science*, 2013. doi:10.1016/j.progpolymsci.2013.07.008. http://www.sciencedirect.com/science/article/pii/S0079670013000889.)

They are strongly influenced by molecular weight and the number of chain defects. Amorphous α, β, and γ polymorphs have a density of 1.68, 1.92, 1.97 and 1.93 g cm^{-3}, respectively. The typical density of commercial PVDF products is in the range of 1.75–1.78 g cm^{-3}, reflecting a degree of crystallinity around 40%. The melt density of a PVDF homopolymer is about 1.45–1.48 g cm^{-3} at 230°C and 1.0 bar (Drobny 2009). The excellent combination of properties and processability of PVDF mean that it is available in a wide range of melt viscosities as powders and pellets to fulfill typical fabrication requirements (Cui et al. 2013).

8.3.3 PVDF as Membrane Materials

PVDF has received great attention as a membrane material since 1980s with regard to its outstanding properties such as high mechanical strength, thermal stability, chemical resistance, and high hydrophobicity. Remarkable progress has been made in the fabrications of PVDF membranes with high performance, and PVDF membranes have been extensively applied in ultrafiltration and microfiltration for general separation purposes, and are currently being explored as potential candidates in the applications of membrane contactor and MD (Liu et al. 2011).

Recently, PVDF has become a more popular material to produce hydrophobic membranes through phase inversion processes, mainly for membrane contactor and MD applications. It is preferred to other more hydrophobic polymers, such as polypropylene and polytetrafluoroethylene, because of its excellent combination of properties and its solubility in common organic solvents. Furthermore, the excellent thermal stability of PVDF has made it interesting as a membrane material in a wide range of industrial applications. In addition, unlike other crystalline polymers, PVDF exhibits thermodynamic compatibility with other polymers, such as poly(methyl methacrylate) (PMMA), over a wide range of blend compositions, which can be useful in the fabrication of membrane with desired properties. PVDF can be further chemically modified to obtain specific functions. In addition, it can be cross-linked when subjected to electron beam radiation or gamma radiation.

There are a few other fluorocopolymers, besides PVDF, which can be used as membrane materials, such as poly(vinylidenefluoride-co-hexafluoropropylene) and poly(vinylidene difluoride-co-chlorotrifluoroethylene). The chemical structure of these copolymers is presented in Figure 8.1.

8.3.4 Organic Solvents for PVDF

As will be discussed in Section 8.4, the solubility of a polymer in a solvent is a necessary condition for a polymer to be a good material for membrane preparation through phase inversion. The solvent plays a crucial role in determining the ultimate membrane performance and properties. With the proper solvent selection, high polymer chain mobility can be maintained, which is directly influenced by both polymer–solvent and polymer–polymer interactions. Therefore, a uniform distribution of polymer in a solvent is a necessity. In contrast, poor solvents result in the aggregation of polymer molecules. The identified good solvents of PVDF that are mainly used in the immersion precipitation method are listed in Table 8.1.

TABLE 8.1
Common Solvents for PVDF

Solvents	Boiling Point (°C)
N,N-Dimethylacetamide (DMA)	165
N,N-Dimethylformamide (DMF)	153
Dimethylsulfoxide (DMSO)	189
Hexamethyl phosphoramide (HMPA)	232.5
N-Methyl-2-pyrrolidone (NMP)	202
Tetramethylurea (TMU)	176.5
Triethyl phosphate (TEP)	215.5
Trimethyl phosphate (TMP)	197.2
Acetone (Ac)	56.1
Tetrahydrofuran (THF)	65

Source: Bottino, A. et al., *Journal of Membrane Science*, 57(1), 1–20, 1991. doi:10.1016/S0376-7388(00)81159-X; Wienk, I.M. et al., *Journal of Membrane Science*, 113(2), 361–371, 1996. doi:10.1016/0376-7388(95)00256-1.

8.4 MEMBRANE PREPARATION THROUGH PHASE INVERSION

The synthesis of polymeric membranes is mostly achieved through the phase inversion process. Phase inversion is one of the most versatile and economical processes used to develop polymeric membranes. In this process, a membrane is prepared from a thermodynamically stable solution, in its most basic form consisting of a polymer and a solvent. This polymer solution is often called the *casting solution*, and is sometimes also called *dope*. The casting solution is cast onto a supporting material, which is typically nonwoven, but sometimes also woven, resulting in a thin liquid-cast–polymer film, further referred to as *cast film*, which is later transformed into a solid membrane. The phase transformation from liquid to solid is induced by disturbing the stability of polymer solution.

In this section, the phase inversion process will be discussed briefly. A more detailed explanation of the various phase inversion techniques can be found elsewhere (Mulder 1996).

The precipitation of liquid polymer solution to form a solid membrane can be achieved in several ways:

1. Thermal gelation: Casting solution is cast hot. Polymer precipitation is induced by lowering the temperature of the cast film. As it cools, the polymer precipitates, and the solution separates into a polymer matrix phase containing dispersed pores filled with solvent. After the polymer solution is cast, the volatile solvent evaporates and the cast film is enriched in the nonvolatile solvent, thus inducing polymer precipitation to form the membrane structure. The membranes formed using this process typically have large pores.

2. Solvent evaporation: A mixture of solvents, with one of them being volatile, is used to form the casting solution. After casting, the volatile solvent evaporates, thus changing the polymer-film–solution composition, which causes precipitation. In the simplest form, a polymer is dissolved in a two-component solvent mixture consisting of a volatile solvent, in which the polymer is readily soluble, and a less volatile nonsolvent, typically water or an alcohol.

3. Water vapor adsorption: The cast film is placed under a humid atmosphere. As water is typically a strong nonsolvent, water adsorption into the cast film causes polymer precipitation. In its strictest form, this process is the opposite of the solvent evaporation. Instead of introducing precipitation by evaporating a volatile solvent, the polymer film is exposed under a humid environment allowing water vapor to be adsorbed onto the cast film.

4. Immersion precipitation: Here, the cast film is immersed in a nonsolvent bath (normally water). Absorption of water and loss of solvent cause the cast film to rapidly precipitate from the top surface down.

In reality, a combination of the aforementioned processes is normally applied. For instance:

- The solvent evaporation process can be combined with immersion precipitation process. After allowing for evaporation for some time, the film is immersed in a nonsolvent. This way, porosity and pore size of the membrane can be controlled. At shorter evaporation times, membranes with fine micropores can be achieved and vice versa under prolonged evaporation times.
- Instead of using a nonvolatile solvent, a volatile solvent can be exposed to a humid atmosphere where solvent evaporation and water vapor adsorption occur simultaneously.
- Time gaps between casting and immersion in the immersion precipitation allows volatile solvent to evaporate or water vapor to be adsorbed before immersion into the nonsolvent.
- In many cases, the temperature of the nonsolvent in the immersion bath is also controlled, thus enabling a combination of immersion precipitation and thermal gelation.

8.5 RATIONALIZATION OF IMMERSION PRECIPITATION PROCESS

Among the four types of phase inversion processes, immersion precipitation remains by far the most important membrane preparation technique, both for dense and porous membranes. This technique is also mostly used to prepare PVDF membranes for MD application. Due to the importance of this technique, this section covers it in detail.

Immersion precipitation in its most simple form is conducted in the following way. A polymer solution consisting of polymer and a solvent is cast as a thin film upon a support and then immersed into a nonsolvent bath. The solvent diffuses into the

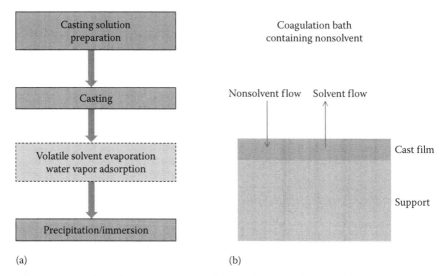

FIGURE 8.2 Schematic representation of (a) basic immersion precipitation process and (b) film bath interface during phase separation.

coagulation bath and, at the same time, the nonsolvent diffuses into the cast film. After a given period of time, the exchange of solvent and nonsolvent has proceeded so far that the solution becomes thermodynamically unstable and phase separation (demixing) starts. Finally, the solid polymeric film is obtained with an asymmetric or symmetric structure. A schematic representation of the process is shown in Figure 8.2.

Development of the immersion precipitation technique proceeds in two approaches. Due to the complexity and difficulties of the process, most of the industrial users have taken an empirical approach based on trial and error. This resulted in several rules-of-thumb to guide membrane producers on how to prepare a desirable immersion precipitation membrane. On the other hand, some theories on membrane formation based on fundamental studies of the precipitation process have also been developed. However, most of them are complex and cannot very well explain the formation mechanisms in great detail. The most convenient way to rationalize the immersion precipitation process is using phase diagrams (Strathmann et al. 1971; Mulder 1996; Baker 2012), as extensively exemplified in this chapter. They can be used to evaluate the two most important aspects of the process, namely, its thermodynamics and kinetics.

8.5.1 THERMODYNAMICS ASPECTS

To understand the mechanism of membrane formation from the three basic components (polymer, solvent, and nonsolvent), the behavior of a mixture of these components is presented in a three-component phase diagram. A typical three-component phase diagram for the components used to prepare membranes in an isothermal process is shown in Figure 8.3. The corners of the triangle represent the three pure components: polymer, solvent, and nonsolvent, typically water. The regions within the triangle represent mixtures of the three components. The diagram has two

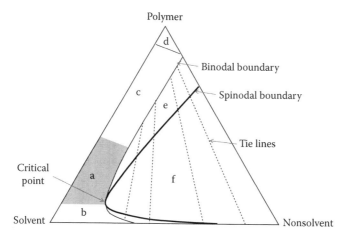

FIGURE 8.3 Schematic of the three-component phase diagram often used to rationalize the immersion precipitation of membranes prepared by phase inversion. (a) one-phase stable region, which typically represents the initial casting solution composition; (b) one-phase stable region but cannot be used as initial composition to prepare membrane; (c) one-phase gel region; (d) glassy region; (e) metastable region; and (f) unstable region.

principal regions: a one-phase region, in which all components are miscible and a two-phase region, in which the system separates into a polymer-rich phase and a liquid polymer-poor phase.

With reference to Figure 8.3, depending on polymer concentration in the casting solution, polymer/solvent composition can be classified into different regions: a liquid polymer solution (zones a and b on Figure 8.3), a polymer gel region (c), and a glassy solid polymer region (d). The typical polymer solution for membrane preparation is in liquid and viscous form at low polymer concentration (a). If the polymer concentration is increased, the viscosity of the solution increases rapidly, reaching such high values that the system can be regarded as a gel. If the casting solution contains more than 90% polymer, the swollen polymer gel may become rigid, so that the polymer chains can no longer rotate, thus it becomes a solid polymer glass (d).

During the following polymer precipitation step, the cast film loses some of its solvent content and gains some nonsolvent material. Thus, the cast film composition moves from the one-phase stable solution region (a) to the two-phase unstable region (f). In between stable and unstable regions, there exists a metastable region (e), where the composition is thermodynamically stable, but will not normally precipitate unless it is well-nucleated. As more solvent leaves the cast film and nonsolvent enters, the film composition enters the metastable region by crossing the binodal boundary that separate the one phase and the metastable regions, where liquid–liquid demixing occurs. Further, as solvent and nonsolvent exchange proceeds, the film composition crosses into another region of the phase diagram, in which a one-phase solution is always thermodynamically unstable. The boundary between the metastable and unstable regions is called the *spinodal boundary*. In this region, the polymer solution spontaneously separates into two phases with their compositions linked by a tie line. The phase separation can be very fast through a spinodal decomposition or slower through nucleation and growth mechanisms, depending on the path of the composition change in the phase diagram (Mulder 1996; Baker 2012).

Depending on the side from which the critical point is approached, two types of nucleation can be distinguished (Nunes and Inoue 1996). When demixing starts somewhere below the critical point, nucleation and growth of a polymer-rich phase occur in the polymer-poor phase, which is not practical in membrane preparation (Ismail and Yean 2003; Sun et al. 2013). The growth of polymer-rich nuclei prevents the formation of an interconnected and integrated polymer matrix, resulting in powdery agglomerates with low-integrity (see Figure 8.4). On the other hand, when demixing starts above the critical point, nucleation of the polymer-poor phase occurs. These nuclei can grow further until a surrounding continuous phase solidifies, through either crystallization, gelation, or vitrification (Mulder 1996). The polymer-lean phase will form the porous structure, whereas the polymer-rich phase will result in the solid matrix surrounding the pores of the membrane. Occasionally, nucleated droplets of polymer-lean phase will grow into macrovoids if the diffusion flow of solvents from the surrounding polymer solution into the nuclei is greater than the diffusion flow of the nonsolvent from the nuclei to the surrounding polymer solution (Ismail and Yean 2003). A graphical illustration of a solid polymer formation is shown in Figure 8.4.

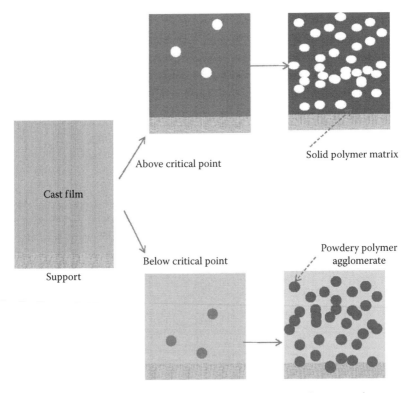

FIGURE 8.4 An illustration of polymer-rich and polymer-poor phase growth, as a result of different paths for entering the metastable region in the phase diagram. The round shape represent the nucleus of polymer-rich phase (black) and polymer-lean phase (white).

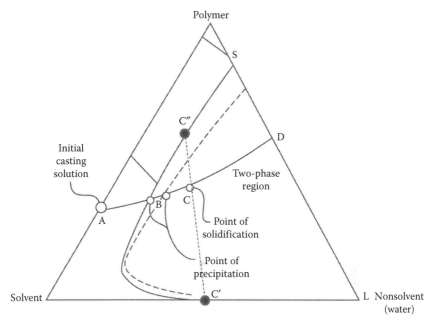

FIGURE 8.5   The path through the three-component phase diagram from the initial casting solution (A) to the final membrane (D). (Data from Strathmann, H. et al., *Journal of Applied Polymer Science*, 15, 811–828, 1971. doi:10.1002/app.1971.070150404.)

To understand the evolution of film composition and how phase separation takes place, the evolution path of a cast film during immersion precipitation from starting casting solution to solid membrane is shown in Figure 8.5, which demonstrates how the composition changes during different stages of immersion precipitation. The path starts at a point representing the original casting solution (A) and finishes at a point representing the composition of the final membrane (D). At point (B), as the solvent and nonsolvent are miscible with each other, the casting solution moves from a composition in the one-phase region to a composition in the two-phase region by losing solvent and gaining nonsolvent, thus crossing the binodal boundary. This brings the casting solution into the metastable two-phase region, where it starts to demix. Polymer solution compositions in this region are thermodynamically unstable, but will not precipitate unless well-nucleated. At point (C), as more solvent and nonsolvent exchange, the composition crosses into another region of the phase diagram in which the solution is always thermodynamically unstable. In this region, the polymer solution spontaneously separates into two phases (C″ and C′), both having an equal chemical potential, with compositions linked by a tie-line. Points C″ and C′ represent the compositions of polymer-rich and polymer-poor phases, respectively. At composition D, the two phases are in equilibrium: a solid (polymer-rich) phase, which forms the matrix of the final membrane, represented by point S, and a liquid (polymer-poor) phase, which constitutes the membrane pores filled with precipitant, represented by point L. Position of composition D on the line S–L determines the overall porosity of the membrane.

8.5.2 KINETIC ASPECTS

The precipitation path in Figure 8.5 is shown as a single line representing the average composition of the whole membrane. This is actually a deliberate simplification of what actually occurs, in which a considerably thick polymer film is represented by a single point. When the cast film is immersed into the nonsolvent bath, the top surface starts to precipitate rapidly and the two phases formed on precipitation do not have time to form a fine microporous structure. Therefore, the rate and the path of precipitation through the phase diagram differ across the film thickness and with time, as illustrated in Figure 8.6. The top surface layer then acts as a barrier that slows the subsequent solvent–nonsolvent exchange in the layers beneath, thus increasing the porosity at the bottom of the surface layer. At time t_2, for example, a few seconds after the precipitation process has initiated, the top surface of the polymer film has almost completely precipitated, and the composition of this surface layer is close to the polymer nonsolvent axis. On the other hand, at the bottom surface of the film where precipitation has just initiated, the composition is close to that of the original casting solution. Therefore, the morphology of the membranes along their cross-section is different, as also shown in Figure 8.6.

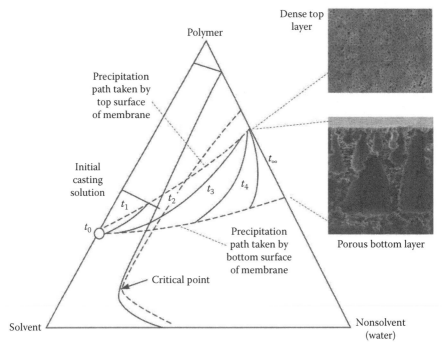

FIGURE 8.6 The surface layer of water-precipitated membranes precipitates faster than the underlying layer. The precipitation pathway is best represented by the movement of a line through the three-component phase diagram. SEM images were taken from a membrane prepared in PVDF/DMF/water system (own data). (Data from Baker, R.W., *Membrane Technology and Applications, 3rd Edition*, Wiley, Chichester, 2012. http://eu.wiley.com/WileyCDA/WileyTitle/productCd-0470743727.html.)

An important feature of membranes fabricated through immersion precipitation is their asymmetric structure; a relatively denser top *skin* on top of a more porous support layer. The skin layer provides the selectivity, such as size exclusion in microfiltration, whereas the porous support layer contributes to membrane strength. To rationalize the formation of this asymmetric structure of the final membrane, the kinetic aspect of immersion precipitation has to be considered. Formation kinetics is strongly related to the exchange rate of solvent and nonsolvent and the nature of phase separation (demixing).

In order to identify the type of demixing process and how it occurs, it is necessary to know the exact local composition at a given instant. Although it is very difficult to determine this composition experimentally, many attempts have been proposed to describe this phenomenon. These attempts mainly consider the outflow diffusion of solvent and the inflow diffusion of nonsolvent that determine the composition changes for various polymer–solvent–nonsolvent systems (Cohen et al. 1979; Altena et al. 1985; Patsis et al. 1990; Kim et al. 1996). A considerable correlation between the thickness of polymer film and the resulting membrane complicates the system. It affects in particular the degree of vertical membrane shrinkage and eventually the thickness of membrane matrix, as well as membrane porosity. This implies that the top skin moves during the formation process as a result of the difference between the inflow of nonsolvent to the polymer film and the outflow of the solvent to the coagulation bath.

The rate of solvent and nonsolvent exchange depends on the miscibility of the two, and the affinity of the nonsolvent for the polymer. Based on this property, two typical demixing paths exist, namely, instantaneous and delayed demixings, as illustrated in Figure 8.7. Occurrence of demixing can be followed during immersion by light transmission experiments (van't Hof et al. 1992). Instantaneous demixing means that the membrane is formed immediately after immersion as a result of spinodal decomposition (Hao and Wang 2003), whereas it takes some time before the ultimate membrane is formed in the case of delayed demixing. Membranes formed

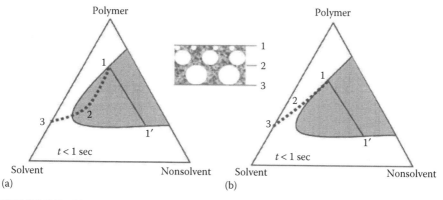

FIGURE 8.7 Phase diagram showing instantaneous (a) and delayed (b) demixings (composition present at the top, middle, and bottom of the film indicated by 1, 2, and 3, respectively). (Data from Vandezande, P. et al., *Chemical Society Reviews*, 37, 365–405, 2008. doi:10.1039/B610848M.)

by instantaneous demixing have a porous top layer and an open-cell macrovoid-like or sponge-like support layer. A relatively denser skin layer forms at the initial stage of the polymer precipitation, and the thickness of this skin layer grows with time (Patsis et al. 1990). Such membranes generally show size exclusion capabilities and are used in microfiltration, ultrafiltration, and MD processes. Therefore, to obtain a membrane with an open pore structure, the immersion precipitation parameters are normally optimized toward instantaneous demixing. Membranes formed by delayed demixing tend to have a dense skin, symmetric structure, and are appropriate for uses in gas separation, pervaporation, nanofiltration and reverse osmosis (Shimizu et al. 2002).

8.6   FACTORS AFFECTING PVDF MEMBRANE FABRICATION

This section discusses the progress and recent findings of PVDF membrane developments through immersion precipitation, in particular for MD application. In general, phase inversion parameters are optimized in order to achieve the criteria discussed in Section 8.2. The effects of different fabrication parameters are comprehensively discussed. It is interesting to note that all fabrication parameters have a profound influence on the final membrane morphology and performance. The fabricated MD membrane does not follow the dictate of a single predominant parameter, rather it is the interplay of all fabrication parameters and their delicate balance, which brings about the most suitable morphology, crystallinity, porosity, PSD, LEP, mechanical strength, and eventually MD performance.

8.6.1   Phase Inversion Parameters

8.6.1.1   Effect of Exposure Time between Casting and Precipitation

As discussed in Section 8.4, phase separation can be achieved by a combined process. An exposure time after casting and before immersion is often introduced in immersion precipitation (Figure 8.2). This time interval can either be used to evaporate a volatile solvent or to adsorb water vapor from air in a humid atmosphere. This step can significantly affect the composition path of polymer film on a phase diagram, even before the immersion step starts.

A good example to demonstrate the occurrence of combined solvent evaporation and water adsorption during time gaps between casting and immersion was reported by Li et al. (2011) in a PVDF/triethylphosphate (TEP) and dimethyl acetamide (DMA)/water system. In this system, water vapor adsorption was dominant due to the relatively humid atmosphere surrounding the cast film (RH = 60%). At short exposure time (below 15 min), a delayed liquid–liquid demixing was observed. As the exposure time increased, the liquid–liquid demixing was accelerated as inferred from the rapid decrease of light transmittance of the coagulation bath (Figure 8.8). At a very long exposure time, the light transmittance decreased even before the immersion. Liquid–liquid demixing occurred on the top of cast polymer film through nucleation and growth, resulting in the formation of cellular and very open surface pores. As a result, a more porous membrane was obtained at higher preimmersion exposure time (Figure 8.8).

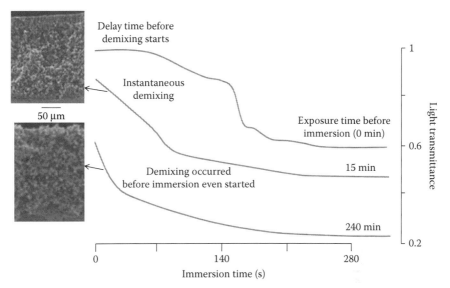

FIGURE 8.8 The effect of exposure time under humid atmosphere (RH = 60%) before immersion in coagulation bath of the PVDF/(TEP/DMA)/water system with 15% PVDF concentration. SEM images show the impact on the resulting membrane structure. (Data from Li, Q. et al., *Polymers for Advanced Technologies*, 22, 520–531, 2011. doi:10.1002/pat.1549.)

The impact of cast film exposure between casting and immersion can be rationalized using a phase diagram. In the case of volatile solvent evaporation, this evaporation shifts the cast film composition toward the polymer corner on the diagram (especially at the top of cast polymer film) and reduces the solvation character (polymer chain mobility) of the polymer in the solvent. This allows for a shift in demixing from liquid–liquid to solid–liquid through gelation, crystallization, and/or vitrification, especially for crystalline and semicrystalline polymers such as PVDF. When the cast film is immersed in the nonsolvent bath, the polymer-rich phase becomes even more concentrated and the polymer-lean phase tends to be leaner. This leads to a denser skin layer as well as smaller pores and lower porosity in the sublayer (Figure 8.9, path 1). On the other hand, when preimmersion evaporation occurs in water vapor, adsorption of water vapor shifts the cast film composition, especially at the top layer, closer to the binodal line, which then accelerates the liquid–liquid demixing during the immersion step (Figure 8.9, path 2). The advantage of manipulating the preimmersion exposure time and surrounding humidity is through the ability to manipulate the cast film locally at the top skin layer, which mostly determines the performance of the resulting membrane, especially in MD.

8.6.1.2 Effect of Solvent Choice

The solvent plays an important role in determining the ultimate membrane properties and performance. The most important criterion for solvent selection is its ability to dissolve the polymer fully. The commonly used organic solvents to prepare PVDF membranes through immersion precipitation are dimethyl formamide (DMF) (Gugliuzza and Drioli 2009), DMA (Lin et al. 2003; Yeow et al. 2005; Yan et al. 2006),

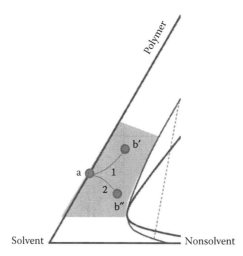

FIGURE 8.9 The change of composition across the cast film thickness from bottom (a) to top (b), after volatile solvent evaporation (path 1) and after nonsolvent (water) adsorption (path 2).

N-methylpyrrolidone (NMP) (Elashmawi 2008), dimethyl sulfoxide (DMSO), hexamethylphosphoramide (HMPA), tetramethylurea (TMU), trimethyl phosphate (TMP), and TEP (Bottino et al. 1991; Lin et al. 2006b; Shi et al. 2008). In addition, combinations of these solvents are also employed, such as TMP–DMA, TEP–DMA, tricresyl phosphate–DMA and tri-*n*-butyl phosphate–DMA (Li et al. 2010).

To understand how the solvent selection affects the PVDF membrane properties, the phase diagram of PVDF/solvent/water system of different solvents is presented in Figure 8.10. In general, the stronger the solvent, the closer the binodal line is to the solvent corner. This means that the addition of smaller amounts of nonsolvent is required to induce liquid–liquid demixing in a polymer/strong-solvent system compared to a polymer/weak-solvent system. It is worth noting that the strength of solvent is judged not only from the binodal boundary but also from the solubility parameter. Therefore, in some cases, the general role mentioned earlier is not fully applicable.

In most polymers, the impact of solvent/nonsolvent pair selection on the resulting membrane structure and performances can be explained based on their mutual affinity (miscibility) (Mulder 1996). In general, if their miscibility increases, more solvent is required in the nonsolvent bath to affect demixing. In other words, higher solvent/nonsolvent mutual affinity leads to delayed demixing and vice versa. However, such phenomenon is not clearly observed in PVDF polymers. Fortunately, an extensive study by Bottino et al. (1991) can be used as a basis to understand the effect of solvent selection on the structure and overall performance of PVDF membranes. The structure and performance of PVDF membranes prepared from PVDF/various solvents/water system can be correlated with the solvent/nonsolvent diffusivity (Figure 8.11). At the very first moment after immersion, but before top skin formation occurs, solvent/nonsolvent exchange starts before PVDF diffusion becomes important. In this condition, the polymer can be considered practically stable. Therefore, the top

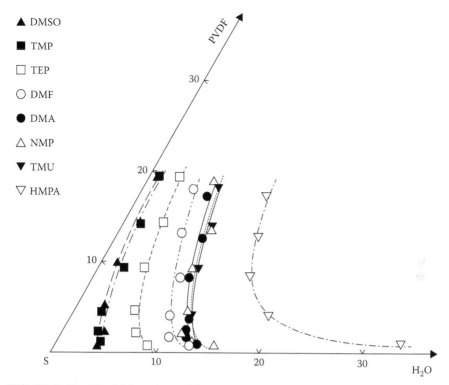

FIGURE 8.10 Binodal boundary of PVDF/solvent/water. (Data from Bottino, A. et al., *Journal of Membrane Science*, 57, 1–20, 1991. doi:10.1016/S0376-7388(00)81159-X.)

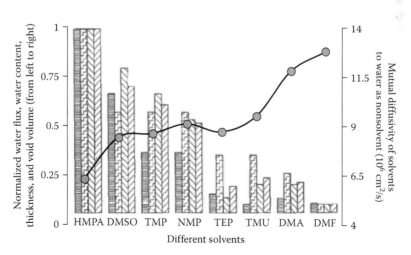

FIGURE 8.11 The relation between structures and the performance of PVDF membranes prepared using different solvents (with water as the nonsolvent) and the mutual diffusivity of solvent/water. (Adapted from Bottino, A. et al., *Journal of Membrane Science*, 57, 1–20, 1991. doi:10.1016/S0376-7388(00)81159-X.)

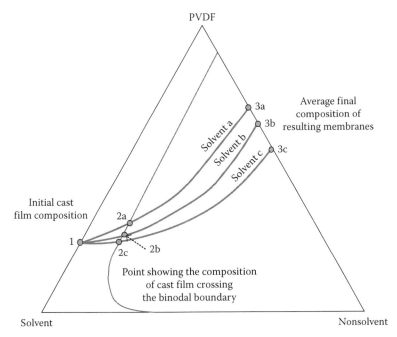

FIGURE 8.12 An illustration showing the change of the composition pathway of PVDF solution prepared from different solvents (a, b, and c with decreasing solvent/nonsolvent mutual diffusivity).

skin formation and membrane structure depend essentially on solvent/nonsolvent diffusion. Under these circumstances, also proven by several solvent/nonsolvent diffusion models, when the solvent/nonsolvent diffusivity increases, the composition pathway on the phase diagram of the cast polymer film enters the binodal boundary at higher polymer concentration. This results in the formation of denser, less permeable, low porosity PVDF membranes, as illustrated in Figure 8.12.

The effect of solvent strength is demonstrated by the cross-sectional structure of PVDF membrane, prepared from PVDF/solvent/water system, shown in Figure 8.13. When TEP (weak solvent) is used as the solvent, a symmetrical and sponge-like structure with no cavities through the whole thickness was obtained. Because TEP is a weak solvent for PVDF, the spinodal boundary is further from the TEP–PVDF line in phase diagram (Figure 8.10), a major presence of water is required to induce phase separation. Therefore, liquid–liquid demixing occurs at a later stage (delayed) and the diffusive exchange of the solvent/nonsolvent across the thickness of the membrane can still be prolonged, thus macrovoids cannot develop. The weak affinity of TEP/water also favors the formation of a sponge-like structure. On the other hand, when NMP (strong solvent) is used, irregular macrovoids are observed just underneath the top skin layer. This structure suggests the formation of a thin skin layer at an early stage of immersion (instantaneous demixing). This thin layer restricts further exchange of NMP/water, thus allowing sufficient time for macrovoid formation.

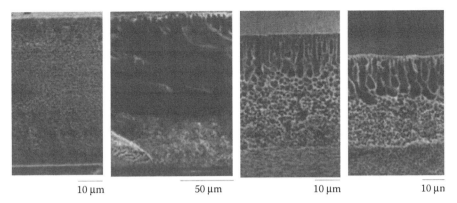

10 µm   50 µm   10 µm   10 µn

FIGURE 8.13 Cross-sectional structure of PVDF membranes prepared using water as the nonsolvent and TEP, NMP, DMF, and DMA as solvent (from left to right). (Data from Yeow, M.L. et al. *Journal of Applied Polymer Science*, 92, 1782–1789, 2004. doi:10.1002/app.20141.)

Solvent selection was also proven to affect the resulting structure in temperature-induced phase separation (TIPS) (Lloyd et al. 1990; Gu et al. 2006; Lu and Li 2009). Diluents such as phthalates promote the formation of irregular fuzzy structures (Lloyd et al. 1990). PVDF membranes prepared from dimethyl phthalate (DMP) as a solvent showed larger spherulite structures compared to that of membranes prepared from DBP, or mixtures of DMP with dioctyl sebacate and dioctyl adipate. It can be ascribed from these differences that the degree of polymer–solvents interactions influences the extent of PVDF crystallization (Lloyd et al. 1990; Gu et al. 2006). Using different types of solvents, the crystallization temperature of PVDF during TIPS can also be changed. A more complex mechanism is indeed expected for a combined solvent system. For instance, in the PVDF/DMAc/TEP system, TEP (weaker solvent) acts as a latent solvent in the immersion precipitation process (Liu et al. 2012).

8.6.1.3 Effect of PVDF Grade

Insignificant thermodynamic effect, but a more profound kinetic effect, were observed when using different grades of PVDF from Kynar™ in PVDF/NMP/water system (Sun et al. 2013). The binodal lines for all tested PVDFs almost overlapped, suggesting no significant differences between the different PVDF grades. However, their interactions with NMP were different, as observed from the different slopes of solvent evaporation rate just before immersion when exposed to air, which indicates differences in solvent diffusion rates, presumably affected by the difference in structure. Unfortunately, it is impossible to analyze what particular structure offered this effect because the chemical structures of the tested PVDFs were not available. Only limited physical properties were provided by the manufacturer.

It has been demonstrated that the skin layer properties with various crystalline or amorphous microstructures determined the surface pore structure and roughness (Buonomenna et al. 2007a). However, limited explanations are available to rationalize these properties, suggesting the need for more intensive studies on polymorphism in porous PVDF membranes.

8.6.1.4   Effect of PVDF Concentration

Increasing the PVDF concentration in the cast solution drastically increases the solution viscosity, which directly affects the structure and performance of the resulting membranes. In fact, this limits the maximum PVDF concentrations capable of producing one liquid-phase PVDF/solvent mixture to 20–30 wt% PVDF, depending on the PVDF grade and the solvent used (see Figures 8.3 and 8.4). Therefore, viscosity and concentration of the polymer solution are key factors that control the morphological structure of the produced membrane. This is further complicated by variation in temperature that directly affects the solution viscosity and, as a result, the thermodynamics and kinetics of the phase inversion. In addition, as the PVDF solution is cast on a woven/nonwoven support fabric, viscosity plays a crucial role in the peeling strength of the precipitated cast layer (Huo et al. 2009). Depending on the type and chemistry of the support fabric, too low PVDF concentration will deteriorate the structure of polymer film matrix because the solution may fully penetrate the support fabric.

Increasing the PVDF concentration in the initial polymer solution leads to even higher polymer concentration at the cast film/nonsolvent interface, due to the nonsolvent/solvent exchange process during the immersion process. This increased polymer concentration results in lower membrane porosity as well as surface contact angle (CA), as CA is also affected by the surface porosity and not only polymer chemistry (Ahmad and Ramli 2013). The polymer chains tend to align more closely as a result of the formation of a denser skin. This closer structure of the membrane with high polymer concentration increases the thickness of the top layer. It also leads to a reduction of both number and size of macrovoids/cavities, while keeping the structure asymmetric (Bottino et al. 1991; Ortiz de Zárate et al. 1995; Tomaszewska 1996; Ahmad and Ramli 2013). Typical cross-sectional SEM images of membranes prepared from different PVDF concentrations are shown in Figure 8.14.

The structure across the membrane thickness is also affected by PVDF concentration in the casting solution, in conjunction with changing the coagulation bath composition. In a study by Liu et al. (2012), undesired morphological characters such as the finger-like pores or macrovoids could be avoided by manipulating the PVDF concentration and coagulation bath composition, without the assistance of pore formers. Furthermore, all their membranes exhibited a bicontinuous structure (Liu et al. 2012). At low concentration (10 wt%), bicontinuous interconnected pores for the whole cross section with spherical crystals emerging in the cross section were obtained. At 15% PVDF, the membranes exhibited a porous structure with interconnected pores.

8.6.1.5   Effect of Coagulation Bath Composition and Temperature

The interaction between the solvent and the nonsolvent significantly affects the demixing process. Therefore, the coagulation medium is one of the key elements in determining the sequence of phase separation in the immersion precipitation process. As explained in Section 8.2, it is well known that water acts as a strong nonsolvent. Thus, the presence of water in the coagulation medium during immersion precipitation often leads to a rapid liquid–liquid demixing process and consequently PVDF membranes with asymmetric structure and finger-like voids are formed.

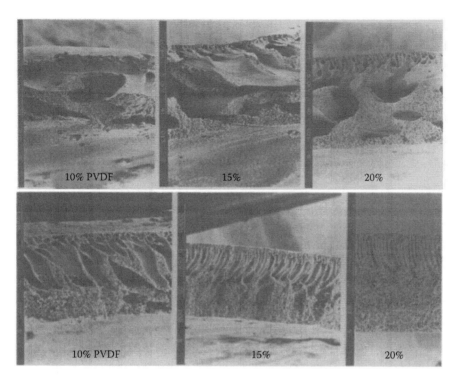

FIGURE 8.14 Typical cross section of PVDF membranes prepared from PVDF/NMP/water (upper) and PVDF/TMP/water (lower). (Adapted from Bottino, A. et al., *Journal of Membrane Science*, 57, 1–20, 1991. doi:10.1016/S0376-7388(00)81159-X.)

Like the solvent, the effect of nonsolvent in coagulation bath can also be explained using a phase diagram. As shown in Figure 8.15, the binodal boundary of PVDF/NMP/nonsolvents system is also a function of nonsolvent. The closer the spinodal line to the solvent–polymer line, the stronger is the nonsolvent. This means that only a small amount of nonsolvent is needed to disturb the polymer/solvent stability and induce liquid–liquid demixing. For the four nonsolvents shown in Figure 8.15, the strength of nonsolvents follows the order of water > methanol > ethanol > isopropanol.

Weaker nonsolvents with a lower solubility parameter other than water, such as ethanol, are sometimes used. These weaker nonsolvents lead to the formation of a denser membrane (Albrecht et al. 2001; Young and Chen 1995). Systems with a rapid phase inversion rate (strong nonsolvent) tend to form macrovoids with finger-like structures, whereas systems with a slow phase inversion rate yield sponge-like structures (Young and Chen 1995). In addition, polymer crystallinity can be affected by the choice of nonsolvent, as reported by Buonomenna et al. (2007b) for the (DMA/water and DMA/C1–C8 alcohols) system.

As an example of weak nonsolvent, the membrane structure prepared using 1-octanol as the nonsolvent in PVDF/DMF system is compared with PVDF/DMF/water system in Figure 8.16. An asymmetric structure consisting of dense thin top layer was formed using water as the nonsolvent. In contrast, a symmetric structure

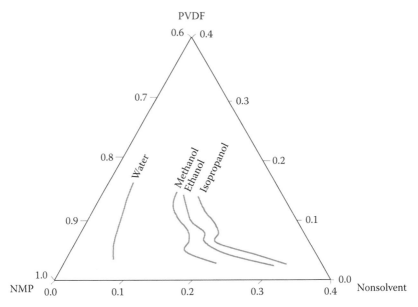

FIGURE 8.15 Phase diagram of ternary PVDF/NMP/nonsolvent systems at 25°C. (Data from Sukitpaneenit, P. and T.-S. Chung. *Journal of Membrane Science*, 340, 192–205, 2009. doi:10.1016/j.memsci.2009.05.029.)

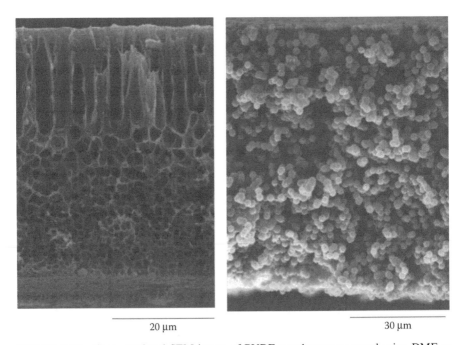

FIGURE 8.16 Cross-sectional SEM image of PVDF membranes prepared using DMF as the solvent and water (left) and 1-octanol (right) as the nonsolvent. (Data Young, T.-H. et al. *Polymer* 40, 5315–5323, 1999. doi:10.1016/S0032-3861(98)00747-2.)

with a uniform cross section packed with almost identical spherical particles was obtained when replacing water with the softer 1-octanol as the nonsolvent.

Addition of solvent into the nonsolvent coagulation bath leads to delayed demixing and eventually results in the formation of membranes with sponge-like structure or even spherulite formation (Young et al. 1999; Lin et al. 2006a). The presence of solvent in the coagulation bath limits the outflow diffusion of solvent in the cast film. However, the PVDF concentration at film interface also decreases, which induces a more porous structure. Therefore, two opposing effects occur: delayed demixing, which tends to produce membranes with a dense top layer, and low polymer concentration at the film interface, which tends to produce a more open top layer. The maximum amount of solvent that can be added is limited by the position of the binodal boundary on the phase diagram (see Figure 8.10).

An example of how the PVDF membrane properties can be engineered through the addition of soft solvent in a dual-bath coagulation system was presented by Thürmer et al. (2012). As shown in Figure 8.17, when using a single bath with water

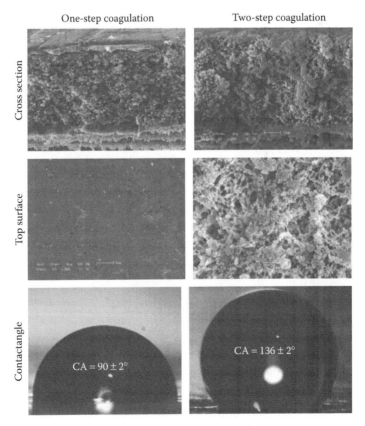

FIGURE 8.17 SEM images and CA measurements of PVDF membranes (20%) in DMF coagulated using a single bath containing water (left) and a two-stage coagulation bath first in ethanol for two seconds followed by water (right). (Data from Thürmer, M.B. et al., *Materials Research* 15, 884–890, 2012. doi:10.1590/S1516-14392012005000115.)

as the nonsolvent, the resulting membrane had a dense top structure, with an asymmetric sponge-like substructure. On the other hand, a porous top structure with globular sublayer was obtained using a dual-bath system; cast film first immersed in ethanol for two seconds, followed by immersion in water in a second bath. The symmetric structure resulting from application of weak solvent also leads to higher hydrophobicity, a desired property for MD (Peng et al. 2005; Li et al. 2011; Thürmer et al. 2012).

Li et al. (2011) applied a two-stage coagulation bath to improve the hydrophobicity and mechanical properties of PVDF membranes made using a PVDF/(TEP/DMA)/water system. In the first coagulation bath, the cast film was immersed in (TEP/DMA)/water nonsolvent system with a fixed ratio of 60/40 (TEP/DMA) but using variable amounts in water, for 30 seconds. The phase separation was completed in the second bath containing only water. As shown in Figure 8.18, higher amounts of (TEP/DMA) in the first bath have led to delayed demixing and vice versa. This was attributed to the small concentration gradient across the membrane-bath interface in the first bath, which hindered the solvent–nonsolvent exchange rate for the film immersed at higher TEP/DMA concentration. Therefore, macrovoids disappeared and a porous top layer was obtained. Moreover, the membrane structure was altered. The symmetry of the cross-sectional structure changed from being asymmetric to symmetric by increasing the amount of TEP/DMA in the first coagulation bath. At 60% TEP/DMA in water, the resulting membrane had a sponge structure with

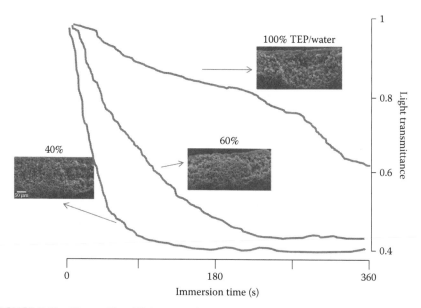

FIGURE 8.18 The profile of light transmittance of the second coagulation bath containing water. The cast films have been immersed in the first bath containing 40%, 60%, and 100% TEP/DMA in the ratio 40:60. The initial PVDF/TEP/DMA ratio of the polymer solution was 15/51/34. (Data from Li, Q. et al., *Polymers for Advanced Technologies*, 22, 520–531, 2011. doi:10.1002/pat.1549.)

interconnected pores throughout the whole thickness with a porous top surface. On the other hand, at 100% TEP/DMA, a symmetric structure with skinless surface was obtained. In addition, the pore radius and porosity of the PVDF membranes increased as the TEP/DMA concentration increased from 0% to 60%.

The application of dual coagulation baths to improve porosity, reduce maximum pore, and increase hydrophobicity was also reported by Ahmad and Ramli (2013). A series of casting films of 15% PVDF in NMP solution were immersed in the first bath containing 100% ethanol for 1 h, followed by immersion in a second bath containing increasing concentration of NMP in water (0%–80%). At higher NMP concentrations in water, a softer coagulation medium, the demixing was further delayed resulting in a lower polymer concentration at the film interface, which governs the porosity of the top layer and prevents the formation of asymmetric structure. This resulted in both increased porosity and hydrophobicity of the PVDF membrane (Peng et al. 2005). It also shifted the PSD of the membrane to a smaller average and maximum pore size.

In a dual coagulation bath system, a 15% PVDF concentration in the cast solution was found to be optimum for achieving favorable characteristics such as smaller pores, high porosity, and more hydrophobicity (Ahmad and Ramli 2013). In another study, with regard to MD performance, 15% PVDF concentration was found to be optimum for both PVDF/DMA/water and PVDF/DMF/water systems (de Zárate et al. 1995). de Zárate et al. (1995) reported that above that concentration, the PVDF membranes become too dense, whereas below that value the membrane matrix seemed to be rather inconsistent with big holes that appeared sporadically on the membrane surface.

In addition to the coagulation bath composition, precipitation temperature is one of the key factors determining the ultimate morphology and crystallinity of PVDF membranes. Variations in fabrication conditions such as the quenching and dissolution temperatures have also been shown to influence the crystalline nature, and thereby the morphology, of the fabricated membranes.

As a semicrystalline polymer, the kinetics of phase separation for PVDF are more difficult to comprehend than other amorphous polymers such as polysulfone (Wienk et al. 1996). The immersion temperature is proven to affect the crystallinity of PVDF membranes. For example, in a PVDF/DMA/water system (12% PVDF concentration), Buonomenna et al. (2007a) found that α crystal type was more dominant than β type at high coagulation bath temperature (60°C) and vice versa at 25°C. They explained the importance of the kinetic aspect. At higher temperature, solution viscosity is lower, which affects the two-phase separation process by increasing the mass transfer between the solution and precipitation bath, favoring the liquid–liquid demixing.

The morphology of the bicontinuous structure also varied with the change of coagulation bath temperature. By lowering the bath temperature from 80°C to 28°C, the casting solution viscosity increased, reflecting the microcosmic transformation of PVDF macromolecule chain conformation (Liu et al. 2012). Actually, the gelation behavior took place at a higher concentration of PVDF in TEP (e.g., 20 wt%), due to the formation of microcrystallites as the connecting sites between amorphous and crystalline zones. Therefore, the metastable PVDF/TEP solution offered a window of opportunity to control the unique membrane structure through both thermal and NIPS.

8.6.1.6   Effect of PVDF Blending with Other Polymers

The morphology of PVDF membrane can be significantly upgraded by blending with other polymers. In one study, the addition of 1% PMMA was found to increase, by 14-fold, the water permeance of the PVDF membrane, without losing its retention capacity (Nunes and Peinemann 1992). In another study, the filtration performance of PVDF membranes could be enhanced 20-fold through the addition of an amphiphilic comb polymer, which yielded a segregation of membrane chemistry across the membrane thickness. The surface coverage by the comb polymer provided hydrophilic surfaces with excellent stability and substantially higher surface porosities than the plain PVDF membranes (Hester and Mayes 2002). However, traditionally, most PVDF blending studies targeted filtration properties such as fouling resistance or improved cleanwater filtration permeance. Introduction of polymer blending, depending on the blend materials, not only influenced the morphology of the resulting membrane but also caused the surface to become hydrophilic, which is undesirable in MD.

Recently, copolymers such as poly(vinylidene fluoride-*co*-hexafluoropropylene) (PVDF-HFP) and poly(vinylidene fluoride-*co*-tetrafluoroethylene) (PVDF-TFE) were used to prepare MD membranes in flat-sheet and hollow-fiber forms using the phase inversion technique (Feng et al. 2004; García-Payo et al. 2009). These copolymers were added for the purpose of enhancing flexibility and processability of the membrane. Membrane surface modification using different technologies such as grafting, coating, or blending fluorinated surface-modifying macromolecules (SMMs) with hydrophilic polymers were also tested for different MD systems and configurations (Khayet et al. 2005; Krajewski et al. 2006; Suk et al. 2006; Jin et al. 2008).

8.6.2   Effects of Additives

Additives used in PVDF membranes synthesis for MD applications are not restricted to pore formers, though the latter are the most common. They encompass all additives used to impart desirable traits to an MD candidate membrane, such as high porosity, hydrophobicity, mechanical strength, antifouling properties, and surface modifications. Pore-forming agents, though diverse in their chemical nature, when present in a polymer blend have a functional role of imparting a porous nature or enhancing the interconnectivity between pores. Pore-forming agents induce multifunctional effects that may be synergistic or detrimental to the properties of the end membrane. In most cases, a pore former can increase solution viscosity or accelerate the phase inversion process, improve the membrane morphology, and enhance the membrane separation as well as its performance.

Additives can be broadly categorized into: (1) polymeric additives, such as poly(vinyl pyrrolidone) (PVP) and poly(ethylene glycol) (PEG); (2) weak nonsolvents (such as glycerol); (3) weak cosolvents (such as ethanol and acetone [Ac]); and (4) low-molecular-weight inorganic salts (such as lithium chloride [LiCl] and lithium perchlorate [LiClO$_4$]) (Yeow et al. 2004). The effects of these additives on the resulting PVDF membrane morphology have been extensively reported. The additives tested to date include PVP (Deshmukh and Li 1998), PEG (Feng et al. 2006), polystyrene sulfonic acid (Uragami et al. 1981), ethanol (Shih et al. 1990), LiCl (Tomaszewska 1996), and LiClO₄ (Yeow et al. 2005).

Addition of PVP and PEG was found to favor macrovoid formation in the fabrication of PVDF membranes. However, difficulties arose in completely washing out the polymer additives from the fabricated membranes. One should also note that polymeric additives exist at different molecular weights, which plays an important role. In contrast to polymeric additives, small molecular inorganic additives can easily diffuse out during the membrane formation and washing process. LiCl is an interesting inorganic additive for membrane fabrication because it interacts strongly with certain solvents and can form complexes with the carbonyl groups in such solvents as DMF, DMA, and NMP through the ion–dipole interaction. The strong LiCl–solvent interaction may result in a more favorable membrane morphology (Lee et al. 2002).

Among the polymeric additives, PVP, being hydrophilic in nature, can enhance the liquid–liquid demixing by allowing the water to intrude into the polymer solution at a faster rate, creating an environment that favors the production of more macrovoids (Wang et al. 1999). Low-molecular-weight PVP tends to create small pores and easily leaches out from the membrane. Most of the high molecular weight PVP remains in the membrane and may block the path of the interconnected voids. In contrast, as an additive, PEG has been reported to improve the pure water flux of membranes and membrane shrinkage behavior after treatment (Uragami et al. 1981). Either the enlargement or suppression of macrovoids can be obtained using the same additive, but with a variation of additive concentration or additive molecular weight. Although some additives have the tendency to form macrovoids, others help in suppressing them, improving the interconnectivity of the pores and resulting in higher porosities in both the top layer and the sublayer (Boom et al. 1992). In case of PVP as an additive, the thermodynamic effect dominates in the entire concentration range used. It significantly increases the permeate flux, whereas LiCl reduces macrovoid formation.

Fontananova et al. (2006) studied the effect of LiCl and PVP on PVDF homopolymer and copolymer and found that additives such as PVP and LiCl increase the thermodynamic miscibility of solvent and nonsolvent, thus enhancing the liquid–liquid demixing. But at the same time, a contradictory kinetic effect comes into play due to the increase in viscosity, which delays the mutual diffusion and slows down the phase inversion effect. At low concentrations of the salt, macrovoids are suppressed, whereas at higher concentrations, macrovoids are enlarged, favoring a higher permeate flux (Kimmerle and Strathmann 1990; Wang et al. 1999). Addition of LiCl in the casting solution remarkably changed the morphology of the PVDF membrane, where the formation of porous structure and larger cavities could be observed and these effects were enhanced with higher LiCl concentrations (Sun et al. 2013). The precipitation rate of the polymer from dope solution upon immersion process becomes relatively higher because of the high tendency of LiCl to dissolve in water. Similar results were obtained by Tomaszewska (1996) in the preparation of PVDF flat-sheet membranes for MD.

When used as an additive, high LiClO₄ concentrations increased the gelation in the LiClO₄/water/DMF/PVDF system (Lin et al. 2003). At a low LiClO₄ content, porous membranes were obtained, whereas macrovoids in the PVDF membrane structure were observed when the membrane was cast from high salt-containing dope. Yeow et al. (2005) investigated the influence of $LiClO_4$ on the morphology of the resulting PVDF flat-sheet membrane. At a relatively low $LiClO_4$ content, an increase in the mean pore size and a more uniform PSD could be obtained, though an excessive additive amount was reported to cause adverse effects.

Recent studies have employed the use of both Ac and phosphoric acid in the PVDF solution (Hou et al. 2012). The membrane prepared from $PVDF/DMAc/Ac/H_3PO_4$ dope had a thinner top skin, more porous sponge-like layer, larger pore size, and narrower PSD, owing to the synergistic effect of Ac and H_3PO_4. The strong interactions among the components of the dope tended to delay the phase separation rate, so the macrovoids became more like tears with wider ends. In addition, spherical nodule aggregates were formed in the polymer-rich phase area, the sponge-like structures changed to be more porous and the membrane porosity was improved. In the case of adding the mixture, because of the synergistic effect of Ac and H_3PO_4, the top skin got thinner, the sponge-like layer obtained more porous network structure and the pore size of membrane surface became larger.

Incorporation of Al_2O_3 and TiO_2 was also found effective to reduce PVDF fouling propensity. However, this approach is more suitable for pressure-driven membrane filtration and less interesting for MD application. The fouling resistance was achieved due to the hydrophilic nature of Al_2O_3 and TiO_2. It reduces surface roughness and CA (Cao et al. 2006; Yan et al. 2006; Oh et al. 2009). Unfortunately, limited data exist to link the presence of these particles in the casting solution to their influence in the immersion precipitation process.

8.6.3 OPTIMIZATION OF IMMERSION PRECIPITATION PARAMETERS

As presented in Sections 8.6.1 and 8.6.2, many parameters can be tuned to obtain desired PVDF membrane properties. In an attempt to find an optimum membrane for MD, the optimization of a few basic phase inversion parameters were investigated by Khayet et al. (2010), namely, PVDF concentration, exposure time, coagulation bath temperature, and additive (PEG). They reported that a set of optimum parameters could be found using a fractional factorial design process, which predicted the variables influence on the direct contact MD (DCMD) membrane flux and its salt rejection coefficient. Their results revealed that increasing the polymer content, the coagulant bath temperature and the solvent precoagulation evaporation time improved the salt rejection coefficient but reduced the DCMD permeate fluxes. In contrast, the increase of the additive PEG concentration in the PVDF–HFP casting solution has the opposite effect.

8.6.4 OTHER PARAMETERS

8.6.4.1 Effect of PVDF Dissolving Temperature

Wang et al. (2009) investigated the effect of PVDF dissolution temperature (50°C–120°C) in a PVDF/DMA/water system containing 15% PVDF and 1% water in the polymer solution. They found that, increasing the dissolution temperature

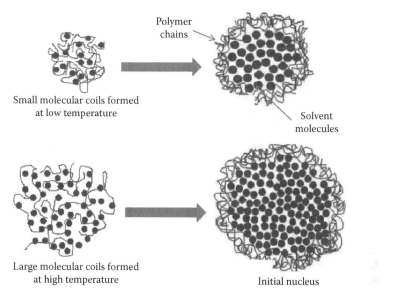

FIGURE 8.19 An illustration of molecular coils formed at low and high PVDF dissolution temperatures, and how they evolve during liquid–liquid demixing through nucleation and growth mechanism. (Data from Wang, X. et al., *Journal of Macromolecular Science, Part B*, 48, 696–709, 2009. doi:10.1080/00222340902958950.)

reduced the membrane porosity, thus the MD flux. Although, after dissolution the polymer was cooled to 25°C, it maintained its state, which eventually increased the size of the cavities across the thickness with the increase in the dissolving temperatures. At higher dissolution temperature, larger molecular coils were formed, which remained in this state for some time even after the solution was cooled (see Figure 8.19). Therefore, the droplets could easily increase their size during liquid–liquid demixing through nucleation and growth mechanisms.

A similar pattern was also observed by Lin et al. (2006a) using 23% PVDF in the nonsolvent octanol. Because softer nonsolvent was used, symmetric membrane with a globular structure across the thickness was obtained. They reported that higher the dissolution temperature, higher the globule size, which suggested the same mechanism shown in Figure 8.19.

8.6.4.2 Rasonic Irradiation

Ultrasonic-assisted phase inversion was proven to be effective in controlling the morphology of PVDF membranes prepared through immersion precipitation, both with and without LiCl as an additive (Tao et al. 2013). By increasing ultrasonic intensity (up to 300 W), morphology of the upper layer of the membrane could be changed dramatically from the original dense skin to regular large voids and then to uniform cellular pores. For the membranes prepared in the presence of LiCl, finger-like pores were suppressed, then aggravated, and the unfavorable cavities in membrane bulk were eliminated. Ultrasonic irradiation also slightly affected the crystalline structure, thermal stability, and tensile elongation. These findings opened the possibility of interfering with the phase inversion process, which was normally restricted to changing only the phase inversion parameters.

8.6.4.3 TiO$_2$ Coating

Super hydrophobic mixed-matrix membrane for MD were fabricated by creating a hierarchical structure with multilevel roughness by depositing TiO$_2$ nanoparticle on microporous PVDF membranes by means of a low temperature hydrothermal process (Razmjou et al. 2012). The TiO$_2$-coated membranes were then fluorosilanized using a perfluor-ododecyl-trichlorosilane. Although both control and modified membranes showed similar fouling behaviors, a significantly higher flux recovery was reported for the modified membrane compared to the control membrane. It is well known that reducing the surface free energy by functionalization with low surface energy materials, particularly fluorosilanes, is the most common method for the generation of super hydrophobic surfaces. Alternative attempts have been made to generate a hierarchical nanostructure surface morphology with multilevel surface roughness in order to modulate surface wettability.

8.6.4.4 Posttreatment

Super-hydrophobic PVDF membrane can be achieved by posttreatment using HYFLON 60XAD, a new family of copolymer of tetrafluoroethylene and 2,2,4-trifluoro-5-trifluorometoxy-1,3-dioxole. The membranes were made using a copolymer of two fluorinated polymers (Gugliuzza and Drioli 2007). HYFLON 60XAD is used to impart high water repellence, high mass transfer, and enhanced mechanical resistance for PVDF membranes as a result of the changes in the spherulite network of the membrane films.

8.6.4.5 Effect of Nonwoven Support

Most fundamental studies on the impact of immersion precipitation factors have cast membranes on glass supports during the synthesis phase. In the reality of full-scale application, PVDF is cast on a woven or a nonwoven fabric. Therefore, a different membrane behavior is expected. Huo et al. (2009) demonstrated that the woven support thickness is closely related to the optimum PVDF concentration. Below 10% PVDF, the polymer solution was too runny, and thus fully penetrated the support. As a result, a poorly integrated membrane structure was formed, with many big holes defects. On the other hand, when an integrated PVDF matrix was formed, lower concentrations yielded a higher peeling strength.

8.7 CONCLUSIONS

PVDF is widely investigated as one of the main materials for membrane fabrication, especially for MD applications. Much research has been done to unravel its morphological intricacies, crystallization behavior, and suitability for applications. Each investigation has revealed the complexity of this unique and versatile fluoropolymer, and added another aspect of fundamental understanding crucial for PVDF application in MD. In this chapter, the fabrication process of PVDF membranes, in particular through immersion precipitation, was discussed. For MD applications, it is required that the membrane matrix has a well interconnected sponge-like cellular structure, with a low nominal pore size and very narrow PSD, to ensure high LEP. It is crucial to eliminate/control the inclusion of macrovoids, as they act as failure points in the MD process. This chapter has demonstrated how phase diagrams can be effectively used to understand the immersion precipitation process. The interplay of different factors affecting the fabrication of PVDF membranes through immersion precipitation was also discussed. Many parameters have significant influence during the immersion precipitation process, they are inter-related with each other, which makes it very difficult, if not impossible, to unravel their individual effects. However, taking into account the number and the progress of research and development of PVDF membranes for MD applications, it is expected that a much better PVDF membrane is soon to be developed.

BIBLIOGRAPHY

Ahmad, A.L. and W.K.W. Ramli. 2013. "Hydrophobic PVDF Membrane via Two-stage Soft Coagulation Bath System for Membrane Gas Absorption of CO_2." *Separation and Purification Technology* 103: 230–240. doi:10.1016/j.seppur.2012.10.032.

Albrecht, W., Th. Weigel, M. Schossig-Tiedemann, K. Kneifel, K.-V. Peinemann, and D. Paul. 2001. "Formation of Hollow Fiber Membranes from Poly(ether Imide) at Wet Phase Inversion Using Binary Mixtures of Solvents for the Preparation of the Dope." *Journal of Membrane Science* 192(1/2): 217–230. doi:10.1016/S0376-7388(01)00504-X.

Alkhudhiri, A., N. Darwish, and N. Hilal. 2012. "Membrane Distillation: A Comprehensive Review." *Desalination* 287: 2–18. doi:10.1016/j.desal.2011.08.027.

Altena, F.W., J. Smid, J.W.A. Van den Berg, J.G. Wijmans, and C.A. Smolders. 1985. "Diffusion of Solvent from a Cast Cellulose Acetate Solution During the Formation of Skinned Membranes." *Polymer* 26(10): 1531–1538. doi:10.1016/0032-3861(85)90089-8.

Ameduri, B. 2009. "From Vinylidene Fluoride (VDF) to the Applications of VDF-Containing Polymers and Copolymers: Recent Developments and Future Trends[†]." *Chemical Reviews* 109(12): 6632–6686. doi:10.1021/cr800187m.

Baker, R.W. 2012. *Membrane Technology and Applications, 3rd Edition*. Chichester: Wiley. http://eu.wiley.com/WileyCDA/WileyTitle/productCd-0470743727.html.

Boom, R.M., I.M. Wienk, Th. van den Boomgaard, and C.A. Smolders. 1992. "Microstructures in Phase Inversion Membranes. Part 2. The Role of a Polymeric Additive." *Journal of Membrane Science* 73(2/3): 277–292. doi:10.1016/0376-7388(92)80135-7.

Bottino, A., G. Camera-Roda, G. Capannelli, and S. Munari. 1991. "The Formation of Microporous Polyvinylidene Difluoride Membranes by Phase Separation." *Journal of Membrane Science* 57(1): 1–20. doi:10.1016/S0376-7388(00)81159-X.

Buonomenna, M.G., P. Macchi, M. Davoli, and E. Drioli. 2007a. "Poly(Vinylidene Fluoride) Membranes by Phase Inversion: The Role the Casting and Coagulation Conditions Play in Their Morphology, Crystalline Structure and Properties." *European Polymer Journal* 43(4): 1557–1572. doi:10.1016/j.eurpolymj.2006.12.033.

Buonomenna, M.G., P. Macchi, M. Davoli, and E. Drioli. 2007b. "Poly(Vinylidene Fluoride) Membranes by Phase Inversion: The Role the Casting and Coagulation Conditions Play in Their Morphology, Crystalline Structure and Properties." *European Polymer Journal* 43(4) (April): 1557–1572. doi:10.1016/j.eurpolymj.2006.12.033.

Cao, X., J. Ma, X. Shi, and Z. Ren. 2006. "Effect of TiO_2 Nanoparticle Size on the Performance of PVDF Membrane." *Applied Surface Science* 253(4): 2003–2010. doi:10.1016/j.apsusc.2006.03.090.

Cohen, C., G.B. Tanny, and S. Prager. 1979. "Diffusion-controlled Formation of Porous Structures in Ternary Polymer Systems." *Journal of Polymer Science: Polymer Physics Edition* 17(3): 477–489. doi:10.1002/pol.1979.180170312.

Cui, Z., E. Drioli, and Y.M. Lee. 2013. "Recent Progress in Fluoropolymers for Membranes." *Progress in Polymer Science*. doi:10.1016/j.progpolymsci.2013.07.008. http://www.sciencedirect.com/science/article/pii/S0079670013000889.

Curcio, E. and E. Drioli. 2005. "Membrane Distillation and Related Operations—A Review." *Separation & Purification Reviews* 34(1): 35–86. doi:10.1081/SPM-200054951.

Deshmukh, S.P. and K. Li. 1998. "Effect of Ethanol Composition in Water Coagulation Bath on Morphology of PVDF Hollow Fibre Membranes." *Journal of Membrane Science* 150(1): 75–85. doi:10.1016/S0376-7388(98)00196-3.

Dillon, D.R., K.K. Tenneti, C.Y. Li, F.K. Ko, I. Sics, and B.S. Hsiao. 2006. "On the Structure and Morphology of Polyvinylidene Fluoride–nanoclay Nanocomposites." *Polymer* 47(5): 1678–1688. doi:10.1016/j.polymer.2006.01.015.

Drobny, J.G. 2009. *Technology of Fluoropolymers*. Boca Raton, FL: CRC Press.

Elashmawi, I.S. 2008. "Effect of LiCl Filler on the Structure and Morphology of PVDF Films." *Materials Chemistry and Physics* 107(1): 96–100. doi:10.1016/j.matchemphys.2007.06.045.

Feng, C., B. Shi, G. Li, and Y. Wu. 2004. "Preparation and Properties of Microporous Membrane from Poly(Vinylidene Fluoride-co-Tetrafluoroethylene) (F2.4) for Membrane Distillation." *Journal of Membrane Science* 237(1/2): 15–24. doi:10.1016/j.memsci.2004.02.007.

Feng, C., R. Wang, B. Shi, G. Li, and Y. Wu. 2006. "Factors Affecting Pore Structure and Performance of Poly(Vinylidene Fluoride-co-Hexafluoro Propylene) Asymmetric Porous Membrane." *Journal of Membrane Science* 277(1/2): 55–64. doi:10.1016/j.memsci.2005.10.009.

Fontananova, E., J.C. Jansen, A. Cristiano, E. Curcio, and E. Drioli. 2006. "Effect of Additives in the Casting Solution on the Formation of PVDF Membranes." *Desalination* 192(1–3): 190–197. doi:10.1016/j.desal.2005.09.021.

García-Payo, M.C., M. Essalhi, and M. Khayet. 2009. "Preparation and Characterization of PVDF–HFP Copolymer Hollow Fiber Membranes for Membrane Distillation." *Desalination* 245(1–3): 469–473. doi:10.1016/j.desal.2009.02.010.

Gilron, J., L. Song, and K.K. Sirkar. 2007. "Design for Cascade of Crossflow Direct Contact Membrane Distillation." *Industrial & Engineering Chemistry Research* 46(8): 2324–2334. doi:10.1021/ie060999k.

Gu, M., J. Zhang, X. Wang, H. Tao, and L. Ge. 2006. "Formation of Poly(Vinylidene Fluoride) (PVDF) Membranes via Thermally Induced Phase Separation." *Desalination* 192(1–3): 160–167. doi:10.1016/j.desal.2005.10.015.

Gugliuzza, A. and E. Drioli. 2007. "PVDF and HYFLON AD Membranes: Ideal Interfaces for Contactor Applications." *Journal of Membrane Science* 300(1/2): 51–62. doi:10.1016/j.memsci.2007.05.004.

Gugliuzza, A. and E. Drioli. 2009. "New Performance of Hydrophobic Fluorinated Porous Membranes Exhibiting Particulate-like Morphology." *Desalination* 240(1–3): 14–20. doi:10.1016/j.desal.2008.07.007.

Guillen-Burrieza, E., R. Thomas, B. Mansoor, D. Johnson, N. Hilal, and Arafat, H. 2013. "Effect of Dry-Out on the Fouling of PVDF and PTFE Membranes Under Conditions Simulating Intermittent Seawater Membrane Distillation (SWMD)." *Journal of Membrane Science* 438: 126–139. doi:10.1016/j.memsci.2013.03.014.

Hao, J.H. and S. Wang. 2003. "Studies on Membrane Formation Mechanism by the Light Transmission Technique. I." *Journal of Applied Polymer Science* 87(2): 174–181. doi:10.1002/app.11244.

Hasegawa, R., Y. Takahashi, Y. Chatani, and H. Tadokoro. 1972. "Crystal Structures of Three Crystalline Forms of Poly(Vinylidene Fluoride)." *Polymer Journal* 3(5): 600–610. doi:10.1295/polymj.3.600.

Hester, J.F. and A.M. Mayes. 2002. "Design and Performance of Foul-Resistant Poly(Vinylidene Fluoride) Membranes Prepared in a Single-Step by Surface Segregation." *Journal of Membrane Science* 202(1/2): 119–135. doi:10.1016/S0376-7388(01)00735-9.

Hou, D., G. Dai, J. Wang, H. Fan, L. Zhang, and Z. Luan. 2012. "Preparation and Characterization of PVDF/nonwoven Fabric Flat-sheet Composite Membranes for Desalination through Direct Contact Membrane Distillation." *Separation and Purification Technology* 101: 1–10. doi:10.1016/j.seppur.2012.08.031.

Huo, R., Z. Gu, K. Zuo, and G. Zhao. 2009. "Preparation and Properties of PVDF-Fabric Composite Membrane for Membrane Distillation." *Desalination* 249(3): 910–913. doi:10.1016/j.desal.2009.06.069.

Ismail, A.F. and L.P. Yean. 2003. "Review on the Development of Defect-free and Ultrathin-skinned Asymmetric Membranes for Gas Separation through Manipulation of Phase Inversion and Rheological Factors." *Journal of Applied Polymer Science* 88(2): 442–451. doi:10.1002/app.11744.

Jin, Z., D.L. Yang, S.H. Zhang, and X.G. Jian. 2008. "Hydrophobic Modification of Poly(phthalazinone Ether Sulfone Ketone) Hollow Fiber Membrane for Vacuum Membrane Distillation." *Journal of Membrane Science* 310(1/2): 20–27. doi:10.1016/j.memsci.2007.10.021.

Khayet, M. 2008. "Membrane Distillation." In *Advanced Membrane Technology and Applications*, edited by Norman N.L., A.G. Fane, W.S.W. Ho, and T. Tsuura, pp. 297–369. Wiley, New York. http://onlinelibrary.wiley.com/doi/10.1002/9780470276280.ch12/summary.

Khayet, M. 2011. "Membranes and Theoretical Modeling of Membrane Distillation: A Review." *Advances in Colloid and Interface Science* 164(1/2): 56–88. doi:10.1016/j.cis.2010.09.005.

Khayet, M., C. Cojocaru, and M.C. García-Payo. 2010. "Experimental Design and Optimization of Asymmetric Flat-sheet Membranes Prepared for Direct Contact Membrane Distillation." *Journal of Membrane Science* 351(1/2): 234–245. doi:10.1016/j.memsci.2010.01.057.

Khayet, M. and T. Matsuura. 2001. "Preparation and Characterization of Polyvinylidene Fluoride Membranes for Membrane Distillation." *Industrial & Engineering Chemistry Research* 40(24): 5710–5718. doi:10.1021/ie010553y.

Khayet, M., J.I. Mengual, and T. Matsuura. 2005. "Porous Hydrophobic/hydrophilic Composite Membranes: Application in Desalination Using Direct Contact Membrane Distillation." *Journal of Membrane Science* 252(1/2): 101–113. doi:10.1016/j.memsci.2004.11.022.

Kim, H.J., R.K. Tyagi, A.E. Fouda, and K. Ionasson. 1996. "The Kinetic Study for Asymmetric Membrane Formation via Phase-Inversion Process." *Journal of Applied Polymer Science* 62(4): 621–629. doi:10.1002/(SICI)1097-4628(19961024) 62:4<621::AID-APP5>3.0.CO;2-V.

Kimmerle, K. and H. Strathmann. 1990. "Analysis of the Structure-determining Process of Phase Inversion Membranes." *Desalination* 79(2–3): 283–302. doi:10.1016/ 0011-9164(90)85012-Y.

Krajewski, S.R., W. Kujawski, M. Bukowska, C. Picard, and A. Larbot. 2006. "Application of Fluoroalkylsilanes (FAS) Grafted Ceramic Membranes in Membrane Distillation Process of NaCl Solutions." *Journal of Membrane Science* 281(1–2): 253–259. doi:10.1016/j.memsci.2006.03.039.

Lee, H.J., J. Won, H. Lee, and Y.S. Kang. 2002. "Solution Properties of Poly(Amic Acid)–NMP Containing LiCl and Their Effects on Membrane Morphologies." *Journal of Membrane Science* 196(2): 267–277. doi:10.1016/S0376-7388(01)00610-X.

Li, B. and K.K. Sirkar. 2004. "Novel Membrane and Device for Direct Contact Membrane Distillation-Based Desalination Process." *Industrial & Engineering Chemistry Research* 43(17): 5300–5309. doi:10.1021/ie030871s.

Li, Q., Z.-L. Xu, and M. Liu. 2011. "Preparation and Characterization of PVDF Microporous Membrane with Highly Hydrophobic Surface." *Polymers for Advanced Technologies* 22(5): 520–531. doi:10.1002/pat.1549.

Li, Q., Z.-L. Xu, and L.Y. Yu. 2010. "Effects of Mixed Solvents and PVDF Types on Performances of PVDF Microporous Membranes." *Journal of Applied Polymer Science* 115(4): 2277–2287. doi:10.1002/app.31324.

Lin, D.-J., K. Beltsios, T.-H. Young, T.-S. Jeng, and L.-P. Cheng. 2006a. "Strong Effect of Precursor Preparation on the Morphology of Semicrystalline Phase Inversion Poly(vinylidene Fluoride) Membranes." *Journal of Membrane Science* 274(1–2): 64–72. doi:10.1016/j.memsci.2005.07.043.

Lin, D.-J., C.-L. Chang, F.-M. Huang, and L.-P. Cheng. 2003. "Effect of Salt Additive on the Formation of Microporous Poly(Vinylidene Fluoride) Membranes by Phase Inversion from LiClO4/Water/DMF/PVDF System." *Polymer* 44(2): 413–422. doi:10.1016/S0032-3861(02)00731-0.

Lin, D.-J., C.-L. Chang, C.-K. Lee, and L.-P. Cheng. 2006b. "Preparation and Characterization of Microporous PVDF/PMMA Composite Membranes by Phase Inversion in Water/DMSO Solutions." *European Polymer Journal* 42(10): 2407–2418. doi:10.1016/j.eurpolymj.2006.05.008.

Liu, F., N.A. Hashim, Y. Liu, M.R.M. Abed, and K. Li. 2011. "Progress in the Production and Modification of PVDF Membranes." *Journal of Membrane Science* 375(1/2): 1–27. doi:10.1016/j.memsci.2011.03.014.

Liu, F., M.-M. Tao, and L.-X. Xue. 2012. "PVDF Membranes with Inter-Connected Pores Prepared via a Nat-ips Process." *Desalination* 298: 99–105. doi:10.1016/j.desal.2012.05.016.

Lloyd, D.R., K.E. Kinzer, and H.S. Tseng. 1990. "Microporous Membrane Formation via Thermally Induced Phase Separation. I. Solid-liquid Phase Separation." *Journal of Membrane Science* 52(3): 239–261. doi:10.1016/S0376-7388(00)85130-3.

Lu, X. and X. Li. 2009. "Preparation of Polyvinylidene Fluoride Membrane via a Thermally Induced Phase Separation Using a Mixed Diluent." *Journal of Applied Polymer Science* 114(2): 1213–1219. doi:10.1002/app.30184.

Mulder, M. 1996. *Basic Principles of Membrane Technology*. Dordrecht, the Netherlands: Kluwer.

Nunes, S.P. and T. Inoue. 1996. "Evidence for Spinodal Decomposition and Nucleation and Growth Mechanisms during Membrane Formation." *Journal of Membrane Science* 111(1): 93–103. doi:10.1016/0376-7388(95)00281-2.

Nunes, S.P. and K.V. Peinemann. 1992. "Ultrafiltration Membranes from PVDF/PMMA Blends." *Journal of Membrane Science* 73 (1): 25–35. doi:10.1016/0376-7388(92)80183-K.

Oh, S.J., N. Kim, and Y.T. Lee. 2009. "Preparation and Characterization of PVDF/TiO$_2$ Organic–Inorganic Composite Membranes for Fouling Resistance Improvement." *Journal of Membrane Science* 345(1/2): 13–20. doi:10.1016/j.memsci.2009.08.003.

de Zárate, J.M.O., L. Peña, and J.I. Mengual. 1995. "Characterization of Membrane Distillation Membranes Prepared by Phase Inversion." *Desalination* 100(1–3): 139–148. doi:10.1016/0011-9164(96)00015-X.

Patsis, A.V., E.H. Henriques, and H.L. Frisch. 1990. "Interdiffusion in Complex Polymer Systems Used in the Formation of Microporous Coatings." *Journal of Polymer Science Part B: Polymer Physics* 28(13): 2681–2689. doi:10.1002/polb.1990.090281314.

Peng, M., H. Li, L. Wu, Q. Zheng, Y. Chen, and W. Gu. 2005. "Porous Poly(Vinylidene Fluoride) Membrane with Highly Hydrophobic Surface." *Journal of Applied Polymer Science* 98(3): 1358–1363. doi:10.1002/app.22303.

Razmjou, A., E. Arifin, G. Dong, J. Mansouri, and V. Chen. 2012. "Superhydrophobic Modification of TiO₂ Nanocomposite PVDF Membranes for Applications in Membrane Distillation." *Journal of Membrane Science* 415–416: 850–863. doi:10.1016/j.memsci.2012.06.004.

Saffarini, R.B., B. Mansoor, R. Thomas, and H.A. Arafat. 2013. "Effect of Temperature-Dependent Microstructure Evolution on Pore Wetting in PTFE Membranes Under Membrane Distillation Conditions." *Journal of Membrane Science* 429: 282–294. doi:10.1016/j.memsci.2012.11.049.

Saffarini, R.B., E.K. Summers, H.A. Arafat, and J.H. Lienhard. 2012a. "Economic Evaluation of Stand-alone Solar Powered Membrane Distillation Systems." *Desalination* 299: 55–62. doi:10.1016/j.desal.2012.05.017.

Saffarini, R.B., E.K. Summers, H.A. Arafat, and J.H. Lienhard. 2012b. "Technical Evaluation of Stand-alone Solar Powered Membrane Distillation Systems." *Desalination* 286: 332–341. doi:10.1016/j.desal.2011.11.044.

Shi, L., R. Wang, Y. Cao, D.T. Liang, and J.H. Tay. 2008. "Effect of Additives on the Fabrication of Poly(Vinylidene Fluoride-co-Hexafluropropylene) (PVDF-HFP) Asymmetric Microporous Hollow Fiber Membranes." *Journal of Membrane Science* 315(1/2): 195–204. doi:10.1016/j.memsci.2008.02.035.

Shih, H.C., Y.S. Yeh, and H. Yasuda. 1990. "Morphology of Microporous Poly(Vinylidene Fluoride) Membranes Studied by Gas Permeation and Scanning Electron Microscopy." *Journal of Membrane Science* 50(3): 299–317. doi:10.1016/S0376-7388(00)80627-4.

Shimizu, H., H. Kawakami, and S. Nagaoka. 2002. "Membrane Formation Mechanism and Permeation Properties of a Novel Porous Polyimide Membrane." *Polymers for Advanced Technologies* 13(5): 370–380. doi:10.1002/pat.201.

Souhaimi, M.K. and T. Matsuura. 2011. *Membrane Distillation: Principles and Applications.* Amsterdam, the Netherlands: Elsevier.

Strathmann, H., P. Scheible, and R.W. Baker. 1971. "A Rationale for the Preparation of Loeb-Sourirajan-Type Cellulose Acetate Membranes." *Journal of Applied Polymer Science* 15(4): 811–828. doi:10.1002/app.1971.070150404.

Suk, D.E., T. Matsuura, H.B. Park, and Y. M. Lee. 2006. "Synthesis of a New Type of Surface Modifying Macromolecules (nSMM) and Characterization and Testing of nSMM Blended Membranes for Membrane Distillation." *Journal of Membrane Science* 277(1/2): 177–185. doi:10.1016/j.memsci.2005.10.027.

Suk, D.E., G. Pleizier, Y. Deslandes, and T. Matsuura. 2002. "Effects of Surface Modifying Macromolecule (SMM) on the Properties of Polyethersulfone Membranes." *Desalination* 149(1–3): 303–307. doi:10.1016/S0011-9164(02)00817-2.

Sukitpaneenit, P. and T.-S. Chung. 2009. "Molecular Elucidation of Morphology and Mechanical Properties of PVDF Hollow Fiber Membranes from Aspects of Phase Inversion, Crystallization and Rheology." *Journal of Membrane Science* 340(1/2): 192–205. doi:10.1016/j.memsci.2009.05.029.

Sun, A.C., W. Kosar, Y. Zhang, and X. Feng. 2013. "A Study of Thermodynamics and Kinetics Pertinent to Formation of PVDF Membranes by Phase Inversion." *Desalination* 309: 156–164. doi:10.1016/j.desal.2012.10.005.

Tao, M.-M., F. Liu, and L.-X. Xue. 2013. "Poly(Vinylidene Fluoride) Membranes by an Ultrasound Assisted Phase Inversion Method." *Ultrasonics Sonochemistry* 20(1): 232–238. doi:10.1016/j.ultsonch.2012.08.013.

Thürmer, M.B., P. Poletto, M. Marcolin, J. Duarte, and M. Zeni. 2012. "Effect of Non-solvents Used in the Coagulation Bath on Morphology of PVDF Membranes." *Materials Research* 15(6): 884–890. doi:10.1590/S1516-14392012005000115.

Tomaszewska, M. 1996. "Preparation and Properties of Flat-Sheet Membranes from Poly(Vinylidene Fluoride) for Membrane Distillation." *Desalination* 104(1–2): 1–11. doi:10.1016/0011-9164(96)00020-3.

Uragami, T., Y. Naito, and M. Sugihara. 1981. "Studies on Synthesis and Permeability of Special Polymer Membranes." *Polymer Bulletin* 4(10): 617–622. doi:10.1007/BF00256290.

Vandezande, P., L.E.M. Gevers, and Vankelecom, I.F.J. 2008. "Solvent Resistant Nanofiltration: Separating on a Molecular Level." *Chemical Society Reviews* 37(2): 365–405. doi:10.1039/B610848M.

van't Hof, J.A., A.J. Reuvers, R.M. Boom, H.H.M. Rolevink, and C.A. Smolders. 1992. "Preparation of Asymmetric Gas Separation Membranes with High Selectivity by a Dual-Bath Coagulation Method." *Journal of Membrane Science* 70(1): 17–30. doi:10.1016/0376-7388(92)80076-V.

Wang, D., K. Li, and W.K. Teo. 1999. "Preparation and Characterization of Polyvinylidene Fluoride (PVDF) Hollow Fiber Membranes." *Journal of Membrane Science* 163(2): 211–220. doi:10.1016/S0376-7388(99)00181-7.

Wang, X., X. Wang, L. Zhang, Q. An, and H. Chen. 2009. "Morphology and Formation Mechanism of Poly(Vinylidene Fluoride) Membranes Prepared with Immerse Precipitation: Effect of Dissolving Temperature." *Journal of Macromolecular Science, Part B* 48(4): 696–709. doi:10.1080/00222340902958950.

Wienk, I.M., R.M. Boom, M.A.M. Beerlage, A.M.W. Bulte, C.A. Smolders, and H. Strathmann. 1996. "Recent Advances in the Formation of Phase Inversion Membranes Made from Amorphous or Semi-Crystalline Polymers." *Journal of Membrane Science* 113(2): 361–371. doi:10.1016/0376-7388(95)00256-1.

Yan, L., Y.S. Li, C.B. Xiang, and S. Xianda. 2006. "Effect of Nano-Sized Al_2O_3-particle Addition on PVDF Ultrafiltration Membrane Performance." *Journal of Membrane Science* 276(1/2): 162–167. doi:10.1016/j.memsci.2005.09.044.

Yeow, M.L., Y.T. Liu, and K. Li. 2004. "Morphological Study of Poly(Vinylidene Fluoride) Asymmetric Membranes: Effects of the Solvent, Additive, and Dope Temperature." *Journal of Applied Polymer Science* 92(3): 1782–1789. doi:10.1002/app.20141.

Yeow, M.L., Y. Liu, and K. Li. 2005. "Preparation of Porous PVDF Hollow Fibre Membrane via a Phase Inversion Method Using Lithium Perchlorate ($LiClO_4$) as an Additive." *Journal of Membrane Science* 258(1/2): 16–22. doi:10.1016/j.memsci.2005.01.015.

Young, T.-H. and L.-W. Chen. 1995. "Pore Formation Mechanism of Membranes from Phase Inversion Process." *Desalination* 103(3): 233–247. doi:10.1016/0011-9164(95)00076-3.

Young, T.-H., L.-P. Cheng, D.-J. Lin, L. Fane, and W.-Y. Chuang. 1999. "Mechanisms of PVDF Membrane Formation by Immersion-Precipitation in Soft (1-Octanol) and Harsh (Water) Nonsolvents." *Polymer* 40(19): 5315–5323. doi:10.1016/S0032-3861(98)00747-2.

9 Membrane Contactor for Carbon Dioxide Absorption and Stripping

Rosmawati Naim, Masoud Rahbari-Sisakht, and Ahmad Fauzi Ismail

CONTENTS

9.1 Introduction 285
9.2 Basic Principles 286
9.3 Mass Transfer Resistance 288
9.4 Membrane Characteristics...290
    9.4.1 Polymer Membrane...290
    9.4.2 Membrane Wetting 293
        9.4.2.1 The Potential of Membrane Wetting..............................293
        9.4.2.2 Membrane Wetting Prevention 294
    9.4.3 Membrane Structure and Surface Modification 295
9.5 Highlights on CO_2 Absorption and Stripping Performance 299
9.6 Liquid Absorbent for Membrane Contactor 301
9.7 Recent Development and Future Direction of Membrane Contactor..........304
    9.7.1 Recent Development 304
    9.7.2 Future Direction..305
9.8 Conclusion 309
References...309

9.1 INTRODUCTION

Many chemically processed materials occur as a mixture of different components in gas, liquid, or solid phases. In order to remove one or more components from its original mixtures, it must be contacted with another phase. There are several conventional and established methods used in the industry to separate them, which includes distillation, leaching, absorption, adsorption, ion exchange, chromatography, crystallization, and extraction [1]. These established processes incorporate a variety of geometrical and flow configurations such as packed columns, spray columns, and bubble-tray columns in various phases, which include gas–liquid, liquid–liquid, vapor–liquid, or solid–liquid phases.

For example, in industrial gases and hydrocarbon gas processing, a particular chemical absorption process has employed alkanolamine solution to separate and recover CO_2 and H_2S gases from exhaust gases. However, these existing methods have several drawbacks such as higher desorption energy consumption, size and weight of equipment restrictions, high vapor losses, corrosion, and the formation of degradation products [2]. Therefore, new methods have been developed to separate these gases with less energy requirements, higher purity of permeate gases, and efficient operation over extended periods. Membrane contactors are one of the potential technologies seen as reliable for replacing these conventional systems. In contrast to the conventional systems, such as packed tower, distillation column, scrubbers, and strippers, membrane contactors offer several advantages, including higher interfacial area per unit volume, no flooding or entrainment, flexibility and compactness, reduced weight and size, no moving parts, independence from gas and flow rates, and ease of scale-up due to modular design.

The membrane contactor technology has been used since 1980, where it was first employed in blood oxygenation. Since then, it has been further implemented into various areas of interests, such as liquid/liquid extraction, dense gas extraction, semiconductors, flue gas separation [3], olefin/paraffin separation [4,5], water treatment [6], fermentation, and enzymatic transformation [7], to name a few. The main characteristic of the contactors is the membrane itself, which does not present any particular selectivity for the species to be transferred and only acts as a barrier between the involved phases, avoiding their mixing. This membrane is typically an hydrophobic microporous membrane, which is able to permeate one phase while blocking the other phase at the surface pore, as the interface is established and mass transfer occurs simply by diffusion [8]. Owing to the hydrophobic attributes, liquid on the other side will remain in the membrane pores provided pressure is slightly higher than the gas side.

Membrane contactors provide a continuous process for contacting two different phases in which one of the phases must be a fluid. Whether using a flat-sheet, hollow-fiber, or spiral-wound type, the membrane acts as a separator for two interfaces as it has two sides compared to conventional separation processes, which involve only one interface in a two-phase system. Therefore, it allows the formation of an immobilized phase interface between the two phases participating in the separation process [9]. Generally, there are five different classes of contacting operations: gas–liquid, liquid–liquid, supercritical fluid–liquid, liquid–solid, and contactors as reactors [10]. The most commonly used operation in industry are gas–liquid also known as vapor–liquid, liquid–liquid, and supercritical fluid–liquid. Each class of system has its own modes of operation but in this study, emphasis will be focused on the gas–liquid contacting systems. Table 9.1 describes the membrane contactor in summary.

9.2 BASIC PRINCIPLES

In principal, the ability of a membrane contactor to remove dissolved gases from an aqueous solution, by absorption or stripping mode, involves a mass transfer process between two phases. According to Henry's law, the amount of gas that will dissolve

TABLE 9.1

Membrane Contactor in Summary

Common Membrane Material	Common Liquid Absorbent	Membrane Contactor Categories/Applications	Advantages	Disadvantages
Polyvinylidene fluoride (PVDF) Polypropylene (PP) Polytetrafluoroethylene (PTFE) Polyetherimide (PEI) Polysulfone (PSf) Poly-4-methyl-1-pentene	*Gas–liquid phase* • Distilled water • Sodium hydroxide • Potassium hydroxide • Potassium carbonate • Amine groups • MEA, DEA, and MDEA • Piperazine • AMP • Amino acid salts (e.g., sodium glycinate, etc.)	**1. Gas–liquid phase** • Acid gas removal, flue gas separation, oxygenation of aqueous solution, degassing of water/aqueous solution for ultrapure water production, carbonation of aqueous solutions/beverages, ammonia recovery, stripping of volatile organic compounds (VOCs), and humidification/dehumidification of airstreams **2. Liquid–liquid phase** • Liquid/liquid extraction, membrane distillation, pervaporation, olefin/paraffin separation, water treatment, fermentation and enzymatic, ammonia removal from water, semiconductors, and pharmaceutical, metallurgical processes **3. Supercritical fluid–liquid** • Extraction of ethanol and acetone from aqueous solution. It is also known as porocritical extraction, which utilizes liquid phase at both sides of the membrane. The aqueous solution maintained at a high pressure and extraction solvent is a near-critical or a supercritical fluid **4. Contactors as reactors** • A combination of reaction with separation to increase conversion. Commonly used when a reaction involves some form of catalyst; there are two main types of membrane reactors, the inert membrane reactor and the catalytic membrane reactor • Commonly used in dehydrogenation reactions (e.g., dehydrogenation of ethane)	• No flooding at high flow rates • No entrainment or foaming • Volume reduction • High interfacial surface area • Operational flexibility • Economical—compact nature • Linear scale up—modularity • Low weight and compact equipment	• Membrane wetting • Capillary condensation • Chemical reaction with liquid absorbent

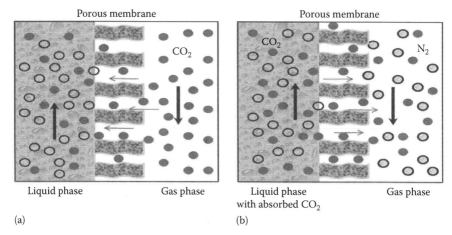

FIGURE 9.1 Schematic representation of (a) CO_2 absorption process and (b) CO_2 stripping mechanism through a gas–liquid membrane contactor.

into a water/solvent at equilibrium is proportional to its partial pressure in the vapor phase that is in contact with the water/solvent:

$$p = Hx \tag{9.1}$$

where:

p is gas partial pressure

H is Henry's law coefficient as a function of water temperature

x is the concentration of dissolved solute at equilibrium

At normal temperature and pressure, (25°C at 1 atm) water usually contains dissolved oxygen, nitrogen, and traces of CO_2 and other gases. If the partial pressure of the gas in contact with water is reduced, the amount of gas dissolved in the water will be reduced correspondingly. This reduction of partial pressure can be controlled by reducing the total pressure of the gas phase by applying a vacuum condition or sweep gas on the gas side, or by altering the concentration of gases in contact with water [11]. Modifying the pressure on either sides will create a driving force to move gases from the liquid phase into the gas phase or vice versa. Figure 9.1 illustrates the mechanism of absorption and stripping processes based on Henry's law.

9.3 MASS TRANSFER RESISTANCE

The mass transfer resistance theory in membrane contactors can be presented as a resistance in electric circuit. In Figure 9.2, the mass transfer resistance in the membrane contactor consists of resistance from the gas phase, membrane, and the liquid phase. The gas and liquid phases contribute to the overall resistance because of the formation of boundary layers close to the membrane surface. This implies that the concentration of the bulk of the two phases is different from its concentration at the membrane surfaces.

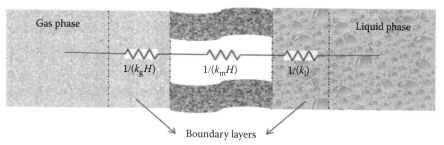

FIGURE 9.2 Resistance to the mass transport offered by the gas, membrane, and the liquid in a membrane contactor system.

The resistance offered by the membrane with liquid-filled pores will be different (generally higher) from the gas-filled pores, due to the different effective diffusion coefficients. There are two absorption processes operating: it is either by physical or chemical absorption. To predict the overall mass transfer coefficient (K_{ov}), the gas, membrane, liquid mass transfer coefficient, and Henry's law constant must be known. The mass transfer coefficient is a function of the system geometry, fluid properties, and flow velocity [1]. It can be expressed by a resistance in series model [9]:

$$\frac{1}{K_{ov}d_i} = \frac{1}{E\,k_l\,d_i} + \frac{1}{H\,k_m\,d_{ln}} + \frac{1}{H\,k_g\,d_o} \tag{9.2}$$

where:

K_{ov} is the overall mass transfer coefficient

k_l, k_m, and k_g are the liquid, membrane, and gas side mass transfer coefficient (m/s), respectively

H is represented by Henry's constant

E is the enhancement factor that describes the influence of a chemical reaction on the mass transfer rate

Meanwhile, d_o, d_i, and d_{ln} are the outer, inner, and log mean diameter of the fiber

In the case when pure CO_2 gas is used as the feed gas, the mass transfer resistance in the membrane phase can be neglected. In this case, according to the Fick's law, which is usually used to describe the mass transfer resistance in the membrane phase, the membrane thickness and tortuosity has little effect on mass transfer [12]. Leveque's correlation is a common equation used to describe the tube side mass transfer. Because the liquid absorbent flow inside the hollow fiber is typically laminar, the mass transfer coefficient in the tube can be estimated from the Leveque's correlation [13]:

$$Sh = \frac{k_l\,d_i}{D_L} = 1.62\left(\frac{d_i}{L}\,Re\,Sc\right)^{0.33} \tag{9.3}$$

where:

D_L is the diffusion coefficient of CO_2 in the liquid phase (m²/s)

L is the effective length of membrane (m)

For porous membrane with nonwetted mode of operation, k_m can be described using Fick's law as follows:

$$k_m = \frac{D_{g,eff}\varepsilon}{\delta\tau} \tag{9.4}$$

where:
$D_{g,eff}$ is the effective diffusion coefficient of gas in the gas-filled membrane pores
δ, ε, and τ are thickness, porosity, and tortuosity of the membrane, respectively

Using the Yang–Cussler [14] correlation, which is valid for fluid flows at low Reynolds number, the gas side mass transfer can be estimated from

$$Sh = \frac{k_g d_h}{D_{i,g}} = 1.25\left(Re\frac{d_h}{L}\right)^{0.93}(Sc)^{0.33} \tag{9.5}$$

where:
d_h is hydraulic diameter (m)
L is active length of hollow-fiber membrane (m)

9.4 MEMBRANE CHARACTERISTICS

9.4.1 POLYMER MEMBRANE

In industrial processing, common materials used in the membrane separation processes are polymers, where they offer several advantages, including superior chemical and heat resistance, solubility in most of organic solvents as well as ease of production and handling. Polypropylene (PP), polytetrafluoroethylene (PTFE), polyvinylidene fluoride (PVDF), and polyetherimide (PEI) are among the prominent polymers extensively used in membrane contactors. Some of these membrane materials such as PP and PTFE have been commercialized and utilized in beverage carbonation, gas dehydration, flue gas removal, and ammonia recovery. The commercial producers for these membranes are Membrana-Charlotte (NC, USA), Hoechst Celanese Corporation (NC, USA), and W.L. Gore and Co (DE, USA). Some of the commercialized membranes such as teflon (PTFE) have been successfully proven to resist wetting; however, due to the unavailability of small-size fiber diameters and the high price of PTFE, the manufacturing cost will make CO_2 removal using this membrane not economically attractive in comparison to conventional absorption processes. Table 9.2 illustrates the common types of membrane that are commercially available.

Generally, there are still some obstacles that may hinder the efficient and reliable operation of membrane in contactor systems. For example, membrane wetting due to capillary condensation of water vapor in the membrane pores [15], pores enlargement due to reaction between the solvent and membrane material [16], and also membrane deterioration due to long-term operation. Therefore, much effort has been applied to improve the performance of existing membranes in gas–liquid membrane contactors. For the absorption process, operation at ambient temperature is sufficient to obtain high absorption flux, but for stripping, higher temperature is needed to ensure the

TABLE 9.2
Commercially Available Membrane Material for Membrane Contactors

Products	Base Polymers	Manufacturers
Liqui-Cel	Polypropylene	Membrana, NC, USA
Accurel Q3	Polypropylene	Membrana, NC, USA
GORE-TECH	e-PTFE	Markel, PA, USA
PTFE	PTFE membrane	W. L. Gore and Co., DE, USA
Celgard X10	Polypropylene/poly(4-methyl-1-pentene)	Hoechst Celanese Corporation, NC, USA
PVDF	Polyvinylidenefluoride	Memcor, NSW, Australia.
		Tianjin Motian Membrane Engineering Technology Co. Ltd., Tianjin, China
Lestosil	Polydimethylsiloxane (PDMS)-based membrane	NPO 'Polimersintez', Vladimir, Russia
Solef	Polyvinylidene fluoride	Solvay, Tavaux, France
PP	Polypropylene	Tianjin Blue Cross Membrane Technology Co., Tianjin, China

desorption performance is at an accepted level. Hence, membrane materials that can withstand such conditions are desperately in demand. Incorporation of membrane contactors in stripping or regeneration processes has not been discussed extensively in the open literature. However, some research has been reported on the absorption/stripping process using different polymer materials and liquid absorbents. Kosaraju et al. [17] used hydrophobic poly(4-methyl-1-pentene) (PMP) in absorption/stripping with the novel solvent polyamidoamine (PAMAM). Mansourizadeh and Fauzi [18] and Mansourizadeh [19] reported on hollow-fiber PVDF membrane for a CO_2 absorption/stripping process by distilled water.

Meanwhile, Simioni et al. [20] employed flat-sheet PTFE and PES membrane for CO_2 stripping from potassium carbonate solvent. In addition, Khaisri et al. [21] undertook a comparison study of purchased PTFE, PP, and PVDF membrane in terms of membrane resistance and absorption performance. These researchers identified that PVDF has lower contact angle and higher membrane resistance compared to PP membrane, but in terms of absorption flux, PVDF membrane surpassed the former membrane. This can be correlated to the membrane pore size, porosity, and structure of PVDF membrane. Even though PTFE membrane in the study of Khaisri et al. [21] produced higher contact angles and low membrane resistance, from an economic point of view, PVDF is the preferable choice as the price was 96% lower compared to commercially available PTFE membrane.

Rangwala [22] studied a PP membrane (Hoechst Celanese Corporation) in different module sizes with aqueous NaOH and aqueous diethanolamine (DEA) as liquid absorbent. Measurement of membrane mass transfer indicated that the PP fiber was partially wetted by both the aqueous solutions despite the fact that the membrane should be nonwetted by solution. Barbe et al. [16] studied the surface morphology of PP membrane, Celgard 2500 and Accurel 1E-PP, and found that both membranes had undergone changes in porosity, pore size, and pore length with exposure to water for

72 h. This was made possible by the high lateral forces exerted by the water on the surrounding fibrils in the larger pores, which contributed to the enlargement of pores.

Khaisri et al. [23] used PTFE membrane (Markel Corporation) in a long-term CO_2 desorption from aqueous monoethanolamine (MEA) solution for 200 h at 100°C. Membranes of different porosities of 23% and 40% were tested and the result showed that the desorption performance of membranes with 40% porosity reduced continuously due to the intrusion of MEA into the membrane pore. PTFE membrane is known as chemical and temperature resistant, due to the doubling of pore size and porosity of membrane that contributes to a high potential of wetting. Pressure drop along the module may aid this condition. Khaisri et al. [24] concluded that the membrane with low porosity and small pore size would be appropriate to be employed in a membrane contactor in terms of flux stability performance.

From the discussion above, it is clearly seen that choosing the right membrane material is crucial in order to have high CO_2 absorption and stripping flux performance, with the ability of the membrane to withstand long hours of operation and high temperature conditions, especially in stripping. Of the many polymers that have been applied in membrane processes, PVDF is one of the materials that have been extensively used in membrane contactors. It is a notable polymer used in membrane contactors due to its high hydrophobicity and superb ability in resisting chemical and mechanical failures even in acidic conditions. It has been applied in several other process applications such as membrane distillation [25], liquid/liquid extraction, dense gas extraction, semiconductors, flue gas separation [3], olefin/paraffin separation [5], and water treatment [26]. Tan et al. [27] applied polyvinylidenefluoride hollow-fiber membrane contactor (PVDF HFMC) for ammonia stripping from water; meanwhile, Park et al. [28] studied NO_2 absorption using polyvinylidenefluoride hollow-fiber membrane (PVDF HFM) with several alkaline solutions and water. Lin et al. [29] studied PVDF and PP membrane for CO_2 absorption using mixed absorbent. In addition, Mansourizadeh et al. [30] used PVDF hollow-fiber membranes with the addition of a LiCl pore former and other potential additives to produce a micropores structure with higher CO_2 flux and lower mass transfer resistance.

Although polymeric membranes have received much attention from researchers in the membrane contactor area, the reliability of the membrane to withstand wetting after long operating hours and at high temperature is still doubtful. An attempt was made by Koonaphapdeelert et al. [31], who employed ceramic hollow-fiber membrane for CO_2 stripping from MEA at high temperatures up to 100°C. It is interesting to have such a system that does not require a cooling down procedure for flue gas emission because the system already operates at a similar temperature. However, the brittleness of the membrane is still a problem to overcome when the process involves material handling in a harsh working environment. In addition, Zhang and Wang [32] have developed highly hydrophobic organic–inorganic composite hollow-fiber membrane by incorporating a fluorinated silica ($fSiO_2$) inorganic layer on the PEI organic substrate. The membrane showed significant increase in contact angle measurement and performed with reasonable stability throughout the 31 days long-term operation using 2 M sodium taurinate. The membrane showed remarkable chemical compatibility after long-term exposure to liquid absorbent, as the hydrophobicity of the membrane remained constant and much higher than the original PEI substrate. Therefore, much work is required to improve membrane properties and performance, so that it is durable for operation in harsh operating conditions with resilience over long-term operation.

9.4.2 MEMBRANE WETTING

Membrane wetting can be considered as the main problem in membrane contactor applications. Wetting increases the membrane mass transfer resistance and reduce the mass transfer efficiency of the system. This occurs when the liquid absorbent enters the membrane pores, over a prolonged period of operation time resulting in gradual membrane wetting. The membrane wetting can be classified into three different modes: nonwetted, partially wetted, and completely wetted. In the nonwetted modes, the membrane pores are completely filled with gas phase over a period of time. Meanwhile, in the partially or completely wetted mode, the membrane pores are partially or completely filled with the liquid absorbent over time. According to Mavroudi et al. [33], when 13% of the membrane pores were liquid-filled, the mass transfer resistance of the membrane can increase to more than 98%. Meanwhile, from a simulation by Wang et al. [34], it was estimated that if 5% of the membrane pores were wetted, the overall mass transfer coefficient might be reduced by 20%. It was also claimed that the absorption rate in wetted mode was six times lower than those in nonwetted mode. In order to retain the best performance of membrane in CO_2 absorption and stripping, it is suggested that the maximum acceptable percentage of membrane wetting is about 40% [21].

9.4.2.1 The Potential of Membrane Wetting

Membrane wetting can occur in several ways and varies according to the mutual interaction of the liquid absorbent and the membrane materials. Dealing with physical absorbents is easier than chemical absorbents because the problem of wetting is less significant. However, the CO_2-absorption capacity of a physical absorbent is lower than that of a chemical solvent. Therefore, experimental work has been focusing on the wetting of chemical absorbents on the membrane material. Wang et al. [35] experimented with Celgard X40-200, Celgard X50-215, and PP membrane in aqueous diethanolamine solution for 3, 10, and 30 days. Both fibers were found wetted by the diethanolamine solution, which was observed through the morphology changes in scanning electron microscopy analysis. Pore structure and surface roughness of the membrane was also altered significantly as detected by atomic force microscopy (AFM) and X-ray photoelectron spectroscopy (XPS) analysis. It was possible that the chemical reactions between the membrane and diethanolamine solution had decreased the surface hydrophobicity (reduction of contact angle), which led to the deterioration of membrane matrixes. In addition, a similar study on the Celgard membrane for CO_2 absorption in diethanolamine solution experienced the same wetting problem indicated by the reduction of absorption flux, changes in membrane morphology, and the reduction in the overall mass transfer coefficient [34].

In another study, Lv et al. [36] studied PP membrane in prolonged contact with various liquid absorbents, including MEA, monodiethanolamine (MDEA), and deionized water, for 90 days. The results showed that diffusion of liquid absorbent into the membrane pores caused the membrane to swell, thus decreasing the surface hydrophobicity significantly. However, it was suggested that the wetting problem can be overcome by increasing the hydrophobicity of the membrane surfaces. Study on the effects of membrane porosity on desorption performance was performed by Khaisri et al. [23] using two different modules with porosity of 23% and 40%. Although the results showed that the module with high membrane porosity (40%) produced a high desorption rate, the performance dropped when the module was used for a long operation time due to the wetting problem. PTFE membrane used in their study is one of the most chemical-resistant polymers available in the market. However, it was suggested that the effects of increasing temperature and possible reaction with aqueous MEA solution might have contributed to the continuous decline of desorption flux. This observation was also supported by Lu et al. [37], who reported that high membrane porosity significantly promoted pore wetting compared to the wetting of a membrane with lower porosity.

The liquid absorbents used may be compatible with the membrane material; however, the harsh operating condition may affect the physical appearance and performance of the membrane. This is even experienced by commercial membrane where it tends to react with liquid absorbents such as sodium hydroxide and amine solution, especially when contacted for long hours and exposed to high temperature conditions [38]. By exposing PP membrane in diethanolamine aqueous solution for 3 and 10 days, Wang et al. [35] reported that not only was the pore structure of the membrane altered, but the surface roughness of the membrane surface was also increased. This is in agreement with a study reported by Rangwala [22] where it was speculated that partial wetting of PP membranes could be possible when amine aqueous solution was applied.

9.4.2.2 Membrane Wetting Prevention

In most applications, the interface between two phases in a membrane contactor lies on one side of the membrane surface. This also implies that one phase usually occupies the porous region of the membrane. However, this situation is quite different from *gas membranes* where the gas phase fills the pores of the membrane, whereas desorbing and absorbing solutions are in contact with both sides of the membrane [39]. Under these conditions, the gas or liquid in the fiber lumen cannot be displaced with the gas or fluid outside the shell side until the pressure difference between the external phase and the phase in the pores is greater than the breakthrough pressure (P). The relationship of pore size, contact angle for liquid and membrane material, surface tension of liquid absorbent, and pressure difference between gas and liquid are governed by the Young–Laplace equation

$$P = \frac{-2\sigma\cos\theta}{r} \tag{9.6}$$

where:
P is the breakthrough pressure
σ is the surface tension of the liquid absorbent
θ is the contact angle for liquid and membrane material
r is the pore radius (m) of the microporous membrane

Breakthrough pressure can be viewed as the success of a long-term wetting resistance provided that the pressure is retained at a higher value. In order to achieve this, the liquid absorbent used should have a higher surface tension, lower maximum pore size, and high contact angle. Taking into consideration the value of θ, when θ = 0, the membrane pores will be totally wetted by the liquid. Meanwhile, if the contact angle is formed around (30° < 0 < 89°) the pores will be *partially wetted* and for 90° and above, the substrate is nonwetting. Various contact angles between liquid and polymer were observed depending on the affinity between liquid and polymer. In addition, surface roughness can affect contact angle and breakthrough pressure [40].

In order to minimize or prevent membrane from wetting, researchers have recently focused on the screening of liquid absorbents and membrane materials, namely, mixing various types of liquid absorbents [2,41]; surface modifying membrane through grafting [42]; incorporating pore-forming agent [43]; and inorganic filler on the membrane surface [32]. Polymer with high surface energy is prone to wetting. To avoid this, polymer should have a low surface energy and the smallest possible maximum pore size. Wetting pressure is inversely proportional to membrane pore size [44], where higher pressure is needed for lower pore size (0.1 μm) compared with lower pressure needed for higher pore size (10 μm). The surface energy of polymers and contact angles with water are exemplified in Table 9.3.

Dindore et al. [41] provides an extensive study on membrane-solvent combination for CO_2 removal in gas–liquid membrane contactors. They highlighted that the possibility of membrane wetting is primarily governed by the membrane pore size, surface tension of the liquid and the mutual interactions of liquid and membrane materials signified by contact angle. Dindore et al. [41] concluded that the combination of PP membrane and propylene carbonate as liquid absorbents is the best option for further research. However, the study was limited to PTFE membrane and no comparison was made with the other types of membranes.

9.4.3 MEMBRANE STRUCTURE AND SURFACE MODIFICATION

Many attempts have been made to modify membrane microstructure by controlling the phase inversion process, changing the spinning parameters, and adding

TABLE 9.3
Surface Energy of Polymer Materials

Polymer	Critical Surface Tension (mN/m) or (mJ/m²)	Water Contact Angle (°)
Polytetrafluoroethylene	19.4	109.2
Polytrifluoroethylene	23.9	–
Polypropylene	30.5	102.1
Polyvinylidene fluoride	31.6	89
Polyethylene	33.2	–
PDMS	20.1	107.2

Source: M. Mulder, *Basic Principles of Membrane*, Kluwer Academic Publishers, Dordrecht, the Netherlands, 1991.

nonsolvent additives such as LiCl, PVP, and glycerol. Manipulation of fabrication parameters may also create good membrane structure that can produce high gas permeability, high absorption, stripping flux as well as high wetting resistance. The structure of the fabricated membrane has significant influence on the membrane contactor performance, especially in long-term operation. The combined structure of finger-like and sponge-like formation can be considered as ideal, as this provides an easy gas channeling for mass transfer processes and the ability to handle high intrusion pressure of liquid absorbent.

In the fabrication of microporous hollow-fiber membrane, the development of membrane pores can be correlated to thermodynamic and kinetic effects of the polymer dope. When additives were introduced into the polymer dope, two significant effects transpired. First, instantaneous demixing would occur, which was due to the thermodynamic enhancement of phase separation by reducing the miscibility of the solution dope with the nonsolvent. Second, the increased viscosity of the solution would cause kinetic hindrance against the phase separation process, which resulted in a delay of solution demixing [45].

Mansourizadeh and Ismail [46] studied various nonsolvent additives in polysulfone hollow-fiber membrane for CO_2 absorption. Various membrane microstructures were obtained with increases in dope viscosity and reduced precipitation rate. Addition of glycerol and utilization of 95:5 bore fluid composition solvent over water, produced combined sponge-like and finger-like structure that increased the membrane wetting pressure and adversely affected the gas permeability. This structure, however, contributed to high CO_2 absorption flux (1.09×10^{-3} mol/m²s), which can be related to the small pore size, high surface porosity, and high water entry pressure of the membrane. Bakeri et al. [47] reported on hydrophilic PEI hollow-fiber membrane with different polymer concentrations used in CO_2 absorption tests. By applying water as bore fluid, finger-like macrovoid structure was produced and the thickness of the inner skin layer increased gradually as the dope PEI concentration increased from 10 to 15 wt%. At low polymer concentration (10 wt%), the CO_2 absorption flux was the highest, compared to other polymer concentrations, but had a low breakthrough pressure (<1 bar) and restricted the performance of the membranes.

In another study, Ghasem et al. [48] fabricated PVDF membrane through thermally induced phase separation at different extrusion temperatures of 150°C–170°C. The results revealed that at a high extrusion temperature of 170°C, the membrane had high porosity, high strength, high water contact angle, and high removal efficiency. From SEM analysis, it was observed that the formation of dense layer and spherulites particles might be the cause of high liquid entry pressure (>20 bar) and high contact angle (>130°) (Figure 9.3).

It was reported that the morphology and surface properties of membrane affect its performance in separation processes [49,50]. Surface properties such as surface porosity, pore size, surface roughness, surface charge, and hydrophobicity affect the performance of membrane in separation processes such as ultrafiltration, nanofiltration, membrane distillation, and membrane contactors. For example, it was reported that the surface hydrophobicity of membrane had great effect on membrane fouling in ultrafiltration processes used to separate humic acids from water [51]. Furthermore,

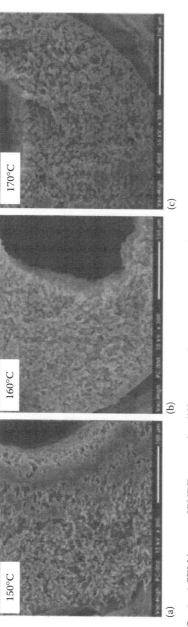

FIGURE 9.3   (a–c) SEM images of PVDF prepared at different extrusion temperature. (Data from N. Ghasem et al., *Sep. Purif. Technol.*, 99, 91–103, 2012.)

the surface hydrophobicity of membrane had significant effect on the performance of membrane in membrane distillation and membrane contactor processes, as the penetration of liquid into membrane pores reduced the performance drastically. Various processes have been reported to change the hydrophobicity/hydrophilicity of membranes; these include UV-assisted grafting, plasma treatment, and physical adsorption. In general, the surface modification processes can be divided into three categories: (1) surface modification by physical methods, (2) surface modification by chemical methods, and (3) bulk modification.

One interesting method for surface modification of membrane is the blending of surface modifying macromolecules (SMMs) into casting or spinning solution. SMMs are macromolecules with an amphipathic structure, the main chain consists of a polyurea or polyurethane polymer (hydrophilic part), which is end-capped with two low polarity fluorine-based polymer (oligomer) chains (hydrophobic part). During membrane fabrication processes, because of incompatibility of SMM with the base polymer, SMM migrates to the membrane–air interface; accumulates on the surface; and makes nanoscale aggregates [52], which change the surface properties of membrane. It should be noted that because of the amphipathic structure of SMM, the migration of SMM to the membrane–water interface is also possible, when water is used as bore fluid in hollow-fiber fabrication process. Migration of SMM to the membrane surface decreases the interfacial energy of system and stabilizes it.

Rahbari-Sisakht et al. [53] fabricated novel surface-modified PVDF hollow-fiber membrane for CO_2 absorption in a hollow-fiber membrane contactor system. The surface-modified PVDF membrane showed larger pore size, higher effective surface porosity, contact angle, and porosity but lower critical water entry pressure compared to the PVDF hollow-fiber membrane without SMM. Performance of the surface-modified membrane in contactor applications for CO_2 absorption through distilled water as absorbent was studied. The results showed that the surface-modified PVDF membrane had higher performance compared to control PVDF membranes. With the membrane prepared from SMM in the spinning dope, a maximum CO_2 flux of 7.7×10^{-4} mol/m^{2}s was achieved at 300 ml/min of absorbent flow rate, which was almost 93% more than the other membrane. In a long-term stability study, CO_2 flux was decreased by only 7.7% using surface-modified PVDF membrane during 150 h of operation.

Rahbari-Sisakht et al. [54] investigated the effect of novel surface modifying macromolecules (nSMMs) on the morphology and performance of PSf hollow-fiber membrane for CO_2 absorption. The performance of surface-modified membrane in contactor application for CO_2 absorption through distilled water as absorbent was studied. The results show that surface-modified membrane had higher performance compared to plain polysulfone membranes. With the membrane prepared from SMM in the spinning dope, a maximum CO_2 flux of 5.8×10^{-4} mol/m^{2}s was achieved at 300 ml/min of absorbent flow rate, which was almost 76% more than the other membrane. In a long-term stability study, the initial flux reduction was found to be about 18% after 50 h of operation of the surface-modified membrane.

9.5   HIGHLIGHTS ON CO_2 ABSORPTION AND STRIPPING PERFORMANCE

The design for a gas–liquid membrane contactor system is based on whether the absorption process is a physical dissolution or a chemical reaction. In a process where physical or chemical absorption of a gas, for example, CO_2 absorption in water or aqueous solution, the regeneration process can be done by stripping or desorbing the sweep gas at increased temperature or decreased pressure. For stripping CO_2 from liquid absorbent, gas can be recovered simply by flowing inert gas (e.g., N_2 or He) or by air stripping. According to research by Cabassud et al. [55] on disinfecting spring water and enhancing the pH value, they used air stripping rather than nitrogen due to economic considerations. In addition, they conducted several comparison tests that led to no significant variation of data when using either air or nitrogen as sweep gas.

In order to be feasible for CO_2 scrubbing, membrane contactors should possess a novel membrane that has high CO_2 transport rates and can prevent the loss of volatile amines from their aqueous solutions. To date, limited data has been reported discussing details of stripping operations and their performance using membrane contactors, despite the fact that stripping units are responsible for the major cost component in impurity removal processes and their contribution to high-energy consumption (Tobeisen et al. 2005). However, a few studies on CO_2 stripping from aqueous amine solution based on conventional plate or packed columns have been reported [50–53]. A pilot-scale membrane contactor hybrid system with a conventional stripper system was developed by Yeon et al. [3]. They compared the performance of PVDF hollow-fiber membrane using aqueous solution of MEA and triethanolamine (TEA) with a conventional packed column to recover CO_2 from flue gas. The CO_2-absorption rate of the contactor was 2.7 times higher than that of a packed column and the system successfully operated for 80 h, while maintaining a CO_2 removal efficiency of above 90%. An economical evaluation in terms of electric power consumption for both systems was carried out, and it was found that the electrical power consumption of the membrane contactor system was two times lower than the packed column. This is because of the high CO_2-removal efficiency of the membrane contactor system and the decrease in stripping temperature used for the regeneration of mixed MEA–TEA absorbent.

Recently, Mansourizadeh and Ismail [56] reported on the CO_2 stripping from water using PVDF hollow-fiber membrane, incorporating glycerol in the polymer dope formulation. They obtained 3.0×10^{-8} mol/m²s of stripping flux compared to the absorption flux of 8.5×10^{-4} mol/m²s. This work revealed that the CO_2-stripping efficiency increased as the concentration of the inlet liquid to the stripper increased. However, considerable decrease in CO_2-stripping efficiency was experienced as the inlet liquid flow rate increased from 50 to 200 ml/min, indicating less contact time between gas and liquid phase and insufficient time for the mass transfer between phases to occur. In addition, the accuracy of the data remains an issue as the water tends to vaporize at elevated temperatures of the stripping process.

TABLE 9.4
Research on CO_2 Absorption–Stripping Processes Using Gas–Liquid Membrane Contactors

Researchers	Membrane Types	Applications	Liquid Absorbents	CO_2 Stripping Flux (mol/m²s)
Kosaraju et al. [17]	Celgard polypropylene (HFM)	CO_2 absorption and stripping	Dendrimer polyamidoamine (PAMAM)	Not reported
Mansourizadeh and Ismail [54]	PVDF	CO_2 absorption and stripping	Distilled water	3.0×10^{-8} mol/m²s (stripping) 8.5×10^{-4} mol/m²s (absorption)
Koonaphapdeelert et al. [31]	Ceramic HFM	CO_2 stripping	Aqueous monoethanolamine (MEA)	Not reported
Khaisri et al. [23]	PTFE (Markel Corp., Glen Allen, VA) (HFM)	CO_2 stripping	Monoethanolamine (MEA)	5.0×10^{-4} [a] 3.0×10^{-3} [b]
Simioni et al. [20,58]	PTFE (Sartorius Stedim Biotech, Aubagne, France) (FSM)	CO_2 stripping	Potassium carbonate	1.33×10^{-2}
	PES with polyester support (Pall, USA) (FSM)		Potassium carbonate	2.00×10^{-2}
	Plasma nylon membrane (FSM)			Not reported
Naim et al. [59]	PVDF + LiCl (HFM)	CO_2 stripping	Aqueous diethanolamine (DEA)	1.6×10^{-2}
Mansourizadeh et al. [56]	PVDF	CO_2 absorption and stripping	Distilled water	8.5×10^{-4} (absorption) 3.0×10^{-8} (stripping)
Naim et al. [43]	PVDF + PEG-400 (HFM)	CO_2 stripping	Aqueous diethanolamine	4.0×10^{-2}
Naim and Ismail [60]	PEI	CO_2 stripping	Aqueous diethanolamine	6.0×10^{-3}
Rahbari-Sisakht et al. [61]	PVDF + surface-modified macromolecule (SMM)	CO_2 stripping	Aqueous diethanolamine	3.0×10^{-4}
Rahbari-Sisakht et al. [62]	Polysulfone	CO_2 stripping	Distilled water	4.0×10^{-5}

[a] 23% membrane porosity.

[b] 40% membrane porosity.

HFM, hollow-fiber membrane; FSM, flat-sheet membrane; M, molarity; T, temperature; V_l, liquid velocity; Q_l, gas velocity; PP, polypropylene; PVDF, polyvinylidene fluoride; PTFE, polytetrafluoroethylene; PMP, poly(4-methyl-1-pentene).

Some interesting results have been observed by researchers when investigating the performance of hollow-fiber membrane for CO_2 desorption. Astarita and Savage [57] revealed that the theory of absorption can also be applied to desorption with only mild restrictions. However, this is more complicated compared to modeling absorption, as one should take into consideration the reversibility of chemical reaction in order to employ the chemical thermodynamics variables into the calculations. The conditions highlighted are as follows: (1) mass transfer must lie within the diffusion and fast reaction regimes; (2) only single overall reaction can be considered; (3) the total capacity of the liquid for dissolved gas greatly exceeds its purely physical (Henry's law) capacity; and (4) the gas phase composition is linearly related to the concentration of dissolved but unreacted gas.

Kosaraju et al. [17] studied CO_2 absorption–stripping behavior of nonvolatile absorbent using the novel dendrimer PAMAM for wetting properties and employed hydrophobic PMP for over 75 days of operation with different flow rates of MEA absorbent. Their findings showed that the absorbent produces excellent performance without wetting of the porous PP fibers by the dendrimer for 55 days. On the contrary, PMP membrane pores were found to be MEA permeated due to the lower density of the polymer and its procession of large micropores. Nevertheless, no detailed discussion was made on the performance of the stripping unit.

In another study, the potential of an hollow-fiber ceramic membrane for CO_2 stripping from aqueous MEA solution at high temperatures up to 100°C was evaluated by Koonaphapdeelert et al. [31]. Flooding tests on the ceramic membrane proved that their membranes are immune from hydrodynamic problems and the module performances in terms of height of transfer unit (HTU) and carbonation ratio difference were described. Remarkable high mass-transfer efficiency was achieved with high surface area per unit volume and lower HTU when compared to conventional packing. Nonetheless, the mass-transfer coefficient model applied in their work had constantly overestimated the experimental value. Therefore, the employment of better mass-transfer models should be focused on, and efforts to reduce such deviation in the future should be undertaken. Table 9.4 summarizes some of the absorption and stripping performances reported in the open literature.

9.6   LIQUID ABSORBENT FOR MEMBRANE CONTACTOR

In membrane contactor processes, various types of aqueous liquid have been employed such as pure water, aqueous solution of NaOH, KOH, amine solution, and amino acid salts. Each of the absorbent has its own specialties that define a selective process application. Li and Chen [63] conducted a study on the selection of liquid absorbent in a membrane contactor, in which they highlighted criteria for choosing the chemical solvent to be implemented in membrane contactor. The criteria included high reactivity with CO_2, liquids with low surface tension, good chemical compatibility with membrane material, regenerability, low vapor pressure, and good thermal stability. Because any liquid that has surface tension lower than the critical surface tension of the polymers may wet the membrane spontaneously, the solvents must have a substantially higher surface tension than the critical surface tension values of the polymer. A low critical surface tension means that the surface has a low energy per unit area.

A great deal of research is focusing on the development and enhancement of new absorbents that can perform at high temperature and are compatible with membrane materials [2,17,49–52]. Chakma [64] discussed strategies for improving energy efficiency that were based on chemical solvents by optimizing process design and implementing mixed solvents that can reduce regeneration energy as much as 30%. An ideal liquid absorbent for contactors application should have good compatibility with membrane material, be noncorrosive or less corrosive, fast absorption and stripping rates with high surface tension, and regenerability (less regeneration energy). These characteristics reflect the performance and efficiency achieved by membrane contactors when minimal wetting occurs in the membrane structure. There are several common liquid absorbents employed in a gas–liquid membrane contactor such as distilled water, aqueous alkaline solution ($NaOH$, Na_2SO_3, etc.), conventional amine solution (DEA, MEA, and MDEA), and other enhanced absorbents such as 2-amino-2-methyl-1-propanol (AMP) promoted with piperzine [29].

In choosing the right liquid absorbent for a large scale of CO_2 removal in an existing power plant, reactive absorption is considered as the only quick and efficient technology that can be adopted in the industry. For that reason, post-combustion capture from power plant flue gases may be the best option to reduce CO_2 emissions. Mangalapally et al. [65] elaborated on a pilot plant experimental study of post-combustion CO_2 capture using reactive liquid absorbent MEA and their newly invented solvents. Because the utilization of MEA solution had contributed to the major drawback of high-energy requirement for solvent regeneration, Mangalapally et al. [65] decided to develop novel solvents, namely, CASTOR 1 and CASTOR 2, a blend of amine groups. Performance of the invented solvents, in comparison with standard MEA solution, in terms of regeneration energy with variation flow rates was evaluated on a gas-fired absorption/desorption pilot plant. The results showed that at lower solvent flows, CASTOR 2 produced promising results and for higher column height, CASTOR 1 and CASTOR 2 provided lower regeneration energy compared to MEA.

Some lab-scale testing had employed distilled water as a liquid absorbent; however, the absorption flux achieved was still far lower than that of CO_2-reactive solutions such as amine groups. Although CO_2 stripping using distilled water is acceptable, the operating stripping condition at higher temperature may result in inconsistency of data measurement because CO_2 absorbed in water can easily be released into the atmosphere at increasing temperatures. Shim et al. [66] tested novel absorbents, KoSol-1 (KS-1) and KoSol-2 (KS-2), at pilot plant scale for post-combustion CO_2 capture. They showed that their modified absorbent had a faster absorption rate, higher absorption capacity, and faster stripping rate than MEA solution. In addition, they revealed that it is possible to reduce the operation energy by more than 50% compared to using MEA. Unfortunately, for the post-combustion process, by employing a MEA solution, high energy consumption is required in the stripper to break the strong bond and nonsteric repulsion between MEA solution and CO_2. Meanwhile, Park et al. [67] utilized AMP as liquid absorbent with PTFE hollow-fiber membrane. It was reported that the outlet concentration of CO_2 decreased as AMP concentration increased while the gas flow rates decreased and the maximum range of inlet concentration of CO_2 was 40%. In addition, the use of sterically hindered amines, such as AMP or Mitsubishi Heavy Industry (KS-1, KS-2) provides another option, which has an exceptionally low corrosive nature and, unlike MEA, does not require a corrosion inhibitor [68].

Al-Saffar et al. [69] studied a porous and nonporous membrane (Celgard X-10 and Dow Corning silicone membrane) with water and aqueous DEA solution as liquid absorbent. They stated that their work on the CO_2 absorption rate using lower DEA concentration had much greater CO_2 absorption than that reported by Li and Teo [70], who used higher concentrations of aqueous NaOH. This can be explained given that CO_2 reacts directly with amine groups to form a carbamate ion, whose reaction is faster than the hydration reaction [71]. Performance of a hollow-fiber membrane contactor made of PP and hydrophobic PMP polymer with nonvolatile absorbent solution, dendrimer PAMAM, was studied by Kosaraju et al. [17]. The CO_2 absorption–stripping behavior was observed by wetting properties and running tests over 75 days with different flow rates of MEA absorbent. Their findings exhibit excellent absorbent performance with no wetting on PP fibers by the dendrimer for 55 days. On the contrary, PMP membrane pores were found to be MEA permeated due to the lower density of the polymer and having large micropores. Nevertheless, no detailed descriptions were explained on stripping unit.

Meanwhile, Rongwong et al. [72] compared absorption performance of various liquid absorbents, including water, MEA, DEA, and AMP. It was found that absorption flux increases in the order of MEA > AMP > DEA > water, which can be explained by the different reactivity of absorbent toward CO_2. It was emphasized that membrane wetting might occur in long-term absorption processes using mixed absorbents, but adding organic salt (sodium glycinate) in the aqueous solution can prevent the liquid intrusion into membrane pores. Mahajani and Danckwerts [73] studied the rate of CO_2 desorption in a 30% potash solution at 100°C used with a variety of amine groups (MEA, DEA, and TEA) as promoting agents. The outcome of their work indicated that DEA was the most effective amine for increasing the rate of desorption. Yan et al. [74] highlighted that using potassium glycinate for absorbing CO_2 from flue gas, this can reduce the potential of membrane wetting due to the higher surface tension compared to MEA and MDEA solutions. In pilot plant scales, they tested PP hollow-fiber membrane with aqueous potassium glycinate running continuously for 40 h and yet the systems were still able to maintain a removal efficiency of 90%. However, this new liquid absorbent has the disadvantage of reducing the mass transfer rate up to 50°C when operated at high temperatures. This condition can be correlated to the decrease in CO_2 solubility with increasing temperature and was supported by the work of Song et al. [75]. Table 9.5 highlights some of the research on the liquid absorbent in membrane contactors.

TABLE 9.5

Research on Liquid Absorbent Employed in Membrane Contactor Systems

Researchers	Liquid Absorbents	Membrane Materials	Findings
Ma'mun et al. [2]	Single and mixed amine-based absorbents	Not mentioned	2-(2-aminoethyl-amino)ethanol (AEEA) is a good absorbent for CO_2 capture from low pressure gases
Dindore et al. [76]	Propylene carbonate	Polypropylene HFM	Suitable combination for bulk CO_2 removal
Kumar et al. [77]	Potassium salt of taurine	Polyolefin HFM	Does not wet polyolefin membrane
Song et al. [75]	16 aqueous amino acids and blending with piperazine	Not mentioned	Smaller distances between amino, carboxyl groups, and bulkier substituted groups: resulted faster desorption rates, increased cyclic capacities, and slower initial absorption rates
			Adding a small amount of piperzine enhanced initial desorption and absorption rates
Lu et al. [78]	Activated MDEA and MDEA	Polypropylene HFM	CO_2 absorption performance of activated MDEA was better than MDEA
Bottino et al. [79]	MEA	PTFE and PP	High-removal efficiency
Todorovic et al. [80]	Sodium carboxymethylcellulose (CMC)	Polypropylene HFM	Increase concentration of CMC, increase liquid mass transfer resistance
Zhang et al. [81]	MDEA + PG	Polypropylene HFM	Hybrid solution of MDEA + PG was better than MDEA + MEA
Lu et al. [82]	Composite amino acid-based solution (glycin salt + piperazine)	Polypropylene HFM	Performance of the composite solution is better than single glycin salt solution

9.7 RECENT DEVELOPMENT AND FUTURE DIRECTION OF MEMBRANE CONTACTOR

9.7.1 RECENT DEVELOPMENT

Among the commercially available membrane in the market, Liqui-Cel membrane, which is made from PP polymer, is extensively used in membrane contactor systems for pilot plant test. For the last 10 years, membrane contactors have been used in the diverse areas of microelectronics, pharmaceutical, power, food and beverage, industrial photographic, ink, and analytical markets [83]. Among European countries, the Netherlands is actively committed to installing and testing membrane contactor systems for capturing CO_2 from flue gas (pre- and post-combustion capture).

Its efforts can be seen through numerous collaborations between Netherlands Organisation for Applied Scientific Research (TNO) and other industrial partners. Table 9.6 summarizes the current and ongoing field tests on membrane contactor systems for various applications.

In South Korea, Korean Electric Power Research Institute (KEPRI) had set up Liqui-Cel membrane contactor systems at Wolsung nuclear power plant for the removal of dissolved oxygen in the cooling unit to overcome pipeline and carbon steel ball corrosion in the system. Meanwhile, Aker Solutions had installed a pilot plant membrane contactor for natural gas dehydration at the Chevron Texaco gas plant in Texas, and CO_2 capture from flue gas at Kansai Nanko power plant in Osaka, Japan. The systems operated smoothly without any difficulties and only a slight deviation from the experimental and simulated data ($0 \sim 20\%$). Due to the modular and compact design of the membrane contactors, Aker Solutions (Norway) had developed a compact membrane contactor system for offshore application based on PTFE membrane. Although the weight and space for the absorber and stripper were tremendously reduced, the cost saved from reducing the equipment size will be offset by the cost of the membrane.

Hollow-fiber membrane using PP (Accurel) was also developed as a novel extraction device for plutonium extraction by Gupta et al. [84]. A membrane contactor for gas humidification was studied by Usachov et al. [85], utilizing polydimethylsiloxane-based membranes (Lestosil). Having the experience of operating a membrane contactor system as an absorber and stripper in various Kvaerner projects, Herzog and Pedersen [86] shared the successful development of a membrane contactor system as a viable commercial technology using PTFE membrane. They highlighted several factors in terms of economic and technological hurdles for commercializing the membrane contactor, including government incentives, marketability, collaboration, start-up problems, flexibility, focus, and novelty.

Recent work by Timofeev [87] reported on the use of a hybrid membrane contactor system for creating semibreathing air, particularly for medical institutions, to improve the respiratory system and ease the rapid recovery of the patient after surgery. Zhou et al. [88] used polyether etherketone polymer with surface modification through chemical grafting for syngas CO_2 separation (absorption and desorption). The developed membrane possesses extreme hydrophobicity and very high water entry pressure, up to 41 bars. Similar membrane was also implemented in pilot plant tests in Midwest Generation's Joliet Power Station using a hybrid membrane/absorption process for postcombustion CO_2 capture. In addition, Chakma [64] identified that in order to increase the efficiency of the CO_2 capture processes, 80% of the total energy needed for regeneration should be reduced. Therefore, much attention should be given to the reliability of the membrane contactor in stripping more than for the absorption process.

9.7.2 FUTURE DIRECTION

Membrane contactor technology, due to its attractive benefits, is proved to potentially substitute conventional distillations and packed columns in the separation industry. Although various experimental studies have been conducted at the laboratory and pilot scales to analyze the feasibility of the system, further advancement in

TABLE 9.6

Field Test/Commercial Pilot Plant Operation of Membrane Contactor Systems

Company and Collaborator	Location	Application	Membrane Material/ Liquid Absorbent	Performance/Design Operation	Reference
Korean Electric Power Research Institute (KEPRI), South Korea	Wolsung nuclear power plant, South Korea	Oxygen removal from water	Liqui-Cel membrane	–	Gabelman and Hwang [89]
Aker Solutions	Kansai Nanko power plant in Osaka, Japan	CO_2 removal from flue gas	Liqui-Cel membrane	–	Herzog and Falk-Pedersen [86]
Aker Solutions	Chevron Texaco gas plant in Texas	Natural gas dehydration	Liqui-Cel membrane	–	
Aker Solutions	TNO, GKSS in Hamburg, and at Gore in Munich	Natural gas treatment	PTFE membrane	–	
Aker Solutions	Offshore and on shore installation	Gas turbine exhaust treatment	GORE-TEX® ePTFE	Under commercialization	
Aker Solutions—from Herzog and Pedersen	Gas terminal pilot plant (Aberdeen, Scotland)	Removal of exhaust gas	GORE-TEX® ePTFE membrane/aqueous MDEA	Pressure regulation (problem for high pressure of natural gas). Exhaust gas flow rate: 2610 kg/h, 85% (195 kg/h) of CO_2 separated from exhaust gas	
Kvaerner & W.L. Gore Co.	State oil gas terminal, Norway		GORE-TEX® ePTFE	Pressure: 88 bar(g), gas flow rate: 5000 Nm³/h, liquid flow rate: 5 m³/h, CO_2 content reduced: 6% to 3.5%	
Gas Technology Institute & Duke Energy Field Service	Marla compressor station, Kersey, CO	Gas dehydration	Special coated membrane produced by W.L. Gore	Need to maintain lean glycol (10°C–15°C) to prevent condensation in membrane unit	Meyer and King [90]

					Reference
Liquid Purification Engineering International Co. Ltd.	Power plant in Thailand	Removal of CO_2 from DI water system	Liqui-Cel membrane	—	Klaassen et al. [91]
Ammonia Gas Absorption Technology project (AGATE)	Ammonia production plant	Ammonia recovery	Polypropylene membrane	—	Klaassen et al. [92]
TNO Science and Industry	KoSa Netherlands BV in Vlissingen, the Netherlands	Pertraction process	Liqui-Cel membrane	—	
TNO and partners	AVEBE, the Netherlands	Sulfur dioxide (SO_2) removal from flue gas at potato starch production plant	Polyolefin membrane/CORAL liquid absorbent (mixture of amino acids and alkaline salts)	—	
Aliachem	Pardubice, Czech Republic	Ammonia recovery at dye intermediates production plant	Polypropylene/water	Capacity to absorb ammonia: 50 kg/hr, ammonia recovery: >20 wt%, reduced 99.9 wt% of ammonia to the environment	
TNO and E.ON Benelux	Maasvlakte, Rotterdam	Removal of CO_2 from flue gas (coal-fired power plant)	Polyolefin membrane/CORAL liquid	More energy efficient and compactness, low overall cost per tonne of CO_2 captured, capacity: 70–250 kg CO_2/hour (max. 1250 Nm^3 incoming flue gas/hr)	
TNO	Pilot plant trial (feasibility study)	CO_2 removal from exhaust gas	Polypropylene membrane (Accurel Q3/2-Membrana)	CO_2 concentration: 3.5%, gas flow rate: 25 m^3/hr, CO_2 removal: 80%, liquid flow rate: 20–100 l/hr	
Pepsi-Cola Bottling Company, Charleston, WV	Pepsi-Cola beverage plant, West Virginia	Beverage carbonation	Membrane made by Hoechst Celanese Corporation	—	Mackey and Mojonnier [93]

several areas is needed before full-scale industrial technology can be implemented. To accelerate the success of this technology in the near future, the following challenges and recommendations need to be tackled:

- Numerous researchers have studied the potential of commercial membrane such as PTFE and PP membrane for long-term stability performance. Some of these commercial membranes are reliable in other applications; however, when contacted with conventional liquid absorbent such as aqueous amine solution in a membrane contactor system, the membranes became susceptible to wetting and lost their initial microstructure. This was unexpected as the commercial membranes have high hydrophobicity properties and high chemical tolerance. Fundamental study on the wetting mechanism is required as the detail of the process is still ambiguous.
- The critical area of research that should be addressed is the membrane wetting. Wetting of membrane pores with liquid absorbent should be eliminated or minimized to ensure efficient mass transfer in the membrane contactor, especially during long hours of operation. Compatibility of the membrane materials and liquid absorbent can be assessed by performing membrane–liquid screening processes, using a highly hydrophobic membrane and the modification of membrane surface. It is worth mentioning that the simple and fast method to analyze the membrane–liquid compatibility is by standard immersion tests at different temperatures. In addition, membrane properties such as contact angle, liquid entry pressure, and flux should remain intact before and after long contact with the liquid absorbent.
- Initial studies on membrane contactors focused on the absorption process, but this has been extended to solvent regeneration. Only a few research studies have reported their outcomes and comprehensive studies are required in this area because the operation involves high operating temperatures. This is because when operated at higher temperature, both liquid absorbent and membrane properties may be altered. This will affect the membrane stability, thus the membrane becomes susceptible to wetting.
- Not much work has been reported in the literature on the combination of both absorption and desorption processes for CO_2 removal. This is appropriate as the real application consists of both processes running at the same time. Thus, future studies should use the real conditions of gas mixtures, operating pressure, and temperature in order to compare membrane contactor performance with that of conventional methods.
- Comparative study in terms of economic points of view should be undertaken to have a clear view on the practicability of membrane contactor systems before they can be further developed to the full-scale level. Thus, capital expenditures and rate of investment estimation for future pilot plant implementation of this technology should be determined.

9.8   CONCLUSION

Membrane contactors have been extensively studied for various applications in the separation industries. The major findings from research studies and pilot plant projects on membrane contactor systems have been reviewed and discussed in this chapter. This technology has proven beneficial by offering a combined membrane separation and absorption process under one roof. Although some doubts may arise from the reported literature, the authors believe that membrane contactors have the potential to perform better than conventional separation columns. The key to success of this technology is to overcome the membrane wetting problem, which can lower the efficiency of the mass transfer process and reduces membrane life span. Thus, enormous effort is required to solve this issue by focusing on membrane–liquid compatibility at different operating conditions and in long-term stability studies. In addition, much work should be examined stripping and the regeneration process as the regeneration cost contributes significant amounts of money to the total operating cost. Finally, the fundamental understanding of the membrane–liquid correlation with the wetting mechanism should be explored in order to have a detailed view of the membrane contactor process and to ensure the successful implementation of the system at a larger scale.

REFERENCES

1. C.J. Geankoplis, *Transport Processes and Separation Process Principles*, 4th Edition, Pearson Prentice Hall, Upper Saddle River, NJ, 2003.
2. S. Ma'mun, H.F. Svendsen, K.A. Hoff, O. Juliussen, Selection of new absorbents for carbon dioxide capture, *Energy Convers. Manag.* 48 (2007) 251–258.
3. S.-H. Yeon, K.-S. Lee, B. Sea, Y.-I. Park, K.H. Lee, Application of pilot-scale membrane contactor hybrid system for removal of carbon dioxide from flue gas, *J. Memb. Sci.* 257 (2005) 156–160.
4. K. Nymeijer, T. Visser, R. Assen, M. Wessling, Super selective membranes in gas-liquid membrane contactors for olefin/paraffin separation, *J. Memb. Sci.* 232 (2004) 107–114.
5. D.C. Nymeijer, T. Visser, R. Assen, M. Wessling, Composite hollow fiber gas–liquid membrane contactors for olefin/paraffin separation, *Sep. Purif. Tech.* 37 (2004) 209–220.
6. S. Jiahui, F. Xuliang, H. Yiliang, J. Qiang, Emergency membrane contactor based absorption system for ammonia leaks in water treatment plants, *J. Environ. Sci.* 20 (2008) 1189–1194.
7. G.D. Bothun, B.L. Knutson, H.J. Strobel, S.E. Nokes, Mass transfer in hollow fiber membrane contactor extraction using compressed solvents, *J. Memb. Sci.* 227 (2003) 183–196.
8. A. Criscuoli, E. Drioli, Membrane contactore for gaseous stream treatments, in: A.K. Pabby, S.S.H. Rizvi, A.M. Sastre (Eds.), *Handbook of Membrane Separations: Chemical, Pharmaceutical, Food, and Biotechnological Applications*, CRC Press, Boca Raton, FL, 2008: pp. 1041–1055.
9. K.K. Sirkar, Other new membrane processes, in: W.S.W. Ho, K.K. Sirkar (Eds.), *Membrane Handbook*, Chapman & Hall, New York, 1992: pp. 885–899.
10. A.S. Kovvali, K.K. Sirkar, Membrane contactors: Recent developments, in: D. Bhattacharyya, D.A. Butterfield (Eds.), New Insights into Membrane Science and Technology: Polymeric and Biofunctional Membranes, Elsevier B.V., Amsterdam, the Netherlands, 2003: pp. 147–164.

11. F. Wiesler, Membrane contactors: An introduction to the technology, *Ultrapure Water* (1996) May/June, 27–31.

12. W. Zhang, J. Li, G. Chen, W. You, Y. Jiang, W. Sun, Experimental study of mass transfer in membrane absorption process sing membranes with different porosities, *Ind. Eng. Chem. Res.* 49 (2010) 6641–6648.

13. H. Kreulen, C.A. Smolders, G.F. Versteeg, W.P.M. van Swaaij, Microporous hollow fibre membrane modules as gas- liquid contactors. Part 1. Physical mass transfer processes A specific application : Mass transfer in highly viscous liquids, *J. Memb. Sci.* 78 (1993) 197–216.

14. M.-C. Yang, E.L. Cussler, Designing hollow-fiber contactors, *AIChE J.* 32 (1986) 1910–1916.

15. N. Nishikawa, M. Ishibashi, H. Ohta, N. Akutsu, H. Matsumoto, T. Kamata et al., CO2 removal by hollow fiber gas-liquid contactor, *Energy Convers. Manag.* 36 (1995) 415–418.

16. A.M. Barbe, P.A. Hogan, R.A. Johnson, Surface morphology changes during initial usage of hydrophobic, microporous polypropylene membranes, *J. Memb. Sci.* 172 (2000) 149–156.

17. P. Kosaraju, A.S. Kovvali, A. Korikov, K.K. Sirkar, Hollow fiber membrane contactor Based CO2 absorption-stripping using novel solvents and membranes, *Ind. Eng. Chem. Res.* 44 (2005) 1250–1258.

18. A. Mansourizadeh, A. Fauzi, A developed asymmetric PVDF hollow fiber membrane structure for CO2 absorption, *Int. J. Greenh. Gas Control* 5 (2011) 374–380.

19. A. Mansourizadeh, Experimental study of CO_2 absorption/stripping via PVDF hollow fiber membrane contactor, *Chem. Eng. Res. Des.* 90 (2012) 555–562.

20. M. Simioni, S.E. Kentish, G.W. Stevens, Membrane stripping: Desorption of carbon dioxide from alkali solvents, *J. Memb. Sci.* 378 (2011) 18–27.

21. S. Khaisri, D. deMontigny, P. Tontiwachwuthikul, R. Jiraratananon, Comparing membrane resistance and absorption performance of three different membranes in a gas absorption membrane contactor, *Sep. Purif. Technol.* 65 (2009) 290–297.

22. H. Rangwala, Absorption of carbon dioxide into aqueous solutions using hollow fiber membrane contactors, *J. Memb. Sci.* 112 (1996) 229–240.

23. S. Khaisri, D. deMontigny, P. Tontiwachwuthikul, R. Jiraratananon, CO_2 stripping from monoethanolamine using a membrane contactor, *J. Memb. Sci.* 376 (2011) 110–118.

24. S. Khaisri, D. deMontigny, P. Tontiwachwuthikul, R. Jiraratananon, Membrane contacting process for CO_2 desorption, *Energy Procedia* 4 (2011) 688–692.

25. A. Bottino, G. Capannelli, A. Comite, Novel porous poly (vinylidene fluoride) membranes for membrane distillation, *Desalination* 183 (2005) 375–382.

26. J. Shao, H. Liu, Y. He, Boiler feed water deoxygenation using hollow fiber membrane contactor, *Desalination* 234 (2008) 370–377.

27. X. Tan, S.P. Tan, W.K. Teo, K. Li, Polyvinylidene fluoride (PVDF) hollow fibre membranes for ammonia removal from water, *J. Memb. Sci.* 271 (2006) 59–68.

28. H.H. Park, B.R. Deshwal, H.D. Jo, W.K. Choi, I.W. Kim, H.K. Lee, Absorption of nitrogen dioxide by PVDF hollow fiber membranes in a G–L contactor, *Desalination* 243 (2009) 52–64.

29. S.-H. Lin, C.-F. Hsieh, M.-H. Li, K.-L. Tung, Determination of mass transfer resistance during absorption of carbon dioxide by mixed absorbents in PVDF and PP membrane contactor, *Desalination* 249 (2009) 647–653.

30. A. Mansourizadeh, A.F. Ismail, M.S. Abdullah, B.C. Ng, Preparation of polyvinylidene fluoride hollow fiber membranes for CO2 absorption using phase-inversion promoter additives, *J. Memb. Sci.* 355 (2010) 200–207.

31. S. Koonaphapdeelert, Z. Wu, K. Li, Carbon dioxide stripping in ceramic hollow fibre membrane contactors, *Chem. Eng. Sci.* 64 (2009) 1–8.

32. Y. Zhang, R. Wang, Fabrication of novel polyetherimide-fluorinated silica organic–inorganic composite hollow fiber membranes intended for membrane contactor application, *J. Memb. Sci.* 443 (2013) 170–180.

33. M. Mavroudi, S.P. Kaldis, G.P. Sakellaropoulos, A study of mass transfer resistance in membrane gas—liquid contacting processes, *J. Memb. Sci.* 272 (2006) 103–115.

34. R. Wang, H.Y. Zhang, P.H.M. Feron, D.T. Liang, Influence of membrane wetting on CO_2 capture in microporous hollow fiber membrane contactors, *Sep. Purif. Tech.* 46 (2005) 33–40.

35. R. Wang, D.F. Li, C. Zhou, M. Liu, D.T. Liang, Impact of DEA solutions with and without CO_2 loading on porous polypropylene membranes intended for use as contactors, *J. Memb. Sci.* 229 (2004) 147–157.

36. Y. Lv, X. Yu, S.-T. Tu, J. Yan, E. Dahlquist, Wetting of polypropylene hollow fiber membrane contactors, *J. Memb. Sci.* 362 (2010) 444–452.

37. J.-G. Lu, Y.-F. Zheng, M.-D. Cheng, Wetting mechanism in mass transfer process of hydrophobic membrane gas absorption, *J. Memb. Sci.* 308 (2008) 180–190.

38. N.A. Hashim, Y. Liu, K. Li, Stability of PVDF hollow fibre membranes in sodium hydroxide aqueous solution, *Chem. Eng. Sci.* 66 (2011) 1565–1575.

39. B.W. Reed, M.J. Semmens, E.L. Cussler, Membrane contactors, in: S.A. Noble, R.D. Stern (Eds.), Membrane Separations Technology: Principles and Applications, Elsevier B.V., Amsterdam, the Netherlands 1995: pp. 467–498.

40. D. Myers, *Surfaces, Interfaces and Colloids. Principles and Applications.* VCH Publishers, New York, 1991.

41. V.Y. Dindore, D.W.F. Brilman, F.H. Geuzebroek, G.F. Versteeg, Membrane–solvent selection for CO_2 removal using membrane gas–liquid contactors, *Sep. Purif. Technol.* 40 (2004) 133–145.

42. S. Wongchitphimon, R. Wang, R. Jiraratananon, Surface modification of polyvinylidene fluoride-co-hexafluoropropylene (PVDF–HFP) hollow fiber membrane for membrane gas absorption, *J. Memb. Sci.* 381 (2011) 183–191.

43. R. Naim, A.F. Ismail, A. Mansourizadeh, Effect of non-solvent additives on the structure and performance of PVDF hollow fiber membrane contactor for CO_2 stripping, *J. Memb. Sci.* 423–424 (2012) 503–513.

44. M. Mulder, *Basic Principles of Membrane*, Kluwer Academic Publishers, Dordrecht, the Netherlands, 1991.

45. I.M. Wienk, R.M. Boom, M.A.M. Beerlage, A.M.W. Bulte, C.A. Smolders, H. Strathmann, Recent advances in the formation of phase inversion membranes made from amorphous or semi-crystalline polymers, *J. Memb. Sci.* 113 (1996) 361–371.

46. A. Mansourizadeh, A.F. Ismail, Effect of additives on the structure and performance of polysulfone hollow fiber membranes for CO_2 absorption, *J. Memb. Sci.* 348 (2010) 260–267.

47. G. Bakeri, A.F. Ismail, M. Shariaty-Niassar, T. Matsuura, Effect of polymer concentration on the structure and performance of polyetherimide hollow fiber membranes, *J. Memb. Sci.* 363 (2010) 103–111.

48. N. Ghasem, M. Al-Marzouqi, N.A. Rahim, Effect of polymer extrusion temperature on poly(vinylidene fluoride) hollow fiber membranes: Properties and performance used as gas–liquid membrane contactor for CO_2 absorption, *Sep. Purif. Technol.* 99 (2012) 91–103.

49. M. Khayet, M.N.A. Seman, N. Hilal, Response surface modeling and optimization of composite nanofiltration modified membranes, *J. Memb. Sci.* 349 (2010) 113–122.

50. M. Khayet, G. Chowdhury, T. Matsuura, Surface modification of polyvinylidene fluoride pervaporation membrnaes, *AIChE J.* 48 (2002) 2843–2843.

51. L. Zhang, G. Chowdhury, C. Feng, T. Matsuura, R. Narbaitz, Effect of surface-modifying macromolecules and membrane morphology on fouling of polyethersulfone ultrafiltration membranes, *J. Appl. Poly. Sci.* 88 (2003) 3132–3138.

52. V.A. Pham, J.P. Santerre, T. Matsuura, R. Narbaitz, Application of surface modifiying macromolecules in polyethersulfone membranes: Influence on PES surface chemistry and physical properties, *J. Appl. Poly. Sci.* 73 (1999) 1363–1378.

53. M. Rahbari-Sisakht, A.F. Ismail, D. Rana, T. Matsuura, A novel surface modified poly-vinylidene fluoride hollow fiber membrane contactor for CO_2 absorption, *J. Memb. Sci.* 415–416 (2012) 221–228.

54. M. Rahbari-Sisakht, A.F. Ismail, D. Rana, T. Matsuura, Effect of novel surface modifying macromolecules on morphology and performance of Polysulfone hollow fiber mem-brane contactor for CO_2 absorption, *Sep. Purif. Technol.* 99 (2012) 61–68.

55. C. Cabassud, C. Burgaud, J.-M. Espenan, Spring water treatment with ultrafiltration and stripping, *Desalination* 137 (2001) 123–131.

56. A. Mansourizadeh, A.F. Ismail, CO_2 stripping from water through porous PVDF hollow fiber membrane contactor, *Desalination* 273 (2011) 386–390.

57. G. Astarita, D.W. Savage, Theory of chemical desorption, *Chem. Eng. Sci.* 35 (1980) 649–656.

58. R.H. Weiland, M. Rawal, R.G. Rice, Stripping of carbon dioxide from monoethanol-amine solutions in a packed column, *AIChE J.* 28 (1982) 963–973.

59. R. Naim, A.F. Ismail, A. Mansourizadeh, Preparation of microporous PVDF hollow fiber membrane contactors for CO_2 stripping from diethanolamine solution, *J. Memb. Sci.* 392–393 (2012) 29–37.

60. R. Naim, A.F. Ismail, Effect of polymer concentration on the structure and performance of PEI hollow fiber membrane contactor for CO_2 stripping, *J. Hazard. Mater.* 250–251C (2013) 354–361.

61. M. Rahbari-Sisakht, A.F. Ismail, D. Rana, T. Matsuura, Carbon dioxide stripping from diethanolamine solution through porous surface modified PVDF hollow fiber membrane contactor, *J. Memb. Sci.* 427 (2013) 270–275.

62. M. Rahbari-Sisakht, A.F. Ismail, D. Rana, T. Matsuura, D. Emadzadeh, Carbon dioxide stripping from water through porous polysulfone hollow fiber membrane contactor, *Sep. Purif. Technol.* 108 (2013) 119–123.

63. J.-L. Li, B.-H. Chen, Review of CO_2 absorption using chemical solvents in hollow fiber membrane contactors, *Sep. Purif. Tech.* 41 (2005) 109–122.

64. A. Chakma, CO_2 capture processes—Opportunities for improved energy efficiencies, *Energy Convers. Manag.* 38 (1997) 551–556.

65. H.P. Mangalapally, R. Notz, S. Hoch, N. Asprion, G. Sieder, H. Garcia et al., Pilot plant experimental studies of post combustion CO2 capture by reactive absorption with MEA and new solvents, *Energy Procedia* 1 (2009) 963–970.

66. J.-G. Shim, J.-H. Kim, J.H. Lee, K.-R. Jang, Highly efficient absorbents for post-combustion CO2 capture, *Energy Procedia* 1 (2009) 779–782.

67. S.-W. Park, D.-S. Suh, K.-S. Hwang, H. Kumazawa, Gas absorption of carbon dioxide in a hollow fiber contained liquid membrane absorber, *Korean J. Chem. Eng.* 14 (1997) 285–291.

68. A.A. Olajire, CO_2 capture and separation technologies for end-of-pipe applications—A review, *Energy* 35 (2010) 2610–2628.

69. H.B. Al-Saffar, B. Ozturk, R. Hughes, A comparison of porous and non-porous gas-liquid membrane contactors for gas separation, *Trans IChemE* 75 (1997) 685–692.

70. K. Li, W.K. Teo, An ultrathin skinned hollow fibre module for gas absorption at elevated pressures, *Trans IChemE* 74 (1996) 856–862.

71. T.R. Carey, J.E. Hermes, G.T. Rochelle, A model of acid gas absorption/stripping using methyldiethanolamine with added acid, *Gas Sep. Purif.* 5 (1991) 95–109.

72. W. Rongwong, R. Jiraratananon, S. Atchariyawut, Experimental study on membrane wetting in gas–liquid membrane contacting process for CO_2 absorption by single and mixed absorbents, *Sep. Purif. Tech.* 69 (2009) 118–125.

73. V.V. Mahajani, P.V. Danckwerts, The stripping of CO_2 from amine-promoted potach solutions at 100°C, *Chem. Eng. Sci.* 38 (1983) 321–327.

74. S.-P. Yan, M.-X. Fang, W.-F. Zhang, S.-Y. Wang, Z.-K. Xu, Z.-Y. Luo et al., Experimental study on the separation of CO_2 from flue gas using hollow fiber membrane contactors without wetting, *Fuel Process. Tech.* 88 (2007) 501–511.

75. H.-J. Song, S. Park, H. Kim, A. Gaur, J.-W. Park, S.-J. Lee, Carbon dioxide absorption characteristics of aqueous amino acid salt solutions, *Int. J. Greenh. Gas Control* 11 (2012) 64–72.

76. V.Y. Dindore, D.W.F. Brilman, P.H.M. Feron, G.F. Versteeg, CO_2 absorption at elevated pressures using a hollow fiber membrane contactor, *J. Memb. Sci.* 235 (2004) 99–109.

77. P.S. Kumar, J.A. Hogendoorn, P.H.M. Feron, G.F. Versteeg, New absorption liquids for the removal of CO_2 from dilute gas streams using membrane contactors, *Chem. Eng. Sci.* 57 (2002) 1639–1651.

78. J. Lu, L. Wang, X. Sun, J. Li, X. Liu, Absorption of CO_2 into aqueous solutions of methyldiethanolamine and activated methyldiethanolamine from a gas mixture in a hollow fiber contactor, *Ind. Eng. Chem. Res.* 44 (2005) 9230–9238.

79. A. Bottino, G. Capannelli, A. Comite, R. Firpo, R. Di Felice, P. Pinacci, Separation of carbon dioxide from flue gases using membrane contactors, *Desalination* 200 (2006) 609–611.

80. J. Todorovic, D.M. Krstic, G.N. Vatai, M.N. Tekic, Gas absorption in a hollow-fiber membrane contactor with pseudo-plastic liquid as an absorbent, *Desalination* 193 (2006) 286–290.

81. W. Zhang, Q. Wang, M. Fang, Z. Luo, K. Cen, Experimental study on the absorption of CO2 from flue gas by aqueous solutions of methyldiethanolamine + potassium glycinate in hollow fiber membrane contactor, *IEEE Xplore* (2010) 23–26.

82. J.G. Lu, Y.-F. Zheng, M.-D. Cheng, Membrane contactor for CO2 absorption applying amino-acid salt solutions, *Desalination* 249 (2009) 498–502.

83. Z. Guoliang, Y. Zhihong, S. Haimin, M. Qin, Novel membrane contactors used in waste gas/liquid separation, *Recent Pat. Eng.* 3 (2009) 18–24.

84. S. Gupta, N.S. Rathore, J.V. Sonawane, A.K. Pabby, R.R. Singh, A.K. Venugopalan et al., Hollow fiber membrane contactor: Novel extraction device for plutonium extraction, *BARC Newsl.* (2003) 181–189.

85. V. Usachov, V. Teplyakov, A. Okunev, N. Laguntsov, Membrane contactor air conditioning system: Experience and prospects, *Sep. Purif. Technol.* 57 (2007) 502–506.

86. H. Herzog, O. Falk-Pedersen, The Kvaerner membrane contactor: Lessons from a case study in how to reduce capture costs, *Proc. 5th Int. Conf. Greenh. Gas Control Technol.* (2001) 121–125.

87. D.V. Timofeev, Hybrid membrane contactor system for creating semi-breathing air, *J. Phys. Conf. Ser.* 345 (2012) 1–3.

88. J. Zhou, H. Meyer, B. Bikson, Pre-combustion carbon capture by a nanoporous, superhydrophobic membrane contactor process, *CPR Meet. Prod. Cond. High Sulfur Biogas Fuel Cell-Preliminary Des.* 1–29 (2010).

89. A. Gabelman, S.-T. Hwang, Hollow Fiber Membrane Contactors, *J. Memb. Sci.* 159 (1999) 61–106.

90. H.S. Meyer, R. Palla, S.I. King, Field tests support reliability of membrane gas/liquid contactor, *GasTIPS*, Summer (2002) 37–40.

91. R. Klaassen, P. Feron, A. Jansen, Membrane contactor applications, *Desalination* 224 (2008) 81–87.

92. R. Klaassen, P.H.M. Feron, A.E. Jansen, Membrane contactors in industrial applications, *Chem. Eng. Res. Des.* 83 (2005) 234–246.

93. J. Mackey, J. Mojonnier, CO_2 Injection Using Membrane Technology, *Bev-Plants 8th Int. Conf. Oper. Technol. Adv. Beverage Plants Warehouses*, March 21–23, 1995: pp. 1–22.

94. D.A. Glasscock, J.E. Critchfield, G.T. Rochelle, CO_2 absorption/desorption in mixtures of methyldiethanolamine with monoethanolamine or diethanolamine, *Chem. Eng. Sci.* 46 (1991) 2829–2845.

95. I. Alatiqi, M.F. Sabri, W. Bouhamra, E. Alper, Steady-state rate-based modelling for CO2/amine absorption-desorption systems, *Gas Sep. Purif.* 8 (1994) 3–11.

96. M. Simioni, S.E. Kentish, G.W. Stevens, Polymeric alternatives to teflon for membrane stripping, *Energy Procedia* 4 (2011) 659–665.

97. U.E. Aronu, H.F. Svendsen, K.A. Hoff, Investigation of amine amino acid salts for carbon dioxide absorption, *Int. J. Greenh. Gas Control* 4 (2010) 771–775.

Section II

Fabrication Processes
for Inorganic Membrane

10 Microstructured Ceramic Hollow-Fiber Membranes
Development and Application

Zhentao Wu, Benjamin F.K. Kingsbury, and Kang Li

CONTENTS

10.1   Introduction 318
10.2   Viscous Fingering-Induced Phase Inversion Process—A Brief History.... 319
10.3   State-of-the-Art Progress.. 320
      10.3.1   Introduction.. 320
            10.3.1.1   Single-Layer Ceramic Hollow-Fiber Membranes 320
            10.3.1.2   Multilayer Ceramic Hollow-Fiber Membranes 321
      10.3.2   Parameters for Morphology Control 321
            10.3.2.1   Single-Layer Ceramic Hollow-Fiber Membranes 321
            10.3.2.2   Multilayer Ceramic Hollow-Fiber Membranes 323
      10.3.3   Sintering 323
            10.3.3.1   Single-Layer Hollow-Fiber Membranes 324
            10.3.3.2   Co-Sintering of Multilayer Hollow-Fiber Membranes.... 324
10.4   Examples... 325
      10.4.1   Asymmetric Alumina Hollow-Fiber Membranes for
            Wastewater Treatment 326
      10.4.2   A Controlled Sintering Process for a More Permeable
            Alumina Membrane 330
      10.4.3   Single-Step Formation of Triple-Layer Membranes
            for Microtubular SOFC 335
            10.4.3.1   Compositions of Each Layer 336
            10.4.3.2   Morphology and Microstructure 337
            10.4.3.3   Mechanical Strength and Gas-Tightness
                   of Electrolyte... 339
            10.4.3.4   Cell Performance... 340
      10.4.4   A New Generation of Catalytic Convertor Substrate................... 341
            10.4.4.1   Fabrication Process 342
            10.4.4.2   Reduced Substrate Volume 342

        10.4.4.3   Reduced Catalyst Loading 343
        10.4.4.4   Reduced Pressure Drop...343
10.5   Conclusions..343
Acknowledgment 344
References..344

10.1   INTRODUCTION

Membrane separation is relatively new to conventional separation and purification techniques. Due to its unique advantages, especially low energy consumptions, membrane separation has gained an important role in industrial processes and has been used in a broad range of applications. Over the whole historical development of membranes, ceramic membranes have been tagged as costly products, mainly due to the multistep fabricating processes associated with high energy penalty, and are subsequently only beneficial for applications where the operating conditions such as high temperatures, strong acid, or base environment preclude the use of existing polymeric membranes.

In order to obtain high selectivity and permeability of ceramic membranes, fabrication techniques rely largely on forming a thinner separation layer with smaller pore size or narrower pore size distribution on a more porous, normally multilayer substrate with a gradient pore structure, which usually needs a string of techniques involving repeated coating and sintering. The surface of ceramic membranes, especially microfiltration (MF) to ultrafiltration (UF) ranges, with a good packed pore structure allows them to be further used as an ideal substrate or support for the formation of other inorganic membranes with materials such as zeolite, metal organic framework, or graphene for molecular separation. Apart from the improved microstructures, such as a thinner separation layer on a more porous substrate, high geometric surface area (GSA) of the membrane is another dimension that membrane scientists and engineers keep pursuing. An increase of ceramic membrane area per unit volume, that is, from a planar design to a tubular configuration with monolithic or honeycomb-like structures, together with high permeation fluxes could contribute to a strong and sustainable growth of ceramic membranes, strengthening its competitiveness against polymeric membranes.

Besides the abovementioned advantages, ceramic hollow-fiber membranes unveil further superiorities over other ceramic membranes. In addition to a substantially higher surface area/volume ratio due to the smaller radius, the unique microchannels inside the membrane wall play an important role in lowering mass transfer resistance and elevate the accessible geometric and specific surface areas. Such a structure is formed due to the concurrence of several process phenomena during the membrane fabrication, which is different from the conventional ram-extrusion-based process where only a simple symmetric structure can be delivered.

This chapter will first introduce fundamental principles in fabricating microstructured ceramic hollow-fiber membranes. Different aspects of the processing parameters in relation to microstructures of the resultant membranes will then be outlined. Finally, examples of using the ceramic hollow-fiber membranes with the designed structures in a variety of applications are given.

10.2 VISCOUS FINGERING-INDUCED PHASE INVERSION PROCESS—A BRIEF HISTORY

Phase inversion, also known as phase separation or polymer precipitation, has been widely used for fabricating polymeric and mixed matrix membranes. This process can be described as a phase separation process of a homogeneous polymer solution induced by either temperature change, immersing in a nonsolvent bath (wet process), or exposing it to a nonsolvent atmosphere (dry process) [1]. In all phase separation processes, a liquid polymer solution finally turns into two phases: a solid polymer-rich phase building up the membrane matrix and a liquid polymer-poor phase forming the membrane pores [2].

Early development of ceramic hollow-fiber membranes, which proved the technical feasibility of using a phase inversion process, was largely based on the methodologies from the polymer counterparts, except for the steps that ceramic particles need to be uniformly dispersed inside a polymeric binder solution, and high temperature sintering is necessary to consolidate the membrane microstructures and to create their mechanical strength. Formation of the two basic structures, sponge-like and finger-like, was considered as a result of different polymer precipitation speeds, due to the solvent–nonsolvent exchange [3], which is also similar to that used for the fabrication of some polymeric membranes. At this stage, capillary membranes prepared through ram-extrusion-based processes were also called hollow-fiber membranes, and their most significant difference from the membrane fabricated through phase inversion process is the lack of finger-like structure. Apart from fabrication techniques, various ceramic materials have been employed, and the hollow-fiber membranes developed were applied for gas separation [4–18], substrate [19–24], membrane reactors [19,25–31], and membrane contactors [32–34]. The innovative idea of a single-step fabrication of multilayer ceramic hollow-fiber membranes, through a process involving phase inversion and the concept of co-extrusion and co-sintering, also emerged recently [35].

Viscous fingering, a well-known phenomenon that occurs at the interface between fluids with different viscosities at the first moment of mixing, was later used to explain the formation of finger-like structures inside ceramic hollow-fiber membranes [36–38], in addition to the solvent–nonsolvent-induced phase inversion process that leads to the precipitation of polymeric binder phase. As a result, impacts of adjusting fabricating parameters, such as suspension viscosity, air-gap, extrusion rate, and internal coagulant flow rate on membrane morphologies can be evaluated in a more scientific way. Unlike the polymeric membranes where the finger-like void was always considered to negatively impact on the resultant membrane performances, in ceramic membranes, it is treated as a positive factor in lowering mass transfer resistance [39,40] or enlarging GSA for catalyst deposition [30,41]. Such a change in interpreting the roles of finger-like structures infiltrates into different applications, and contributes to generating a number of new concepts, such as considering each finger-like channel with a catalyst wash-coat as a microreactor [27–29]. Meanwhile, more progress in developing high-quality dual-layer ceramic hollow-fiber membranes has also been achieved. The dual-layer membranes have been used as a compact membrane reactor [42,43], or for microtubular solid oxide fuel cells (SOFC) [44–48].

To the best of our knowledge, further innovative development in fabricating more advanced ceramic hollow-fiber membranes, such as triple-layer hollow fibers using the viscous fingering-induced phase inversion process for important applications, is possible in a sustainable way, especially toward integrating multiple microstructured hollow fibers into a functional unit for separation, reaction, and energy applications. Meanwhile, fundamental studies focusing on the density and dimensions of micro-channels are currently being carried out.

10.3 STATE-OF-THE-ART PROGRESS

10.3.1 INTRODUCTION

10.3.1.1 Single-Layer Ceramic Hollow-Fiber Membranes

Single-layer ceramic hollow-fiber membranes fabricated through the viscous fingering-induced phase inversion process normally consist of two basic structures: a finger-like and a sponge-like. Different morphologies can be formed as shown in Table 10.1, assuming the top and bottom surfaces of the schematic drawings in the first column as the exterior and interior surfaces of a hollow-fiber membrane, respectively. For most of the morphologies in Table 10.1, the proportions between the finger-like structure and the sponge-like structure can be adjusted. Further details are given as follows:

TABLE 10.1
Fundamental Characteristics of Different Major Morphologies

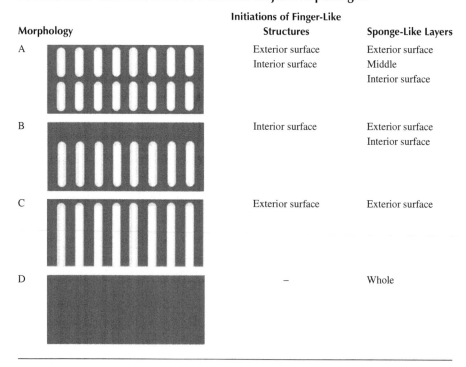

Morphology	Initiations of Finger-Like Structures	Sponge-Like Layers
A	Exterior surface Interior surface	Exterior surface Middle Interior surface
B	Interior surface	Exterior surface Interior surface
C	Exterior surface	Exterior surface
D	–	Whole

With regard to Morphology A, the viscous fingering is initiated from both the exterior and interior surfaces, leaving two sponge-like skin layers next to each surface and a third sponge-like layer between the two finger-like *layers*. It should be noted that, besides these three sponge-like layers, two sponge-like skin layers in interior and exterior surfaces and one middle sponge-like layer, the walls between finger-like structures are also sponge-like structures, formed during the growth of the finger-like voids. While for Morphology B, viscous fingering is initiated from the interior surface, with only one sponge-like skin layer formed in the interior surface, leaving the exterior surface with a sponge-like layer having an adjustable thickness. In some special cases, such viscous fingering can reach and break through the exterior surface, forming finger-like channels. Morphology C with only one sponge-like skin layer in the exterior surface can be formed by initiating viscous fingering from the exterior surface. Because a weak non-solvent is used as an internal coagulant, the fingers can penetrate through the interior surface, forming the finger-like channels. For Morphology D, no finger-like structure can be formed due to an excessively high initial viscosity and/or fast changes in the viscosity of the spinning suspension, which is further discussed in the Section 10.3.2.1.

10.3.1.2 Multilayer Ceramic Hollow-Fiber Membranes

Apart from single-layer membranes, the use of the viscous fingering-induced phase inversion process allows a single-step formation of multilayer hollow-fiber membranes with flexible control of thickness and morphology of each layer. To date, dual- and triple-layer membranes have been fabricated, and the same methodology can be transferred toward membranes made of more layers, although the subsequent sintering and application of such membranes may need further considerations.

Similar to the single-layer one, multilayer hollow-fiber membranes also consist of the two basic structures, a finger-like and a sponge-like. And the major morphologies listed in Table 10.1 can be achieved in the multilayer membranes as well, with the finger-like structure presented in one layer or two layers with voids or channels penetrating through the interface between the two adjacent layers, or forming a possible secondary finger-like structure from the interface.

10.3.2 Parameters for Morphology Control

To deliver various microstructured ceramic hollow-fiber membranes, a number of technical parameters need to be considered, such as suspension rheology, suspension composition and viscosity, air-gap, extrusion rate of suspension, and flow rate of internal coagulant, which have been systematically investigated previously [36–38]. This section addresses how to use these technical parameters to form the membrane morphologies listed in Table 10.1.

10.3.2.1 Single-Layer Ceramic Hollow-Fiber Membranes

1. Morphology A: Viscous fingering initiated from both the exterior and interior surfaces and creates two separate sponge-like skin layers on each surface and a third thicker sponge-like layer between the two finger-like layers.

   In this case, both the internal and external coagulants need to be in contact with the interior and exterior surfaces at the same time (i.e., wet spinning, a zero air-gap), allowing the viscous fingering to take place from both the membrane surfaces. Normally, an external coagulation bath supplies a sufficient amount of *fresh* nonsolvent, leading to a slightly denser exterior sponge-like skin layer due to the instant polymer binder precipitation. Meanwhile, as a result of the limited dimensions of the lumen (a couple of millimeters), the actual internal coagulant or bore liquid is a mixture of solvent from suspension and nonsolvent, the internal coagulant. The polymer precipitation on the interior surface is thus slightly slower than the exterior counterpart, forming a more porous interior sponge-like skin layer. Because the solvent in suspension diffuses out in two directions, finger-like structures initiated from the two surfaces cannot join together, leaving a third thicker sponge-like layer between the two finger-like layers.

If the internal and external coagulants are of the same fluid, increasing the air-gap is an efficient way of obtaining a thicker interior finger-like layer, because it delays the initiation of the exterior finger-like structure. Use of a higher internal coagulant flow rate is also helpful for such a purpose, although it may increase the pressure inside the lumen.

2. Morphology B: Viscous fingering initiated from the interior surface only and develops radially toward the exterior surface, leaving an interior sponge-like skin layer and a thicker exterior sponge-like layer.

This morphology can derive from Morphology A, by increasing the air-gap to a scale that no exterior finger-like structure can be formed when the precursor fiber reaches the external coagulation bath. This means that, within this air-gap the out-diffusion of the solvent from the interior surface causes a significant increase of the suspension viscosity at the exterior surface, exceeding the upper limit for initiating the finger-like structure from viscous fingering when the outer surface is in contact with external coagulant. Although due to the fact that the actual internal coagulant is still a mixture of solvent (from suspension) and nonsolvent, the interior sponge-like skin layer is normally more porous than the external thicker sponge-like layer. In addition to larger air-gaps and higher bore liquid flow rates, adjusting the composition of the spinning suspension, such as adding a small amount of nonsolvent (i.e., water) into the spinning suspension, allows the formation of shorter finger-like voids, and subsequently eliminates finger-like voids when water contents reach 8% [36], that is, symmetric membranes of Morphology D in Table 10.1.

3. Morphology C: Viscous fingering initiated from the exterior surface and penetrated through the membrane wall, leaving a plurality of openings on the interior surface.

A zero air-gap is normally used for this morphology, and in contrast from the other morphologies, a bore liquid, which is a solvent, needs to be used to prevent polymer precipitation at the interior surface. This allows sustainable development of the finger-like channels from the exterior surface, until they penetrate through the interior surface. Solvents similar to the one used in the suspension can be employed directly as the bore liquid for this morphology, although sometimes it may partially eliminate readily formed finger-like structure when it diffuses outward into the external coagulation

bath, and partially dissolves the interior surface. Therefore, the strength of the solvent used as the bore liquid needs to be carefully controlled.
4. Morphology D: No viscous fingering takes place with a fully sponge-like structure throughout the membrane.

This morphology is very similar to the ceramic membranes fabricated through a ram-extrusion-based process. Solvent–nonsolvent exchange still proceeds, but the interface between the spinning suspension and nonsolvent (coagulant) are stable enough, so that no viscous fingering can be initiated. The detailed information on this morphology can be found elsewhere [36–38].

10.3.2.2 Multilayer Ceramic Hollow-Fiber Membranes

Besides microstructuring single-layer ceramic hollow-fiber membranes, the viscous fingering-induced phase inversion process is also useful in developing multilayer membranes in one step, by co-extruding a number of layers at the same time. This is quite meaningful to reducing the fabrication costs and generating new applications of ceramic membranes. Currently, the major uses of multilayer ceramic hollow-fiber membranes (dual and triple layer) include catalytic membrane reactors [42,43,49] and SOFC [44,47,50], in which each layer is composed of different ceramic materials for specifically defined functions.

Adjusting extrusion rate of the suspension for each layer is normally enough for controlling the thickness of a layer. The fundamental principles and technical parameters introduced above for single-layer hollow-fiber membranes apply to the multilayer membranes as well with the following additional factors:

1. The overall extrusion speed, which is different from the extrusion rate of a suspension, is largely determined by the layer with the highest thickness, the quickest polymer precipitation, or the highest initial viscosity.
2. Adjusting extrusion rate of the other layers is normally based on the layer *mastering* the overall extrusion speed.
3. Adhesion between layers is normally very good, when the same polymer binder is used for all layers. Precipitated polymer thus forms a continuous phase throughout the precursor membranes.
4. If finger-like voids or channels need to penetrate through the interface between two adjacent layers, initial viscosity of the suspension for the layer where the finger-like structure initiates is suggested to be comparable or higher than that of the next layer.
5. If different solvents are employed for each layer, interdiffusion of the solvents at the interface between the two adjacent layers may affect adhesion and development of finger-like structure.
6. For greater uniformity of the multilayer membranes, difference in the compositions of the suspensions for each layer should not be too significant.

10.3.3 SINTERING

The viscous fingering-induced phase inversion process integrates microstructuring and continuous fabrication of ceramic hollow-fiber membranes in one step, whereas a high temperature sintering is still necessary to finalize membrane microstructure and deliver appropriate mechanical strength.

10.3.3.1   Single-Layer Hollow-Fiber Membranes

Conventional sintering of ceramic membranes consists of three major steps: initial, intermediate, and final sintering. The parameters associated with the stages of sintering for polycrystalline solids [51] are listed in Table 10.2.

In contrast to the ceramic membranes fabricated through a ram-extrusion-based technique, there is a significant amount of polymeric phase inside nascent microstructured ceramic hollow-fiber membranes. As a result, an additional calcination step for removing the polymeric phase smoothly is necessary, prior to the initial stage of sintering (Table 10.2) [52]. However, the final pore structure, especially the pore size and porosity of the sponge-like structure that directs many applications of such membranes, is still determined by conventional sintering parameters. For example, improved mechanical strength normally achieved at high sintering temperatures and/or long dwelling time is at the expense of membrane porosity.

Very recently, an innovative idea of precipitated polymeric binder as a pore structure *stabilizer* has been exploited [53]. Because the polymeric binder precipitates between ceramic particles inside a precursor membrane, removing it in a controlled way, such as underdesigned temperature profile and atmosphere, can affect the interaction between ceramic particles at high temperatures. This allows the formation of a strong grain boundary without losing membrane porosity, thus leading to a highly porous membrane with great mechanical strength, even if high sintering temperatures are used. Moreover, the precipitated polymeric binder is a continuous phase inside precursor membranes, and it functions as a pore structure stabilizer to alleviate over the densification of ceramics at high temperatures. As a result, it should be distinguished from a pore former that is widely used for retaining membrane porosity.

10.3.3.2   Co-Sintering of Multilayer Hollow-Fiber Membranes

A viscous fingering-induced phase inversion process has allowed a single-step formation of nascent multilayer membranes with controlled structures and great adhesion

---

TABLE 10.2
Parameters Associated with the Stages of Sintering Ceramics

Stage	Typical Microstructural Feature	Relative Density Range	Idealized Model
Initial	Rapid inter-particle neck growth	≤ 0.65	Two monosize spheres in contact
Intermediate	Equilibrium pore shape with continuous porosity	0.65–0.90	Tetrakaidecahedron with cylindrical pores of the same radius along the edges
Final	Equilibrium pore shape with isolated porosity	≥ 0.90	Tetrakaidecahedron with spherical monosize pores at the corners

Source: Rahaman, M.N., *Ceramic Processing and Sintering*, Marcel Dekker, New York, 2003.

between layers. This is uniquely different from the existing multistep technique routes. In addition to this, a successful co-sintering of such membranes relies more on the nature of ceramic materials involved in each layer. The technical factors that need to be considered include the following:

10.3.3.2.1 Material Composition

Because each layer is of a specific function, such as separation and catalysis, appropriate ceramic materials need to be selected first. In most cases, a mixture of ceramics gives more flexibility in fine-tuning layer functions, improving adhesion between layers, and matching sintering behaviors and thermal expansion coefficients between layers.

10.3.3.2.2 Adhesion between Layers

After removing the polymeric binder during sintering, strong adhesion between the layers is built up by forming grain boundaries of ceramic materials. From this viewpoint, a ceramic mixture seems to be a more feasible choice. For example, one layer is composed of A and B, and the adjacent layer is made of B and C, the adhesion between these two layers can be formed by the continuous phase of B at the interface.

10.3.3.2.3 Sintering Behavior

Ceramic materials undergo densification at elevated temperatures. A ceramic material *shrinks* by how much and at what a rate at a certain temperature, namely, sintering behavior needs to be considered. If the sintering behaviors of two adjacent layers are too different from each other, the membrane will not survive the co-sintering. And currently, adjusting material composition of each layer is the dominating way in matching the sintering behaviors.

10.3.3.2.4 Co-Sintering Profile

Even for a ceramic mixture, sintering behavior can still not be matched perfectly. As a result, co-sintering profiles can sometimes be designed to alleviate the impacts from slightly mismatched sintering behaviors. For example, if there is a major difference in sintering behaviors at the intermediate stage (Table 10.2), a quick heating-up rate can be considered for a shorter time window at this sintering stage.

10.3.3.2.5 Interactions between Layers

Each layer is formed by *packing* ceramic particles together. The multilayer membranes are more likely to survive the co-sintering if the outer layer *shrinks* more and/or faster than the inner one by *compressing* the layers together.

Of course, for different applications, different membrane materials are used. Actual implementation of these technical factors needs to be adjusted from case to case.

10.4 EXAMPLES

As a result of the wide range of ceramic materials, membrane morphologies, and microstructures, the microstructured ceramic hollow-fiber membranes have been used in various applications, which have been extensively reviewed [52,54,55].

In this section, four examples of different applications are outlined. Each example addresses different aspects of the potential in using microstructured ceramic hollow-fiber membranes for separation and chemical reaction.

10.4.1 ASYMMETRIC ALUMINA HOLLOW-FIBER MEMBRANES FOR WASTEWATER TREATMENT

Use of membrane techniques for wastewater treatment has been commercialized for several decades. Ceramic membranes always play an important role, especially when the operating conditions are too harsh for polymeric membranes. From an engineering viewpoint, packing more membrane area into a unit of a smaller size is constantly pursued. Ceramic membranes of a hollow-fiber configuration are no doubt a great choice due to their substantially higher surface area/volume ratio.

Commercialized ceramic membranes can be of different configurations, from disc/flat-sheet to multichannel tubes. The membrane microstructure is always very similar, that is, a thin separation layer made of finer particles supported onto a more porous multilayer substrate with a gradient pore structure, as shown in Figure 10.1. The only layer involved in separation is normally the thinnest layer with the finest pore size. The thickest layer, which consists of big particles for low resistance to permeates, provides the mechanical strength, whereas the layers in between are needed for a uniform formation of the top separation layer. Normally, different techniques are used for each individual layer, which results in soaring expenditures with the number of layers involved.

In contrast, the finger-like microvoids or channels inside microstructured ceramic hollow-fiber membranes significantly reduce mass transfer resistance, and the thin sponge-like skin layer can function as a separation layer. Thus, a single-step viscous

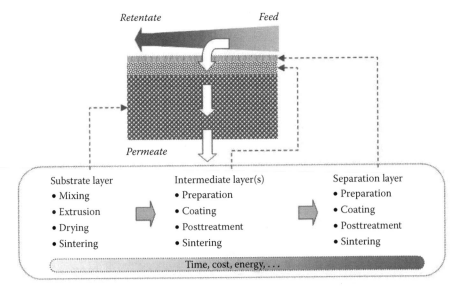

FIGURE 10.1 Schematic diagram of multilayer structure of ceramic membranes.

fingering-induced phase inversion process can deliver membranes suitable for the purpose of wastewater treatment, such as microfiltration (MF) and UF. And from this viewpoint, Morphology C in Table 10.1 can be selected using alumina (1 μm, alpha, 99.9% metals basis, surface area 6–8 m²/g) as the membrane material.

Using a solvent-based bore liquid allows the viscous fingering to penetrate through the whole cross section, leaving a highly porous inner surface with little resistance to the water permeation and a sponge-like skin layer for separation, as shown in Figure 10.2.

Based on the bubble point method, the pore size of the exterior separation layer is at approximately 0.1 μm (Figure 10.3), which is ideal for MF. The effects of sintering temperature on the pore structure of this separation layer are not significant.

Due to the complicated pore structure of such asymmetric membranes, mercury intrusion is always used to distinguish between finger-like and sponge-like structures. As can be seen in Figure 10.4, finger-like structure shrinks significantly with the increasing sintering temperatures, and the size of the *openings* on the interior surface (Figure 10.2c) is reduced from around 16.6 to 9.6 μm when the sintering temperature is elevated from 1204°C–1455°C.

A second peak in the mercury intrusion results represents the pore structure of the sponge-like structure, including the separation layer and walls between the finger-like microchannels. For mercury intrusion method, mercury would fill larger pores,

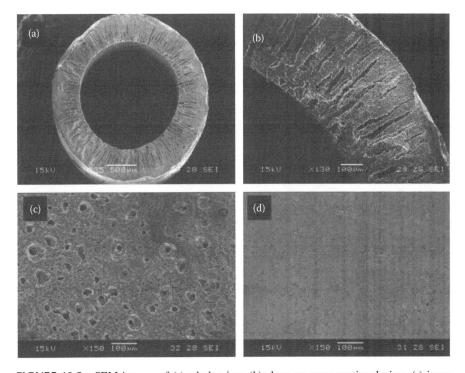

FIGURE 10.2 SEM images of (a) whole view, (b) close-up cross-sectional view, (c) inner surface, and (d) outer surface of the membrane sintered at 1342°C. (Data from Lee, M. et al., *Journal of Membrane Science*, 461, 39–48, 2014.)

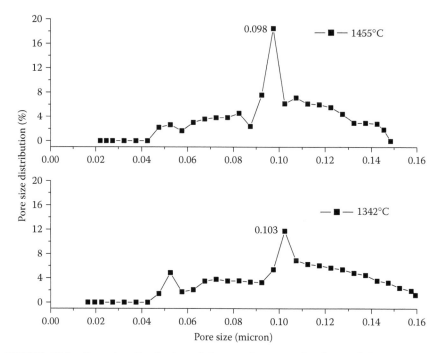

FIGURE 10.3 Pore size distribution of the exterior separation layer of the membranes sintered at 1342°C and 1455°C (bubble point method). (Data from Lee, M. et al., *Journal of Membrane Science*, 461, 39–48, 2014.)

finger-like channels in this case, before penetrating into smaller pores in sponge-like structures. Due to the fact that the pore size of the sponge-like structure in Figure 10.4 is much larger than the separation layer (Figure 10.3), it is reasonable to predict that there is a gradient pore structure along the radial direction of the sponge-like structure. Moreover, the porosity of such sponge-like structure drops quickly with the increasing temperatures (reduced peak area), whereas the pore size slightly increases at higher temperatures, which is in line with previous studies [36,37].

Apart from pore structure, sintering temperature also affects the mechanical properties of the membrane, as shown in Figure 10.5 where three-point bending test results are presented. Ceramic densification at high temperatures contributes to the improved mechanical strength. Although due to the small dimensions and thinner membrane wall, mechanical strength, especially fracture loading, of each individual hollow-fiber membrane of this type cannot compete with commercialized ceramic membranes. It should be noted that a bundle of such membranes co-sintered as one ceramic unit (see Section 10.4.4) will lead to a substantial increase in mechanical strength, together with an highly integrated membrane area when certain designs of the membrane module is achieved.

A higher sintering temperature normally results in more ceramic densification, leading to changes in membrane pore structure (Figure 10.4), mechanical properties (Figure 10.5), and membrane dimensions (outer diameter [OD] of the hollow fiber membrane at various sintering temperatures) with a clear trend. However, the change

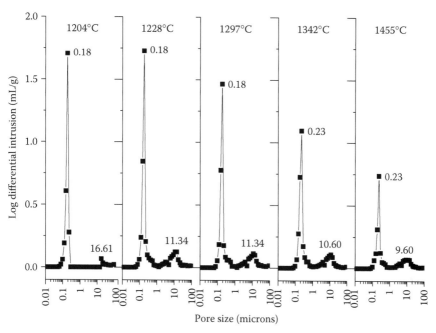

FIGURE 10.4 Mercury intrusion of the membranes sintered at different temperatures. (Data from Lee, M. et al., *Journal of Membrane Science*, 461, 39–48, 2014.)

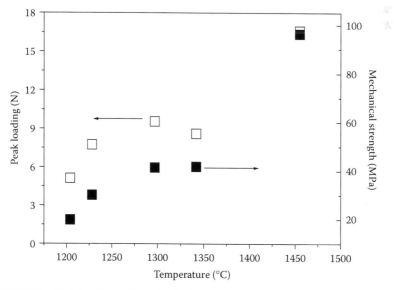

FIGURE 10.5 Peak loading and mechanical strength of the membrane sintered at different temperatures. (Data from Lee, M. et al., *Journal of Membrane Science*, 461, 39–48, 2014.)

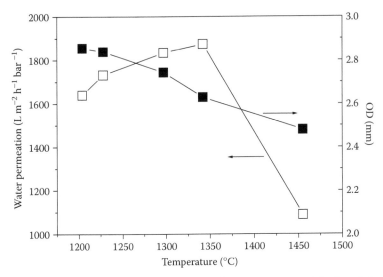

FIGURE 10.6 Pure water permeation (transmembrane pressure = 1 bar) and the outer diameters of the membranes sintered at different temperatures. (Data from Lee, M. et al., *Journal of Membrane Science*, 461, 39–48, 2014.)

of pure water permeation shows a slightly different trend in response to elevated sintering temperatures (Figure 10.6). This may indicate that the densification of the exterior sponge-like skin layer, such as pore structure (Figure 10.3) and its thickness is different from the other parts of the membrane, as well as the highly asymmetric membrane structure (Figure 10.2).

10.4.2 A CONTROLLED SINTERING PROCESS FOR A MORE PERMEABLE ALUMINA MEMBRANE

After phase inversion, precursor membranes consist of ceramic particles surrounded by a continuous and uniformly precipitated polymeric binder, no matter what kind of membrane morphology. Polyethersulfone (PESf) is one of the most widely used polymeric binders for fabricating microstructured ceramic hollow-fiber membranes. Its function was first considered as a binder connecting ceramic particles in precursor membranes, before being further attributed with another important role in structuring the microchannels during the concurrence of phase inversion and viscous-fingering phenomenon. However, PESf has been little linked to the sintering process, as it is fully burnt off before the actual sintering of ceramic particles occurs, namely, normal sintering.

As stated, the uniformly precipitated PESf phase can be used as a pore structure *stabilizer* in a controlled sintering process, in order to balance the trade-off between membrane porosity and mechanical properties in a normal sintering process.

Instead of being burnt off directly in a normal sintering process, the PESf phase is first thermally treated in static air, followed by converting it into carbon in an oxygen-free environment, together with the sintering of alumina particles at high temperatures (Table 10.3). The drop in membrane porosity is thus not solely a factor of temperature due to the presence of carbon around alumina particles. Appropriate thermal treatment allows control over the amount of carbon left during the following steps, which can be used to balance the retained membrane porosity and final membrane strength.

In this example, a microstructured alumina hollow-fiber membrane with Morphology B (Table 10.1) is used, and its morphology and microstructure after normal sintering is shown in Figure 10.7. Because the actual bore liquid for the membranes of this type is a mixture of internal coagulant (distilled water) and solvent, the polymer precipitation is slower than the exterior surface, leading to a more porous interior surface, which is retained after normal sintering.

The feasibility of using precipitated PESf as a pore structure stabilizer in the controlled sintering process can be proven by thermogravimetric analysis (TGA), as shown in Figure 10.8. The significant weight loss of PESf in air indicates a reasonable temperature window for controlling the quantity of stabilizer around alumina particles

TABLE 10.3
Sintering Parameters for Normal Sintering and Controlled Sintering

Temperature (°C)		Rate (°C/min)	Dwelling Time (min)	Atmosphere	Sintering Type
From	To				
RT	600	2	–	Static air	Normal sintering
600	600	–	120		
600	1450	5	–		
1450	1450	–	240		
1450	RT	3	–		
RT	400–600	5	–	Static air	Controlled
400–600	RT	5	–	(first step)	sintering
RT	1450	5	–	N$_2$	
1450	1450	–	240	(second step)	
1450	RT	3	–		
RT	800	5	–	Static air	
800	800	–	120	(third step)	
800	RT	3	–		

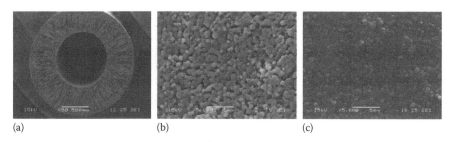

(a)   (b)   (c)

FIGURE 10.7 SEM images of the alumina hollow-fiber membrane (normal sintering) (a) whole view, (b) interior surface, and (c) exterior surface. (Data from Wu, Z.T. et al., *Journal of Membrane Science*, 446, 286–293, 2013.)

(first step). Although in an oxygen-free atmosphere, an adjustable amount of carbon (<40%) can be maintained (second step) for good membrane porosity. This type of carbon can then be removed in air (third step). At the end of the first step, there is a significant change in the colors of the precursor membranes (Figure 10.9), indicating the sensitivity to control by thermal-treatment temperatures.

Mercury intrusion is then used to characterize and compare the change in the pore structure of the membranes sintered normally and in a controlled way. Normally, mercury tends to penetrate into the larger pores of a sample, before further filling

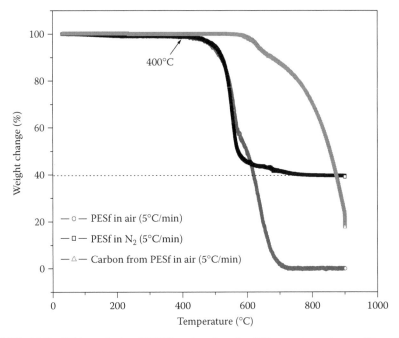

FIGURE 10.8 TGA analysis of PESf and carbon in different atmospheres. (Data from Wu, Z.T. et al., *Journal of Membrane Science*, 446, 286–293, 2013.)

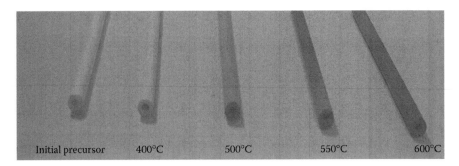

FIGURE 10.9 Photographic image of precursor fibers thermally treated at different temperatures (first step of controlled sintering). (Data from Wu, Z.T. et al., *Journal of Membrane Science*, 446, 286–293, 2013.)

smaller ones at a higher pressure. Due to the slightly more porous interior surface (Figure 10.7b), mercury should start penetrating into the normally sintered membrane from the interior surface, before filling in the finger-like microvoids, forming the peak at approximately 0.68 μm, as shown in Figure 10.10. The second peak, at approximately 0.18 μm, represents the rest of the sponge-like structure.

In terms of the membranes of the same initial morphology but sintered in a controlled way, the thermal-treatment temperature (first step, 400°C–600°C) affects the final membrane pore structure in two ways: (1) the interior sponge-like skin layer becomes denser at a higher thermal-treatment temperature and (2) pore size of the rest of the sponge-like structure is bigger at relatively low thermal-treatment temperatures (400°C–500°C), and is comparable to the normally sintered membrane at high thermal-treatment temperatures (550°C–600°C). Although the reason for a denser interior sponge-like skin layer in controlled sintering is not yet clear, a more porous sponge-like structure with a pore size much bigger than 0.18 μm (normal sintering) demonstrates the effectiveness of suppressed ceramic densification, due to the presence of such pore structure stabilizer. Apart from the pore size, the porosity of the sponge-like structure thermally treated between 400°C and 500°C is significantly higher, especially those below 450°C as shown in Figure 10.11.

The controlled sintered membranes are thus more permeable than the normally sintered counterpart, in terms of higher water permeation fluxes shown in Figure 10.12.

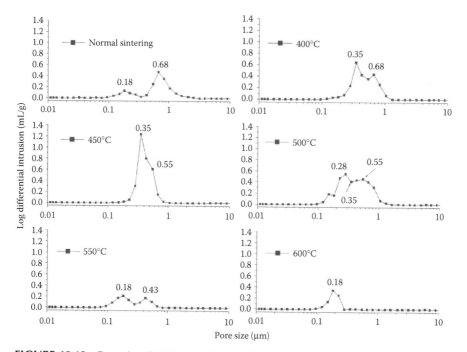

FIGURE 10.10 Pore size distribution of normally sintered membranes and controlled sintered membranes with various pretreatment temperatures. (Data from Wu, Z.T. et al., *Journal of Membrane Science*, 446, 286–293, 2013.)

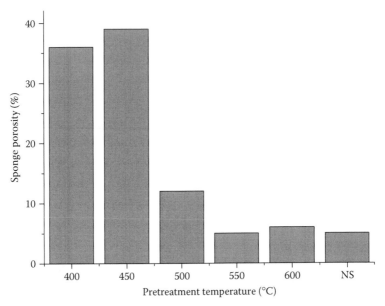

FIGURE 10.11 Porosity of sponge-like structure of normally sintered membrane and controlled sintered membranes with various pretreatment temperatures. (Data from Wu, Z.T. et al., *Journal of Membrane Science*, 446, 286–293, 2013.)

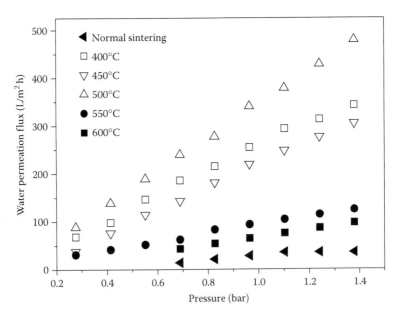

FIGURE 10.12 Pressure dependence of water permeation flux of normally sintered and controlled sintered membranes at various pretreatment temperatures. (Data from Wu, Z.T. et al., *Journal of Membrane Science*, 446, 286–293, 2013.)

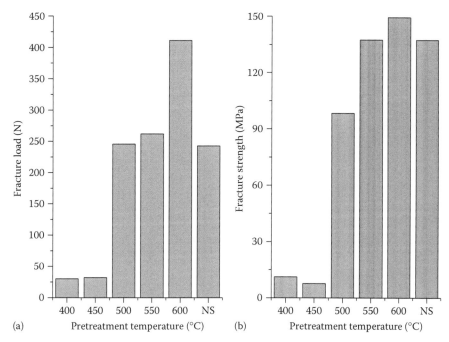

FIGURE 10.13 Mechanical evaluation of normally sintered and controlled sintered membranes at various pretreatment temperatures (a) fracture load and (b) strength. (Data from Wu, Z.T. et al., *Journal of Membrane Science*, 446, 286–293, 2013.)

Furthermore, low thermal-treatment temperatures (400°C–500°C) contribute to bigger pore size (Figure 10.10) and higher porosity (Figure 10.11) of the sponge-like structure, leading to an improved water permeation flux of up to approximately 13 times (at 1.38 bar). Although the water permeation flux is codetermined by several factors, such as pore size, porosity, and thickness of a membrane/separation layer, it is very clear that using PESf as the source of a structure stabilizer in controlled sintering is efficient for a more permeable ceramic membrane.

With regard to mechanical properties, as shown in Figure 10.13, for diametrical compression tests, the more permeable membranes (at lower thermal-treatment temperatures) are significantly weaker than those thermally treated at higher temperatures (first step) that are comparable with the normally sintered membrane. It should be noted that stronger membranes can be obtained by further increasing the sintering temperatures, due to the formation of strengthened grain boundaries. Meanwhile, the presence of stabilizer (carbon) between grains significantly suppresses the grain growth, retaining high porosity of a membrane. This is also uniquely different from normal sintering.

10.4.3 SINGLE-STEP FORMATION OF TRIPLE-LAYER MEMBRANES FOR MICROTUBULAR SOFC

In addition to ceramic membranes commercially available for filtration, a multilayer membrane structure can be found in other applications such as catalytic membrane reactors and SOFC. Taking SOFC, a promising technique for a sustainable generation of clean energy, as an example, each single cell is composed of a cathode, electrolyte, and anode. More components such as interconnect and current collector are needed for a stack or system. Conventional multistep processes are normally expensive and time consuming, which make it extremely challenging to develop a microtubular SOFC, which is outstanding in rapid start-up/shut-down, high power density, good cycling performance, and thermal shock resistance.

The viscous fingering-induced phase inversion process incorporating coextrusion and co-sintering is perfect for the applications of this type. Besides a single-step formation of a multilayer hollow fiber with controlled morphology and microstructure, which significantly reduces fabrication costs and contributes to improved performance, there is more flexibility in designing the functions of each layer. For example, collecting current efficiently from the electrode located inside the lumen of microtubular SOFC is challenging, due to the very limited dimensions. Inserting metal mesh or wires works for lab-scaled research, while technology innovations are highly demanded for developing efficient stacks or systems. If the most inner layer of the multi-layer hollow fibers can function as a current collector, technique challenge of this type can thus be solved.

Microtubular SOFC is normally anode supported. Finger-like microvoids are efficient in reducing resistance for fuel diffusion, resulting in improved cell performance due to the lowered concentration polarization resistance [48]. However, the presence of finger-like microvoids also leads to lower conductivity, reduced mechanical strength, and less triple-phase-boundary (TPB) of the anode, limiting further optimization of cell performance, unless there is still possibility to upgrade the other cell components. As a result, this example introduces the development of an anode/ anodic functional layer (AFL)/electrolyte triple-layer hollow fiber in one step, and the importance of controlling the thickness of AFL in affecting final cell performance.

10.4.3.1 Compositions of Each Layer

Besides its function as a cell component, the composition of each layer is also determined by its sintering behavior, in order for the membrane to survive the co-sintering. As can be seen in Figure 10.14a, the final shrinkages (dL/L_0) of electrolyte $Ce_{0.9}Gd_{0.1}O_{1.95}$ (CGO), AFL (60%CGO–40%NiO), and anode (40%CGO–60%NiO) are approximately 17.5%, 12.5%, and 11.0%, respectively. This means that, during the co-sintering, the outer layer shrinks more and *presses* the inner layer to form a denser structure. If this does not happen, for instance, the inner layer shrinks more and *drags* the outer layer; thus the whole membrane can easily collapse during co-sintering.

In terms of sintering rates (Figure 10.14b), CGO starts to shrink at a lower temperature with a higher rate than NiO. The temperature gap of the highest sintering rate of CGO and NiO is approximately 180°C, which indicates that the two phases *hinder* the sintering of each other. This is why the AFL and anode have a *dual-peak* shape for the sintering rate curve. Moreover, the layer with more CGO starts to shrink at a lower temperature and a higher rate, and the highest sintering rate increases at a lower temperature as well.

A greater uniformity in sintering behaviors of the three layers can be achieved by adjusting the material composition, but at the expense of cell performance. As a result, a fast heating rate, for example 15°C/min, is needed for co-sintering such hollow fibers.

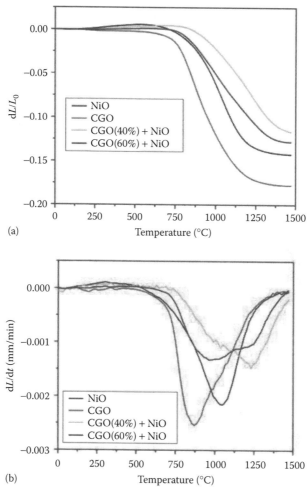

FIGURE 10.14 Sintering behaviours of materials for each layer: (a) sintering curves (heating rate: 5°C/min) and (b) sintering rate curves. (Data from Li, T. et al., *Journal of Membrane Science*, 449, 1–8, 2014.)

10.4.3.2 Morphology and Microstructure

Morphology and microstructure of successfully co-sintering electrolyte/AFL/anode triple-layer hollow fibers are shown in Figure 10.15. The anode is of a typical asymmetric structure, supporting a sponge-like intermediate AFL and exterior electrolyte. Varying co-extrusion parameters, such as the extrusion rate of AFL suspension, will not affect the uniformity of the three layers. Due to similar material composition between the AFL and anode, scanning electron microscopy in back-scattered electron (BSE) mode can be used to distinguish the two layers.

A faster extrusion of the AFL suspension (1–5 ml/min) leads to gradually reduced thicknesses of anode and electrolyte (Figure 10.16a), due to the elongation within the

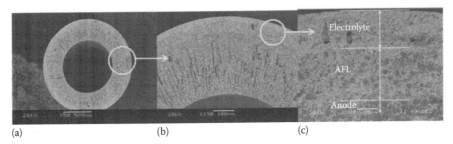

(a)   (b)   (c)

FIGURE 10.15   SEM images in BSE mode) of the sintered triple-layer hollow fiber: (a) whole view, (b) cross section, and (c) a higher magnification of cross section. (Data from Li, T. et al., *Journal of Membrane Science*, 449, 1–8, 2014.)

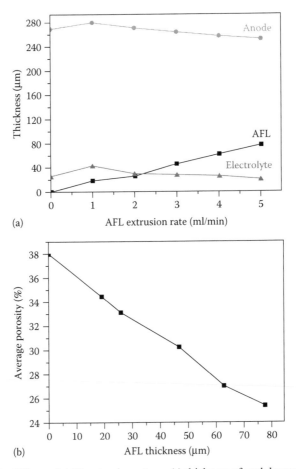

FIGURE 10.16   Effects of AFL extrusion rate on (a) thickness of each layer and (b) average porosity of AFL and anode. (Data from Li, T. et al., *Journal of Membrane Science*, 449, 1–8, 2014.)

air-gap (Morphology B in Table 10.1). When the extrusion rate of the AFL is zero, it is an electrolyte/anode dual-layer hollow fiber, without AFL in between. After reducing NiO into Ni, both AFL and anode are very porous. Due to a higher concentration of CGO in AFL, the overall porosity of AFL and anode (helium pycnometer) keeps reducing with the increasing AFL thickness. The mean size of NiO particle (12–22 μm) is approximately two orders of magnitude greater than CGO particle (0.1–0.4 μm), which means that the higher concentration of CGO still contributes to more TPB, when compared with the anode.

10.4.3.3 Mechanical Strength and Gas-Tightness of Electrolyte

The triple-layer hollow fibers become stronger with increased AFL thickness (Figure 10.17a), due to the fact that, in contrast to the anode, there is more CGO, which is the major contribution of mechanical strength, in AFL. Figure 10.17a further indicates that a fast heating rate is efficient in suppressing the negative effects of slightly mismatched sintering behavior (Figure 10.14). If not, the formation of microcracks will have impact on the mechanical properties of the co-sintered hollow

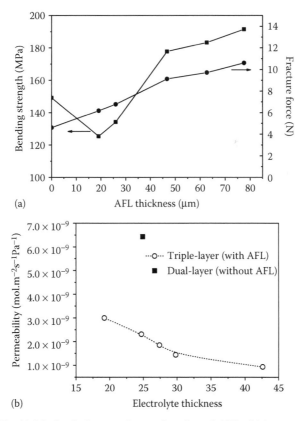

FIGURE 10.17 (a) Mechanical strength as a function of AFL thickness and (b) electrolyte gas-tightness as a function of electrolyte thickness. (Data from Li, T. et al., *Journal of Membrane Science*, 449, 1–8, 2014.)

fibers. In contrast to the dual-layer counterpart (without AFL), the triple-layer hollow fiber, with an AFL of around 20 μm, has a higher fracture force. But due to the bigger dimensions, its bending strength is still smaller.

With the increasing AFL thickness, the electrolyte layer is thinner (Figure 10.16a), leading to a slightly reduced gas-tightness (Figure 10.17b). And when compared with the dual-layer counterpart of the same electrolyte thickness (approximately 25 μm), the gas-tightness of the triple-layer hollow fiber is still significantly improved. This indicates that AFL acting as an intermediate layer contributes to a better-matched sintering behavior between electrolyte and anode.

10.4.3.4 Cell Performance

After coating LSCF ($La_{0.6}Sr_{0.4}Co_{0.2}Fe_{0.8}O_{3-\delta}$) based composite cathode, dual- and triple-layer hollow fibers, with the dimensions listed in Table 10.4, were used for investigating the effects of AFL thickness on cell performance.

As can be seen in Figure 10.18a, the open-circuit voltage (OCV) of the cells with AFL is approximately 0.85 V, slightly higher than the one without AFL (approximately 0.8 V). This is suggested to be due to the more gas-tight electrolyte with better-matched co-sintering behavior in the presence of intermediate AFL layer. For the cells with AFL, power density keeps decreasing with the increasing AFL thickness. Moreover, the correct control over the thickness of AFL always contributes to improved power density when compared with the one without AFL, and the highest power density is increased by approximately 36% when AFL is around 16.9 μm (ER1). Meanwhile, AFL cannot be too thick, as the cell performance is slightly worse than ER0 when the AFL is around 52.7 μm.

In terms of the function of AFL, a thicker AFL provides more TPB, contributing to less activation polarization resistance. On the other hand, it increases concentration polarization due to increased diffusion resistance. All these are clearly shown in Figure 10.18b, in which the AFL slightly increases the ohmic resistance, and a thicker AFL shows a higher polarization resistance. When the polarization resistance is higher than the cell without AFL, due to the trade-off between lowered activation polarization resistance and increased concentration polarization resistance as a function of AFL thickness, power density of the corresponding cell is worse than the one without AFL.

All these clearly demonstrate the importance of controlling AFL in affecting cell performance.

TABLE 10.4
Dimensions of Dual- and Triple-Layer Hollow Fibers for Microtubular SOFC

Extrusion Rate of AFL (ml/min)	OD (μm)	ID (μm)	AFL (μm)	Anode (μm)	Electrolyte (μm)
0	1465.2	883.0	/	275.8	15.3
1	1518.9	880.9	16.9	282.4	14.6
2	1540.2	878.3	32.6	284.4	13.9
3	1571.3	883.7	52.7	276.9	14.2

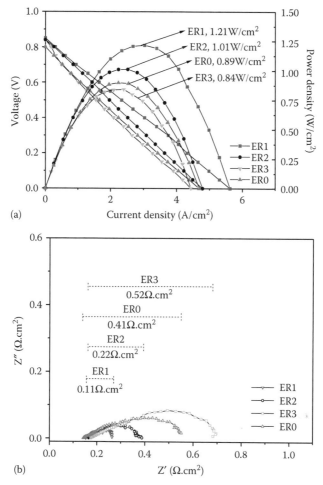

FIGURE 10.18 Effects of AFL thickness on (a) OCV and power density and (b) electrochemical impedance (open-circuit). (Data from Li, T. et al., *Journal of Power Sources*, 999–1005, 2015. DOI:10.1016/j.jpowsour.2014.10.004.)

10.4.4 A New Generation of Catalytic Convertor Substrate

Catalytic converters have been used for automotive exhaust gas treatment since 1975, and government regulation of emissions has driven advances. A catalytic converter consists of numerous channels, through which the exhaust gasses pass, coated with a catalytically active layer containing platinum group metals (PGMs), which is designed to increase the available surface area of the catalyst. The most important property of a ceramic substrate is GSA. By increasing the GSA, the thickness of the catalytic coating can be reduced, thereby reducing the diffusion limitations within the coating. The GSA of substrates has doubled over the last 40 years, from 2500 m^2/m^3 to around 4500 m^2/m^3. A technological limit has been reached and there have been no significant improvements over the last 10 years, as significantly higher GSA cannot be achieved using current manufacturing techniques. In addition, current substrate designs rely on reducing the diameter of the channels through which the exhaust gasses flow to achieve increases in GSA, which leads to increased exhaust system backpressure, which negatively impacts engine performance. Therefore, there is a trade-off between high GSA (high exhaust gas conversion) and low exhaust system backpressure, which cannot be avoided for current substrate technology.

In contrast, the microstructured ceramic hollow fibers (Morphology C of Table 10.1) are not limited by the trade-off of this type, as their major GSA comes from a plurality of open microchannels inside the fiber wall. This means that the microstructured hollow fibers with a relatively bigger inner diameter (ID) can still supply great GSA for the purpose of catalyst coating. Furthermore, the sponge-like structure among the open microchannels provides a huge specific surface area, and can sometimes be ideal for direct catalyst deposition. A number of additional outstanding advantages of using microstructured ceramic hollow fibers as a new-generation catalytic convertor substrate are introduced below in Sections 10.4.4.2 through 10.4.4.4.

10.4.4.1   Fabrication Process

Ceramic substrates with advanced microstructures have been developed with GSA values in excess of 30,000 m^2/m^3, representing a seven- to eightfold increase and a step change in ceramic substrate technology. Figure 10.19a shows a ceramic substrate recently developed. The large GSA is a result of the structure within the channel walls (Figure 10.19b) that are accessible to the exhaust gasses from the surface of the channels (Figure 10.19c).

The manufacturing process comprises of continuous extrusion of ceramic/polymer/solvent systems combined with phase inversion. This versatile technique is continuous, scalable, and cost-effective and does not require high-tech or expensive equipment. With an understanding of the fundamental processes involved in the formation of microstructures in ceramic substrates, the morphology can be easily controlled. The control over substrate structure can be achieved by varying simple parameters during the extrusion process, rather than developing new techniques or redesigning or modifying equipment.

10.4.4.2   Reduced Substrate Volume

The high GSA achieved allows the volume of the substrate to be reduced. By reducing the volume, the substrate can be positioned closer to the engine manifold and the

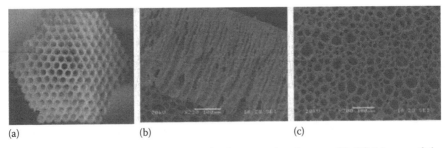

(a)   (b)   (c)

FIGURE 10.19   (a) Front view photograph of a ceramic substrate, (b) SEM image of the cross section of the channel wall, and (c) SEM image of the channel wall surface. (Data from Kingsbury, B. et al., SAE Technical Paper 2014-01-2805, 2014.)

catalyst will reach effective operating temperature more quickly. Before the catalyst reaches its operating temperature, 50% of HC, CO, and NOx emissions occur, so the adoption will significantly reduce emissions during testing cycles. Reduced substrate volume will also increase the design freedom available to OEMs. In a modern car, space within the engine bay is at a premium and is a major advantage. An additional benefit of the manufacturing process is the ability to fabricate substrates with a wider variety of shapes, which is not achievable with the current technology. A substrate could be designed to fit in a bend within an exhaust system, reducing the distance between the exhaust manifold and the substrate, leading to reduced emissions and effective use of space within the engine bay.

10.4.4.3   Reduced Catalyst Loading

PGM cost accounts for over 60% of the cost of the catalytic converter (average of £155 per vehicle in Europe) unit as a whole, so it is essential that the quantity of PGM is minimized. Conversion efficiency is improved by increasing catalyst loading. However, the benefit of increased catalyst loading becomes substantially reduced as the quantity is increased, due to diffusion limitations within the catalytic coating. To meet emissions regulations excessive costs result from high catalyst loading to achieve very small benefits in conversion. Increasing the GSA reduces the quantity of catalyst needed and increases the contact between the catalyst and the exhaust gas. Substrate and catalytic coating development has focused on reducing diffusion limitations to maximize catalytic activity and reduce catalyst cost. The reduction in catalyst loading achieved by increasing GSA is the most important factor in converter design, due to the high cost of PGM catalysts. The innovative microstructure inside the hollow fibers offers an eightfold increase in GSA allowing for optimal dispersion within the substrate and diffusion limitations can be all but eliminated. Additional benefits include significantly less or slower aging.

10.4.4.4   Reduced Pressure Drop

For current ceramic substrates, GSA is increased by reducing the diameter of the channels through which the exhaust gasses flow, leading to an increase in the resistance to gas flow, reducing engine power output and increasing fuel consumption and CO_2 emissions. Preparing substrates with advanced microstructures allows large channel diameters to be achieved while maintaining extremely high GSA, overcoming the trade-off between high GSA and backpressure. A reduction of 70% in pressure drop compared to a commercial 900 cpsi ceramic substrate can be achieved. This large reduction in pressure drop leads to substantial reductions in fuel consumption and improvements in engine power output.

10.5   CONCLUSIONS

The viscous fingering-induced phase inversion process introduced in this chapter is uniquely different from existing techniques in fabricating ceramic hollow-fiber membranes. Apart from a wide range of ceramic materials that can be employed, this technique combines great flexibilities in designing the ceramic membranes, such as *microstructuring* ceramic membranes and assembling multilayers and hollow-fiber membrane formation into a single-step process. This leads to great potentials in improving the membrane performance, due to the lowered mass transfer resistance, the enlarged accessible area, and reduced membrane fabrication costs, which is still challenging in large-scale ceramic membrane processes. Potential engineering applications of such microstructured ceramic hollow-fiber membranes include water filtration, microtubular SOFC, and automotive emissions control, after assembling specified single-channel hollow-fiber membranes into a module or unit. Meanwhile, a microstructured ceramic hollow-fiber or tubular membrane consisting of multiple channels is coming into shape. Besides greater mechanical properties, it provides more characteristics favorable for engineering applications. All these contribute to form a platform technique, delivering new ceramic membrane products for different applications in near future.

ACKNOWLEDGMENT

The authors gratefully acknowledge the research funding provided by Engineering and Physical Sciences Research Council (EPSRC) in the United Kingdom (grant nos: EP/I010947/1 and EP/G012679/1).

REFERENCES

1. Nunes, S.P. and K.V. Peinemann (eds.), Membrane preparation, in *Membrane Technology*. 2001, Wiley-VCH Verlag GmbH, Germany. pp. 6–11.
2. Baker, R.W. (eds.), Membranes and modules, in *Membrane Technology and Applications*. 2004, Wiley, New York. pp. 89–160.
3. Tan, X.Y., S.M. Liu, and K. Li, Preparation and characterization of inorganic hollow fiber membranes. *Journal of Membrane Science*, 2001. **188**(1): 87–95.
4. Tan, X.Y. and K. Li, Modeling of air separation in a LSCF hollow-fiber membrane module. *AIChE Journal*, 2002. **48**(7): 1469–1477.
5. Tan, X.Y., N.T. Yang, and K. Li, Modeling of a SrCe0.95Yb0.05O3-alpha hollow fibre membrane reactor for methane coupling. *Chinese Journal of Chemical Engineering*, 2003. **11**(3): 289–296.
6. Liu, S.M. and G.R. Gavalas, Preparation of oxygen ion conducting ceramic hollow-fiber membranes. *Industrial & Engineering Chemistry Research*, 2005. **44**(20): 7633–7637.
7. Liu, S.M. and G.R. Gavalas, Oxygen selective ceramic hollow fiber membranes. *Journal of Membrane Science*, 2005. **246**(1): 103–108.
8. Liu, Y.T. and K. Li, Preparation of SrCe0.95Yb0.05O3-alpha hollow fibre membranes: Study on sintering processes. *Journal of Membrane Science*, 2005. **259**(1/2): 47–54.
9. Tan, X.Y., Y.T. Liu, and K. Li, Mixed conducting ceramic hollow-fiber membranes for air separation. *AIChE Journal*, 2005. **51**(7): 1991–2000.
10. Tan, X.Y., Y.T. Liu, and K. Li, Preparation of LSCF ceramic hollow-fiber membranes for oxygen production by a phase-inversion/sintering technique. *Industrial & Engineering Chemistry Research*, 2005. **44**(1): 61–66.
11. Li, K., X.Y. Tan, and Y.T. Liu, Single-step fabrication of ceramic hollow fibers for oxygen permeation. *Journal of Membrane Science*, 2006. **272**(1/2): 1–5.
12. Liu, S. et al., Ba0.5Sr0.5Co0.8Fe0.2O3-delta ceramic hollow-fiber membranes for oxygen permeation. *AIChE Journal*, 2006. **52**(10): 3452–3461.
13. Liu, Y.T., X.Y. Tan, and K. Li, SrCe0.95Yb0.05O3-alpha (SCYb) hollow fibre membrane: Preparation, characterization and performance. *Journal of Membrane Science*, 2006. **283**(1/2): 380–385.

14. Liu, Y.T., X.Y. Tan, and K. Li, SrCe0.95Yb0.05O3-alpha hollow-fiber membrane and its property in proton conduction. *AIChE Journal*, 2006. **52**(4): 1577–1585.
15. Thursfield, A. and I.S. Metcalfe, Air separation using a catalytically modified mixed conducting ceramic hollow fibre membrane module. *Journal of Membrane Science*, 2007. **288**(1/2): 175–187.
16. Liu, H. et al., Novel dual structured mixed conducting ceramic hollow fibre membranes. *Separation and Purification Technology*, 2008. **63**(1): 243–247.
17. Tan, X.Y. et al., Enhancement of oxygen permeation through La0.6Sr0.4Co0.2Fe0.8O3-delta hollow fibre membranes by surface modifications. *Journal of Membrane Science*, 2008. **324**(1/2): 128–135.
18. Wang, H.H., C. Tablet, and J. Caro, Oxygen production at low temperature using dense perovskite hollow fiber membranes. *Journal of Membrane Science*, 2008. **322**(1): 214–217.
19. Garcia-Garcia, F.R. et al., Hollow fibre membrane reactors for high H-2 yields in the WGS reaction. *Journal of Membrane Science*, 2012. **405**: 30–37.
20. Israni, S.H., B.K.R. Nair, and M.P. Harold, Hydrogen generation and purification in a composite Pd hollow fiber membrane reactor: Experiments and modeling. *Catalysis Today*, 2009. **139**(4): 299–311.
21. Kilgus, M. et al., Palladium coated ceramic hollow fibre membranes for hydrogen separation. *Desalination*, 2006. **200**(1–3): 95–96.
22. Medrano, J.A. et al., Two-zone fluidized bed reactor (TZFBR) with palladium membrane for catalytic propane dehydrogenation: Experimental performance assessment. *Industrial & Engineering Chemistry Research*, 2013. **52**(10): 3723–3731.
23. Pan, X.L. et al., Pd/ceramic hollow fibers for H-2 separation. *Separation and Purification Technology*, 2003. **32**(1–3): 265–270.
24. Pan, Y.C., B. Wang, and Z.P. Lai, Synthesis of ceramic hollow fiber supported zeolitic imidazolate framework-8 (ZIF-8) membranes with high hydrogen permeability. *Journal of Membrane Science*, 2012. **421**: 292–298.
25. Caro, J. et al., Perowskit hollow fibre membranes for the catalytic partial oxidation of methane to synthesis gas. *Chemie Ingenieur Technik*, 2007. **79**(6): 831–842.
26. Garcia-Garcia, F.R. et al., Asymmetric ceramic hollow fibres applied in heterogeneous catalytic gas phase reactions. *Catalysis Today*, 2012. **193**(1): 20–30.
27. Garcia-Garcia, F.R. and K. Li, New catalytic reactors prepared from symmetric and asymmetric ceramic hollow fibres. *Applied Catalysis A: General*, 2013. **456**: 1–10.
28. Garcia-Garcia, F.R. et al., Catalytic hollow fibre membrane micro-reactor: High purity H-2 production by WGS reaction. *Catalysis Today*, 2011. **171**(1): 281–289.
29. Garcia-Garcia, F.R. et al., Dry reforming of methane using Pd-based membrane reactors fabricated from different substrates. *Journal of Membrane Science*, 2013. **435**: 218–225.
30. Gbenedio, E. et al., A multifunctional Pd/alumina hollow fibre membrane reactor for propane dehydrogenation. *Catalysis Today*, 2010. **156**(3/4): 93–99.
31. Tan, X., Z. Pang, and S. Liu, Catalytic perovskite hollow fibre membrane reactors for methane oxidative coupling. *Journal of Membrane Science*, 2007. **302**(1/2): 109–114.
32. Faiz, R. et al., Separation of Olefin/Paraffin Gas Mixtures Using Ceramic Hollow Fiber Membrane Contactors. *Industrial & Engineering Chemistry Research*, 2013. **52**(23): 7918–7929.
33. Koonaphapdeelert, S. et al., Solvent distillation by ceramic hollow fibre membrane contactors. *Journal of Membrane Science*, 2008. **314**(1/2): 58–66.
34. Koonaphapdeelert, S., Z.T. Wu, and K. Li, Carbon dioxide stripping in ceramic hollow fibre membrane contactors. *Chemical Engineering Science*, 2009. **64**(1): 1–8.
35. de Jong, J. et al., Towards single step production of multi-layer inorganic hollow fibers. *Journal of Membrane Science*, 2004. **239**(2): 265–269.
36. Kingsbury, B.F.K. and K. Li, A morphological study of ceramic hollow fibre membranes. *Journal of Membrane Science*, 2009. **328**(1/2): 134–140.

37. Kingsbury, B.F.K., Z.T. Wu, and K. Li, A morphological study of ceramic hollow fibre membranes: A perspective on multifunctional catalytic membrane reactors. *Catalysis Today*, 2010. **156**(3/4): 306–315.

38. Wang, B. and Z.P. Lai, Finger-like voids induced by viscous fingering during phase inversion of alumina/PES/NMP suspensions. *Journal of Membrane Science*, 2012. **405**: 275–283.

39. Zydorczak, B., Z.T. Wu, and K. Li, Fabrication of ultrathin La0.6Sr0.4Co0.2Fe0.8O3-delta hollow fibre membranes for oxygen permeation. *Chemical Engineering Science*, 2009. **64**(21): 4383–4388.

40. Kanawka, K. et al., Microstructure and performance investigation of a solid oxide fuel cells based on highly asymmetric YSZ microtubular electrolytes. *Industrial & Engineering Chemistry Research*, 2010. **49**(13): 6062–6068.

41. Wu, Z.T. et al., A Novel Inorganic Hollow Fiber Membrane Reactor for Catalytic Dehydrogenation of Propane. *AIChE Journal*, 2009. **55**(9): 2389–2398.

42. Wu, Z.T. et al., Effects of separation layer thickness on oxygen permeation and mechanical strength of DL-HFMR-ScSZ. *Journal of Membrane Science*, 2012. **415**: 229–236.

43. Wu, Z.T., B. Wang, and K. Li, A novel dual-layer ceramic hollow fibre membrane reactor for methane conversion. *Journal of Membrane Science*, 2010. **352**(1/2): 63–70.

44. Droushiotis, N. et al., Fabrication by Co-extrusion and electrochemical characterization of micro-tubular hollow fibre solid oxide fuel cells. *Electrochemistry Communications*, 2010. **12**(6): 792–795.

45. Othman, M.H.D. et al., Electrolyte thickness control and its effect on electrolyte/anode dual-layer hollow fibres for micro-tubular solid oxide fuel cells. *Journal of Membrane Science*, 2010. **365**(1/2): 382–388.

46. Othman, M.H.D. et al., High-Performance, Anode-Supported, Microtubular SOFC Prepared from Single-Step-Fabricated, Dual-Layer Hollow Fibers. *Advanced Materials*, 2011. **23**(21): 2480.

47. Othman, M.H.D. et al., Novel fabrication technique of hollow fibre support for micro-tubular solid oxide fuel cells. *Journal of Power Sources*, 2011. **196**(11): 5035–5044.

48. Othman, M.H.D. et al., Dual-layer hollow fibres with different anode structures for micro-tubular solid oxide fuel cells. *Journal of Power Sources*, 2012. **205**: 272–280.

49. Wu, Z.T., B. Wang, and K. Li, Functional LSM-ScSZ/NiO-ScSZ dual-layer hollow fibres for partial oxidation of methane. *International Journal of Hydrogen Energy*, 2011. **36**(9): 5334–5341.

50. Li, T., Z.T. Wu, and K. Li, Single-step fabrication and characterisations of triple-layer ceramic hollow fibres for micro-tubular solid oxide fuel cells (SOFCs). *Journal of Membrane Science*, 2014. **449**: 1–8.

51. Rahaman, M.N., *Ceramic Processing and Sintering*. 2003, Marcel Dekker, New York.

52. Li, K., *Ceramic Membranes for Separation and Reaction*. 2007, Wiley.

53. Wu, Z.T. et al., A controlled sintering process for more permeable ceramic hollow fibre membranes. *Journal of Membrane Science*, 2013. **446**: 286–293.

54. Wu, Z.T. et al., *Advanced Membrane Science and Technology for Sustainable Energy and Environmental Applications*. 2011, Woodhead Publishing Series in Energy, Philadelphia, PA.

55. Kingsbury, B.F.K., Z.T. Wu, and K. Li, *Membranes for Membrane Reactors: Preparation, Optimization and Selection*. 2011, Wiley, New York.

56. Lee, M. et al., Micro-structured alumina hollow fibre membranes—Potential applications in wastewater treatment. *Journal of Membrane Science*, 2014. **461**: 39–48.

57. Kingsbury, B. et al., *Advanced ceramic substrate with ordered and designed microstructure for applications in automotive catalysis*, SAE Technical Paper 2014-01-2805, 2014, doi:10.4271/2014-01-2805.

11 Ceramic Hollow-Fiber Support through a Phase Inversion-Based Extrusion/Sintering Technique for High-Temperature Energy Conversion Systems

Mohd Hafiz Dzarfan Othman,
Mukhlis A. Rahman, Kang Li, Juhana Jaafar,
Hasrinah Hasbullah, and Ahmad Fauzi Ismail

CONTENTS

11.1   Introduction 348
11.2   Fabrication of Ceramic Hollow-Fiber Support through a Phase
         Inversion-Based Extrusion/Sintering Technique 350
         11.2.1   Preparation of a Spinning Suspension 350
         11.2.2   Extrusion of Ceramic Hollow-Fiber Precursor 351
         11.2.3   Sintering Process 354
11.3   Applications of Ceramic Hollow-Fiber Membrane in
         High-Temperature Energy Conversion Systems 356
         11.3.1   Application I: Hollow-Fiber Membrane Microreactor 356
                     11.3.1.1   Background 356
                     11.3.1.2   Fabrication of Hollow-Fiber Support for Microreactors 357
                     11.3.1.3   Morphology of Hollow-Fiber Support for
                                     Microreactor 359
                     11.3.1.4   Deposition of Membrane onto the Hollow-Fiber
                                     Microreactor 360
                     11.3.1.5   Impregnation of Catalyst into Microreactor Pore
                                     Structures 362
                     11.3.1.6   Membrane Microreactor Performance 365

      11.3.2   Application II: Microtubular SOFCs..366
            11.3.2.1 Background 366
            11.3.2.2 Single-Step Fabrication of Electrolyte/Anode
                          Dual-Layer Hollow Fiber..367
            11.3.2.3 Morphology and Physical Properties of Dual-Layer
                          Hollow Fibers...369
            11.3.2.4 Development of a Complete SOFC 372
            11.3.2.5 Fuel Cell Performances...374
11.4 Conclusion...378
References...378

11.1 INTRODUCTION

Over the years, ceramic membrane has demonstrated many advantages compared to polymeric membrane due to its superior thermal stability, excellent chemical resistance, and great mechanical strength. Figure 11.1 shows the timeline for the development of ceramic membrane technology over the last 80 years. Despite examples such as the development and implementation of ceramic membranes for uranium enrichment during the Manhattan project in 1942, and later by the French Atomic Agency in the 1950s, it was not until the 1980s that ceramic membrane research started to develop in earnest and ceramic membranes were applied to more conventional applications such as filtration and gas separation. The increased use of ceramic membranes in recent years is driven by their advantages over traditional technologies.

Similar to polymeric membranes, there are three main types of ceramic membrane geometries, disk/flat-sheet, tubular, and hollow-fiber, which are normally

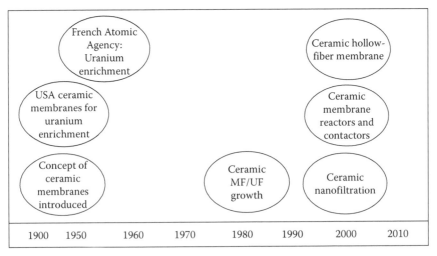

FIGURE 11.1 Progression of ceramic membrane technology from its inception in the 1940s to the present day. (Data from B.F.K. Kingsbury, A morphological study of ceramic hollow fibre membranes: A perspective on multifunctional catalytic membrane reactors, PhD Thesis, Imperial College London, London, 2010.)

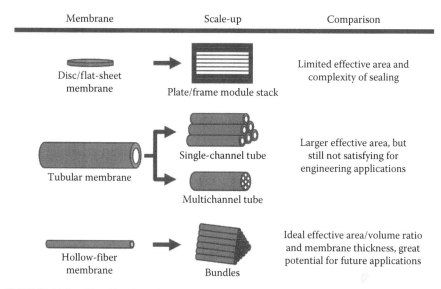

Membrane	Scale-up	Comparison

Disc/flat-sheet membrane → Plate/frame module stack — Limited effective area and complexity of sealing

Tubular membrane → Single-channel tube / Multichannel tube — Larger effective area, but still not satisfying for engineering applications

Hollow-fiber membrane → Bundles — Ideal effective area/volume ratio and membrane thickness, great potential for future applications

FIGURE 11.2 Classification of ceramic membrane support in view of geometrical configurations. (Data from Z. Wu, Dual-layer functional ceramic hollow fibre membranes for partial oxidation of methane, PhD Thesis, Imperial College London, London, 2012.)

fabricated as a support for various applications, as shown in Figure 11.2. Although suffering from the limited surface area and the substantial challenge of sealing at elevated temperature, flat-sheet-type membranes are still receiving a lot of attention due to their ease of fabrication and quality control. In addition, the supports of this type of membrane are ideal for many applications because parameters such as temperature, surface area, and membrane thickness can be easily determined.

In an effort to enhance the surface area of ceramic support, tubular membranes have been developed using an extrusion process followed by high temperature sintering. A module consisting of a number of single-channel tubes packed together or taking the form of a multichannel tubular configuration contributes to an increase in the surface area/volume ratio by approximately 250 m^2 m^{-3}. Membrane sealing can be achieved outside the high temperature zone and is much easier than that of disk/flat-sheet counterparts. However, these membranes normally have a symmetric structure due to the membrane fabrication process. Membrane thickness is, in this case, equal to that of tube wall, which has to be of a certain high level to give a reasonable mechanical strength. In contrast, hollow-fiber membranes with higher surface area/volume ratios of up to 3000 m^2 m^{-3} show a number of advantages over the planar and tubular counterparts. In addition to significantly higher surface area, easy membrane sealing, and great membrane uniformity for large-scale production, the effective membrane thickness of hollow-fiber membranes can be reduced to a much lower level of approximately 200 μm.

Ceramic hollow-fiber support can be fabricated by a phase inversion-based extrusion/sintering technique [3], which allows a more flexible control over the membrane macro/microstructures by adjusting fabricating parameters such as air-gap, extrusion rate, internal coagulant composition, and the amount of nonsolvent additive in the spinning suspensions [4]. Such unique structural diversity delivers considerable possibilities of further improving oxygen separation performance, although the mechanical property of these membranes turns out to be the concern for promising engineering applications.

11.2 FABRICATION OF CERAMIC HOLLOW-FIBER SUPPORT THROUGH A PHASE INVERSION-BASED EXTRUSION/SINTERING TECHNIQUE

Early development of ceramic hollow fibers using a phase inversion-based extrusion/sintering technique was applied in gas separation and ultrafiltration by Okubo et al. [5]. This technique was further established by Tan et al. [6] and Liu et al. [7]. Since then, the phase inversion-based extrusion/sintering technique has been widely used to prepare ceramic hollow fibers for different applications. There are basically three main steps in preparing ceramic hollow fibers using a phase inversion-based extrusion/sintering technique: (1) preparation of a spinning suspension, (2) extrusion of ceramic hollow fiber precursors, and (3) finally sintering [3]. Each step plays a specific and crucial part in the production of the desired ceramic hollow fibers, as listed in Table 11.1.

11.2.1 Preparation of a Spinning Suspension

The spinning suspension is a viscous mixture composed of ceramic particles, solvent, binder, and additives, each of which has different influences on the suspension. The particle size, distribution, and shape are important factors to consider when selecting the ceramic powders, as they could affect the dispersion of ceramic particles in the suspension, and the properties of the produced fiber. For the binders, most of them are long-chain polymers that dissolve in the solvent. It should be noted that the binder must be fully burned away without leaving ash or tar during the sintering process. The interaction between the binder and the solvent is very important, as it determines the rheology of the suspension and affects the solidification process of the hollow fiber. Solvents used in the dry-jet wet extrusion technique must exhibit a high exchange rate with nonsolvents, a term that is used to refer to the coagulant. The exchange rate between the solvent and the nonsolvent has an effect on the cross-sectional structure of the hollow fiber. Other additives such as dispersants or lubricants are often added to improve the particle dispersion in the suspension. The dispersion depends on the ability of the dispersant to break the surface interaction between particles to allow them to remain separate. The amount of additive used should be as small as possible to achieve a high volume fraction of ceramic particles in the spinning dope.

The typical process to prepare a homogeneous spinning suspension involves several steps, which are as follows: (1) mixing a dispersant with a selected solvent in a container; (2) adding ceramic particles, which must be well dispersed by a sieve; (3) adding milling balls to homogenize the particle suspension; (4) adding polymer binders and additive; and (5) degassing the suspension dope before the extrusion process. A typical spinning suspension may contain approximately 50–70 wt% of

TABLE 11.1

Steps of the Phase Inversion-Based Extrusion/Sintering Technique, Their Factors, and Influences on the Resultant Hollow-Fiber Properties

Steps	Factors	Influences
Preparation of spinning suspension	• Concentration of compounds • Particle size/distribution • Particle/binder ratio • Solvent types • Dispersant • Additives • Mixing speed (shear stress) • Mixing temperature	• Homogeneous • Viscosity • Particle packing density • Rheology • Particle dispersion
Extrusion of ceramic hollow-fiber precursors	• Suspension viscosity/homogeneous • Particle size/distribution • Extrusion speed • Air gap • Environment conditions (humidity/temperature) • Spinneret dimension/configuration • Internal/external coagulant (solvent%, flow rate, and temperature) • Suspension temperature	• Uniformity (bore shape) • Particle packing density • Surface morphology (smooth/rough) • Precursor morphology (dense/porous) • Precursor structure (symmetric/asymmetric) • Precursor thickness • Precursor dimension • Precursor length
Sintering process	• Particle characteristics (thermal expansion) • Particle size/distribution • Sintering profile • Sintering placement (vertical/horizontal) • Environmental gas	• Membrane dimension • Membrane morphology (dense/porous) • Cracking/defect • Grain size • Porosity • Pore size • Mechanical strength

ceramic material, 20–40 wt% of solvent, 5–7 wt% of binder, and a very small amount of dispersant (~0.1–1.5 wt%). Ceramic material is by far the main constituent of the suspension, and it is important that the spinning suspension is mixed sufficiently, so that ceramic particles are completely surrounded by the binder. The suspension should also have a considerable viscosity for the ease of fiber forming, whereas at the same time playing an important role in varying the hollow-fiber structure [4].

11.2.2 EXTRUSION OF CERAMIC HOLLOW-FIBER PRECURSOR

The second step in the dry-jet wet extrusion/sintering method is the extrusion of hollow-fiber precursors. The hollow-fiber precursor is the terminology used to refer to the extruded hollow fiber before sintering, which consists of ceramic material, polymer binder, and dispersant. Prior to the extrusion process, it is essential to degas the prepared spinning suspension to prevent the incorporation of gas bubbles into the hollow-fiber structure, which reduces its integrity and uniformity. The spinning suspension is then extruded through the opening of a tube-in-orifice spinneret to form a fiber precursor. At the same time, the nonsolvent internal coagulant is pumped through the center of the spinneret, causing the precursor to be in the form of a hollow and longitudinal fiber. The fiber will then travel into the nonsolvent external coagulation bath where it forms a coil.

Formation of the hollow-fiber precursor results from the exchange between the solvent and the nonsolvent (coagulant), which induces the precipitation of polymer in the suspension and hence consolidates the ceramic material [8]. This process is known as phase inversion [9], and the basic concept of this process can be explained using the ternary phase diagram of a polymeric system, which involves the polymer, solvent, and nonsolvent, as shown in Figure 11.3.

The entire phase inversion process of a polymeric solution is represented by the path from A to D. The original polymeric solution is at point A, where no precipitation agent (nonsolvent) is present in the solution. After the immersion of the polymeric solution into a nonsolvent coagulation bath, the solvent diffuses out of the polymer solution, whereas the nonsolvent diffuses into the solution. In the case when the solvent flux is higher than the nonsolvent flux, the polymer concentration at the interface would increase, and at some point, the polymer starts to precipitate (as represented by point B). The continuous replacement of the solvent by the nonsolvent would result in the solidification of the polymer-rich phase (point C). Further solvent/nonsolvent exchange would cause shrinkage of the polymer-rich phase and finally reach point D, where the two phases (solid and liquid) are in equilibrium. A solid (polymer-rich) phase that forms the membrane structure is represented by point S and a liquid (polymer-poor) phase that constitutes the membrane pores filled with nonsolvent is represented by point L.

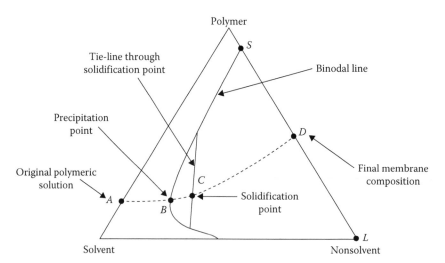

FIGURE 11.3 Schematic ternary phase diagram of a polymer/solvent/nonsolvent during polymeric membrane formation. (Data from H. Strathmann and K. Kock, *Desalination*, 21, 241–255, 1977.)

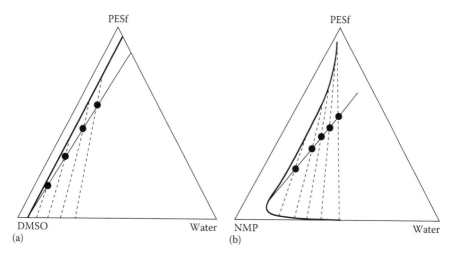

FIGURE 11.4 Phase diagrams for (a) PESf/DMSO/water and (b) PESf/NMP/water systems. Thick line, binodal line; thin line, spinodal line; dashed lines, tie-lines; dot, solidification point. (Reproduced from B.F. Barton and A.J. McHugh, *Journal of Polymer Science Part B: Polymer Physics*, 37, 1449–1460, 1999; L. Zeman and G. Tkacik, *Journal of Membrane Science*, 36, 119–140, 1988.)

It should be noted that the type of solvent, polymer binder or nonsolvent, could affect the precipitation mechanism in the phase inversion process. As an example, the *N*-methyl-2-pyrrolidone (NMP) and dimethyl sulfoxide (DMSO) systems possess dissimilar precipitation lines (Figure 11.4), where the precipitation point for the polyethersulfone (PESf)/DMSO/water system is much closer to the original casting solution (0% water) line than that of the PESf/NMP/water system. Because the rates of solvent/nonsolvent exchange for both systems are not very much different, the time to reach the precipitation point for PESf/DMSO/water system is much shorter than that for the PESf/NMP/water system. As the macrostructure of the membrane is largely determined at the precipitation point, PESf/DMSO/water and PESf/NMP/water systems would yield different macrostructures.

Great effort has been made to both control and understand the formation mechanism for a wide range of structures observed in polymeric membrane fabrication [13,14]. However, due to the large differences between polymeric and ceramic systems, in particular the low polymer concentration, this information is of limited use during ceramic membrane preparation. In fact, only two morphologies have so far been observed in ceramic systems, namely, finger-like voids and a sponge-like structure. Examples of both these structures are given in Figure 11.5.

Hydrodynamically unstable viscous fingering is a well-known phenomenon that occurs at the interface between fluids with different viscosities in the first moments of mixing and has been applied here to explain the formation of finger-like voids in ceramic membrane precursors [16]. When the suspension is in contact with the nonsolvent, a steep concentration gradient results in solvent/nonsolvent exchange, a rapid increase in local viscosity, and finally precipitation of the polymer phase. However, due to instabilities at the interface between the suspension and the precipitant,

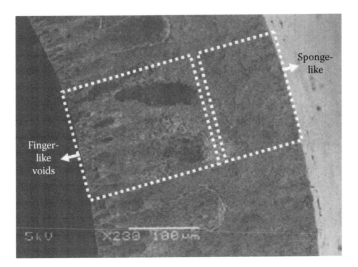

FIGURE 11.5 Scanning electron microscopy (SEM) image of the macrostructure of a ceramic hollow-fiber precursor. (Data from S. Liu and K. Li, *Journal of Membrane Science*, 218, 269–277, 2003.)

there is a tendency for viscous fingering to occur, initiating the formation of finger-like voids. Under normal circumstances, a stable interface would be established between the two phases of differing viscosities; however, due to the presence of invertible polymer binder, a rapid viscosity increase followed by polymer precipitation retains the viscous-fingering structure. The relative thickness of finger-like and sponge-like regions greatly affects the membrane or membrane support properties, including mechanical strength and permeation flux. The versatility of ceramic hollow fibers means it is essential that fiber morphology can be controlled, so that it may be tailored for specific applications [16,17].

11.2.3 SINTERING PROCESS

The hollow-fiber precursors, which normally comprise of ceramic particles, binder, and dispersant, are heat-treated in a furnace to develop the desired microstructure and properties. Ideally, this treatment, commonly known as *sintering*, is the process by which a powder compact is transformed into a strong and dense ceramic body without melting the powder. As the sintering process is required for a ceramic hollow-fiber precursor, which has been prepared from a dry-jet wet extrusion, this method is often called the *combined dry-jet wet extrusion and sintering technique*.

Sintering of the hollow-fiber precursors can be divided into three major stages: pre-sintering, thermolysis, and final sintering. Pre-sintering is a means of removing residual liquid that may remain after the formation of the fiber precursor and any moisture adsorbed from the atmosphere during transporting and setting. The adsorbed moisture may persist in the precursor up to 200°C [17]. It is important to increase the temperature slowly, as expanding vapor within the lattice may cause cracks and fractures. Thermolysis is a process that burns out the organic components such as binder and dispersant and occurs as the temperature is increased to approximately 800°C. It is an important step as incomplete binder removal and uncontrolled thermolysis may introduce defects in the hollow fiber, which may then impair the performance of the hollow fiber.

Final sintering is defined as a transforming process of ceramic particles into a strong and dense ceramic body, and it does not start to densify until the temperature reaches approximately 60%–80% of the melting point of the fiber material. The final sintering process usually involves three stages: initial stage, intermediate stage, and final stage, which are shown in Figure 11.6.

The initial stage consists of the rearrangement of ceramic particles and neck growth at contacting points between particles. The initial stage lasts until the radius of the grown neck reaches 40%–50% of the particle radius. In the intermediate stage, the grain boundaries start to develop. Ceramic particles are bound together and pore channels are formed along the grain edges. This stage normally covers the major part of the sintering process and leads to a major shrinkage of the ceramic membrane. When pores continue shrinking and some of them pinch off and become isolated at the grain corners, it is considered that the final stage has started. Growth of grains mainly occurs in this stage and the pores are gradually eliminated until the membrane turns into a fully dense structure. However, in most of the cases, the macrostructure of the fiber precursor formed during the phase inversion process is

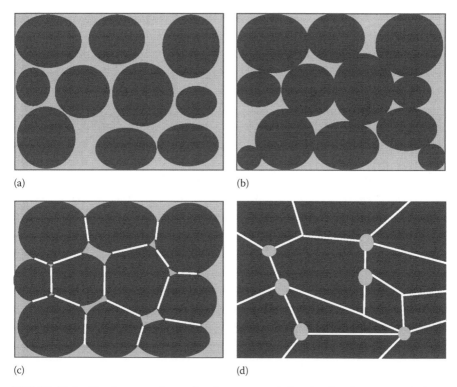

(a)   (b)

(c)   (d)

FIGURE 11.6 Development of ceramic microstructure during the sintering process: (a) loose particles; (b) initial stage; (c) intermediate stage; and (d) final stage.

retained during sintering. Finger-like voids above a certain size cannot be eliminated, although at elevated sintering temperature sponge-like regions will be densified and eventually become gas tight for some ceramic materials [16,17].

11.3 APPLICATIONS OF CERAMIC HOLLOW-FIBER MEMBRANE IN HIGH-TEMPERATURE ENERGY CONVERSION SYSTEMS

Ceramic hollow-fiber membrane has been used widely in various separation applications, such as wastewater treatment, oil recovery, membrane contactors, and gas separation. Generally, the performance of ceramic hollow-fiber membrane surpasses its polymeric counterpart in separating the desired product. Ceramic membrane possesses high mechanical strength and high melting points rendering them suitable to be used under extreme conditions such as higher temperature and pressure. Similar to polymeric hollow-fiber membranes, phase inversion process has been used to fabricate ceramic hollow-fiber precursors. This fabrication technique is rather simple, which enables it to be used directly for large-scale production of ceramic hollow-fiber membrane.

Recently, ceramic hollow-fiber membrane has been used in energy-conversion applications, which require a combination of reaction and separation processes. The next subtopics focus on the development of membrane microreactors and solid oxide fuel cells (SOFCs) using ceramic hollow-fiber support fabricated through a phase inversion-based extrusion technique. These are examples of energy conversion applications for ceramic hollow-fiber membrane. Both applications require the combination of chemical reactions in either generating hydrogen or electricity followed by the separation of the product to the permeate stream. Therefore, these dual-mode processes require an advanced ceramic hollow-fiber support that integrates reaction zones and a site for depositing a selective layer or membrane in a single unit.

11.3.1 APPLICATION I: HOLLOW-FIBER MEMBRANE MICROREACTOR

11.3.1.1 Background

Internal combustion engines that use gasoline as a fuel have been widely used in transportation. However, there is an increasing demand for renewable and environmental-friendly fuel due to the depletion of fossil fuels and the uncontrollable release of greenhouse gases such as CO_2. To alleviate such problems related to excessive release of CO_2, bioethanol is regarded as an alternative fuel to replace gasoline as a fuel for the internal combustion engine. In comparison with fossil fuel, ethanol can be generated from renewable sources, and the release of CO_2 from this biofuel is less severe. The release of CO_2 from ethanol will be compensated by the growth of plants/feedstock used in the production of ethanol [18]. Unlike fossil fuel, there will be a cycle of production and consumption of CO_2 if ethanol is consumed as a fuel.

Although ethanol can be used directly in the internal combustion engine, as implemented in Brazil, its performance in this conventional technology is still lower compared with that of gasoline or even the fuel cell system [19]. As an alternative, it can be used as a hydrogen carrier for the on-board hydrogen generation in the transportation application. A number of attempts have been made to develop a compact reformer that can yield hydrogen for the polymer electrolyte membrane fuel cell (PEMFC) used in transportation [20–22]. To use this approach, palladium/ silver (Pd/Ag) membrane can be incorporated to the reformer and used as a separation unit to produce pure hydrogen [23,24]. Incorporation of Pd/Ag membrane to the reformer is made because pure hydrogen can increase the performance of fuel cell due to the absence of CO_2 in fuel stream. Chellapa et al. from MesoFuel, Inc developed MesoChannel™ membrane reformers, which can use ethanol as the fuel to obtain pure hydrogen for the PEMFC [23]. The system can be operated under pressures ranging from 4 to 6 bar and at temperatures ranging from 575°C to 625°C. The CO levels can be reduced down to 5 ppm, which enable the hydrogen stream to be connected directly to the PEMFC. Recently, Falco developed a process scheme and carried out modeling on a four-tube-and-shell membrane reactor using 2D mathematical models in the production of hydrogen through ethanol steam reforming (ESR) for the PEMFC [24]. The result of his simulation study was later validated by experimental data. Using Pd/Ag membrane fabricated on tubular tubes with 150 cm length and 6 cm outer diameter, the system produced 64.7 NL min⁻¹ of hydrogen, which is equal to a 4 kW FC feedstock requirement. Hydrogen recovery of the system is 67% of total H_2 produced through the ESR.

11.3.1.2 Fabrication of Hollow-Fiber Support for Microreactors

To develop a reformer, which can include an inorganic membrane used as a separation unit, an exceptional substrate design is required, so that the reaction and separation units can be incorporated and shaped into a compact design. Asymmetric ceramic hollow fiber can be a very promising choice to be used as a substrate for this application. The catalytic hollow-fiber membrane microreactor (CHFMMR) was developed successfully using asymmetric yttria-stabilized zirconia (YSZ) hollow fiber prepared by a phase inversion-based extrusion technique followed by a sintering process at higher temperatures to produce hydrogen from the ESR. This novel design integrates the ESR catalyst in a conical microchannel and the Pd/Ag membrane on the outer surface of YSZ hollow fiber.

Incorporation of the ESR catalyst and hydrogen selective membrane into a single unit of YSZ hollow fiber also gives an advantage for the on-board hydrogen generation to be shaped into a more compact and efficient design suitable for vehicular applications [25]. Figure 11.7a shows a schematic diagram of a single CHFMMR that consists of a YSZ hollow fiber, in which the catalyst was impregnated into the finger-like region and the Pd/Ag membrane was platted on its outer surface. The Pd/Ag membrane, which has high permeability and infinite selectivity to hydrogen, enables the separation process to be carried out efficiently. The separation occurs from the reaction zone to produce pure hydrogen in the shell side whereas CO_2 will be left in the lumen as illustrated in Figure 11.7b.

YSZ has been chosen as a substrate for the development of the CHFMMR due to its high mechanical strength and competitive price compared to other ceramic materials, even though it has been mostly used in electrochemical applications such as SOFCs, oxygen sensors, and high temperature pH sensors [26]. The precursor of the YSZ hollow fiber is sintered at 1300°C, which is lower than the sintering temperature

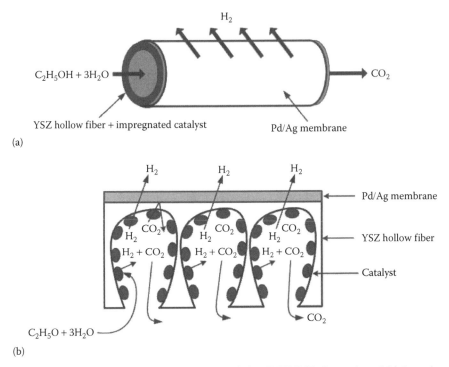

FIGURE 11.7 (a) Schematic representation of the CHFMMR for on-board high purity hydrogen production using the ESR reaction. (b) The reactants enter the conical microchannels in which the ESR takes place. H_2 will be separated using the Pd/Ag membrane, whereas CO_2 will be retained in the lumen. (Data from M.A. Rahman et al., *Journal of Membrane Sciences*, 390–391, 68–75, 2012.)

of ceramic material. Although high sintering temperatures enable the densification of YSZ particles, which results in an increase of mechanical strength, the sintering process is still carried out at lower temperatures due to the following reasons. YSZ hollow fibers sintered at temperatures below 1400°C are likely to have a porous structure, which is necessary for the catalyst impregnation process [27]. In addition, low sintering temperatures preserve the hydroxyl group on a ceramic substrate, which is necessary for the catalyst adhesion. Hydroxyl group on the surface of ceramic material will diminish when the sintering process is carried out at elevated temperatures. Typically, high sintering temperature is used to obtain a gas-tight layer on the outer surface suitable for the application of microtubular fuel cell [28]. This is not practical for the development of the CHFMMR, which requires a porous sponge-like region to enhance the diffusion of reaction products. As reported by Wei et al., the mechanical strength of YSZ hollow fiber sintered at 1300°C is 211 MPa [26].

In this work, YSZ hollow fiber substrate was prepared using a phase inversion-based extrusion/sintering technique. Arlacel P135 (1.3 wt%) was dissolved in the NMP solution before the addition of YSZ powders. The suspension was then rolled with 20 mm agate milling balls for 48 h and the milling continued for further 48 h after the addition of polymer binder. The suspension was degassed to remove air bubbles before being transferred into a stainless container. The tube-in-orifice spinneret with an outer diameter of 3 mm and an inner diameter of 1.2 mm was used. The deionized water was used as the internal coagulant at a flow rate of 3 cm^3 min^{-1} and an air gap ranging from 30 to 40 cm was used during the spinning process. The YSZ hollow-fiber precursors were left in the external coagulation bath overnight to enable the completion of the phase-inversion process before being cut and dried at room temperature. YSZ hollow fibers were obtained at sintering temperatures of 1300°C by sintering its precursor using a tubular furnace.

11.3.1.3 Morphology of Hollow-Fiber Support for Microreactor

The outer diameter of the YSZ hollow fiber is approximately 1 mm, as shown in Figure 11.8. Smaller diameters of YSZ hollow fiber can be obtained if the air gap between the bottom of spinneret and the surface of coagulation bath is increased during the spinning process. The high air gap enables the YSZ precursor to be stretched due to the action of gravitational force as it is being formed. The cross section of YSZ hollow fibers consists of finger-like structures that occupy about 80% of the cross section of YSZ hollow fiber, as shown in Figure 11.8b and c. However, the rest

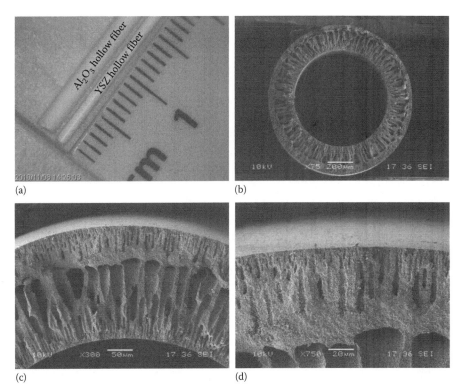

(a)   (b)

(c)   (d)

FIGURE 11.8 Images of the YSZ hollow fibers used to develop the CHFMMR. (a) Photographic images of a set of bare YSZ hollow fibers prepared using the phase inversion-based extrusion technique followed by a sintering process. (b–d) SEM images of the YSZ hollow fiber at different magnifications.

of the cross section consists of a sponge-like region and small finger-like structures originating from the outer surface. Presence of large voids in the finger-like structures enables the ESR catalyst to be distributed uniformly and enhances the efficiency of catalytic processes. The outer surface of the YSZ hollow fiber that consists of a thin-skin layer and a small finger-like region can be observed in Figure 11.8d. The thickness of the outer thin-skin layer, which is approximately 6.0 μm, together with its small and narrow pore size distribution ($D_p = 0.1$ μm), enables this substrate to be used for the deposition of Pd/Ag membrane. In addition, the small finger-like region situated between the thin-skin layer and the large finger-like region enhances the diffusion of both reactants and products across the YSZ hollow fiber.

11.3.1.4 Deposition of Membrane onto the Hollow-Fiber Microreactor

Pd/Ag membranes are plated onto the outer surface of YSZ hollow fibers using the electroless plating (ELP) technique. The plating process of palladium on the YSZ hollow fibers is initially carried out to obtain a total thickness of 4 μm. This process is followed by a silver plating process to obtain a 1 μm-thick silver layer on the Pd membrane. An annealing process in a hydrogen stream at 400°C for 24 h is carried out later to enable the diffusion of silver in the palladium matrix after the plating process. The annealing process is undertaken to facilitate the diffusion of silver in the matrix of Pd membrane to produce a robust Pd/Ag membrane. The Pd/Ag membrane is fabricated prior to the impregnation of the ESR catalyst. Precaution is taken to avoid the possibility of catalyst dissolution in the plating solution due to the presence of ammonium hydroxide solution and edetic acid [29].

Figure 11.9a shows the Pd/Ag membrane on the outer surface of a YSZ hollow fiber. The actual thickness of this membrane is approximately 5 μm, and the red line on Figure 11.9a represents the actual top surface of the Pd/Ag membrane. The appearance of an uneven surface on the Pd/Ag surface is illustrated in Figure 11.9b. It is expected that the deposition of Pd takes place initially on the seeded Pd nuclei and continues to grow as the plating process proceeds, which contributes to the rough surface of Pd/Ag membrane.

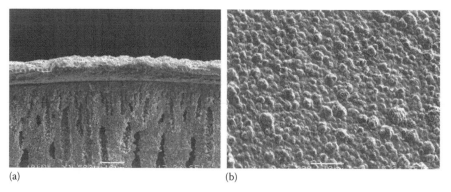

(a)   (b)

FIGURE 11.9 SEM images of (a) the Pd/Ag membrane on the outer surface of YSZ hollow fiber and (b) the top surface of Pd/Ag membrane. (Data from M.A. Rahman et al., *Journal of Membrane Sciences*, 390–391, 68–75, 2012.)

Energy dispersive X-ray spectroscopy (EDS) analysis is used to characterize the Pd/Ag membrane on the outer surface of the CHFMMR. Figure 11.10a shows the SEM image of the edge of YSZ hollow fiber that consists of a Pd/Ag membrane and the 10 wt% NiO/MgO-CeO$_2$ catalyst on its inner surface. This figure shows that the Pd/Ag membrane remains intact, although it has been exposed to a catalyst solution during the catalyst impregnation process. Figure 11.10b shows a dense distribution of zirconium that represents the cross section of the YSZ hollow fibers. Figure 11.10c and d shows the distribution of palladium and silver across the membrane, respectively, after an annealing process at 400°C for 24 h in an H$_2$ atmosphere. Uniform distributions of palladium and silver suggest that a uniform Pd/Ag alloy has been

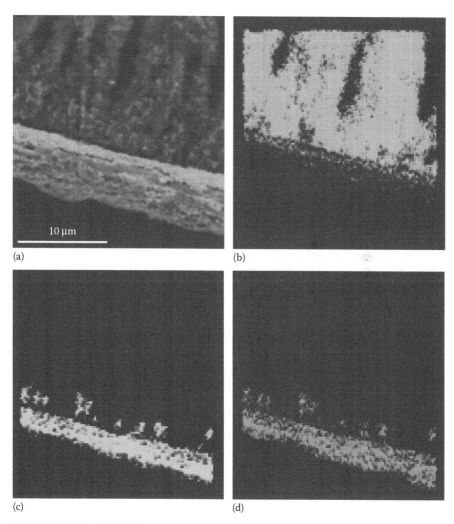

(a)   (b)

(c)   (d)

FIGURE 11.10   (a) SEM image that shows the outer surface of the CHFMMR. Distribution of (b) zirconium that represents the YSZ hollow fiber and (c) palladium and (d) silver ions on the Pd/Ag membrane.

produced through the annealing process. It is interesting to note that the distribution of silver across this metallic membrane is less intense than its matrix due to the lower concentration of silver used during the plating process. Figure 11.10c and d also shows that palladium and silver ions can be found in the thin-skin layer of YSZ hollow fiber. This suggest that during the deposition of the Pd/Ad membrane plating solutions can penetrate into the outer layer of YSZ hollow fiber. The *Pd/Ag roots* formed during the plating process may improve the adhesion of the Pd/Ad membrane, thus avoiding a delamination process at high temperatures.

11.3.1.5   Impregnation of Catalyst into Microreactor Pore Structures

In the development of the CHFMMR, a single-step impregnation process was used to impregnate the ESR catalyst into the inner surface of YSZ hollow fiber. Preparation of the ESR catalyst and the impregnation process into YSZ hollow fibers have been carried out using the sol-gel Pechini method [30]. In this approach, the YSZ hollow fiber, with Pd/Ag membrane on its outer surface, was wrapped with the polytetrafluoroethylene (PTFE) tape to prevent a direct contact between the Pd/Ag membranes and the catalyst solution. A homogeneous catalyst solution was injected into the lumen of YSZ hollow fiber using a glass pipette, and this process was repeated a number of times. The wrapped YSZ hollow fiber was put into a container and a small amount of catalyst solution was added at one end of the YSZ hollow fiber to maintain a capillary force into the finger-like structure. The YSZ hollow fibers were then dried in an oven at 60°C for 24 h and further dried at 115°C to complete the polymerization of a polymeric resin precursor. The catalyst solution turned into xerogel, which adsorbed onto the inner surface of alumina hollow-fiber substrates.

After the drying process, the YSZ hollow fiber with Pd/Ag membrane was unwrapped and assembled into a tubular ceramic module. Oxidation of the xerogel formed during the drying process was carried out in a tubular furnace. The temperature was increased from room temperature to 400°C at a rate of 5°C and held for 1 h. An air flow was introduced into the lumen to complete the oxidation process, and argon was introduced on the outer surface of the Pd/Ag membrane throughout this process to prevent oxidation of the Pd/Ag membrane, which may cause the formation of pin-holes.

X-ray diffraction (XRD) analysis is used to study the presence of NiO and CeO_2 obtained in this catalyst. Figure 11.11a shows the XRD spectrum of fresh catalyst that has broad peaks, which represents a low calcination temperature. This condition was used to burn off organic compounds formed during the drying process and to facilitate an oxidation process. The XRD analysis was also carried out on the YSZ hollow fiber before and after impregnation with the catalyst. Figure 11.11b and c shows high intensity peaks that signify the YSZ hollow fiber was sintered at a high temperature, which facilitates the growth of a large particle grain. It can be deduced that XRD analysis should not be used to examine the presence of the ESR catalyst because the high intensity peaks of the YSZ hollow fiber may saturate the analysis and hide the presence of NiO and CeO_2 peaks.

Distribution of the catalyst in the CHFMMR in this work is examined using the SEM–EDS technique. This analysis was carried out on a similar sample to that shown in Figure 11.10, and the dispersions of cerium and nickel ions on the cross section of YSZ hollow fiber is shown in Figure 11.12. These SEM–EDS images

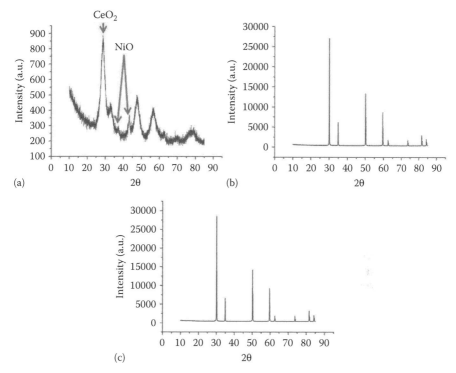

FIGURE 11.11   (a) XRD spectrum of fresh 10 wt%NiO/MgO–CeO₂ catalyst prepared using the sol-gel Pechini method after a calcination process at 400°C for 1 h. The XRD spectra for the YSZ hollow fiber (b) before and (c) after impregnation with the catalyst.

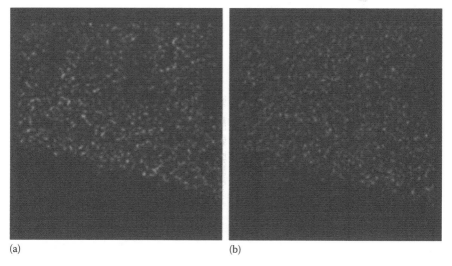

(a)   (b)

FIGURE 11.12   SEM–EDS showing the distribution of (a) cerium and (b) nickel on the cross section of YSZ hollow fiber.

show a uniform distribution of cerium and nickel ions in the YSZ hollow fiber when the impregnation of the catalyst is carried out using the sol-gel Pechini method. This impregnation technique enables the catalyst to be dispersed uniformly even at sponge-like regions adjacent to the Pd/Ag membrane. It is expected that the catalyst solution can easily reach other parts of the substrate cross section as well. The catalyst impregnation process starts from the lumen of the YSZ hollow fiber and the catalyst solution diffuses through the narrow opening of the conical microchannels under capillary action and diffuses to the outer layer where it is then halted by the dense Pd/Ag membrane.

Permeation tests were carried out using pure N_2 on YSZ hollow fiber before and after impregnation with the catalyst to study the effect of impregnation processes on the permeability of the YSZ hollow fibers and the results are shown in Figure 11.13. The permeation data shows that the flux of N_2 through the YSZ hollow fiber impregnated with catalyst is similar to that of a bare YSZ hollow fiber substrate. This result suggests that the presence of the catalyst has a negligible effect on the flux of the YSZ hollow fiber. This behavior may be associated with low catalyst loading in the YSZ hollow fibers, which is only $1.3 \pm 0.12\%$. The catalyst can reach the inner surface of YSZ hollow fiber as shown in Figure 11.12a and b; however, the adhesion process is rather difficult due to the nature of this ceramic, which has limited hydroxyl groups on its surface [26]. It is believed that the catalyst amount can be increased if the YSZ hollow fiber undergoes a pretreatment using alkaline solution to restore the hydroxyl groups on its surface [26,31]. An increase in the catalyst amount may not affect the flux of this substrate even after the catalyst impregnation process, as long as the sol-gel Pechini method is used to disperse the catalyst in YSZ

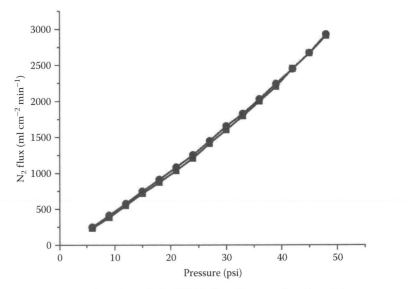

FIGURE 11.13 Fluxes of N_2 through the YSZ hollow fiber as a function of the transmembrane pressure before (●) and after (■) a catalyst deposition process. (Data from M.A. Rahman et al., *Journal of Membrane Sciences*, 390–391, 68–75, 2012.)

hollow fibers. The sol-gel Pechini method tends to produce loose catalyst particles, which are unlikely to block the flow of reactants in the narrow channels of the YSZ hollow fibers [30].

11.3.1.6 Membrane Microreactor Performance

Performance of a single unit of the CHFMMR, which has 2.5 mg of ESR catalyst, was tested and the results obtained were compared with a fixed-bed reactor and two units of the catalytic hollow fiber microreactor (CHFMR). CHFMR has similar feature to the CHFMMR, but it has no Pd/Ag membrane on its outer surface. The fixed-bed reactor and CHFMR use 5.7 and 5.0 mg of ESR catalyst, respectively. As seen from Figure 11.14, the amount of H_2 produced increased from 0.6 cm^3 min^{-1} at 350°C to 3.9 cm^3 min^{-1} at 510°C when the ESR is performed in the CHFMMR, whereas H_2 production increased slightly in the fixed-bed reactor and the CHFMR with increases from 0.2 to 1.0 cm^3 min^{-1} and 0.1 to 2.0 cm^3 min^{-1}, respectively. Improvement in the production of H_2 can be associated with the incorporation of Pd/Ag membrane adjacent to the conical microchannels in the finger-like structure. Pd/Ag membrane that has high permeability and infinite selectivity to H_2 enables only H_2 to be separated from the reaction zone. Removal of H_2 from the reaction zone shifts thermodynamic equilibria for most intermediate reactions in the ESR toward product formation leading to a rapid ethanol conversion and enabling H_2 with a high purity to be produced in the shell-side.

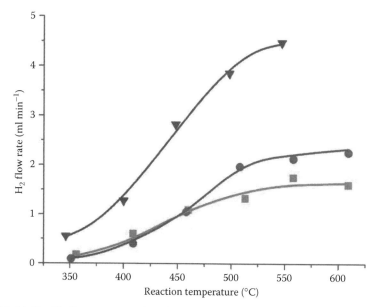

FIGURE 11.14 H_2 flow rate as a function of the reaction temperature obtained in (■) a fixed–bed reactor (GHSV = 590,000 cm^3 g^{-1} hr^{-1}), (●) CHFMR (GHSV = 672,000 cm^3 g^{-1} hr^{-1}), and CHFMMR (GHSV = 720,000 cm^3 g^{-1} hr^{-1}) with a sweep gas of (▼) 60 cm^3 min^{-1}. (Data from M.A. Rahman et al., *Journal of Membrane Sciences*, 390–391, 68–75, 2012.)

11.3.2 APPLICATION II: MICROTUBULAR SOFCs

11.3.2.1 Background

Apart from the promising use as a substrate for membrane microreactors, ceramic hollow fiber is also suitable to be employed as a support for SOFCs. SOFCs have been considered as a promising electricity-generation technology because of their high efficiency in converting chemical energy to electrical power [32]. SOFCs exhibit various advantages over other types of fuel cells. Use of a solid electrolyte, instead of liquid, makes SOFC easier to maintain, as this reduces corrosion and catalyst wetting problems. A major advantage of SOFCs is that hydrocarbons can be used directly as fuels without the need for further transformation. High operating temperature ensures that fuels will reform internally and then oxidized rapidly without the need for any specialized and expensive catalysts. Lack of high-activity catalysts, which can be easily poisoned, also gives SOFCs a high tolerance to impurities. SOFCs can achieve the highest total energy efficiency among all types of fuel cells. In addition, generated heat can be easily utilized in a simple and economical way.

Operation of an SOFC involves two primary electrode reactions: (1) oxidation of fuel at the anode and (2) reduction of oxidant at the cathode, as shown in Figure 11.15. The electrolyte is usually made of an oxygen-ion-conducting material, with oxygen vacancies in the crystal structure allowing the diffusion of oxygen ions at high temperatures. At the cathode, oxygen gas *captures* electrons and is reduced to oxygen ions. The oxygen ions move through the electrolyte and then react with hydrogen gas at the anode to release electrons and form water as the product. The electrons flow from the anode to the cathode through an external circuit where the current is generated.

To date, two types of SOFC designs, planar and tubular SOFCs, have been widely studied. Although planar SOFCs in a highly compact configuration show high power density, this design is challenged by the sophisticated sealing near the edges of the cell because only few sealants can be employed at the high temperature operation [33]. In contrast, the tubular geometry is more reliable and can be operated in the absence of high-temperature sealants. However, it is difficult to achieve high power

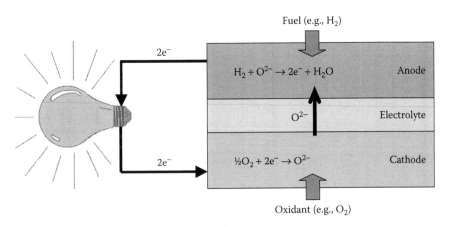

FIGURE 11.15   Operation principles of SOFC.

output due to a less efficient cell packing. In order to improve the volumetric power output, the diameter of tubular designs needs to be reduced from a few centimeters (e.g., Siemens–Westing cell) to the order of 1 mm, namely, microtubular SOFCs, as they are often referred to in the literature. The design of this type can significantly increase the surface area of electrodes, improve mechanical properties, and more importantly, speed up the start-up and the shut-down processes, and as a result, the microtubular SOFCs have attracted increasing interests in recent years [16,17,34–38].

Besides the progress in SOFC design, a development in the fabrication process is critically important to ensure the success of the cell. Previously, the fabrication of microtubular SOFC could only be achieved through multiple-step processes [16,17,34–38]. A support layer, for example, an anode tube, was first prepared and pre-sintered to provide mechanical strength to the fuel cell. The electrolyte layer was then deposited and sintered prior to the final coating of cathode layer. Each step involved at least one high-temperature heat treatment, making the cell fabrication time consuming and costly, with unstable control over cell quality.

For a more economical fabrication of microtubular SOFC with more reliability and flexibility in quality control, an advanced dry-jet wet extrusion technique, that is, a phase inversion-based co-extrusion process, followed by co-sintering and reduction processes was employed to fabricate a novel electrolyte/anode dual-layer hollow fiber. Using the co-extrusion technique, one of the layers has to be thick in order to provide mechanical strength to the fiber, and in this design, the anode is chosen to be the thick layer due to the much lower ohmic losses (as shown in Figure 11.16). Use of co-extrusion has many advantages over conventional dry-jet wet extrusion methods such as simplified fabrication and better control over the membrane structure. Furthermore, the risk of defects formation can be reduced and at the same time greater adhesion between the layers can be achieved.

11.3.2.2   Single-Step Fabrication of Electrolyte/Anode Dual-Layer Hollow Fiber

Studies on the fabrication and use of polymeric dual-layer hollow fiber membranes through phase inversion-based co-extrusion began in the late 1970s for hemodialysis [39]. In the late 1980s, Yanagimoto invented dual-layer asymmetric flat-sheet and hollow-fiber membranes to improve the antifouling properties of polymeric membranes used for water purification [40,41]. The patent disclosures about the procedures of fabricating polymeric multilayer hollow fibers for gas separation were made by Du Pont de

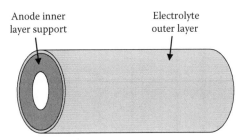

FIGURE 11.16   Schematic drawing of electrolyte/anode dual-layer hollow fiber.

Nemours [42]. Although there has been extensive investigation on polymeric dual-layer hollow fibers in the last 10 years [43–47], very limited research has been involved in the fabrication using ceramic materials [48–50], due to a number of challenges in the fabrication processes, such as different sintering behavior between the layers.

In the preparation of the electrolyte/anode dual-layer hollow fibers for this work, two ceramic suspensions were prepared separately. Suspension of the anode inner layer was composed of 60 wt% of nickel oxide (NiO) and 40 wt% of cerium gadolinium oxide (CGO), whereas the one for the electrolyte outer layer contained 100% CGO powder. The suspension compositions for the fabrication of four different dual-layer hollow fibers are listed in Table 11.2. Ceramic powders were first mixed with DMSO and Arlacel P135 as the solvent and dispersant, respectively, and stirred for 48 h. After this time, PESf pellets as the polymer binder were slowly added into the mixtures with stirring at ~300 rpm. The mixing continued for an extra 48 h to obtain homogeneous spinning suspension.

The dual-layer hollow-fiber precursors were prepared by a phase inversion-based co-extrusion technique. Prior to the co-extrusion, both spinning suspensions were degassed while stirring at room temperature to fully remove the air trapped inside the suspensions. Both spinning suspensions were then loaded into two stainless steel containers and forced simultaneously through a triple-orifice spinneret. The co-extrusion process is represented schematically in Figure 11.17.

TABLE 11.2

Compositions of the Spinning Suspension for Dual-Layer Hollow Fibers

	Composition (wt%)					
Layer	**NiO**	**CGO**	**PESf**	**Arlacel P135**	**DMSO**	**Viscosity at 50 s^{-1} (cP)**
Anode	42.00	28.00	7.00	0.12	22.88	32500
Electrolyte	–	64.00	6.40	0.12	29.48	14900

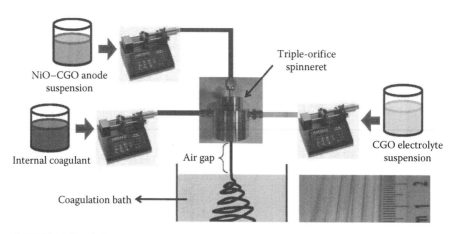

FIGURE 11.17 Schematic diagram of the phase inversion-based co-extrusion process of the dual-layer hollow-fiber precursor.

TABLE 11.3

Co-Extrusion Conditions of Dual-Layer Hollow Fibers with Different Electrolyte Thicknesses

Sample	Air Gap (cm)	Internal Coagulant Flow Rate (cm³ min⁻¹)	Inner Layer Extrusion Rate (cm³ min⁻¹)	Outer Layer Extrusion Rate (cm³ min⁻¹)	Electrolyte Thickness (μm)
R-0.5				0.5	10
R-1				1	19
R-2				2	35
R-3	20	10	7	3	53
R-4				4	65
R-5				5	70

Source: M.H.D. Othman et al., *Journal of Membrane Science*, 365, 382–388, 2010.

The co-extrusion conditions are summarized in Table 11.3, in which the samples are named according to the extrusion rate of the electrolyte outer layer. The hollow-fiber precursors were then co-sintered by heating in the air at 1500°C for 12 h in a tubular furnace. The co-sintered dual-layer CGO/NiO–CGO hollow fibers were reduced to CGO/Ni–CGO dual-layer hollow fibers at 550°C using pure hydrogen for 2.5 h.

11.3.2.3 Morphology and Physical Properties of Dual-Layer Hollow Fibers

Morphology of the reduced hollow fibers was examined using SEM. The hollow fibers were snapped in order to obtain clear cross-sectional fracture. Figure 11.18 shows the SEM images of the electrolyte/anode dual-layer hollow fibers with different electrolyte thicknesses co-sintered at 1500°C for 12 h followed by a reduction at 550°C for 2.5 h. The anode structure of the hollow fibers is composed of short finger-like voids originating from the inner surface with the rest of the anode occupied by a sponge-like structure. Formation of this kind of asymmetric structure was explained in detail by Kingsbury and Li [3] and in Section 11.2.2. Each type of structure (i.e., finger-like voids and sponge-like structures) has its own effect on the anode characteristics.

FIGURE 11.18 SEM images of the overall view of the reduced dual-layer hollow fibers consisting of different electrolyte thicknesses. The images were taken in backscattered electrons (BSE) mode. (Data from M.H.D. Othman et al., *Journal of Membrane Science*, 365, 382–388, 2010.)

As an example, finger-like voids in the anode are believed to have a strong effect on the permeation of fuel, as longer finger-like voids allow gas to diffuse much easier through the anode. However, the presence of such macrovoids is less advantageous in terms of both mechanical strength and electrical conductivity of the anode. Based on this understanding, it is thus crucial to control the growth of the finger-like voids in the anode support, and in this fabrication work, the relative length of the finger-like voids was controlled to be around 35% of the overall anode thickness.

Variation in outer electrolyte thickness of the dual-layer hollow fibers is achieved by simply adjusting the corresponding extrusion rate, without affecting the structure of the anode inner layer. Viscosity of the spinning suspension for the outer layer should not be too high in order to ensure a smooth flow of the suspension and a full and uniform coverage, especially when the extrusion rate is very low. As shown in Figure 11.19, dual-layer hollow fibers with different electrolyte thicknesses were successfully produced when the extrusion rate of the outer layer was varied from 0.5 to 5 cm^3 min^{-1}. The fiber R-5 that extruded from 5 cm^3 min^{-1} produced the thickest electrolyte layer, whereas the thinnest electrolyte generated for the R-0.5 prepared from 0.5 cm^3 min^{-1} of extrusion rate (Figure 11.20).

Figure 11.19 also shows that all the six batches of hollow fibers have good adhesion between the inner and outer layers, as all of them are free from delamination problems. Furthermore, the electrolyte outer layer has very uniform thickness on the anode inner layer and covers the whole length of hollow-fiber surface and thus shows another advantage of co-extrusion as a deposition technique of a thin layer on the hollow-fiber support.

Table 11.3 lists the measured thickness of electrolyte of fabricated samples. The data of the table shows that the extrusion rate significantly affects the thickness of the outer layer, as the thickness of the electrolyte outer layer varies from 10 to 70 μm when the extrusion rate was controlled from 0.5 to 5 cm^3 min^{-1}. In addition, the electrolyte layer did not fully cover the anode when the extrusion rate was lower than 0.5 cm^3 min^{-1}. These results demonstrate that co-extrusion is a suitable technique for producing thin and uniform electrolyte layers, comparable with other conventional and complex depositing techniques such as dip-coating and chemical vapor deposition.

The mechanical property of the dual-layer hollow fibers with different electrolyte thicknesses together with the anode single-layer hollow fiber was investigated by a three-point bending test using a tensile tester with a load cell of 1 kN, as shown in Figure 11.20. As expected, the dual-layer hollow fiber with the thinnest electrolyte layer (10 μm for R-0.5) showed the lowest bending strength of about 143 MPa, which is still about 30% higher than the single-layer hollow fiber. The bending strength of dual-layer hollow fibers gradually increased to 146, 151, 157, 163, and 166 MPa with the increase of thickness of the electrolyte outer layer from 19 μm (R-1), 35 μm (R-2), 53 μm (R-3), 65 μm (R-4), and 70 μm (R-5), respectively. This can be explained by the difference in the porosity between anode and electrolyte of the hollow fibers. In order to obtain a high efficiency for the anode layer, a 60 wt% of NiO was used and after the reduction process, this became a mechanically weak Ni network due to the interconnected pores structure in the anode layer. Thus, the deposition of a thicker dense electrolyte layer on the anode support improves the mechanical strength of the hollow fiber.

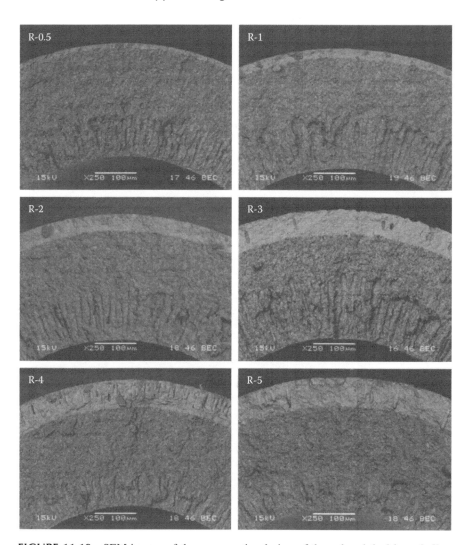

FIGURE 11.19 SEM images of the cross-sectional view of the reduced dual-layer hollow fibers consisting of different electrolyte thicknesses. The images were taken in backscattered electrons (BSE) mode. (Data from M.H.D. Othman et al., *Journal of Membrane Science*, 365, 382–388, 2010.)

In order to investigate the effect of the electrolyte thickness on the gas-tightness property of the developed dual-layer hollow fibers, a gas-tightness test was performed. As seen in Figure 11.21, the nitrogen permeability of the hollow fiber consisting of 10 μm electrolyte (R-1) is 11.7×10^{-9} mol m^{-2} s^{-1} Pa^{-1}. This value decreases to 6.6×10^{-9} mol m^{-2} s^{-1} Pa^{-1} for the fiber consisting of 19 μm electrolyte (R-1) and is further reduced to 4.5×10^{-9}, 2.9×10^{-9}, 2.1×10^{-9}, and 1.1×10^{-9} mol m^{-2} s^{-1} Pa^{-1} when the electrolyte thickness is increased to 35 μm (R-2), 53 μm (R-3), 65 μm (R-4), and 70 μm (R-5), respectively. This indicates that the thicker the dense layer, the better the gas-tightness property of the electrolyte is. Without

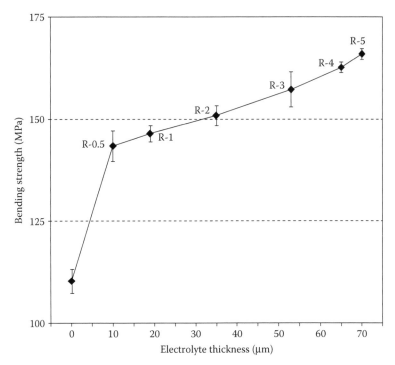

FIGURE 11.20 Bending strength of the reduced dual-layer hollow fibers measured at room temperature as a function of electrolyte thickness (number of samples = 5). (Data from M.H.D. Othman et al., *Journal of Membrane Science*, 365, 382–388, 2010.)

electrolyte outer layer, the permeability of the nitrogen gas through the hollow fiber is 1081×10^{-9} mol m^{-2} s^{-1} Pa^{-1}, which is more than two orders of magnitude higher than the permeability of the fiber R-0.5 with the thinnest electrolyte layer.

Although fiber R-0.5 showed the lowest mechanical strength and gas-tightness property among the reported dual-layer hollow fibers, the electrochemical measurements were carried out using the fuel cell that was prepared from this fiber, as it was anticipated that a thinner electrolyte has a lower resistance. The fibers R-1, R-2, and R-4 were also tested in order to study the effect of electrolyte thickness on the electrochemical performance.

11.3.2.4 Development of a Complete SOFC

Prior to the electrochemical measurements, multilayers of cathode were deposited on the electrolyte of the dual-layer hollow fibers. Two cathode material slurries were prepared by mixing the powders with ethylene glycol in ratio 1:1 in weight; one was a mixture of 50 wt% lanthanum strontium cobalt ferrite (LSCF) and 50 wt% CGO, whereas another one was 100 wt% LSCF. The co-sintered dual-layer hollow fibers were then cut into 50 mm lengths, and the outer layer surface was then covered with masking tape, leaving 10 mm uncovered at the center of the fibers. The cathode multilayers were deposited using a brush painting technique onto the hollow fibers; the first and second layers were the mixture LSCF–CGO, and the third layer was

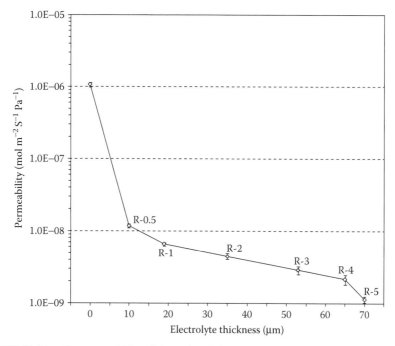

FIGURE 11.21 Gas permeability of the reduced dual-layer hollow fibers measured at room temperature as a function of electrolyte thickness, with the initial pressure of 40 psig. This graph is used to describe the gas-tightness property of the electrolyte layer (number of samples = 3). (Data from M.H.D. Othman et al., *Journal of Membrane Science*, 365, 382–388, 2010.)

100 wt% LSCF. Each layer was left to dry in air for 1 h before the next layer was applied. After the third layer dried, the hollow fibers were sintered at 1200°C for 5 h to form a complete microtubular SOFC with 10 mm length cathode.

Morphology of the sintered cathode layer was observed using SEM and the image is shown in Figure 11.22. As seen, the obtained cathode of approximately 40 μm is

FIGURE 11.22 SEM image of cathode layer prepared using brush-painting technique and sintered at 1200°C for 5 h. The images were taken in secondary electron imaging (SEI) mode. (Data from M.H.D. Othman et al., *Journal of Power Sources*, 205, 272–280, 2012.)

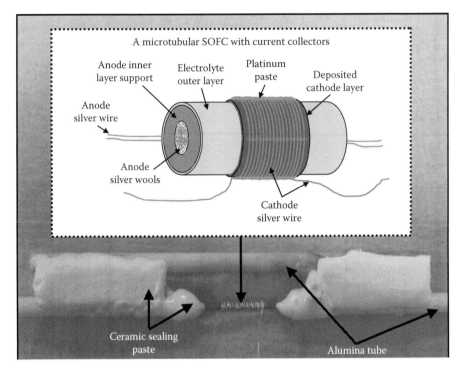

FIGURE 11.23 Experimental apparatus of the single microtubular SOFC reactor.

porous and uniform, which greatly facilitates gas transport and oxidant reduction reaction in the cathode. The cathode layer is also in good contact with the electrolyte surface of the dual-layer hollow fibers. Furthermore, although the cathode is made of three layers, no obvious boundaries between the layers can be observed, indicating a very low resistance for the transfer of electrons and oxygen ions.

The resultant cell was then fixed in a gas-tight alumina tube (Multilab Ceramics, UK) using a ceramic sealing paste, Ceramabond (Aremco, USA), after applying current collectors, as shown in Figure 11.23. The current collection was made by wrapping silver wire around the cathode layer, and conductive platinum paste was used to reduce potential losses between the wire and the cathode surface. Silver wool was packed inside the fiber lumen for the anode current collection, producing excellent electrical contact onto the anode wall. The complete cell reactor was inserted in the center of a tube furnace. An additional thermocouple was placed close to the cell in the furnace in order to measure the temperature of the cell surface. The silver wires attached to the anode and the cathode were then connected to a potentiostat/galvanostat and a booster.

11.3.2.5 Fuel Cell Performances

Prior to the fuel cell test, the anode of the cell was preconditioned in 5 cm^3 min^{-1} of hydrogen (saturated with water vapor of 0.12 cm^3 min^{-1} at 20°C, 1 atm) at 550°C for 2.5 h in order to reduce NiO–CGO to Ni–CGO. Hydrogen of 15 cm^3 min^{-1} (saturated with water vapor of 0.35 cm^3 min^{-1} using a bubbling cylinder at 20°C, 1 atm) and air of

40 cm^3 min^{-1} (20°C, 1 atm) were then flowed through the anode and cathode, respectively, in a counter-flow arrangement. The current-voltage (I-V) characterization was recorded using cycling voltammetry at galvanostatic mode (software Nova 1.5, Autolab) at operating temperature of 550°C–600°C, and a current step of approximately 1.1 s. The AC impedance spectra were measured on the same electrochemical workstation (0.01 Hz–100 kHz) with signal amplitude of 10 mV under open-circuit conditions and was controlled by FRA 4.9 (Frequency Response Analyzer connection, Autolab).

Figure 11.24a shows cell voltages and power densities as a function of current density for the single cell of microtubular SOFC that was prepared from R-0.5, R-1, R-2, and R-4, with 15 cm^3 min^{-1} of hydrogen (saturated with water vapor of 0.35 cm^3 min^{-1} at 20°C, 1 atm) flowing through the anode and 40 cm^3 min^{-1} (20°C, 1 atm) of air flowing to the cathode. The operation temperature was set at 600°C because the CGO electrolyte has very favorable ionic conductivity at temperatures as low as 600°C [53]; however, higher temperature would cause severe reduction in CGO, which leads to high electronic conductivity [54]. The measured open-circuit voltage (OCV) showed an increase from 0.77 to 0.93 V with the increase in the electrolyte thickness of the dual-layer hollow fibers from 10 to 65 µm. As the thicker electrolyte shows better gas-tightness property (as shown in Figure 11.21), it reduces the possibility for the fuel or oxidant to transport toward the opposite electrode and results in higher OCV. The low OCV value of the fiber R-0.5 is not only due to the gas tightness issue but also probably because of the minor electronic current leakage across the electronic conductive path in the CGO electrolyte [55], resulting from the reduction of Ce^{4+} to Ce^{3+} at the reducing atmosphere.

The maximum power densities achieved were 0.52, 0.62, 0.77, and 1.11 W cm^{-2} for the cells R-4, R-2, R-1, and R-0.5, respectively, which increased with the decrease in the electrolyte layer thickness. In the thinner electrolyte, the traveling distance of the oxygen ion from the cathode side through the electrolyte to the anode side could be shortened, and thus reduces the ohmic resistance of the cell (this will be discussed further using impedance data in the following paragraph). In order to observe a precise relationship between the maximum power density and the electrolyte thickness, the maximum power density of each cell was normalized with their electrolyte thickness and the result is shown in Figure 11.24b. From this figure, it is seen that the electrolyte thickness-normalized maximum power density increased exponentially with the decrease of electrolyte thickness. One of the possible reasons to explain this trend is that the reduction of electrolyte thickness not only reduces the ohmic resistance but also enhances the rate of the reactions in both electrodes, due to the more effective oxygen ion transfer from the cathode to the anode. These results justify the need of reducing the electrolyte thickness due to the significant impact on the performance of the cell. In comparison with previous studies, the maximum power density obtained for cell R-0.5 (1.11 W cm^{-2}) is still lower than the ones produced by conventional ram-extrusion method with similar electrolyte thickness (10 µm), which were reported to be 1.29 W cm^{-2} at 600°C [55] and 1.31 W cm^{-2} at 550°C [56,57], respectively. The difference can be explained by the impedance data shown below.

Figure 11.25a shows impedance data for the cells of microtubular SOFCs at 600°C, which were used to investigate the effect of electrolyte thickness on the ohmic area

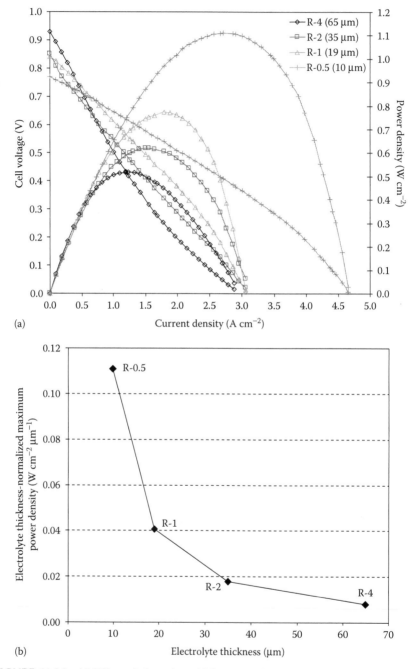

(a)

(b)

FIGURE 11.24 (a) Effect of electrolyte thickness on the voltages and power densities of the cells. The cell operated at temperature of 600°C using hydrogen flow rate of 15 cm^3 min^{-1} (20°C, 1 atm) and air flow rate of 40 cm^3 min^{-1} (20°C, 1 atm); (b) the electrolyte thickness-normalized maximum power density of the cells as a function of electrolyte thickness. (Data from M.H.D. Othman et al., *Journal of Membrane Science*, 365, 382–388, 2010.)

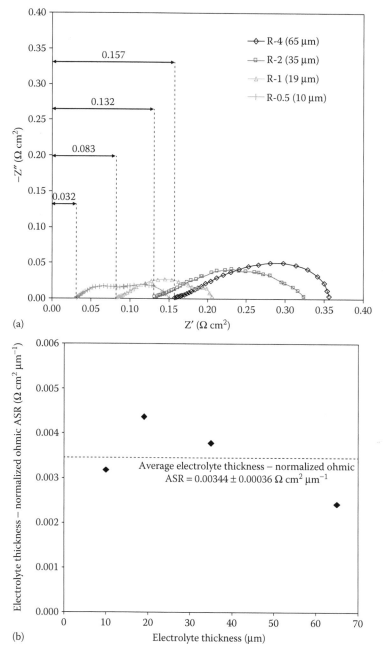

FIGURE 11.25 (a) Effect of electrolyte thickness on impedance spectra of cells measured under open-circuit condition at temperature range of 550°C–600°C using signal amplitude of 10 mV at frequencies of 100 kHz–0.01 Hz and hydrogen flow rate of 15 cm³ min⁻¹ (20°C, 1 atm) and air flow rate of 40 cm³ min⁻¹ (20°C, 1 atm). (b) The electrolyte thickness-normalized ohmic ASR of the cells as a function of electrolyte thickness. (Data from M.H.D. Othman et al., *Journal of Membrane Science*, 365, 382–388, 2010.)

specific resistance (ASR). The high-frequency intercept on the real impedance axis represents the value of ohmic ASR in the cell, which is generated from the ionic resistance in the electrolyte layer, both ionic electronic resistances in the electrodes, and the contact resistance from the interfaces and current collectors [58]. From this figure, it is seen that ohmic ASR of the cells were reduced with the decrease in the electrolyte thickness of the cells. Assuming each cell had the same ohmic resistances in their electrodes, interfaces, and current collectors because of their similar material compositions, structures, and fabrication techniques, the ionic resistance in the electrolyte layer was decreased significantly by the reduction in the electrolyte thickness. This is also one of the major reasons for the higher power output of thinner electrolyte layer that is shown in Figure 11.24a.

In order to prove that the major cause of the difference in the ohmic ASR in this work is the electrolyte thickness, the ohmic ASR of each cell was normalized with the electrolyte thickness and the trend is shown in Figure 11.25b. This graph demonstrates that the electrolyte thickness-normalized ohmic ASR of each cell was still in a close range, with the average value of $0.00344 \pm 0.00036 \ \Omega \ cm^2 \ \mu m^{-1}$. This also indicates that the ohmic ASR was almost directly proportional to the electrolyte thickness for the cells that consisted of similar microstructure.

11.4 CONCLUSION

This chapter has discussed the fabrication of ceramic hollow fibers through a phase inversion-based extrusion/sintering technique and the use of such fibers as a support for two types of energy conversion systems: membrane microreactors and microtubular SOFCs. Compared with the other conventional extrusion techniques such as plastic mass ram extrusion, this phase inversion-based extrusion shows better control over the internal macrostructure of the hollow fiber. The versatility of phase inversion-based extrusion is further demonstrated by the preparation of dual-layer hollow fiber supports in a single step using co-extrusion/co-sintering. Use of this technique significantly simplifies the fabrication process when compared to the single-layer extrusion methods, while still allowing full control over the hollow-fiber structures. The generic advantages of the design of dual-layer hollow fiber and the single-step membrane fabrication through a phase inversion-based co-extrusion process can be transferred to the development of other inorganic membranes with more advanced structures, expanding and speeding up the development of membrane separation and membrane reactor processes.

REFERENCES

1. B.F.K. Kingsbury, A morphological study of ceramic hollow fibre membranes: A perspective on multifunctional catalytic membrane reactors, PhD Thesis, Imperial College London, London, 2010.
2. Z. Wu, Dual-layer functional ceramic hollow fibre membranes for partial oxidation of methane, PhD Thesis, Imperial College London, London, 2012.
3. K. Li, *Ceramic Membranes for Separation and Reaction*, Wiley, Chichester, 2007.
4. B.F.K. Kingsbury and K. Li, A morphological study of ceramic hollow fibre membranes, *Journal of Membrane Science* 328(1/2) (2009) 134–140.

5. T. Okubo, K. Haruta, K. Kusakabe, S. Morooka, H. Anzai, and S. Akiyama, Preparation of a sol-gel derived thin membrane on a porous ceramic hollow fiber by the filtration technique, *Journal of Membrane Science* 59(1) (1991) 73–80.

6. X. Tan, S. Liu, and K. Li, Preparation and characterization of inorganic hollow fiber membranes, *Journal of Membrane Science* 188(1) (2001) 87–95.

7. S. Liu, K. Li, and R. Hughes, Preparation of porous aluminium oxide (Al_2O_3) hollow fibre membranes by a combined phase-inversion and sintering method, *Ceramics International* 29(8) (2003) 875–881.

8. R.W. Baker, *Membrane Technology and Applications*, Wiley, Chichester, 2004.

9. S.I.D.N. Loeb and S.R.I.N. Sourirajan, Sea water demineralization by means of an osmotic membrane, in *Saline Water Conversion—II* (Ed: R.F. Gould), Advances in Chemistry, vol 38, American Chemical Society, Washington, DC, 1963, pp. 117–132.

10. H. Strathmann and K. Kock, The formation mechanism of phase inversion membranes, *Desalination* 21(3) (1977) 241–255.

11. B.F. Barton and A.J. McHugh, Kinetics of thermally induced phase separation in ternary polymer solutions. I. Modeling of phase separation dynamics, *Journal of Polymer Science Part B: Polymer Physics* 37(13) (1999) 1449–1460.

12. L. Zeman and G. Tkacik, Thermodynamic analysis of a membrane-forming system water/N-methyl-2-pyrrolidone/polyethersulfone, *Journal of Membrane Science* 36 (1988) 119–140.

13. Z.S. Li, Investigation of the dynamics membrane formation by of poly(ether sulfone) membrane formation by precipitation immersion, *Journal of Polymer Science Part B: Polymer Physics* 43(5) (2005) 498–510.

14. J. Barzin and B. Sadatnia, Theoretical phase diagram calculation and membrane morphology evaluation for water/solvent/polyethersulfone systems, *Polymer* 48(6) (2007) 1620–1631.

15. S. Liu and K. Li, Preparation TiO_2/Al_2O_3 composite hollow fibre membranes, *Journal of Membrane Science* 218(1/2) (2003) 269–277.

16. N. Droushiotis, U. Doraswami, K. Kanawka, G.H. Kelsall, and K. Li, Characterization of NiO-yttria stabilised zirconia (YSZ) hollow fibres for use as SOFC anodes, *Solid State Ionics* 180(17–19) (2009) 1091–1099.

17. N. Yang, X. Tan, and Z. Ma, A phase inversion/sintering process to fabricate nickel/yttria-stabilized zirconia hollow fibers as the anode support for micro-tubular solid oxide fuel cells, *Journal of Power Sources* 183(1) (2008) 14–19.

18. L. F. Wang, C. Saricks, and D. Santini, *Effects of Fuel Ethanol Use on Fuel-Cycle Energy and Greenhouse Gas Emissions*, Center for Transportation Research Argonne National Laboratory; United States Department of Energy, Argonne National Laboratory, Lemont, IL, 1999.

19. L.R. Lynd, Overview and evaluation of fuel ethanol from cellulosic biomass: Technology, economics, the environment, and policy, *Annual Review of Energy and the Environment* 21 (1996) 403–465.

20. B. Höhlein, M. Boe, J. Bøgild-Hansen, P. Bröckerhoff, G. Colsman, B. Emonts, R. Menzer, and E. Riedel, Hydrogen from methanol for fuel cells in mobile systems: Development of a compact reformer, *Journal of Power Sources* 61(1/2) (1996) 143–147.

21. B. Emonts, J.B. Hansen, S.L. Jorgensen, B. Hohlein, and R. Peters, Compact methanol reformer test for fuel-cell powered light-duty vehicles, *Journal of Power Sources* 71(1/2) (1998) 288–293.

22. L.J. Pettersson and R. Westerholm, State of the art of multi-fuel reformers for fuel cell vehicles: Problem identification and research needs, *International Journal of Hydrogen Energy* 26(3) (2001) 243–264.

23. A.S. Chellappa, T.R. Vencill, and D.C. Lamont, Pure hydrogen from multiple fuels in membrane reformers, *Abstracts of Papers of the American Chemical Society* 228 (2004) 168.

24. M. De Falco, Ethanol membrane reformer and PEMFC system for automotive application, *Fuel* 90(2) (2011) 739–747.

25. M.A. Rahman, F.R. García-García, and K. Li, The development of catalytic hollow fibre membrane microreactor as a microreformer unit for the automotive application, *Journal of Membrane Sciences* 390–391 (2012) 68–75.

26. C.C. Wei, Yttria stabilised zirconia (YSZ) and their applications, PhD Thesis, Imperial College London, London, 2009.

27. C.C. Wei, O.Y. Chen, Y. Liu, and K. Li, Ceramic asymmetric hollow fibre membranes—One step fabrication process, *Journal of Membrane Science* 320(1/2) (2008) 191–197.

28. M.H.D. Othman, Z.T. Wu, N. Droushiotis, U. Doraswami, G. Kelsall, and K. Li, Single-step fabrication and characterisations of electrolyte/anode dual-layer hollow fibres for microtubular solid oxide fuel cells, *Journal of Membrane Science* 351(1/2) (2010) 196–204.

29. E.Y. Nevskaya, I.G. Gorichev, S.B. Safronov, B.E. Zaitsev, A.M. Kutepov, and A.D. Izotov, Interaction between copper(II) oxide and aqueous ammonia in the presence of ethylenediaminetetraacetic acid, *Theoretical Foundations of Chemical Engineering* 35(5) (2001) 503–509.

30. M. Kakihana, "Sol-Gel" preparation of high temperature superconducting oxides, *Journal of Sol-Gel Science and Technology* 6(1) (1996) 7–55.

31. C.C. Wei and K. Li, Preparation and Characterization of a Robust and Hydrophobic Ceramic Membrane via an Improved Surface Grafting Technique, *Industrial & Engineering Chemistry Research* 48(7) (2009) 3446–3452.

32. S.C. Singhal and K. Kendall, Introduction to SOFCs, in *High-Temperature Solid Oxide Fuel Cells: Fundamentals, Design and Applications* (Eds: S.C. Singhal and K. Kendall), Elsevier, Oxford, 2003, pp. 1–22.

33. S. Weil, C. Coyle, J. Hardy, J. Kim, and G.G. Xia, Alternative planar SOFC sealing concepts, *Fuel Cells Bulletin* 2004(5) (2004) 11–16.

34. Y. Funahashi, T. Shimamori, T. Suzuki, Y. Fujishiro, and M. Awano, Fabrication and characterization of components for cube shaped micro tubular SOFC bundle, *Journal of Power Sources* 163(2) (2007) 731–736.

35. R. Campana, R.I. Merino, A. Larrea, I. Villarreal, and V.M. Orera, Fabrication, electrochemical characterization and thermal cycling of anode supported microtubular SOFC, *Journal of Power Sources* 192(1) (2009) 120–125.

36. T. Suzuki, Y. Funahashi, T. Yamaguchi, Y. Fujishiro, and M. Awano, Effect of anode microstructure on the performance of micro tubular SOFCs, *Solid State Ionics* 180(6–8) (2009) 546–549.

37. C.C. Wei and K. Li, Yttria-stabilized zirconia (YSZ)-based hollow fibre solid oxide fuel cells, *Industrial and Engineering Chemistry Research* 47(5) (2008) 1506–1512.

38. C. Yang, W. Li, S. Zhang, L. Bi, R. Peng, C. Chen, and W. Liu, Fabrication and characterization of an anode-supported hollow fibre SOFC, *Journal of Power Sources* 187(1) (2009) 90–92.

39. W. Henne, G. Dunweg, W. Schmitz, R. Pohle, and F. Lawitzki, Method of producing dialyzing membrane, US Patent 4164437, 1979.

40. T. Yanagimoto, Manufacture of ultrafiltration membranes, Japanese Patent 62019205, 1987.

41. T. Yanagimoto, Method for manufacture of hollow-fibre porous membranes, Japanese Patent 63092712, 1988.

42. O.M. Ekiner, R.A. Hayes, and P. Manos, Novel multicomponent fluid separation membranes, US Patent 5085676, 1992.

43. D. Wang, K. Li, and W.K. Teo, Preparation of annular hollow fibre membranes, *Journal of Membrane Science* 166(1) (2000) 31–39.

44. Y. Li, C. Cao, T.S. Chung, and K.P. Pramoda, Fabrication of dual-layer polyethersulfone (PES) hollow fiber membranes with an ultrathin dense-selective layer for gas separation, *Journal of Membrane Science* 245(1/2) (2004) 53–60.

45. L. Jiang, T.S. Chung, D.F. Li, C. Cao, and S. Kulprathipanja, Fabrication of Matrimid/ polyethersulfone dual-layer hollow fiber membranes for gas separation, *Journal of Membrane Science* 240(1/2) (2004) 91–103.

46. Y. Li and T.S. Chung, Exploration of highly sulfonated polyethersulfone (SPES) as a membrane material with the aid of dual-layer hollow fiber fabrication technology for protein separation, *Journal of Membrane Science* 309(1/2) (2008) 45–55.

47. N. Widjojo, T.S. Chung, and S. Kulprathipanja, The fabrication of hollow fiber membranes with double-layer mixed-matrix materials for gas separation, *Journal of Membrane Science* 325(1) (2008) 326–335.

48. J. de Jong, N.E. Benes, G.H. Koops, and M. Wessling, Towards single step production of multi-layer inorganic hollow fibers, *Journal of Membrane Science* 239(2) (2004) 265–269.

49. Z. Wu, B. Wang, and K. Li, A novel dual-layer ceramic hollow fibre membrane reactor for methane conversion, *Journal of Membrane Science* 352(1/2) (2010) 63–70.

50. K. Kanawka, M.H.D. Othman, Z. Wu, N. Droushiotis, G. Kelsall, and K. Li, A dual layer Ni/Ni-YSZ hollow fibre for micro-tubular SOFC anode support with a current collector, *Electrochemistry Communications* 13(1) (2011) 93–95.

51. M.H.D. Othman, N. Droushiotis, Z. Wu, K. Kanawka, G. Kelsall, and K. Li, Electrolyte thickness control and its effect on electrolyte/anode dual-layer hollow fibres for micro-tubular solid oxide fuel cells, *Journal of Membrane Science* 365 (2010) 382–388.

52. M.H.D. Othman, N. Droushiotis, Z. Wu, G. Kelsall, K. Li, Dual-layer hollow fibres with different anode structures for micro-tubular solid oxide fuel cells, *Journal of Power Sources* 205 (2012) 272–280.

53. S. El-Houte and M. El-Sayed Ali, EMF measurements on gadolinia doped ceria, *Ceramics International* 13(4) (1987) 243–246.

54. B.C.H. Steele, Appraisal of $Ce_{1-y}Gd_yO_{2-y/2}$ electrolytes for IT-SOFC operation at 500°C, *Solid State Ionics* 129(1/4) (2000) 95–110.

55. T. Suzuki, Y. Funahashi, T. Yamaguchi, Y. Fujishiro, and M. Awano, Development of fabrication/integration technology for micro tubular SOFCs, in *Micro Fuel Cells— Principles and Applications* (Ed: T.S. Zhao), Elsevier, Oxford, 2009, pp. 141–177.

56. F. Calise, G. Restuccia, and N. Sammes, Experimental analysis of micro-tubular solid oxide fuel cell fed by hydrogen, *Journal of Power Sources* 195(4) (2010) 1163–1170.

57. Y.W. Sin, K. Galloway, B. Roy, N.M. Sammes, J.H. Song, T. Suzuki, and M. Awano, The properties and performance of micro-tubular (less than 2.0 mm O.D.) anode supported solid oxide fuel cell (SOFC), *International Journal of Hydrogen Energy* 36(2) (2011) 1882–1889.

58. M. Mogensen and P. Hendriksen, Testing of electrodes, cells and short stacks, in *High-Temperature Solid Oxide Fuel Cells: Fundamentals, Design and Applications* (Eds: S.C. Singhal and K. Kendall), Elsevier, Oxford, 2003, pp. 261–289.

12 Development of Large-Scale Industrial Applications of Novel Membrane Materials

Carbon Nanotubes, Aquaporins, Nanofibers, Graphene, and Metal-Organic Frameworks

Kailash Chandra Khulbe, Chaoyang Feng,
Ahmad Fauzi Ismail, and Takeshi Matsuura

CONTENTS

12.1 Introduction 384
12.2 CNT Membrane 385
    12.2.1 Preparation of CNT Membranes 385
    12.2.2 Properties and Applications of CNT Membranes 386
    12.2.3 Summary of CNT Membranes 390
12.3 AQP Membrane 390
    12.3.1 Preparation of AQP Membrane 391
    12.3.2 Properties and Applications of AQP 392
    12.3.3 Summary 394
12.4 Nanofibers 395
    12.4.1 Preparation of Nanofibers 395
        12.4.1.1 Electrospun Nanofibrous Membranes with a Single
                Component 395
        12.4.1.2 Preparation of Nanofibers with Side-by-Side
                Components 396
    12.4.2 Properties and Applications of Nanofibers 401
        12.4.2.1 Application of Nanofibers in RO, UF, NF, and MF 402

          12.4.2.2   Application of Nanofibers MD....................................403
          12.4.2.3   Application of Nanofibers in Fuel Cells.......................405
          12.4.2.4   Application of Nanofibers for Adsorption...................405
          12.4.2.5   Application of Nanofibers for Catalysts......................406
          12.4.2.6   Other Applications of Nanofibers 407
          12.4.2.7   Application of Hollow Nanofibers for Membrane
                     Separation Processes 408
     12.4.3   Summary of Application of Nanofibers....................................408
12.5   Graphene Membrane 411
     12.5.1   Graphene Preparation 411
     12.5.2   Application and Properties of Graphene 412
     12.5.3   Summary of Graphene Membrane Application 415
12.6   MOF Membranes... 415
     12.6.1   MOF Membranes Preparation 416
          12.6.1.1   The Direct Growth/Deposition of MOF Thin Films
                     on Solid Substrate 416
          12.6.1.2   Colloidal Deposition Method for MOF Membranes.... 417
          12.6.1.3   LBL or Liquid Phase Epitaxy of Surface Grown,
                     Crystalline MOF Multilayers 417
          12.6.1.4   Electrochemical Synthesis of MOF Membranes 417
          12.6.1.5   Gel-Layer Synthesis of MOF Membranes 418
          12.6.1.6   Evaporation-Induced Crystallization of MOF
                     Membranes...418
          12.6.1.7   Other Methods for Fabrication of MOF Membranes......418
     12.6.2   Characterization and Properties of MOFs.................... 419
     12.6.3   Application of MOF Membranes.............................. 419
     12.6.4   Summary of Application of MOF Membranes.........................424
12.7   Conclusions..425
References...426

12.1   INTRODUCTION

Traditionally, separation membranes are mostly dense polymeric films where advanced chemistry is used to control the surface properties of the films produced. A wide range of polymers and production techniques have been used, resulting in a great diversity in the structure and function of separation membranes tailored to a wide variety of applications. Membrane separation is usually described in terms of pore/solute size, pore/solute charge, and dielectric effects, coupled with diffusion or convective flow. Occasionally, more complex partitioning and transport mechanisms are used; however, more synthetic membranes may be broadly described as polymer sheets containing micron to nanometer-sized holes. Polymeric membranes are costly, with short life, and need maintenance during the process. However, there is a continuing quest for membranes with improved performance to provide better separations at even lower energy demands.

In the last 15 years, novel membrane materials such as carbon nanotubes (CNTs), aquaporins (AQPs), nanofibers, graphene, and metal-organic frameworks (MOFs) have been discovered and attempts were made to develop the corresponding membranes. These materials opened new frontiers for the separation of systems such as gases, liquids, nanoparticles, and viruses from water, desalination, and the environment. In this chapter, these materials and their uses as membranes are discussed.

12.2 CNT MEMBRANE

Iijima [1], in 1991, reported the first detailed transmission electron microscopy images of arc-grown multiwalled CNTs (MWCNTs). The single-walled CNTs (SWCNTs) were reported later on [2]. CNTs are nanoscale cylinders of graphene with exceptional properties such as high mechanical strength, high aspect ratio, and large specific surface area. To exploit these properties for membranes, macroscopic structures need to be designed with controlled porosity and pore size.

CNTs are allotropes of carbon with a cylindrical nanostructure. Nanotubes have been constructed with length to diameter ratio up to 132,000,000:1, significantly larger than that for any other material [3]. These cylindrical carbon molecules have unusual properties, which are valuable for nanotechnology, electronics, optics, and other fields of material science and technology [4]. Nanotubes are categorized as single-walled nanotubes (SWNTs) and multiwalled nanotubes (MWNTs). Individual nanotubes naturally align themselves into *ropes* held together by van der Waals forces.

SWNTs have a diameter of close to 1 nm, with a tube length that can be many millions of times longer. The structure of an SWNT can be conceptualized by wrapping a one-atom-thick layer of graphite called *graphene* into a seamless cylinder. MWNT consists of multiple rolled layers (concentric tubes) of graphene. Double-walled CNTs (DWCNTs) form a special class of nanotubes because their morphology and properties are similar to those of SWNT, but their resistance to chemicals is significantly improved.

12.2.1 PREPARATION OF CNT MEMBRANES

Primarily, there are four approaches to the synthesis of membranes based on CNTs[5]:

1. Deposition of carbonaceous materials inside preexisting-ordered porous membranes, such as anodized alumina, also known as template-synthesized CNT membranes
2. Membranes based on the interstices between nanotubes in a vertical array of CNTs, subsequently referred to as the dense-array outer-wall CNT membrane
3. Encapsulation of as-grown vertically aligned CNTs (VA-CNT) by a space-filling inert polymer or ceramic matrix followed by opening up the CNT tips using plasma chemistry or the open-ended CNT membrane
4. Membranes composed of nanotubes as fillers in a polymer matrix, also known as mixed-matrix membranes (MMMs)

12.2.2   PROPERTIES AND APPLICATIONS OF CNT MEMBRANES

Chen and Hill [6] used classical phonon theory for the collective motion of water molecules inside the nanotube to formulate the basic equations of motion for water diffusing through a CNT. On making a *sufficiently long* hypothesis, the average water flow time can be deduced analytically. Furthermore, on incorporating the external forces at the tube ends, it showed that water is virtually incompressible for external forces up to 3 pN. On determining the variation of the water flow time under random fluctuations in the presence of the external forces, it was observed that the random effect diminishes as the external force increases. This outcome could open up a precise engineering approach for using such nanotube membranes in numerous applications.

Pilates et al. [7] discussed the first systematic approach for investigating the internal configuration of template-based CNT arrays in detail. Key findings were made for the customized optimization of the resulting nanotube membranes for a variety of applications, including separations, nanofluidics and nanoreactors, biological capturing and purification, and controlled drug delivery and release.

CNTs also offer an exciting opportunity to mimic natural protein channels due to the following factors: (1) a mechanism for dramatically enhanced fluid flow, (2) the ability to place *gatekeeper* chemistry at the entrance to pores, and (3) being electrically conductive to localize electric field or perform electrochemical transformations [8]. The transport mechanisms through CNT membranes are primarily (1) ionic diffusion near bulk expectation, (2) gas flow enhanced 1–2 orders of magnitude primarily due to specular reflection, and (3) fluid flow 4–5 orders of magnitude faster than conventional materials due to a nearly ideal slip-boundary interface. CNTs can be applied in water purification, energy generation, and bioseparations. Rana et al. [9] fabricated VA-CNTs over self-ordered nanoporous alumina films and studied their surface properties. It was reported that the surface had super hydrophobic character. Kang et al. [10] demonstrated the fabrication and detailed characterization of nanocomposite membranes containing a high loading (up to ~40 vol.%) of aluminosilicate SWNTs. The transport properties of the poly(vinyl alcohol) (PVA)/SWNT membranes were investigated experimentally by ethanol/water mixture pervaporation measurements, computationally by grand canonical Monte Carlo and molecular dynamics, and by a macroscopic transport model for anisotropic permeation through nanotube–polymer composite membranes. The nanocomposite membranes substantially enhanced the water flux throughout with increasing SWNT volume fraction, with a cost of a moderate reduction of the water/ethanol selectivity. The model was parameterized purely from molecular simulation data with no fitted parameters, and showed reasonably good agreement with the experimental permeability data.

Sears et al. [11] discussed the fabrication and application of two types of CNT-based membranes, namely, (1) Bucky papers (BPs) and (2) isoporous CNT membranes. Both of these membranes have distinctively different structures and porosity. BP membranes are comprised of randomly entangled CNTs that are fabricated by a relatively simple process involving vacuum filtration. BPs are typically formed by first purifying the CNTs and dispersing them in a suitable solvent. Once a well-dispersed solution is achieved, it is filtered through a porous support, which captures the CNTs to form an optically opaque CNT BP. If the BP is thick enough, it can be peeled off the support filter intact. In BP, the deposited nanotubes, which are like nanofibers, work as a membrane. Figure 12.1 shows an SEM image of the surface of an aligned CNT BP. CNTs could be fabricated into BP structures as self-supporting membranes, and their hydrophobic property can be used for water desalination by membrane distillation (MD). CNT BPs possess very interesting properties such as natural hydrophobicity, high porosity, and very high specific surface area, making them promising candidates for separation applications such as MD. Thus, the properties of BP depend on the type of CNTs used and their pretreatment, purification, and dispersion. BP can be applied for applications such as direct contact MD (DMCD), capacitive deionization, and filtration of particles, including bacteria and viruses. On the other hand, isoporous CNT membranes use CNTs as pores across an otherwise impermeable matrix material. A handful of groups have published different approaches to isoporous CNT membrane construction with promising permeance results. Despite the smaller CNT diameter, gas permeances equal to or higher than that of commercial polycarbonate membranes with cylindrical, 10 nm diameter pores, were reported. This is made possible, in part, by a higher CNT pore density compared to polycarbonate membranes. However, as demonstrated by Holt et al. [12], flow enhancement due to the atomically smooth and hydrophobic CNT surface may also play a large role for CNT pore diameters less than ~2 nm. Several groups have also demonstrated fast liquid flow through the CNT interior, 2–3 orders of magnitude greater than that predicted by conventional theory, and seem to confirm theoretical predictions. These isoporous membranes are, therefore, of great interest for nanofiltration (NF) membranes with both high flux and high selectivity.

Liu et al. [13] showed that windowed-CNTs can efficiently separate the CO_2/CH_4 mixture using molecular dynamics simulation. Four CO_2/CH_4 mixtures with 10%, 30%, 50%, and 80% CO_2 were investigated as a function of applied pressure from 80 to 180 bar. In all simulated conditions, only CO_2 permeation was observed. On the other hand, CH_4 was completely rejected by the nitrogen-functionalized windows or pores

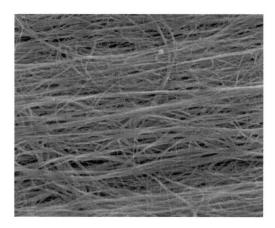

FIGURE 12.1   SEM image of the surface of an aligned carbon nanotube BP.

on the nanotube wall in the accessible time scale, while maintaining a fast diffusion rate along the tube. The estimated time-dependent CO_2 permeance ranges from 10^7 to 10^5 GPU (gas permeation unit), compared with around 100 GPU for typical polymeric membranes. It was suggested that a windowed-CNT can be used as a highly efficient medium, configurable in hollow-fiber-like modules, for removing CO_2 from natural gas.

Liu et al. [14] functionalized DWCNTs as artificial water channel proteins. For the first time, molecular dynamics simulations showed that the bilayer structure of DWCNTs is advantageous for CNT-based transmembrane channels. Shielding of the amphiphilic outer layer could guarantee biocompatibility of the synthetic channel and protect the inner tube (functional part) from disturbance of the membrane environment. This novel design could promote more sophisticated nanobiodevices, which could function in a bioenvironment with high biocompatibility.

Wu et al. [15] studied the CNT membranes with inner diameter (ID) ranging from 1.5 to 7 nm for enhanced electro-osmotic flow. After functionalization through electrochemical diazonium grafting and carbodiimide coupling reaction, it was found that neutral caffeine molecules can be efficiently pumped through electro-osmosis. An electro-osmotic velocity as high as 0.16 cm s^{-1} V^{-1} was observed. Power efficiencies were improved 25–110-fold compared to related nanoporous materials, which has important applications in chemical separations and compact medical devices. Nearly ideal electro-osmotic flow was seen in the case where the mobile cation diameter nearly matched the ID of the SWCNT, resulting in a condition of using one ion to pump one neutral molecule at equivalent concentrations.

Hinds et al. [16] fabricated an array of aligned CNTs incorporated across a polymer film to form a well-ordered nanoporous membrane structure. The measured nitrogen permeance was consistent with the flux calculated by Knudsen diffusion through nanometer-scale tubes of the observed microstructure. Data on $Ru(NH_3)_6^{3+}$ transport across the membrane in aqueous solution also indicated transport through aligned CNT cores of the observed microstructure. The lengths of the nanotubes within the polymer film were reduced by selective electrochemical oxidation, allowing for tunable pore lengths. Oxidative trimming processes resulted in carboxylate end groups that were readily functionalized at the entrance to each CNT inner core. Membranes with CNT tips that were functionalized with biotin showed a reduction in $Ru(NH_3)_6^{3+}$ flux by a factor of 15 when bound with streptavidin, thereby demonstrating the ability to gate molecular transport through CNT cores for potential applications in chemical separations and sensing.

New classes of MWCNTs/carbon nanocomposite thin films were introduced by Tseng et al. [17]. These were prepared by incorporating MWCNTs into polyimide (PI) precursor solution. The carbon films were obtained in only one coating step by spin-coating on a microporous alumina substrate and carbonization at 773 K. The MWCNTs/carbon nanocomposite thin film exhibited an ideal CO_2 flux that was 8656.6 Barrer, and a separation factor for CO_2/N_2 of 4.1 at room temperature and 1 atm feed pressure was achieved. It was 2–4 times higher than that of pure carbon membrane prepared by the same procedure and conditions.

Holt et al. [12] have demonstrated an impressive capability to produce dense parallel arrangements of DWCNTs, with both sides open in a supported membrane structure, and they reported that methane molecules were translating inside a CNT.

Verweij et al. [18] prepared 2–6 mm thin membranes consisting of DWCNTs, all aligned perpendicular to the apparent membrane surface. These tubes are open at both ends and the space between the tubes is filled with dense Si_3N_4. Pure gas and water fluxes were measured at room temperature with the application of a small pressure difference. Interpretation of the results led to the conclusion that the membranes showed much higher fluxes than what was estimated from Knudsen gas diffusion and Poiseuille viscous flow models. The membranes have straight-channel morphology with a narrow pore-size distribution and exceptionally smooth pore walls. The unusual geometry and surface properties make it difficult to compare the properties of membrane with common membranes, but it is certain that the mass transport in the aligned DWCNTs is fast indeed.

Yu et al. [19] fabricated high-density, VA-CNT membranes. The CNT arrays were prepared by chemical vapor deposition (CVD), and the arrays were collapsed into dense membranes by capillary forces due to solvent evaporation. The average space between the CNTs after shrinkage was ~3 nm, which was comparable to the pore size of the CNTs. Thus, the interstitial pores between CNTs were not sealed, and gas permeated through both CNTs and interstitial pores. NF of gold nanoparticles and N_2 adsorption indicated the pore diameters were approximately 3 nm. Gas permeances, based on total membrane area, were 1–4 orders of magnitude higher than VA-CNT membranes in the literature, and gas permeabilities were 4–7 orders of magnitude higher than the literature values. Gas permeances were approximately 450 times more than those predicted for Knudsen diffusion, and ideal selectivities were similar to or higher than Knudsen selectivities. These membranes separated a larger molecule (triisopropyl orthoformate) from a smaller molecule (*n*-hexane) during pervaporation, possibly due to the preferential adsorption, which indicates separation potential for liquid mixtures.

Khan et al. [20] reported the gas transport behavior of MMMs, which were prepared from MWCNTs and dispersed within polymers of intrinsic microporosity (PIMs) matrix. The MWCNTs were chemically functionalized with poly(ethylene glycol) (PEG) for a better dispersion in the polymer matrix. Gas permeation measurements showed that MMMs incorporated with pristine or functionalized MWCNTs (f-MWCNTs) exhibited improved gas-separation performance compared to pure PIMs. The f-MWCNTs MMM showed better performance in terms of permeance and selectivity in comparison to pristine MWCNTs. The gas permeances of the derived MMMs are increased to approximately 50% without sacrificing the selectivity at 2 wt% of f-MWCNTs' loading. Addition of f-MWCNTs inside the polymer matrix also improved the long-term gas transport stability of MMMs in comparison with PIMs. The high permeance, selectivity, and long-term stability of the fabricated MMMs suggest that the reported approach can be utilized in practical gas-separation technology.

Ahn et al. [21] discussed CNT-based membranes, which are an emerging technology for water purification system applications. This article reviewed the appropriate manufacturing methods for CNT membranes and speculated on their performances. Diverse types of CNT membranes were discussed; however, there are no commercial products available. Future applications of integrated CNT membrane system were also outlined. Table 12.1 shows applications of CNT membranes as well as future applications, as suggested by Majumder and Ajyan [5].

TABLE 12.1

Current Status and Future Applications of CNT Membranes

Type of CNT Membrane	Pore Size (nm)	Application Demonstrated	Future Applications
Template-synthesized	20	Water-vapor/O_2 separation	Nanofluidic interconnects in microfluidic platforms
Dense-array outer-wall	200	Electro-osmotic transport	
	20–100	Heavy hydrocarbon (liquid) separation	UF membranes
		Bacteria separation	Electrochemical membrane reactors
		Voltage-controlled transport	
		Protein separation	
		Compression-modulated transport	
Open-ended	1–2	Nanoparticle separation	Desalination
		Small-molecule separation	Recovery of homogeneous catalysts
		Voltage-gated transport for transdermal drug delivery	Dialysis
			Recovery
			Stochastic sensing
Mixed-matrix	Nonporous	Gas separation	Gas separation
		Pervaporation	Pervaporation

Source: Majumder, M. and Ajayan, P.M., Carbon nanotube membranes: A new frontier in membrane science. In: E. Drioli and L. Giorno (Eds.), *Comprehensive Membrane Science and Engineering*, vol 1, Oxford: Academic Press, pp. 291–310, 2010.

12.2.3 SUMMARY OF CNT MEMBRANES

CNT membranes open a new door for membranes used for separation purposes. These membranes can be used for water separation, gas separation, pervaporation, fuel cells, membrane reactors, and medical science. However, the application of CNT membranes needs a lot of research for its development.

The safety of the CNT membrane is an important issue to be resolved. Concerns have been raised about the environmental and health impacts of CNT' particles that may be released from CNT membranes during operation [22,23]. However, there is little chance that CNT membranes will be able to pass through a wholly integrated membrane element. After much effort for regulating nanomaterials, no safety guidelines for the use of CNT membranes in the water purification process have been adopted so far. Thus, an interdisciplinary approach should be used to establish a policy for CNT membrane technology.

12.3 AQP MEMBRANE

AQPs are selective membrane channel proteins found in the lipid bilayer of living cells that work to transport water across the cell membrane. AQPs accomplish this task while excluding any unwanted ions or other polar molecules, making them a perfect model for the formation of low-energy water filtration systems [24]. Bowen [25] discussed how we could learn from biological membranes in the development of membranes that benefit from biomimetics to achieve better selectivity and higher permeability. Biomimetic membranes are designed to mimic the highly selective transport of water or solutes across cell membranes. One promising biomimetic membrane technology employs AQPs to regulate the flow of water, providing increased permeability, and near-perfect solute rejection. AQPs act as water channels that allow water molecules to pass through a single file, whereas the transport of ions, protons, and hydroxyl ions is abrogated by an electrostatic-tuning mechanism of the channel interior. Small molecules also have restricted passage because their electrochemical properties do not *fit*. The result is that only water molecules, nothing else, can pass through AQP water pores. AQP membranes are suggested to be 100 times more permeable than commercial reverse osmosis (RO) membranes. Johns Hopkins University's Peter Agre won a Nobel Prize in chemistry, together with Roderick MacKinnon, for the AQP research.

12.3.1 Preparation of AQP Membrane

Until now, most recombinant AQPs have been expressed only in lab-scale quantities for screening, functional, regulatory, or structural studies [26]. One of the main obstacles in protein production is that membrane protein over expression *in vivo* is hampered by their complex structure, hydrophobic transmembrane regions, host toxicity, and the time-consuming and low-efficiency refolding steps required. Recent developments of high expression systems may, however, provide insights into how large-scale AQP production may be realized. These include *Escherichia coli*, *Saccharomyces cerevisiae*, *Pichia pastoris*, and baculovirus/insect cell-based systems. For review, see the articles written by Altramura and Calamita [27] and Tang et al. [26]. A new approach for fabricating AQP-based biomimetic membrane was developed by Zhao et al. [28], which involves embedding AQP-containing proteoliposomes or proteopolymersomes in a cross-linked polyamide matrix. A microporous substrate was first soaked in an aqueous solution of *m*-phenylenediamine, which also contained a given amount of AQP-containing vesicles (Figure 12.2). Soaked substrates were then exposed to a trimesoyl chloride solution to form an interfacial polymerized polyamide rejection layer, where the vesicles were dispersed in the thin rejection layer. In this design, the AQP-containing vesicles provide preferential water paths through the polyamide layer, and thus significantly enhance the membrane water permeability. Due to its simple fabrication procedure, this technique can be easily scaled up to produce large membrane area, as authors claimed.

Table 12.2 summarizes the existing approach of preparing AQP-based biomimetic membranes.

A company actually named Aquaporin has developed a novel biometric water membrane, based on AQP water channels. Though it is still in the developmental phase, the AQP membrane can extract water with extremely high initial salt-rejection rates of 99.99987% and initial water fluxes of ~10 kg m^{-2} h^{-1}, at osmotic gradients of 30 bar in draw solutions, with a membrane design very suitable for industrial up-scaling [29].

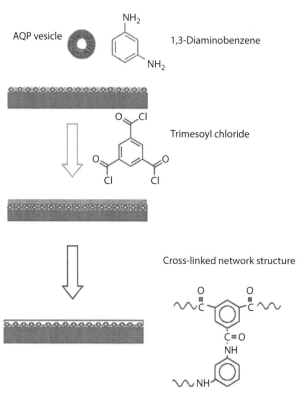

FIGURE 12.2 Preparation of AQP-based biomimetic membranes by interfacial polymerization. The polymerization results in a thin matrix (light gray, O-O-O-O-O-O) with AQP forming the active layer supported by a microporous substrate.

12.3.2 Properties and Applications of AQP

Biological membranes have excellent water-transport characteristics, with certain membranes able to regulate permeability over a wide range [30]. Kumar et al. [30] discussed the idea of incorporating AQP properties into desalination membranes.

The AQPs are a family of small membrane-spanning proteins (monomer size around 30 kDa [kilodalton]) that are expressed at plasma membranes in many cell types involved in fluid transport. AQPs appear to assemble in membranes as homo-tetramers, in which each monomer, consisting of six membrane-spanning α-helical domains with cytoplasmically oriented amino and carboxyl termini, contains a distinct water pore. Medium-resolution structural analysis indicated that the six titled helical segments form a barrel surrounding a central pore-like region that contains additional protein density. Several of the mammalian AQPs (e.g., AQP1, APQ2, AQP4, and AQP5) appear to be highly selective for the passage of water, whereas others, termed *aquaglyceroporins*, also transport glycerol (e.g., AQP3 and AQP8) and even larger [31].

Borgnia et al. [32] revealed that the high water permeability characteristic of mammalian red blood cell membranes is caused by AQP1. This channel freely permits the movement of water across the cell membrane, but it is not permeated by other

TABLE 12.2
Examples of Biomimetic Membrane Designed for Water Reuse and Desalination

Approach	WP (L.m^{-2} h^{-1} bar^{-1})	R_{NaCl} (%)	P_{Max} (bar)
Charged lipid mixture vesicles deposition onto NF membranes	0.83	n.d.	10
Vesicle fusion facilitated by hydraulic pressure on hydrophilic NF membranes coated with positively charged lipids	3.6 ± 0.2	35 ± 8	1
Membranes across multiple micron scale apertures either as free-standing lipid or polymer membranes	n.d.	n.d.	n.d.
Membranes across multiple micron scale apertures and stabilized by hydrogel encapsulation	12–40	n.d.	2
Aquaporin containing polymersomes on methacrylate functionalized CA membrane	34.2 ± 6.9	32.9 ± 9.1	5
Detergent-stabilized His-tagged aquaporin added to monolayers with nickel-chelating lipids	n.d.	n.d.	n.d.
Proteopolymersome deposition onto gold-functionalized PC track etched substrates	n.d.[a]	n.d.[a]	n.d.
Interfacial polymerization method with embedded proteoliposomes	4 ± 0.4	96.3 ± 1.2	14

Source: Tang, C.Y. et al., *Desalination*, 308, 34–40, 2013.

Note: Performance data are presented as water permeability (WP), NaCl rejection (R_{NaCl}), and maximal external pressure applied (P_{Max}) when operated in RO, CA (cellulose acetates), and PC (polycarbonate).

[a] RO tests were not performed: Based on FO (forward osmosis) tests, a WP of 16.4 Lm^{-2}h^{-1}, and a salt flux of 6.6 gm^{-2}h^{-1} were obtained for membrane prepared with a protein-to-polymer molar ratio 1:100, with 0.3 M sucrose as draw and 200 ppm NaCl as feed.

small, uncharged molecules or charged solutes. Figure 12.3 shows the structural basis of water-specific transport through the AQP1 water channel. AQP1 is a tetramer with each subunit containing an aqueous pore likened to an hourglass formed by obversely arranged tandem repeats. Cryoelectron microscopy of reconstituted AQP1 membrane crystals has revealed the three-dimensional structure at 3Å–6Å. Ten mammalian AQPs have been identified in water-permeable tissues and fall into two groupings. Orthodox AQPs are water-selective and include AQP2, a vasopressin-regulated water channel in the renal collecting duct, in addition to AQP0, AQP4, and AQP5. Multifunctional aquaglyceroporins such as APQ3, AQP7, and AQP9 allow the permeation of water, glycerol, and some other solutes. AQPs are being identified in numerous other species, including amphibia, insects, plants, and microbials. Recent work has suggested that, in addition to water movement, AQPs might transport other physiologically important molecules across membranes, including CO_2, H_2O_2, NH_3/NH_4^+, boron, and silicon, and therefore may be involved in a number

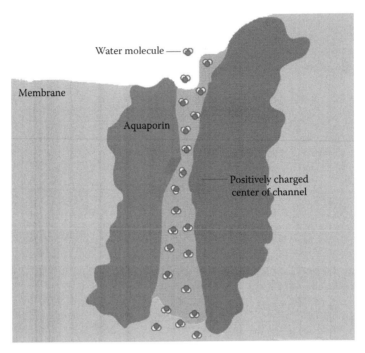

FIGURE 12.3 Structural basis of water-specific transport through the AQP1 water channel. (With the permission of Water Technology Markets, ewlish@globalwaterintel.com.)

of fundamental processes in plants, such as nutrient acquisition, photosynthesis, and stress responses. However, measuring transport through AQP channels directly and teasing apart a direct role in transport of these molecules and possible indirect effects associated with water has proven to be exceptionally difficult [33].

Thus, concluding points to this section are as follows: (1) AQPs are membrane channels that have conserved structure and facilitate the transport of water or small neutral solutes (e.g., urea, boric acid, and silicic acid) or gases (e.g., ammonia and carbon dioxide); (2) AQPs exhibit a high isoform multiplicity that reflects distinct transport specificities and subcellular localization; (3) AQP transport activity can be regulated by multiple mechanisms, including regulation of transcript or protein abundance, subcellular trafficking, or grating by phosphorylation or cytosolic protons; (4) AQPs play a central role in plant–water relations; and (5) multiple integrated roles of AQPs in carbon and nitrogen assimilation and micronutrient uptake are being uncovered [34].

12.3.3 SUMMARY

Perhaps the most challenging part of biomimetic membrane development is to understand the interaction between the membrane and its support, particularly when this support is also porous, and thus can support mass transport across the membrane [35]. With embedded AQPs, the support must bear pressure up to 10 bar and allow water flux >100 L m^{-2} h^{-1}. Therefore, the development of the AQP membrane is closely linked to the simultaneous development of suitable porous support.

One difficulty in commercializing biomimetic membrane is to make them as stable and strong as polymeric membranes, so they are able to withstand the operating pressures and repeated fouling and cleaning events expected to be encountered in most full-scale applications.

12.4 NANOFIBERS

Nanoscale materials can be designed to exhibit novel and significantly improved physical and chemical properties. Polymer nanofibers are an important class of nanomaterials, which have attracted increasing attention in the last 10 years because of their high surface-to-mass (or volume) ratio and special characteristics attractive for advanced application [36]. Burger et al. [37] wrote a review on nanofibrous materials and their applications. The review of Balamurugan et al. [38] discussed the trends in nanofibrous membranes and their suitability for air and water filtration, and the authors mainly presented the application of nanofibers in textile industry, air cleaning in hospitals and other domains, and environmental applications using microfiltration (MF) and ultrafiltration (UF).

More than 100 polymers, both synthetic and natural, have been successfully electrospun into nanofibers, mostly from polymer solutions, as any polymer may be electrospun into nanofibers, provided that the polymer molecular weight is sufficiently high and the solvent can be evaporated in the time during the jet transit period, over a distance between the spinneret and the collector. Standard polymers successfully electrospun into nanofibers include polyacrylonitrile (PAN), poly(ethylene oxide) (PEO), poly(ethylene terephthalate) (PET), polystyrene (PS), poly(vinyl chloride) (PVC), Nylon 6, PVA, poly(ε-caprolactone), Kevlar (poly[p-phenylene terephthalamide]), poly(vinylidene fluoride) (PVDF), polybenzimidazole, polyurethanes (PUs), polycarbonates, polysulfones, poly(vinyl phenol) (PVP), and many others [36,37]. Electrospinning has also been used to produce nanofibers from natural biomacromolecules, including cellulose [either electrospun from cellulose acetate (CA) with subsequent hydrolysis or directly electrospun from cellulose solutions in N,N-dimethylacetamide with lithium chloride], collagen and gelatin, modified chitin, chitosan, and DNA.

12.4.1 Preparation of Nanofibers

There are a number of processing techniques to fabricate nanofibers, which include electrospinning, drawing, template synthesis, phase separation, interfacial polymerization, self-assembly, and so on [39,40]. Among these, electrospinning process is a very effective method to make porous hydrophobic membranes.

12.4.1.1 Electrospun Nanofibrous Membranes with a Single Component

The electrospinning process is similar to the conventional spinning method with the solution passing through a spinneret. However, instead of using air or mechanical devices to create the extrusion force, a high voltage is applied to the solution so

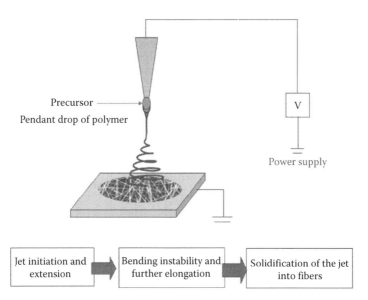

Precursor

Pendant drop of polymer

V

Power supply

| Jet initiation and extension | → | Bending instability and further elongation | → | Solidification of the jet into fibers |

FIGURE 12.4   Schematic diagram of the electrospinning process.

that the solution is charged, thereby creating a repulsive force. At a critical voltage, the repulsive force overcomes the surface tension of the solution and a jet will erupt from the tip of the spinneret. Unlike conventional spinning, the jet is only stable near the tip of the spinneret, after which the jet will undergo bending instability. As the charged jet accelerates toward regions of lower potential, the entanglements of the polymer chain will prevent the jet from breaking up while the solvent evaporates resulting in fiber formation. A grounded plate is used to collect the fibers. Figure 12.4 is a schematic of the electrospinning process.

12.4.1.2   Preparation of Nanofibers with Side-by-Side Components

Gupta and Wilkes [41] described simultaneous electrospinning of two polymer solutions in a side-by-side fashion for two bicomponent polymer systems, namely, PVC/segmented PU (PVC/Estanew) and PVC/PVDF. The schematic diagram of the equipment is shown in Figure 12.5.

The two syringes that contain the polymer solutions lie in a side-by-side fashion. A common syringe pump (KDS 100, KD Scientific Inc., Holliston, MA) controls the flow rate of the two polymer solutions. The platinum electrodes dipped in each of these solutions are connected in parallel to the high voltage DC supply (Spellman CZE 1000R). The free ends of the Teflon needles attached to the syringes are adhered together. The internal diameter of the Teflon needle is 0.7 mm with a wall thickness of about 0.2 mm. The length of the Teflon capillary is about 6 cm. When the syringe pump is turned on, the two polymer solutions flow outward through each of the Teflon needles until they come in contact at the tip of the needles. The grounded target used for collecting the solidified polymer filaments was a steel wire (wire diameter ~0.5 mm) mesh of count 20 × 20 (20 steel wires per 1″ each in the horizontal and vertical axes).

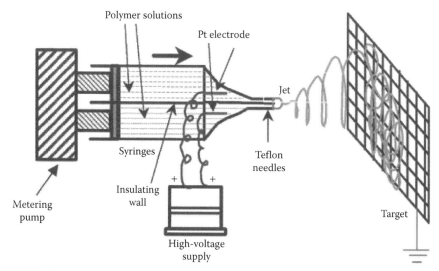

FIGURE 12.5   Schematic representation of the bicomponent fiber electrospinning setup.

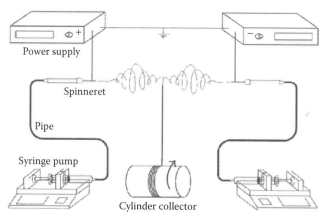

FIGURE 12.6   Schematic diagram of a dual-opposite-spinneret electrospinning (DOSE) apparatus.

Well-aligned and uniform side-by-side bicomponent fibers were produced through dual-opposite-spinneret electrospinning (DOSE). Side-by-side TiO_2/SnO_2 nanofibers were obtained after calcining as-spun fibers [42].

The DOSE apparatus is illustrated in Figure 12.6. Electrospinning solution was loaded into the syringe and pumped by the syringe pump. A flat-tipped stainless steel syringe needle was used as the spinneret. Two spinnerets were assembled horizontally in opposite directions, and each was connected to a separate high-voltage power supply. A rotating cylinder covered with aluminum foil was used as a collector. From the DOSE apparatus, it was easy to collect well-aligned, uniform, and side-by-side electrospun fibers.

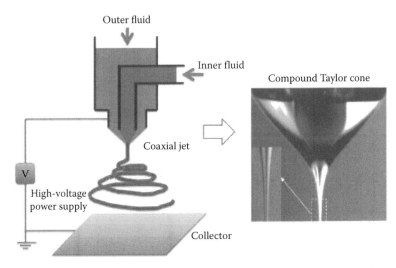

FIGURE 12.7 Basic setup for coaxial electrospinning and fabrication process of common core–shell nanofibers.

Various nanofibers with different secondary nanostructures have also been fabricated. Especially, core–shell nanofibers draw more and more interest and attention, because of the great potential applications and prospects for nanochannel, nanocapsule, and small encapsulating devices. Coaxial electrospinning is an effective, fast, and controlled technique to construct core–shell nanostructures into nanofibers [36,43–45], which will lead to the fabrication of hollow nanofibers.

Recently, advances in the technique of electrospinning have allowed this method to be used to directly fabricate core–sheath and hollow nanofibers of composites, ceramics, and polymers, with diameters ranging from 20 nm to 1 μm [43,44,46].

Li and Xia [43] prepared hollow nanofibers with walls made of inorganic/polymer composites or ceramics by letting two immiscible liquids flow through a two capillary spinneret, followed by selective removal of the cores. Figure 12.7 shows a basic setup for coaxial electrospinning and the fabrication process of common core–shell nanofibers. It is similar to the conventional one except for the use of a spinneret containing two coaxial capillaries. The core solution is fed into the inner spinneret (mineral oil), whereas the shell solution, polymeric solution that may include a polymer, a catalyst, a solvent, and a sol-gel precursor, is transported into the outer one. Coaxial fibers are collected on the metal substrate under high-voltage conditions. A too high or too low speed of inner fluid is unfavorable. An appropriate injecting speed and the ratio of inner–outer fluid speed should be considered and investigated [47].

The multifluidic compound-jet electrospinning is presented in Figure 12.8. Several metallic capillaries with outer diameter (OD) of 0.4 mm and ID of 0.2 mm were arranged at the several vertexes of an equilateral triangle. Then the bundle of capillaries was inserted into a plastic syringe (OD = 3.5 mm, ID = 2.0 mm) with gaps between individual inner capillaries and outer syringe. Two immiscible viscous liquids were

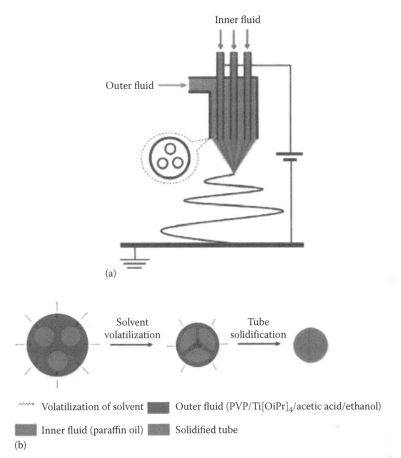

FIGURE 12.8   Multichannel coaxial electrospinning system (a) and multichannel fiber shaping (b).

fed separately to the three inner capillaries and outer syringe in an appropriate flow rate. A 20% polyvinylpyrrolidone (PVPD) ethanol solution served as outer liquid, whereas a nondissolution paraffin oil was chosen as inner liquid. Then a high-voltage generator between three inner metallic capillaries and a metallic plate coated with a piece of aluminum foil, which acted as counter electrode, provided the drive and control for the electrospinning. The inner is for immiscible paraffin oil and outer for $Ti(OiPr)_4$ solution, and the both issued out separately from individual capillaries. With a suitable high-voltage application, a whipping compound fluid jet is formed under the spinneret and then a fibrous membrane is collected on the aluminum foil.

Srivastava et al. [48] demonstrated a microfluidic approach to fabricate hollow and core/sheath nanofibers by electrospinning. They successfully fabricated hollow PVP + titania (TiO_2) composite and core/sheath polypyrrole/PVP nanofibers in the order of 100 and 250 nm, respectively, using elastomeric microfluidic devices.

Multilayered hollow nanofiber was also fabricated by Ge et al. [49] through the combination of the electrostatic layer-by-layer (LbL) and electrospinning methods.

It was observed that the formation of multilayered polyethylene films on PS fiber surface was strongly influenced by the number of repetitions of deposition and the pH value.

McCann et al. [50] described the use of electrospinning in the fabrication of polymeric, ceramic, and composite nanofibers with core–sheath, hollow-fiber, or porous structure, as well as the efforts made to improve their morphological homogeneity, functionality, and device performance. Figure 12.9a shows the schematic diagram for the fabrication of core–sheath nanofibers with a coaxial spinneret, which was used by McCann et al. [50]

Wang et al. [51] used the coaxial technique to fabricate poly(bis[p-methylphenoxy])phosphazene (PMPPh)/PAN nanofibrous membranes. Under the optimized electrospinning conditions of a 15 wt% concentration and 0.2 mL/h flow rate for both PMPPh and PAN, nanofibers of diameter 150 ± 20 nm and a clear core/sheath structure were obtained.

In another report by Theron et al. [52], the alignment of individual fibers on a tapered and grounded wheel-like bobbin was enabled by an electrostatic field-assisted assembly technique. The bobbin was able to wind a continuously as-spun nanofiber at its tip-like edge. This alignment approach has resulted in PEO-based nanofibers,

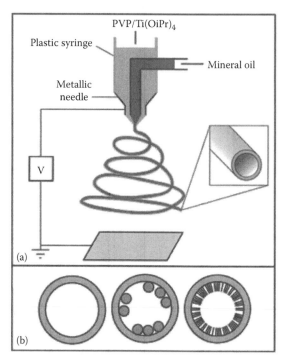

FIGURE 12.9 (a) Schematic diagram for the fabrication of core–sheath nanofibers with a coaxial spinneret. The mineral oil is fed through a silica capillary, whereas the outer sheath layer contains a sol-gel precursor and poly(vinyl pyrrolidone). (b) With subsequent removal of the oil core, hollow nanofibers can be generated. In addition, colloidal particles and long-chain silanes can be added to the mineral oil to generate hollow nanofibers with functionalized interiors.

with diameters ranging from 100 to 300 nm. They were also able to successfully take off from the wheel a braid of aligned nanofibers with a diameter of 5–10 μm and density of about 100 nanofibers μm^{-2} and a length of about 10 cm. Theron et al. [52] also reported that the length of the individual nanofiber was up to hundreds of μm. Later, Inai et al. [53] used the same method as that of Theron et al. [52] with some modification and obtained aligned single nanofibers of poly(L-lactide). They could collect single nanofibers with diameters ranging from 600 to 900 nm and lengths that were sufficient for a tensile strength test with 2 cm gauge length.

A novel nanofibrous membrane was fabricated by Franco et al. [54] through electrospinning composed of polyvinyl phosphoric acid and PVA. Chemical cross-linking was done by immersion in methanol and methanol/GA. Visual examination and dimensional analyses showed that heat treatment produced discoloration on the membrane surface and chemical cross-linking reduced membrane dimensions. Tensile strength and strain of cross-linked membranes improved compared to noncross-linked counterpart. Swelling and degradation were also depending on cross-linking condition. Biocompatibility was observed to be more favorable in heat-treated membranes.

Yang et al. [55] demonstrated successfully how to make a highly aligned and continually packed electrospun core/shell PEG/poly(L-lactic acid) fibers by balancing the electrostatic force, the solution viscosity, and the weight of the jetted fibers. The diameter of the electrospun hollow fibers was in the range of tens of micrometers with a wall thickness of a few micrometers.

12.4.2   PROPERTIES AND APPLICATIONS OF NANOFIBERS

Electrospun nanofibrous membranes possess several attractive qualities, such as high porosity, pore sizes ranging from tens of nanometer to several micrometers, interconnected open pore structure, and high permeability of gases [56]. Moreover, the higher porosity of the nanofibrous membranes enables them to have a large surface area per unit volume of membrane accessible to the liquid filtration applications such as MD. The electrospinning technique is the only one that allows the production of continuous polymeric nanofibers and provides numerous opportunities to manipulate and control surface area, fiber diameter, the porosity of the nanofibrous layer, fiber density as well as basis weight (fiber weight per area).

Characterization of nanofibers using different techniques has shown that the behavior of nanofibers during the morphological characterization is specific and significantly different compared to the rigid porous polymers. Therefore, the morphological characterization of nanofibrous materials requires a complex approach and evaluation of the results of various methods [57].

Zhou et al. [58] studied the electrospun CA nanofibrous membranes prepared from a CA solution in dichloromethane, formic acid, acetic acid, and trifluoroacetic acid (TFA). SEM results revealed that the as-spun CA nanofibers using TFA as solvent were continuous and smooth, with the fiber diameter ranging from 100 to 300 nm. The nanofibrous membranes exhibited high surface area and high porosity compared with the common filter paper, which reached 2.020×10^7 m^{-1} and 87%, respectively. Also, the membranes displayed very high water permeability and a good hydrolytic stability.

Electrospun nanofibrous membranes (ENMs) were found to be applicable for particle separation from water. Thus, they can be used for the treatment of wastewater prior to the treatment by UF, NF, and RO [56,59–72]. Gopal et al. [73] demonstrated the PVDF ENM applicability in particulate removal. Characterization of these electrospun membranes revealed that they have similar properties to that of conventional MF membranes. The electrospun membranes were used to separate 1, 5, and 10 μm PS particles. The electrospun membranes were successful in rejecting more than 90% of the microparticles from solution. It was suggested by Gopal et al. that nanofibrous membranes have potential for the pretreatment of water prior to RO or as prefilters to minimize fouling and contamination prior to UF or NF.

Another interesting water purification where ENMs can be used is membrane adsorption whereby the separation of chiral isomers or the removal of various hazardous substances can be done [74–77]. Guibo et al. [78] reported that the thermally treated nanofibrous membranes prepared from polyamide showed a high efficiency for filtration.

12.4.2.1 Application of Nanofibers in RO, UF, NF, and MF

Chu et al. [79] patented a membrane including a coating layer that was applied onto cellulose nanofibers produced from oxidized cellulose microfibers by immersing in GA and incorporating polyacrylic acid. They claimed that the electrospun membrane can be used in MF, whereas the nanofibrous membranes can be used in UF, NF, and RO after chemical modification and coating.

For molecular filtration, Chen et al. [80] fabricated carbonaceous nanofibrous (CNF) membranes functionalized by beta-cyclodextrins (CNF-β-CD membrane). The membrane showed a remarkable capability to function as an ideal molecular filter through complexation of phenolphthalein molecules with the CD molecules grafted on the CNFs. As a typical dye pollutant, fuchsin acid could also be effectively removed from the solution through such a membrane. Engineering the surface of this CNF membrane may be used for other applications such as chiral separation and drug delivery.

Ma et al. [71] used a new class of ultrafine polysaccharide nanofibrous membranes for water purification by UF. It was reported by Ma et al. that polysaccharide membranes with three kinds of barrier layer exhibited very high permeation flux (10 times increase) and high rejection rate (>99.5%) compared with commercial UF membrane for the separation of oil/water emulsions. In these membranes, ultrafine polysaccharide nanofibers had pores of about 20 nm diameter.

Chen and Liang et al. [81] fabricated a new kind of free-standing CNF membrane, which was capable of the filtration and separation of nanoparticles with different sizes from solution. Fabricated CNF membranes were very flexible and mechanically robust enough for filtration under a high applied pressure without any damage. Free-standing CNF membranes were fabricated through a solvent-evaporation-induced self-assembly process. A wool-like homogeneous suspension was obtained by vigorous magnetic stirring of the CNF-50 in ethanol for several hours. After casting the suspension onto a Teflon substrate and drying at ambient temperature, a brown paper-like material was found, which could be easily detached from the substrate without cracking. The size and shape of the membrane were determined by the Teflon substrate used for casting. Thus, the CNF membrane obtained was completely free-standing and mechanically robust enough for filtration operation, even under a high applied pressure, without any damage. The highly porous structure and hydrophilic properties of the fibrous membranes facilitate the permeation of water through the membranes and a high flux can be achieved. More importantly, these CNF membranes have very narrow pore size distributions and showed excellent size-selective rejection properties.

12.4.2.2 Application of Nanofibers MD

It is expected that MD by ENMs can be applied for many purposes such as wastewater treatment, food processing, and the treatment of pharmaceutical products.

Jafar and Sarbatly [82] suggested that the desalination of geothermal water (groundwater, temperature 60°C) can be done by MD using nanofibrous membranes. Khayat et al. [83] used copolymer (PVDF-*co*-F6PP) or homopolymer (PVDF) nanofibrous or nanostructured flat membranes for DCMD. The membranes are produced by means of electrospinning with solvents having different affinities with the polymer or copolymer used, thereby allowing the thickness of the membrane, the diameter of the nanofibers, and the pore size and the porosity of the membranes (empty space between nanofibers) to be controlled. The resulting membranes can be used in DMCD for different purposes, such as saltwater treatment, wastewater treatment, concentration of pharmaceutical products and foods, and the production of distilled or ultrapure water.

Rošic et al. [84] discussed the fundamental aspects of the electrospinning process and the properties of nanofibers, as well as highlighted the enormous potential of nanofibers as drug-delivery systems and tissue scaffolds. As ENMs have high porosity, interconnected open-pore structure, tailorable membrane thickness, and high surface hydrophobicity, they are suitable for MD.

Shih [85] prepared two composite nanofibrous membranes of PVDF and poly(vinylidene fluoride-*co*-hexafluoropropylene) (PVDF-HFP) by electrospinning to be employed in a DCMD. Using SEM observations, a porosity analyzer technique, and contact angle measurement, it was found that the nanofibrous membrane with an average fiber diameter of 170 nm was the best membrane to be applied in DCMD systems. It was observed that over a period of 12 h, the permeate flux of the PVDF-HFP composite membrane was 4.28 kg m^{-2} h^{-1}, which was significantly higher than the PVDF membrane and even higher than the PTFE commercial membrane.

Feng [86] and Feng et al. [87] developed novel nanofibrous membranes for seawater desalination by air-gap MD. The PVDF nanofibrous membranes were characterized by SEM, AFM, and DSC, measurement of LEPw (liquid entry pressure of water), equilibrium contact angle, and particle separation. It was found that the pore size of the PVDF nanofibrous membrane was around 1.5 μm. The equilibrium contact angle of some nanofibrous membranes was above 120°. Feng et al. [86] attempted for the first time to use ENMs for desalination by MD. PDVF nanofibrous membrane could produce potable water (NaCl concentration <280 ppm) from a saline water of NaCl concentration 6 wt% by air-gap MD (AGMD). This new approach may eventually enable the MD process to compete with conventional seawater desalination processes such as distillation and RO. It was reported that NaCl rejection ranges from 98.7% to 99.9%, depending on the feed NaCl concentration. In addition, NaCl concentration in the permeate was found to be between 110 and 280 ppm, which is below the salt concentration limit for the drinking water. Salt-rejection rate of 99.9% for the feed containing 22% NaCl was also reported. The membrane performance was practically unchanged after almost two months of operational period. Prince et al. [88] prepared PVDF–clay nanocomposite membranes by electrospinning a solution in acetone and then used the fabricated membrane for MD.

Composite nanofibrous membranes were prepared by electrospinning and thermal treatment from PVDF-tetramethyl orthosilicate (TMOS) blend solutions by Kim et al. [89]. The contents of TMOS in PVDF were 10, 20, and 40 wt%, and the membranes were designated as TMOS10, TMOS20, and TMOS40, respectively, based on the weight of PVDF. Thermal treatment on these composite membranes resulted in the enormous enhancement of the mechanical properties and hydrophobicity. When the concentration of TMOS in the solution increased, the average diameter of nanofibers reduced, which was 403 ± 63 nm for PVDF, 343 ± 59 nm for TMOS10, 265 ± 42 nm for TMOS20, and 240 ± 35 nm for TMOS40, respectively. X-ray diffraction (XRD) results revealed that the crystal structure of PVDF in the composite membranes transformed from α-phase to β-phase due to the formation of silica particles by the thermal treatment. As the porosity of membranes was apparently improved by the introduction of the electrospinning method as compared with the phase separation process, they concluded that the electrospinning process is useful for the preparation of composite nanofibrous membranes for MD.

Feng [86] attempted ethanol separation from water by vacuum MD (VMD) using a PVDF nanofibrous membrane. VMD experiments were conducted at ethanol feed concentrations of 20, 40, 60, and 80 wt% and at feed temperatures of 40°C, 50°C, and 60°C. It was observed that the flux increased with an increase in ethanol concentration in the feed. The flux also increased with the feed temperature. The feed concentration and temperature combined increased the flux from less than 1.5 kg m^{-2} h^{-1} to over 3.3 kg m^{-2} h^{-1} at room temperature.

It was observed that the ethanol concentration in the permeate was nearly constant for a given feed ethanol concentration, regardless of the feed temperature. Table 12.3 summarizes the permeate ethanol concentration as a function of the feed ethanol concentration. Comparing the ethanol concentration in the vapor phase at V/L equilibrium, permeate ethanol concentrations are significantly lower. The separation factors were from 1.42 to 2.45.

Feng et al. [90] tried the removal of chloroform from water by means of gas stripping MD (GSMD) technique using electrospun PVDF nanofibrous membrane. It was shown that the process was feasible. The overall mass transfer coefficient of chloroform through the nanofibrous membrane was found to be 2×10^{-5} m s^{-1} at room temperature, which was more than the highest value obtained earlier for a hollow-fiber air-stripping system. It was concluded that volatile organic compounds (VOCs) could be removed by the membrane gas stripping using electrospun PVDF nanofibrous membrane due to the high surface hydrophobicity and appropriate pore sizes of the membrane.

TABLE 12.3

Ethanol Concentration in the Permeate versus Feed Ethanol Concentration and Its Comparison with Vapor/Liquid Equilibrium

Feed Ethanol Concentration (wt%)	Permeate Ethanol Concentration (wt%)	Vapor/Liquid Equilibrium[a] (wt% alcohol)
20	38	67.5
40	37	77.3
60	80	81.6
80	85	87.1

Source: [a] Perry, R.H. and Green, D.W., *Perry's Chemical Engineering Handbook*, 4th Edition, McGraw-Hill, New York, 1987.

12.4.2.3   Application of Nanofibers in Fuel Cells

Nanofibers proved to be promising materials as a proton exchange membrane and have potential application in fuel cells [91]. Lee et al. [92] fabricated a new type of proton-conducting fuel cell membrane. The membrane was composed of a 3D interconnected network of Nafion (persulfonic acid) nanofibers that was embedded in an uncharged and inert polymer matrix. Membranes were made with an average fiber diameter in the range of 161–730 nm, where the nanofiber mat occupied 60%–80% of the membrane value. The resulting membranes exhibited high proton conductivity (0.06–0.08 S/cm at 25°C in water) and improved mechanical properties, as compared to a recast homogeneous Nafion film.

Thavasi et al. [93] highlighted the potential and application of electrospun nanofibrous materials for solving critical energy and environmental issues. Electrospinning allows the production of nanofibers from various materials, organic and inorganic, in different configurations and assemblies. It is highly beneficial for energy devices. Inorganic materials, especially metal oxides, can be synthesized, but electrospining improves the conducting and ceramic properties of the material. Excitonic solar cells fabricated with aligned nanofibrous metal oxide electrodes provide higher solar-energy conversion energy. Fuel cells fabricated with nanofibrous electrodes provide uniform dispersion of catalysts, which increases the electrocatalytic activity to obtain higher chemical-electric energy conversion efficiency. The nanofibers used in filtration membranes for environmental remediation minimize the pressure drop, which provides better efficiency than conventional fiber mats.

12.4.2.4   Application of Nanofibers for Adsorption

Sang et al. [94] used nanofibrous membrane prepared from chloridized PVC by high-voltage electrospinning for the removal of divalent metal cations (Cu^{2+}, Cd^{2+}, and Pb^{2+}) from the simulated groundwater. To obtain the best heavy metal removal, several experimental methods were investigated, including static adsorption, direct filtration, soil-addition filtration, diatomic-addition filtration, and micellar-enhanced filtration (MEF). The experimental results revealed that the rejection of copper in the simulated groundwater by MEF can reach more than 73%, the rejection of lead more than 82%, and the rejection of cadmium more than 91%. It was indicated the nanofibrous membranes can be used for the treatment of groundwater containing Cu^{2+}, Pb^{2+}, and Cd^{2+} with high efficiency.

Haider and Park [95] studied the metal adsorbability of chitosan electrospun nanofiber (~235 nm in diameter) mats in an aqueous solution. The chitosan nanofiber mats, which were neutralized with potassium carbonate, showed good erosion stability in water and high adsorption affinity for metal ions in an aqueous solution. The equilibrium adsorption capacities for Cu(II) and Pb(II) were 485.44 mg g^{-1} (2.85 mmol g^{-1}) and 263.155 mg g^{-1} (0.79 mol g^{-1}), respectively. The Cu(II) adsorption data were ~6 and ~11 times higher than the reported highest values of chitosan microsphere (80.71 mg g^{-1}) [94] and the plain chitosan (45.20 mg g^{-1}) [96], respectively. It was suggested by the authors that the chitosan electrospun nanofiber mats can be applied to filter out (or neutralize) toxic metal ions and microbes without losing their original properties such as biocompatibility, bioactivity, nonantigenicity, and nontoxicity.

Menkhaus et al. [97] functionalized electrospun nanofibrous membranes (CA) with 3D monolayers through atom transfer radical polymerization (ATRP). It was noted that functionalized nanolayers immobilized onto electrospun regenerated nanofibers have an ultra-high capacity as adsorbent with chemically selective adsorption sites, 50 times higher than current commercial membrane adsorption systems and over 12 times higher than packed bed resins. It was suggested by Menkhaus et al. [97] that different base matrix materials can be utilized, such as natural polymers, synthetic polymers, and inorganic materials, along with various forms of chemical functionality such as ion-exchange, affinity, and hydrophilicity.

Vu et al. [98] synthesized PVP/TiO_2 nanofiber by electrospinning. The electrospun nanofibers were immersed in an aqueous CH_3COOH (pH = 5) solution to produce the crystalline TiO_2 nanofibers. Behavior of As(III) adsorption on TiO_2 samples was studied. It was noticed that the surface area and pore volume play an important role in the adsorption of As(III) onto the TiO_2 nanofibers. The amorphous TiO_2 nanofibers have a higher adsorption capacity and rate than the crystalline TiO_2 nanofibers due to their higher surface area and pore volume. Because the amorphous TiO_2 nanofibers have a continuously long nanofibrous shape, high adsorption capacity, and fast adsorption rate, they are promising materials for large-scale engineering applications for the removal of As(III) in water.

12.4.2.5 Application of Nanofibers for Catalysts

Nanofibrous membrane-supported TiO_2 was used as a catalyst for the oxidation of benzene to phenol in a microreactor at ambient conditions through *in situ* production of hydroxyl radicals as oxidant [99]. Reaction conditions were optimized and the performance of the microreactor was compared with the conventional laboratory scale reaction, in which hydrogen peroxide was used as oxidant. The microreactor gave a better yield of 14% for phenol compared to 0.14% in the conventional laboratory scale reaction. Reaction conditions such as reaction time, reaction pH, and applied potential were optimized. With optimized reaction conditions, selectivity of >37% and >88% conversion of benzene were obtained.

Formo et al. [100] functionalized the surfaces of TiO_2 (both anatase and rutile) and ZrO_2 nanofibrous membranes with Pt, Pd, and Rh nanoparticles. These membranes were used as a catalytic system for cross-coupling reactions in a continuous flow reactor. Contrary to the conventional setup for an organic synthesis, a continuous flow system has advantages such as short reaction time and no need for separation. The membrane-based catalytic system can also be fully regenerated for reuse. Zhang et al. [101] studied the photocatalytic efficiencies of TiO_2 continuous and porous nanofibrous membranes, and suggested that this membrane has potential applications for environmental purifications.

12.4.2.6   Other Applications of Nanofibers

A simple method for the formation of molecularly imprinted rhodamine B (RhB) was developed through electrospinning by Liu et al. [102]. RhB molecularly imprinted microspheres were produced by precipitation polymerization using RhB, acrylamide, ethylene glycol dimethacrylate, azobisisobutyronitrile, and acetonitrile as template, functional monomer, cross-linking agent, initiator, and porogen, respectively. Then molecularly imprinted membranes (MIMs) were prepared through electrospinning with PET as the matrix polymer. High pressure liquid chromatography (HPLC) analysis showed that in the optimized conditions of separation and enrichment, the recovery rate could reach 97.8%–117.1%, with relative standard deviations ($n = 3$) of 1.36%–2.19% in employing MIMs to the RhB simulated water samples. It was revealed that the imprinted polymer exhibited higher affinity for RhB compared to non-MIMs and molecularly imprinted particles.

Sueyoshi et al. [103] used chitin nanofibers for chiral separation. The membrane showed separation ability by adopting concentration gradient as a driving force for membrane transport. In simple words, the chitin nanofibrous membrane selectively transported the D-isomer of glutamic acid (Glu), phenylalanine (Phe), and lysine (Lys) from the corresponding racemic amino acid mixtures.

Xu et al. [104] used electrospun polysulfone nanofiber successfully in the advanced treatment of secondary biotreatment sewage. Electrospun CA nanofibrous membranes have the possibility of being new materials for potential filters [64]. Membrane adsorption for the purification of biopharmaceutical therapeutics, wastewater treatments, and chemical separations has shown great promise as an alternative to traditional packed-bed adsorption. Gibson et al. [105] showed that electrospun fiber coating produces an exceptionally lightweight multifunctional membrane for protective clothing, with applications that exhibit high breathability, elasticity, and filtration efficiency. It is also reported that electrospun PVDF nanofibrous membranes can be modified by plasma-induced grafting of acrylic acid to improve their wettability [106]. Xu et al. [107] electrospun a solution of soyprotein isolate (SPI) and PEO mixtures in 1,1,1,3,3,3-hexafluoro-2-propanol (HFIP) and nanowoven nanofibrous membranes were prepared. The diameters of most of the nanofibers were in the range of 200–300 nm. SPI and PEO showed high compatibility in the fiber, and SPI was homogeneously dispersed at nanoscale. Crystallization of SPI and PEO in the fiber was significantly different from that of their pure forms. All the nanofibrous membranes showed super hydrophilicity, and thus these membranes can be used in filtration and biomedical applications.

12.4.2.7 Application of Hollow Nanofibers for Membrane Separation Processes

Coaxial electrospinning has been extensively exploited as a simple technique to generate core–sheath and hollow nanofibers. These materials have been used, however, mostly for field emitter display and magnetic media, biocatalyst, controlled release, and adsorbent materials. No one has so far attempted to utilize the wall of hollow nanofibers as the membrane for separation processes. One of the obvious advantages of hollow nanofibrous membrane is its extremely large surface/volume ratio. According to a preliminary calculation based on fiber diameter of 1000 nm (1 μm) and fiber length of 10 cm, about 300, hollow nanofibers are necessary in a bundle to make an effective membrane area of 1 cm^2, which is the minimum for separation experiments. The cross-sectional area occupied by the hollow nanofibers is only 2×10^{-6} cm^2. Accordingly, the module size of hollow nanofibers will be extremely small [108].

Development of the following two techniques is the most essential for the construction of hollow nanofiber modules: (1) preparation of hollow nanofibers and (2) alignment of many hollow nanofibers, which are sufficient for making a hollow nanofiber bundle.

Regarding the second technique, there are several papers in which the method of aligning nanofibers is presented. For example, Lee et al. [109] showed the method to align nanofibers by collecting the electrospun nanofibers on a rotating drum of 4 cm radius and 10 cm length with a surface linear velocity of 2.6 m/s.

For making a module, nanofibers should be co-electrospun using a polymer solution (sheath) and a mineral oil (core) from the spinneret constructed by the method modified by Inai et al. [53]. A braid of nanofibers of a few centimeter length is removed from the rotating wheel and inserted into a stainless capillary tube, and both ends of the nanofiber braid are potted to the tube by epoxy glue. If the open ends of the nanofibers are covered by solidified glue, the glue should be sanded to make the nanofiber ends open. Permeation and separation experiments of gas mixtures and solutions can be conducted using the hollow nanofiber modules so constructed. The hollow nanofibrous membrane module may be more suitable for gas permeation experiments considering the high pressure drop caused by the liquid stream in the lumen side of the hollow nanofiber.

12.4.3 Summary of Application of Nanofibers

ENMs can contribute to MD, an emerging desalination and water-treatment technology. Nanofibrous membranes have exhibited great advantages over conventional media in environmental applications, such as air and water filtration, even though there remain a large number of challenges, such as mass production schemes of high-quality nanofiber-based composites and the selection of suitable materials and of appropriate chemistry to introduce the desired functionality to meet specific needs [109].

All techniques that have been developed for conventional polymeric or composite asymmetric membranes seem to be applicable for ENMs. Applications of the following techniques are considered to be worth exploring to investigate the potential of ENMs for water treatment.

1. It is shown that ENMs can be used as prefilters prior to water treatment by UF or RO. This has to be further investigated.
2. Pore filling of ENMs is worth investigating in the future by grafting polymers of desired properties. It has potential for the development of temperature and pH sensitive smart membranes and membranes for vapor separations with limited swelling of the polymer matrix.
3. High surface hydrophobicity of ENMs makes them useful not only for MD but also for membrane contactors.

Although the initial intensive work in the field of nanofibers has been completed, a lot of questions need to be answered. The biggest challenge is a complete understanding of the electrospinning mechanism. In order to control the properties, orientation, and mass production of the nanofibers, it is necessary to understand quantitatively how electrospinning transforms the fluid solution through a millimeter-sized needle into solid fibers having diameters that are 4–5 orders smaller. The next problem to solve is the process efficiency and repeatability in electrospinning. Applications of ENMs for different systems are summarized in Table 12.4.

TABLE 12.4
Examples of ENMs Application

Polymer/Inorganic Material	Process	Area of Application	References
Polyvinylidene (PVDF)	Filtration	Pretreatment of water prior RO, NF, and UF	[73]
Polysulfone	Filtration	Prefilters for particulate removal	[56]
Carbonized nanofibrous (precursor PAN)	Filtration	Removal of disinfection by-products from water	[76]
Nylon 6	Water treatment	Filtration (to remove micron/micron particles)	[59]
Polyethersulfone	Prefiltration	Water and other liquid separation	[60]
Fluorinated copolyimide	Filtration	Water treatment	[61]
Cellulose/polyacrylonitile (PAN)	Filtration	Bacteria and virus removal	[62]
Poly(vinyl alcohol) (PVA) cross-linked with glutaraldehyde	Filtration	Oil/water emulsion filtration	[63]
PAN/chitosan	Ultrafiltration and nanofiltration	Water filtration	[64]
PVA/PVA hydrogel	Ultrafiltration	Oil/water emulsion	[65]
UV cured PVA	Ultrafiltration	Oil/water emulsion	[66]
Polyamide/PAN	Nanofiltration	Water treatment	[67]
Polyethersulfone	Filtration	Water treatment	[68]
PVA/PAN	Ultrafiltration	Oil/water emulsion	[69]
Interfacial polymerization of piperazine (PIP)	Ultrafiltration	Water treatment	[70]

(Continued)

TABLE 12.4 (*Continued*)
Examples of ENMs Application

Polymer/Inorganic Material	Process	Area of Application	References
Polysaccharide (cellulose and chitin)	Ultrafiltration	Water purification and oil/water emulsion	[71]
PVDF-*co*-F6PP, homopolymer PVDF	DCMD	Water treatment, ultrapure water, etc.	[82]
PVDF and PVDF-HFP	DCMD	Water treatment, etc.	[84]
PVDF	AGMD	Desalination, etc.	[86,87]
PVDF/clay	MD	Desalination	[87]
PVDF/TMOS	MD	Separation	[88]
PVDF	VMD	Ethanol/water	[85]
PVDF	GSMD	VOCs removal from water	[89]
PAN	Nanofiltration	Salt removal	[72]
Polystyrene/beta-cyclodextrin	Nanofiltration	Removal of organic compounds from water	[74]
CA	Filtration	Filtration of water	[58]
Polyamide	Filtration	Water treatment	[78]
PVDF	Particle removal	Water treatment	[73]
Oxidized cellulose/GA, polyacrylic acid	Filtration	UF, NF, and RO	[79]
Carbon	Filtration	Water treatment	[80]
Functionalized carbonaceous nanofiber membrane	Filtration	Chiral separation and drug delivery	[80]
CA (molecularly imprinted with glutamic acid)	Adsorption	Chiral separation	[75]
PVO/TiO$_2$	Adsorption	Removal of As(III) from water	[97]
Chitosan	Adsorption	Removal of Cu(II) and Pb(II) from water	[94]
Cellulose acetate	Heavy metal ion adsorption	Water treatment	[77]
Nanofiber/TiO$_2$	Catalyst	Oxidation of benzene to phenol	[98]
TiO$_2$	Photocatalytic	Environmental purification	[100]
TiO$_2$, ZrO$_2$/Pt, Pd, Rh	Catalytic system	Cross-coupling reactions in a continuous flow reactor	[99]
Nafion (persulfuric acid)	Proton conductivity	Fuel cell	[90]
Organic/inorganic materials	Energy device	Solar cell, etc.	[92]
Polyvinyl chloride	Adsorption	Removal of Cu, Cd, and Pb ions from groundwater	[93]
Chitosan	Adsorption	Removal of Cu, Pb ions from water, etc.	[94,95]
Functionalized CA	Adsorption	Metallic ion adsorption	[96]
PVP-TiO$_2$	Adsorption	As(III) from water, etc.	[97]
PET/RhB	HPLC	Analysis	[101]
Chitin	Separation	Chiral separation (isomer)	[102]
Polysulfone	Biotreatment	Sewage, etc.	[103]

12.5   GRAPHENE MEMBRANE

Graphene is one of the most exciting materials being investigated today, not only out of academic curiosity but also for its potential applications. It is a substance composed of pure carbon, with atoms arranged in a regular hexagonal pattern similar to graphite, but in a one atom thick sheet. It is very light, with a square meter sheet weighing only 0.77 mg. It is an allotrope of carbon whose structure is a single planar sheet of sp^2-bonded carbon atoms, which are densely packed in a honeycomb crystal lattice [110]. Geim and Novoselov received the Nobel Prize for their groundbreaking experiments regarding the two-dimensional (2D) material graphene [111]. Graphene is most easily visualized as an atomic-scale chicken wire (Figure 12.10) made of carbon atoms and their bonds. In other words, graphene, a chemically stable and an electrically conductive layer of one atom in thickness, is a single layer of graphite, which is at the ultimate limit [112]. The crystalline or *flake form of graphite* consists of many graphene sheets stacked together. The carbon–carbon bond length in graphene is about 0.142 nm. Graphene sheets stack to form graphite with an interplanar spacing of 0.335 nm. Graphene represents the first truly 2D atomic crystal. It has a unique atomic structure that gives it remarkable mechanical and thermal properties. Graphene is the basic structural element of some carbon allotropes, including graphite, charcoal, CNTs, and fullerene. It can also be considered as an indefinitely large aromatic molecule, the limiting case of the family of flat polycyclic aromatic hydrocarbons.

12.5.1   GRAPHENE PREPARATION

Graphene sheets can be prepared by different methods: (1) mechanical cleavage of graphite using adhesive tapes, (2) chemical reduction of graphite oxide, (3) epitaxial growth (EG) of graphene by thermal graphitization of SiC, (4) CVD of hydrocarbon gases on transition metals, (5) exfoliation of graphite by sonication, (6) quenching,

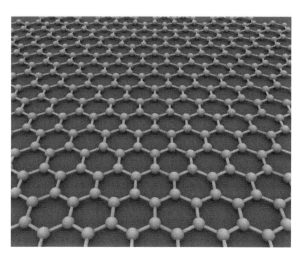

FIGURE 12.10   Graphene is an atomic-scale honeycomb lattice made of carbon atoms.

and (7) wet ball milling. An alternative method using SiC as carbon source for graphene sheet preparation has also been demonstrated. In this technique, the catalyst for SiC graphitization was an Ni film deposited on an SiC single-crystal substrate or SiC amorphous thin film on an SiO_2 substrate [113]. As a robust yet flexible membrane, graphene provides infinite possibilities for the modification or functionalization of its carbon backbone [114]. At present, the conventional method for preparing high-quality graphene film is CVD. However, a very promising route for the bulk production of graphene sheets can be chemical reduction and dispersion of graphene in aqueous solutions [115].

Graphene oxide (GO) is a new material that outperforms many other paper-like materials in stiffness and strength. GO paper is a free-standing carbon-based membrane material made by flow-directed assembly of individual GO sheets. Its combination of macroscopic flexibility and stiffness is a result of a unique interlocking-tile arrangement of the nanoscale GO sheets. GO is a layered material consisting of hydrophilic oxygenated graphene sheets (GO sheets) bearing oxygen functional groups on their basal planes and edges. GO-based thin films have been fabricated through solvent-casting methods [116].

12.5.2 APPLICATION AND PROPERTIES OF GRAPHENE

Graphene has displayed a variety of intriguing properties including high electron mobility at room temperature (250,000 cm^2 Vs^{-1}), exceptional thermal conductivity (5000 W m^{-1} K^{-1}), and superior mechanical properties with a Young's modulus of 1 TPa (terapascal, 1 terapascal = 10^{12} pascals) [116]. Thus, graphene sheets offer extraordinary electronic, thermal, and mechanical properties and are expected to find a variety of applications, such as in sensors, batteries, supercapacitors, hydrogen storage systems, and as reinforcement fillers of nanocomposites [118, 117]. Various polymers and nanoparticles, including metal, metal oxide, and semiconductor composites, have been developed based on the unique properties of graphene. It possesses similar mechanical properties as CNTs but has superior electrical and thermal properties and a larger surface area (2620 $m^2 g^{-1}$) because of its 2D crystal structure.

Ortolani et al. [119] used high-resolution transmission electron microscopy successfully to provide a complete nanoscale geometrical and physical picture of the 3D structure of various wrinkle and fold configurations of monolayer graphene. O'Hern et al. [120] fabricated graphene composite membranes with nominal areas more than 25 mm^2 through transferring a single layer of CVD graphene onto a porous polycarbonate substrate. A combination of pressure-driven and diffusive-transport measurements provided evidence of size-selective transport of molecules through the membrane, which was attributed to the low-frequency occurrence of intrinsic 1–15 nm diameter pores in the CVD graphene. O'Hern et al.'s results presented the first step toward the realization of practical membranes that use graphene as the selective material. It was claimed that in future these membranes could be used to filter tiny harmful compounds from water or in controlled drug delivery. All this will be possible only if researchers find a way to engineer graphene with holes of the exact size needed. Hauser and Schwerdtfeger suggested that there are a lot of chemical methods that can be used to modify graphene pores, so it is a platform technology for a new class of membranes [121]. Further promising applications include membranes that filter microscopic contaminants from water, or that separate specific types of molecules from biological samples. Karnik et al. [122] observed the actual holes in the graphene membrane, characterizing the material with a high-powered electron microscope. They found that the pores ranged in size from about 1 to 12 nm.

Ruiz-Vargas et al. [123] studied the image grain boundaries and ripples in graphene membranes obtained by CVD, using atomic force microscopy (AFM). Nanoindentation measurements revealed that out-of-plane ripples effectively soften the in-plane stiffness of graphene. Moreover, grain boundaries significantly decrease the breaking strength of membranes. It was revealed that two graphene membranes brought together formed membranes with higher resistance to breaking. On measuring the adhesion energy of graphene sheets with a silicon oxide substrate through a pressurized blister test, an adhesion energy of 0.45 ± 0.02 J m^{-2} for monolayer graphene and 0.31 ± 0.03 J m^{-2} for samples containing two to five graphene sheets are reported [124]. These values are larger than the adhesion energies measured in typical micromechanical structures and are comparable to solid–liquid adhesion energies. This is due to the extreme flexibility of graphene, which allows it to conform to the topography of even the smoothest substrate, thus making its interaction with the substrate more liquid-like than solid-like.

Bunch et al. [125] demonstrated that a monolayer graphene membrane was impermeable to standard gases, including helium, and by applying a pressure difference across the membrane measured both the elastic constants and the mass of a single layer of graphene. The pressurized graphene membrane was suggested to be the world's thinnest balloon and provided a unique separation barrier between two distinct regions that was only one atom thick.

Porous graphene membranes have been suggested for the separation of hydrogen from methane, separation of helium from other noble gases and methane, selective passage of ions, characterization of DNA, filtration of water, and the separation of nitrogen from hydrogen. It was also reported that graphene pores are capable of separating fermionic He from bosonic He [126]. Surface adsorption effects have a significant influence on the gas permeability, especially at low temperature in graphene membranes [127]. Hauser and Schwerdtfeger [126] discussed that nanoporous graphene membranes can be CH_4-selective for gas purification by adjusting pore sizes in the graphene membrane.

Membranes act as selective barriers and play an important role in processes such as cellular compartmentalization and industrial-scale chemical and gas purification. The ideal membrane should be as thin as possible to maximize flux, mechanically robust to prevent fracture, and have well-defined pore sizes to increase selectivity. Graphene is an excellent starting point for developing size-selective membranes because of its atomic thickness, high mechanical strength, relative inertness, and impermeability to all standard gases. However, pores that can exclude larger molecules but allow smaller molecules to pass through would have to be introduced into the material [128]. Koeing et al. [128] showed that ultraviolet-induced oxidative etching can create pores in micrometer-sized graphene membranes, and the resulting membranes can be used as molecular sieves. A pressurized blister test and mechanical resonance were used to measure the transport of a range of gases (H_2, CO_2,

Ar, N_2, CH_4, and SF_6) through the pores. The experimentally measured leak rate, separation factors, and Raman spectrum agree well with models based on effusion through a small number of ångstrom-sized pores.

Du et al. [127] found that the porous graphene membranes could be used to separate hydrogen and nitrogen molecules. In order to gain a series of pore sizes, they drilled carbon atoms from the graphene. The selectivity and permeability could be controlled by creating nanopores with different shapes and sizes. The mechanisms of hydrogen and nitrogen permeation through the porous graphene are different. The selectivity of the porous graphene largely results from the different permeation mechanism of these molecules.

Schrier [129] also demonstrated that graphene could be made selectively permeable by the introduction of pores. Schrier proposed an economical means of separating He from the other noble gases and alkanes present in natural gases using tailored graphene. It was also demonstrated by Schrier and McLain [130] that isotope separation can be done using graphene membranes.

Permeation through nanometer pore materials has been attracting unwavering interest due to fundamental differences in governing mechanisms at macroscopic and molecular scales, the importance for filtration and separation techniques, and the crucial role played by selective molecular transport through cellular membranes. Nair et al. [131] reported that submicron-thick membranes made from GO are completely impermeable to liquids, vapors, and gases but allow unimpeded permeation of water; H_2O permeates through the membranes at least 10^{10} times faster than He. Nair et al. suggested that there was a nearly frictionless flow of a monolayer of water through the 2D capillaries formed by closely spaced graphene sheets. Diffusion of other molecules is blocked by the water that clogs the capillaries and by their reversible narrowing in low humidity. Despite being only one atom thick, graphene is believed to be impermeable to all gases and liquids. This makes it tempting to exploit this material as a barrier film. Because of the ways graphene can currently be mass produced, films made from GO present a particularly interesting candidate. Using this graphene derivative, it is possible to make laminates, which are a collection of micron-sized graphene crystals forming an interlocked layered structure.

At present, RO accounts for nearly half of the world's installed desalination capacity, and works by driving salt (feed) water across a semipermeable membrane under high pressure. The membrane allows water molecule to pass through, but not salt ions. Though this technique is energy efficient, it struggles with membrane fouling and slow water transport. Cohen-Tanugi and Grossman [132] calculated that graphene with subnanometer pores can surpass the water permeability of current polymeric RO membrane by 2–3 orders of magnitude, making it a high performance membrane for water desalination. Wang and Karnik [133] discussed the *Cohen-Tanugi and Grossman* suggestion and concluded that "Graphene promises water desalination at throughputs much higher than state-of-the art membranes" (p.552). It was suggested also that chemical functionalization of the pores of graphene with hydrogen will increase water selectivity, whereas functionalization with hydroxyl group will increase the speed of water transport. Due to the subnanometer pores in the graphene membrane, it is a promising RO membrane. Suk and Aluru [134] indicated that the graphene membrane can be used as an ultra-efficient water transporter, compared to thin CNT membranes, whenever the diameter is larger than 0.8 nm.

The results with graphene membranes for water processing are promising, but it still needs more work. More work is needed to achieve monodispersed pore sizes and the desired chemical functionalization of graphene membranes. Perhaps the greatest difficulty will be the scalable manufacturing of the large graphene membranes with subnanometer pores with a narrow size distribution, while maintaining the structural integrity of the graphene and keeping costs low. It is also necessary for graphene membranes to be robust over time, and avoid excessive fouling under operating conditions. Also, membrane modules must be designed to rapidly remove the salt on the feed side at high flow rates.

12.5.3   Summary of Graphene Membrane Application

Graphene is an ideal material for next generation membranes due to its atom scale thickness, remarkable mechanical strength, and potential for size selective transport through nanometer-scale holes in its lattice. Arguably, in the near future, graphene membranes will take the place of polymeric membranes for filtration (RO, UF, NF, and MD) and gas separation. Although there are some problems for desalination using graphene membranes, they will make clean water more accessible around the world due to their novel and tunable transport properties.

12.6   MOF MEMBRANES

At present, the membrane market for gas separation is dominated by polymeric membranes due to their low cost of production and because they exhibit high gas fluxes and mechanical flexibility. However, in general, polymer membranes have short lifetimes, low thermal and chemical stabilities, and low selectivities. Robeson [135] defined an upper boundary for polymer membranes delineating a limit on their selectivity/permeability performance. Although this limit has been adjusted since its inception, it still indicates that there is a limiting trade-off between the selectivity and permeability of a polymer membrane. In addition, some of these challenges make polymer membranes generally limited to the separation of noncondensable gases. Condensable gas separation such as olefin/paraffin and butane isomer separations is an important area where membrane technology can expand [136]. Microporous membranes with pore apertures below the nanolevel can exhibit size selectivity by serving as a molecular sieve, which is promising for overcoming Robeson's *upper boundary* limits in membrane-based gas separation. Zeolites, PIMs, metal oxides, and active carbon [137] are the typical materials used. MOFs or 3D porous coordination polymers (PCPs) have attracted tremendous attention over the past few years. MOFs are a relatively new class of hybrid materials. As hybrid organic–inorganic material, MOFs consist of metal cations or cationic oxide clusters that are linked by organic molecules, thus forming a crystalline network. MOFs are often also termed *coordination polymers*. MOFs have the potential to answer some of the challenges facing researchers producing materials for gas-separation membranes. They offer a large variety in structure, pore size, and functionality, together with very large surface area. Therefore, numerous applications are predictable, particularly in gas separation and gas storage [138]. Due to their unique properties, MOFs are also excellent candidates for gas-separation membranes [139]. Thus, by chemical functionalization of the organic linkers in the structures, MOFs afford easy control over pore size and chemical/physical properties making MOFs attractive for membrane-based gas separation.

12.6.1 MOF MEMBRANES PREPARATION

There is a similarity between zeolites and MOF membrane preparation. For the synthesis of an MOF membrane, all the tools such as seeding, microwave heating, ceramic porous supports, and intergrowth-supporting additives, which are used for zeolite membranes, can be applied. Thus, the synthesis of MOFs can be achieved in two ways:

1. Hydrothermal methods
2. Solvothermal methods

Fabrication of thin films of crystalline framework materials follows one of the following two approaches: *in situ* growth and *secondary or seeded* growth. MOFs are normally assembled by bridging organic ligands that stay intact during the synthesis. The templating is usually achieved using organic anions.

For the fabrication of well-defined and highly porous coatings a number of new or modified methods for the deposition of MOF thin films have been developed. Different routes are used for the growth of MOF thin films on solid substrate. There are many other techniques, including chemical modification of the support surfaces with self-assembled monolayers (SAM) [140,141].

12.6.1.1 The Direct Growth/Deposition of MOF Thin Films on Solid Substrate

The growth of MOF thin films on solid substrate is based on the conventional MOF synthesis scheme using solvothermal methods. For example, in the case of MOF-5, the conventional scheme is as follows. The reactants, $Zn(NO_3)_2.4H_2O$, and terephthalic acid, are mixed together and the mixture is kept at elevated temperatures for an extended period (105°C, 1 atm). After a period of three days, the reaction product consists of particles. When aiming at a rigid deposition of the MOF particles on a substrate, the most naïve approach would be to simply immerse a substrate during MOF formation. Microwave heating is also used for the preparation of MOF thin films. Microwave-induced thermal deposition is a novel method for easy preparation of MOF thin films on porous substrates. The first MOF membranes using microwave-induced rapid seeding were reported in 2009 by the Lai and Jeong groups [142,143]. Yoo et al. [142] used this method for the preparation of multilayer film (MOF-5/zeolite silicate-1 film). Substrates, nanoporous anodized alumina discs, coated with various conductive thin films were placed vertically in vials containing MOF-5 precursor solution. MOF-5 crystals were grown under microwave irradiation. The fundamental basis of this method is to create the high temperatures needed for the MOF formation only in the vicinity of the surface. As a result the reaction does not proceed within the solution but only in a thin layer at the interface to the substrate heated by the microwave radiation. The most important instrumental parameter in microwave heating is the power of irradiation and the question of how it is created, delivered, and controlled by the reaction. This depends mainly on the device obtained. In general, reactions occur between a few seconds and some minutes, with times rarely exceeding 1 h. At present, microwave heating seems the most promising way to achieve short reaction times and small crystallite sizes, both crucial for applications in devices [144].

12.6.1.2 Colloidal Deposition Method for MOF Membranes

Horcajada et al. [145] used this method for the fabrication of porous iron carboxylate MOF MIL-89. For this particular MOF, the iron(III) acetate, transmuconic acid, sodium hydroxide, deionized water, and methanol were first mixed to yield an orange gel. The gel was aged at 100°C for 3 days in a Teflon-lined Parr bomb. This process yielded monolithic gels and xerogels consisting of MIL-89 crystallites embedded in ethanol. Thin layers of this material can be deposited on solid substrate by a simple dipping procedure. When removing the substrate from the gel, a film was left on the surface. After drying, a film of MOF crystallites remained on the surface.

12.6.1.3 LBL or Liquid Phase Epitaxy of Surface Grown, Crystalline MOF Multilayers

Shekhah et al. [146] used LBL or liquid phase epitaxy to fabricate porous MOF thin films. The two components, copper (II) acetate (Cu[ac]$_2$) and 1,3,5-benzenetricarboxylic acid (H$_3$btc), were dissolved in ethanol and the substrate was immersed into each solution in a cyclic way. Each immersion was followed by rinsing with ethanol. By starting with Cu(ac)$_2$, a linear increase of thickness of the deposited HKUST-1 layer with the number of (alternating) immersion cycles in Cu(ac)$_2$ and H$_3$btc was observed. The thickness of deposited film was monitored by AFM, IR spectroscopy and X-ray–photoelectron spectroscopy.

12.6.1.4 Electrochemical Synthesis of MOF Membranes

This technique was used by researchers at BASF [147,148] for the preparation of thin films of HKUST-1 on copper substrate. For the electrochemical synthesis of powdered MOF-5, a copper electrode was immersed into a solution of the organic MOF building block, benzenedicarboxylic acid (bdc) in the case of MOF-5. By applying an appropriately biased electrical voltage, the electrochemical oxidation of Cu atoms led to the dissolution of Cu^{2+} metal ions and subsequently to the formation of crystallites in the vicinity of the electrode surface. The continuous supply of more Cu^{2+} leads to a continuous growth of the crystallites.

12.6.1.5   Gel-Layer Synthesis of MOF Membranes

Schoedel et al. [149] used this technique to synthesize oriented MOF thin films on modified Au-substrate. Modification of Au substrate was achieved by growing a COOH- or OH-terminated alkanethiolate-based SAM. Subsequently, the substrate was coated with a layer of a PEO gel, which served as a storage medium for the metal-containing reactants. Formation of the MOF thin films was then induced by pouring solutions of H_3btc (benzene-tricarboxylic acid) or NH_2bdc (amino-benzenedicarboxylic acid) on the top of gel layer. The thickness of the MOF thin films deposited using the gel layer method could be varied by adjusting the concentration of the metal-containing reactant within the gel.

12.6.1.6   Evaporation-Induced Crystallization of MOF Membranes

Ameloot et al. [150] used this method for producing thin layers of MOF crystallites on a variety of different substrates. This technique is based on first forming a clear precursor solution (mother liquor), which does not contain crystallites, and using the appropriate solvents that stabilize the MOF constituents. As a result, the formation of small nuclei is avoided. Slow evaporation of solvents leads to crystallization of small MOF crystallites.

12.6.1.7   Other Methods for Fabrication of MOF Membranes

Lu and Zhu [151] developed a method based on a liquid–liquid interfacial coordination mechanism for the synthesis of free-standing MOF membranes. MOF precursors, zinc nitrate [$(Zn(NO_3)_2$] and terephthalic acid (TPA or H_2bdc [benzene dicarboxylic acid]), as well as catalyst triethylamine (TEA), were dissolved in two immiscible solvents, namely, dimethylformamide (DMF) and hexane. A region of reactant concentrations critical to membrane formation was identified; a free-standing membrane could only be found in the region of high Zn/TPA and low TEA concentration. SEM results revealed that the top layer was particulate, whereas the bottom layer had a sheet-like morphology, which was further identified by XRD data as 3D $Zn_4O(bdc)_3$ (also known as MOF-5) and 2D Znbdc DMF (MOF-2) for the top and bottom, respectively. MOF-2 was the dominant material in the overall composition. Nitrogen adsorption tests showed an average Langmuir surface area of 709 $m^2\,g^{-1}$ for the membrane, which demonstrated its potential for gas separation applications.

Ben et al. reported a convenient method for the synthesis of polymer-supported and large-scale free-standing MOF membranes [152]. First, a poly(methylmethacrylate) (PMMA) was spin-coated on a template substrate, which could be any solid surface of metal or plastic. Then the PMMA surface was hydrolyzed by concentrated sulfuric acid and converted into PMAA. After that, the PMMA–PMAA coated substrate was immersed into MOF precursor solution in an autoclave for a suitable reaction time. For the further preparation of a free-standing MOF membrane, the as-synthesized MOF membrane can be separated from the substrate by dissolving the PMMA–PMAA in chloroform. Thus, an intact free-standing MOF membrane with manifold size and shape, and a thickness from hundreds of nanometers to hundreds of micrometer was obtained. Ben et al. claimed that this MOF membrane fabrication method is simple and convenient and can be readily applied to a variety of other material compositions to produce functional membranes with diverse micropore structures. Thus, MOF membrane technology is opening a lot of opportunities for the development of new functional nanodevices.

However, there are some challenges associated with fabricating films of MOF materials, including poor substrate–film interactions, moisture sensitivity, and thermal/mechanical instability, as even nanometer-scale cracks and defects can affect the performance of a membrane for gas separation.

12.6.2 CHARACTERIZATION AND PROPERTIES OF MOFs

MOFs are hybrid organic–inorganic nanoporous materials that exhibit regular crystalline lattices with well-defined pore structures. Chemical functionalization of the organic linkers in the structured MOFs affords facile control over pore size and chemical/physical properties, making MOFs attractive for membrane-based gas separation. However, there are some challenges associated with fabricating films of MOF materials, including poor substrate–film interactions, moisture sensitivity, and thermal/mechanical instability, as even nanometer-scale cracks and defects can affect the performance of a membrane for gas separation. The adsorption affinities of MOFs can be tailored by introducing functional groups, which strongly interact with specific molecular species [153,154]. Li et al. [153] wrote a review on the selective gas adsorption and separation of MOFs and noted that the influence of the mixed gas adsorption selectivity on the total membrane selectivity can be roughly estimated by the simple relationship *membrane selectivity~adsorption selectivity × diffusion selectivity* [154]. Furthermore, gas separations in MOFs, including the molecular sieving effect, kinetic separation, the quantum sieving effect for H_2/D_2 separation, and MOF-based membranes were also summarized. Silva et al. [155] demonstrated the characterization of MFI zeolite membranes (crystallizing a zeolite layer on the internal face of macroporous tubular support of α-Al_2O_3) by means of permeability determination of near critical and supercritical CO_2. They developed a new experimental system using a transient methodology to evaluate the permeance (and potential selectivity) of dense gases, liquids, and supercritical fluids through microporous membranes. From the experiments, flux and permeance of CO_2 at high pressure conditions were calculated and the obtained values were coherent with the mass transport properties through micropores. However, the experimental permeability cannot be described by classical mechanisms; Knudsen likes diffusion or viscous flow.

12.6.3 APPLICATION OF MOF MEMBRANES

MOFs are an emerging class of nanoporous materials comprising of metal centers connected by various organic linkers to create one-dimensional, 2D, and 3D porous structures with tunable pore volumes, surface areas, and chemical properties. Due to their porosity, large inner surface area, tunable pore sizes, and topologies, MOF membranes lead to versatile architectures and promising applications. In separation processes, MOF membranes can be applied in ion exchange, gas adsorption, and storage (in particular, hydrogen and methane gas), liquid separation processes, drug delivery, sensor technology, heterogeneous catalysis, hosts for metal colloids or nanoparticles or polymerization reactions, pollutant sequestration, microelectronics, luminescence, nonlinear optics, and magnetism [156–158].

Several attempts have prepared gas sieving MOF membranes by growing MOFs as polycrystalline films on top of macroporous supports. These attempts include protopical, large-pored MOFs such as HKUST-1, MOF-5, and Mil-53, as well as membranes of the small-pored ZIF family (zeolitic imidazolate framework) [159]. Large-pored MOFs exhibit pore sizes that are much larger than the kinetic diameter of common light gas molecules (e.g., H_2, N_2, CH_4, and CO_2). Thus, a sharp molecular sieve separation by steric or size exclusion was not observed, whereas ZIF membrane led to successful separation of H_2, for example. Thus, it is difficult to obtain an efficient gas separation using large-pored MOFs. It was demonstrated by Banerjee et al. [160] that by functionalizing isoreticular imidazolate frameworks, pore sizes can be controlled for favorable gas separation. Further, MOFs with polar functionalities have the potential to emerge as important materials for natural gas purification and landfill gas separation.

Li et al. [161] fabricated an ultramicroporous zeolite imidazolate framework, ZIF-7 membrane, by synthesizing it on a porous alumina support using a microwave-assisted secondary growth technique. The solution used for seeding was prepaid by dispersing ZIF-7 nanoseeds into polyethyleneimine (PEI) solution. As its pore dimensions approached the size of H_2, a high H_2 selectivity was obtained without any sophisticated pore size engineering to target H_2/CO_2 separation. The membrane was thermally stable for use at elevated temperature (ZIF-7 is stable at least up to 500°C). At 165°C the activated ZIF-7 membrane showed a H_2 permeance of about 7×10^{-9} mol m^{-2} s^{-1} Pa^{-1} and a H_2/N_2 separation factor of 8.

Ranjan and Tsapatsis [162] prepared preferentially oriented and well-intergrown films of microporous MOF on porous alpha-alumina discs through a seeded growth technique. PEI polymer was used with the expectation of enhanced attachment of seeds through H-bonding. A solvent mixture consisting of methanol and water (1:5) was used for solvothermal growth. $Cu(NO_3)_2.3H_2O$ (0.072 g) and 4,4′-(hexafluoroisopropylidine)-bis(benzoic acid) (0.366 g, excess) was added to the methanol–water solvent mixture (15 mL), and solvothermal growth was carried out at 150°C for 12 h. The crystals obtained after solvothermal growth were crushed into submicrometer-sized crystals. Seed layers of these particles were deposited manually and by secondary growth on PEI coated support. The orientation was such that the pores were preferentially aligned along the membrane thickness and the membrane exhibited ideal selectivity for H_2/N_2 as high as 23.

Cao et al. [163] for the first time introduced a continuous and intergrown Cu-BTC membrane prepared on novel potassium hexatitanate support through *in situ* growth. This kind of support can adsorb Cu^{2+} in a moderate acid environment and is more suitable for the growth of Cu^{2+}-containing MOF membranes than other traditional supports, such as porous alumina support. Compared to other MOF membranes, the Cu-BTC membrane showed a higher selectivity for He over CO_2, N_2, and CH_4. Thus, this kind of membrane may have potential applications for helium recovery.

Keskin and Sholl [164] presented the first results by examining the ability of MOFs to act as gas separation membranes. The calculations were based on atomistic simulations using rigid MOFs. Characterizing the properties of these membranes using mixed gas feeds rather than single-component gases is crucial. The results for MOF-5 for single component predicted that MOF-5 would show strong ideal selectivities for CH_4 in CO_2/CH_4 mixtures.

Huang's group (Ningbo Institute of Materials and Technology & Engineering, Chinese Academy of Sciences) [165] developed a hydrogen-selective ZIF-90 membrane as a high gas-separation membrane. Through the condensation reaction between the free aldehyde of ZIF-90 and the amino group of organosilica, both narrowing of pores and the sealing of intercrystalline defects of the polycrystalline ZIF-90 layer took place, and thus the gas-separation selectivity was enhanced. For binary mixtures at 225°C and 1 bar, the mixture separation factors of H_2/CO_2, H_2/CH_4, H_2/C_2H_6, and H_2/C_3H_8 were found to be 20.1, 70.5, 250, and 458, respectively, and a relatively high H_2 permeance of about 2.85×10^{-7} mol.m^{-2}s^{-1}Pa^{-1} could be obtained due to the avoidance of pore blocking.

Bétard et al. [159] fabricated an MOF membrane by stepwise deposition of reactants. Two-pillared layered MOFs with the general formula $[Cu_2L_2P]_n$ (L = dicarboxylate linker, P = pillaring ligand) were selected. For this demonstration, they selected the nonpolar $[Cu_2(ndc)_2(dabco)]_n$ (**1**: ndc = 1,4-naphthalene dicarboxylate; dabco = 1,4-diazabicyclo(2.2.2)octane and the polar $[Cu_2(BME\text{-}bdc)_2(dabco)]_n$) (**2**: BME-bdc = 2,5-bis(2-methoxyethoxy)-1,4-benzene dicarboxylate). The framework structure of **1** is shown in Figure 12.11.

Structure **2** is quite similar to **1** by just replacing ndc by BME-bdc (see Figure 12.12).

Performances of both membranes were evaluated in gas-separation experiments of CO_2/CH_4 (50/50) mixtures using a modified Wicke–Kallenbach technique. The

FIGURE 12.11   Structure of $[Cu_{2L}L_2P]_n$ MOFs (here $\{Cu_2(ndc)_2(dabco)]_n$) (**1**) with linker L = ndc(1,4-naphtalenedicarboxylate) and ligand P = dabco(1,4-diazabicyclo(2.2.2)octane seen in [1 0 0] direction.

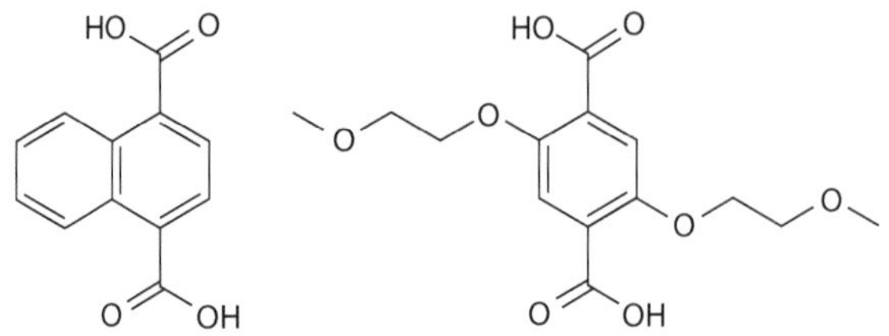

FIGURE 12.12 Dicarboxylic acid linkers used in this work; left, H_2ndc (ndc: 1,4-naphtalene dicarboxylate) and right, H_2BME-bdc (BME-bdc: 2,5-bis(2-methoxyethoxy)-1,4-benzene dicarboxylate).

separation-active MOF layer was located inside the macroporous support at a depth of size in the μm scale range. The microstructures of the MOF-based membranes resemble foam with the intergrown lamellae as transport-selective membrane. Proof of principle was given by the fact that the functionalization of the linker can induce CO_2 membrane selectivity; CO_2/CH_4 mixtures were separated with an anti-Knudsen separation factor of 4–5 in favor of CO_2.

High-quality, thin (14 μm) MOF-5 membrane was prepared by Zhao et al. [166] through a secondary growth method. The MOF-5 membrane was composed of one to two layers of MOF-5 crystals of about 5–20 μm in size. The MOF-5 membranes were permselective for CO_2 over N_2 or H_2 under experimental conditions. The MOF-5 membranes showed a separation factor for CO_2/H_2 of close to 5 with a feed CO_2 composition of 82% and a separation factor for CO_2/N_2 greater than 60 with a feed CO_2 composition of 88% at 445 kPa and 298 K.

For the first time, a homochiral MOF membrane was reported by Wang et al. [167] for the enantioselective separation of chiral compounds, especially chiral drug intermediates. In this study, a homochiral MOF material [Zn_2(bdc)(L-lac)(dmf)] (DMF) (ZnBLD) was used for the preparation of an MOF separation membrane through the solvothermal reaction of a metal cation, a chiral ligand, and an organic connector. Zn-BLD exhibits preferential adsorption ability to (S)-methyl phenyl sulfoxide (S-MPS) over R-MPS. They claimed that this membrane will allow potential development of a new, sustainable, and highly efficient chiral separation technique.

Nan et al. [168] used seeding method (step by step) to prepare HKUST-1 (known as Cu_3(btc)$_2$) on porous α-alumina disc. The H_2 selectivity with respect to CH_4, N_2, and CO_2 was studied. For the single-component permeation, the ideal selectivities of the membrane for H_2/CH_4, H_2/N_2, and H_2/CO_2 were 2.9, 3.7, and 5.1, respectively. On the other hand, binary separation factors of HKUST-1 membrane for H_2/CH_4, H_2/N_2, and H_2/CO_2 were 3, 3.7, and 4.7, respectively. Jeong's group's [169] reported ideal selectivities for H_2/CH_4, H_2/N_2, and H_2/CO_2 at room temperature were about 2.4, 3.7, and 3.5, respectively, for the HKUST-1 membrane fabricated by the secondary growth method. HKUST-1 crystal has pore sizes around 9Å, which is much larger than the kinetic diameters of the studied gases. It was concluded that HKUST-1 membrane has no obvious advantages for H_2 separation, and its potential application could lie in the separation of larger substances, such as protein and aromatic hydrocarbons.

MMMs consist of an inorganic or inorganic–organic hybrid material in the form of micro or nanoparticles (discrete phase) incorporated into a polymeric matrix. Use of two materials with different flux and selectivity provides the possibility to design a better gas separation membrane, allowing the synergistic combination of polymers, with easy processability and the superior gas-separation performance of inorganic materials.

For the future, a new class of membrane called MMM with MOFs, which combine the advantages of polymers and MOFs [170], is promising. In general, MMMs with zeolites face difficulties including expensive synthesis of defect-free crystals with long preparation time and a limited range of zeolite structures with a discontinuous pore size, little chemical tailorability, and poor zeolite interface. Synthesis of MOFs with various physical/chemical properties is relatively easier than that of zeolites, and the MOF/polymer interface can be controlled by varying the affinity of organic linkers to the polymer matrix. Also, the surface of MOFs can be modified by functionalization for favorable interaction with the polymer. MOFs, in general, offer greater pore volumes and weigh less than zeolites. Thus, for a given mass loading, an MMM with MOF will affect the membrane behavior significantly greater than an MMM with a zeolite. A significant number of MMMs with MOFs are reported in the literature [136]. Adams et al. [171] synthesized an MOF of copper and terephthalic acid (CuTPA), and incorporated it into a PVAc [poly(vinyl acetate)] matrix to form an MMM membrane with MOFs. Helium gas transport properties of these CuTPA MMMs showed improvements over the pure polymer gas transport properties.

Rosi's group introduced metal-adeninate biometal-organic frameworks (Bio-MOFs), which are appealing candidates for molecular gas applications because of their permanent microporosity, high surface areas, chemical stability, and exceptional CO_2 adsorption capacities, due to the presence of basic biomolecule building units [172]. In particular, $Zn_8(ad)_4(BPDC)_6O.2Me_2NH_2$ (where ad = adeninate, BPDC = biphenyldicarboxylate) denoted as Bio-MOF-1, a 3D porous MOF with infinite-adeninate columnar building units, which are incorporated through biphenyldicarboxylate linkers, is an attractive material with great potential for CO_2 separation.

Bohrman and Carreon [173] demonstrated the preparation of continuous and reproducible BIO-MOF-1 $\{Zn_8(ad)_4(BPDC)_6 \cdot 2Me_2NH_2, 8DMF, 11H_2O\}$, which was a rigid, permanently porous (~around 700 m^2 g^{-1} surface area) [174] membrane supported on porous stainless steel tubes. Membranes were prepared through secondary seeded growth. These membranes displayed high CO_2 permeances and separation ability for CO_2/CH_4 gas mixtures. The observed CO_2/CH_4 selectivities were above the Knudsen selectivity, indicating that the separation was promoted by preferential CO_2 adsorption over CH_4. This preferential adsorption was attributed to the presence of adeninate amino basic sites present in the Bio-MOF-1 structure. These results indicate the feasibility of the development of a novel type of membrane that could be promising for diverse molecular gas separations. Nik et al. [175] synthesized glassy PI, 6FDA-ODA that was mixed with several as-synthesized MOFs fillers at 25 wt% content to fabricate CO_2/CH_4 gas separation MMMs. The experimental results showed that the presence of -NH_2 functional groups in MOF structure could lead to creating rigidified polymer at the interface of the filler and polymer matrix, and therefore decrease the permeability, while increasing the selectivity.

Influence of three different MOFs in MMMs for binary gas mixture separation was studied by Basu et al. [176]. Both dense and asymmetric MOFs-containing membranes were prepared by solvent evaporation and phase inversion, respectively, with Matrimid® as a base polymer and three different MOFs: {$Cu_3(BTC)_2$}, ZIF-8, and MIL-53(Al) as fillers. Dense membranes and asymmetric membranes for all three studied MOFs showed improvement in CO_2/CH_4 and CO_2/N_2 selectivity and permeance as compared to the unfilled reference membrane.

12.6.4 SUMMARY OF APPLICATION OF MOF MEMBRANES

During the last seven years, MOF membranes have been developed and tested for gas separation. There is, so far, no major industrial gas separation that is using inorganic membranes. MOF membranes are promising candidates for the shape-selective separation of a gas mixture by molecular sieving. In addition to molecular sieving based on molecular exclusion, MOF membrane can separate a gas mixture by the interplay of mixed gas adsorption and diffusion. MOF membranes show the effect of framework flexibility, which facilitates module construction but prohibits a sharp cut-off [177]. An application of MMMs, which consist of MOF nanoparticles in a standard polymer, in industrial gas separation is predicted for the near future.

MOF is a relatively new class of crystalline porous materials consisting of metal clusters connected by organic ligands. By choosing appropriate metal centers and organic ligands, the pore size and surface properties of MOFs can be tuned to a great extent. This structural flexibility opens the door for the application of MOFs in gas separation, gas storage, adsorption, catalysis, sensors, and so forth. However, to make a thin film of MOFs is still challenging. There are two major challenges in engineering MOFs: (1) *growing to position* and (2) *growing in shape*. It is difficult to direct MOF growth into a specific location. Due to their crystalline rigidity, it is also difficult to process MOFs into certain shapes by postsynthetic methods. These difficulties have significantly limited MOF applications in separation and filtration areas, where membrane-like shapes are desired. Compared to the broad range of separations carried out by pure polymer membranes, examples with MOF-MMMs are still quite limited.

MOFs membranes can be used in the following processes where gases are involved.

1. Gas separation: MOFs permit certain molecules to pass through their pores based on kinetic diameter and size. This is especially useful for separating carbon dioxide.
2. Gas purification: MOFs have strong chemisorptions between its odor-generating, electron-rich molecules and the framework permitting passing of the gas through the MOF.
3. Gas storage: MOFs can store carbon monoxide, carbon dioxide, oxygen, and methane because of their high adsorption enthalpies.
4. Heterogeneous catalysis: Due to their size and shape selectivity and accessible bulk volume, selective gas permeation for the catalytic reaction is enabled.

TABLE 12.5
Application of MOFs for Gas Separation

MOF Membranes	Single Gas Permeation/Separation of Gases	References
ZIF-8	Ethane/methane	[154]
ZIF-8	H_2/CH_4	[159]
ZIF-8	CO_2/CH_4	[157]
MOF-5	H_2, CH_4, N_2, CO_2, and SF_6	[143]
MOF-5	CO_2/H_2, CO/N_2	[166]
HKUST-1	H_2/CO_2, H_2/CH_4, H_2/N_2	[168]
ZIF-90	H_2/CO_2, H_2/CH_4, H_2/C_2H_6, and H_2/C_3H_8	[165]
ZIF-7, -8, = 22, -90	Molecular sieving of H_2	[177]
ZIF-9	CO_2/CH_4 and CO_2/N_2	[158]
ZIF-7	H_2/CO_2	[161]
MIL-53	H_2, CH_4, N_2, and CO_2	[176]
$[Cu_2L_2P]_n$	CO_2/CH_4	[158]
Bio-MOF membranes		
Bio-MOFs	CO_2/CH_4	[172]
Bio-MOF-1	CO_2/CH_4	[169]
Bio-MOF-1	CO_2/CH_4	[173]
MMM-containing MOFs		
Functionalized MOF containing MMMs	CO_2/CH_4	[175]
Matrmid®/[$Cu_3(BTC)_2$], ZIF-8, and MIL-53(Al)	CO_2/CH_4 and CO_2/N_2	[174]

It is challenging to develop MOF membranes with high gas-separation performance due to the heterogeneous nucleation and the very poor growth of MOF crystals on a porous surface. MMM with MOFs, which combines the advantages of polymers and MOFs, is a promising membrane for future. Table 12.5 shows the applications of MOFs for gas separation.

12.7 CONCLUSIONS

Among the five topics chosen in this chapter as novel membranes and membrane materials, three of them, namely, CNTs, AQPs, and graphene, were reported only recently as new membrane materials for water treatment, especially for seawater desalination, in journals of high impact factors such as *Nature* and *Science*. They promise orders of magnitude higher water flux than conventional polymeric membranes with almost complete rejection of salt, which have indeed been confirmed by small-scale laboratory experiments. Currently, attempts are being made worldwide to make these membranes practical for commercial purposes. In fact, one of them, AQP membrane, has already been commercialized for desalination.

There are, however, a number of obstacles to be overcome from the engineering viewpoint. Production of sizable membrane area at low cost and the ability of the membrane to bend and wind for installation in a narrow module space are the major technical challenges that have not yet been resolved for any of these novel membranes and membrane materials. This situation is very similar to carbon and zeolite molecular sieve membranes for which excellent gas separation and permeation results have been recorded but none went into large-scale commercial applications.

The answers to these challenges are probably given by the fabrication of MMMs, in which the novel membrane materials are loaded in a host of polymeric membranes as filler nanoparticles. Thus, high processibilty of polymeric membranes is combined with high separation performance of filler materials. Usually, separation performance of MMMs is much improved compared with the host polymeric membrane, but the improvement is not as spectacular as expected from the novel membrane materials. A flux that is orders of magnitude higher than the host polymeric membrane has never been achieved by MMMs.

Despite the above challenges and difficulties, graphene membrane seems the closest to practical applications because films of sizable area with flexibility and high mechanical strength can be fabricated from graphene or graphene-related materials relatively easily.

The situation of MOF membrane is similar to the above-mentioned three materials, and much research for the practical utilization of MOFs is currently under way.

ENMs are unique among the chosen novel membranes and membrane materials. Industrial mass production of ENMs is possible. They are flexible, mechanically strong enough, and have uniquely high porosities. Their pore sizes are, however, in a range of fractions of micrometer to micrometers. Thus, ENMs can be used for MF, membrane adsorption, membrane contactor, and MD. However, for ENMs to be used in UF, NF, RO, and gas separation, substantial pore size reduction is required, which can be accomplished by *in situ* polymerization, dip-coating, and grafting. This is the focus of recent research worldwide. Hollow ENMs can also be fabricated. Nobody has attempted to utilize hollow ENMs as separation membranes. This would be very interesting because of the extremely high surface and volume ratio of hollow ENMs, much higher than those of hollow fibers.

REFERENCES

1. Iijima, S., 1991, Helical microtubules of graphitic carbon, *Nature* 354:56–58.
2. Iijima, S., Ichihashi, T., 1993, Single-shell carbon nanotubes of 1-nm diameter, *Nature* 363:603–605.
3. Sur, U.K., 2011, Carbon nanotube radio. In: S. Bianco (Ed.), *Carbon Nanotubes—From Research to Applications,* p. 24. http://www.asknature.org/product/c70bbf7b3ad4097b961d686cca053ea.
4. Wang, X., Li, Q., Xie, J., Jin, Z., Wang, J., Li, Y., Jiang, K., Fan, S., 2009, Fabrication of ultra-long and electrically uniform single-walled carbon nanotubes on clean substrates, *Nano Lett* 9(9): 3137–3141.
5. Majumder, M., Ajayan, P.M., 2010, Carbon nanotube membranes: A new frontier in membrane science. In: E. Drioli and L. Giorno (Eds.), *Comprehensive Membrane Science and Engineering*, volume 1, pp. 291–310, Oxford: Academic Press.
6. Chen, Y., Hill, J.M., 2011, A mechanical model for single-file transport of water through carbon nanotube membranes, *J Memb Sci* 372(1/2):1157–1160.
7. Pilatos, G., Vermisoglou, E.C., Romanos, G.E., Karanikolos, G.N., Boukos, N., Likodimos, V., Kanellopoulos, N.K., 2010, A closer look inside nanotubes: Pore structure evaluation of anodized alumina templated carbon nanotube membranes through adsorption and permeability studies, *Adv Funct Mater* 20(15):2500–2510.

8. Hinds, B., 2010, Dramatic transport properties of carbon nanotube membranes for a robust protein channel mimetic platform, *Curr Opin Solid State Mater Sci* 16:1–9.
9. Rana, K., Kucukayan-Dogu, G., Bengu, E., 2012, Growth of vertically aligned carbon nanotubes over self-ordered nano-porous alumina films and their surface properties, *Appl Surf Sci* 18(1):7112–7114.
10. Kang, D.Y., Tong, H.M., Zang, J., Choudhary, R.P., Sholl, D.S., Beckham, H.W., Jones, C.W., Nair, S., 2012, Single-walled aluminosilicate nanotube/poly(vinyl alcohol) nanocomposite membranes, *ACS Appl Mater Interfaces* 4:965–976.
11. Sears, K., Dumée, L., Schűtz, J., She, M., Huyuh, C., Hawkins, S., Duke, M., Gray, S., 2010, Recent developments in carbon nanotube membranes for water purification and gas separation, *Materials* 3:127–149.
12. Holt, J.K., Park, H.G., Wang, Y., Staderman, M., Artyukhin, A.B., Grigoropoulos, C.P., Noy, A., Bakajin, O., 2006, Fast mass transport through sub-2-nanometer carbon nanotubes. *Science* 312:1034–1037.
13. Liu, H., Cooper, V.R., Dai, S., Jiang, D., 2012, Windowed carbon nanotubes for efficient CO_2 removal from natural gas, *J Phys Chem Lett* 3:3343–3347.
14. Liu, B., Li, X., Li, B., Xu, B., Zhao, Y., 2009, Carbon nanotube based artificial water channel protein: Membrane perturbation and water transportation, *Nano Lett* 9(4):1386–1394.
15. Wu, J., Gerstandt, K., Majumder, M., Zhan, X., Hinds, B.J., 2011, Highly efficient electroosmotic flow through functionalized carbon nanotube, *Nanoscale* 3:3321–3328.
16. Hinds, B.J., Chopra, N., Rantell, T., Andrews, R., Gavals, V., Bachas, L.G., 2004, Aligned multiwalled carbon nanotube membranes, *Science* 303(5654):62–65.
17. Tseng, H.H., Kumar, I.A., Weng, T.H., Lu, C.Y., Wey, M.Y., 2009, Preparation and characterization of carbon molecular sieve membranes for gas separation-the effect of incorporated multi-wall carbon nanotubes, *Desalination* 240:40–45.
18. Verweij, H., Schillo, M.C., Li, J., 2007, Fast mass transport through carbon nanotube membrane, *Small* 3(12):1996–2004.
19. Yu, M., Funke, H.H., Falconer, J.L., Noble, R.D., 2009. High density, vertically-aligned carbon nanotube membranes, *Nano Lett* 9(1):225–229.
20. Khan, M.M., Filiz, V., Bengtson, G., Shishatskiy, S., Rahman, M., Abetz, V., 2012, Functionalized carbon nanotubes mixed matrix membranes of polymers of intrinsic microporosity for gas separation, *Nanoscale Res Lett* 7:504–515.
21. Ahn, C.H., Baek, Y., Lee, C., Kim, S.O., Kim, S., Lee, S., Kim, S.H., Bae, S.K., Park, J., Yoon, J., 2012, Carbon nanotube-based membranes: Fabrication and application to desalination, *J Ind Eng Chem* 18:1551–1559.
22. Lam, C., James, J.T., McCluskey, R., Arepalli, S., Hunter, R.L., 2006, A review of carbon nano tube toxicity and assessment of potential occupational and environmental health risks, *CRC Crit Rev Toxicol* 36:189–217.
23. Kostarelos, K., 2008, The long and short of carbon nanotube toxicity, *Nat Biotechnol* 26:774–776.
24. Preston, G.M., Carroll, T.P., Guggino, W.B., Agre, P., 1992, Appearance of water channels in *Xenopus* oocytes expressing red cell CHIP28 protein, *Science* 256(5055): 385–387.
25. Bowen, W.R., 2006, Biometric separation—learning from the early development of biological membranes, *Desalination* 199: 225–227.
26. Tang, C.Y., Zhao, Y., Wang, R., Hélix-Nielsen, C., Fane, A.G., 2013, Desalination by biomimetic aquaporin membranes: Review of status and prospects, *Desalination* 308:34–40.
27. Altramura, N., Calamita, G., 2012, Systems for production of proteins for biomimetic membrane devices. In: C. Hélix-Nielsen (Ed.). *Biomimetic Membranes for Sensor and Separation Applications*, Springer, Dordrecht, the Netherlands, 2012, pp. 233–250.

28. Zhao, Y., Qiu, C., Li, X., Vararattanavech, A., Shen, W., Torres, J., Hélix-Nielsen et al. 2012, Synthesis of robust and high-performance aquaporin-based biomimetic membranes by interfacial polymerization-membrane preparation and RO performance characterization, *J Memb Sci* 423–424:422–428.

29. Aquaporin-PolyNano, 2013, Updated by M.P. Jaensen on October 9, 2013. http://www.polynano.org/Parttners/Aquaporin.aspx.

30. Kumar, M., Grzelakowski, M., Zilles, J., Clark, M., Meier, W., 2007, Highly permeable polymeric membranes based on the incorporation of the functional water channel protein Aquaporin Z, *Proc Natl Acad Sci USA* 104:225–227.

31. Verkman, A.S., Mitra, A.K., 2000, Structure and function of auaporin water channels, *Am J Physiol Renal Physiol* 278(1):F13–F28.

32. Borgnia, M., Nielsen, S., Engel, A., Agre, P., 1999, Cellular and molecular biology of the aquaporin water channels, *Annu Rev Biochem* 68:425–458.

33. Eckardt, N.A., 2008, Aquaporins and chloroplast membrane permeability, *Plant Cell* 20(3):499.

34. Maurel, C., Verdoucq, L., Luu, D.T., Santoni, V., 2008, Plant aquaporins: Membrane channels with multiple integrated functions, *Ann Rev Plant Biol* 59:595–624.

35. Reimhult, E., Kumar, K., 2008, Membrane biosensor platforms using nano- and microporous supports, *Trends Biotechnol* 26(2):82–89.

36. Khayet, M., Matsuura, T., *Membrane Distillation Principles and Applications*, Elsevier, England. 2011.

37. Burger, C., Hsiao, B.S., Chu, B., 2006, Nanofibrous materials and their applications, *Annu Rev Mater Res* 36:333–368.

38. Balamurugan, R., Sundarrajan, S., Ramakrishna, S., 2011, Recent trend in nanofibers and their suitability for air and water filtrations, *Membranes* 1:232–248.

39. Huang, Z.M., Zhang, Y., Kotaki, Z., Ramakrishna, S., 2003, A review on polymer nanofibers by electrospinning and their applications in nanocomposites, *Compos Sci Technol* 63:2223–2253.

40. Feng, C., Khulbe, K.C., Matsuura, T., 2010, Recent progress in the preparation, characterization, and applications of nanofibers and nanofiber membranes via electrospunning/interfacial polymerization, *J Appl Poly Sci* 115:56–776.

41. Gupta, P., Wilkes, G.L., 2003, Some investigations on the fiber formation by utilizing a side-by-side bicomponent electrospinning approach, *Polymer* 44:6353–6359.

42. Xu, F., Li, L., Cui, X., 2012, Fabrication of aligned side-by-side TiO$_2$/SnO$_2$ nanofibers via dual-opposite-spinneret electrospinning (DOSE), *J Nanomater* Volume 2012, Article ID 575926, 5 pages, doi:10.1155/2012/575926.

43. Li, D., Xia, Y., 2004, Direct fabrication of composite and ceramic hollow nanofibers by electrospinning, *Nano Lett* 4(5):933–938.

44. Zhang, Z., Huang, Z.-M., Xu, X., Lim, C.T., Ramakrishna, S., 2004, Preparation of core-shell structured PCL-r-Gelatin Bi-component nanofibers by coaxial electrospinning, *Chem Mater* 16:3406–3409.

45. Xia, Y., Li, D., 2009, Electrospinning of fine hollow fibers, US 7,575,707 B2.

46. Yu, J.H., Fridrikh, S.V., Rutledge, G.C., 2004, Production of submicrometer fibers by two fluid electrospinning, *Adv Mater* 16:1562–1566.

47. Li, F., Zho, Y., Song, Y., 2010, Core-shell nanofibers: Nano channel and capsule by coaxial electrospinning. In: A. Kumar (Ed.). *Nanofibers*, chapter 22, InTech, Rijeka, Croatia, pp. 420–438.

48. Srivastava, Y., Loscertales, I., Marquez, M., Thorsen, T., 2007, Electrospinning of hollow and core/sheath nanofibers using a microfluidic manifold, *Microfluid Nanofluidics* 4(3):245–250.

49. Ge, L., Wang, X., Tu, Z., Pan, C., Wang, C., Gu, Z., 2007, Fabrication of multilayered hollow nanofibers and estimates Young's modulus, *Jpn J Appl Phys* 46(10A):6790–6795.

50. McCann, J.T., Li, D., Xia, Y., 2005, Electrospinning of nanofibers with core-sheath, hollow or porous structure, *J Mater Chem* 15:735–738.

51. Wang, S.G., Jiang, X., Chen, P.C., Yu, A.G., Huang, X.J., 2012, Preparation of coaxial – electrospun poly{bis(p-methylphenoxy)}phosphazene nanofiber membrane for enzyme immobilization, *J Mol Sci* 13:14136–14148.

52. Theron, A., Zussman, E., Yarin, A.L., 2001, Electrostatic field-assisted alignment of electrospun nanofibers, *Nanotechnol* 12:384–390.

53. Inai, R., Kotaki, M., Ramakrishna, S., 2005, Structure and properties of electrospun PLLA single nanofibers, *Nanotechnol* 16:208–213.

54. Franco, R.A., Min, Y.K., Yang, H.M., Lee, B.Y., 2012, On stabilization of PVPA/PVA electrospun nanofiber membrane and its effect on material properties and biocompatibility, *J Nanomater* 2012(2012) Article ID 393042, 9 pages, doi: 10.1155/2012/393042.

55. Yang, J.C., Lee, S.Y., Tseng, W.C., Shu, Y.C., Lu, J.C., Shie, H.S., Chen, C.C., 2012, Formation of highly aligned, single-layered, hollow fibrous assemblies and the fabrication of large pieces of PLLA membranes, *Macromol Mater Eng* 297:115–122.

56. Gopal, R., Kaur, S., Feng, C.Y., Chan, C., Ramakrishna, S., Tabe, S., Matsuura, T., 2007, Electrospun nanofibrous polysulfone membranes as pre-filters: Particulate removal, *J Memb Sci* 289:210–219.

57. Širc, J., Hobzová, Kostina, N., Munzarová, M., Juklíčková, M., Lhotka, M., Kubinová, Š., Zajícová, A., Michálek., J., 2012, Morphological characterization of nanofibrs: Methods and application in practice, *J Nanomater* Volume 2012, Article ID 327369, 14 pages, doi:1155/2012/327369.

58. Zhou, W., He, J., Cui, S., Gao, W., 2011, Studies of electrospun cellulose acetate nanofibrous membranes, *The Open Mater Sci J* 5:51–55.

59. Aussawasathien, D., Teerawattananon, C., Vongachariya, A., 2006, Separation of micronto submicron particles from water: Electrospun nylon-6 nanofibrous membranes as pre-filters, *J Memb Sci* 315(1/2):11–19.

60. Homaeigohar, S.S., Buhr, K., Ebert, K., 2010, Polyethersulfone electron nanofibrous composite membrane for liquid filtration, *J Memb Sci* 365:68–77.

61. Karube, Y. Kawakami, H. 2010, Fabrication of well-aligned electrospun nanofibrous membrane based on fluorinated polyimide, *Polym Adv Technol* 21(12) 861–866.

62. Sato, A., Wang, R., Ma, H., Hsiao, B.S., Chu, B., 2011, Novel nanofibrous scaffolds for water filtration with bacteria and virus removal capability, *J Electron Microscopy* 60(3) 201–209.

63. Wang, X., Chen, X., Yoon, K., Fang. D., Hsiao, B.S., Chu. B., 2005, High flux filtration medium based on nanofibrous substrate with hydrophilic nanocomposite coating, *Environ Sci Technol* 39:7684–7691.

64. Yoon, K., Kim. K., Wang, X., Fang, D., Hsiao, B.S., Chu, B., 2006, High flux ultrafiltration membranes based on electrospun nanofibrous PAN scaffolds and chitosan coating, *Polymer* 47:2434–2441.

65. Wang, X., Fang, K., Yoon, D., Hsiao, B.S., Chu, B., 2006, High performance ultrafiltration composite membrane based on poly(vinyl alcohol) hydrogel coating on crosslinked nanofibrous poly(vinylalcohol) scaffold, *J Memb Sci* 278(1/2):261–268.

66. Tang, Z., Wei, J., Yung, L., Ji, Ma, H., Qiu, C., Yoon, K., Wan, F., Feng, D., Hsiao, B.S., Chu, B., 2009, UV-cured poly(vinyl alcohol) ultrafiltration membrane based on electrospun nanofiber scaffolds, *J Memb Sci* 328(1/2):1–5.

67. Yoon, K., Hsiao, B., Chu, B., 2009, High flux nanofiltration membranes based on interfacially polymerized polyamide barrier layer on polyacrylonitrile scaffolds, *J Memb Sci* 326(2):484–492.

68. Tang, Z., Qiu, C., McCutcheon, J.R., Yoon, K., Ma, H., Fang, D., Lee, E., Kopp, C., Hsiao, B.S., Chu, B., 2009, Design and fabrication of electrospun polyethersulfone nanofibrous scaffold for high-flux nanofiltration membranes, *J Polym Sci Part B: Polym Phys* 47:2288–2300.

69. Wang, X., Zhang, K., Yang, Y., Wang, L., Zhou, Z., Zhu, M., Hsiao, B.S., Chu, B., 2010, Development of hydrophilic barrier layer on nanofibrous substrate as composite membrane in a facile route, *J Memb Sci* 356:110–116.
70. Yung, L., Ma, H., Wang, X., Yoon, K., Wang, R., Hsiao, B.S., Chu, B., 2010 Fabrication of thin-film nanofibrous composite membranes by interfacial polymerization using ionic liquids as additives, *J Memb Sci* 365(1/2):52–58.
71. Ma, H.Y., Burger, C., Hsiao, B.S., Chu, B., 2011, Ultrafine polysaccharide nanofibrous membranes for water purification, *Biomacromolecules* 12(4):970–976.
72. Kaur, S., Barhate, R., Sundarrajan, S., Matsuura, T., Ramakrishna, S., 2011, Hot pressing of electrospun membrane and its influence on separation performance on thin film composite nanofiltration membrane, *Desalination* 279:201–209.
73. Gopal, R., Kaur, S., Ma, Z., Chan, C., Ramakrishna, S., Matsuura, T., 2006, Electrospun nanofibrous filtration membrane, *J Memb Sci* 281:581–586.
74. Uyar, T., Havelund, R., Nur, Y., Hacaloglu, J., Besenbacher, F., Kingshott, P., 2009, Molecular filters based on cyclodextrin functionalized electrospun fibers, *J Memb Sci* 332:120–137.
75. Sueyoshi, Y., Fukushima, C., Yoshikawa, M., 2010, Molecularly imprinted nanofiber membranes from cellulose acetate aimed for chiral separation, *J Memb Sci* 357(1/2):90–97.
76. Gurdev, S., Rana, D., Matsuura, T., Ramakrishna, S., Narbaitz, R.M., Tabe, S., 2010, Removal of disinfection byproducts from water by carbonized electrospun nanofibrous membranes, *Sep Purif Techno* 74:202–212.
77. Tian, Y., Wu, M., Liu, R., Li, Y., Wang, D., Tan, J., Wu, R., Huang, Y., 2011, Electrospun membrane of cellulose acetate for heavy metal ion adsorption in water treatment, *Carbohydr Polym* 83:743–748.
78. Guibo, Y., Qing, Z., Yahong, Z., Yin, Y., Yumin, Y., 2013, The electrospun polyamide 6 nanofiber membrane used as a high efficiency filter materials: Filtration potential, thermal treatment, and their continuous production, *J Appl Polym Sci* 128(2):1061–1069.
79. Chu, B., Hsiao, B., Ma, H., 2011, High flux high efficiency nanofiber membranes and methods of production thereof, U.S. 20110198282.
80. Chen, P., Liang, H.W., Lv, X.H., Zhu, H.Z., Yao, H.B., Yu, S.H., 2011, Carbonaceous nanofiber membrane functionalized by beta-cyclodextrins for molecular filtration, *ACS Nano* 5(7):5928–5935.
81. Liang, H.W., Wang, L., Chen, P.Y., Lin, H.T., Chen, L.F., He, D., Yu, S.H., 2010, Carbonaceous nanofiber membranes for selective filtration and separation of nanoparticles, *Adv Mater* 22:4691–4695.
82. Jafar, S., Sarbatly, R., 2012, Geothermal water desalination by using nanofiber membrane, *Int. Conf. on Chemical, Environmental and Biological Science*, Penang, Malaysia, February 11–12, pp. 46–50.
83. Khayet, M., Payo, G., Carmen, A.S., Carmen, M., 2011, Nanostructured flat membranes for direct contact membrane distillation. WO/2011/117443.
84. Rošic, R., Pelipenko, J., Kristl, J., Kocbek, P., Baumgartner, S., 2012, Properties, Engineering and applications of polymeric nanofibers: Current research and future advances, *Chem Biochem Eng Q* 26(4):417–425.
85. Shih, J.H., 2011, A study of composite nanofiber membrane applied in seawater desalination by membrane distillation, Master's Thesis, National Taiwan University of Science and Technology, Taipei City, Taiwan.
86. Feng, C.Y., 2009, Development of novel nanofiber membranes for seawater desalination by air-gap membrane distillation, PhD Thesis, University of Ottawa, Ottawa, Canada, 2009.
87. Feng. C., Khulbe, K.C., Matsuura, T., Gopal, R., Kaur, S., Ramakrishna, S., Khayet, M., 2008, Production of drinking water from saline water by air-gap membrane distillation using polyvinylidene fluoride nanofiber membrane, *J Memb Sci* 311:1–6.

88. Prince, J.A., Singh, G., Rana, D., Matsuura, T., Anbharasi, V., Shanmugasundaram, T.S, 2012, Preparation and characterization of highly hydrophobic poly(vinylidene fluoride)-clay nanocomposite nanofiber membranes (PVDF-clay NNMs) for desalination using direct contact membrane distillation, *J Memb Sci* 397–398:80–86.

89. Kim, Y.J., Ahn, C.H., Choi, M.O., 2010, Effect of thermal treatment on the characteristics of electrospun PVDF-silica composite nanofibrous membrane, *European Polym J* 46:1957–1965.

90. Feng, C., Khulbe, K.C., Tabe, S., 2012, Volatile organic compound removal by membrane gas stripping using electro spun nanofiber membrane, *Desalination* 287:98–102.

91. Tamura, T., Kawakami, H., 2010, Aligned electrospun nanofiber composite membranes for fuel cell electrolytes, *Nano Lett* 10:1324–1328.

92. Lee, K.M., Choi, J., Wycisk, R., Pintauro, P.N., Mather, P., 2009, Nafion Nnanofiber membranes, *ECS Trans* 25(1):1451–1458.

93. Thavasi, V., Singh, G., Ramakrishna, S., 2008, Electrospun nanofibers in energy and environmental S applications, *Energy Environ Sci* 1:205–221.

94. Sang, Y., Li, F., Gu, Q., Liang, C., Chen, J., 2008, Heavy metal-contaminated ground water treatment by a novel nanofiber membrane, *Desalination* 223:349–360.

95. Haider, S., Park, S.Y., 2009, Preparation of the electrospun chitosan nanofibers and their applications to the adsorption of Cu(II) and PB(II) ions from aqueous solution, *J Memb Sci* 328:90–96.

96. Ngah, W.S.W., Endud, C.S., Mayanar, R., 2002, Removal of copper(II) ions from aqueous solution onto chitosan and cross-linked chitosan beads, *React Funct Polym* 50:182–190.

97. Menkhaus, T.J., Varadaraju, H., Zhang, L., Scchneiderman, S., Bjustrom, S., Liu, L., Fong, H., 2010, Electrospun nanofiber membranes surface functionalized with 3-dimensional nanolayers as an innovative adsorption medium with ultra-high capacity and throughout, *Chem Commun* 46:3720–3722.

98. Vu, D., Li, X., Li, Z., Wang, C., 2013, Phase structure effects of electrospun TiO_2 nanofiber membranes on As(III) adsorption, *J Chem Eng Data* 58:71–77.

99. Basheer, C., 2013, Nanofiber-Membrane-supported TiO_2 was used as a catalyst for oxidation of benzene to phenol, *J Chemistry* Article ID 562305, 7 pages, doi:10.1155/2013/562305.

100. Formo, E., Yavuz, M.S., Lee, E.P., Lane. L., Xia, Y., 2009, Functionalization of electrospun ceramic nanofibre membranes with noble-metal nanostructures for catalytic applications, *J Mater Chem* 19:3878–3882.

101. Zhang, X., Xu, S., Han, G., 2009, Fabrication and photocatalytic activity of TiO_2 nanofiber membrane, *Mater Lett* 63:1761–1763.

102. Liu, H., Lei, X., Zhai, Y., Li, L., 2012, Electrospin nanofiber membranes containing molecularly imprinted polymer (MIP) for Rhodamine B (RhB), *Adv Chem Eng Sci* 2:266–274.

103. Sueyoshi, Y., Hasimoto, T., Yoshikawa, M., Ifuku, S., 2012, Chitin nanofiber membranes for chiral separation, *Sustain Agric Res* 1(1):42.

104. Xu, Z., Gu, Q., Hu, H., Li, F., 2008, A novel electrospun polysulfone fiber membrane: Application to advanced treatment of secondary biotreatment sewage, *Environ Technol* 29(1):13–21.

105. Gibson, P., Schreuder-Gibson, H., Rivin, D., 2001, Transport properties of porous membranes based on electrospun nanofibers, *Colloids Surf A* 187–188:469–481.

106. Huang, F.L., Wang, Q.Q., Wei, Q.F., Gao, W.D., Shou, H.Y., Jiang, S.D., 2010, Dynamic wettability and contact angles of poly(vinylidene fluoride) nanofiber membranes grafted with acrylic acid, *Express Polym Letters* 4(9):557–558.

107. Xu, X. Jiang, L., Zhou, Z., Wu, X., Wang, Y., 2012, Preparation and properties of electrospun soy protein isolate/polyethylene oxide nanofiber membranes, *ACS Appl Mater Interfaces* 4(8):4331–4337.

108. Feng, C., Khulbe, K.C., Matsuura, T., Tabe, S., Ismail, A.F., 2013, Preparation and characterization of electro-spun nanofiber membranes and their possible applications in water treatment, *Sep Puri Tech* 102:118–135.

109. Lee, C.H., Shin, H.J., Cho, I.H., Kang, Y.M., Kim, A.I., Park, K.-D., Shin, J.-W., 2005, Nanofiber alignment and direction of mechanical strain affect the ECM production of human ACL fibroblast, *Biomaterials* 26:1261–1270.

110. Geim, A.K., Novoselov, K.S., 2007, The rise of graphene, *Nature Mater* 6(3):183–191.

111. Nobelprize.org, The Nobel Prize in Physics, October 5, 2010. Royal Swedish Academy of Sciences, Stockholm, Sweden.

112. Meyer, J.C., Geim, A.K., Katsnelson, M.I., Novoselov, K.S., Booth, T.J., Roth, S., 2007, The structure of suspended graphene sheets *Nature* 446(7131):60–63.

113. Li, C., Li, D., Yang, J., Zeng, X., Yuan, W., 2011, Preparation of single- and few-layer graphene sheets using Co deposition on SiC substrate. *J Nanomater* 2011:1–6.

114. Zhu, B.Y., Murali, S., Cai, W., Li, X., Suk, J.W., Potts, J.R., Rooff, R.S., 2010, Graphene and graphene oxide: Synthesis, Properties, and applications, *Adv Mater* 22:3906–3924.

115. Hassan, K., Sandberg, M.O., Nur, O., Willander, M., 2011, Polycation stabilization of graphene suspensions, *Nanoscale Res Lett* 6(1):493.

116. Dikin, D.A., Stankovich, S., Zimney, E.J., Piner, R.D., Dommett, G.H.B., Evmenenko, G., Nguyen, S.T., Ruoff, R.S., 2007, Preparation and characterization of graphene oxide paper, *Nature* 448:457–460.

117. Singh, V., Joung, D., Zhai, L., Das, S., Khondaker, S.I., Seal, S., 2011, Graphene based Materials: Past, present and future. *Prog Mater Sci* 56:1178–1271.

118. Soldano, C., Mahmood, A., Dujardin, E., 2010, Production, properties and potential of graphene. *Carbon* 48(8):2127–2150.

119. Ortolani, L., Cadelano, E., Veronese, G.P., Boschi, C.D.E., Snoeck, E., Colombo, L., Morandi, V., 2012, Folded graphene membranes: Mapping curvature at the nanoscale, *Nano Lett* 12:5207–5212.

120. O'Hern, S.C., Stewart, C.A., Boutilier, M.S.H., Idrobo, J.C., Bhaviripudi, S., Das, S.K., Kong, J., Laoui, T., Atieh, M., Karnik, R., 2012, Selective molecular transport through intrinsic defects in a single layer of CVD graphene, *ACS Nano* 6(11):10130–10138.

121. Chu, J., 2012, Tiny pores in graphene could give rise to membranes, *MIT News*, October 23.

122. Karnik, R., O'Hern, S.C., Idrobo, J.C., 2012, Tiny pores in graphene could give rise to membranes, *MIT News*, October 22.

123. Ruiz-Vargas, C.S., Zhuang, H.L., Huang, P.Y., van der Zande, A.M., Garg, S., McEuen, P.L., Muller, D.A., Hennig, R.G., Park, J., 2011, Softened elastic response and unzipping in chemical vapor deposition graphene membranes, *Nano Lett* 11(6):2259–2263.

124. Koenig, S.P., Boddeti, N.G., Dunn, M.L., Bunch, J.S., 2011, Ultra strong adhesion of graphene membranes, *Nature Nanotechnol* 6:543–546.

125. Bunch, J.S., Verbridge, S.S., Alden, J.S, van der Zande, A.M., Parpia, J.M., Craighead, H.G., McEuen, P.L., 2008, Impermeable atomic membranes from graphite sheets, *Nano Lett* 8(8):2458–2462.

126. Hauser, A.W., Schwerdtfeger, P., 2012, Methane-selective nanoporous graphene membranes for gas purification, *Phys Chem Chem Phys* 14:13292–13298.

127. Du, H., Li, J., Zhang, J., Su, G., Li, X., Zhao, Y., 2011, Separation of hydrogen and nitrogen gases with porous graphene membrane, *J Phys Chem C* 115:23261–23266.

128. Koenig, S.P., Wang, L., Pellegrino, J., Bunch, S., 2012, Selective molecular sieving through porous graphene, *Nature Nanotechnol* 7:728–732.

129. Schrier, J., 2010, Helium separation using porous graphene membranes, *J Phys Chem Lett* 1:2284–2287.

130. Schrier, J., McClain, J., 2012, Thermally-driven isotope separation across nanoporous graphene, *Chem Phys Lett* 521:118–124.
131. Nair, R.R., Wu, H.A., Jayaram, P.N., Grigorieval, I.V., Geim, A.K., 2012, Unimpeded permeation of water through helium-leak-tight graphene-based membranes, *Nature Nanotechnol* 2:728–732.
132. Cohen-Tanugi, D., Grossman, J.C., 2012, Water desalination across nanoporous graphene, *Nano Lett* 12:3602–3608.
133. Wang, E.N., Karnik, R., 2012, Graphene cleans up water, *Nature Nanotechnol* 7:552–554.
134. Suk, M.E., Aluru, N.R., 2010, Water transport through ultrathin graphene, *J Phys Chem Lett* 1:1590–1594.
135. Robeson, L.M., 2008, The upperbound revisited, *J Memb Sci* 320(1/2):390–400.
136. Shah, M., McCarthy, M.C., Sachdeva, S., Lee, A.K., Jeong, H.K., 2012, Current status of metal-organic framework membranes for gas separations: Promises and challenges, *Ind Eng Chem Res* 51:2179–2199.
137. Lin, Y.S., Kumakiri, I., Nair, B.N., Alsyouri, H., 2002, Microporous inorganic membranes, *Sep Purif Method* 31:229–379.
138. Meek, S.T., Greathouse, J.A., Allendorf, M.D., 2011, Metal-organic frameworks: A rapidly growing class of versatile nanoporous materials, *Adv Mater* 23:249.
139. Gascon, J., Kapteijn, F., 2010, Metal-organic framework membranes-high potential, bright future? *Angew Chem Int Ed* 49:1530–1532.
140. Zacher, D., Shekhah, O., Woll, C., Fischer, R.A., 2009, Thin films of metal-organic frameworks, *Chem Soc Rev* 38(5):1418–1429.
141. Shekhah, O., Liu, J., Fischer, R.A., Wöll, Ch., 2011, MOF thin films: Existing and future applications, *Chem Soc Rev* 40:1081–1106.
142. Yoo, Y., Lai, Z.P., Jeong, H.K., 2009, Fabrication of MOF-5 membranes using microwave induced rapid seeding and solvothermal secondary growth, *Microporous Mesoporous Mater* 123(1–3):100–106.
143. Liu, Y.Y., Ng, Z.F., Khan, E.A., Jeong, H.K., Ching, C.B., Lai, Z.P., 2009, Synthesis of continuous MoF-5 membranes on porous alpha-alumina substrate, *Microporous Mesoporous Mater* 118(1–3):296–301.
144. Klinowski, J., Almeida Paz, F.A., Paz, Silva, P., Rocha, J., 2011, Microwave-assisted synthesis of metal-organic frameworks, *Dalton Trans* 40:321–330.
145. Horcajada, P., Serre, C., Grosso, D., Boissiere, C., Perruchas, S., Sanchez, C., Ferey, G., 2009, Colloidal route for preparing optical thin films of nanoporous metal–organic frameworks, *Adv Mater* 21(19):1931–1935.
146. Shekhah, O., Wang, H., Kowarik, S., Schreiber, F., Paulus, M., Tolan, M., Sternemann, C. et al., 2007, Step-by-step route for the synthesis of metal-organic frameworks, *J Am Chem Soc* 129(49):15118–15119.
147. Muller, U., Pütter, H., Hesses, M., Wessel, H., Schubert, M., Huff, J., Guzmann, M., 2005, Method for electrochemical production of a crystalline porous metal organic skeleton material, WO/2005/049892.
148. Muller, U., Pütter, H., Hesses, M., Wessel, H., Schubert, M., Huff, J., Guzmann, M., 2005, Method for electrochemical production of a crystalline porous metal organic skeleton material, US 2007/0227898.
149. Schoedel, A., Scherb, C., Bein, T., 2010, Oriented nanoscale films of metal-organic frameworks by room-temperature gel-layer synthesis, *Angew Chem Int Ed* 49(40):7225–7228.
150. Ameloot, R., Gobechiya, E., Ujii, H., Martens, J.A., Hofkens, J., Alaerts, L., Sels, B.F., De Vos, D.E., 2010, Direct patterning of oriented metal-organic framework crystals via control over crystallization kinetics in clear precursor solution, *Adv Mater* 22(24):2685–2688.
151. Lu, H., Zhu, S., 2013, Interfacial synthesis of free-standing metal-organic framework membranes, *Eur J Inorg Chem* 2013(8):1294–1300.

152. Ben, T., Lu, C., Pei, C., Xu, S., Qiu, S., 2012, Polymer supported and free standing metal-organic framework membrane, *Chem Eur J* 18:10250–10253.

153. Li, J.R., Kuppler, R.J., Zhou, H.C., 2009, Selective gas adsorption and separation in metal–organic frameworks, *Chem Soc Rev* 38:1477–1504.

154. Bux, H., Chmelik, C., van Baten, J.M., Krishna, R., Caro, J., 2010, Novel MOF-membrane for molecular sieving predicted by IR-diffusion studies and molecular modeling, *Adv Mater* 22(42):4741–4743.

155. Silva, L., Plaza, A., Romero, J., Sanchez, J., Rios, G.M., 2008, Characterization of MFI membranes by means of permeability determination of near critical and supercritical CO_2, *J Chil Chem Soc* 53(1):1415–1421.

156. Jeazet, H.B.T., Staudt, C., Janiak, C., 2012, Metal-organic frameworks in mixed-matrix membranes for gas separation, *Dalton Trans* 41:14003–1427.

157. Venna, S.R., Carreon, M.A., 2010, Highly permeable zeolite imidazolate framework-8 membranes for CO_2/CH_4 separation, *J Am Chem Soc* 132(1):76–78.

158. Bae, T.H., Lee, S., Qiu, W., Koros, W.J., Jones, C.W., Naire, S., 2010, A high-performance gas-separation membrane containing submicrometer-sized metal–organic framework crystals, *Angew Chem Int Ed* 49:9863–9866.

159. Bétard, A., Bux, H., Henke, S., Zacher, D., Caro, J., Fischer, R.A., 2012, Fabrication of a CO_2-selective membrane by stepwise liquid-phase deposition of an alkylether functionalized pillared-layered metal-organic framework $\{Cu_2L_2P\}_n$ on a macroporous support, *Microporous Mesoporous Mater* 150:76–82.

160. Banerjee, R., Furukawa, H., Britt, D., Knobler, C., O'Keeffe, M., Yaghi, O.M., 2009, Control of pore size and functionality in isoreticular zeolitic imidazolate frameworks and their carbon dioxide selective capture properties, *J Am Chem Soc* 131:3875–3877.

161. Li., Y.S., Liang, F.Y., Bux, H., Feldhoff, A., Yang, W.S., Caro, J., 2010, Molecular sieve membrane: Supported metal-organic framework with high hydrogen selectivity, *Angew Chem Int Ed* 49:548–551.

162. Ranjan, R., Tsapatsis, M., 2009, Microporous metal organic framework membrane on porous support using the seeded growth method, *Chem Mater* 21:4920–4924.

163. Cao, F., Zhang, C., Xiao, Y., Huang, H., Zhang, W., Liu, D., Zhong, C., Yang, Q., Yang, Z., Lu, X., 2012, Helium recovery by a Cu-BTC metal-organic-framework membrane, *Ind Eng Chem Res* 51:11274–11278.

164. Keskin, S., Sholl, D.S., 2007, Screening metal-organic framework materials for membrane-based methane/carbon dioxide separations, *J Phys Chem C* 111:14055–14059.

165. Huang, A., Wang, N., Kong, C., Caro, J., 2012, Organosilica-functionalized zeolite imidazolate framework ZIF-90 membrane with high gas separation membrane, *Angewandte Chemie* 51(2):10551–10555.

166. Zhao, Z., Ma, X., Kasik, A., Li, Z., Lin, Y.S., 2013, Gas separation properties of metal organic framework (MOF-5) membranes, *Ind Eng Chem Res* 52:1102–1108.

167. Wang, W., Dong, X., Nan, J., Jin, W., Hu, Z., Chen, Y., Jiang, J., 2012, A homochiral metal-organic framework membrane for enantioselective separation, *Chem Commun* 48:7022–7024.

168. Nan, J., Dong, X., Wang, W., Jin, W., Xu, N., 2011, Step by step procedure for preparing HKUST-1 membrane, *Langmuir* 27:4309–4312.

169. Guerrero, V.V., Yoo, Y., McCarthy, M.C., Jeong, H.K., 2010, *J Matter Chem* 20:3938–3943.

170. Wang, Z., Cohen, S.M., 2009, Postsynthetic modification of metal organic frameworks, *Chem Soc Rev* 38(5):1315–1329.

171. Adams, R., Carson, C., Ward, J., Tannenbaum, R., Koros, W., 2010, Metal organic framework mixed matrix membrane for gas separation, *Microporous Mesoporous Mater* 131(1–3):13–20.

172. An, J., Shade, C.M., Chengelis-Czegan, D.A., Petoud, S., Rosi, N.L., 2011, Zinc-adeninate meta-organic framework for aqueous encapsulation and sensitization of near-infrared and visible emitting lanthanide cations, *J Am Chem Soc* 133:1220–1223.

173. Bohrman, J.A., Carreon, M.A., 2012, Synthesis and CO_2/CH_4 separation performance of bio-MOF-1 membranes, *Chem Commun* 48:5130–5132.

174. An, J., Geib, R., Rosi, N.L., 2009, Cation-triggered drug release from a porous zinc-adeninate metal-organic framework, *J Am Chem Soc* 131:8376–8377.

175. Nik, O.G., Chen, X.Y., Kaliaguine, S., 2012, Functionalized metal organic framework-polyimide mixed matrix membranes for CO_2/CH_4 separation, *J Memb Sci* 413–414:48–61.

176. Basu, S., Cano-Odena, A., Vankelecom, I.F.J., 2011, MOF containing mixed-matrix-membranes for CO_2/CH_4 and CO_2 binary gas mixture separations, *Sep Puri Technol* 81:31–40.

177. Caro, J., 2011, Are MOF membranes better in gas separation than those made of zeolites? *Curr Opin Chem Eng* 1:77–83.

13 Pd-Based Membranes and Membrane Reactors for Hydrogen Production

Silvano Tosti

CONTENTS

13.1   Introduction 437
13.2   Hydrogen/Metal Interaction 438
      13.2.1   Properties of Hydrogenated Pd Alloys 438
      13.2.2   Stability of Composite Pd Membranes 442
13.3   Hydrogen Mass Transfer through Metals 446
      13.3.1   Hydrogen Solubility..446
      13.3.2   Hydrogen Diffusion 447
      13.3.3   Hydrogen Permeation 448
13.4   Fabrication of Metal Membranes...452
      13.4.1   Pd-based Self-Supported Membranes..453
      13.4.2   Pd-based Composite Membranes..457
      13.4.3   Other Metal Membranes...460
13.5   Applications: Pd-Membrane Reactor..464
13.6   Conclusions..480
References..480

13.1   INTRODUCTION

Long-term exploitation of fossil fuels is considered to be unsustainable due to their limited availability and the increasing emissions of greenhouse gases that are considered to be responsible for environmental pollution and climate change. Among the future energy scenarios, the use of hydrogen is proposed as an energy vector of renewable sources such as wind and solar energy [1–3]. However, at present, for large-scale use of hydrogen, the development of reliable technologies for its separation, transport, and storage is necessary. In the recent decades, membrane technologies have been successfully applied in the hydrogen separation processes. The main advantages with respect to traditional separation processes are that the membranes permit continuous operation and energy saving. Furthermore, intrinsic modularity and easy scale-up of membrane systems make their integration with other separation processes practicable [4–8].

For hydrogen separation, several kinds of Palladium (Pd)-alloy membranes have been studied [9]. In industrial applications, the need for reducing costs leads to the development of membranes made of thin metal layers, mainly Pd-composite membranes, where the thin Pd-alloy films are covered over ceramic or metal supports. These membranes are characterized by high hydrogen permeance but not complete selectivity and, therefore, cannot be used for producing ultrapure hydrogen. Conversely, dense self-supported membranes exhibit infinite hydrogen selectivity with permeance slightly lower than the composite ones.

This chapter deals with the manufacturing of Pd-Ag membranes with particular focus on the self-supported types produced through cold-rolling and diffusion welding [10–12]. The design and fabrication of both Pd membranes and membrane modules is strongly influenced by the chemical–physical properties of the Pd alloys that are, therefore, discussed in detail. Particularly, the hydrogen uploading involves a significant strain of the Pd-Ag alloy, which is responsible for the stability of the thin metal layers covered over the porous supports and can produce mechanical stresses in the permeator tubes fixed to the membrane modules.

Finally, specific design and mechanical configurations of Pd-based membrane reactors as well as their main applications for producing hydrogen through dehydrogenation reactions such as water gas shift and reforming of hydrocarbons and alcohols are described.

13.2 HYDROGEN/METAL INTERACTION

Hydrogen interaction with metals can significantly affect their chemical–physical properties. The Pd/H system is of interest for several applications such as catalysts for hydrogenation/dehydrogenation reactions and materials for hydrogen storage. In particular, membranes made of Pd alloys have been extensively studied for hydrogen separation because of their high permeability, easy activation, and reduced poisoning when operating at high temperature [4,13]. Hydrogenated Pd alloys may especially modify the metal properties. Knowledge of this behavior has to be taken into consideration for the optimum design and manufacturing of membranes and membrane modules.

13.2.1 Properties of Hydrogenated Pd Alloys

Pd alloys are preferred to pure palladium, which exhibits inadequate properties for membrane applications. Pd lattice can absorb large amounts of hydrogen with consequent embrittlement of the metal. In fact, at atmospheric pressure and below 300°C, hydrogen in Pd co-exists under two hydride forms, phases α and β, with different lattice parameters, 0.3894 and 0.4025 nm, respectively. Transition between these two phases produces cyclic lattice strains that, at macroscopic level, are responsible for the embrittlement of the metal [4,14–16].

Alloying of Pd introduced for reducing the embrittlement due to hydrogenation is also effective in increasing hydrogen permeability, mechanical strength, and other features important for membrane technologies. Commercial Pd-Ag alloys of silver content 20–25 wt% are largely used to produce membranes for hydrogen separation.

In fact, Pd-Ag alloy exhibits reduced hydrogen embrittlement and, particularly, silver composition in the range 20–25 wt% maximizes hydrogen permeability, mechanical strength, and electrical resistivity. As described in the following paragraphs, silver atoms have high mobility in the metal lattice and silver addition to Pd also makes feasible the joining through diffusion welding of the Pd-Ag foils.

By adding Ag to Pd, the α/β miscibility gap, the co-existence of two hydride phases, reduces, thus reducing the hydrogen embrittlement. In particular, at room temperature, Pd alloys with silver content over 20%–30% present one hydride phase [17]. Addition of silver to Pd also increases the hydrogen solubility that has a maximum for the silver content of commercial Pd-Ag alloys (silver 20%–30%) as shown in Figure 13.1 [18]. Hydrogen solubility is related to permeability, which also exhibits maximum for the Pd-Ag alloys of commercial grade. Figure 13.2 reports the hydrogen permeability of Pd alloys versus the content of the alloying element [4]. Apart from Pd-Y, Pd-Ag has the highest permeability among the Pd alloys. However, Pd-Y alloys are seldom used because of their modest workability.

Mechanical properties are very important for the reliability of both self-supported and composite Pd membranes. Alloying of Pd increases the mechanical strength, and, similar to other chemical–physical properties, the tensile strength of the Pd-Ag alloy, both in the worked and annealed state, is also maximum at a silver content around 25%–30%, as shown in Figure 13.3 [19]. A similar behavior is reported for the hardness versus silver content of the Pd alloy [19].

Recent designs of membrane modules use the direct ohmic heating of the Pd membranes. Electrical resistance of the metal membranes is very low and, therefore,

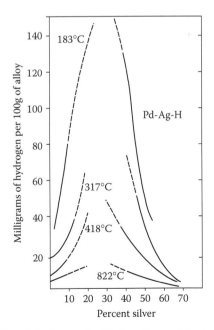

FIGURE 13.1 Solubility of hydrogen in Pd-Ag alloys at 1 atm versus silver content. (Reprinted from A.G. Knapton, *Platin. Met. Rev.*, 21, 44–50, 1977.)

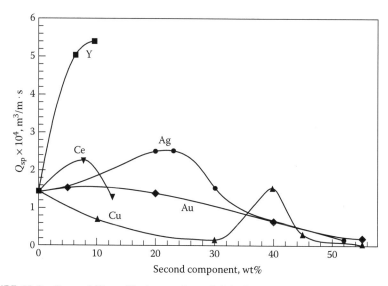

FIGURE 13.2   Permeability of hydrogen through Pd alloys at 350°C. (From J. Shu et al., Catalytic palladium-based membrane reactors: A review. *Canadian Journal of Chemical Engineering.* 1991. 69. 1036–1060. Copyright Wiley-VCH Verlag GmbH & Co. KGaA. Reproduced with permission.)

such systems need to be supplied by high currents. In view of this, it is important to use alloys with slightly higher electrical resistivity. As shown in Figure 13.4, the electrical resistivity of Pd-Ag alloys is also maximum in the range of silver content 20–40 wt% [20].

Another important feature to be considered for the design of membranes and membrane modules is the strain of the hydrogenated Pd-Ag alloy. In fact, high hydrogen uploading of this alloy is responsible for a significant macroscopic expansion of the permeators. The dilatometric behavior of hydrogenated commercial Pd-Ag alloy was described by Fort and Harris [21]. The expansion/contraction of both hydrogenated and nonhydrogenated Pd-Ag alloy (Ag 25%) is shown in Figure 13.5. Below 100°C, the hydrogenated alloy expands by following the same trend of the nonhydrogenated metal. Afterward, the hydrogenated alloy shrinks much more than the nonhydrogenated alloy up to about 300°C. Over this temperature most of the hydrogen is released from the lattice and then the hydrogenated alloy behaves like the nonhydrogenated one.

It is noteworthy to compare the strain ($\varepsilon_{H/Pd-Ag}$) of hydrogenated and nonhydrogenated Pd-Ag (ε_{Pd-Ag}) by increasing the temperature from room temperature to about 300°C. Hydrogenated Pd alloy contracts its length by about 1.5% ($\varepsilon_{H/Pd-Ag} \approx -0.015$), whereas the nonhydrogenated alloy elongates by about 0.3% ($\varepsilon_{Pd-Ag} \approx 0.003$). Therefore, it results with $\varepsilon_{H/Pd-Ag} \approx -5 \times \varepsilon_{Pd-Ag}$. Such behavior has to be taken into consideration for the design of reliable membrane modules; this will be discussed in detail in Section 13.2.2.

Measurements of Fort and Harris have been confirmed by recent studies, which tested Pd-Ag permeators in the form of tubes [22,23].

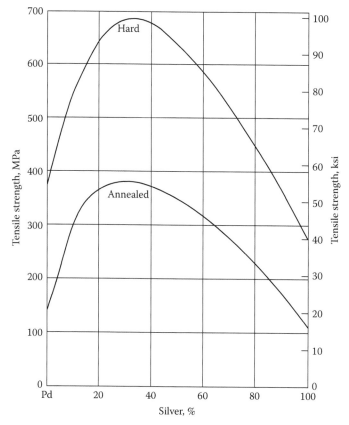

FIGURE 13.3 Tensile strength of Pd-Ag alloys versus silver content. (Reprinted with permission from ASM International, *ASM Handbook, Metals Handbook: Volume 2, Properties and Selection: Nonferrous Alloys and Special-Purpose Materials*, Materials Park, OH, 2000. www.asminternational.org. All rights reserved.)

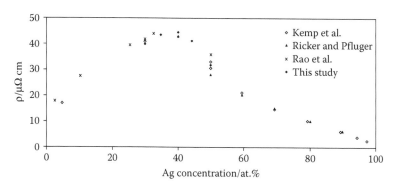

FIGURE 13.4 Electrical resistivity of Pd-Ag alloys versus silver concentration at room temperature. (Reprinted with permission from S. Arajs et al., *Phys. Rev. B*, 15, 2429–2431, 1977. Copyright 1977 the America Physical Society. http://link.aps.org/doi/10.1103/PhysRevB.15.2429.)

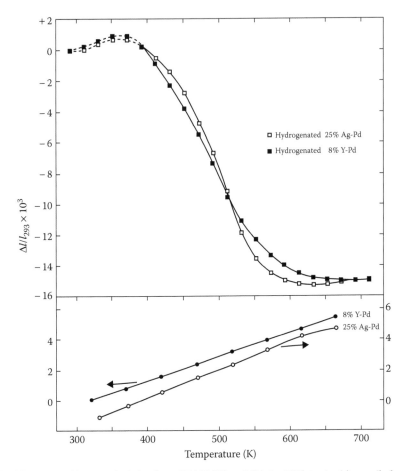

FIGURE 13.5 Dilatometric behavior of Pd-Y 8% and Pd-Ag 25% in the binary (below in the graph) and hydrogenated (above in the graph) forms. The expansion or contraction is expressed as the ratio of the change in length (Δl) to the room temperature length ($l_{293\ K}$). (Reprinted from *Journal of the Less-Common Metals*, 41, D. Fort, I.R. Harris, The physical properties of some palladium alloy hydrogen diffusion membrane materials, 313–327, Copyright 1975, with permission from Elsevier.)

13.2.2 STABILITY OF COMPOSITE PD MEMBRANES

In order to reduce costs, composite membranes made of thin Pd-alloy films deposited over porous supports have been developed. The thickness of the metal films affects the permeance, selectivity, and stability of the composite membranes. The relationship between the metal layer thickness of composite membranes and their cost, permeance, and selectivity can be clearly defined. Conversely, in order to establish the effect of the metal layer thickness over the membrane stability, it is necessary to consider the dilatometric behavior of the hydrogenated Pd-Ag alloy. Such a behavior is responsible for the shear stresses at the interface metal/support that can involve the peeling of the metal layer.

During normal operation Pd-based membranes are hydrogenated only at high temperature (typically over 300°C). On the contrary, with the purpose of simulating malfunctions, such as failure of control systems, stability tests have to be carried out with Pd-Ag hydrogenated at low temperatures (ambient temperature); this analysis considers thermal and hydrogenation cycling of a composite membrane also at low temperature. In fact, maximum stresses at the metal/support interface occur when the Pd-Ag is cooled from 300°C to ambient temperature under a hydrogen atmosphere or vice versa and hydrogenated Pd-Ag is heated from ambient temperature to 300°C. The metal layer expands under the effect of the hydrogen uploading whereas the support contracts or vice versa.

The scheme of Figure 13.6 represents the forces acting on the metal layer and the support of a composite membrane in the case of thermal and hydrogenation cycling. Expansion of the Pd-Ag layer when hydrogenated is constrained by its adhesion to the support. Compression stress of Pd-Ag layer is calculated as follows:

$$\sigma = E_{Pd\text{-}Ag}\left(\varepsilon_{H/Pd\text{-}Ag} - \varepsilon_{sup}\right) \tag{13.1}$$

where:

σ is the compression stress (Pa)

$E_{Pd\text{-}Ag}$ is the Young's modulus of Pd-Ag (Pa)

ε_{sup} is the strain of the porous support

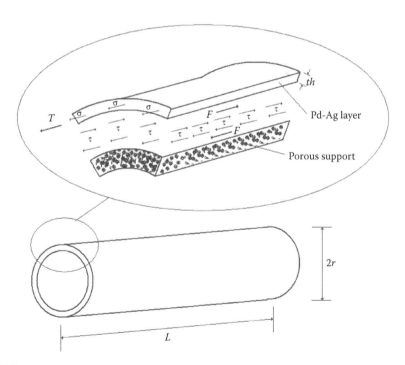

FIGURE 13.6   Schematic representation of a composite membrane (below) and scheme of the stresses (τ) at the interface between metal layer and porous support (above).

This compression stress works over the annular cross section of the thin metal layer of thickness th (m). The resulting axial force F (N) is

$$F \cong \sigma 2\pi r\,th = E_{\text{Pd-Ag}}\left(\varepsilon_{\text{H/Pd-Ag}} - \varepsilon_{\text{sup}}\right)2\pi r\,th \tag{13.2}$$

This traction force is equilibrated by the shear force T (N) acting at the interface metal/support along all the porous tube of length L (m) and external radius r (m). Such a shear force T can be calculated by considering the shear stress τ (Pa) at the interface metal/support as follows:

$$T = \tau 2\pi r L \tag{13.3}$$

By posing the equilibrium condition $F = T$, it results as follows:

$$\tau = E_{\text{Pd-Ag}}\left(\varepsilon_{\text{H/Pd-Ag}} - \varepsilon_{\text{sup}}\right)th/L \tag{13.4}$$

This expression establishes that the stability of a composite membrane is affected, through the shear stresses at the interface metal/support, by the thickness of the metal layer and the length of the membrane tube as well as by the mechanical properties of the Pd-Ag alloy (Young's modulus and strain).

With the purpose of establishing a criterion of stability of composite Pd membranes based on the thickness of the metal layer, an assessment of the shear stresses can be performed for the two cases of ceramic and metal supports. The thermal strain of the ceramic support is assumed to be negligible, $\varepsilon_{\text{cer}} \approx 0$ (i.e., infinite stiffness of the ceramic material), whereas the thermal strain of the metal support is postulated to be equal to that of nonhydrogenated Pd-Ag, $\varepsilon_{\text{met}} \approx \varepsilon_{\text{Pd-Ag}}$.

In case of ceramic support, the difference between the Pd-Ag and ceramic strains is

$$\Delta\varepsilon = \varepsilon_{\text{H/Pd-Ag}} - \varepsilon_{\text{sup}} = \varepsilon_{\text{H/Pd-Ag}} - \varepsilon_{\text{cer}} \text{ that is } \Delta\varepsilon \approx \varepsilon_{\text{H/Pd-Ag}} \tag{13.5}$$

In case of metal support, the differential strain is

$$\Delta\varepsilon = \varepsilon_{\text{H/Pd-Ag}} - \varepsilon_{\text{sup}} = \varepsilon_{\text{H/Pd-Ag}} - \varepsilon_{\text{met}} \text{ that is, } \Delta\varepsilon \approx \varepsilon_{\text{H/Pd-Ag}} - \varepsilon_{\text{Pd-Ag}} = 1.2 \times \varepsilon_{\text{H/Pd-Ag}} \tag{13.6}$$

First of all, this analysis demonstrates that the kind of the support (ceramic or metal) does not significantly affect the stability of the composite membrane. Accordingly, the following calculations will be performed for the case of ceramic support for which $\Delta\varepsilon \approx \varepsilon_{\text{H/Pd-Ag}}$.

Using in Equation 13.4 the values of strain of hydrogenated Pd-Ag obtained from the literature [21], the shear stress at the metal layer/support interface versus temperature is obtained as a function of the temperature for different Pd-Ag thickness [22] (Figure 13.7). Shear stress values in the range 5–10 MPa (50–100 kg cm^{-2}) are deemed to be a limit for the adhesion of the metal layer to the support. In view of this, composite Pd membranes with metal film thickness over 5 mm could exhibit a poor durability under hydrogenation cycling from room temperature to 200°C–300°C.

From this stress analysis, we can conclude that for exhibiting a good stability, a composite membrane has to contain a thin metal layer. On the contrary, to achieve

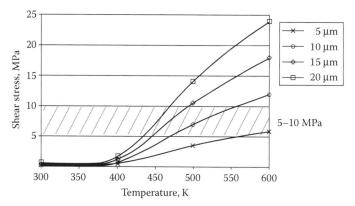

FIGURE 13.7 Shear stress at the interface Pd-Ag/ceramic support versus the temperature for different thicknesses of the metal layer. (Reprinted from *Journal of Membrane Science*, 196, S. Tosti et al., Sputtered, electroless, and rolled palladium-ceramic membranes, 241–249, Copyright 2002, with permission from Elsevier.)

high selectivity, the thickness of the metal layer should be large enough to block all the pores of the support. That is, the thickness of the metal layer has to be larger than the pores of the support of a composite membrane, but these pores cannot be too small to avoid large mass transfer resistance through the support.

For a composite membrane made of thin metal layers, durability and selectivity can only be partially satisfied at the same time. In fact, the pores of the support cannot be too small, to avoid large mass transfer resistance, and the thickness of the metal layer has to be larger than the pore size, to have good selectivity, but not too large so as shear stress will compromise its adhesion over the support.

Figure 13.8 shows a composite Pd obtained by covering through electroless plating a porous ceramic support with a Pd-Ag layer of thickness 10–15 μm membrane after thermal and hydrogenation cycling.

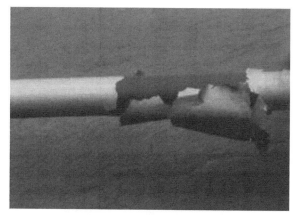

FIGURE 13.8 Peeling of the Pd-Ag layer deposited through electroless plating over a ceramic porous support of external diameter 10 mm.

13.3 HYDROGEN MASS TRANSFER THROUGH METALS

Hydrogen can permeate selectively dense metal layers, and, accordingly, this phenomenon is exploited when metal membranes are applied for separating ultrapure hydrogen from gas mixtures [24–27]. The hydrogen mass transfer through dense metal membranes includes several transport mechanisms, which are discussed in the following sections.

13.3.1 HYDROGEN SOLUBILITY

The hydrogen solubility is a measure of the amount of hydrogen present in the metal lattice; it is controlled by the kinetics of the reactions that occur over the metal surface. When hydrogen gas comes in contact with a metal, two hydrogen fluxes are established through the metal surfaces. Figure 13.9 represents the ideal case in which a piece of metal is put in a container under a hydrogen atmosphere at constant temperature and pressure. The hydrogen flux entering the metal surface is proportional to the hydrogen partial pressure in the gas phase

$$J_{in} = Ka p_{H2} \tag{13.7}$$

where:
J is the flux of hydrogen (molecules), mol m^{-2} s^{-1}
Ka is the rate of adsorption, mol m^{-2} s^{-1} Pa^{-1}
p_{H2} is the hydrogen partial pressure in the gas, Pa

The hydrogen flux leaving the metal is due to the recombination of two protons in one molecule and, then, it is proportional to the square of the hydrogen concentration in the metal lattice

$$J_{out} = Ka'[H]^2 \tag{13.8}$$

where:
$[H]$ is the hydrogen concentration into the metal, mol m^{-3}
Ka is the rate of desorption from the metal surface, m^4 mol^{-1} s^{-1}

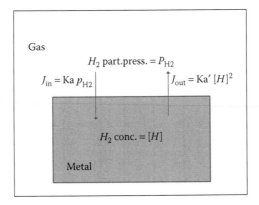

FIGURE 13.9 Hydrogen solubility into metals with hydrogen fluxes entering and leaving the metal surface.

The total hydrogen flux J is

$$J = J_{in} - J_{out} \tag{13.9}$$

By combining Equations 13.7 through 13.9 and introducing the constant Ke = Ka'/Ka, the square of hydrogen concentration in the metal can be calculated as follows:

$$[H]^2 = \frac{p_{H2up} - (J/Ka)}{Ke} \tag{13.10}$$

The square of hydrogen concentration in the metal is proportional through the ratio of the adsorption/desorption constants to the hydrogen partial pressure in the gas phase minus a term (J_{tot}/Ka) representing the pressure drop due to surface effects. The term $1/Ka$ can be defined as a mass transfer resistance due to the surface reactions.

It is noteworthy to consider the previous expression under steady-state conditions. In this case, the hydrogen in the metal lattice is in equilibrium with that in the gas phase and, therefore, $J = 0$. By calculating the square root of Equation 13.10, the well-known Sieverts' law results as follows:

$$[H] = \frac{1}{Ke^{0.5}} p_{H2}{}^{0.5} \tag{13.11}$$

This law establishes at a given temperature the relationship between the concentration of hydrogen (protons) into the metal lattice in equilibrium with the hydrogen (molecules) in the gas phase.

In fact, the solubility coefficient S (mol m^{-1} Pa$^{-0.5}$) is related to the constant Ke (m^6 Pa mol^{-2}) through the following equation:

$$S = \frac{1}{Ke^{0.5}} \tag{13.12}$$

Its dependence on the temperature is expressed by an Arrhenius-based law as follows:

$$S = S_0 e^{-\frac{E_s}{RT}} \tag{13.13}$$

where:
S_0 is the solubility pre-exponential factor, m^2 s^{-1}
E_s is the activation energy of the solubility, J mol^{-1}
RT is the gas constant, 8.314 J mol^{-1} K^{-1}

13.3.2 HYDROGEN DIFFUSION

A gradient of hydrogen concentration in the metal lattice produces a flux according to Fick's second law, which, in the one-dimensional case, is as follows:

$$\frac{\partial [H]}{\partial t} = \frac{\partial}{\partial x}\left(D\frac{\partial [H]}{\partial x}\right) = D\frac{\partial^2 [H]}{\partial x^2} \tag{13.14}$$

where:

  D is the diffusion coefficient, m² s⁻¹

  t is the time, s

  x is the spatial abscissa, m

Under equilibrium ($t = > \infty$) and other general hypotheses (i.e., D = constant), the integration of Equation 13.14 for a metal layer of thickness th (m) leads to Fick's first law:

$$J = D \frac{\left[H_{up}\right] - \left[H_{out}\right]}{th} \tag{13.15}$$

where the indexes *up* and *down* represent the zones at high (upstream) and low (downstream) hydrogen concentration, respectively.

An Arrhenius law is also applied to the diffusion coefficient to account for its dependence on temperature:

$$D = D_0 e^{-\frac{E_D}{RT}} \tag{13.16}$$

where:

  D_0 is the diffusivity pre-exponential factor, m² s⁻¹

  E_D is the activation energy of the diffusion, J mol⁻¹

13.3.3 HYDROGEN PERMEATION

When the surfaces of a dense metal layer are exposed to hydrogen at different partial pressures, a mass transfer of hydrogen (permeation) occurs from the side at higher pressure (upstream) to that at lower one (downstream). In this analysis, five main transport mechanisms are recognized for the permeation:

- Adsorption of the molecular hydrogen on the first metal surface (upstream surface)
- Dissociation of the hydrogen into two protons at the first metal surface (upstream surface)
- Diffusion of the protons through the metal lattice
- Recombination of the two protons at the opposite side of the metal wall (downstream surface)
- Desorption of the molecular hydrogen from the metal surface (downstream surface)

Consider the case of a dense metal wall, as represented schematically in Figure 13.10. Hydrogen at high pressure (p_{H2up}) enters the left surface of the metal and leaves the right surface at low pressure (p_{H2down}). The hydrogen concentration entering the metal in proximity of the high and low pressure surfaces is $[H_{up}]$ and $[H_{down}]$, respectively.

Using Equation 13.10 and considering the hydrogen flux moving from left to right as positive, for the square of the hydrogen concentration at high and low pressure the following two expressions can be written as follows:

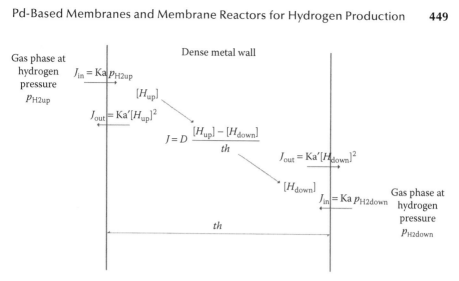

FIGURE 13.10   Scheme of the hydrogen permeation through a dense metal wall.

$$\left[H_{up}\right]^2 = \frac{p_{H2up} - \left(J/Ka\right)}{Ke} \tag{13.17}$$

$$\left[H_{down}\right]^2 = \frac{p_{H2down} + \left(J/Ka\right)}{Ke} \tag{13.18}$$

Equations 13.17 and 13.18 are obtained under the hypothesis, generally verified, that the adsorption/desorption constants (Ka, Ka′, and then Ke) are equal for the two metal surfaces. Different values of these constants could be due to the diverse status of the surfaces with presence of impurities, blanketing of the membrane surface, and variation of the rugosity. Accordingly, the gradient of the hydrogen concentration in the metal wall is obtained by elaborating Equation 13.17 with 13.18 as follows:

$$\left[H_{up}\right] - \left[H_{down}\right] = \frac{\left[p_{H2up} - \left(J/Ka\right)\right] - \left[p_{H2down} + \left(J/Ka\right)\right]}{Ke\left(\left[H_{up}\right] + \left[H_{down}\right]\right)} \tag{13.19}$$

Fick's first law can be applied to the hydrogen concentration gradient by combining Equations 13.15 and 13.19 as follows:

$$J = \frac{D}{th} = \frac{\left[p_{H2up} - \left(J/Ka\right)\right] - \left[p_{H2down} + \left(J/Ka\right)\right]}{Ke\left(\left[H_{up}\right] + \left[H_{down}\right]\right)} \tag{13.20}$$

By simplifying and introducing at the denominator the square root of the expressions 13.17 and 13.18, we can write the following equation, where the solubility coefficient defined in Equation 13.12 has been also considered:

$$J = \frac{p_{H2up} - p_{H2down}}{\left(2/Kd\right) + \left(th/SD\right)\left[p_{H2up} - \left(J/Ka\right)\right]^{0.5} + \left[p_{H2down} + \left(J/Ka\right)\right]^{0.5}} \tag{13.21}$$

The product $S \times D$ is defined as the permeability coefficient Pe (mol s^{-1} m^{-1} Pa$^{-0.5}$). According to Equations 13.13 and 13.16, its dependence from the temperature is

$$Pe = Pe_0 e^{-\frac{E_p}{RT}} \tag{13.22}$$

where:
Pe_0 is the permeability pre-exponential factor (mol m^{-1} s^{-1} Pa$^{-0.5}$)
E_p is the activation energy of permeability (J mol^{-1})

Clearly, it results that $Pe_0 = D_0 \times S_0$ and $E_P = E_D + E_S$.
By introducing the permeability coefficient in Equation 13.21, it is described as follows:

$$J = \frac{p_{\text{H2up}} - p_{\text{H2down}}}{(2/\text{Ka}) + (th/Pe)\left[p_{\text{H2up}} - (J/\text{Ka})\right]^{0.5} + \left[p_{\text{H2down}} + (J/\text{Ka})\right]^{0.5}} \tag{13.23}$$

As previously discussed in Section 13.3.3, 1/Ka accounts for the surface mass transfer resistance that, usually, is assumed to be negligible for the membranes of thickness larger than 0.1 mm [28]. However, even in case of thick membranes, surface effects could occur and affect significantly the permeation as a consequence of the presence of some gases such as CO, CO_2, and methane contaminating the metal surface [29]. In practice, when the term $1/Ka$ is negligible, the permeation occurs under a diffusion-controlled regime; the surface reactions of recombination/dissociation are fast and the kinetics of diffusion of hydrogen through the membrane is the controlling step. In this case, Equation 13.28 reduces to the following well-known expression:

$$J = \frac{Pe}{th}\left(p_{\text{H2up}}^{0.5} - p_{\text{H2down}}^{0.5}\right) \tag{13.24}$$

This equation is usually reported as a consequence of the Sieverts' law described by the expression 13.11 and, then, it is also called *Sieverts' law of permeation*.
Through algebraic passages, Equation 13.23 can be rewritten as follows:

$$J = \frac{Pe}{th}\left\{\left[p_{\text{H2up}} - \left(\frac{J}{\text{Ka}}\right)\right]^{0.5} - \left[p_{\text{H2down}} + \left(\frac{J}{\text{Ka}}\right)\right]^{0.5}\right\} \tag{13.25}$$

A comparison of Equations 13.24 and 13.25 identifies that the term $1/Ka$ is a mass transfer resistance R_S (m^2 s Pa mol^{-1}) taking into account surface phenomena or *wall effects*, leading to deviations from Equation 13.24 [30–32]. In fact, the term J/Kd (Pa) introduces a pressure drop reducing the permeation driving force of the Sieverts' law of permeation. In other words, the expression 13.24 can be derived from Equation 13.25 by modifying the driving force for the surface effects as shown in Figure 13.11.
Expression 13.25 is applied for calculating the values of permeability and surface resistance of dense Pd-Ag permeators in recent experiments carried out in a large temperature (200°C–450°C) and pressure range (200–800 kPa) [33,34].

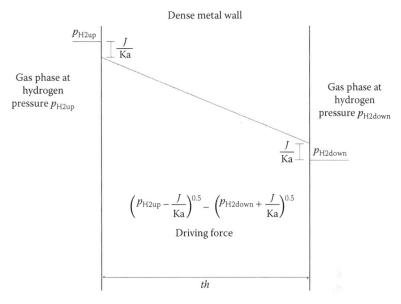

Dense metal wall

Driving force

$$\left(p_{H2up} - \frac{J}{Ka}\right)^{0.5} - \left(p_{H2down} + \frac{J}{Ka}\right)^{0.5}$$

FIGURE 13.11 Permeation driving force of Sieverts' expression in the presence of surface effects.

Equation 13.23 can be modified for the case of a diffusion-controlled regime. This case occurs when the diffusion through the metal lattice is fast compared to the adsorption/desorption. The term *th/D* of Equation 13.21 can be neglected for very thin membranes and/or high diffusion coefficients and it can be written as follows:

$$J = \frac{Kd}{2}\left(p_{H2up} - p_{H2\,down}\right) \tag{13.26}$$

In this case, there is a linear dependence of the hydrogen flow rate versus the hydrogen partial pressures difference and the pressure exponent is equal to 1.

Equation 13.23 that is obtained for the case of a dense metal layer can be modified by including other mass transfer resistances as follows:

$$J = \frac{p_{H2up} - p_{H2\,down}}{2R_S + \left(th/Pe\right)\left[\left(p_{H2up} - JR_S\right)^{0.5} + \left(p_{H2\,down} + JR_S\right)^{0.5}\right] + R_p + R_F} \tag{13.27}$$

where:
R_S is the surface resistance previously defined
R_F and R_p (m² s Pa mol⁻¹) are the mass transfer resistances due to gas-film and porous support (for Pd-composite membranes), respectively

This general expression can be simplified. In fact, many authors report a Sieverts' law modified by introducing a generic pressure exponent *n*, which is in the range 0.5–1, as follows:

$$J = \frac{Pe}{th}\left(p_{H2up}^{n} - p_{H2\,down}^{n} \right) \tag{13.28}$$

where:
the permeability coefficient is given in mol s^{-1} m^{-1} Pa^{-n}

For a membrane made of several layers of different materials, such as the case of a composite membrane, just one permeability coefficient cannot be used. In fact, hydrogen permeability is a parameter strictly related to each material and, for a composite membrane, the permeance Φ (mol s^{-1} m^{-2} Pa^{-n}) is therefore introduced. In this way, the expression 13.28 takes the following form:

$$J = \Phi\left(p_{H2up}^{n} - p_{H2\,down}^{n} \right) \tag{13.29}$$

Assessment of the permeance for a composite membrane has to take into consideration the permeability of all layers, their mass transfer mechanism, and thickness. An example of the calculation of permeance of dense metal composite membranes is reported by Tosti [35].

In practice, the value of the pressure exponent found in the literature for the expressions 13.28 and 13.29 may depend on the controlling mass transfer mechanism (diffusion or adsorption/desorption) through the metal layer. The pressure exponent may also be influenced from the surface effects, presence of contaminants, porous support, and gas-film resistance.

For example, we can consider a Pd-ceramic composite membrane consisting of a ceramic porous support covered by a Pd-alloy layer. Hydrogen permeation through the metal follows the Sieverts' law ($n = 0.5$), whereas the hydrogen transport through the porous ceramic support is ruled by the Darcy or Knudsen laws ($n = 1$) [36–38]. For such a membrane, it results in $0.5 < n < 1$, and the deviations from the Sieverts' law are due to the porous support.

13.4   FABRICATION OF METAL MEMBRANES

Hydrogen can be separated through metal membranes made of Pd alloys. In Section 13.2.2, it was discussed that the selectivity as well as the stability of these membranes is strongly related to the thickness of the metal layers and the properties of the hydrogenated Pd-Ag alloy. In particular, dense self-supported membranes exhibit complete selectivity and are used for producing ultrapure hydrogen. Thin-walled Pd-Ag permeator tubes have been produced in order to reduce the costs and increase the hydrogen permeance. Further increase of permeance can be attained using composite membranes made of very thin metal layers deposited over porous supports. However, for these membranes, high selectivity and good stability cannot be obtained at the same time. Examples of composite membranes using metal supports (supported metal membranes) or layers of different metals (laminated membranes) have also been studied.

13.4.1   Pd-based Self-Supported Membranes

Thin-walled Pd-Ag tubes have been produced through cold-rolling and diffusion welding of metal foils [9]. These permeators of wall thickness 50–60 μm have at 400°C a rupture pressure of about 1.7 MPa (see Table 13.1) and are used in hydrogen separation processes with operating pressure up to 0.2–0.3 MPa [39]. These dense and defect-free metal membranes exhibit complete hydrogen selectivity as well as chemical–physical stability as demonstrated in long-term tests [40]. In particular, at 300°C–350°C with a transmembrane pressure of 200 kPa, about 3 Nm^3 m^{-2} h^{-1} of hydrogen can be separated through these membranes.

Thin metal foils can be obtained by cold-working, a process of plastic deformation occurring below 30% of the metal melting absolute temperature. Usually, cold-working processes such as pressing, drawing, and rolling are carried out at ambient temperature. In particular, cold-rolling can operate at high working speeds and, therefore, it is used industrially for reducing the thickness of metal foils. The characteristics of the rolling mills depend on the mechanical properties of the material and the required rolling speed and thickness reduction. The rolls' diameter especially influences the minimum thickness achievable; the metal sheet is the thinnest after rolling when the diameter of the rolls is at their smallest. However, rolls of small diameter can produce curved metal foils because of their bending, as shown in Figure 13.12 [41]. Production of Pd-Ag foils of thickness 50 μm has been carried out through a four-high rolling mill, in which two support rolls of large diameter avoid the bending of the smaller working rolls. The scheme and picture of the four-high rolling mill used at ENEA Frascati laboratories are reported in Figures 13.13 and 13.14, respectively.

Cold-rolling modifies the mechanical properties of the metal; both hardness and tensile strength increase because the dislocations movement in the metal lattice is constrained by the presence of grain boundaries and by the occurrence of new sources of dislocation, which requires higher stresses in order to be moved [42,43].

TABLE 13.1
Calculated Rupture Pressure, Pr, for 10 mm Diameter Tube Permeators

	Pr (MPa), annealed		Pr (MPa), hard	
	Tamb	400°C	Tamb	400°C
Pd, 50 μm	1.50	0.67	3.78	1.68
Pd, 70 μm	2.10	0.93	5.29	2.35
Pd-Ag, 50 μm	3.79	1.68	6.84	3.04
Pd-Ag, 70 μm	5.30	2.36	9.58	4.25

Source: *Int. J Hydrogen Energy*, 25, S. Tosti et al., Rolled thin Pd and Pd-Ag membranes for hydrogen separation and production, 319–325, Copyright 2000, with permission from Elsevier.

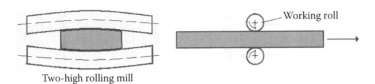

Two-high rolling mill

FIGURE 13.12   Rolls bending of a two-high rolling mill; the worked metal foils are curved. (From Tosti, S.: Metallic membranes prepared by cold rolling and diffusion welding, in *Membranes for Membrane Reactors: Preparation, Optimization and Selection*, A. Basile and F. Gallucci, eds., 155–167, 2011. Copyright Wiley-VCH Verlag GmbH & Co. KGaA. Reproduced with permission.)

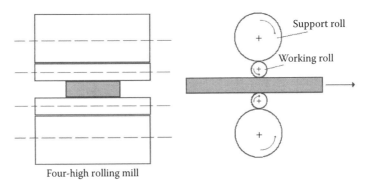

Four-high rolling mill

FIGURE 13.13   Scheme of a four-high rolling mill. (From Tosti, S.: Metallic membranes prepared by cold rolling and diffusion welding, in *Membranes for Membrane Reactors: Preparation, Optimization and Selection*, A. Basile and F. Gallucci, eds., 155–167, 2011. Copyright Wiley-VCH Verlag GmbH & Co. KGaA. Reproduced with permission.)

FIGURE 13.14   The four-high rolling mills at ENEA Frascati laboratories. (From Tosti, S.: Metallic membranes prepared by cold rolling and diffusion welding, in *Membranes for Membrane Reactors: Preparation, Optimization and Selection*, A. Basile and F. Gallucci, eds., 155–167, 2011. Copyright Wiley-VCH Verlag GmbH & Co. KGaA. Reproduced with permission.)

Through heat treatment (i.e., annealing), dislocation is removed and the mechanical properties can be recovered.

When cold-worked, the hardness of commercial Pd-Ag moves from about 90 to 180 HB (Brinell hardness) and the tensile strength from about 380 to 680 MPa. During rolling, the measurement of the hardness verifies the need for annealing the material. Annealing has been carried out at 800°C–1000°C per 1–2 h under a controlled atmosphere (Ar with 5% of hydrogen or vacuum) for avoiding the oxidation of the Pd-Ag alloy. When annealing under vacuum, the modification of the composition of the Pd alloy due to the evaporation of silver has to be considered.

After cold-rolling, the Pd-Ag foils are joined to form permeator tubes through diffusion welding [10]. Other welding techniques have been tested, such as brazing and tungsten inert gas (TIG) welding. However, brazing may introduce metal impurities, contaminating the Pd-Ag alloy, thus reducing its hydrogen permeability and resistance to embrittlement. TIG welding involved thermal stresses zones of the Pd-Ag membranes where defects such as cracks and microholes occur during thermal and hydrogenation cycling [39]. Figure 13.15 shows details of a defect produced after hydrogenating a Pd-Ag membrane tube TIG welded.

Diffusion welding is a technique used to join metals; it consists of pressing at temperatures 50%–75% of the melting point of the metal parts to be joined [44,45]. Usually, the load applied provokes no macroscopic deformation of the material. At high temperature, the metal atoms diffuse through the lattice, thus joining the parts pressed together.

Diffusion welding process is controlled by temperature and time through Fick's second law (13.14) described in Section 13.3.2 for the case of hydrogen diffusion in a metal lattice.

Integration of expression 13.14 introduces the *length of diffusion* λ (m), a parameter indicating the depth achieved at the time t by a solute that diffuses through the metal lattice with diffusivity D as follows:

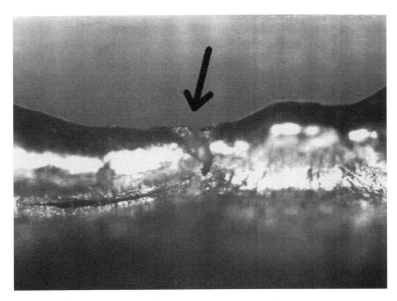

FIGURE 13.15 Details of an arc-welded tube after thermal and hydrogenation cycling (100×). (With kind permission from Springer Science + Business Media: *Journal of Material Science*, Diffusion bonding of Pd-Ag membranes, 39, 2004, 3041–3046, S. Tosti, L. Bettinali.)

$$\lambda = \sqrt{4Dt} \tag{13.30}$$

Metals characterized by high diffusion coefficients allow the diffusion welding at relatively low temperatures for short times. Another important technological feature of adding silver to palladium is given by the high diffusivity of silver that makes easier the diffusion welding. In fact, the diffusion coefficient of silver is as follows:

$$D = D_0 \exp\left(\frac{-E_d}{RT}\right) \tag{13.31}$$

with $D_0 = 6{,}7\ 10\text{--}5\ \mathrm{m^2\ s^{-1}}$ and $E_d = 45{,}2\ \mathrm{kcal\ mol^{-1}}$ [46].

Through Equation 13.30, the diffusion length of silver is calculated versus time for different temperatures, as presented in Figure 13.16. For instance, Pd-Ag foils of thickness 50 μm can be joined through diffusion welding when the diffusion length is about 100 μm; the thickness of two metal foils overlapped. For such a case, it is observed that less than 1 h at 1000°C is needed.

Metal foils are joined through diffusion welding by pressing them together through specially designed devices [10,47]. In the first apparatus developed at ENEA Frascati laboratories, the overlapped limbs of the thin Pd-Ag sheet wrapped around an alumina bar are compressed in the device shown in Figure 13.17.

The cross section of the welded zone is shown in Figure 13.18a, whereas the scheme of the overlapped limbs of the Pd-Ag foils is shown in Figure 13.18b.

In an alternative method, a thermomechanical press applies the pressure needed to weld the metal limbs of the Pd-Ag foil under the heat treatment. In fact, this press,

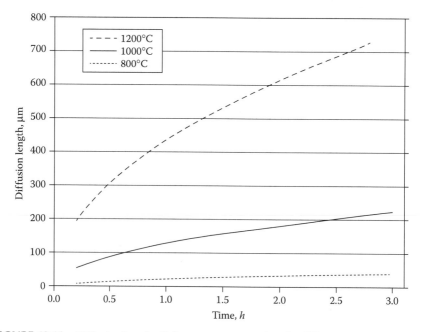

FIGURE 13.16   Diffusion length of silver atoms versus time for different temperatures.

FIGURE 13.17   Picture and schematic representation of the diffusion welding device. (From Tosti, S.: Metallic membranes prepared by cold rolling and diffusion welding, in *Membranes for Membrane Reactors: Preparation, Optimization and Selection*, A. Basile and F. Gallucci, eds., 155–167, 2011. Copyright Wiley-VCH Verlag GmbH & Co. KGaA. Reproduced with permission.)

shown in Figure 13.19, is composed of two stainless steel plates pressed together by threatened bars (screws) made of a special metal alloy (INVAR). Such an alloy is characterized by a negligible thermal expansion coefficient, and under the heat treatment, it compresses the steel plates and the other parts, in particular, the overlapped limbs of the Pd-Ag foil. The permeator tube is manufactured by brazing to the Pd-Ag membrane two stainless steel tube ends, providing the mechanical stiffness needed for its tight connection to the membrane module; see Figure 13.20.

13.4.2   Pd-based Composite Membranes

In order to reduce the costs and increase the hydrogen permeance, thin Pd-based layers are deposited over porous supports made of glass, ceramic, or metal. Both physical and chemical (reactive) techniques are adopted for preparing the Pd coatings. Physical methods consist of physical vapor deposition [48,49], spray pyrolysis [50], sputtering [11,40,51], co-condensation [52], and electroplating [53–55]. Among the reactive deposition methods, the most important are chemical vapor deposition [48,56], electrochemical vapor deposition, sol-gel technique, and electroless plating [11,57–59].

The Pd-composite membranes with ceramic supports are usually composed of three parts, a Pd-based layer, the ceramic porous support, and an intermediate ceramic layer of small porosity, which has the function of improving the adhesion of the metal layer over the support. Apte et al. covered a bilayer ceramic porous support

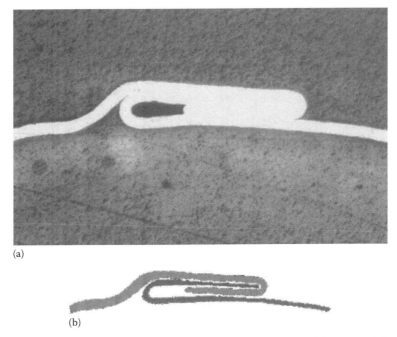

(a)

(b)

FIGURE 13.18 (a) Cross section of the diffusion bonding joint (32×); (b) scheme of the overlapping method used for the diffusion bonding joint. (With kind permission from Springer Science + Business Media: *Journal of Material Science*, Diffusion bonding of Pd-Ag membranes, 39, 2004, 3041–3046, S. Tosti, L. Bettinali.)

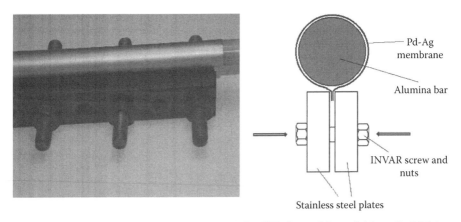

FIGURE 13.19 Thermomechanical press used for diffusion welding of thin-walled Pd-Ag tubes. (From Tosti, S.: Metallic membranes prepared by cold rolling and diffusion welding, in *Membranes for Membrane Reactors: Preparation, Optimization and Selection*, A. Basile and F. Gallucci, eds., 155–167, 2011. Copyright Wiley-VCH Verlag GmbH & Co. KGaA. Reproduced with permission.)

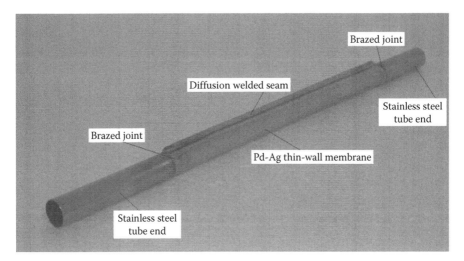

FIGURE 13.20 Thin-wall permeator; the Pd-Ag membrane tube is joined with two stainless steel tube ends by brazing. (From Tosti, S.: Metallic membranes prepared by cold rolling and diffusion welding, in *Membranes for Membrane Reactors: Preparation, Optimization and Selection*, A. Basile and F. Gallucci, eds., 155–167, 2011. Copyright Wiley-VCH Verlag GmbH & Co. KGaA. Reproduced with permission.)

with Pd-alloy film through electroless plating [60]. It is noteworthy to consider that the presence of the intermediate ceramic layer with small pores could significantly affect the hydrogen permeance. A modified electroless plating procedure ensuring high permeance was developed by Hou et al. by filling the ceramic porous support with $Al(OH)_3$ that decomposes to porous alumina Al_2O_3 [61].

Composite membrane fabricated using metal supports can also use an intermediate layer for reducing the thermal mismatching between the support and the Pd-alloy film. For instance, ceramic layers have been positioned in between stainless steel porous supports and Pd-Ag and Pd films [62]. The ceramic interlayer also reduces intermetallic diffusion of the supports into the active Pd-based layers that could contaminate the membrane. Intermetallic film made of a porous Pd-Ag layer obtained through continuous electroless plating of alternating Pd and Ag baths has been used in the fabrication of composite Pd membranes, which consist of porous stainless steel plates with a selective Pd layer [63]. These membranes were successfully tested up to 500°C. Their effectiveness as diffusion barriers for Pd membrane deposited over porous stainless steel supports at 400°C was demonstrated by the presence of a zirconia layer [64].

Very thin and defect-free metal foils ($<2\,\mu m$) have been produced by sputtering techniques or other vaporizing methods; the metal is coated over a low adherence material such as glass, oxides, and nitrides, and after the removal of the foils, it can be applied over a porous support [65]. A composite membrane consisting of a Pd-Ag film of $2.2\,\mu m$ coated over a porous stainless steel tube of average pore size $2\,\mu m$ exhibited at 400°C a permeance of $6.6 \times 10^{-3}\,mol\,m^{-2}\,s^{-1}\,Pa^{-0.5}$ and an H_2/N_2 separation factor of 1400 [66].

13.4.3 OTHER METAL MEMBRANES

As described in Section 13.4.1, self-supported membranes consisting of thin-walled Pd-Ag tubes of diameter 10 mm can be produced with a minimum thickness of about 50 µm. In fact, tubes of lower thickness could not work under the operating conditions typical for hydrogen separation processes (300°C–400°C and at least 200–300 kPa). However, further thickness reduction could be beneficial for increasing the permeance and reducing costs. These characteristics can be attained by joining dense Pd-based foils of thickness lower than 50 µm with metal structures that provide them the mechanical strength needed to withstand the operating pressures. In view of this, metal-supported membranes and laminated Pd membranes have been studied [35].

Thin Pd-Ag foils are joined to metal structures such as grids using diffusion welding. Flat metal-supported membranes consisting of thin Pd-Cu foils (25–63 µm) have been joined by diffusion welding to Cu frames with heat treatment at about 300°C [67]. Again, diffusion welding is applied to enclose thin Pd-alloy foils between fine mesh metal fabrics [68]. Presence of a bilateral support permits reducing drastically the thickness of the Pd alloy (10–30 µm), thus increasing the membrane permeance.

Metal-supported membranes in the form of both flat membranes and tubes have also been produced at ENEA Frascati laboratories [35], where Pd-Ag sheets have been joined to flat metal supports through diffusion welding using the thermomechanical press shown in Figure 13.21 [69,70].

Metal supports used for these membranes consisted of both stainless steel grids and nickel-perforated sheets. A Pd-Ag composite membrane fabricated using a stainless steel grid is shown in Figure 13.22. The cross section of this membrane presented in Figure 13.23 shows detail of the joint between the Pd-Ag foil (50 µm thick) and the stainless steel grid.

A nickel-perforated metal support of thickness 210 µm and hole diameter 2.5 mm used for producing a tubular supported membrane with a Pd-Ag foil of thickness 42 µm is presented in Figure 13.24. First, diffusion welding is applied for joining the Pd-Ag foil to the flat nickel support and then for welding the limbs of the supported membrane wrapped around an alumina bar, as described in the scheme of Figure 13.25.

With the aim of cost reduction, several low-cost metals, such as Ni, Nb, V, and Ti, and their alloys have been studied as alternatives to Pd in order to manufacture membranes for hydrogen separation. In fact, V group metals exhibit higher permeability than Pd, whereas other metals such as Ni and Fe could be of practical interest because of their very low cost. Permeability of these metals is shown in Figure 13.26 [35]. Applications of new materials are promising in terms of increased hydrogen permeability. However, the new membranes tested exhibited poor stability and reliability because their high values of permeability are linked with high hydrogen uploading into the metal lattice that causes embrittlement [71].

Furthermore, under the operating conditions of the processes for separating hydrogen, the non-noble metals could react with gases such as oxygen and nitrogen and form surfaces layers that reduce or block the hydrogen permeation. To avoid this problem, a laminated membrane is produced by covering Ni and Nb sheets with Pd-Ag thin foils. The membrane thickness is then reduced by cold-rolling, according to the scheme of Figure 13.27.

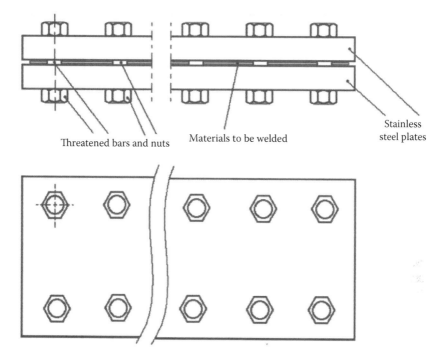

FIGURE 13.21   Schematic representation of the press used for joining the Pd–Ag foils to flat metal supports. (Reprinted from *Int. J Hydrogen Energy*, 28, S. Tosti, Supported and laminated Pd-based metallic membranes, 1455–1464, Copyright 2003, with permission from Elsevier.)

FIGURE 13.22   Composite membrane obtained by welding a Pd–Ag foil to a stainless steel grid. (Reprinted from *Int. J Hydrogen Energy*, 28, S. Tosti, Supported and laminated Pd-based metallic membranes, 1455–1464, Copyright 2003, with permission from Elsevier.)

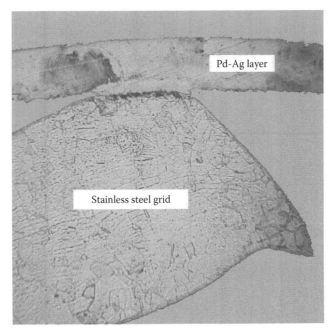

FIGURE 13.23 Cross section of the composite membrane obtained by welding a Pd–Ag foil (50 μm thick) to a stainless, steel grid. (Reprinted from *Int. J Hydrogen Energy*, 28, S. Tosti, Supported and laminated Pd-based metallic membranes, 1455–1464, Copyright 2003, with permission from Elsevier.)

FIGURE 13.24 The external surface of the Ni-supported membrane (support hole diameter of 2.5 mm). (From Tosti, S. et al.: Metal supported and laminated Pd-based membranes, in *Membranes for Membrane Reactors: Preparation, Optimization and Selection*, A. Basile and F. Gallucci, eds. 275–287. 2011. Copyright Wiley-VCH Verlag GmbH & Co. KGaA. Reproduced with permission.)

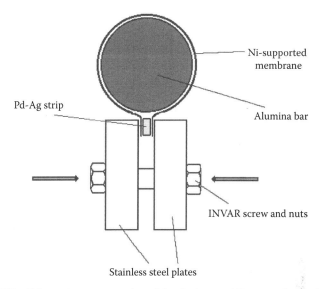

FIGURE 13.25 Schematic representation of the device used for preparing an Ni-supported membrane tube. (From Tosti, S. et al.: Metal supported and laminated Pd-based membranes, in *Membranes for Membrane Reactors: Preparation, Optimization and Selection*, A. Basile and F. Gallucci, eds. 275–287. 2011. Copyright Wiley-VCH Verlag GmbH & Co. KGaA. Reproduced with permission.)

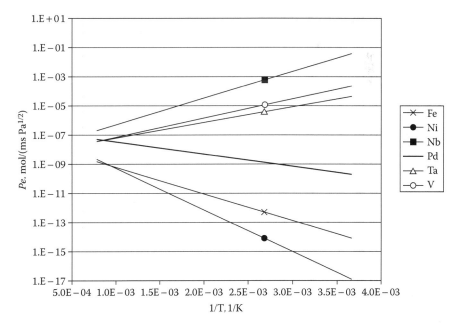

FIGURE 13.26 Hydrogen permeability through metals. (Reprinted from *Int. J Hydrogen Energy*, 28, S. Tosti, Supported and laminated Pd-based metallic membranes, 1455–1464, Copyright 2003, with permission from Elsevier.)

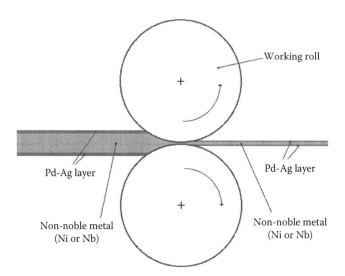

FIGURE 13.27 Schematic representation of the cold-rolling of composite Pd-based membranes. (From Tosti, S. et al.: Metal supported and laminated Pd-based membranes, in *Membranes for Membrane Reactors: Preparation, Optimization and Selection*, A. Basile and F. Gallucci, eds. 275–287. 2011. Copyright Wiley-VCH Verlag GmbH & Co. KGaA. Reproduced with permission.)

The cross section of a flat laminated membrane consisting of an Ni sheet covered over its surfaces by two Pd-Ag foils of thickness 28 μm is shown in the micropho-tography image of Figure 13.28.

This flat membrane has been cold-rolled to reduce its overall thickness to 141 μm, about 127 μm of Ni sheet with two Pd-Ag layers of 7 μm. The permeator tube shown in Figure 13.29 was constructed from this laminated membrane.

Another laminated membrane prepared by covering an Nb sheet (1 mm) with two Pd-Ag foils (25 μm) has been cold-rolled to obtain a composite membrane of overall thickness 128 μm, with about 122 μm of Nb bulk with two very thin Pd-Ag layers of 3 μm. Permeator tube produced with this laminated membrane was tested with hydrogen at 180°C and 200 kPa and exhibited very poor durability. Embrittlement due to the high hydrogen uploading produced a quick failure of this membrane, as shown in Figure 13.30.

13.5 APPLICATIONS: Pd-MEMBRANE REACTOR

Self-supported Pd-Ag membranes are used for producing ultrapure hydrogen. Their stability can be affected by thermal cycling; thus, the design of the membrane module has to take into consideration the significant hydrogenation strain of the Pd-Ag alloy. In fact, the membrane module has to be tightly connected to the Pd-Ag membrane in order to ensure its selectivity, and at the same time has to permit its elongation/contraction due to the thermal and hydrogenation cycles. In the case of a single-tube membrane module, both ends of the Pd-Ag permeator should be tightly fixed to the module, as demonstrated in Figure 13.31. However, when hydrogenated,

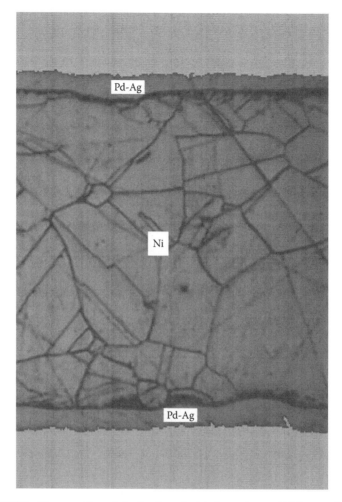

FIGURE 13.28 Cross section of the laminated nickel membrane (thickness of the Pd-Ag layers 28 µm). (Data from Tosti, S. et al.: Metal supported and laminated Pd-based membranes, in *Membranes for Membrane Reactors: Preparation, Optimization and Selection*, A. Basile and F. Gallucci, eds. 275–287. 2011. Copyright Wiley-VCH Verlag GmbH & Co. KGaA. Reproduced with permission.)

the Pd-Ag permeator expands much more than the module, which, thus, applies a compressive force over the thin-walled tube; see Figure 13.32 [72]. For instance, such a configuration could involve a compressive force of about 2600 N for a Pd-Ag tube of wall thickness 50 µm and diameter 10 mm hydrogenated at low temperature. The finger-like (or dead-end) configuration shown in Figure 13.33 permits the free elongation/contraction of a Pd-Ag tube that is tightly fixed to the membrane module at one of its end. In this way, no stress is applied to the permeator tube, thus increasing its stability in presence of hydrogenation [73].

Membrane modules are also produced for membrane reactors, manufacturing devices that combine a fixed bed catalytic reactor with a permselective membrane.

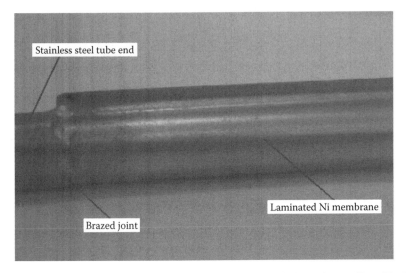

FIGURE 13.29   The permeator tube produced by the laminated Ni membrane. (From Tosti, S. et al.: Metal supported and laminated Pd-based membranes, in *Membranes for Membrane Reactors: Preparation, Optimization and Selection*, A. Basile and F. Gallucci, eds. 275–287. 2011. Copyright Wiley-VCH Verlag GmbH & Co. KGaA. Reproduced with permission.)

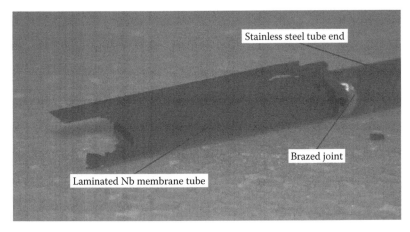

FIGURE 13.30   Failure of a laminated Nb membrane after hydrogenation. (From Tosti, S. et al.: Metal supported and laminated Pd-based membranes, in *Membranes for Membrane Reactors: Preparation, Optimization and Selection*, A. Basile and F. Gallucci, eds. 275–287. 2011. Copyright Wiley-VCH Verlag GmbH & Co. KGaA. Reproduced with permission.)

Hydrogen removal through the selective Pd membrane may promote the conversion of dehydrogenation reactions beyond the thermodynamic equilibrium according to the well-known *shift effect* of the membranes [4,7].

Both single- and multitube Pd-membrane reactors adopting the finger-like configuration were developed at ENEA laboratories [74,75]. Reaction tests verified the

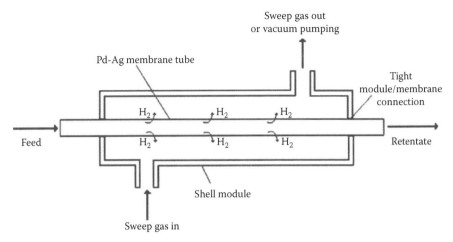

FIGURE 13.31   Schematic representation of a single-tube Pd-membrane reactor with the catalyst located in the lumen side and hydrogen recovery in the shell side. (Reprinted from *Int. J Hydrogen Energy*, 35, S. Tosti, Overview of Pd-based membranes for producing pure hydrogen and state of art at ENEA laboratories, 12650–12659, Copyright 2010, with permission from Elsevier.)

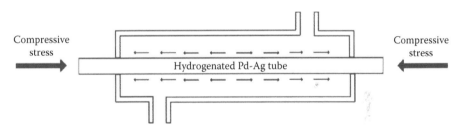

FIGURE 13.32   Schematic representation showing how hydrogen uploading expands the Pd-Ag permeator (arrows are just indicating its expansion); the permeator is then compressed by the module. (Reprinted from *Int. J Hydrogen Energy*, 35, S. Tosti, Overview of Pd-based membranes for producing pure hydrogen and state of art at ENEA laboratories, 12650–12659, Copyright 2010, with permission from Elsevier.)

capability of these Pd-membrane reactors to produce ultrapure hydrogen by attaining reaction conversions and hydrogen yields higher than the traditional reactors. The main dehydrogenation reactions tested concerned water gas shift, methane and ethanol steam and oxidative reforming, acetic acid and methanol steam reforming, and the treatment of liquid biomasses [12,76–88].

A single-tube module used in both permeation and reaction tests consists of a Pd-Ag permeator tube assembled in a finger-like configuration inside a shell made of Pyrex glass. The membrane is heated through a Pt wire coil surrounding the permeator tube; see Figure 13.34 [87]. The temperature is measured by a thermocouple above the surface of the membrane tube.

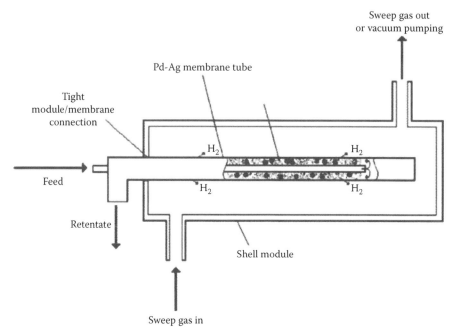

FIGURE 13.33 Schematic representation of a dead-end Pd-membrane reactor. (Reprinted from *Int. J Hydrogen Energy*, 35, S. Tosti, Overview of Pd-based membranes for producing pure hydrogen and state of art at ENEA laboratories, 12650–12659, Copyright 2010, with permission from Elsevier.)

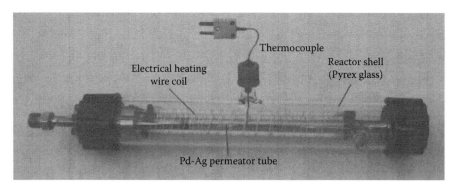

FIGURE 13.34 The Pd-membrane reactor module. (Reprinted from *Int. J Hydrogen Energy*, 38, S. Tosti et al., Pd-based membrane reactors for producing ultra pure hydrogen: Oxidative reforming of bio-ethanol, 701–707, Copyright 2013, with permission from Elsevier.)

As discussed previously, electrical resistivity of Pd-Ag is maximum for a silver content typical of the commercial alloy (20–30 wt%) that can be ohmically heated by relatively low electrical currents. Direct ohmic heating of the Pd-Ag tube was successfully applied with the advantage of reducing the power consumption and shortening the temperature ramping; it is schematically represented in Figure 13.35 and shown in Figure 13.36 [89,90].

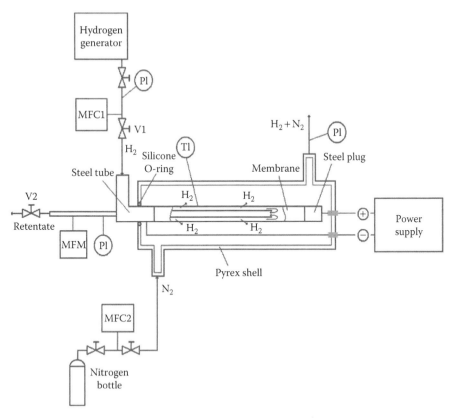

FIGURE 13.35 Schematic representation of a finger-like single-tube membrane module with direct ohmic heating. (Reprinted from *Int. J Hydrogen Energy*, 35, S. Tosti et al., Electrical resistivity, strain and permeability of Pd-Ag membrane tubes, 7796–7802, Copyright 2010, with permission from Elsevier.)

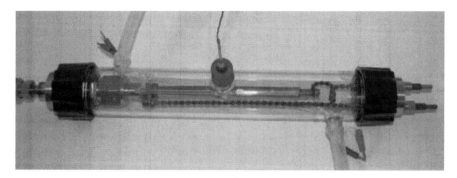

FIGURE 13.36 The finger-like single-tube membrane module with direct ohmic heating.

Thin-walled Pd-Ag tubes (thickness 50 μm) obtained through cold-rolling and diffusion welding have also been used as a hydrogen diffusion cathode of an alkaline water electrolyser, which is schematically represented in Figure 13.37 [91–93]. Such a device is an *electrochemical membrane reactor* where the water electrolysis reaction and the permeation of hydrogen take place simultaneously. In a traditional alkaline electrolyser, hydrogen is produced with a purity of about 99.5%, and for applications where higher purity is required, a further purification step is needed downstream of the electrolyser. Conversely, when a Pd-Ag thin-walled tube is used as a cathode of the electrolyser, more than 50% of the hydrogen produced by electrolysis permeates through the membrane and is then recovered as ultrapure gas. The characteristics of the alkaline electrolyser with the Pd-Ag membranes are reported in Table 13.2.

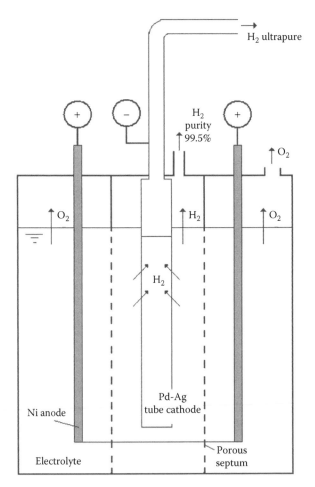

FIGURE 13.37 Schematic representation of the alkaline electrolyser using a Pd-Ag thinwalled tube as cathode. (Reprinted from *Int. J Hydrogen Energy*, 36, A. Pozio et al., Pd-Ag hydrogen diffusion cathode for alkaline water electrolysers, 5211–5217, Copyright 2011, with permission from Elsevier.)

Processing of hydrogen isotopes in the fuel cycle of fusion machines burning deuterium–tritium mixtures needs Pd membranes of infinite hydrogen selectivity and optimum stability in order to meet the requirements of reliability and safety of nuclear plants [72,94,95]. Therefore, these processes use Pd alloy self-supported membranes in finger-like configuration modules. In the final clean-up of tokamak exhausts, tritium has to be removed from water and methane molecules. To do this, at the Tritium Laboratory of Karlsruhe, the membrane reactor PERMCAT, shown in the scheme of Figure 13.38, has been studied [96–98]. The tritiated gases are sent into the reactor shell where a catalyst bed promotes the exchange of tritium (T) with hydrogen (protium), which is fed in a counter-current mode in the membrane lumen. The isotopic exchange reactions are as follows:

$$2H_2 + CT_4 \Leftrightarrow CH_4 + 2T_2 \tag{13.32}$$

$$H_2 + T_2O \Leftrightarrow H_2O + T_2 \tag{13.33}$$

Only the hydrogen isotopes can permeate through the dense Pd-Ag tube, and, thanks to the isotopic exchange over the catalyst, the tritium is transferred from the shell side to the lumen. The decontamination factors required (up to 100) can be attained using long and thin-walled permeator tubes.

For PERMCAT reactors, several design configurations of Pd-Ag membrane tubes of wall thickness 100 µm were proposed [99], as described in Figure 13.39. The first reactor (a) uses a typical finger-like configuration; the tritiated gases are fed in the shell side whereas the hydrogen is sent inside the membrane lumen. In the second reactor (b), a preloaded bellows is joined to the Pd-Ag membrane tube in order to compensate for expansion/contraction. In a third reactor (c), the permeator

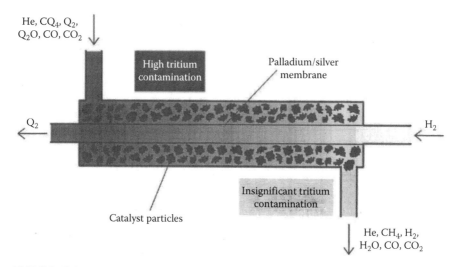

FIGURE 13.38 Schematic representation of a PERMCAT reactor (catalyst in shell side). (Reprinted from *Fusion Engineering and Design*, 49–50, M. Glugla et al., A Permcat reactor for impurity processing in the JET Active Gas Handling System, 817–823, Copyright 2000, with permission from Elsevier.)

is a corrugated Pd-Ag tube working as a bellow-type membrane. Other PERMCAT-kind membrane reactors were also designed and manufactured at ENEA Frascati laboratories [22,100].

The reactor depicted in Figure 13.40 used a thin-walled Pd-Ag tube of length 500 mm [22,41]. The mechanical design and assembly procedures shown in Figure 13.41 permit the avoidance of any combined compressive and bending stress of the permeator tube [12]. As previously described in Section 13.2.1, maximum expansion of hydrogenated Pd-Ag can be estimated to be about 1.5%, that is, 7.5 mm for this tube of length 500 mm. Accordingly, the Pd-Ag tube joined with two stainless steel metal bellows was fabricated shorter than the shell module by 7.5 mm (Figure 13.41a). Then the permeator (plus the two bellows) was welded to the module by applying a traction force T, which stretched the bellows (Figure 13.41b). During operation, the hydrogenated Pd-Ag tube expands and its elongation is compensated by the contraction of the bellows, so that the traction force T reduces to zero. In practice, no compressive stress and bending is applied to the thin wall and long tube (Figure 13.41c), thus ensuring its long life.

The Pd-based membrane reactor shown in Figure 13.42 was designed for water detritiation of the Joint European Torus (JET) housekeeping waste [100–102]. In this device, the direct ohmic heating of the membrane was achieved. As reported in the scheme of Figure 13.43, a bimetallic spring made of Cu and Inconel gave electrical continuity (Cu) and pretensioning of the Pd-Ag tube (Inconel).

In laboratory applications, the scale-up of membrane devices used multitube modules. At ENEA Frascati laboratories, a bundle of 19 Pd-Ag tubes in finger-like configuration was realized for separating up to 6 NL min^{-1} of ultrapure hydrogen. Scheme and picture of the multitube reactor are shown in Figure 13.44 [12].

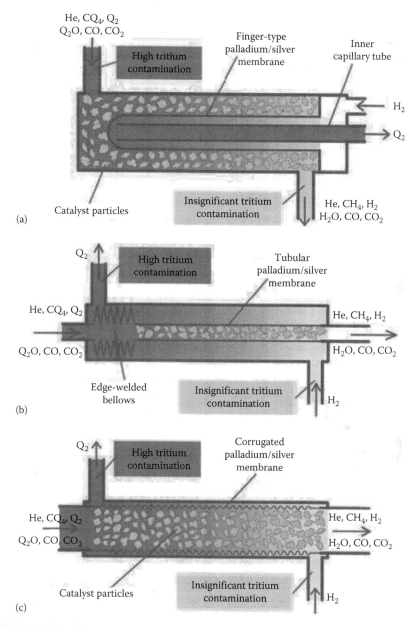

FIGURE 13.39 Diagrams showing three different PERMCAT reactors developed at Tritium Laboratory of Karlsruhe. (Reprinted from *Fusion Engineering and Design*, 82, D. Demange et al., Experimental validation of upgraded designs for PERMCAT reactors considering mechanical behaviour of Pd/Ag membranes under H_2 atmosphere, 2383–2389, Copyright 2007, with permission from Elsevier.)

FIGURE 13.40 ENEA Pd-Ag membrane reactor (PERMCAT-kind): steel shell module (above) and the 500 mm long Pd-Ag tube (below). (From Tosti, S.: Metallic membranes prepared by cold rolling and diffusion welding, in *Membranes for Membrane Reactors: Preparation, Optimization and Selection*, A. Basile and F. Gallucci, eds. 155–167. 2011. Copyright Wiley-VCH Verlag GmbH & Co. KGaA. Reproduced with permission.)

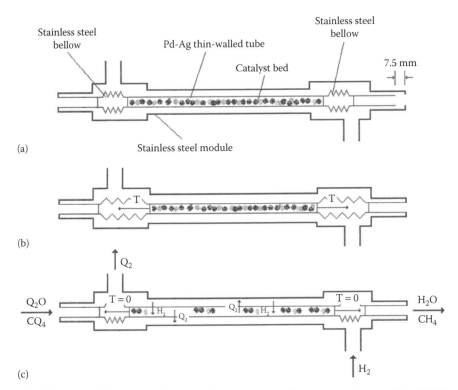

FIGURE 13.41 Diagrams of the assembly procedure and operation scheme of the ENEA Pd-Ag membrane reactor (PERMCAT-like). (Reprinted from *Int. J Hydrogen Energy*, 35, S. Tosti, Overview of Pd-based membranes for producing pure hydrogen and state of art at ENEA laboratories, 12650–12659, Copyright 2010, with permission from Elsevier.)

Steam and oxidative reforming of ethanol and methane were tested through a two-step process, in which the multitube was coupled to a traditional reformer operating at high temperature. The experimental apparatus is shown in Figure 13.45 [82,84,86,103–105]. Such a process maximized the reaction yield and reduced the permeation area required for separating the hydrogen produced [75,106].

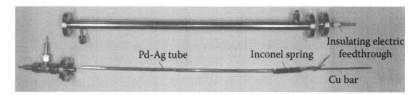

FIGURE 13.42 The Pd-Ag membrane reactor developed for the JET housekeeping waste detritiation (before assembling). (Reprinted from S. Tosti, Membranes and Membrane Reactors for tritium Separation, in *Tritium in Fusion: Production, Uses and Environmental Impact*, S. Tosti and N. Ghirelli, eds., 203–240, Copyright 2013, with permission from Nova Science Publishers, Inc.)

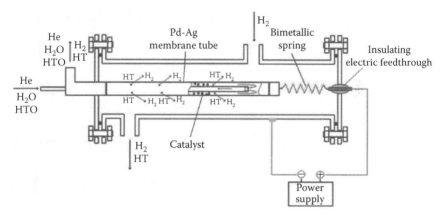

FIGURE 13.43 Schematic representation of the membrane reactor powered by direct ohmic heating and using a bimetallic spring. (Reprinted from *Fusion Engineering and Design*, 86, S. Tosti et al., Design of Pd-based membrane reactor for gas detritiation, 2180–2183, Copyright 2011, with permission from Elsevier.)

Recently, combined methane and ethanol reforming was tested in a two-step process with the purpose of demonstrating its applicability to bioethanol mixtures enriched with methane [84]. Model studies assessed energy balances of the two-step system by verifying their effectiveness [107,108].

Scale-up and especially the reduced size of membrane devices can be addressed by studying compact Pd-membrane modules. A membrane configuration similar to the design of flat-and-frame heat exchangers was developed [109]. Several Pd-Ag thin foils are joined to stainless steel frames using gaskets made of brazing alloy, as shown in the scheme of Figure 13.46. In order to increase their mechanical strength, the Pd-Ag foils can also be supported by stainless steel grids. Figures 13.47 and 13.48 show the permeator before and after the assembly, respectively.

Recent literature reports on other flat-membrane devices [110–114]. Design of these devices takes into consideration that the mass and heat transfer, which occur simultaneously, have to be maximized. For instance, the membrane reactor shown in Figure 13.49 consists of a zone where the heat needed for the dehydrogenation reaction

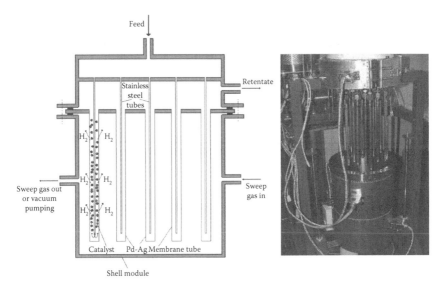

FIGURE 13.44 Multitube membrane module: diagram of the WGS reactor (left) and view of the Pd-Ag permeator tubes (right). (Reprinted from *Int. J Hydrogen Energy*, 35, S. Tosti, Overview of Pd-based membranes for producing pure hydrogen and state of art at ENEA laboratories, 12650–12659, Copyright 2010, with permission from Elsevier.)

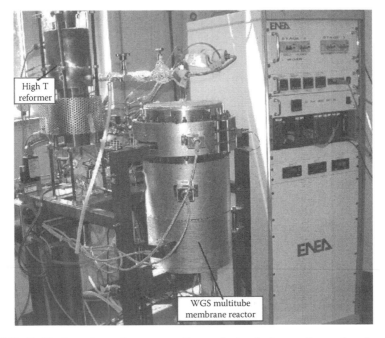

FIGURE 13.45 Experimental apparatus for producing hydrogen from ethanol reforming (two-step process). (Reprinted from *Int. J Hydrogen Energy*, 35, S. Tosti, Overview of Pd-based membranes for producing pure hydrogen and state of art at ENEA laboratories, 12650–12659, Copyright 2010, with permission from Elsevier.)

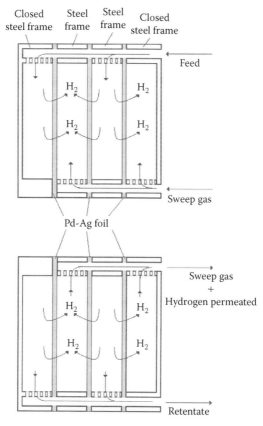

FIGURE 13.46 Diagram of a compact flat-and-frame Pd permeator.

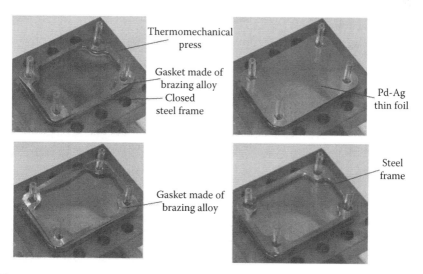

FIGURE 13.47 Assembly of the compact flat-and-frame Pd permeator.

FIGURE 13.48   View of the compact flat-and-frame Pd permeator.

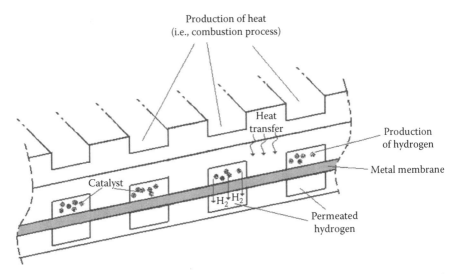

FIGURE 13.49   Flat membrane reactor. (Reprinted from *Int. J Hydrogen Energy*, 35, S. Tosti, Overview of Pd-based membranes for producing pure hydrogen and state of art at ENEA laboratories, 12650–12659, Copyright 2010, with permission from Elsevier.)

is produced through combustion, a second zone wherein the hydrogen is produced over a catalyst, and a third zone where the hydrogen permeated is collected [111].

A similar design is adopted in the reactor shown in Figure 13.50; it uses a membrane made of Pd, V, Cu, or their alloys for separating hydrogen, although the reaction takes place between the heating and the hydrogen zones [112]. Combined steam reforming and partial oxidation of methane are performed in the membrane reactor shown in Figure 13.51; it achieves an optimum temperature profile through the dosed feeding of oxygen (or air) in the zone where a part of the methane fed is burnt to sustain the endothermic reaction (steam reforming of methane) [113].

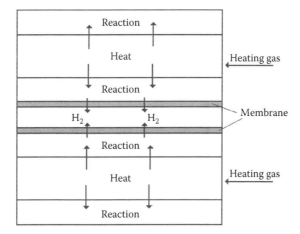

FIGURE 13.50 Diagram of a compact membrane reactor. (Reprinted from *Int. J Hydrogen Energy*, 35, S. Tosti, Overview of Pd-based membranes for producing pure hydrogen and state of art at ENEA laboratories, 12650–12659, Copyright 2010, with permission from Elsevier.)

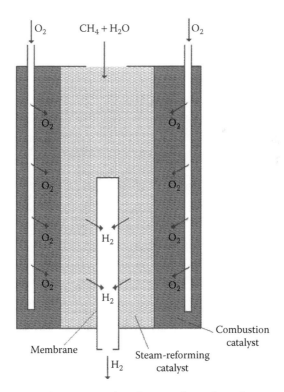

FIGURE 13.51 Schematic representation of an autothermal membrane reactor. (Reprinted from *Int. J Hydrogen Energy*, 35, S. Tosti, Overview of Pd-based membranes for producing pure hydrogen and state of art at ENEA laboratories, 12650–12659, Copyright 2010, with permission from Elsevier.)

13.6 CONCLUSIONS

Fabrication of Pd membranes has to take into consideration both the hydrogen/metal interaction and hydrogen mass-transfer mechanisms through the metal lattice. Particularly, the strong hydrogen/metal interaction is responsible for the change of chemical, physical, and mechanical properties of the Pd alloys used.

The Pd-Ag alloy with 20%–25% silver was widely used and is commercially available for preparing membranes for hydrogen separation. In view of this, the main properties affecting the fabrication of Pd-Ag membranes, including hydrogen solubility and permeability, strain, and the diffusion of silver through the metal lattice are described in detail.

Particularly, it has been discussed how the addition of silver to palladium increases the hydrogen permeability and the mechanical strength of the membranes, reduces its embrittlement under hydrogenation, and facilitates the joining of metal parts through diffusion welding. These features of the Pd-Ag alloys address the fabrication techniques aimed at producing membranes characterized by high hydrogen permeability and chemical physical stability. Diffusion welding has been used for the fabrication of both self-supported thin-walled tubes and composite metal membranes consisting of thin Pd-Ag layers covered over metal structures.

This chapter also discusses how the stability of self-supported Pd membranes needs an appropriate design of the membrane module. The finger-like design of membrane modules in both single- and multitube configuration avoids any mechanical stress of the thin-walled permeators, thus ensuring the durability of the membranes.

Important developments of metal membranes for separating hydrogen are expected from the study of metals alternative to Pd alloys. In fact, at present, the cost of Pd represents the main hurdle for the wide diffusion of Pd-based permeators in the processes for producing hydrogen. In view of this, most of the results already attained in the fabrication technology of Pd membranes and described in this chapter could be extended to new metal membranes. Future work should also be addressed to the membrane module design in order to scale up and especially reduce the size of these devices. An understanding and control of the mass transfer and heat transport phenomena that occur simultaneously in a membrane reactor will also be required for its optimum design.

REFERENCES

1. O. Biccikovci, P. Straka, Production of hydrogen from renewable resources and its effectiveness, *International Journal of Hydrogen Energy*, 37 (2012) 11563–11578.
2. M.F. Orhan, I. Dincer, M.A. Rosen, M. Kanoglu, Integrated hydrogen production options based on renewable and nuclear energy sources, *Renewable and Sustainable Energy Reviews*, 16(8) (2012) 6059–6082.
3. C. Mansilla, J. Louyrette, S. Albou, C. Bourasseau, S. Dautremont, Economic competitiveness of off-peak hydrogen production today—A European comparison, *Energy*, 55 (2013) 996–1001.
4. J. Shu, B.P.A. Grandjean, A. Van Neste, S. Kalaguine, Catalytic palladium-based membrane reactors: A review, *Canadian Journal of Chemical Engineering*, 69 (1991) 1036–1060.

5. E. Drioli, E. Fontananova, Membrane technology and sustainable growth, *Chemical Engineering Research and Design*, 82(12) (2004) 1557–1562.

6. M. Mulder, *Basic Principles of Membrane Technology*, 2nd edn., Boston, MA: Kluwer Academic Publishers, 1991.

7. E. Kikuchi, Membrane reactor application to hydrogen production, *Catalysis Today*, 56 (2000) 97–101.

8. A.G. Dixon, Recent research in catalytic inorganic membrane reactors, *International Journal of Chemical Reactor Engineering*, 1 (2003) R6. http://www.bepress.com/ijcre/vol1/R6.

9. Ø. Hatlevik, S.K. Gade, M.K. Keeling, P.M. Thoen, A.P. Davidson, J.D. Way, Palladium and palladium alloy membranes for hydrogen separation and production: History, fabrication strategies, and current performance, *Separation and Purification Technology*, 73 (2010) 59–64.

10. S. Tosti, L. Bettinali, Diffusion bonding of Pd-Ag membranes, *Journal of Material Science*, 39 (2004) 3041–3046.

11. S. Tosti, L. Bettinali, S. Castelli, F. Sarto, S. Scaglione, V. Violante, Sputtered, electroless, and rolled palladium-ceramic membranes, *Journal of Membrane Science*, 196 (2002) 241–249.

12. S. Tosti, Overview of Pd-based membranes for producing pure hydrogen and state of art at ENEA laboratories, *International Journal of Hydrogen Energy*, 35 (2010) 12650–12659.

13. J. Chabot, J. Lecomte, C. Crumet, J. Sannier, Fuel clean-up system: poisoning of palladium-silver membranes by gaseous impurities, *Fusion Science and Technology*, 14 (1988) 614–618.

14. G.J. Grashoff, C.E. Pilkington, C.W. Corti, The purification of hydrogen—A review of the technology emphasising the current status of palladium membrane diffusion, *Platinum Metals Review*, 27 (1983) 157–168.

15. F.A. Lewis, K. Kandasamy, B. Baranowski, The "uphill" diffusion of hydrogen—strain-gradient-induced effects in palladium alloy membranes, *Platinum Metals Review*, 32 (1988) 22–26.

16. H.P. Hsieh, Inorganic membrane reactors—A review, *AIChE Symposium Series*, 268(85) (1989) 53–67.

17. M.L.H. Wise, J.P.G. Farr, I.R. Harris, X-ray studies of the α/β miscibility gaps of some palladium solid solution-hydrogen systems, *Journal of the Less-Common Metals*, 41 (1975) 115–127.

18. A.G. Knapton, Palladium alloys for hydrogen diffusion membranes—A review of high permeability materials, *Platinum Metals Review*, 21 (1977) 44–50.

19. ASM. *ASM Handbook, Metals Handbook: Volume 2, Properties and Selection: Nonferrous Alloys and Special-Purpose Materials*, Materials Park, OH: ASM International, 2000.

20. S. Arajs, K.V. Rao, Y.D. Yao, W. Teoh, Electrical resistivity of palladium-silver alloys at high temperature, *Physical Review B*, 15(4) (1977) 2429–2431. http://link.aps.org/doi/10.1103/PhysRevB.15.2429.

21. D. Fort, I.R. Harris, The physical properties of some palladium alloy hydrogen diffusion membrane materials, *Journal of the Less-Common Metals*, 41 (1975) 313–327.

22. S. Tosti, L. Bettinali, F. Borgognoni, D.K. Murdoch, Mechanical design of a PERMCAT reactor module, *Fusion Engineering and Design*, 82 (2007) 153–161.

23. D. Demange, S. Welte, M. Glugla, Experimental validation of upgraded designs for PERMCAT reactors considering mechanical behavior of Pd/Ag membranes under H2 atmosphere, *Fusion Engineering and Design*, 82 (2007) 2383–2389.

24. J. Gabitto, C. Tsouris, Hydrogen transport in composite inorganic membranes, *Journal of Membrane Science*, 312 (2008) 132–142.

25. T.B. Flanagan, D. Wang, K.L. Shanahan, Diffusion of H through Pd membranes: Effects of non-ideality, *Journal of Membrane Science*, 306 (2007) 66–74.

26. X. Pan, M. Kilgus, A. Goldbach, Low-temperature H2 and N2 transport through thin Pd66Cu34Hx layers, *Catalysis Today*, 104 (2005) 225–230.

27. S. Tosti, F. Borgognoni, C. Rizzello, V. Violante, Water gas shift reaction via Pd-based membranes, *Asia-Pacific Journal of Chemical Engineering*, 4 (2009) 369–379.

28. T.L. Ward, T. Dao, Model of hydrogen permeation behavior in palladium membranes, *Journal of Membrane Science*, 153 (1999) 211–231.

29. C.V. Miguel, A. Mendes, S. Tosti, L.M. Madeira, Effect of CO and CO2 on H2 permeation through finger-like Pd-Ag membranes, *International Journal of Hydrogen Energy*, 37 (2012) 12680–12687.

30. F. Guazzone, E.E. Engwall, Y.H. Ma, Effects of surface activity, defects and mass transfer n hydrogen permeance and n-value in composite palladium-porous stainless steel membranes, *Catalysis Today*, 118 (2006) 24–31.

31. S. Hara, M. Ishitsuka, H. Suda, M. Mukaida, K. Haraya, Pressure-dependent hydrogen permeability extended for metal membranes not obeying the square-root law, *Journal of Physics and Chemistry B*, 113 (2009) 9795–9801.

32. B.D. Morreale, M.V. Ciocco, R.M. Enick, B.I. Morsi, B.H. Howard, A.V. Cugini, K.S. Rothenberger, The permeability of hydrogen in bulk palladium at elevated temperatures and pressures, *Journal of Membrane Science*, 212 (2003) 87–97.

33. M. Vadrucci, F. Borgognoni, A. Moriani, A. Santucci, S. Tosti, Hydrogen permeation through Pd-Ag membranes: Surface effects and Sieverts' law, *International Journal of Hydrogen Energy*, 38 (2013) 4140–4152.

34. A. Santucci, F. Borgognoni, M. Vadrucci, S. Tosti, Testing of dense Pd-Ag tubes: Effect of pressure and membrane thickness on the hydrogen permeability, *Journal of Membrane Science*, 444 (2013) 378–383.

35. S. Tosti, Supported and laminated Pd-based metallic membranes, *International Journal of Hydrogen Energy*, 28 (2003) 1455–1464.

36. E.A. Mason, A.P. Malinauskas, *Gas Transport in Porous Media: The Dusty-Gas Model*, New York: Elsevier, 1983.

37. K. Keizer, R.J.R. Uhlhorn, R.J. Van Vuren, A.J. Burggraaf, Gas separation mechanisms in microporous modified γ-Al2O3 membranes, *Journal of Membrane Science*, 1988 (39) 285–300.

38. B.S. Massey, *Mechanics of Fluids*, 6th edn., London: Chapman & Hall, 1989.

39. S. Tosti, L. Bettinali, V. Violante, Rolled thin Pd and Pd-Ag membranes for hydrogen separation and production, *International Journal of Hydrogen Energy*, 25 (2000) 319–325.

40. M. Konno, M. Shindo, S. Sugawara, S. Saito, A composite palladium and porous aluminium oxide membrane for hydrogen gas separation, *Journal of Membrane Science*, 37 (1988) 193.

41. S. Tosti, Metallic membranes prepared by cold rolling and diffusion welding, in *Membranes for Membrane Reactors: Preparation, Optimization and Selection*, A. Basile and F. Gallucci, eds., New York: Wiley, 2011, pp. 155–167.

42. J.S. Koeheler, The nature of work hardening, *Physical Review*, 86 (1952) 52–59.

43. M.A. Meyers, A. Mishra, D.J. Benson, Mechanical properties of nanocrystalline materials, *Progress in Materials Science*, 51 (2006) 427–556.

44. S.B. Dunkerton, Diffusion bonding—Process and applications, *Welding and Metal Fabrication*, 59(3) (1991) 132–136.

45. C. Deminet, Method of diffusion bonding, US Patent 4 013 210 (1977).

46. R.C. Weast, M.J. Astle, W.H. Beyer, *Handbook of Chemistry and Physics*, 67th edn., CRC Press, Boca Raton, FL, 1986.

47. S. Tosti, L. Bettinali, D. Lecci, F. Marini, V. Violante, Method of bonding thin foils made of metal alloys selectively permeable to hydrogen, particularly providing membrane devices, and apparatus for carrying out the same, European Patent EP 1184125 (2001)

48. C.F. Powell, J.H. Oxley, J.M. Blocher, *Vapor Deposition*, New York: Wiley, 1966.

49. S. Ilias, R. Govind, Development of high temperature membranes for membrane reactor: An overview, *AIChE Symposium Series*, 268(85) (1989) 18–25.

50. Z.Y. Lia, H. Maedaa, K. Kusakabea, S. Morooka, H. Anzaib, S. Akiyamab, Preparation of palladium-silver alloy membranes for hydrogen separation by the spray pyrolysis method, *Journal of Membrane Science*, 78 (1993) 247.

51. V.M. Gryaznov, O.S. Serebryannikova, Yu.M. Serov, M.M. Ermilova, A.N. Karavanov, A.P. Mischenko, N.V. Orekhova, Preparation and catalysis over palladium composite membranes, *Applied Catalysis A*, 96 (1993) 15.

52. G. Capannelli, A. Bottino, G. Gao, A. Grosso et al., Porous Pt/γ-Al$_2$O$_3$ catalytic membrane reactors prepared using mesitylene solvated Pt atoms, *Catalysis Letters*, 20 (1993) 287–297.

53. F.A. Lowenheim, *Modern Electroplating*, New York: Wiley, 1974, pp. 342–357; 737–747.

54. E.M. Wise, *Palladium—Recovery, Properties and Uses*, New York: Academic Press, 1968.

55. H. Kikuchi, Alloy-plated membranes for hydrogen separation and their manufacture, Japanese Patent No. 88 294 925 (1988).

56. S. Uemiya, M. Koseki, T. Kojima, Preparation of highly permeable membranes for hydrogen separation using a CVD technique, in *Proceedings of the 3rd International Conference on Inorganic Membranes*, Y.H. Ma, ed., Worcester, MA, July 10–14, 1994, p. 545.

57. S. Itoh, R. Govind, Combined oxidation and dehydrogenation in a palladium membrane, *Industrial & Engineering Chemistry Research*, 28 (1989) 1554–1557.

58. S. Uemiya, Y. Kude, K. Sugino, N. Sato, T. Matsuda, E. Kikuchi, A palladium/porous glass composite membrane for hydrogen separation, *Chemistry Letters*, 10 (1988) 1687–1690.

59. S. Uemiya, T. Matsuda, E. Kikuchi, Hydrogen permeable palladium-silver alloy membrane supported on porous ceramics, *Journal of Membrane Science*, 56 (1991) 325.

60. P.S. Apte, J.M. Schwartz, S.W. Callahan, Hydrogen transport membrane fabrication method, WO 2007/058913 A2 (2007).

61. S. Hou, K. Jiang, W. Li, H. Xu, L. Yuan, Metal palladium composite membrane or alloy palladium composite membrane and their preparation methods, WO 2005/065806 (2005).

62. Z. Dardas, Y. She, T.H. Vanderspurt, J. Yamanis, C. Walker, Composite palladium membrane having long-term stability for hydrogen separation, PCT/US/2005/047047 (2005).

63. M.E. Ayturk, I.P. Mardilovich, E.E. Engwall, Y.H. Ma, Synthesis of composite Pd-porous stainless steel (PSS) membranes with a Pd/Ag intermetallic diffusion barrier, *Journal of Membrane Science*, 285 (2006) 385–394.

64. S.K. Gade, M.K. Keeling, D.K. Steele, J.D. Way, P.M. Thoen, High flux Pd membranes deposited on stainless steel supports by electroless plating, *Proceedings of the 9th Conference on Inorganic Membranes*, Lillehammer, Norway, June 25–29, 2006.

65. R. Bredesen, H. Klette, Method of manufacturing thin metal membranes, US6086729 (2000).

66. T.A. Peters, M. Stange, H. Klette, R. Bredesen. High pressure performance of thin Pd-23%Ag/stainless steel composite membranes in water gas shift gas mixtures; influence of dilution, mass transfer and surface effects on the hydrogen flux, *Journal of Membrane Science*, 316 (2008) 119–127.

67. W. Juda, C.W. Krueger, R.T. Bombard, Diffusion-bonded palladium-copper alloy framed membrane for pure hydrogen generators and the like and method of preparing the same, US5904754 (1999).

68. N. Iniotakis, C.-B. von der Decken, H. Fedders, W. Frohling, F. Sernetz, Hydrogen permeation membrane, US Patent 4699637 (1987).

69. S. Tosti, L. Bettinali, D. Lecci, V. Violante, Procedimento di saldatura di strutture metalliche di rinforzo a lamine in lega di palladio per la fabbricazione di membrane composite per la separazione di idrogeno e apparato per la sua realizzazione, Italian Patent Grant 0001323876 (2004).

70. S. Tosti, A. Basile, F. Gallucci, Metal supported and laminated Pd-based membranes, in *Membranes for Membrane Reactors: Preparation, Optimization and Selection*, A. Basile and F. Gallucci, eds., New York: Wiley, 2011, pp. 275–287.

71. A. Santucci, S. Tosti, A. Basile, Alternatives to palladium in membranes for hydrogen separation: Nickel, niobium and vanadium alloys, ceramic supports for metal alloys and porous glass membranes, in *Handbook of Membrane Reactors*, volume 1, A. Basile, ed., Cornwall: Woodhead Publishing Series in Energy, 2013, pp. 183–217.

72. S. Tosti, Membranes and membrane reactors for tritium separation, in *Tritium in Fusion: Production, Uses and Environmental Impact*, S. Tosti and N. Ghirelli, ed., New York: Nova Science Publishers, 2013, pp. 203–240.

73. S. Tosti, A. Basile, L. Bettinali, F. Borgognoni, F. Chiaravalloti, F. Gallucci, Long-term tests of Pd–Ag thin wall permeator tube, *Journal of Membrane Science*, 284 (2006) 393–397.

74. S. Tosti, A. Basile, L. Bettinali, D. Lecci, C. Rizzello, Dispositivo a membrana a fascio tubiero per la produzione di idrogeno ultrapuro, Italian patent n. RM2005A000399 (2005).

75. S. Tosti, A. Basile, L. Bettinali, F. Borgognoni, F. Gallucci, C. Rizzello, Design and process study of Pd membrane reactors, *International Journal of Hydrogen Energy* 33 (2008) 5098–5105.

76. S. Tosti, A. Basile, G. Chiappetta, C. Rizzello, V. Violante. Pd-Ag membrane reactors for water gas shift reaction. *Chemical Engineering Journal*, 93 (2003) 23–30.

77. F. Gallucci, A. Basile, S. Tosti, A. Iulianelli, E. Drioli. Methanol and ethanol steam reforming in membrane reactors: An experimental study. *International Journal of Hydrogen Energy*, 32 (2007) 1201–1210.

78. A. Basile, F. Gallucci, A. Iulianelli, F. Borgognoni, S. Tosti. Acetic acid steam reforming in a Pd–Ag membrane reactor: The effect of the catalytic bed pattern. *Journal of Membrane Science*, 311 (2008) 46–52.

79. A. Basile, A. Parmaliana, S. Tosti, A. Iulianelli, F. Gallucci, C. Espro, J. Spooren. Hydrogen production by methanol steam reforming carried out in membrane reactor on Cu/Zn/Mg-based catalyst, *Catalysis Today*, 137 (2008) 17–22.

80. F. Gallucci, S. Tosti, A. Basile. Pd–Ag tubular membrane reactors for methane dry reforming: A reactive method for CO_2 consumption and H_2 production, *Journal of Membrane Science*, 317 (2008) 96–105.

81. D. Mendes, V. Chibante, J.M. Zheng, S. Tosti, F. Borgognoni, A. Mendes, L.M. Madeira. Enhancing the production of hydrogen via water gas shift reaction using Pd-based membrane reactors. *International Journal of Hydrogen Energy*, 35 (2010) 12596–12608.

82. S. Tosti, F. Borgognoni, A. Santucci. Multi-tube Pd-Ag membrane reactor for pure hydrogen production. *International Journal of Hydrogen Energy*, 35 (2010) 11470–11477.

83. A. Santucci, M.C. Annesini, F. Borgognoni, L. Marrelli, M. Rega, S. Tosti. Oxidative steam reforming of ethanol over a Pt/Al_2O_3 catalyst in a Pd-based membrane reactor. *International Journal of Hydrogen Energy* 36 (2011) 1503–1511.

84. F. Borgognoni, S. Tosti, M. Vadrucci, A. Santucci. Pure hydrogen production in a Pd-Ag multi-membranes module by methane steam reforming. *International Journal of Hydrogen Energy* 36 (2011) 7550–7558.

85. S. Tosti, M. Fabbricino, A. Moriani, G. Agatiello, C. Scudieri, F. Borgognoni, A. Santucci. Pressure effect in ethanol steam reforming via dense Pd-based membranes. *Journal of Membrane Science*, 77 (2011) 65–74.

86. F. Borgognoni, S. Tosti. Pd-Ag multi-membranes module for hydrogen production by methane auto-thermal reforming. *International Journal of Hydrogen Energy*, 37 (2012) 1444–1453.

87. S. Tosti, M. Zerbo, A. Basile, V. Calabrò, F. Borgognoni, A. Santucci. Pd-based membrane reactors for producing ultra pure hydrogen: Oxidative reforming of bio-ethanol. *International Journal of Hydrogen Energy* 38 (2013) 701–707.

88. S. Tosti, C. Accetta, M. Fabbricino, M. Sansovini, L. Pontoni, Reforming of olive mill wastewater through a Pd-membrane reactor, *International Journal of Hydrogen Energy* 38 (2013) 10252–10259.

89. S. Tosti, L. Bettinali, R. Borelli, D. Lecci, F. Marini, Dispositivo a membrana di permeazione per la purificazione di idrogeno, Italian Patent RM2009U000143 (2009).

90. S. Tosti, F. Borgognoni, A. Santucci, Electrical resistivity, strain and permeability of Pd-Ag membrane tubes, *International Journal of Hydrogen Energy* 35 (2010) 7796–7802.

91. A. Pozio, M. De Francesco, Z. Jovanovic, S. Tosti, Pd-Ag hydrogen diffusion cathode for alkaline water electrolysers, *International Journal of Hydrogen Energy*, 36 (2011) 5211–5217.

92. Z. Jovanovic, M. De Francesco, S. Tosti, A. Pozio, Structural modification of Pd-Ag alloy induced by electrolytic hydrogen absorption, *International Journal of Hydrogen Energy* 36 (2011) 7728–7736.

93. A. Pozio, S. Tosti, L. Bettinali, R. Borelli, M. De Francesco, D. Lecci, F. Marini, Elettrolizzatore alcalino con catodo tubolare in Pd-Ag per la produzione di idrogeno ultrapuro, Italian Patent RM2009U000200 (2009).

94. S. Konishi, S. Nishio, K. Tobita, The DEMO design team, DEMO plant design beyond ITER, *Fusion Engineering and Design*, 63–64 (2002) 11–17.

95. M. Glugla, A. Antipenkov, S. Beloglazov, C. Caldwell-Nichols, I.R. Cristescu, I. Cristescu, C. Day, L. Doerr, J.P. Girard, E. Tada, The ITER tritium systems, *Fusion Engineering and Design*, 82 (2007) 472–487.

96. R.-D. Penzhorn, R.D. Rodriguez, M. Glugla, A catalytic plasma exhaust purification system, *Fusion Technology*, 14 (1988) 450–455.

97. M. Glugla, A. Perevezentsev, D. Niyongabo, R.-D. Penzhorn, A. Bell, P. Herrmann, A Permcat reactor for impurity processing in the JET Active Gas Handling System, *Fusion Engineering and Design*, 49–50 (2000) 817–823.

98. B. Bornschein, M. Glugla, K. Gunther, R. Lasser, T.L. Le, K.H. Simon, S. Welte, Tritium tests with a technical PERMCAT for final clean-up of ITER exhaust gases, *Fusion Engineering and Design*, 69 (2003) 51–56.

99. D. Demange, S. Welte, M. Glugla, Experimental validation of upgraded designs for PERMCAT reactors considering mechanical behaviour of Pd/Ag membranes under H2 atmosphere, *Fusion Engineering and Design*, 2007 (82) 2383–2389.

100. S. Tosti, N. Ghirelli, F. Borgognoni, P. Trabuc, A. Santucci, K. Liger, F. Marini, Membrane reactor for the treatment of gases containing tritium, Patent filing PCT/IT2011/000205 (2012).

101. N. Ghirelli, S. Tosti, P. Trabuc, F. Borgognoni, K. Liger, A. Santucci, X. Lefebvre, Process for the detritiation of soft housekeeping waste and plant thereof, Patent filing PCT/IT2011/000211 (2012).

102. S. Tosti, C. Rizzello, F. Borgognoni, N. Ghirelli, A. Santucci, P. Trabuc, Design of Pd-based membrane reactor for gas detritiation, *Fusion Engineering and Design*, 86 (2011) 2180–2183.

103. S. Tosti, R. Borelli, A. Santucci, L. Scuppa, Pd–Ag membranes for auto-thermal ethanol reforming, *Asia-Pacific Journal of Chemical Engineering* 5 (2010) 207–212.

104. F. Borgognoni, S. Tosti, M. Vadrucci, A. Santucci, Combined methane and ethanol reforming for pure hydrogen production through Pd-based membranes, *International Journal of Hydrogen Energy* 38 (2013) 1430–1438.

105. F. Borgognoni, S. Tosti, Multi-tube Pd-Ag membrane module for pure hydrogen production: Comparison of methane steam and oxidative reforming, *International Journal of Hydrogen Energy*, 38 (2013) 8276–8284.

106. S. Tosti, A. Basile, D. Lecci, C. Rizzello, Membrane process for hydrogen production from reforming of organic products, such as hydrocarbons or alcohols, European Patent 1829821 (2006).

107. G. Manzolini, S. Tosti, Hydrogen production from ethanol steam reforming: Energy efficiency analysis of traditional and membrane processes, *International Journal of Hydrogen Energy* 33 (2008) 5571–5582.

108. D. Mendes, S. Tosti, F. Borgognoni, A. Mendes, L.M. Madeira, Integrated analysis of a membrane-based process for hydrogen production from ethanol steam reforming, *Catalysis Today* 156 (2010) 107–117.

109. S. Tosti, L. Bettinali, F. Borgognoni, F. Marini, A. Santucci, A. Basile, Dispositivo compatto a membrana metallica per la produzione di fluidi gassosi, Italian Patent RM2010U000066 (2010).

110. R.E. Buxbaum, Hydrogen generator apparatus, US 2004/0163313 A1 (2004).

111. O. Gorke, P. Pfeifer, K. Schubert, Reactor and method for the production of hydrogen, EP1669323 A1 (2006).

112. Z. Jia, L.A. Stryker, D.E. Decker, Supercritical process, reactor and system for hydrogen production, CA 2648589 A1 (2007).

113. N. Toshiyuki, M. Nohuhiko, Y. Manabu, Permselective membrane type reactor and method for hydrogen production, US2008226544 (A1) (2008).

114. E. Gernot, A. Deschamps, Staged system for producing purified hydrogen from a reaction gas mixture comprising a hydrocarbon compound, US 2008/0311013 A1 (2008).

Section III

Fabrication Processes for
Composite Membrane

14 Current Progress of Nanomaterial/Polymer Mixed-Matrix Membrane for Desalination

Goh Pei Sean, Ng Be Cheer,
and Ahmad Fauzi Ismail

CONTENTS

14.1 Introduction 489
14.2 Types of MMMs for Desalination 491
    14.2.1 RO Membranes 491
    14.2.2 FO Membranes 492
    14.2.3 Nanofiltration 493
14.3 The Technological Needs of MMMs for Desalination 494
14.4 Modification of Nanomaterials and Fabrication of MMMs 496
14.5 Performance Evaluation of MMMs for Desalination 498
14.6 Future Prospective and Concluding Remarks 503
References 504

14.1 INTRODUCTION

Freshwater is currently becoming a scarce commodity and is used unsustainably in the majority of the world's regions, as only 0.014% of the earth's total volume of water is suitable for direct human consumption. It is anticipated that in the next few decades, access to water for drinking, agriculture, and industrial use will become an increasingly crucial challenge for many countries around the world [1]. Improving the effectiveness and efficiency of water-purification technology, to produce clean water and protect the environment in a sustainable manner, is considered by many as perhaps the main challenge of the twenty-first century [2]. Intensive efforts are underway throughout the world to avert this looming crisis with the conservation of the existing limited freshwater supply and the conversion of the abundantly available seawater through various desalination technologies. There is a large interest in the development of economically attractive desalination technologies. Over the years, a number of desalination methods have been developed among which distillation, reverse osmosis (RO), and electrodialysis are the most commonly known and widespread technologies [3]. A common goal for current research is to make these technologies more energy-efficient and cost-effective, for the desalination of both seawater and brackish water. Considering that there is more brackish water than freshwater in the world, it is clear that it is particularly attractive to utilize the large brackish water resources for human consumption and residential use, agriculture, and industry [4].

Membrane-based filtration is the current leading energy-efficient technology for the cleanup and desalination of brackish water, recycled water, and arid seawater. Membrane-based filtration is also capable of removing dangerous impurities, such as arsenic, as well as toxic organic compounds [5]. By technology type, the majority of desalination capacity is membrane based, mainly on RO membranes. Figure 14.1 shows the global market share of the technologies applied for desalination [6]. Thus, it is reported that membrane-based desalination accounts for about 44% of the installed capacity of water desalination in the world [7]. Low pressure or low-energy membranes technology represents a fast-growing niche of membrane industry, and these products will continue to be incorporated into future water and wastewater-treatment applications throughout the world [8].

Rapid growth in nanotechnology has advocated a blooming interest in the environmental applications of nanomaterials particularly to revolutionize the century-old conventional water treatment processes [9]. Availability of different types of nanostructured materials coupled with the improved knowledge and skills on the production of various forms of structurally engineered novel materials has opened new doors in the development of functional membranes. It is anticipated that the use of nanoparticles for water desalination will allow the production of more competitive materials that requires less labor, capital, land, and energy. More desirably, nanoparticles can be easily designed to produce tailor-made particles for specific applications. Currently, desalination membranes based on the incorporation of nano-materials are available on the market, with others either close to market launch or in the process of being developed. Advancement in nanotechnology has introduced the use of MMMs, where nanofillers are dispersed throughout a polymeric matrix [10]. As an attractive material that demonstrates outstanding separation properties, MMM has been the subject of worldwide academic studies conducted by many researchers,

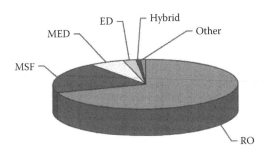

FIGURE 14.1 Global market share of the technologies applied for desalination. MSF, multistage flash; MED, multiple effect distillation; ED, electrodialysis; RO, reverse osmosis. (Data from T. Mezher et al., *Desalination*, 266, 263, 2011.)

especially those related to membrane technology. These types of membranes have generated additional capability for the development of innovative membrane materials for numerous separation processes such as fuel cell applications, gas separation, pervaporation, and water filtration [11–13]. The past decades have witnessed substantial progress and exciting breakthroughs in both the fundamental and application aspects of MMM in various forms of separation. These emerging materials for separation have been traditionally accomplished by incorporating conventional inorganic fillers such as zeolite, carbon molecular sieve, as well as metal oxides and silica nanoparticles in a polymer matrix. Recent advances have shifted toward the introduction of new and novel materials, namely, carbon nanotubes (CNTs), metal organic framework, and clay-layered silicate as potential fillers in the polymer matrix.

Despite the current state-of-the-art in the membrane desalination technology that is exploiting the advantages of nanomaterials, several challenges and hurdles have been encountered in the development of this newly emerging technology that are still remaining as major concerns for the worldwide desalination industries. Substantial uncertainty also remains about the environmental impact of the technology. Therefore, efforts to expand and fully realize the advantages of nanomaterials may face great challenge in the near term. With the major focus placed on the utilization of nanomaterials, this chapter first examines the current desalination membranes in use and then describes the fabrication of nanocomposite membrane incorporating nanomaterials. Through a discussion on the utilization of MMMs for desalination purposes, the performance evaluation related to the application of MMMs in desalination is highlighted.

14.2   TYPES OF MMMs FOR DESALINATION

14.2.1   RO Membranes

Today, RO is the leading desalination technology. It has overtaken conventional thermal technology such as multistage flash (MSF) and is expected to maintain its leadership in the near future though new technologies, such as membrane distillation, electrodialysis, capacitive deionization, and forward osmosis (FO), that have been proposed [14]. Although the demand for drinking water in the world is increasing and regulations on drinking-water quality have become a lot more stringent [15], RO membranes are becoming increasingly popular for water purification applications that require high salt rejection such as brackish and seawater desalination. RO is considered as the simplest and most efficient technique for the purposes of seawater desalination [16], as the process is inherently simple to design and operate compared with many traditional separation processes such as distillation, extraction, ion exchange, and adsorption. The global market for RO components is currently greater than $2.6 billion and is expected to reach $3.7 billion by 2014 [8]. Capital and installation costs for RO systems are highly specific to their applications. Tailored to fit a multiplicity of demands and configurations, the technology remains heavily system and application oriented. Commercial interest in RO technology is increasing globally due to continuous process improvements, which in turn lead to significant cost reductions. These advances include developments in membrane materials and module design, process design, feed pretreatment, energy recovery, and reduction in energy consumption [14]. For membrane desalination, decreasing costs and the production of water with superior quality are among a number of significant reasons why it continues to be the water treatment technology of choice in the world.

Since the start of seawater desalination with RO membranes, the membranes have been extensively developed with the aim of achieving greater water permeability while also maintaining or even enhancing the rejection of salts [17,18]. The most successful RO membranes available commercially in the market are most probably thin-film composites, which typically consist of a three-layer structure. The thin dense active membrane layer of thickness in the order of 100 nm (the *thin film*) is attached to a more open and thicker intermediate layer of about 40 μm thickness, which is attached to an even more open support layer. In addition to the ongoing research into conventional polymeric RO membrane materials, nanotechnology has opened the way to incorporating nanomaterials into RO processes. RO membranes incorporated with various forms of nanomaterials have been proposed to offer attractive permeability characteristics, and many scientists expected that nanotechnology could possibly bring revolutionary advancements to the desalination industry. Currently, thin film nanocomposite (TFN), a new type of nanotechnology-enhanced membrane material, is studied as prospective candidate for desalination purposes. A new fabrication strategy to manufacture TFN membrane through the dispersion of inorganic fillers in the thin polymeric layers has been performed during interfacial polymerization. Several studies have developed methods on how to incorporate zeolites, silica, and silver nanoparticles [19–22]. Many studies have demonstrated that TFN membranes may significantly enhance the membrane properties such as permeability, selectivity, and stability in various membrane separation processes [23,24]. For example, TFN RO membranes have been developed by incorporating pure metal, metal oxide, and zeolite nanoparticles into a polyamide (PA) rejection layer [14]. Unlike other new potential types of desalination membranes, such as those proposed with graphene oxide sheets [25] or CNTs [26] as the active membrane layer, the TFN membrane is already commercially available.

14.2.2 FO MEMBRANES

While RO provides important process performance in desalination, this technology has been criticized as being energy intensive, as the membrane processes are driven by external hydraulic pressures. From a certain viewpoint, the broad application of these membrane processes is severely limited by the worsening global energy crisis [27–29]. In recent years, FO has been explored for desalination, wastewater treatment, and drug release [30,31]. FO is an emerging osmotic process that uses a semipermeable membrane to affect the separation of water from dissolved solutes by an osmotic pressure gradient. Unlike RO, FO does not require high pressure for salt separation, allowing low energy consumption to produce water because FO is a naturally occurring process where water transfers across a selectively permeable membrane driven by an osmotic gradient. The main advantages of using FO are that it operates at low or no hydraulic pressures, it has high rejection of a wide range of contaminants, and it may have lower membrane-fouling propensity than pressure-driven membrane processes [32–34]. As the only pressure involved in the FO process is due to flow resistance in the membrane module, the equipment used is very simple and membrane support is less of a problem. It is an emerging technology that promises to provide an energy-efficient solution in the area of water purification and desalination [35]. Several patents have been awarded for different methods and systems for water desalination by FO [36,37].

Broad application of FO relies on two critical things [38–40]: (1) a good FO membrane, which should be extra-thin, highly porous, and hydrophilic to reduce membrane resistance and increase permeate flux, and (2) ideal draw solutes with high osmotic pressure and easy separation. All the practical experience developing FO as an alternative desalination process has exposed two major limitations: a lack of high-performance membranes and the necessity for an easily separable draw solution. Moreover, when considering seawater desalination, especially when high water recovery is desired, FO can be utilized only if the draw solution can induce a high osmotic pressure. While FO possesses many advantages over the widely used RO, the lack of efficient membranes to reduce internal concentration polarization [40] has resulted in low water flux in the FO process. This is one of the key challenges faced by the FO community [41]. Despite the efforts in advancing FO membranes, only a few membranes such as one made from cellulose have been commercialized because of their high rejection properties and mechanical stability, even though FO has low water flux. Recently, advances in nanotechnology have led to the development of nanostructured materials, which may form the basis for novel FO membranes. New FO membranes have been extensively developed, including nanomaterial composite membranes that are incorporated with nanofiber, zeolite, and CNT [42].

14.2.3 NANOFILTRATION

Nanofiltration (NF) membranes, sometimes called *loose* RO membranes, are porous membranes that exhibit performance between that of RO and UF membranes. Typically, NF involves the separation of monovalent and divalent salts, or organic solutes with molecular weight in the range of 200–1000 g/mol [43,44]. The typically excellent performance of NF membrane, such as high permeate flux, small investment, and low operation cost, brings it wider and larger applications in water softening and dyestuffs desalination [45–47]. Although the core underlying process principles are different, RO and FO have one common feature: membrane forms the heart of the process and hence the membrane performance is directly related to the energy efficiency of the process. Thus, the primary focus of the ongoing research activities rests on membrane materials or, more specifically, improving their physical–chemical behavior [48]. On the other hand, the central element of the NF process rests on the premise that the pressure-driven membrane desalination process can be carried out for any brackish or saltwater by replacing the more energy-consuming RO membranes with more energy-efficient NF membranes. Similar to RO membranes, NF membranes mostly fall into two categories: (1) asymmetric bulk structure, which has a very thin, permselective dense skin layer supported on a more porous sublayer of the same polymer, commonly formed by a phase inversion process induced by an immersion precipitation technique [49] and (2) thin-film composite structure with a thin polymer layer supported by one or more porous layers in different polymers. Interfacial polymerization is the most widely used method to prepare thin-film composite NF membranes. In this regard, interfacially cross-linked aromatic PA on a polyarylsulfone support layer has increasingly been recognized as one of the best processes for the production of these membranes [50–52].

NF membrane serves an appropriate choice for brackish and seawater treatment where very high salt rejection is not necessary or even desirable. Water softening using NF membranes is able to reduce hardness and also remove organics, color, bacteria, and other impurities from raw water. Although RO is necessary for desalinating seawater and treating brackish water, with high levels of dissolved solids, many water supplies do not need such total salt removal. NF partially demineralizes water, removing 10%–90% of dissolved salts, compared with up to 99.5% for RO. However, more frequently, NF membranes have been used in the pretreatment of brackish water for desalination worldwide, due to their remarkable abilities to selectively reject different dissolved salts and provide high water flux with low operating costs and energy consumption [53]. Pretreatment with NF membrane has been identified to assist in improving desalination efficiency, reducing salinity levels, sustaining RO plant performance, and increasing RO membrane lifetime by reducing scaling tendency.

14.3 THE TECHNOLOGICAL NEEDS OF MMMs FOR DESALINATION

Membrane-based desalination, like other desalination technologies, is also not free from some serious concerns. Currently, the main factors that limit the efficiency of the membrane purification technologies include membrane fouling, membrane resistance to the flow, and membrane imperfections, which lead to incomplete rejection or to a drop in the membrane rejection properties over time. A major challenge for membrane technology is the inherent tradeoff between membrane selectivity and permeability. Besides the water permeability and salt rejection, the boron rejection of desalination membranes is also of great importance. In a recent study on the boron rejection of full-scale desalination plants performed for the US Department of the Interior, Bureau of Reclamation [54], it was found that the observed boron rejection in full-scale plants was 65%–80%, which was substantially lower than that measured in lab-scale experiments [55]. The high energy consumption is an important barrier to the wide application of pressure-driven membrane processes. Membrane fouling adds to the energy consumption and the complexity of the process design and operation. Furthermore, it reduces the lifetime of membranes and membrane modules. Generally, fouling is caused by solute adsorbing irreversibly or reversibly onto the surface of the membrane or within the pores of the membrane. It usually causes serious decline in the flux and quality of permeate, ultimately resulting in an increase in the operating pressure with time [9]. Although the term *fouling* can be used to describe both reversible and irreversible solute adsorption, it is the irreversible portion that is most problematic. Irreversible adsorption produces a long-term flux decline that cannot be fully recovered by hydraulically cleaning the membrane. Apart from the cost of energy to run the high-pressure pumps, membrane fouling is an important factor that controls the cost of the water purification unit. Hence, research and development of advanced membranes characterized by high productivity and resistance to fouling may lead to significant energy reduction in the desalination process. Because the maintenance and remediation expenses represent huge implications for the total operating cost, a new generation of membranes with inherent antifouling capability is highly desired. In addition, it was demonstrated that desalted water cost can be reduced if membranes with high permeability are used, even if they exhibit an increase of the salt passage. The water cost reduction can come from either energy savings, because of the lower operating pressures required, or an increase in permeate production, as the plant is running at increased flux and recovery [33].

Undoubtedly, the development of water treatment processes relies much on research progress in advanced materials, in this context, the desalination membrane materials [8]; performance of membrane systems is largely decided by the membrane material. Incorporation of functional nanomaterials into membranes offers a great opportunity to improve the membrane permeability, fouling resistance, and mechanical and thermal stability as well as to render new functions for contaminant degradation and self-cleaning [2]. These nanocomposite MMMs have shown performance exceeding that of existing commercial products based on the standardized polymer chemistry, which were used in typical RO membranes in recent decades, as well as FO membranes developed recently. It is worth saying that nanotechnology is expected to further improve membrane technology and also drive down the high costs of desalination. In this so-called next generation of membrane, high permeate production is achieved by allowing the desalination plant to operate at increased flux and recovery. This is enabled by assembling or incorporating inorganic nanomaterials such as metal and metal oxide nanoparticles, CNTs, and silica into polymeric membranes to bring about tremendous improvement in water flux and alteration of surface properties, potentially related to fouling, while maintaining or slightly increasing the salt rejection [56,57]. With these improvements, less energy is needed to pump water through the membranes. Also, through the correct incorporation of nanomaterials into membranes, the inactivation of irreversibly adhered microorganism can be achieved, making it possible to minimize the fouling [9]. This form of membrane is found to foul more slowly than the conventionally used membrane, as it exhibits stronger repellence toward the particles that might ordinarily be adsorbed on the surface. The result is a water purification process that is just as effective as current methods but more energy-efficient and potentially much less expensive. Initial tests suggest the new membranes have up to twice the productivity, or consume 50% less energy, reducing the total expense of desalinated water by as much as 25% [58,59].

In addition, the incorporation of antibacterial nanomaterials such as silver nanoparticles into the membrane active layer has received a great deal of attention among the scientific community. Generally, nanoparticles show good antibacterial properties arising from their large surface area to volume ratio providing desirable contact with bacterial cell [60,61]. The mechanisms of nanoparticles' antibacterial activity are frequently discussed in the literature using the example of metallic Ag nanoparticles. Incorporation of silver nanoparticles has been proposed as a solution to mitigate the destructive effects caused by the growth of bacteria in water source, as this class of nanomaterial is found to be nonallergic and nontoxic to mammalian living tissues and is environmental friendly [54]. Antibacterial effects of silver nanoparticles are long lasting and the release manner can be well controlled as compared to that of silver ions. For this reason, silver nanoparticles have received more attention for biofouling mitigation [62]. Nanoparticles that possess effective antimicrobial activity such as silver and titanium oxide (TiO_2) have been successfully incorporated into membranes and displayed improved performance compared to the conventionally used polymeric membranes [9]. Testing with feed water containing *Escherichia coli* has shown superior antibiofouling properties, especially with the aid of UV excitation, without compromising the flux and salt rejection performance of the original membrane. No significant loss of TiO_2 nanoparticles from the membrane was observed after a continuous seven-day RO trial [63].

In some other successful examples, zeolite nanoparticles have been incorporated into a polymer matrix to form a thin-film nanocomposite RO membrane and to create a preferential flow path for water molecules, leading to enhanced water transport through the membrane [64,65]. Use of zeolite in the development of TFN for RO was first reported by Hoek and co-workers [66]. Similarly, Jeong et al. [64] prepared a thin-film RO nanocomposite membrane by interfacial *in situ* polymerization on porous polysulfone support, in which NaA zeolite nanoparticles were incorporated into a thin PA film. Introduction of zeolite nanoparticles into a conventional PA RO thin film has enhanced flux to more than double of the conventional membrane with a salt rejection of 99.7%, which is attributed to the smoother and more hydrophilic negatively charged surface. Silica nanoparticles of various sizes have also been incorporated into a PA polymer matrix for RO desalination [67]. Presence of silica nanoparticles was found to remarkably modify the PA network structure, and subsequently the pore structure and transport properties; with only 1–2 wt% of silica, a membrane was fabricated with significantly enhanced flux and salt rejection.

14.4 MODIFICATION OF NANOMATERIALS AND FABRICATION OF MMMs

One of the key parameters for the successful implementation of nanocomposite MMMs is to improve the dispersion of inorganic nanomaterial fillers in polymer matrix through the establishment of specific interaction between the two phases [68]. Unfortunately, the dispersion of nanofiller is hindered by poor interfacial compatibility of the agglomerated nanofiller with the polymer, hence resulting in the formation of unselective voids in the membranes. Weak interaction between these two phases could also disparagingly lead to the leaching of the nanofiller during the membrane cleaning process and after long-term operation. Possibilities of the nanofiller leaching from the MMM would decrease the membrane lifetime and raise concerns with regard to potential environmental and health risks associated with the release of nanomaterials. Various surface modification routes to enhance nanoparticles dispersions in organic solvents or polymer matrices were reviewed by Kango et al. [69].

Various techniques have been developed and applied for the preparation of MMMs. Depending on the formation processes of the nanocomposites, they can be generally divided into two categories. These are direct physical mixing and *in situ* synthesis [70]. Direct physical mixing method has been extensively used to fabricate nanocomposite MMMs mainly due to its convenience in operation, relatively low cost, and suitability for large-scale production. In a typical fabrication route, regardless of the types of inorganic nanomaterials used, the nanofillers and polymer dope are prepared separately and then mixed through solution, emulsion, fusion, or mechanical forces [71,72]. MMMs can also be prepared by directly coating or depositing inorganic particles onto the membrane surface. Nevertheless, direct mixing of polymers with nanofillers has achieved only limited success for most systems due to the difficulties in deciding the space distribution parameter of nanoparticles in or on the polymer matrix. Nanoparticles usually tend to form larger aggregates during mixing, hence greatly diminishing the advantages of their small dimensions. Furthermore, polymer degradation upon melt compounding and phase separation of nanophase from the polymer phase is sometimes detrimental. Various surface treatments of nanoparticles have been adopted in the synthesis procedure, and the compounding conditions, such as temperature and time, shear force, and the configuration of the reactor, can also be adjusted to achieve a good dispersion of nanoparticles in polymer matrices [73]. Also, in some cases, appropriate dispersing agents or compatibilizers are added to improve the particle dispersion and/or miscibility and adhesion between the nanoparticles and the matrix [74].

The *in situ* synthesis approach is widely used to prepare MMMs, and many transition metal sulfide or halide nanoparticles can be readily preloaded within a polymeric phase through the *in situ* synthesis method. Depending on the different starting materials and fabrication processes, *in situ* synthesis can be generally classified as three types, as illustrated in Figure 14.2. Metal ions are preloaded within a polymer matrix to serve as nanoparticle precursors where the ions are expected to distribute uniformly. Then, the precursors are exposed to the appropriate liquid or gas for the *in situ* synthesis of the corresponding nanoparticles [69,75,76]. Tong et al. [77] developed a sol-gel approach to prepare polyimide-TiO$_2$ hybrid films from soluble polyimides and a modified titanium precursor. Rate of the hydrolysis reaction of titanium alkoxide can be controlled using acetic acid as a modifier. Another similar approach employs the monomers of the polymeric host and the target nanofiller as the starting materials [78,79]. Typically, the nanoparticles are first dispersed into the monomers or precursors of the polymeric hosts followed by *in situ* polymerization under desirable conditions, including the addition of appropriate catalyst. Increasing attention has been paid to this method because it allows the synthesis of nanocomposites with desired physical properties. Furthermore, a direct and good dispersion of the nanoparticles into the liquid monomers or precursors reduces the tendency of their agglomeration in the polymer matrix and thereafter improves the interfacial interactions between both phases. Also, nanoparticles and polymers can be prepared simultaneously by blending the precursors of nanoparticles and the monomers of polymers with an initiator in the correct solvent [80,81]. For example, Jeong et al. [64] reported a method to prepare TFN PA RO membrane by dispersing synthesized zeolite-A nanoparticles in trimesoyl chloride (TMC) solution.

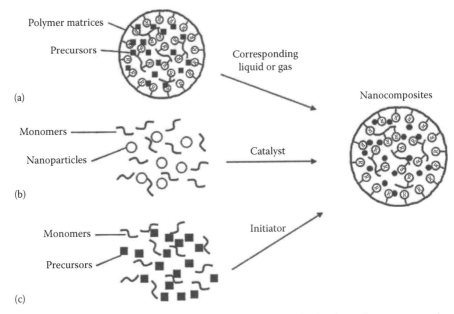

Polymer matrices

Precursors

Corresponding
liquid or gas

(a)

Nanocomposites

Monomers

Nanoparticles

Catalyst

(b)

Monomers

Precursors

Initiator

(c)

FIGURE 14.2 Illustration of *in situ* synthesis process of mixed-matrix nanocomposite membranes: (a) metal ions are preloaded within polymer matrix to serve as nanoparticle precursor; (b) monomers of polymer matrix and the nanofillers as the starting materials; and (c) blending of the nanoparticle precursors and monomers of polymers in solvent.

14.5 PERFORMANCE EVALUATION OF MMMs FOR DESALINATION

The hydrophilic nanozeolite-A was incorporated in a PA layer to modify membrane transport properties, which resulted in membranes with dramatically improved permeability when compared to pure PA membranes [64]. Recently, the effect of particle size and mobile cation on nanozeolite-A was studied by Lind et al. [82,83]. They observed that nanoparticle size may be considered an additional control parameter in designing TFN of RO membranes, and mobile cations play an important role in water permeability with similar salt rejection. In these studies, Linde type-A zeolite particles of varying size were embedded in a thin film in order to enhance the water permeability. The exact mechanism by which the zeolite particles increase the water permeability is still under investigation, although it was hypothesized that the high charge density, superhydrophilicity, and the internal porosity of zeolite molecular sieves provide preferential flow paths for water molecules through the nanocomposite thin films [64]. Alternatively, the increased flux may be due to the low level cross-linking of the TFN [84]. Other possible advantages of the incorporation of zeolite particles in the thin film are reduced fouling and compaction [65]. The commercially available TFN membranes are stated to have a 50%–100% increased water permeability when compared to that of seawater RO membranes, while having a similar salt rejection [17]. TFN consist of a PA layer, and nanosized NaX zeolite was fabricated through interfacial polymerization of TMC and m-phenylenediamine (MPD) monomers over porous polyethersulfone (PES) ultrafiltration support. It was found that the prepared nanocomposite membranes exhibited higher thermal stability and more water permeability than pure PA membranes. Results showed that the addition of nanosized NaX zeolite to the PA membrane led to the improvement of surface properties such as root mean squared roughness, contact angle, solid–liquid interfacial free energy, a decrease in the film thickness, and an increase in pore size and water flux. Nanocomposite membranes formed from a high concentration of monomers in interfacial polymerization exhibited a high water flux and low salt rejection. Excellent membrane performance was observed for the nanocomposite membrane containing about 0.2% (w/v) nanosized NaX zeolite, 0.1% (w/v) TMC, and 2% (w/v) MPD; its flux was 1.8 times higher than that of the neat PA membrane without any change in salt rejection [19]. Also, in a recent study [85], composite RO membranes made from sulfonated poly(arylene ethersulfone) containing amino groups (aPES) and aminated template free zeolite nanoparticle (aTMA) were prepared with the aim of enhancing chlorination resistance and improving membrane performance. Performance of the RO membranes containing aPES and aTMA was evaluated; salt rejection and water flux were 98.8% and 37.8 L/m²h, respectively. Salt rejection decreased by only 12.7% and water flux increased by 2.5 L/m²h after the chlorination test. aPES/aTMA significantly modified the three-dimensional PA network structures and contributed to the high performance because of copolymer chain stiffness due to a high degree of crosslinking in the RO membranes. Therefore, aPES and aTMA, which helped improve water permeability, also protected the active layer structure from degradation and enhanced chlorine resistance of the RO membrane. As depicted in Figure 14.3, intermolecular hydrogen bonding was enhanced by the chemical combination between the aminated nanoparticles and PA copolymer structure. This impeded replacement of hydrogen with chlorine on the amide groups of the aromatic PA membranes. Furthermore, the amino groups on the nanoparticles and TMC combined through the

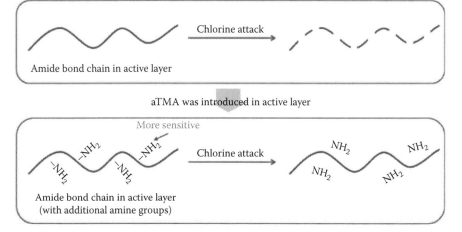

FIGURE 14.3 Diagram showing protection sequence against chlorine attack. (Data from S.G. Kim et al., *Journal of Membrane Science*, 443, 10–18, 2013.)

formation of an amide bond in the RO membrane active layer. These additional amide bonds and unreacted amino groups protect the active layer from chlorine. Due to their higher susceptibility to oxidative chlorination, aromatic primary diamines produce PAs that have resulted in an irreversible reaction at the aromatic nucleus [86]. Free amine groups in aTMA are much more sensitive to oxidation than the amide bonds in the membrane active layer, and this phenomenon explains the high chlorine resistance of the aPES/aTMA membrane.

Hybrid thin-film-composite membrane with TiO_2 nanoparticles was fabricated [63]. TiO_2 nanoparticles imparted an antibiofouling property in the TFC membrane, which has also been reported in other research [87]. A TFN NF membrane was developed through the interfacial incorporation of aminosilanized TiO_2 nanoparticles. PES barrier coating on a porous α-Al_2O_3 ceramic hollow-fiber membrane was employed as the substrate layer [52]. A silane coupling agent containing amino-functional groups was used to modify the surface of TiO_2 nanofillers by reducing their surface energy in order to achieve good dispersion inside the PA skin layer. Figure 14.4 is a schematic

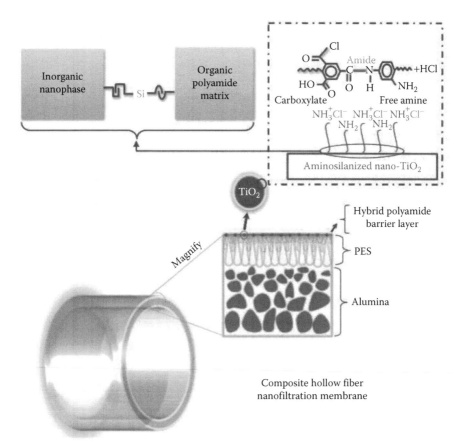

FIGURE 14.4 Schematic representation of the formation of TFN membrane consisting of TiO_2 nanoparticles incorporated into polyamide matrix. (Data from B. Rajaeian et al., *Desalination*, 313, 176–188, 2013.)

representation of the preparation of TFN membrane. The developed membrane has improved permeability as well as selectivity. At low concentration (0.005 wt%), the functionalized TiO_2 nanoparticles improved the salt rejection to 54% as well as water flux to 12.3 L/m^2h. By incorporating a higher concentration of TiO_2 nanoparticles, water flux was increased up to twofold compared with the pure PA membrane with negligible rejection loss. These results demonstrated the competency of using functionalized inorganic nanoparticles to increase the product flux and separation efficiency.

CNT-enhanced TFN membranes were prepared by incorporating the nanotubes into the active layers of membranes used for water treatment [88]. For inclusion into these active layers, a grafting procedure for CNTs was set up to increase their hydrophobicity [89]. Multiwalled CNTs (MWCNTs) grafted with poly(methyl methacrylate) (PMMA) were synthesized through a microemulsion polymerization of methyl methacrylate (MMA) in the presence of acid-modified multiwalled CNTs. Subsequently, PA TFN membranes containing PMMA–MWCNTs were prepared through interfacial polymerization. In such preparation, PMMA was used as a modifier for MWCNTs with an aim of bridging the CNTs with an organic solvent or polymer. It has been previously reported that a covalent bond is formed between PMMA and MWCNTs, which allows the anchoring of PMMA on the CNTs [88]. Schematic representation of the synthesis process of the TFN membrane is shown in Figure 14.5. It has also been identified that a strong interfacial interaction between PMMA and MWCNTs could yield a high solubility in organic solvents. Morphology studies demonstrated that MWCNTs were successfully embedded into the active PA layer [88]. Rejection of Na_2SO_4 was high (99%), and water flux increased by about 62%, which suggested that PMMA–MWCNTs significantly improve selectivity and permeability.

In another similar study, functionalized MWCNT immobilized polyethyleneimine–poly(amide–imide) (PEI–PAI) hollow-fiber membrane was designed and fabricated [35]. PAI hollow fiber was spun through phase inversion, followed by the immobilization of functionalized MWCNTs with vacuum filtration before a chemical posttreatment using PEI was applied to obtain a positively charged selective layer. MWCNT-immobilized PAI membrane surface was then chemically crosslinked with PEI to form a positively charged NF-like selective layer. The resulting membranes were evaluated in an FO process. Membrane prepared using a 0.62 mg/L MWCNTs solution showed a pure water permeability of 4.48 L/m^2h bar with an $MgCl_2$ rejection of 87.8% at 1 bar. This is a 44% enhancement in water permeability without significant deterioration of the salt permeability when compared to membranes without MWCNTs. In addition, FO water flux of membranes prepared using 0.31 and 1.25 mg/L of MWCNTs solutions have been enhanced up to 29% without compromising the solute flux in the active layer, with 0.5 M $MgCl_2$ solution as the draw solution and deionized water as the feed water. It is suggested that the enhancement in the water permeability can be attributed to additional nanochannels between the MWCNTs and the PEI polymer matrix.

Jadav et al. [90] demonstrated that enhanced thermally–chemically stable PA membrane can be achieved, and the membrane performance, in terms of membrane flux and productivity, can be improved by adding a nanoparticle of silica

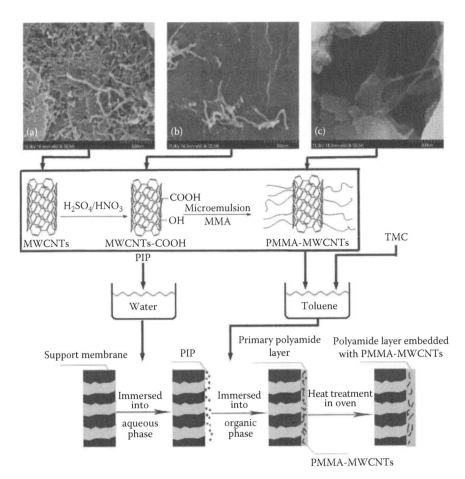

FIGURE 14.5 Schematic representation of TFN membrane synthesis. (Data from D. Baskaran et al., *Angewandte Chemie International Edition*, 43, 2138–2142, 2004.)

into a PA TNF. In a research by Kim et al. [91], composite RO membranes made from sulfonated poly(arylene ether sulfone) containing amino groups (aPES) and hyper-branched aromatic PA-grafted silica (HBP-*g*-silica) were prepared to enhance chlorination resistance and improve membrane performance. As presented in Figure 14.6, the salt rejection and water flux performance of the RO membranes were 96% and 34 L/m²h, respectively. It is worth noting that after the chlorination test, the salt rejection decreased by only 14% and the water flux increased by 4 L/m²h. The aPES/HBP-g-silica significantly modified the three-dimensional PA network structures and contributed to chain stiffness of the copolymer with a high degree of cross-linking in the RO membranes. Consequently, in addition to improving water permeability, the presence of silica nanoparticles protected the active layer structure from degradation and enhanced the chlorine resistance of the RO membrane.

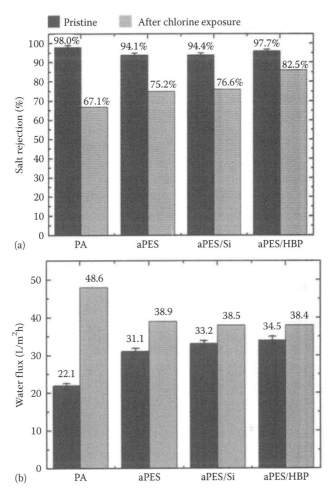

FIGURE 14.6 RO performances and chlorine resistances of composite RO membranes made from sulfonated poly(arylene ether sulfone) containing amino groups (aPES) and hyperbranched aromatic PA-grafted silica (HBP-*g*-silica): (a) salt rejection and (b) water flux. (Data from S.G. Kim et al., *Desalination*, 325, 76–83, 2013.)

14.6 FUTURE PROSPECTIVE AND CONCLUDING REMARKS

Mixed-matrix nanocomposite membranes that can improve membrane performance in desalination have recently attracted considerable interest. Key recent innovations, as well as existing research lines, focus on trying to further reduce the process energy consumption. In addition, much endeavor has concentrated on minimizing the negative effects of scaling and fouling on membranes and obtaining membranes with higher permeate flux. Nanotechnology and its application is one of the rapidly developing sciences. As demand for fresh drinking water continues to increase, nanotechnology can contribute noticeable development and improvement to water treatment process. Advances in nanocomposite membrane technology have led to the development of various nanomaterials, such as zeolite, silica, and metal oxide nanoparticles. Many studies have demonstrated that TFN membranes containing various forms of nanomaterials significantly enhance membrane properties such as permeability, selectivity, and stability in a range of membrane processes. While nanotechnology is leading the way in the development of novel membranes for desalination, there are many fundamental scientific and technical aspects that have to be addressed before the potential benefits may be realized. Development of such membranes is only in the initial stages, and many problems are yet to be overcome. The two major practical challenges are the high cost of nanostructured materials and the difficulty in scaling up nanomembrane manufacturing processes for commercial use. In addition, health and safety issues around the use of nanomaterials have to be addressed in the domestic water industry, particularly with respect to the use of nanoparticles. Extensive and rigorous research is needed to minimize the knowledge gaps in the technical and safety aspects of nanotechnology and dispel public anxiety about these new technologies. Unfortunately most of the recent achievements are still based on laboratory-level tests. Many issues may need to be solved for mass production and application. The majority of methods are confined to scientific research currently due to the high cost, complicated operation procedure, or difficulty in scaling up. Only a few methods are ready for commercial use. In addition, there needs to be further attention paid to studies on long-term fouling. Further, the stability of modifiers should be verified in actual application. Although there is still room to improve our scientific and technological knowledge about nanomaterials in membrane technology, these materials play a key role in the current status and future direction of membrane processes, especially desalination. By addressing the hurdles and challenges encountered during the preparation and processing of these nanocomposite membranes, it is anticipated that this innovative material could offer a new ground for promoting membrane desalination technology.

REFERENCES

1. M. Shatat, M. Worall, S. Riffat. Opportunities for solar water desalination worldwide: Review, *Sustainable Cities and Society* 9 (2013) 67–80.
2. X. Qu, P.J.J. Alvarez, Q. Li. Applications of nanotechnology in water and wastewater treatment, *Water Research* 47 (2013) 3931–3946.
3. M.A. Anderson, A.L. Cudero, J. Palma. Capacitive deionization as an electrochemical means of saving energy and delivering clean water. Comparison to present desalination practices: Will it compete? *Electrochim Acta* 55 (2010) 3845–3856.
4. S. Porada, R. Zhaoa, A. van der Wal, V. Presser, P.M. Biesheuvela. Review on the science and technology of water desalination by capacitive deionization. *Progress in Materials Science* (2013), http://dx.doi.org/10.1016/j.pmatsci.2013.03.005.
5. O. Balcajin, A. Noy, F. Fornasiero, C.P. Grigoropoulos, J.K. Holt, J.B. In, S. Kim, H.G. Park, Nanofluidic carbon nanotube membranes: Applications for water purification and desalination, in: N. Savage et al. (eds.). *Nanotechnology Application for Clean Water*, 77–93, 2009, Norwich: William Andrew Inc.
6. T. Mezher, H. Fath, Z. Abbas, A. Khaled, Techno-economic assessment and environmental impacts of desalination technologies, *Desalination* 266 (2011) 263.
7. C. Fritzmann, J. Löwenberg, T. Wintgens, T. Melin, State-of-the-art of reverse osmosis desalination, *Desalination* 216 (2007) 1–76.
8. M.G. Buonomenna. Nano-enhanced reverse osmosis membranes, *Desalination* 314 (2013) 73–88.

9. A. Matin, Z. Khan, S.M.J. Zaidi, M.C. Boyce, Biofouling in reverse osmosis membranes for seawater desalination: Phenomena and prevention, *Desalination* 281 (2011) 1–16.

10. P.S. Goh, A.F. Ismail, B.C. Ng. Carbon nanotubes for desalination: Performance evaluation and current hurdles, *Desalination* 308 (2013) 2–14.

11. V. Baglio, A.S. Arico, A. Di Blasi, P.L. Antonucci, F. Nannetti, V. Tricoli, V. Antonucci, Zeolite-based composite membranes for high temperature direct methanol fuel cells, *Journal of Applied Electrochemistry* 35 (2005) 207–212.

12. S.A. Hashemifard, A.F. Ismail, T. Matsuura, Mixed matrix membrane incorporated with large pore size halloysite nanotubes (HNTs) as filler for gas separation: Morphological diagram, *Chemical Engineering Journal* 172 (2011) 581–590.

13. G. Liu, F. Xiangli, W. Wei, S. Liu, W. Jin, Improved performance of PDMS/ceramic composite pervaporation membranes by ZSM-5 homogeneously dispersed in PDMS via a surface graft/coating approach, *Chemical Engineering Journal* 174 (2011) 495–503.

14. K.P. Lee, T.C. Arnot, D. Mattia. A review of reverse osmosis membrane materials for desalination—Development to date and future potential, *Journal of Membrane Science* 370 (2011) 1–22.

15. M.A. Shannon, P.W. Bohn, M. Elimelech, J.G. Georgiadis, B.J. Marinas, A.M. Mayes, Science and technology for water purification in the coming decades, *Nature* 452 (2008) 301–310.

16. A.D. Khawaji, I.K. Kutubkhanah, J.-M. Wie. Advanced in seawater desalination technology, *Desalination* 221 (2008) 47–69.

17. B. Hofs, R. Schurer, D.J.H. Harmsen, C. Ceccarelli, E.F. Beerendonk, E.R. Cornelissen. Characterization and performance of a commercial thin film nanocomposite seawater reverse osmosis membrane and comparison with a thin film composite, *Journal of Membrane Science* 446 (2013) 68–78.

18. C.J. Kurth, R. Burk, J. Green, Utilizing nanotechnology to enhance RO membrane performance for seawater desalination, in: *Proceedings of IDA World Congress*, Perth, Australia, 2011, pp. PER11–PER323.

19. M. Fathizadeh, A. Aroujalian, A. Raisi, Effect of added NaX nano-zeolite into polyamide as a top thin layer of membrane on water flux and salt rejection in a reverse osmosis process, *Journal of Membrane Science* 375 (2011) 88–95.

20. M.L. Lind, D.E. Suk, T.V. Nguyen, E.M.V. Hoek, Tailoring the structure of thin film nanocomposite membranes to achieve seawater RO membrane performance, *Environmental Science & Technology* 44 (2010) 8230–8235.

21. P.S. Singh, V.K. Aswal, Characterization of physical structure of silica nanoparticles encapsulated in polymeric structure of polyamide films, *Journal of Colloid and Interface Science* 326 (2008) 176–185.

22. S.Y. Lee, H.J. Kim, R. Patel, S.J. Im, J.H. Kim, B.R. Min, Silver nanoparticles immobilized on thin film composite polyamide membrane: Characterization, nanofiltration, antifouling properties, *Polymers for Advanced Technologies* 18 (2007) 562–568.

23. E.S. Kim, G. Hwang, M.G. El-Din, Y. Liu, Development of nanosilver and multi-walled carbon nanotubes thin-film nanocomposite membrane for enhanced water treatment, *Journal of Membrane Science* 394–395 (2011) 37–48.

24. M. Amini, M. Jahanshahi, A. Rahimpour. Synthesis of novel thin film nanocomposite (TFN) forward osmosis membranes using functionalized multi-walled carbon nanotubes, *Journal of Membrane Science* 435 (2013) 233–241.

25. R.R. Nair, H.A. Wu, P.N. Jayaram, I.V. Grigorieva, A.K. Geim, Unimpeded permeation of water through helium-leak-tight graphene-based membranes, *Science* 27 (2012) 442–444.

26. J.K. Holt, H.G. Park, Y.M. Wang, M. Stadermann, A.B. Artyukhin, C.P. Grigoropoulos, A. Noy, O. Bakajin, Fast mass transport through sub-2-nanometer carbon nanotubes, *Science* 312 (2006) 1034–1037.

27. R.L. McGinnis, M. Elimelech, Global challenges in energy and water supply: The promise of engineered osmosis, *Environmental Science & Technology* 42 (2008) 8625–8629.

28. R. Semiat, Energy issues in desalination processes, *Environmental Science & Technology* 42 (2008) 8193–8201.

29. H. Bai, Z. Liu, D.D. Sun. Highly water soluble and recovered dextran coated Fe_3O_4 magnetic nanoparticles for brackish water desalination, *Separation and Purification Technology* 81 (2011) 392–399.

30. R.W. Holloway, A.E. Childress, K.E. Dennett, T.Y. Cath, Forward osmosis for concentration of anaerobic digester centrate, *Water Research* 41 (2007) 4005–4014.

31. L. Cartinella, T.Y. Cath, M.T. Flynn, G.C. Miller, K.W. Hunter, A.E. Childress, Removal of natural steroid hormones from wastewater using membrane contactor processes, *Environmental Science & Technology* 40 (2006) 7381–7386.

32. Y. Choi, J. Choia, H. Oha, S. Leea, D. Ryook, J. Ha, Toward a combined system of forward osmosis and reverse osmosis for seawater desalination, *Desalination* 247 (2009) 239–246.

33. B. Peñate, L. García-Rodríguez, Current trends and future prospects in the design of seawater reverse osmosis desalination technology, *Desalination* 284 (2012) 1–8.

34. M. Elimelech, Yale constructs forward osmosis desalination pilot plant, *Membrane Technology* 1 (2007) 7–8.

35. K. Goh, L. Setiawan, L. Wei, W. Jiang, R. Wang, Y. Chen. Fabrication of novel functionalized multi-walled carbon nanotube immobilized hollow fiber membranes for enhanced performance in forward osmosis process, *Journal of Membrane Science* 446 (2013) 244–254.

36. T.Y. Cath, A.E. Childress, M. Elimelech, Forward osmosis: Principles, applications, and recent developments, *Desalination* 281 (2006) 70–87.

37. WDR (Water Desalination Report). Company news section 46(3) (2010) 3.

38. Q. Yang, K.Y. Wang, T.S. Chung, Dual-layer hollow fibers with enhanced flux as novel forward osmosis membranes for water production, *Environmental Science & Technology* 43 (2009) 2800–2805.

39. N.Y. Yip, A. Tiraferri, W.A. Phillip, J.D. Schiffman, M. Elimelech, High performance thin-film composite forward osmosis membrane, *Environmental Science & Technology* 44 (2010) 3812–3818.

40. J.R. McCutcheon, R.L. McGinnis, M. Elimelech, Desalination by ammonia–carbon dioxide forward osmosis: Influence of draw and feed solution concentrations on process performance, *Journal of Membrane Science* 278 (2006) 114–123.

41. R. Wang, L. Shi, C.Y. Tang, S. Chou, C. Qiu, A.G. Fane, Characterization of novel forward osmosis hollow fiber membranes, *Journal of Membrane Science* 355 (2010) 158–167.

42. Y.H. Cho, J. Han, S. Han, M.D. Guiver, H.B. Park. Polyamide thin-film composite membranes based on carboxylated polysulfone microporous support membranes for forward osmosis, *Journal of Membrane Science* 445 (2013) 220–227.

43. P. Vandezande, L.E.M. Gevers, I.F.J. Vankelecom, Solvent resistant nanofiltration: Separating on a molecular level, *Chemical Society Reviews* 37 (2008) 365–405.

44. P.S. Zhong, N. Widjojo, T.S. Chung, M. Weber, C. Maletzko. Positively charged nanofiltration (NF) membranes via UV grafting on sulfonated polyphenylenesulfone (sPPSU) for effective removal of textile dyes from wastewater, *Journal of Membrane Science* 417–418, 52–60.

45. A.R. Anim-Mensah, W.B. Krantz, R. Govind, Studies on polymeric nanofiltration-based water softening and the effect of anion properties on the softening process, *European Polymer Journal* 44 (2008) 2244–2252.

46. G. Baumgarten, D. Jakobs, H. Muller. Treatment of AOX-containing waste-water partial flows from pharmaceutical production processes with nanofiltration and reverse osmosis, *Chemie Ingenieur Technik* 76 (2004) 321–325.

47. J. Miao, L.-C. Zhang, H. Lin. A novel kind of thin film composite nanofiltration membrane with sulfated chitosan as the active layer material, *Chemical Engineering Science* 87 (2013) 152–159.
48. S. Sarkar, A.K. SenGupta. A new hybrid ion exchange-nanofiltration (HIX-NF) separation process for energy-efficient desalination: Process concept and laboratory evaluation, *Journal of Membrane Science* 324 (2008) 76–84.
49. R.W. Baker, *Membrane Separation Systems: Recent Developments and Future directions*, Norwich: William Andrew, 1991.
50. Y. Song, P. Sun, L.L. Henry, B. Sun, Mechanisms of structure and performance controlled thin film composite membrane formation via interfacial polymerization process, *Journal of Membrane Science* 251(1/2) (2005) 67–79.
51. L. Li, S. Zhang, X. Zhang, G. Zheng, Polyamide thin film composite membranes prepared from 3, 4′, 5-biphenyl triacyl chloride, 3, 3′, 5, 5′-biphenyl tetraacyl chloride and m-phenylenediamine, *Journal of Membrane Science* 289(1/2) (2007) 258–267.
52. B. Rajaeian, A. Rahimpour, M.O. Tade, S. Liu. Fabrication and characterization of polyamide thin film nanocomposite (TFN) nanofiltration membrane impregnated with TiO_2 nanoparticles, *Desalination* 313 (2013) 176–188.
53. A.A. Abuhabib, A.W. Mohammad, N. Hilal, R.A. Rahman, A.H. Shafie. Nanofiltration membrane modification by UV grafting for salt rejection and fouling resistance improvement for brackish water desalination, *Desalination* 295 (2012) 16–25.
54. J. Kim, M. Wilf, J.-S. Park, J. Brown, *Boron Rejection by Reverse Osmosis Membranes: National Reconnaissance and Mechanism Study*, Denver, CO: Desalination and Water Purification Research (DWPR) Department of the Interior, Bureau of Reclamation, U.S., 2009.
55. K.L. Tu, L.D. Nghiem, A.R. Chivas, Boron removal by reverse osmosis membranes in seawater desalination applications, *Separation and Purification Technology* 75 (2010) 87–101.
56. J.S. Taurozzi, H. Arul, V.Z. Bosak, A.F. Burban, T.C. Voice, M.L. Bruening, V.V. Tarabara, Effect of filler incorporation route on the properties of polysulfone silver nanocomposite membranes of different porosities, *Journal of Membrane Science* 325 (2008) 58–68.
57. J.-F. Li, Z.-L. Xu, H. Yang, L.-Y. Yu, M. Liu, Effect of TiO_2 nanoparticles on the surface morphology and performance of microporous PES membrane, *Applied Surface Science* 255 (2009) 4725–4732.
58. AWWA Membrane Technology Research Committee, Recent advances and research need in membrane fouling, *Journal of AWWA* 97(8) (2005) 79–89.
59. J.B. Li, J.W. Zhu, M.S. Zheng, Morphologies and properties of poly(phthalazinone ether sulfone ketone) matrix ultrafiltration membranes with entrapped TiO_2 nanoparticles, *Journal of Applied Polymer Science* 103(6) (2007) 3623–3629.
60. H. Barani, M. Montazer, N. Samadi, T. Toliyat, In situ synthesis of nano silver/lecithin on wool: Enhancing nanoparticles diffusion, *Colloid and Surface B* 92 (2012) 9–15.
61. M. Moritz, M. Geszke-Moritz. The newest achievements in synthesis, immobilization and practical applications of antibacterial nanoparticles, *Chemical Engineering Journal* 228 (2013) 596–613.
62. J.S. Kim, E. Kuk, K.N. Yu, J. Kim, S.J. Park, H.J. Lee, S.H. Kim et al., Antimicrobial effects of silver nanoparticles, *Nanomedicine* 3 (2007) 95–101.
63. S.H. Kim, S.-Y. Kwak, B.-H. Sohn, T.H. Park, Design of TiO_2 nanoparticle self-assembled aromatic polyamide thin-film-composite (TFC) membrane as an approach to solve biofouling problem, *Journal of Membrane Science* 211 (2003) 157–165.
64. B.H. Jeong, E.M.V. Hoek, Y. Yan, X. Huang, A. Subramani, G. Hurwitz, A.K. Ghosh, A. Jawor, Interfacial polymerization of thin film nanocomposites: A new concept for reverse osmosis membranes, *Journal of Membrane Science* 294 (2007) 1–7.

65. M.T.M. Pendergast, J.M. Nygaard, A.K. Ghosh, E.M.V. Hoek, Using nanocomposite materials technology to understand and control reverse osmosis membrane compaction, *Desalination* 261 (2010) 255–263.

66. E.M.V. Hoek, B.H. Jeong, Y. Yan, Nanocomposite membranes and methods of making and using same. U.S. Application 11/364,885, US 60/660428 (2005).

67. G.L. Jadav, P.S. Singh, Synthesis of novel silica-polyamide nanocomposite membrane with enhanced properties, *Journal of Membrane Science* 328 (2009) 257–267.

68. J.M. Arsuaga, A. Sotto, G. del Rosario, A. Martinez, S. Molina, S.B. Teli, J. de Abajo, Influence of the type, size and distribution of metal oxide particles on the properties of nanocomposite ultrafiltration membranes, *Journal of Membrane Science*, 428 (2013) 131–141.

69. S. Kango, S. Kalia, A. Celli, J. Njuguna, Y. Habibi, R. Kumar, Surface modification of inorganic nanoparticles for development of organic-inorganic nanocomposites: A review, *Progress in Polymer Science* 38 (2013) 1232–1261.

70. J.F. Chen, G.Q. Wang, X.F. Zeng, H.Y. Zhao, D.P. Cao, J. Yun, C.K. Tan, Toughening of polypropylene-ethylene copolymer with nanosized $CaCO_3$ and styrene-butadiene-styrene, *Journal of Applied Polymer Science* 94(2) (2004) 796–802.

71. Q.X. Zhang, Z.Z. Yu, X.L. Xie, Y.W. Mai, Crystallization and impact energy of polypropylene/$CaCO_3$ nanocomposites with nonionic modifier, *Polymer* 45(17) (2004) 5985–5994.

72. D.M. Wu, Q.Y. Meng, Y. Liu, Y.M. Ding, W.H. Chen, H. Xu, D.Y. Ren, *in situ* bubble-stretching dispersion mechanism for additives in polymers, *Journal of Polymer Science Part B: Polymer Physics* 41(10) (2003) 1051–1058.

73. L. Zha, Z. Fang, Polystyrene/$CaCO_3$ composites with different $CaCO_3$ radius and different nano-$CaCO_3$ content-structure and properties, *Polymer Composite* 31 (2010) 1258–1264.

74. Q. Yu, P. Wu, P. Xu, L. Li, T. Liu, L. Zhao, Synthesis of cellulose/titanium dioxide hybrids in supercritical carbon dioxide, *Royal Society of Chemistry* 10 (2008) 1061–1067.

75. S. Ahmad, S. Ahmad, S.A. Agnihotry, Synthesis and characterization of *in situ* prepared poly (methyl methacrylate) nanocomposites, *Bulletin of Materials Science* 30(1) (2007) 31–35.

76. Y. Luo, W. Li, X. Wang, D. Xu, Y. Wang, Preparation and properties of nanocomposites based on poly(lactic acid) and functionalized TiO_2, *Acta Materialia* 57(11) (2009) 3182–3191.

77. Y. Tong, Y. Li, F. Xie, M. Ding, Preparation and characteristics of polyimide-TiO_2 nanocomposite film, *Polymer International* 49 (2000) 1543–1547.

78. E. Tang, G.X. Cheng, X.L. Ma, Preparation of nano-ZnO/PMMA composite particles via grafting of the copolymer onto the surface of zinc oxide nanoparticles, *Powder Technology* 161(3) (2006) 209–214.

79. Z. Wang, Y. Lu, J. Liu, Z. Dang, L. Zhang, W. Wang, Preparation of nano-zinc oxide/EPDM composites with both good thermal conductivity and mechanical properties, *Journal of Applied Polymer Science* 119(2) (2011) 1144–1155.

80. L.A. Utracki, M. Sepehr, E. Boccaleri, Synthetic, layered nanoparticles for polymeric nanocomposites (PNCs), *Polymers for Advanced Technologies* 18 (2007) 1–37.

81. X. Li, D. Wang, G. Cheng, Q. Luo, J. An, Y. Wang, Preparation of polyaniline-modified TiO_2 nanoparticles and their photocatalytic activity under visible light illumination, *Applied Catalysis B: Environmental* 81(3/4) (2008) 267–273.

82. M.L. Lind, A.K. Ghosh, A. Jawor, X. Huang, W. Hou, Y. Yang, E.M.V. Hoek, Influence of zeolite crystal size on zeolite–polyamide thin film nanocomposite membranes, *Langmuir* 25 (2009) 10139–10145.

83. M.L. Lind, B.H. Jeong, A. Subramani, X. Huang, E.M.V. Hoek, Effect of mobile cation on zeolite–polyamide thin film nanocomposite membranes, *Journal of Materials Research* 24 (2009) 1624–1631.

84. M.M. Pendergast, E.M.V. Hoek, A review of water treatment membrane nanotechnologies, *Energy & Environmental Science* 4 (2011) 1946–1971.

85. S.G. Kim, D.H. Hyeon, J.H. Chun, B.-H. Chun, S.H. Kim. Nanocomposite poly(arylene ether sulfone) reverse osmosis membrane containing functional zeolite nanoparticles for seawater desalination, *Journal of Membrane Science* 443 (2013) 10–18.

86. L.F. Liu, S.C. Yu, Y. Zhou, C.J. Gao, Study on a novel polyamide-urea reverse osmosis composite membrane (ICIC–MPD). I. Preparation and characterization of ICIC–MPD membrane, *Journal of Membrane Science* 281 (2006) 88–94.

87. I. Soroko, A. Livingston, Impact of TiO$_2$ nanoparticles on morphology and performance of crosslinked polyimide organic solvent nanofiltration (OSN) membranes, *Journal of Membrane Science* 343 (2009) 189–198.

88. J.N. Shen, C.C. Yu, H.M. Ruan, C.J. Gao, B. van der Bruggen. Preparation and characterization of thin-film nanocomposite membranes embedded with poly(methyl methacrylate) hydrophobic modified multiwalled carbon nanotubes by interfacial polymerization, *Journal of Membrane Science* 442 (2013) 18–26.

89. D. Baskaran, J.W. Mays, M.S. Bratcher, Polymer-grafted multiwalled carbon nanotubes through surface-initiated polymerization, *Angewandte Chemie International Edition* 43 (2004) 2138–2142.

90. G.L. Jadav, V.K. Aswal, P.S. Singh, SANS study to probe nanoparticle dispersion in nanocomposite membranes of aromatic polyamide and functionalized silica nanoparticles, *Journal of Colloid and Interface Science* 351 (2010) 304–314.

91. S.G. Kim, J.H. Chun, B.-H. Chun, S.H. Kim. Preparation, characterization and performance of poly(aylene ether sulfone)/modified silica nanocomposite reverse osmosis membrane for seawater desalination, *Desalination* 325 (2013) 76–83.

15 Fabrication of Polymeric and Composite Membranes

Chun Heng Loh, Yuan Liao,
Laurentia Setiawan, and Rong Wang

CONTENTS

15.1 Introduction 512
15.2 Membrane Preparation through Phase Inversion 513
    15.2.1 Basic Principles 514
        15.2.1.1 Thermodynamics of Polymer Solutions...................... 514
        15.2.1.2 Demixing of Polymer Solutions................................. 516
        15.2.1.3 Phase Diagram and Membrane Formation 519
        15.2.1.4 Influence of Thermodynamics and Kinetics
                  on Membrane Morphology 522
        15.2.1.5 Macrovoid Formation.. 522
    15.2.2 Process of Membrane Fabrication through Immersion
        Precipitation.. 524
    15.2.3 Effects of Dope Formulation on Membrane Formation 525
        15.2.3.1 Polymer 525
        15.2.3.2 Solvent... 526
        15.2.3.3 Additive.. 527
    15.2.4 Effects of Casting/Spinning Conditions on Membrane
        Formation... 529
        15.2.4.1 Exposure to Air.. 529
        15.2.4.2 Composition of Coagulant 529
        15.2.4.3 Coagulant Temperature... 530
        15.2.4.4 Air-Gap 530
        15.2.4.5 Take-Up Speed.. 531
15.3 Preparation of Nanofibrous Membranes through Electrospinning............ 531
    15.3.1 Understanding the Complex Processes Governing
        Electrospinning 532
        15.3.1.1 Onset of Jetting and Rectilinear Jet Development....... 533
        15.3.1.2 Bending Deformation with Looping and Spiraling
                  Trajectories.. 536
        15.3.1.3 Solidification and Deposition on Counter
                  Electrodes/Substrates 537

        15.3.2   Effects of Intrinsic Properties of Polymer Solution..................... 538
                  15.3.2.1   Polymer Concentration and Dope Viscosity............... 538
                  15.3.2.2   Electrical Conductivity 540
                  15.3.2.3   Surface Tension..540
                  15.3.2.4   Solvent..540
        15.3.3   Effects of Operating Conditions 541
                  15.3.3.1   Applied Voltage 541
                  15.3.3.2   Ambient Environment... 541
                  15.3.3.3   Other Operating Parameters 541
15.4   Preparation of Composite Membranes 542
        15.4.1   Interfacial Polymerization 542
                  15.4.1.1   Factors Affecting Interfacial Polymerization 543
                  15.4.1.2   Mixed-Matrix Interfacial Polymerization 543
        15.4.2   Multilayer Polyelectrolyte Deposition 544
                  15.4.2.1   Factors Affecting Multilayer Polyelectrolyte
                              Deposition... 544
        15.4.3   Chemical Cross-Linking 547
        15.4.4   Dual Layer Co-Extrusion/Co-Casting.....................................547
                  15.4.4.1   Fabrication Parameters 548
        15.4.5   Dip-Coating 550
        15.4.6   Chemical Grafting... 551
15.5   Characterizations of Solutions/Membranes... 552
        15.5.1   Polymer Dope Characteristics 552
                  15.5.1.1   Cloud Point 552
                  15.5.1.2   Dope Viscosity and Rheology 552
        15.5.2   Membrane Separation Characteristics..................................... 553
                  15.5.2.1   Fluid Displacement 556
                  15.5.2.2   Fluid Transition... 556
                  15.5.2.3   Microscopic Techniques 557
                  15.5.2.4   Solute Retention 557
15.6   Conclusions.. 558
References.. 558

15.1   INTRODUCTION

Rapid advances in the strategies to design and fabricate synthetic membranes with desired permeability, selectivity, chemo-physical property, multifunctionality, and controlled structure for targeted applications have been seen in the past decades. Common techniques used for porous polymeric membrane fabrication include sintering, stretching, track-etching, phase inversion, and electrospinning. Among these, sintering, stretching, track-etching, and electrospinning are useful in preparing microfiltration (MF) membranes, whereas membranes with a very wide range of pore size covering from MF to ultrafiltration (UF) can be obtained through phase inversion. On the other hand, nonporous membranes used for applications such as nanofiltration (NF), reverse osmosis (RO), and gas separation are normally composite membranes consisting of a porous support layer and a dense skin layer, though some of the nonporous membranes may also have an integral asymmetric structure.

In this chapter, important methods for the preparation of polymeric membranes will be described with focus on nonsolvent-induced phase separation (NIPS) and electrospinning. NIPS is the most commonly used technique for preparing asymmetric porous membranes since its invention pioneered by Loeb–Sourirajan [1]. On the other hand, the preparation of nanofibrous membranes through electrospinning has gained much attention in recent years due to their high orientations, surface area, and porosity. Fundamentals of membrane formation through these two techniques will be illustrated followed by discussion on the various factors affecting the membrane formation process.

By utilizing different materials for the selective and support layers in a membrane, it is possible to optimize each layer individually to achieve the desired characteristics. Preparation of these *composite membranes* is also discussed in this chapter. The composite membranes can be obtained through a single-step method or two-step approaches using the above-mentioned phase-inversion and electrospun membranes as the substrate, followed by applying a top selective layer on the substrate using different materials by various techniques.

It is important to understand the chemistry of the solutions used for membrane preparation and the characteristics of resultant membranes during membrane fabrication. Therefore, a brief introduction of some essential characterization techniques involved in membrane preparation will be included.

15.2 MEMBRANE PREPARATION THROUGH PHASE INVERSION

In 1963, Loeb and Sourirajan successfully developed the first cellulose acetate asymmetric membrane through the phase inversion technique [1]. The developed membrane exhibited higher permeability and separation, thus making the RO separation process possible and practical. Since then, phase inversion has become the most important technique of membrane fabrication for various applications. Advantages of phase inversion over other membrane fabrication techniques include its ability to produce membranes with a wide range of pore size (0.001–10 μm) and the flexibility of material selection.

Phase inversion involves the phase separation (i.e., demixing) of a homogeneous polymer solution followed by the fixation of structure through the solidification of the polymer matrix. Two main approaches that induce the phase separation of a polymer solution are as follows:

1. Thermally induced phase separation (TIPS), in which a polymer solution is prepared at a high temperature and demixing is induced by cooling down the homogenous solution. The simplest system of TIPS involves two components: a polymer and a solvent.
2. NIPS, in which a change in the local composition of the polymer solution is induced by the diffusion of solvent and nonsolvent between the solution and its external environment. Change in composition will lead to demixing and precipitation. The simplest system of NIPS involves three components: a polymer, a solvent, and a nonsolvent.

NIPS of a homogenous polymer solution can be induced by (1) immersion precipitation, immersion of the polymer solution into a nonsolvent bath such that the solvent–nonsolvent exchange occurs; (2) vapor absorption, the absorption of a nonsolvent by the polymer solution when it is subject to a vapor containing nonsolvent until precipitation; and (3) solvent evaporation, the evaporation of a volatile solvent from a polymer solution [2].

In this section, the discussion on membrane preparation will mainly focus on, but not limited to, NIPS by immersion precipitation.

15.2.1 BASIC PRINCIPLES

15.2.1.1 Thermodynamics of Polymer Solutions

Membrane fabrication through phase inversion always involves a thermodynamically stable polymer solution that is subject to demixing. In order to understand the mechanism of membrane formation through phase inversion techniques, thermodynamic principles for mixing and demixing are first introduced. In general, the occurrence of a spontaneous process in a closed system is governed by the change in Gibbs free energy as follows:

$$\Delta G = \Delta H - T \Delta S \qquad (15.1)$$

where:
    ΔG is the free enthalpy change
    ΔH is the change in enthalpy
    ΔS is the change in entropy during the process
    T is the absolute temperature

The described process can be a chemical reaction, or a dissolving process, for example. A process occurs spontaneously if ΔG of the process is negative. For the case of mixing two components at a constant temperature and pressure, the difference in Gibbs free energy of mixing can be expressed as follows:

$$\Delta G_m = \Delta H_m - T \Delta S_m \qquad (15.2)$$

where:
    m refers to mixing

Mixing of two components can occur spontaneously if $\Delta G_m < 0$. In other words, a homogenous solution can be obtained by mixing two components when the Gibbs free energy of the system is reduced by the mixing.

If one of the components that are subject to mixing is polymer, the increase in entropy of mixing, ΔS_m, in the system will be much lower compared with that in the system consisting of only small molecules. Therefore, the change in enthalpy of mixing, ΔH_m, is the important factor that determines whether a mixing is spontaneous. Hildebrand introduced the concept of a solubility parameter that can be used to derive ΔH_m [3]

$$\delta = \left(\frac{C}{v}\right)^{1/2} \qquad (15.3)$$

where:
  δ is the solubility parameter of molecules
  C is the cohesive energy that is equal to the heat of vaporization
  v is the molar volume of the molecules

The term C/v is called the cohesive energy density, which indicates the energy required to remove a molar volume of molecules from the bulk. Considering a binary system, the change in the enthalpy of mixing of two components can, therefore, be expressed as follows [3]:

$$\Delta H_m = v[\delta_1 - \delta_2]^2 \varnothing_1 \varnothing_2 \tag{15.4}$$

where:
  $\varnothing_1$ and $\varnothing_2$ are the volume fractions of the two components, respectively

From here, it can be noted that ΔH_m is minimized (and so for ΔG_m) when the solubility parameters of the two components, δ_1 and δ_2, are similar. As described by the second law of thermodynamics that entropy change of an isolated system is never negative, a small ΔH_m implies that ΔG_m is more likely to be negative ($\Delta G_m \approx -T\Delta S_m$). In other words, the mixing of the two components is more likely to be spontaneous when $\delta_1 \approx \delta_2$.

To improve the accuracy of the solubility parameter, Hansen divided δ into contributions of different intermolecular forces [4] as follows:

$$\delta^2 = \delta_d^2 + \delta_q^2 + \delta_h^2 \tag{15.5}$$

where:
  Subscripts d, q, and h represent the dispersion interaction, dipole–dipole force, and hydrogen bonding components for the solubility parameter, respectively

Difference in solubility parameter between two components in Equation 15.4 is hence expressed as follows:

$$\Delta \delta = \left[(\delta_{d1} - \delta_{d2})^2 + (\delta_{q1} - \delta_{q2})^2 + (\delta_{h1} - \delta_{h2})^2 \right]^{1/2} \tag{15.6}$$

Another model describing the thermodynamics of polymer solutions was introduced by Flory and Huggins [5,6]. In the Flory–Huggins theory, the non-ideality of the polymer solutions due to the large difference in molecular size between the solute and solvent is taken into account. Based on this theory, the Gibbs free energies of binary and ternary systems are expressed in Equations 15.7 and 15.8, respectively [7] as follows:

$$\Delta G_m = RT(n_1 \ln \varnothing_1 + n_2 \ln \varnothing_2 + n_1 \varnothing_2 \chi_{12}) \tag{15.7}$$

$$\Delta G_m = RT \left(n_1 \ln \varnothing_1 + n_2 \ln \varnothing_2 + n_3 \ln \varnothing_3 + n_1 \varnothing_2 \chi_{12} + n_3 \varnothing_1 \chi_{13} + n_3 \varnothing_2 \chi_{23} \right) \tag{15.8}$$

where:

R is the gas constant

n is the number of moles of a component

Subscripts 1, 2, and 3 represent the solvent, polymer, and nonsolvent components, respectively

χ is the Flory–Huggins interaction parameter

χ is a measure of the interaction between the polymer and solvent or nonsolvent, and a positive value reflects a repulsion between the two components and vice versa [8].

Relationship between ΔG_m and χ of a binary system is presented in Figure 15.1. At a low value of χ, the ΔG_m curve is concave and only shows a single minimum point. Mixing is favorable at any composition in such case. With increasing value of χ, an upward convex part starts to appear at the middle region of the ΔG_m curve and the mixing becomes increasingly unfavorable, the reasons for which are discussed in Section 15.2.1.2.

15.2.1.2 Demixing of Polymer Solutions

A homogeneous solution will demix into two or more phases if the demixing process reduces the free energy of the system; the free energy after phase separation, ΔG_{sep}, is smaller than the free energy of the homogeneous system, ΔG_0. Figure 15.2a presents the ΔG_m curve of a binary system having a low χ value. It is seen from the figure that the free energy of such system would increase from ΔG_0 to ΔG_{sep} if a homogeneous solution with an arbitrary composition of $\varnothing_0$ is separated into two phases with (arbitrary) compositions $\varnothing_I$ and $\varnothing_{II}$. In such a case, mixing is favorable at any composition. Therefore, the solution is said to be *stable* and no phase separation will occur.

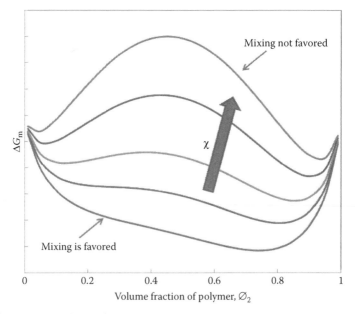

FIGURE 15.1   Dependence of ΔG_m on χ.

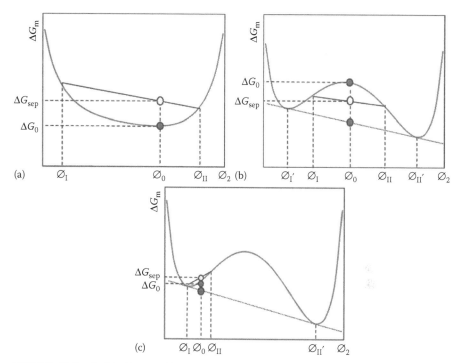

FIGURE 15.2 Stable (a), unstable (b), and metastable (c) states of the solution.

In the case that the χ value of a binary system is high, a mixed solution could be stable, unstable, or metastable depending on its composition. As shown in Figure 15.2b, if the composition of the solution is located at the concave-down section of the ΔG_m curve (i.e., $[\partial^2 \Delta G_m/\partial \varnothing_i^2] < 0$), a small fluctuation in composition from $\varnothing_0$ to $\varnothing_I$ and $\varnothing_{II}$ will lead to a decrease in free energy from ΔG_0 to ΔG_{sep}. Hence, the solution is said to be *unstable* and will spontaneously demix into two phases until the compositions of the phases reach $\varnothing_I'$ and $\varnothing_{II}'$, where a communal tangent line exists and the system is at its most stable state.

Figure 15.2c shows the case for a solution having a composition located outside the concave-down section (i.e., $[\partial^2 \Delta G_m/\partial \varnothing_i^2] > 0$) but in between the two points of the communal tangent line. A small fluctuation in composition from $\varnothing_0$ to $\varnothing_I$ and $\varnothing_{II}$ leads to an increase in free energy from ΔG_0 to ΔG_{sep}, indicating that demixing is not spontaneous. However, the free energy could be minimized when the solution is separated into two phases with compositions $\varnothing_I$ and $\varnothing_{II}'$. A solution with such compositions is in a *metastable* state, and the demixing of the homogeneous solution could only occur when the fluctuation in composition is significantly large.

In equilibrium, the phases in a demixed system should have equal chemical potential [9]. The chemical potential difference of a component after mixing is defined as follows:

$$\Delta\mu_i = \frac{\partial \Delta G_m}{\partial n_i} \tag{15.9}$$

At any composition, the chemical potentials of the pure components are given by the tangent line, as shown in Figure 15.3a. The condition of phase equilibrium in a demixed binary system, which can also be used to determine the binodal, is given by

$$\Delta\mu_i^{I} = \Delta\mu_i^{II} \tag{15.10}$$

where:
$\Delta\mu_i$ is the chemical potential difference of component i
whereas I and II represent the two phases in a system

Binodal denotes the boundary between the stable and metastable states of a solution. Compositions of binodal, $\varnothing_b$, and $\varnothing_b'$ can be determined where the two points are connected by a communal tangent line, as shown in Figure 15.3b. On the other hand, spinodal denotes the boundary between the absolute instability and stable/metastable states of a solution. It is defined as follows:

$$\frac{\partial^2\Delta G_m}{\partial\varnothing_i^{2}} = 0 \tag{15.11}$$

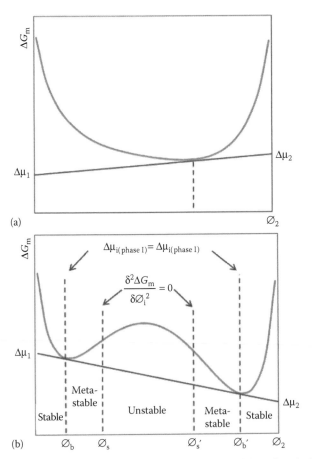

FIGURE 15.3 Chemical potential (a) and stable/metastable/unstable regions in ΔG_m graph (b).

Compositions of spinodal, $\emptyset_s$, and $\emptyset_s'$ are the inflection points of the ΔG_m curve, as shown in Figure 15.3b. A solution is in metastable state when $(\partial^2 \Delta G_m / \partial \emptyset_i^2) > 0$, whereas it is in unstable state when $(\partial^2 \Delta G_m / \partial \emptyset_i^2) < 0$.

15.2.1.3 Phase Diagram and Membrane Formation

Figure 15.4 shows a typical phase diagram for TIPS involving two components. Based on Equations 15.10 and 15.11, the binodal and spinodal curves in the phase diagram are plotted by joining the binodal and spinodal points of ΔG_m curves at different temperatures. When a homogeneous solution is brought into the binodal and spinodal regions, a solution is in metastable and unstable states, respectively, in which the homogeneous solution demixes into two liquid phases. There exists a critical point (A in Figure 15.4) where the binodal and spinodal curves meet each other, and it fulfills both Equation 15.11 and Equation 15.12.

$$\frac{\partial^3 \Delta G_m}{\partial \emptyset_i^3} = 0 \qquad (15.12)$$

In TIPS, a solution remains homogeneous at any composition when the temperature is above the critical point.

In a similar manner, the binodal and spinodal curves in the ternary phase diagram are plotted by joining the binodal and spinodal points of ΔG_m curves at different compositions. A typical ternary phase diagram involving the polymer, solvent, and nonsolvent components is shown in Figure 15.5. In addition to the binodal and spinodal curves that indicate the metastable and unstable regions, the gelation, crystallization/vitrification curves can be included in the phase diagram, as shown in Figures 15.4 and 15.5. These curves represent the boundary where a polymer solution starts to

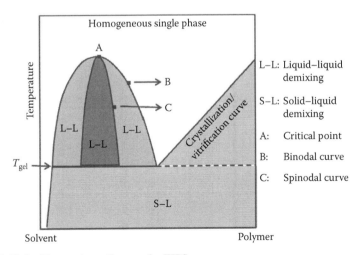

FIGURE 15.4 Binary phase diagram for TIPS.

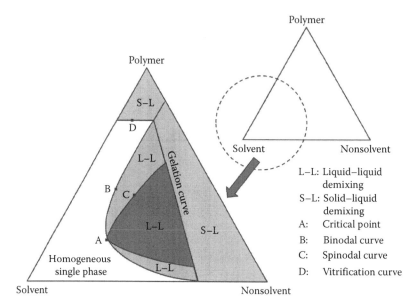

FIGURE 15.5   Ternary phase diagram for NIPS.

become solidified by different mechanisms. Solidification mechanisms of polymer solutions include the following [10]:

1. Crystallization, whereby the polymer chains are immobilized due to the formation of an ordered crystalline phase. Many of the polymers are semi-crystalline, which consist of both an unordered amorphous phase and an ordered crystalline phase.
2. Vitrification, whereby the polymer chains are immobilized when the glass transition temperature is passed. The glass transition temperature of a polymer solution is dependent on the type of polymer and solvent, as well as the concentration of the polymer [11].
3. Gelation, whereby the mobility of polymer chains in a solution is reduced drastically due to the formation of a three-dimensional network by physical cross-linking. Gelation can be induced by the presence of microcrystallites, special interaction between polymer and solvent, and the vitrification of polymer-rich phase during liquid–liquid demixing [12].

Membrane formation is highly dependent on the formation path in the phase diagram for both TIPS and NIPS. Although TIPS and NIPS are induced by different approaches, their mechanisms of phase separation are similar. Therefore, both the phase diagrams for TIPS and NIPS are included in Figure 15.6 to illustrate how the formation path affects the membrane formation, but only the description for NIPS will be included.

At the initial stage of NIPS, a solution initially falls in the single-phase homogeneous region in the ternary phase diagram. For the case of immersion precipitation

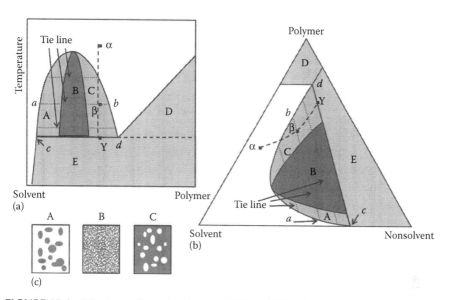

FIGURE 15.6 Membrane formation through TIPS and NIPS illustrated by phase diagram.

or vapor absorption, when the solution is immersed into a nonsolvent (coagulant) bath, the solvent–nonsolvent exchange starts to take place; the solvent within the polymer solution diffuses out into the nonsolvent bath while the nonsolvent diffuses into the polymer solution. As illustrated by path α–β in Figure 15.6 (N), the polymer solution is brought into a metastable or unstable region (region A, B, or C) in the ternary phase diagram after a period of time due to the change in composition. In this region, liquid–liquid demixing occurs and the solution separates into polymer-rich and polymer-lean phases.

Mechanism of liquid–liquid demixing is different for the metastable and unstable regions. For instance, when the solution enters region C through path α–β, the solution is in a metastable state where liquid–liquid demixing is mainly based on the growth of polymer-lean nuclei within a continuous polymer-rich phase (Figure 15.6 (N)). This demixing mechanism for metastable solutions is called *nucleation* and *growth*. Nascent membrane with continuous polymer matrix and dispersed pores is formed in this case. Instead, if the solution enters region A, the demixing is achieved by the nucleation and growth of polymer-rich nuclei. In this case, a structure with low-integrity powdery agglomerates is formed, which is not a practical membrane structure.

On the other hand, when the solution enters region B, which is the unstable region, phase separation takes place without the presence of any nucleus. The solution separates into an interpenetrating network of polymer-rich and polymer-lean phases, as shown in Figure 15.6 (N). This mechanism is called *spinodal decomposition*. The nascent membrane has both interconnected polymer matrix and pores.

Composition of polymer-rich and polymer-lean phases at any stage of the demixing process is determined by the tie lines, whereby two compositions of equal chemical potential are joined together. For instance, the polymer-rich and polymer-lean phases have a composition indicated by points b and a, respectively, when the solution is at a state indicated by β, as shown in Figure 15.6 (N). In addition, the volume fraction of polymer-rich and polymer-lean phases can be determined from phase diagram using the lever rule [13]. For instance, at β *state* in Figure 15.6 (N), the volume fraction of polymer-rich phase is given by

$$X_{\text{polymer-rich}} = \frac{\text{length of } a\beta}{\text{length of } ab} \tag{15.13}$$

where:
$X_{\text{polymer-rich}}$ is the volume fraction of the polymer-rich phase

With further exchange of solvent and nonsolvent, the polymer solution will be eventually brought from β to γ, where the polymer-rich phase is solidified (Figure 15.6 (N)). Structure of the nascent membrane is more or less frozen at this point. In the case of phase inversion by solvent evaporation, it is also possible for solidification to occur before a solution is subject to liquid–liquid demixing. In this case, a homogeneous solution is directly brought into region D in the ternary phase diagram due to the continuous loss of solvent. As a result, the polymer solution undergoes a glass transition and a glassy and dense film is formed.

15.2.1.4 Influence of Thermodynamics and Kinetics on Membrane Morphology

In NIPS by immersion precipitation, the morphology of the resultant membranes is highly controlled by the time interval between the immersion of the dope (polymer solution) into the coagulant and the onset of liquid–liquid demixing. Instantaneous demixing refers to the situation where phase separation occurs almost immediately after immersing a dope into the coagulant. In this case, the formed membranes generally possess a thin and dense skin layer on the top and a substructure with macrovoids. For the case of delayed demixing, after the immersion of a dope into the coagulant, it takes some time before the occurrence of phase separation. Membranes produced by delayed demixing generally exhibit sponge-like structure with a relatively loose top layer [10].

In a thermodynamic context, instantaneous demixing is likely to take place when the initial composition of the homogeneous polymer dope is relatively close to its metastable or unstable state. This is because a little amount of nonsolvent is sufficient to induce phase separation. In addition to thermodynamics, the kinetics of membrane formation, which often refers to the rate of solvent–nonsolvent exchange during the phase inversion, is observed to govern the final membrane structure in many cases [14,15]. If the solvent–nonsolvent exchange rate is slow, delayed demixing will occur even when a homogeneous system is thermodynamically close to its metastable or unstable state and vice versa.

15.2.1.5 Macrovoid Formation

Formation of macrovoids in a membrane is always observed in a system that facilitates instantaneous demixing. Presence of macrovoids is beneficial to the membrane permeability as the tortuosity is reduced. However, mechanical properties of membranes can be compromised by the excessively large macrovoids. Therefore, there is a need to control the macrovoid formation during membrane preparation. Different mechanisms of macrovoid formation have been proposed in the literature. They include interfacial hydrodynamic instability induced by a surface tension gradient [16], osmotic pressure in polymer-lean nucleus [17], shrinkage of polymer matrix [18], and excess intermolecular potential gradients caused by concentration gradients [19].

After immersing a dope into the coagulant bath, phase separation takes place immediately at the interface between the dope and the coagulant due to solvent–nonsolvent exchange. With further exchange of solvent and nonsolvent, the interface between the phase separation region and the homogeneous region gradually migrates inward from the outermost surface. According to McKelvey and Koros, the planar front of the phase separation region moves at a rate inversely proportional to time squared [17]. Composition at a given location in the nascent membrane changes with time. If the composition reaches its metastable state, the nucleation and growth of polymer-lean phase is initiated. When the moving front is far from nucleus, the nucleus grows isotropically due to the in-diffusion of solvent from the surrounding environment until the nucleus wall is solidified, as shown in Figure 15.7a [9]. However, the growth of a nucleus is affected when it is in contact with the moving front. In this case, the rate of solvent–nonsolvent exchange at the back side of the nucleus is dramatically enhanced. Due to osmotic pressure, nonsolvent flows into the nucleus from the back side, which also facilitates the inflow of the solvent from the front side of the nucleus. Because the

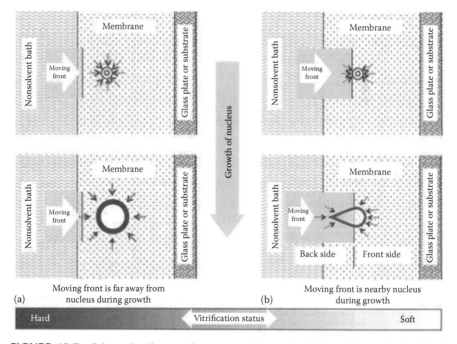

FIGURE 15.7 Schematic diagram for macrovoid growth. (Adapted from D. Li, Dual-layer asymmetric hollow-fiber membranes for gas separation, PhD Thesis, Chemical & Biomolecular Engineering, National University of Singapore, Singapore, 2005.)

back side has more nonsolvent, the wall at the back side of the nucleus becomes more rigid, whereas the front side of the nucleus is still highly plasticized. Therefore, the nucleus grows toward the front side, forming a tear-like shape (Figure 15.7b). Under some conditions, macrovoids are formed when the nuclei are able to grow continuously before the wall is solidified [17].

15.2.2 Process of Membrane Fabrication through Immersion Precipitation

Different devices are used for the preparation of flat-sheet and hollow-fiber membranes. The device used for casting flat-sheet membranes is represented in Figure 15.8. A homogeneous polymer solution is first poured onto a supporting sheet made of, for example, nonwoven material. The dope is then passed through a knife, which is to cast the dope into a certain thickness. The casted dope is exposed to ambient environment for a period of time before it is immersed into a coagulant bath. The exposure time, the properties of the ambient environment, and the temperature of the coagulant can be adjusted to control the membrane formation. Water is the most commonly used coagulant for the preparation of polymeric membranes, but different solutions have also been used. The resultant membrane is then collected for further treatment.

One of the main differences between the preparation of flat-sheet and hollow-fiber membranes is that a bore fluid, an internal coagulant, is needed for hollow-fiber spinning. Because the phase separation takes place from both the outer and inner sides, hollow-fiber spinning is a more complex process than flat-sheet casting. The device used for hollow-fiber spinning is represented in Figure 15.9. As shown in the figure, the polymer dope and bore fluid are simultaneously extruded from a spinneret. Flow rates of the polymer dope and bore fluid can be adjusted by the pumps that are used to push the fluids through the spinneret. The nascent fiber passes through an air-gap before the immersion into the coagulant bath. The resultant fiber is then collected by a drum rotating at a constant speed (take-up speed). Take-up speed must be at least equal to the free-falling speed of the nascent fiber to prevent the fiber from coiling in the coagulant bath.

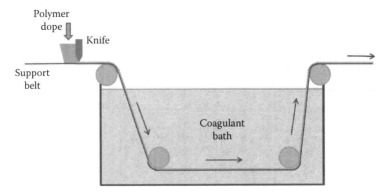

FIGURE 15.8   Schematic diagram for membrane casting.

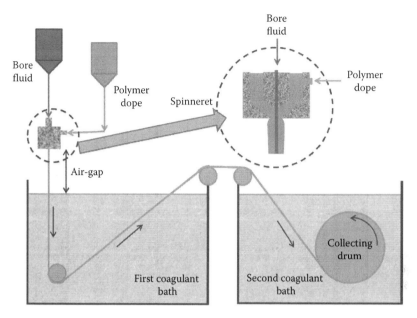

FIGURE 15.9 Schematic diagram for the spinning of hollow-fiber membranes.

After the flat-sheet or hollow-fiber membranes are formed and solidified in the coagulant bath, it is collected and soaked in water for several days to extract any remaining solvent. The membranes can then be used for other treatments or drying. Due to the large surface tension of water, membrane pores tend to collapse during the drying process. Therefore, special drying methods such as solvent exchange and freeze drying are frequently employed to maintain the pore structure [20,21].

15.2.3 EFFECTS OF DOPE FORMULATION ON MEMBRANE FORMATION

In NIPS by immersion precipitation, a homogeneous solution is prepared by dissolving a polymer into a suitable solvent. Additives serving various functions are also often included in the solution. The types of polymer, solvent, and additive, as well as their concentration in the dope, are highly influential on the properties of final membranes, which are discussed in the following sections.

15.2.3.1 Polymer

For MF and UF, membrane selectivity is mainly determined by the pore size of membranes rather than the properties of membrane material itself. However, considering the operating environment and the fouling issue, the selection of polymers as the membrane material is crucial for the practical application of the membranes. In general, polymers used for membrane fabrication through NIPS need to be easily dissolved in some common solvents, resistant to chemical, thermal, and mechanical stresses, and less susceptible to fouling. Common polymers used in NIPS by immersion precipitation and their properties are listed in Table 15.1.

TABLE 15.1

Commonly Used Polymers in NIPS by Immersion Precipitation

Polymers	Crystallinity	Hydrophilicity	Glass Transition Temperature (°C)	Chemical Stability
Cellulose ester	Semicrystalline	Hydrophilic	Varies	Low to moderate
Polysulfone	Amorphous	Moderate	195	High
Polyethersulfone	Semicrystalline	Moderate	230	High
Polyacrylonitrile	Semicrystalline	Hydrophilic	100	Moderate to high
Poly(vinylidene fluoride)	Semicrystalline	Hydrophobic	−35	High
Polyetherimide	Amorphous	Moderate	216	Moderate to high
Polyimide	Amorphous	Moderate	Varies	Moderate to high
Polyether ether ketones	Semicrystalline	Moderate	143	Moderate to high

Concentration of polymer in a dope can significantly affect the properties of the resultant membranes. During phase inversion, a homogeneous dope containing a high amount of polymer tends to separate into two phases with a higher volume fraction of polymer-rich phase according to the lever rule as mentioned previously (see Section 15.2.1.3). Therefore, membranes with a denser structure are formed. As a result, the resultant membranes exhibit lower permeability, higher selectivity, and higher mechanical strength [22,23].

Viscosity of a dope, which is highly related to the kinetics of membrane formation, increases significantly with increasing polymer concentration in the dope. This is because chain entanglement is significantly enhanced with the presence of a larger amount of macromolecules [24]. For hollow-fiber spinning, the dope viscosity should be high enough to ensure a stable and continuous flow from the spinneret during the spinning process.

15.2.3.2 Solvent

A polymer must be dissolved in a suitable solvent to obtain a homogeneous solution for membrane fabrication. Relative affinity between a polymer and a solvent can be assessed by the difference between their solubility parameters, $\Delta\delta$, as calculated by Equation 15.6. Theoretically, a smaller value of $\Delta\delta$ reveals a stronger solubility of a particular solvent to the polymer.

Membrane structure has been observed to be affected by the choice of solvent. In a study preparing poly(vinylidene fluoride) (PVDF) membranes, eight solvents including N,N-dimethylacetamide (DMAc), N,N-dimethylformamide (DMF), dimethylsulfoxide, N-methyl-2-pyrrolidone (NMP), hexamethylphosphoramide, tetramethylurea, triethyl phosphate (TEP), and trimethyl phosphate were employed and the resultant membranes were compared [25]. It was found that the membrane porosity and hence the water permeability of the membranes were dependent on the mutual diffusivity of solvent and nonsolvent (water). It was claimed that the formation path in the ternary phase diagram led to entry into the demixing region at a higher polymer concentration when the solvent–nonsolvent diffusivity was high. As a result, denser membranes were formed at high solvent–nonsolvent diffusivity.

Influence of solvent on macrovoid formation was studied by researchers. For instance, PVDF and polyethersulfone (PES) membranes prepared using DMAc as the solvent exhibit smaller macrovoids than that using NMP as the solvent [23,25,26]. Barzin and Sadatnia proposed that in a PES/DMAc/water system, the polymer-rich phase vitrified earlier than that in PES/NMP/water system. As a result, macrovoid growth in the former system was suppressed [26].

15.2.3.3 Additive

Use of additives is one of the most effective means to control membrane formation in NIPS. Depending on the type of additive used, the presence of an additive in a polymer dope may cause changes in dope properties such as thermodynamics, kinetics, and hydrophilicity. As a result, various effects on membrane formation have been demonstrated using different additives. Common additives used in NIPS can be categorized as polymeric additives (e.g., polyethylene glycol [PEG]), organic molecules (e.g., ethylene glycol), inorganic molecules (e.g., lithium chloride), weak co-solvents (e.g., ethanol), and strong nonsolvent (e.g., water).

Polymeric additives such as PEG and poly(vinyl pyrrolidone) are widely used as the pore-forming additives in membrane fabrication. These additives not only enhance the hydrophilicity but also induce thermodynamic instability of the dopes. Therefore, instantaneous demixing is promoted, and hence membranes with high porosity are formed [27,28]. In recent years, polymeric additives with an amphiphilic nature such as block copolymers have received much attention due to their unique behavior during membrane formation. Amphiphilic additives tend to segregate to the membrane–water interface due to the presence of a hydrophilic segment, whereas the hydrophobic segment of the additives firmly anchors in the polymer matrix [29,30]. Pluronic copolymers, which are commercially available triblock copolymers of poly(ethylene oxide) (PEO) and poly(propylene oxide) (PPO), have been observed to possess both pore-forming and surface-modifying abilities in PES membrane fabrication [31]. Presence of Pluronic copolymers might also narrow down the membrane pore size, which enhances membrane selectivity, as illustrated in Figure 15.10 [32]. In addition, it has been reported that Pluronic copolymers have an extremely strong pore-forming ability in PVDF membrane formation due to the interaction between PVDF macromolecules and the PPO segments in the copolymers [33].

Inorganic salts such as lithium chloride are well known for their ability to enhance the dope viscosity due to their strong interaction with some solvents such as NMP, DMAc, and DMF [34,35]. Wang and Ma have demonstrated how the kinetics of membrane formation affects macrovoid growth in PVDF membranes using a mixture of inorganic salts, ferrous chloride, and hydroxylamine hydrochloride as the additive [36]. With increasing additive concentration, the hydrophilicity of the dope is enhanced, and hence solvent–nonsolvent exchange becomes faster. This facilitates instantaneous demixing, which produces membranes with larger macrovoids. However, when the additive concentration exceeds a certain amount, the dope viscosity becomes so high that the in-diffusion of nonsolvent is highly restricted. Therefore, delayed demixing takes place and the macrovoid size in membranes

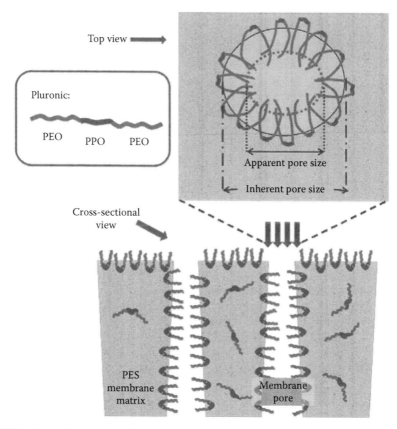

FIGURE 15.10   Illustration of PEO brush layer formation on the surface of membrane internal pores due to the surface segregation of Pluronic.

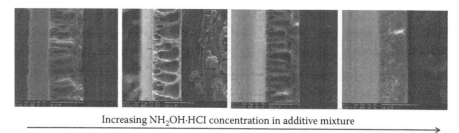

FIGURE 15.11   Effect of inorganic salt concentration on macrovoid formation. (Adapted from Z. Wang, J. Ma, *Desalination*, 286, 69–79, 2012.)

becomes smaller. The membranes become fully sponge-like finally at a high additive concentration, as shown in Figure 15.11 [36].

Water is a strong nonsolvent to many polymer solutions, but it is possible to prepare a homogeneous polymer dope containing a small amount of water. Presence of water brings the dope closer to its metastable or unstable state, and thus facilitates instantaneous demixing during membrane formation. However, when the amount of water added in a dope is large enough to bring the system close to its gelation point, the macrovoid growth is suppressed because nuclei are unable to grow in this case [12].

15.2.4 EFFECTS OF CASTING/SPINNING CONDITIONS ON MEMBRANE FORMATION

In addition to dope formulation, one can control membrane morphology by adjusting various parameters during membrane casting or spinning. Casting and spinning parameters that have significant influence on membrane formation include the exposure time to air, types and temperature of coagulant, air-gap, flow rates of dope and bore fluid, and take-up speed.

15.2.4.1 Exposure to Air

A homogeneous dope can be exposed to air before it is immersed into the coagulant bath. During the exposure period, the volatile solvent or co-solvent in the dope is gradually evaporated, and hence the dope becomes concentrated and highly viscous. This in turn facilitates delayed demixing during phase separation. For instance, Ong et al. prepared PVDF membranes with a dense skin layer, fully sponge structure, and good mechanical strength after evaporating acetone from the casted dope for 20 min before immersing it into the coagulant [37]. Susanto et al. added triethylene glycol (TEG) into a PES dope and exposed the dope to the air for different durations [38]. Water vapor in the ambient environment was absorbed into the dope during the exposure period due to the hygroscopic nature of TEG. Thus, the system was already brought to the metastable condition before it was immersed into the coagulant. This resulted in membranes with large pores as well as skinless and sponge structure, which enhanced the water permeability significantly.

15.2.4.2 Composition of Coagulant

Composition of coagulant has significant effect on membrane morphology. Water is considered a strong coagulant because a little amount of water is sufficient to bring a polymer solution into a metastable or unstable state. In contrast, alcohols and mixtures of solvent and nonsolvent are considered weak coagulants (or so-called soft coagulants). A relatively large amount of weak coagulant is required to be present in a solution for phase separation to take place. Strong coagulants facilitate instantaneous demixing during the phase inversion, leading to the formation of a membrane with a dense skin layer and macrovoids underneath, whereas soft coagulants cause delayed demixing in the dope solution, resulting in the formation of porous and sponge-like nascent membrane with the absence of a dense skin layer [34,39].

In hollow-fiber spinning process, it may not be economically practical to use an organic solvent as the coagulant because the volume of the external coagulant required is very large. However, the use of weak nonsolvent as the bore fluid is a convenient way to eliminate the inner skin of hollow fibers [24,40]. As shown in Figure 15.12b, macrovoids grow only from the shell side but not from the lumen side of a fiber when a weak coagulant is used as the bore fluid. The inner surface of the fiber also becomes loose with big pores (Figure 15.12d) compared with the fiber prepared using a strong coagulant as the bore fluid (Figure 15.12c). Nevertheless, the bore fluid must not be too weak or else the lumen of hollow fibers will not be formed.

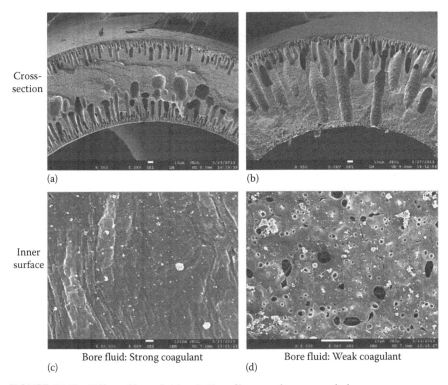

Cross-section

Inner surface

| (a) | (b) |
| (c) Bore fluid: Strong coagulant | (d) Bore fluid: Weak coagulant |

FIGURE 15.12   Effect of bore fluid on hollow-fiber membrane morphology.

15.2.4.3   Coagulant Temperature

Effect of coagulant temperature on membrane morphology has been found to be contradictory. On the one hand, a polymer solution is more thermodynamically stable at an elevated temperature; thus the onset of demixing is delayed. On the other hand, the rate of solvent–nonsolvent exchange rate is enhanced at a higher temperature, which in turn promotes instantaneous demixing. As a result, different observations have been reported in the literature. Choi et al. and Wongchitphimon et al. observed smaller surface pores and macrovoids in PVDF membranes made at a higher coagulant temperature [15,23]. Tang et al. and Chou et al. reported that larger surface pores and macrovoids are formed in PVDF and cellulose acetate membranes, respectively, when the coagulant temperature is higher [41,42]. It is important to determine the controlling mechanism that causes the different results.

15.2.4.4   Air-Gap

Corrugation in the lumen of hollow fibers has frequently been observed during the spinning process, as illustrated in Figure 15.13 [43]. Although the proposed mechanisms for this phenomenon are diverse, it is believed that the corrugation is caused by inward radial forces due to the presence of some instability in the system [28,43]. Degree of corrugation in the inner contour can be alleviated with increasing air-gap, as shown in Figure 15.13. The reasons might be due to (1) the release of stress in the

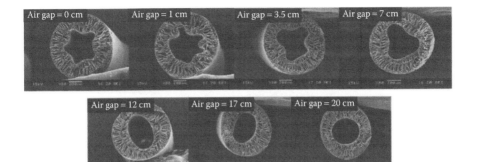

FIGURE 15.13 Effect of air-gap on the inner contour of hollow fiber membranes. (Adapted from S. Bonyadi et al., *Journal of Membrane Science*, 299, 200–210, 2007.)

nascent fiber during the air-gap period and (2) the formation of a more rigid inner skin before the nascent fiber entering the coagulant bath because the duration for phase separation at the lumen side is longer at a higher air-gap.

15.2.4.5 Take-Up Speed

If the take-up speed of spinning is higher than the free-falling speed of the nascent fiber, tensile stress will be induced, which affects the membrane properties. An increase in the surface pore size has been reported at a higher take-up speed, which might be attributed to the merging of pores or expansion of defects under the elongation stress [44,45]. However, contradictory findings have also been reported that membranes with smaller surface pores are obtained when the take-up speed is higher [46,47]. The reasons might be due to (1) changes in dope properties by elongation stress; (2) increase in polymer concentration caused by out-diffusion of solvent, which is induced by rapid shrinkage of nascent fiber; and (3) occurrence of spinodal decomposition as the elongation stress induces thermodynamic instability of the dope solution.

15.3 PREPARATION OF NANOFIBROUS MEMBRANES THROUGH ELECTROSPINNING

Nanofibrous membranes with high orientation, surface area, and porosity have great potential for various applications. Among the methods of producing nanofibrous membranes such as melt fibrillation and gas jet techniques, the electrospinning method has gained much attention due to its low cost and relatively high production rate [48,49]. Differing from the mechanism of membrane formation through conventional NIPS, as discussed in Section 15.2, the formation of nanofibrous membranes through electrospinning is a self-assembly process induced by electric charges. A typical electrospinning setup is composed of three basic parts: a high voltage supply, a capillary (including polymer solution syringes, syringe pump, and spinnerets), and a grounded metal collector, as shown in Figure 15.14 [50]. During an electrospinning process, the precursor solution (polymer solution) is extruded from a spinneret and a small droplet is formed at the tip of the spinneret. In the presence of an electric field, the droplet is charged, causing the formation of a solution jet,

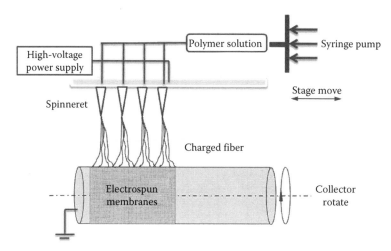

FIGURE 15.14 Schematic illustration of the basic setup for electrospinning. (Adapted from B. Sun et al., *Progress in Polymer Science*.)

which is then extruded from the droplet. The extrusion of the charged jet starts in uniform filaments, followed by vigorous whipping or splitting motion due to electrically driven bending instabilities [51]. Finally, the jet forms a continuous fiber that is deposited and collected on counter electrodes/substrates.

The principle of electrospinning was first illustrated by Formhals in the 1930s, and the first patent related to electrospinning was filed in the United States in 1902 [52,53]. However, considerable attention on the electrospinning process was not gained until the 1990s. Within the last decade, electrospinning of polymers has become a globally recognized technology for the preparation of polymeric nanofibers with diameters down to a few nanometers. Number of journal publications related to electrospinning has increased dramatically in recent years (Figure 15.15).

At the present time, significant progress has been achieved in not only understanding the complex electrospinning processes but also controlling fiber formation by changing operating parameters and materials. These achievements have in turn made electrospinning technologies possible for various applications, including membrane fabrication, tissue engineering, drug delivery, energy storage, and defense/security [54–57]. Many companies such as eSpin Technologies and NanoTechnics are seeking to reap the unique advantages offered by electrospinning, whereas other companies such as Donaldson and Freudenberg have incorporated electrospun fibers in their commercialized filtration products within the last two decades. In this section, the details of the electrospinning process and parameters affecting the electrospinning including the intrinsic properties of the polymer solution, operating conditions, as well as the temperature and humidity of ambient environment are discussed.

15.3.1 Understanding the Complex Processes Governing Electrospinning

During electrospinning, polymeric nanofibrous membranes are formed by the creation and elongation of an electrified fluid jet. When the charged fluid jet is extruded with

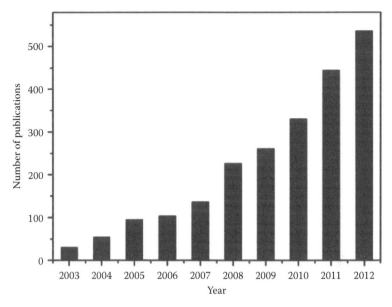

FIGURE 15.15 The annual number of publications on the subject of electrospinning as provided by the search engine of Science Direct Scholar.

an accelerated high speed under the electric field, the pathway of the jet is controlled by various forces, including electrostatic force, drag force, gravity, Coulombic repulsion force, surface tension, and viscoelastic force, as shown in Figure 15.16. In order to control the morphology and structure of nanofibrous membranes, it is necessary to quantitatively analyze how the interactions between electrical forces and surface tension affect the formation of jets that finally solidify into nanofibers. Recently, theoretical and experimental studies have demonstrated that the electrospinning process is generally composed of the following three stages, as shown in Figure 15.17 [58]: (1) onset of jetting and rectilinear jet development, (2) bending deformation with looping and spiraling trajectories, and (3) solidification and deposition on counter electrodes/substrates. These three stages of electrospinning are discussed in detail in the following sections.

15.3.1.1 Onset of Jetting and Rectilinear Jet Development

During electrospinning, an electrical potential difference, measured in volts, is applied between a droplet of the polymer fluid, which is initially suspended at an orifice due to surface tension and viscoelastic stresses, and an electrically conducting grounded collector. Evolution of the shape of a fluid drop in electrospinning process is illustrated in Figure 15.18 [59]. The time zero is set at the time when the first jet appeared (Figure 15.12d), which is after the supply of an electrical field for 28 ms. With an increase of applied voltage, the shape of the droplet at the orifice is gradually transformed into a conical shape, as shown from Figure 15.18a-c. At a critical point where the tip of the droplet becomes sharp, a jet is emanated, as shown in Figure 15.18d. The conical-shaped droplet at the point of emanation is called the *Taylor cone* [60]. Milliseconds after the jet emanation, the conical shape of the droplet becomes

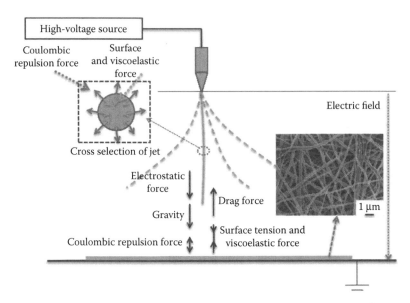

FIGURE 15.16 Schematic diagram illustrating the possible mechanism of nanofiber formation during electrospinning process.

rounder (Figure 15.18f), while a rapidly elongating and thinning jet with electrical charges continuously flows from the droplet. This cone shape is stable as long as an appropriate amount of polymer solution is continuously extruded from the orifice. The critical voltage where the tip of the droplet overcomes the surface tension and generates a jet can be calculated by Taylor's calculation [61]:

$$V_c^2 = \frac{h^2}{l^2}\left[\ln\left(\frac{2l}{r_0}\right) - 1.5\right](1.3\pi r_0 \gamma)(0.09) \tag{15.14}$$

where:
V_c is the critical voltage
h is the distance from the orifice to the collecting screen
l is the length of the liquid column
r_0 is the orifice outer radius
γ is the surface tension of the solution

Before the onset of the first bending instability, the jet of the polymer solution will follow the nearly straight electric field lines for a certain distance away from the tip [62]. Critical length of the straight jet can be predicted by applying Cauchy's inequality [63]:

$$L_c = \frac{4kQ^3}{\pi\rho^2 I^2}\left[\left(\frac{\pi\rho kE}{2\sigma Q}\right)^{2/3} - r_0^{-2}\right] \tag{15.15}$$

where:
L_c is the critical length of the straight jet
Q is the flow rate

σ is the surface charge density
k is the dimensionless conductivity
E is the applied electric field
I is the current passing through the jet
ρ is the liquid density
r_0 is the initial radius of the jet

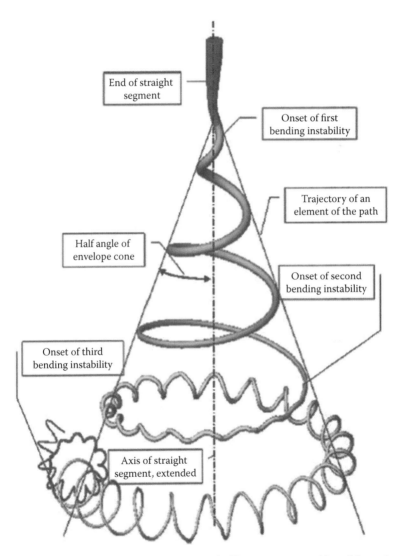

FIGURE 15.17 A diagram showing the prototypical instantaneous position of the pathway of an electrospinning jet. (Adapted from H.R. Darrell, H. Fong, E. Zussman, S. Koombhongse, W. Kataphinan, Nanofiber manufacturing: Toward better process control. in: D.H. Reneker and H. Fong (Ed.), *Polymeric Nanofibers*, American Chemical Society, Washington, DC, 2006, Vol. 198, pp. 7–20.)

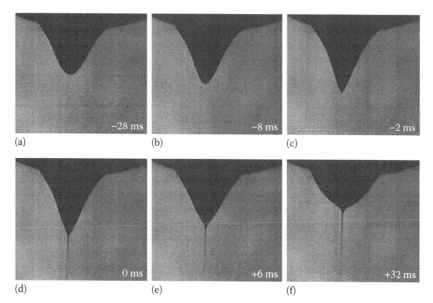

FIGURE 15.18 Evolution of the shape of a fluid drop in electrospinning process. (Adapted from H. Fong, H.R. Darrell, *Electrospinning and the Formation of Nanofibers*, Hanser, Cincinnati, OH, 2000.)

15.3.1.2 Bending Deformation with Looping and Spiraling Trajectories

As shown in Figure 15.17, after the straight segment, the jet is elongated, which is driven by the electrical force induced by the charges carried with the jet, followed by a series of successively bending coils having turns of increased radius [62,64,65]. It was demonstrated that both the bending/stretching of the jet at high frequencies and the reduction of the jet diameter from micrometers to nanometers are mainly caused by instabilities such as Rayleigh, axisymmetric, and bending [66,67]. The classical Rayleigh instability, which is dominated by surface tension and is suppressed at high electric field, is given as follows:

$$\left(\varepsilon_I - \varepsilon_O\right)E^2 + \frac{4\pi^2\sigma^2}{e_O} = \frac{2\pi\gamma}{r} \tag{15.16}$$

where:

ε_I and ε_O are the dielectric constants inside and outside of the jet, respectively
r is the radius of the jet

On the other hand, both axisymmetric and bending/whipping instabilities are caused by an electrically driven force due to the fluctuations in the dipolar component of the charge distribution and are essentially independent of the surface tension of the polymer solution. Axisymmetric instability often occurs at a higher electric field than Rayleigh instability whereas bending or whipping instability is not axisymmetric.

Frame 1   Frame 6

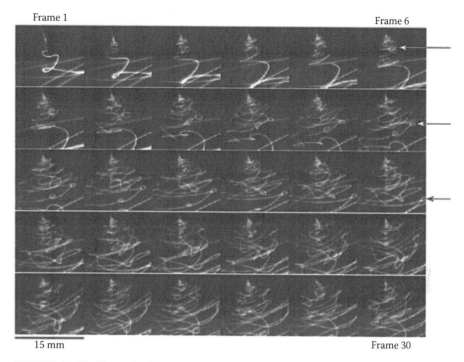

15 mm   Frame 30

FIGURE 15.19 Frame-by-frame images that show the evolution of the bending instability on a jet of 15% polycaprolactone in acetone. (Adapted from D.H. Reneker et al., *Polymer*, 43, 6785–6794, 2002.)

Bending instability of electrospun nanofibers was demonstrated by the images of electrospinning polycaprolactone, as shown in Figure 15.19 [62].

In addition to the above-mentioned instabilities, other characteristic instabilities such as the branching and formation of physical beads have been observed. Branches occur more frequently in a highly concentrated and viscous solution as well as at a higher electric field, whereas beaded nanofibers are formed when a cylindrical fluid jet is broken up into droplets due to the occurrence of capillary instability at smaller electrical charge per unit length [68,69].

15.3.1.3 Solidification and Deposition on Counter Electrodes/Substrates

During the process of nanofiber bending and elongation, the solvent in the fibers evaporates simultaneously, which consequently produces dry polymer fibers on counter electrodes/substrates. A discretized form of an equation accounting for solvent evaporation and polymer solidification has been developed to calculate the jet paths during the course of nonlinear bending instability which leads to the formation of nanofibers [65]. Because of the longitudinal compressive force from jet impingement on a solid flat surface, the buckling instability may occur. This results in bending fibrous structures with sinuous folding and overlapping, which have been observed on a hard flat surface [70].

Collection method for electrospinning process is critical for the morphology and structure of resultant nanofibers. Electrospinning systems using various target electrodes were reported, as shown in Figure 15.20 [71]:

a. A one-dimensional nanofiber was achieved using an optical chopper motor.
b. Residual solvent was removed in a coagulation bath, which forces fiber to crystallize in the coagulation bath.
c. The charged jet was concentrated on the sharp edge of the collector because the electrical field is increased by the sharp edge points.
d. When the rotating speed of the cylindrical collector was fixed at an appropriate number, the fiber possesses the best alignment.
e. The cylindrical grounded copper wire frames were used to further improve the fiber alignment.
f. Double-edge steel bladed in line was developed to obtain highly aligned fibers.
g. The effects of rectangular frame collector materials on prepared fiber were studied.
h. In order to enhance membrane strength by fiber patterning and orientation, metal wire conducting grids were used.
i. The electrospinning system with a copper ring was used to investigate the effects of jet concentration.

In addition, the fabrication of flat-sheet fiber mat by delivering the polymer solution to a metal capillary in the center of the top plate at a constant flow rate was reported [66]. Fiber deposition can also be regulated by controlling the motion of the target mandrel to collect fibers with linear and parallel arrays [72]. It was presented that the electric field used to macroscopically align polymer nanofibers could be used to align polymer chains parallel to the fiber axis [73,74]. In summary, modifying the collectors is an effective way to improve electrospinning and hence expands the possible applications of as-spun nanofibers.

15.3.2 Effects of Intrinsic Properties of Polymer Solution

In recent years, the effects of various parameters on the electrospinning of nanofibers have been widely studied [50,75–78]. One of the parameters that significantly influence the formation of nanofibers is the intrinsic properties of the solution, which includes viscosity, elasticity, and electrical conductivity of the solution, as well as polarity and surface tension of the solvent. These parameters will be briefly discussed in the following sections.

15.3.2.1 Polymer Concentration and Dope Viscosity

It is well known that the diameter of nanofibers, as well as the structure and morphology of electrospun membranes, are influenced by the concentration and viscosity of polymer solutions [79–82]. Polymer concentration highly affects the viscosity and surface tension of the solution, which in turn controls the formation of nanofibers. For instance, a solution with low polymer concentration (i.e., low viscosity)

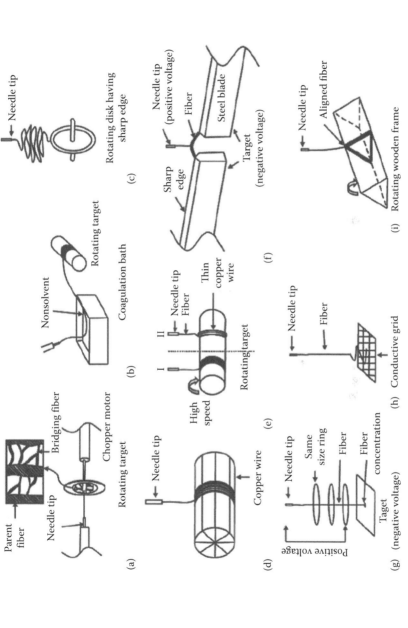

FIGURE 15.20 Schematics of electrospinning systems for various target electrodes. (Adapted from S. Park et al., *Polymer International*, 56, 1361–1366, 2007.)

forms bead-on-string fibers, whereas a solution with higher polymer concentration (i.e., higher viscosity) tends to change the shape of the beads from spherical to spindle-like and thus produce uniform fibers [83,84]. On the other hand, with further enhancement of viscosity, the diameter of nanofibers is increased. Therefore, during electrospinning process the concentration of polymer solution should be optimized to obtain nanofibers with desired structure.

15.3.2.2 Electrical Conductivity

Electrical conductivity of a polymer solution affects its rheological behavior and thus considerably influences the spinnability of the solution [85,86]. Electrical conductivity of a polymer solution is mainly determined by the type of polymer and solvent, as well as the concentration of ionizable salts [83]. During electrospinning, highly conductive solutions tend to produce fibers with significantly reduced diameters, dramatic bending, and a broad diameter distribution [54,55]. Furthermore, the instability of the Taylor cone could be enhanced by increasing dope conductivity, which in turn produces more microsized droplets and dense nets [87]. Addition of ions in the polymer solution could improve conductivity dramatically, and it has been reported that the ions with smaller atomic radius could impose a stronger elongation force on the jet due to their higher charge density and higher mobility [54,88–90]. Addition of formic acid into the polymer solution could also increase the dope conductivity due to the high dielectric constant of formic acid, thus favoring the formation of thinner fibers and dense nano nets [46,91].

15.3.2.3 Surface Tension

It has been observed that the surface tension of the dope, which can be adjusted by adding surfactants, is likely to play an important role in regulating the morphology of nanofibers [89,92–94]. Due to the lower surface tension, nano nets are more regular and uniform, which is attributed to the stable jets encountering fewer perturbations. However, if the concentration of surfactants is too high, the nanofiber morphology may show defects, as the surfactants may self-assemble to form colloidal aggregates.

15.3.2.4 Solvent

As the nanofibers are formed by solvent evaporation during flight in the high electric field, the type of solvent is one of the primary factors that influence fiber morphologies. Effects of various kinds of solvent on the formation of fibrous membranes have been investigated [89,91]. It was suggested that the polymer solution prepared by a good solvent with high solubility was relatively hard to be electrospun compared with that prepared by a lower-solubility solvent [95]. Impacts of the quality of solvent on the electrospinning have been studied qualitatively based on model systems comprising of 28 different solvents as well as mixtures of them with nonpolar biocompatible polymers. It was found that good solvents tended to facilitate electrospraying whereas solvents with partial solubility tended to facilitate stable electrospinning of nanofibers [95]. Spinnability-solubility maps are also constructed for an easy selection of solvents for electrospinning [95].

In summary, it is illustrated that the formation of nanofibers is highly influenced by the variation in intrinsic properties of polymer solution. Therefore, polymer solution formulation is critical in the successful preparation of nanofiber membranes by electrospinning.

15.3.3   EFFECTS OF OPERATING CONDITIONS

15.3.3.1   Applied Voltage

Operating conditions of electrospinning also play significant roles in determining the morphologies and structures of electrospun membranes. The strength of applied electric field is an important element that determines not only the amount of charge in the droplet located on the tip of spinneret but also the magnitude of electrostatic force [92]. It was reported that an increase in applied voltage could favor the formation of thinner fibers and completely split nano nets [87]. The average diameter of nanofibers was decreased and the area density of nano nets was sharply increased with increasing applied voltage, which was attributed to the increase of bead defect density [96].

15.3.3.2   Ambient Environment

In addition to the above-mentioned parameters, ambient temperature and humidity have influences on the prepared nanofibers. It was demonstrated that at a lower temperature, the average diameter of nanofibers decreased as the evaporation rate of solvent was decelerated, and thus there was more time for the nanofiber to elongate and solidify. In contrast, at a higher temperature, the polymer chains had less freedom to move and forced the jet to solidify faster and consequently produced thicker nanofibers [91,97].

On the other hand, according to recent reports, the variation of relative humidity can also control the formation of nanofiber membranes. It was observed that the fibers fabricated at higher humidity would possess a thinner and light fiber-sticking structure compared with that obtained at lower humidity [87,91,92,96].

15.3.3.3   Other Operating Parameters

Distance between the tip and collector can also affect the prepared nanofibers as the nanofibers need a minimum distance to give sufficient time for solvent evaporation. Recently it was demonstrated that if the distance is either too close or too far, beads tend to form [98]. In addition, a decrease in orifice size was found to decrease the fiber diameter [99], and the diameter of nanofibers was increased with increasing dope flow rate [75].

In summary, experimental and theoretical investigations have been carried out to understand the effects of spinning parameters on nanofiber membrane morphologies and structures. Nevertheless, there is still a lack of study on the possible synergistic effects of different parameters such as various polymer solutions under different operating conditions. Further studies are required to gain more comprehensive understanding of the electrospinning process and its control.

15.4 PREPARATION OF COMPOSITE MEMBRANES

Generally, a membrane used for separation consists of a highly selective active layer supported by a porous layer providing adequate mechanical strength to the membrane. There are various methods for the preparation of the support, as discussed in Sections 15.2 and 15.3. In this section, a number of methods used to construct an active selective layer are discussed. The active selective layer can be composed based on the composite membrane concept. In a composite membrane, the active layer and support layer are fabricated separately from different materials. Normally, the support/nonselective layer is formed in the first step followed by the creation of a thin active layer on top of the support surface, for flat-sheet membranes, or on the outer/inner surface for hollow-fiber membranes. Advantages of composite membranes include the following: (1) active layer can be developed easily, and (2) each layer, active or support, can be optimized and customized specifically according to application [100,101]. However, making composite membranes is normally more expensive and time consuming as compared with integral asymmetric membrane preparation [102]. In addition, the risk of introducing defects in each layer increases with increasing numbers of processing steps [10].

Active selective layer of the composite membrane can be developed by several techniques, such as interfacial polymerization, multilayer polyelectrolyte deposition, chemical cross-linking, dual layer co-extrusion/co-casting, dip-coating and UV-photo-grafting, and plasma.

15.4.1 INTERFACIAL POLYMERIZATION

Formation of polyamide thin film by interfacial polymerization is the most commonly used technique to prepare high-performance RO-like and NF-like active layers. Interfacial polymerization is a polymerization of two very reactive monomers occurring at the interface of two immiscible solvents. One monomer should be able to dissolve in an organic solvent whereas the other monomer dissolves in an inorganic/aqueous solvent. The two solvents should be immiscible to each other. There are two types of interfacial polymerization: (1) one phase is dispersed as tiny droplets in the other phase using high-speed stirring to produce micro/nano-capsules or micro/nano-spheres for applications such as drug delivery and dye encapsulation [103], and (2) polymerization forms a thin film enveloping a surface of a support continuously, which is typically applied for membrane fabrication [104,105].

As shown in Figure 15.21a, the interfacial polymerization of polyamide occurs between amine groups ($-NH_2$) from the aqueous phase of an amine solution and carboxyl groups ($-COOH$) from the organic phase of acyl halides solution, which are linked together by peptide bonds or amide links [102]. Several types of monomers and prepolymers, such as piperazine, N,N'-diaminopiperazine, and m-phenylenediamine for amine solutions [106,107], and trimesoyl chloride, sebacoyl chloride, and iso-phthloyl chloride for acyl halides solutions [105], can be used. As an example, Figure 15.21b shows the structure of polyamide formed by the polymerization of trimesoyl chloride and piperazine [108]. Polymerization reaction occurs in the organic phase due to highly unfavorable partition coefficients of acyl halides, which limit

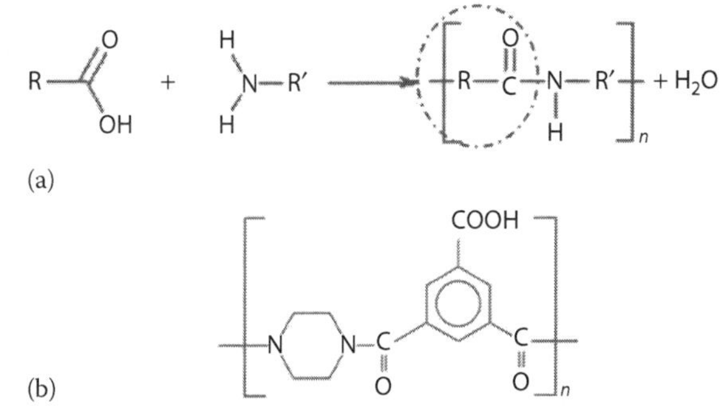

(a)

(b)

FIGURE 15.21 (a) Reaction of polyamide from monomers and (b) polyamide forming from trimesoyl chloride and piperazine. (Adapted from J. Liu et al., *Separation and Purification Technology*, 58, 53–60, 2007.)

their availability in the aqueous solution. Thus, the amine groups have to diffuse across the water–organic interface in order to make a reactive contact [102].

15.4.1.1 Factors Affecting Interfacial Polymerization

Polarity of the amide groups in polyamide results in a material that is very sensitive to water. Adhesion of the polyamide active layer to the support layer can be weakened by preferential hydrogen bonding interactions between the amide groups and the water. Amide groups are also susceptible to oxidation, making polyamide vulnerable to degradation by chlorine disinfectant agents used in wastewater treatment processes [109,110].

There are three important factors in the preparation of a coherent film to achieve a satisfactory membrane performance. First is the ratio of monomer concentrations, aqueous phase to organic phase. For lower ratios, a thin film was obtained at the expense of a lower rejection. For a higher ratio, on the other hand, the rejection is not further improved whereas pure water permeability (PWP) slightly decreased. Second is the contact time between the aqueous and organic phases. Film thickness increases as the contact time increases, which is indicated by a decrease in PWP [111]. When organic phase monomers are not able to diffuse through the interface any more, the thickness remains constant. The last factor is the time for the removal of the excess aqueous solution before contacting with the organic phase. Several methods can be employed to remove the excess aqueous solution such as using a solvent of the organic phase for rinsing or nitrogen purging [106,112].

15.4.1.2 Mixed-Matrix Interfacial Polymerization

Mixed-matrix interfacial polymerization has been developed to embed nanoparticles throughout the polyamide thin film layer. This concept aims to improve the membrane performances. Super-hydrophilic zeolite nanoparticles are used to enhance the water permeation while maintaining high salts rejection [113]. A similar approach has been developed to fabricate aquaporin-based biomimetic membranes that have superior separation performance [114].

15.4.2 Multilayer Polyelectrolyte Deposition

Polyelectrolyte is described as a polymer carrying (an) electrolyte(s) group in its repeating units. Polyelectrolyte presents charge property when it dissociates in water or an aqueous solution. The technique of multilayer polyelectrolyte deposition on membrane surfaces is based on the electrostatic interaction of oppositely charged molecules. A polyelectrolyte membrane can be fabricated by sequential deposition of aqueous polyelectrolyte solutions on a porous substrate, as illustrated in Figure 15.22 [115].

Multilayer polyelectrolyte deposition is simple and flexible for application in developing membrane selective layers with high selectivity and very thin skin thickness (in the range of nm). Structure and functionality of the selective layer can be varied for specific applications depending on the molecular structure and charge density of the polyelectrolytes used. Figures 15.23 [101,116–120] and 15.24 [115,121–125] show the commonly used polyanions (negatively charged) and polycations (positively charged), respectively. Applications of polyelectrolyte membranes include forward osmosis [115,126,127], NF [128], ion exchange [129], pervaporation [130,131], and gas separation [132,133].

15.4.2.1 Factors Affecting Multilayer Polyelectrolyte Deposition

The pH and ionic strength of the polyelectrolyte solution play an important role in the formation of an excellent film of multilayer polyelectrolytes. The solution pH affects the dissociation of the electrolyte in the aqueous solution, especially for weak polyelectrolytes. If both porous substrate and the polyelectrolyte are negatively charged at a high pH and vice versa, the optimum pH used should be in between the iso-electric point of the substrate and the polymer, as illustrated in Figure 15.25. Thus, the substrate and the electrolyte will carry opposite charges [134]. An ultraviolet/ozone (UV/O$_3$)-cleaned porous alumina with surface pore diameter of 0.02 μm is attractive as a substrate for polyelectrolyte membranes due to its positive charges [135].

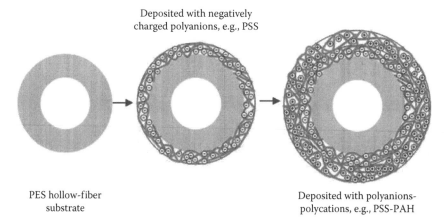

Deposited with negatively
charged polyanions, e.g., PSS

PES hollow-fiber
substrate

Deposited with polyanions-
polycations, e.g., PSS-PAH

FIGURE 15.22 Schematic drawing of multilayer polyelectrolyte deposition on the outer surface of the hollow-fiber membrane. (Adapted from C. Liu et al., *Desalination*, 308, 147–153, 2013.)

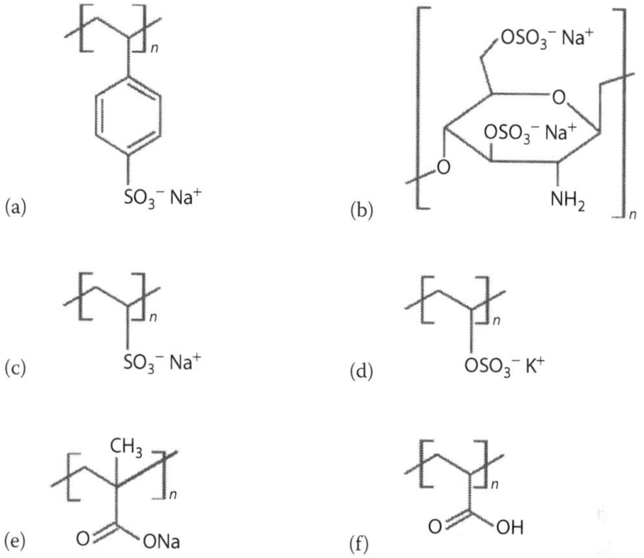

FIGURE 15.23 Commonly used polyanions for the development of active-selective layer: (a) poly(styrene sulfonate) (PSS) sodium salt (Data from L. Setiawan et al., *Desalination*, 312, 99–106, 2013.); (b) sulfated chitosan (Data from J. Miao et al., *Desalination*, 181, 173–183, 2005.); (c) poly(vinyl sulfonic acid) (PVSu) sodium salt (Data from B. Tieke et al., *Advances in Colloid and Interface Science*, 116, 121–131, 2005.); (d) poly(vinyl sulfate) (PVS) potassium salt (Data from B. Tieke et al., *European Physical Journal E*, 5, 29–39, 2001.); (e) poly(methacrylic acid) (PMA) soldium salt (Data from D. Saeki et al., *Journal of Membrane Science*, 447, 128–133, 2013.); and (f) poly(acrylic acid) (PAA). (Data from W. Shan et al. *Journal of Membrane Science*, 349, 268–278, 2010.)

Plasma-treated/hydrolized polyacrylonitrile and cellulose acetate, on the other hand, are negatively charged [126,127]. In addition, PES is rather appealing to be used as a supporting material although it is fairly neutral. Hence, the attachment of polyelectrolyte layer is based on hydrophobic interaction [115].

Ionic strength of the polyelectrolyte solution can be increased by adding salts. At high ionic strength, the electrostatic repulsion of the polymer chain decreases resulting in a coil conformation. Therefore, it increases the thickness of the individual layer [136]. However, in the environment of extremely high salt concentration, only a small amount of polyelectrolyte can be adsorbed by the substrate due to the competition with the smaller charged ions from the salts [134]. In addition to the effort of improving the stability of the polyelectrolytes layers, chemical cross-linking could be employed to enhance the layers' stability. For a high ionic strength feed solution, the membrane ability to reject salts is weakened due to the distinctive behavior of polyelectrolyte. Therefore, at the end of polyelectrolyte deposition, glutaraldehyde could be used as the cross-linker [126,127].

Charge density of a polyelectrolyte pair depends on their molecular structures. It can be defined as the ratio of total ion pairs to the number of carbon atoms in each repeat unit. For example, a pair of PAH/PSS has total ion pair of 1 and carbon

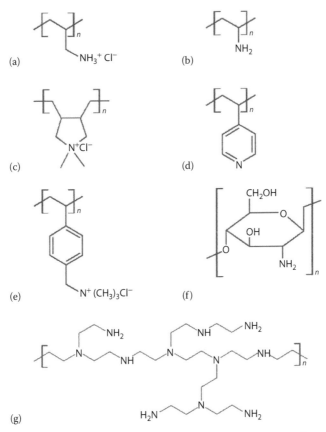

FIGURE 15.24 Commonly used polycations for the development of active-selective layer: (a) poly(allylamine hydrochloride) (PAH) (Data from C. Liu et al., *Desalination*, 308, 147–153, 2013.); (b) polyvinylamine (PVA) (Data from K. Hoffmann, B. Tieke, *Journal of Membrane Science*, 341, 261–267, 2009.); (c) poly(diallyldimethyl ammonium chloride) (PDADMAC) (Data from W. Shan et al. *Journal of Membrane Science*, 349, 268–278, 2010.); (d) poly(4-vinylpyridine) (P4VP) (Data from G. Zhang et al., *Langmuir*, 26, 4782–4789, 2009.); (e) poly(4-vinylbenzyl trimethyl ammonium chloride) (PVTAC) (Data from R.V. Klitzing, B. Tieke, Polyelectrolyte membranes, in: M. Schmidt (Ed.), Polyelectrolytes with Defined Molecular Architecture I, Advances in Polymer Science, 2004, pp. 177–210.); (f) chitosan (Data from M.L. Bruening et al., *Langmuir*, 24, 7663–7673, 2008.); and (g) Polyethylenimine. (Data from J. Wang et al. *Journal of Membrane Science*, 337, 200–207, 2009.)

atoms of 3 for PAH and 8 for PSS (total carbon atoms are 11). So PAH/PSS has a charge density of 0.09. High charge density enhances the membrane separation performance [136,137].

Incorporation of nanoparticles such as silver on the active layer of the membrane may enhance the antifouling or antibacterial properties of the membrane. The layered structure of multilayer polyelectrolyte assembly could improve the stability of the nanoparticles on the membrane surface [128].

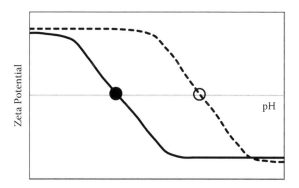

FIGURE 15.25　Plot of pH and zeta potential of polyelectrolyte (——) and substrate (---); ●, iso-electric point of polyelectrolyte; and ○, iso-electric point of substrate.

15.4.3　CHEMICAL CROSS-LINKING

Some membrane materials have characteristic functional groups that are rather reactive. By applying a suitable cross-linker, covalent bonds can be formed to link the functional groups. There is a significant scope through the use of chemical cross-linking in membrane fabrication processes, as summarized in Table 15.2 [100,101,138–146]. One common approach is to improve the mechanical strength of the membrane and to reduce membrane swelling. Another approach is to improve the permselectivity of specific solutes and maintain high permeability depending on the applications. Using this method, the surface and/or membrane pore size can be altered, for example, from UF to NF type skin, and the pore size distribution can be uniformly tuned. Furthermore, the cross-linked complexes may carry charges when an electrolyte compound is used as the cross-linker. Cross-linking degree and the charge density are influenced by several parameters, including reaction time and temperature, cross-linking medium/solvent, molecular structure, and the concentration of the cross-linker. Cross-linking reaction can be confirmed by fourier transform infrared spectrometry using the attenuated total reflection method [100,145,147].

15.4.4　DUAL LAYER CO-EXTRUSION/CO-CASTING

Development of composite dual layer membranes is attractive as beneficial properties of two or more polymeric materials can be combined to suit for various applications accordingly. The material cost of high-performance polymer can be reduced and the polymer having exceptional selectivity with poor mechanical strength can be strengthened by combining them with inexpensive and robust polymers as the support layer [148,149]. Emerging applications of dual layer membranes include forward osmosis, gas separation, and NF membranes, which are composed of a thin dense selective layer supported by a porous polymer matrix [101,104,150–152], and direct contact membrane distillation, which requires a combination of a thin hydrophobic layer to prevent wetting and a hydrophilic layer to enhance water permeability [153].

TABLE 15.2

Chemical Cross-Linking Performed on Various Membranes Using Various Cross-Linkers

Membranes/Cross-Linkers	Approaches	Applications	References
Polyimide/*p*-xylenediamine	Improves chemical and plasticization resistance	Gas separation	[138]
Poly(dimethyl siloxane) (PDMS)/ hydrosilane	Reduces the swelling of PDMS in organic solvent	Nanofiltration	[139]
Polyimide/ethylene diamine (diamines), ethyleneglycol (diols)	Reduces the swelling of polyimide membranes	Pervaporation: ethanol dehydration	[140]
Polybenzimidazole/*p*-xylene dichloride	Improves separation performance and membrane mechanical strength	Pervaporation: acetone dehydration	[141]
Pyridine based aromatic polyether/ phosphoric acid	Improves membranes mechanical strength at high temperatures	Fuel cell	[142]
Poly(vinyl alcohol) (PVA)/ poly(acrylamide-*co*-diallyl dimethylammonium chloride)	Increases alkaline resistance and stability of hydroxyl group	Fuel cell	[143]
Polyimide 6FDA-durene/ benzylamine, 1,3-diaminobenzene	Improves the permselectivity and maintains high permeability	Gas separation	[144]
Polyimide (P84)/polyethylenimine	Develops positively charged membrane	Nanofiltration for industrial wastewater	[145]
Chitosan/toluene diisocyanate	Develops positively charged membrane	Nanofiltration: water softening	[146]
Torlon poly(amide-imide) (PAI)/ polyethylenimine	Develops positively charged NF-like skin	Nanofiltration, forward osmosis	[100,101]

Principles of dual-layer membrane fabrication for hollow-fiber [152,154] and flat-sheet [148,150] membranes are similar; they are prepared by a single-step co-extrusion or casting of two different polymer solutions based on NIPS. Simultaneous formation of the dual-layer structure can be done using a triple-orifice spinneret for hollow-fiber membranes and co-casting using a double-blade casting machine for flat-sheet membranes, as illustrated in Figure 15.26 [150,154].

15.4.4.1 Fabrication Parameters

Fabrication of dual-layer membranes either in hollow-fiber or flat-sheet configuration is rather complex as many parameters are involved, which control the thermodynamic property and the phase inversion kinetics to obtain good lamination between the two layers as well as regular and uniform membrane cross-sectional morphology. Fabrication parameters can be classified into the chemistry of polymer solutions and the operating conditions. The chemistry of polymer solutions depends on the polymer concentration and type, the solvents affinity to the polymer or coagulant, and

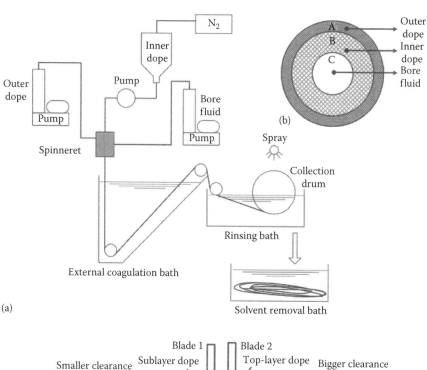

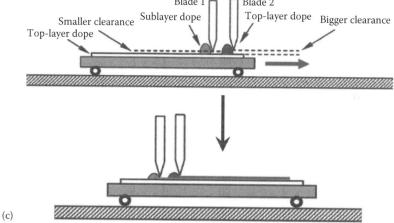

FIGURE 15.26 (a) Schematic diagram of a dual-layer hollow-fiber spinning process; (b) cross section of triple-orifice spinneret (Adapted from L. Setiawan et al., *Journal of Membrane Science*, 423–424, 73–84, 2012.); and (c) fabrication process of a dual-layer flat-sheet membrane using a double-blade casting machine. (Adapted from S.A. Hashemifard et al. *Journal of Membrane Science*, 375, 258–267, 2011.)

the variety and concentration of nonsolvent additives (or pore formers) [150,154]. Operating conditions include air-gap and evaporation time, which are applicable for hollow-fiber spinning and flat-sheet casting, respectively, the composition of the coagulant, the temperatures of the coagulant and polymer solutions, as well as the operating temperature [20,148,154–156].

Delamination of the two layers occurs when the coagulant (i.e., water) is accumulated on the interface between the outer and inner layers for hollow fibers or on the top and sublayers for flat-sheet membranes [154]. For dual-layer hollow-fiber membranes, a higher diffusion rate of water in the outer layer than that in the inner layer leads to an accumulation of water at the interface of outer and inner layers. On the other hand, when the inner layer has a high water penetration rate, its expansion occurs, leading to an irregularity of inner contour [154]. Main challenge in the fabrication of dual-layer flat-sheet membranes is the spreading of the top and sublayer dope solutions during the casting process. In order to address this challenge, the viscosity of the sublayer dope solution should be higher than that of the top layer. Thus, the shear stress excreted by the top layer is unable to drag and sweep the sublayer [150].

15.4.5 DIP-COATING

Polymer or organic materials can be applied to coat the membrane surface by a dip-coating method. The polymer commonly used as coating material should have some special properties such as being hydrophilic and negatively charged, and easy to attach on the support layer. This kind of polymers can be made based on sulfonation reactions such as sulfonated PES (SPES) and sulfonated poly(ether ether ketone) (SPEEK). Coating layer might enhance the performance of the support layer such as improved stability and separation properties. Several parameters need to be considered in selecting the coating polymer such as the strength and stability of the polymer, capability of forming a film, solubility in solvents, cost, and the possibility to cross-link [157]. Figure 15.27 illustrates the dip-coating process [10]. In basic terms, dip-coating process consists of three stages: (1) immersing dry membrane into a coating solution, (2) allowing the coating material to have interaction with the substrate, and (3) drying the membrane.

SPES was utilized as the selective layer of NF hollow-fiber membrane by dip-coating because of its ion exchange (capacity of 0.8 meq/g) and antifouling abilities. SPES carries negative charges due to the sulfonic acid group in the main chain. The major drawback of this polymer is that it can swell easily in water. Once the polymer is dried, the structure of the membrane becomes brittle [148]. Moreover, NF hollow-fiber membrane has been fabricated from PES as the substrate followed by the dip-coating of SPEEK as the selective layer. The thickness of the coating layer mostly depends on the viscosity of the coating solution, which is influenced by temperature,

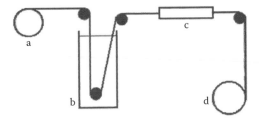

FIGURE 15.27 Schematic illustration of dip-coating: (a) dry porous support hollow fiber membrane, (b) coating bath, (c) oven, (d) hollow-fiber composite membrane. (Adapted from M. Mulder, *Basic Principles of Membrane Technology*, 2nd edn., Kluwer Academic, Dordrecht, the Netherlands/Boston, MA, 1996.)

concentration of the solution, and additives. At a lower concentration, the viscosity of the solution is also low, and, therefore, the coating solution will penetrate to the substrate pores [158].

15.4.6 Chemical Grafting

A membrane surface can be chemically modified by grafting by either grafting another polymer to the surface or growing a new polymer on the surface. The grafted polymer promotes the base polymer characteristics to be more hydrophilic and have antifouling property and solute selectivity. There are several methods to generate the active sites that can lead to the initiation of a graft polymerization, such as UV, plasma, and ion-beam irradiation [159–161].

UV photo-grafting is used to modify the active surface of membranes that are applied for wastewater treatment of textile industries because polyamide membranes cannot be applied here. The outer surface of polysulfone UF hollow-fiber membrane has been modified by UV-photo-grafting using sodium p-styrene sulfonate as the monomer, N,N'-methylenebisacrylamide as the cross-linker, and 4-hydroxybenzophenone as the photo-initiator. Chemical bonds in the porous support polysulfone membrane that is photo-reactive can be cleaved to create radical sites using a UV-photo-grafting technique. Free radical polymerization will occur at these sites in the presence of a vinyl monomer, and thus the polymer chains are grafted to the membrane surface by covalent bonds. The UV-photo-grafting setup is illustrated in Figure 15.28 [162]. At first, the support layer hollow fibers, wetted by water, are dipped in an aqueous monomer solution. Then, the fibers pass through two UV polychromatic lamps [163].

Graft polymerization of methacrylic acid monomer could increase the hydrophilicity and impart negative charges on the membrane surface. It has been used to remove endocrine disrupting chemicals and pharmaceuticals active compounds [159]. In addition, surface grafting using redox initiation has been developed, which offers simplicity to the process. The reaction can be performed in an aqueous media at room temperature without an external activation [161]. However, redox initiation has relatively slow kinetics that requires a high concentration of monomer [164].

Plasma-induced grafting of PEG followed by dip-coating into TiO_2 solution could effectively modify the hydrophobic surface of PVDF membrane to enhance

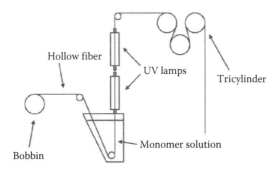

FIGURE 15.28 Continuous UV-photo-grafting setup for the fabrication of hollow-fiber membranes. (Adapted from S. Béquet et al., *Desalination*, 144, 9–14, 2002.)

the anti-oil fouling property for membrane distillation application. The hydrophilic modified membrane exhibits a positive interfacial free energy with oil in water. This result indicates the improvement of the anti-oil fouling property of the PVDF membrane, which thus enhances the membrane distillation performance [165].

15.5 CHARACTERIZATIONS OF SOLUTIONS/MEMBRANES

15.5.1 POLYMER DOPE CHARACTERISTICS

Information on polymer dope characteristics such as the amount of nonsolvent needed to induce phase separation and viscosity is important during membrane preparation. These characteristics that highly influence the thermodynamics and kinetics in membrane formation are useful for adjusting the dope formula during membrane fabrication.

15.5.1.1 Cloud Point

Cloud point of a polymer dope is simply a temperature or composition where the dope becomes no longer homogeneous and turns cloudy due to phase separation. In NIPS, cloud points are usually determined by a titrimetric method that examines the amount of nonsolvent required to make the dope cloudy [166,167]. In this method, the droplets of nonsolvent are slowly added into a polymer dope until permanent turbidity is visually observed in the dope. Frequently, repeating cycles of heating and cooling are involved due to the occurrence of local gelation upon nonsolvent addition.

By obtaining the cloud points of a series of dopes containing different concentration of polymer and solvent, a cloud point curve can be drawn on a ternary phase diagram, as shown in Figure 15.29 [167]. It is believed that a cloud point curve may not always represent the liquid–liquid demixing boundary of a system. Instead, a cloudy dope may also be indicative for the occurrence of crystallization in a system containing a semicrystalline polymer. For instance, when determining the cloud point curve of a PVDF system, the cloud point curves obtained from the titration of semicrystalline PVDF homopolymer solution and amorphous PVDF terpolymer solution represent the crystallization and liquid–liquid demixing boundaries, respectively (Figure 15.29).

15.5.1.2 Dope Viscosity and Rheology

Viscosity of polymer dope is one of the determining parameters on membrane structure because it can highly influence the rate of solvent out-diffusion and nonsolvent in-diffusion during the membrane formation process. On the other hand, the dope rheology provides information on the flowing behavior of the polymer under applied stresses. Dope rheology is particularly important in hollow-fiber membrane fabrication as the extrusion of the polymer dope through a spinneret involves high shear stresses. The shear stress can induce molecular chain orientation and die swell phenomenon, which is closely related to the rheological properties of the dope, and affect the structure and performance of resultant membranes [168,169]. Viscosity and rheological properties of polymer dopes can be measured by a rheometer.

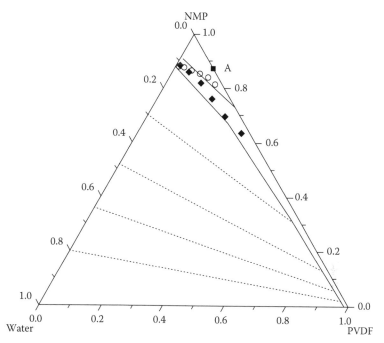

FIGURE 15.29 Isothermal phase diagram of PVDF/NMP/water system: (···) tie line, (♦) liquid–liquid demixing point, (○) gelation or crystallization point, (♦), and binodal line, (○). (Adapted from D.J. Lin et al., *Journal of Polymer Science, Part B: Polymer Physics*, 42, 830–842, 2004.)

15.5.2 MEMBRANE SEPARATION CHARACTERISTICS

Because the prepared membranes will eventually be used for separation purposes, the membrane characteristics relating to separation performance need to be evaluated. The two most important separation parameters are the permeability and selectivity of the membranes. During practical applications, there are many factors affecting the permeability of membranes such as fouling and concentration polarization. Therefore, PWP of a membrane is often provided as one of the standard membrane characteristics, as it eliminates the effects of fouling and concentration polarization. PWP is directly proportional to the transmembrane pressure according to Darcy's Law, but membrane compaction at a high pressure may cause the actual permeability to deviate from the theoretical one [170].

For porous membranes, the selectivity of membranes is governed by the size of membrane pores and pore size distribution. There is no standard technique for determining the membrane pore size, and its distribution as every existing technique has its advantages and limitations. The commonly used techniques for membrane pore characterization involve the displacement or phase transition of a fluid filling the pores. In addition, membrane surface pores can be directly observed through microscopic techniques. Instead of pore size and pore size distribution, it is also common to determine the solute rejection of membrane, which is sometimes more indicative of actual membrane performance. These commonly used techniques are briefly discussed as follows and summarized in Table 15.3.

TABLE 15.3
Commonly Used Techniques for Membrane Pore Characterization

Techniques	Description	Principle	Range	Advantages	Limitations
a. Bubble point	Measures the minimum pressure required to force air to flow through a liquid-filled membrane	Young–Laplace equation	20 nm–100 μm	Simple operation; measures only active pores	Measures only the maximum pore
b. Mercury porosimetry	Measures the volume of mercury forced into membrane pores at increasing pressure	Young–Laplace equation	5 nm–10 μm	Simple operation	Does not distinguish between dead-end pores and interconnecting pores; expensive and requires high pressure
c. Gas–liquid displacement	Measures the gas flow through a liquid-filled membrane at increasing pressure	Young–Laplace equation	20 nm–100 μm	Measures only active pores	Requires dedicated equipment; results might be affected by membrane swelling
d. Liquid–liquid displacement	Liquid filled in membrane pores is forced out by another liquid at increasing pressure	Young–Laplace equation	5–100 nm	Measures only active pores; requires lower pressure as interfacial tension between two liquids can be very low	Requires dedicated equipment; results might be affected by membrane swelling
e. Thermoporometry	Measures freezing point depression of liquid at different pore size by changing temperature	Gibbs–Thomson equation	2–50 nm	Membrane drying can be avoided as wet membrane is used	Requires dedicated equipment; dead-end pores are also evaluated; accuraracy may be limited by calorimetric analysis

Method	Description	Principle	Range	Advantage	Disadvantage
f. Gas adsorption–desorption	Measures the amount of gas adsorbed (desorbed) and condensed (evaporated) in pores at increasing (decreasing) pressure	Kelvin equation	2–100 nm	Simple operation	Does not distinguish between dead-end pores and interconnecting pores
g. Permporometry	Vapor is condensed or filled liquid is evaporated inside pores with changing pressure; gas flux through open pores is measured	Kelvin equation	2–100 nm	Measures only active pores	Sensitive to experiment conditions; correction has to be made to account for the presence of a thin layer of adsorbed gas
h. Evapoporometry	Filled volatile liquid in pores is evaporated over time; mass change over time is measured and relates to vapor pressure	Kelvin equation	2–100 nm	High accuracy; not sensitive to experiment conditions	Requires membrane with larger porosity; dead-end pores are also evaluated
i. Microscopy	Observes membrane morphology directly	Scanning electron	Down–1 nm	Direct observation; provides information on pore geometry	Characterization is usually done over a very small sample area
j. Solute retention	Determines the retention of solute in percentage or MWCO	Physical separation	MW: 1–1000 k	Simple and indicative of membrane performance	Rejection behaviors are affected by type of solutes and operating conditions

15.5.2.1 Fluid Displacement

When measuring membrane pore size using the fluid displacement method, the membrane pores are first filled by a fluid, which can be either air or a wetting liquid. The fluid inside the pores is then displaced by another immiscible fluid under applied pressure. The principle of the fluid displacement method is based on Young–Laplace equation [171]:

$$\Delta P = \frac{2\gamma \cos \theta}{r_{\mathrm{p}}} \tag{15.17}$$

where:
r_{p} is the membrane pore radius
γ is the surface tension between two fluids
θ is the contact angle
ΔP is the applied pressure difference

This equation states that the pressure required to force a fluid to penetrate into a pore is proportional to the surface tension and inversely proportional to the pore radius. Techniques used to determine membrane pore size or pore size distribution include bubble point, mercury porosimetry, gas–liquid displacement, and liquid–liquid displacement (Table 15.3a–d) [171–173].

15.5.2.2 Fluid Transition

Due to the fact that freezing point temperature and vapor pressure of a substance vary with interfacial curvature in a pore, the determination of pore size and pore size distribution can be achieved by observing solid–liquid or liquid–vapor transition in pores. Thermoporometry is a technique involving the measurement of freezing point depression of a liquid in membrane pores using a sensitive differential scanning calorimetry (Table 15.3e). The principle of thermoporometry is described by Gibbs–Thomson equation [174]:

$$\frac{\Delta T}{T_0} = \frac{2\gamma v}{\Delta H_{\mathrm{f}} r_{\mathrm{p}}} \tag{15.18}$$

where:
ΔT is the freezing point depression
T_0 is the original freezing point
γ is the surface tension between liquid–solid interface
v is the molar volume of liquid
ΔH_{f} is the enthalpy of fusion.

The gas adsorption–desorption method involves the analysis of adsorption–desorption isotherms of an inert gas as a function of partial pressure (Table 15.3f) [175].

The principle is based on the change in vapor pressure due to the curvature of the liquid–vapor interface in a capillary pore, as given by Kelvin equation [176]:

$$\ln\frac{P}{P_0} = -\frac{2\gamma v}{r_p RT}\cos\theta \tag{15.19}$$

where:
P and P_0 are the vapor pressure and original vapor pressure, respectively

Because the gas adsorption–desorption method measures both dead-end and interconnecting pores, another variation of this method, namely, permporometry, is also introduced (Table 15.3g) [177]. Instead of measuring the volume of gas adsorbed/desorbed, the gas flux through open pores is measured in permporometry. Evapoporometry, a relatively new technique involving the evaporation of volatile liquid from membrane pores, is also developed based on the Kelvin equation (Table 15.3h) [176].

15.5.2.3 Microscopic Techniques

Microscopic instruments such as scanning electron microscopy (SEM) and field emission SEM (FESEM) are useful for direct observation on the morphology of membrane surface and cross section (Table 15.3i). Use of microscopic methods for characterizing membrane pores provides information on pore geometry, which is difficult to obtain by other characterization techniques. Resolution of SEM and FESEM can be down to 1 nm [178]. Therefore, the size, shape, and distribution of pores on MF and UF membranes can be visualized under SEM and FESEM.

15.5.2.4 Solute Retention

Determining solute retention is a straightforward way to characterize the selectivity of a membrane as retention is directly related to membrane performance. Molecular weight cut-off (MWCO) is the most commonly used term to express the retention of a membrane, particularly for UF membranes. MWCO is defined as the molecular weight of a solute where the retention is 90%. Both linear (PEG) and branched (dextran) molecules have been frequently used as the solute to determine MWCO of membranes, and the result is usually different for the different solutes used (Table 15.3j).

By correlating the molecular weight of the solute and membrane pore radius, it is possible to obtain more information related to the membrane pores if a suitable model fitting is used. For instance, the correlation between pore radius and molecular weight of dextran was given by an empirical formula [179]:

$$r_p = 0.33M^{0.46} \tag{15.20}$$

where:
M is the molecular weight of dextran

Information obtained in Equation 15.20 can be fitted to a mathematical model to obtain the mean pore size and pore size distribution. Various mathematical models have been developed based on different distribution functions and transport theories, as described in detail in the literature [179–182].

15.6 CONCLUSIONS

NIPS has become a widely used membrane preparation technique since its invention. It provides the flexibility of material selection, and membranes with a wide range of pore size can be produced. On the other hand, a relatively new technique, electrospinning, has also gained much attention in membrane preparation due to the unique nanofiber structure of the resultant membranes that exhibits relatively high orientation, surface area, and porosity. Membrane preparation through NIPS and electrospinning involves many affecting parameters, including dope formulation and process conditions. Therefore, experimental studies on the synergetic effects of various parameters are as important as the understanding of the basic membrane formation principles for successful membrane fabrication.

Membranes prepared through NIPS and electrospinning can be further modified to make composite membranes. With a functional top layer and a supportive substrate, the preparation of composite membranes through different modification techniques allows wider applications.

Characterization of polymer solutions used in membrane preparation can provide useful information on adjusting dope formulation. On the other hand, the characterization of membrane separation properties can be carried out by various techniques. Each pore size-determining technique has its advantages and limitations; thus. one should select a suitable technique based on the requirements on applications.

REFERENCES

1. S. Loeb, S. Sourirajan, *Sea Water Demineralization by Means of an Osmotic Membrane*, Saline Water Conversion—II, American Chemical Society, Washington, DC, 1963, pp. 117–132.
2. J. Ren, R. Wang, Preparation of polymeric membranes, in: L.K. Wang, J.P. Chen, Y.-T. Hung, N.K. Shammas (Eds.), *Membrane and Desalination Technologies*, Humana Press, Inc., Totowa, NJ, 2011.
3. J.H. Hildebrand, R.L. Scott, The Solubility of Nonelectrolytes, Reinhold Pub. Corp., New York, 1950.
4. C.M. Hansen, The universality of the solubility parameter, *Industrial & Engineering Chemistry Product Research and Development*, 8 (1969) 2–11.
5. P.J. Flory, Thermodynamics of high polymer solutions, *The Journal of Chemical Physics*, 9 (1941) 660–661.
6. M.L. Huggins, Solutions of long chain compounds, *The Journal of Chemical Physics*, 9 (1941) 440.
7. P.J. Flory, *Principles of Polymer Chemistry*, Cornell University Press, Ithaca, NY, 1953.
8. E.A. Men'shikov, A.V. Bol'Shakova, I.V. Yaminskii, Determination of the Flory-Huggins parameter for a pair of polymer units from AFM data for thin films of block copolymers, *Protection of Metals and Physical Chemistry of Surfaces*, 45 (2009) 295–299.
9. D. Li, Dual-layer asymmetric hollow-fiber membranes for gas separation,: PhD Thesis, Chemical & Biomolecular Engineering, National University of Singapore, Singapore, 2005.

10. M. Mulder, *Basic Principles of Membrane Technology*, 2nd edn., Kluwer Academic, Dordrecht, the Netherlands/Boston, MA, 1996.
11. S.G. Li, T. van den Boomgaard, C.A. Smolders, H. Strathmann, Physical gelation of amorphous polymers in a mixture of solvent and nonsolvent, *Macromolecules*, 29 (1996) 2053–2059.
12. K.-Y. Lin, D.-M. Wang, J.-Y. Lai, Nonsolvent-induced gelation and its effect on membrane morphology, *Macromolecules*, 35 (2002) 6697–6706.
13. W.F. Smith, J. Hashemi, *Foundations of Materials Science and Engineering*, 5th edn., McGraw-Hill, Dubuque, IA, 2010.
14. M. Khayet, C.Y. Feng, K.C. Khulbe, T. Matsuura, Preparation and characterization of polyvinylidene fluoride hollow fiber membranes for ultrafiltration, *Polymer*, 43 (2002) 3879–3890.
15. S. Wongchitphimon, R. Wang, R. Jiraratananon, L. Shi, C.H. Loh, Effect of polyethylene glycol (PEG) as an additive on the fabrication of polyvinylidene fluoride-co-hexafluoropropylene (PVDF-HFP) asymmetric microporous hollow fiber membranes, *Journal of Membrane Science*, 369 (2011) 329–338.
16. R. Matz, The structure of cellulose acetate membranes. 1. The development of porous structures in anisotropic membranes, *Desalination*, 10 (1972) 1–15.
17. S.A. McKelvey, W.J. Koros, Phase separation, vitrification, and the manifestation of macrovoids in polymeric asymmetric membranes, *Journal of Membrane Science*, 112 (1996) 29–39.
18. H. Strathmann, K. Kock, The formation mechanism of phase inversion membranes, *Desalination*, 21 (1977) 241–255.
19. R.J. Ray, W.B. Krantz, R.L. Sani, Linear stability theory model for finger formation in asymmetric membranes, *Journal of Membrane Science*, 23 (1985) 155–182.
20. L. Jiang, T.-S. Chung, D.F. Li, C. Cao, S. Kulprathipanja, Fabrication of Matrimid/polyethersulfone dual-layer hollow fiber membranes for gas separation, *Journal of Membrane Science*, 240 (2004) 91–103.
21. M.M. Teoh, T.-S. Chung, Y.S. Yeo, Dual-layer PVDF/PTFE composite hollow fibers with a thin macrovoid-free selective layer for water production via membrane distillation, *Chemical Engineering Journal*, 171 (2011) 684–691.
22. M.C. García-Payo, M. Essalhi, M. Khayet, Preparation and characterization of PVDF-HFP copolymer hollow fiber membranes for membrane distillation, *Desalination*, 245 (2009) 469–473.
23. S.H. Choi, F. Tasselli, J.C. Jansen, G. Barbieri, E. Drioli, Effect of the preparation conditions on the formation of asymmetric poly(vinylidene fluoride) hollow fibre membranes with a dense skin, *European Polymer Journal*, 46 (2010) 1713–1725.
24. H. Ohya, S. Shiki, H. Kawakami, Fabrication study of polysulfone hollow-fiber microfiltration membranes: Optimal dope viscosity for nucleation and growth, *Journal of Membrane Science*, 326 (2009) 293–302.
25. A. Bottino, G. Camera-Roda, G. Capannelli, S. Munari, The formation of microporous polyvinylidene difluoride membranes by phase separation, *Journal of Membrane Science*, 57 (1991) 1–20.
26. J. Barzin, B. Sadatnia, Correlation between macrovoid formation and the ternary phase diagram for polyethersulfone membranes prepared from two nearly similar solvents, *Journal of Membrane Science*, 325 (2008) 92–97.
27. C. Feng, R. Wang, B. Shi, G. Li, Y. Wu, Factors affecting pore structure and performance of poly(vinylidene fluoride-co-hexafluoro propylene) asymmetric porous membrane, *Journal of Membrane Science*, 277 (2006) 55–64.
28. L. Shi, R. Wang, Y. Cao, C. Feng, D.T. Liang, J.H. Tay, Fabrication of poly(vinylidene fluoride-co-hexafluoropropylene) (PVDF-HFP) asymmetric microporous hollow fiber membranes, *Journal of Membrane Science*, 305 (2007) 215–225.

29. J.F. Hester, P. Banerjee, A.M. Mayes, Preparation of protein-resistant surfaces on poly(vinylidene fluoride) membranes via surface segregation, *Macromolecules*, 32 (1999) 1643–1650.
30. Y.Q. Wang, T. Wang, Y.L. Su, F.B. Peng, H. Wu, Z.Y. Jiang, Remarkable reduction of irreversible fouling and improvement of the permeation properties of poly(ether sulfone) ultrafiltration membranes by blending with pluronic F127, *Langmuir*, 21 (2005) 11856–11862.
31. W. Zhao, Y. Su, C. Li, Q. Shi, X. Ning, Z. Jiang, Fabrication of antifouling polyethersulfone ultrafiltration membranes using Pluronic F127 as both surface modifier and pore-forming agent, *Journal of Membrane Science*, 318 (2008) 405–412.
32. C.H. Loh, R. Wang, L. Shi, A.G. Fane, Fabrication of high performance polyethersulfone UF hollow fiber membranes using amphiphilic Pluronic block copolymers as pore-forming additives, *Journal of Membrane Science*, 380 (2011) 114–123.
33. C.H. Loh, R. Wang, Insight into the role of amphiphilic pluronic block copolymer as pore-forming additive in PVDF membrane formation, *Journal of Membrane Science*, 446 (2013) 492–503.
34. J. Kong, K. Li, Preparation of PVDF hollow-fiber membranes via immersion precipitation, *Journal of Applied Polymer Science*, 81 (2001) 1643–1653.
35. A. Idris, I. Ahmed, M.A. Limin, Influence of lithium chloride, lithium bromide and lithium fluoride additives on performance of polyethersulfone membranes and its application in the treatment of palm oil mill effluent, *Desalination*, 250 (2010) 805–809.
36. Z. Wang, J. Ma, The role of nonsolvent in-diffusion velocity in determining polymeric membrane morphology, *Desalination*, 286 (2012) 69–79.
37. Y.K. Ong, N. Widjojo, T.S. Chung, Fundamentals of semi-crystalline poly(vinylidene fluoride) membrane formation and its prospects for biofuel (ethanol and acetone) separation via pervaporation, *Journal of Membrane Science*, 378 (2011) 149–162.
38. H. Susanto, N. Stahra, M. Ulbricht, High performance polyethersulfone microfiltration membranes having high flux and stable hydrophilic property, *Journal of Membrane Science*, 342 (2009) 153–164.
39. P. Sukitpaneenit, T.S. Chung, Molecular elucidation of morphology and mechanical properties of PVDF hollow fiber membranes from aspects of phase inversion, crystallization and rheology, *Journal of Membrane Science*, 340 (2009) 192–205.
40. S. Atchariyawut, C. Feng, R. Wang, R. Jiratananon, D.T. Liang, Effect of membrane structure on mass-transfer in the membrane gas-liquid contacting process using microporous PVDF hollow fibers, *Journal of Membrane Science*, 285 (2006) 272–281.
41. Y. Tang, N. Li, A. Liu, S. Ding, C. Yi, H. Liu, Effect of spinning conditions on the structure and performance of hydrophobic PVDF hollow fiber membranes for membrane distillation, *Desalination*, 287 (2012) 326–339.
42. W.-L. Chou, D.-G. Yu, M.-C. Yang, C.-H. Jou, Effect of molecular weight and concentration of PEG additives on morphology and permeation performance of cellulose acetate hollow fibers, *Separation and Purification Technology*, 57 (2007) 209–219.
43. S. Bonyadi, T.S. Chung, W.B. Krantz, Investigation of corrugation phenomenon in the inner contour of hollow fibers during the non-solvent induced phase-separation process, *Journal of Membrane Science*, 299 (2007) 200–210.
44. W.L. Chou, M.C. Yang, Effect of take-up speed on physical properties and permeation performance of cellulose acetate hollow fibers, *Journal of Membrane Science*, 250 (2005) 259–267.
45. X. Shen, Y. Zhao, L. Chen, X. Feng, D. Yang, Q. Zhang, D. Su, Structure and performance of temperature-sensitive poly(vinylidene fluoride) hollow fiber membrane fabricated at different take-up speeds, *Polymer Engineering and Science*, 53 (2013) 571–579.
46. S.P. Sun, K.Y. Wang, D. Rajarathnam, T.A. Hatton, T.S. Chung, Polyamide-imide Nanofiltration hollow fiber membranes with elongation-induced nano-pore evolution, *AIChE Journal*, 56 (2010) 1481–1494.

47. P. Sukitpaneenit, T.S. Chung, Molecular design of the morphology and pore size of PVDF hollow fiber membranes for ethanol-water separation employing the modified pore-flow concept, *Journal of Membrane Science*, 374 (2011) 67–82.
48. S. Iwamoto, A.N. Nakagaito, H. Yano, Nano-fibrillation of pulp fibers for the processing of transparent nanocomposites, *Applied Physics A: Materials Science & Processing*, 89 (2007) 461–466.
49. Y. Lin, Y. Yao, X. Yang, N. Wei, X. Li, P. Gong, R. Li, D. Wu, Preparation of poly(ether sulfone) nanofibers by gas-jet/electrospinning, *Journal of Applied Polymer Science*, 107 (2008) 909–917.
50. B. Sun, Y.Z. Long, H.D. Zhang, M.M. Li, J.L. Duvail, X.Y. Jiang, H.L. Yin, Advances in three-dimensional nanofibrous macrostructures via electrospinning, *Progress in Polymer Science*, 39 (2014) 862–890.
51. D.H. Reneker, A.L. Yarin, Electrospinning jets and polymer nanofibers, *Polymer*, 49 (2008) 2387–2425.
52. A. Formhals, Process and apparatus for preparing artificial threads, US Patent No. 1975504, 1934.
53. W.J. Morton, Method of dispersing fluids, US Patent No. 705691, 1902.
54. Y. Liao, R. Wang, M. Tian, C. Qiu, A.G. Fane, Fabrication of polyvinylidene fluoride (PVDF) nanofiber membranes by electro-spinning for direct contact membrane distillation, *Journal of Membrane Science*, 425–426 (2013) 30–39.
55. T.J. Sill, H.A. von Recum, Electrospinning: Applications in drug delivery and tissue engineering, *Biomaterials*, 29 (2008) 1989–2006.
56. C. Chen, K. Liu, H. Wang, W. Liu, H. Zhang, Morphology and performances of electrospun polyethylene glycol/poly (dl-lactide) phase change ultrafine fibers for thermal energy storage, *Solar Energy Materials and Solar Cells*, 117 (2013) 372–381.
57. S. Ramakrishna, K. Fujihara, W.-E. Teo, T. Yong, Z. Ma, R. Ramaseshan, Electrospun nanofibers: Solving global issues, *Materials Today*, 9 (2006) 40–50.
58. H.R. Darrell, H. Fong, E. Zussman, S. Koombhongse, W. Kataphinan, Nanofiber manufacturing: Toward better process control. in: D.H. Reneker and H. Fong (Ed.), *Polymeric Nanofibers*, American Chemical Society, Washington, DC, 2006, Vol. 198, pp. 7–20.
59. H. Fong, H.R. Darrell, *Electrospinning and the Formation of Nanofibers*, Hanser, Cincinnati, OH, 2000.
60. G. Taylor, Disintegration of water drops in an electric field, *Proceedings of the Royal Society of London. Series A, Mathematical and Physical Sciences*, 280 (1964) 383–397.
61. G. Taylor, Electrically driven jets, *Proceedings of the Royal Society of London. Series A, Mathematical and Physical Sciences*, 313 (1969) 453–475.
62. D.H. Reneker, W. Kataphinan, A. Theron, E. Zussman, A.L. Yarin, Nanofiber garlands of polycaprolactone by electrospinning, *Polymer*, 43 (2002) 6785–6794.
63. J.-H. He, Y. Wu, W.-W. Zuo, Critical length of straight jet in electrospinning, *Polymer*, 46 (2005) 12637–12640.
64. D.H. Reneker, A.L. Yarin, H. Fong, S. Koombhongse, Bending instability of electrically charged liquid jets of polymer solutions in electrospinning, *Journal of Applied Physics*, 87 (2000) 4531.
65. A.L. Yarin, S. Koombhongse, D.H. Reneker, Bending instability in electrospinning of nanofibers, *Journal of Applied Physics*, 89 (2001) 3018.
66. Y.M. Shin, M.M. Hohman, M.P. Brenner, G.C. Rutledge, Experimental characterization of electrospinning: The electrically forced jet and instabilities, *Polymer*, 42 (2001) 09955–09967.
67. Y.M. Shin, M.M. Hohman, M.P. Brenner, G.C. Rutledge, Electrospinning: A whipping fluid jet generates submicron polymer fibers, *Applied Physics Letters*, 78 (2001) 1149.
68. H. Fong, I. Chun, D.H. Reneker, Beaded nanofibers formed during electrospinning, *Polymer*, 40 (1999) 4585–4592.

69. A.L. Huebner, H.N. Chu, Instability and breakup of charged liquid jets, *Journal of Fluid Mechanics*, 49 (1971) 361–372.

70. T. Han, D.H. Reneker, A.L. Yarin, Buckling of jets in electrospinning, *Polymer*, 48 (2007) 6064–6076.

71. S. Park, K. Park, H. Yoon, J. Son, T. Min, G. Kim, Apparatus for preparing electrospun nanofibers: Designing an electrospinning process for nanofiber fabrication, *Polymer International*, 56 (2007) 1361–1366.

72. J.A. Matthews, G.E. Wnek, D.G. Simpson, G.L. Bowlin, Electrospinning of collagen nanofibers, *Biomacromolecules*, 3 (2002) 232–238.

73. M.V. Kakade, S. Givens, K. Gardner, K.H. Lee, D.B. Chase, J.F. Rabolt, Electric field induced orientation of polymer chains in macroscopically aligned electrospun polymer nanofibers, *Journal of the American Chemical Society*, 129 (2007) 2777–2782.

74. D. Li, Y. Wang, Y. Xia, Electrospinning of polymeric and ceramic nanofibers as uniaxially aligned arrays, *Nano Letters*, 3 (2003) 1167–1171.

75. S. Agarwal, A. Greiner, J.H. Wendorff, Functional materials by electrospinning of polymers, *Progress in Polymer Science*, 38 (2013) 963–991.

76. X. Wang, B. Ding, G. Sun, M. Wang, J. Yu, Electro-spinning/netting: A strategy for the fabrication of three-dimensional polymer nano-fiber/nets, *Progress in Materials Science*, 58 (2013) 1173–1243.

77. D. Li, J.T. McCann, Y. Xia, M. Marquez, Electrospinning: A simple and versatile technique for producing ceramic nanofibers and nanotubes, *Journal of the American Ceramic Society*, 89 (2006) 1861–1869.

78. D. Li, Y. Xia, Electrospinning of nanofibers: Reinventing the wheel? *Advanced Materials*, 16 (2004) 1151–1170.

79. J.M. Deitzel, J. Kleinmeyer, D. Harris, N.C.B. Tan, The effect of processing variables on the morphology of electrospun nanofibers and textiles, *Polymer*, 42 (2001) 261–272.

80. Z.-M. Huang, Y.Z. Zhang, M. Kotaki, S. Ramakrishna, A review on polymer nanofibers by electrospinning and their applications in nanocomposites, *Composites Science and Technology*, 63 (2003) 2223–2253.

81. Y.J. Ryu, H.Y. Kim, K.H. Lee, H.C. Park, D.R. Lee, Transport properties of electrospun nylon 6 nonwoven mats, *European Polymer Journal*, 39 (2003) 1883–1889.

82. K.H. Lee, H.Y. Kim, H.J. Bang, Y.H. Jung, S.G. Lee, The change of bead morphology formed on electrospun polystyrene fibers, *Polymer*, 44 (2003) 4029–4034.

83. N. Bhardwaj, S.C. Kundu, Electrospinning: A fascinating fiber fabrication technique, *Biotechnology Advances*, 28 (2010) 325–347.

84. D.H. Reneker, I. Chun, Nanometre diameter fibres of polymer, produced by electrospinning, *Nanotechnology*, 7 (1996) 216.

85. T. Subbiah, G.S. Bhat, R.W. Tock, S. Parameswaran, S.S. Ramkumar, Electrospinning of nanofibers, *Journal of Applied Polymer Science*, 96 (2005) 557–569.

86. G.C. Rutledge, S.V. Fridrikh, Formation of fibers by electrospinning, *Advanced Drug Delivery Reviews*, 59 (2007) 1384–1391.

87. X. Wang, B. Ding, J. Yu, Y. Si, S. Yang, G. Sun, Electro-netting: Fabrication of two-dimensional nano-nets for highly sensitive trimethylamine sensing, *Nanoscale*, 3 (2011) 911–915.

88. N.A.M. Barakat, M.A. Kanjwal, F.A. Sheikh, H.Y. Kim, Spider-net within the N6, PVA and PU electrospun nanofiber mats using salt addition: Novel strategy in the electrospinning process, *Polymer*, 50 (2009) 4389–4396.

89. S. Yang, X. Wang, B. Ding, J. Yu, J. Qian, G. Sun, Controllable fabrication of soap-bubble-like structured polyacrylic acid nano-nets via electro-netting, *Nanoscale*, 3 (2011) 564–568.

90. X. Zong, K. Kim, D. Fang, S. Ran, B.S. Hsiao, B. Chu, Structure and process relationship of electrospun bioabsorbable nanofiber membranes, *Polymer*, 43 (2002) 4403–4412.

91. X. Wang, B. Ding, J. Yu, J. Yang, Large-scale fabrication of two-dimensional spider-web-like gelatin nano-nets via electro-netting, *Colloids and Surfaces B: Biointerfaces*, 86 (2011) 345–352.

92. J. Hu, X. Wang, B. Ding, J. Lin, J. Yu, G. Sun, One-step electro-spinning/netting technique for controllably preparing polyurethane nano-fiber/net, *Macromolecular Rapid Communications*, 32 (2011) 1729–1734.

93. S. Talwar, A.S. Krishnan, J.P. Hinestroza, B. Pourdeyhimi, S.A. Khan, Electrospun nanofibers with associative polymer–surfactant systems, *Macromolecules*, 43 (2010) 7650–7656.

94. T. Lin, H. Wang, H. Wang, X. Wang, The charge effect of cationic surfactants on the elimination of fibre beads in the electrospinning of polystyrene, *Nanotechnology*, 15 (2004) 1375.

95. C.J. Luo, M. Nangrejo, M. Edirisinghe, A novel method of selecting solvents for polymer electrospinning, *Polymer*, 51 (2010) 1654–1662.

96. B. Ding, C. Li, Y. Miyauchi, O. Kuwaki, S. Shiratori, Formation of novel 2D polymer nanowebs via electrospinning, *Nanotechnology*, 17 (2006) 3685.

97. N. Wang, X. Wang, B. Ding, J. Yu, G. Sun, Tunable fabrication of three-dimensional polyamide-66 nano-fiber/nets for high efficiency fine particulate filtration, *Journal of Materials Chemistry*, 22 (2012) 1445–1452.

98. X. Zhang, M.R. Reagan, D.L. Kaplan, Electrospun silk biomaterial scaffolds for regenerative medicine, *Advanced Drug Delivery Reviews*, 61 (2009) 988–1006.

99. D.S. Katti, K.W. Robinson, F.K. Ko, C.T. Laurencin, Bioresorbable nanofiber-based systems for wound healing and drug delivery: Optimization of fabrication parameters, *Journal of Biomedical Materials Research Part B: Applied Biomaterials*, 70B (2004) 286–296.

100. L. Setiawan, R. Wang, K. Li, A.G. Fane, Fabrication of novel poly(amide-imide) forward osmosis hollow fiber membranes with a positively charged nanofiltration-like selective layer, *Journal of Membrane Science*, 369 (2011) 196–205.

101. L. Setiawan, R. Wang, S. Tan, L. Shi, A.G. Fane, Fabrication of poly(amide-imide)-polyethersulfone dual layer hollow fiber membranes applied in forward osmosis by combined polyelectrolyte cross-linking and depositions, *Desalination*, 312 (2013) 99–106.

102. R.J. Petersen, Composite reverse osmosis and nanofiltration membranes, *Journal of Membrane Science*, 83 (1993) 81–150.

103. F. Gaudin, N. Sintes-Zydowicz, Correlation between the polymerization kinetics and the chemical structure of poly(urethane–urea) nanocapsule membrane obtained by interfacial step polymerization in miniemulsion, *Colloids and Surfaces A: Physicochemical and Engineering Aspects*, 415 (2012) 328–342.

104. S. Chou, R. Wang, L. Shi, Q. She, C. Tang, A.G. Fane, Thin-film composite hollow fiber membranes for pressure retarded osmosis (PRO) process with high power density, *Journal of Membrane Science*, 389 (2012) 25–33.

105. S. Veríssimo, K.V. Peinemann, J. Bordado, Thin-film composite hollow fiber membranes: An optimized manufacturing method, *Journal of Membrane Science*, 264 (2005) 48–55.

106. S. Veríssimo, K.V. Peinemann, J. Bordado, New composite hollow fiber membrane for nanofiltration, *Desalination*, 184 (2005) 1–11.

107. J. Wei, C. Qiu, C.Y. Tang, R. Wang, A.G. Fane, Synthesis and characterization of flat-sheet thin film composite forward osmosis membranes, *Journal of Membrane Science*, 372 (2011) 292–302.

108. J. Liu, Z. Xu, X. Li, Y. Zhang, Y. Zhou, Z. Wang, X. Wang, An improved process to prepare high separation performance PA/PVDF hollow fiber composite nanofiltration membranes, *Separation and Purification Technology*, 58 (2007) 53–60.

109. P.R. Buch, D.J. Mohan, A.V.R. Reddy, Preparation, characterization and chlorine stability of aromatic-cycloaliphatic polyamide thin film composite membranes, *Journal of Membrane Science*, 309 (2008) 36–44.

110. Y. Yang, X. Jian, D. Yang, S. Zhang, L. Zou, Poly(phthalazinone ether sulfone ketone) (PPESK) hollow fiber asymmetric nanofiltration membranes: Preparation, morphologies and properties, *Journal of Membrane Science*, 270 (2006) 1–12.

111. G.-Y. Chai, W.B. Krantz, Formation and characterization of polyamide membranes via interfacial polymerization, *Journal of Membrane Science*, 93 (1994) 175–192.

112. F. Yang, S. Zhang, D. Yang, X. Jian, Preparation and characterization of polypiperazine amide/PPESK hollow fiber composite nanofiltration membrane, *Journal of Membrane Science*, 301 (2007) 85–92.

113. B.-H. Jeong, E.M.V. Hoek, Y. Yan, A. Subramani, X. Huang, G. Hurwitz, A.K. Ghosh, A. Jawor, Interfacial polymerization of thin film nanocomposites: A new concept for reverse osmosis membranes, *Journal of Membrane Science*, 294 (2007) 1–7.

114. Y. Zhao, C. Qiu, X. Li, A. Vararattanavech, W. Shen, J. Torres, C. Hélix-Nielsen, R. Wang, X. Hu, A.G. Fane, C.Y. Tang, Synthesis of robust and high-performance aquaporin-based biomimetic membranes by interfacial polymerization-membrane preparation and RO performance characterization, *Journal of Membrane Science*, 423–424 (2012) 422–428.

115. C. Liu, W. Fang, S. Chou, L. Shi, A.G. Fane, R. Wang, Fabrication of layer-by-layer assembled FO hollow fiber membranes and their performances using low concentration draw solutions, *Desalination*, 308 (2013) 147–153.

116. J. Miao, G.H. Chen, C.J. Gao, A novel kind of amphoteric composite nanofiltration membrane prepared from sulfated chitosan (SCS), *Desalination*, 181 (2005) 173–183.

117. B. Tieke, A. Toutianoush, W. Jin, Selective transport of ions and molecules across layer-by-layer assembled membranes of polyelectrolytes, p-sulfonato-calix[n]arenes and Prussian Blue-type complex salts, *Advances in Colloid and Interface Science*, 116 (2005) 121–131.

118. B. Tieke, F. Van Ackern, L. Krasemann, A. Toutianoush, Ultrathin self-assembled polyelectrolyte multilayer membranes, *European Physical Journal E*, 5 (2001) 29–39.

119. D. Saeki, M. Imanishi, Y. Ohmukai, T. Maruyama, H. Matsuyama, Stabilization of layer-by-layer assembled nanofiltration membranes by crosslinking via amide bond formation and siloxane bond formation, *Journal of Membrane Science*, 447 (2013) 128–133.

120. W. Shan, P. Bacchin, P. Aimar, M.L. Bruening, V.V. Tarabara, Polyelectrolyte multilayer films as backflushable nanofiltration membranes with tunable hydrophilicity and surface charge, *Journal of Membrane Science*, 349 (2010) 268–278.

121. K. Hoffmann, B. Tieke, Layer-by-layer assembled membranes containing hexacyclenhexaacetic acid and polyethyleneimine N-acetic acid and their ion selective permeation behaviour, *Journal of Membrane Science*, 341 (2009) 261–267.

122. G. Zhang, Z. Ruan, S. Ji, Z. Liu, Construction of metal-ligand-coordinated multilayers and their selective separation behavior, *Langmuir*, 26 (2009) 4782–4789.

123. R.V. Klitzing, B. Tieke, Polyelectrolyte membranes, in: M. Schmidt (Ed.), *Polyelectrolytes with Defined Molecular Architecture I*, Advances in Polymer Science, Vol. 165, 2004, Springer, Berlin; Heidelberg, Germany, pp. 177–210.

124. M.L. Bruening, D.M. Dotzauer, P. Jain, L. Ouyang, G.L. Baker, Creation of functional membranes using polyelectrolyte multilayers and polymer brushes, *Langmuir*, 24 (2008) 7663–7673.

125. J. Wang, Y. Yao, Z. Yue, J. Economy, Preparation of polyelectrolyte multilayer films consisting of sulfonated poly (ether ether ketone) alternating with selected anionic layers, *Journal of Membrane Science*, 337 (2009) 200–207.

126. S. Qi, W. Li, Y. Zhao, N. Ma, J. Wei, T.W. Chin, C.Y. Tang, Influence of the properties of layer-by-layer active layers on forward osmosis performance, *Journal of Membrane Science*, 423–424 (2012) 536–542.

127. C. Qiu, S. Qi, C.Y. Tang, Synthesis of high flux forward osmosis membranes by chemically crosslinked layer-by-layer polyelectrolytes, *Journal of Membrane Science*, 381 (2011) 74–80.

128. X. Liu, S. Qi, Y. Li, L. Yang, B. Cao, C.Y. Tang, Synthesis and characterization of novel antibacterial silver nanocomposite nanofiltration and forward osmosis membranes based on layer-by-layer assembly, *Water Research*, 47 (2013) 3081–3092.

129. G. Liu, D.M. Dotzauer, M.L. Bruening, Ion-exchange membranes prepared using layer-by-layer polyelectrolyte deposition, *Journal of Membrane Science*, 354 (2010) 198–205.

130. P. Zhang, J. Qian, Y. Yang, Q. An, X. Liu, Z. Gui, Polyelectrolyte layer-by-layer self-assembly enhanced by electric field and their multilayer membranes for separating isopropanol–water mixtures, *Journal of Membrane Science*, 320 (2008) 73–77.

131. G. Zhang, X. Gao, S. Ji, Z. Liu, One-step dynamic assembly of polyelectrolyte complex membranes, *Materials Science and Engineering C*, 29 (2009) 1877–1884.

132. F. Van Ackern, L. Krasemann, B. Tieke, Ultrathin membranes for gas separation and pervaporation prepared upon electrostatic self-assembly of polyelectrolytes, *Thin Solid Films*, 327–329 (1998) 762–766.

133. L. Krasemann, B. Tieke, Composite membranes with ultrathin separation layer prepared by self-assembly of polyelectrolytes, *Materials Science and Engineering C*, 8–9 (1999) 513–518.

134. P.R. Van Tassel, Polyelectrolyte adsorption and layer-by-layer assembly: Electrochemical control, *Current Opinion in Colloid & Interface Science*, 17 (2012) 106–113.

135. M.D. Miller, M.L. Bruening, Controlling the nanofiltration properties of multilayer polyelectrolyte membranes through variation of film composition, *Langmuir*, 20 (2004) 11545–11551.

136. L. Krasemann, B. Tieke, Selective ion transport across self-assembled alternating multilayers of cationic and anionic polyelectrolytes, *Langmuir*, 16 (2000) 287–290.

137. L. Ouyang, R. Malaisamy, M.L. Bruening, Multilayer polyelectrolyte films as nanofiltration membranes for separating monovalent and divalent cations, *Journal of Membrane Science*, 310 (2008) 76–84.

138. Y. Liu, T.-S. Chung, R. Wang, D.F. Li, M.L. Chng, Chemical cross-linking modification of polyimide/poly(ether sulfone) dual-layer hollow-fiber membranes for gas separation, *Industrial & Engineering Chemistry Research*, 42 (2003) 1190–1195.

139. N. Stafie, D.F. Stamatialis, M. Wessling, Effect of PDMS cross-linking degree on the permeation performance of PAN/PDMS composite nanofiltration membranes, *Separation and Purification Technology*, 45 (2005) 220–231.

140. N.L. Le, Y. Wang, T.-S. Chung, Synthesis, cross-linking modifications of 6FDA-NDA/DABA polyimide membranes for ethanol dehydration via pervaporation, *Journal of Membrane Science*, 415–416 (2012) 109–121.

141. G.M. Shi, Y. Wang, T.-S. Chung, Dual-layer PBI/P84 hollow fibers for pervaporation dehydration of acetone, *AIChE Journal*, 58 (2012) 1133–1145.

142. K.D. Papadimitriou, M. Geormezi, S.G. Neophytides, J.K. Kallitsis, Covalent cross-linking in phosphoric acid of pyridine based aromatic polyethers bearing side double bonds for use in high temperature polymer electrolyte membrane fuelcells, *Journal of Membrane Science*, 433 (2013) 1–9.

143. J. Qiao, J. Fu, L. Liu, Y. Liu, J. Sheng, Highly stable hydroxyl anion conducting membranes poly(vinyl alcohol)/poly(acrylamide-co-diallyldimethylammonium chloride) (PVA/PAADDA) for alkaline fuel cells: Effect of cross-linking, *International Journal of Hydrogen Energy*, 37 (2012) 4580–4589.

144. C.E. Powell, X.J. Duthie, S.E. Kentish, G.G. Qiao, G.W. Stevens, Reversible diamine cross-linking of polyimide membranes, *Journal of Membrane Science*, 291 (2007) 199–209.

145. C. Ba, J. Langer, J. Economy, Chemical modification of P84 copolyimide membranes by polyethylenimine for nanofiltration, *Journal of Membrane Science*, 327 (2009) 49–58.

146. R. Huang, G. Chen, B. Yang, C. Gao, Positively charged composite nanofiltration membrane from quaternized chitosan by toluene diisocyanate cross-linking, *Separation and Purification Technology*, 61 (2008) 424–429.

147. R. Huang, G. Chen, M. Sun, C. Gao, Preparation and characterization of quaterinized chitosan/poly(acrylonitrile) composite nanofiltration membrane from anhydride mixture cross-linking, *Separation and Purification Technology*, 58 (2008) 393–399.

148. T. He, M.H.V. Mulder, H. Strathmann, M. Wessling, Preparation of composite hollow fiber membranes: Co-extrusion of hydrophilic coatings onto porous hydrophobic support structures, *Journal of Membrane Science*, 207 (2002) 143–156.

149. R.X. Liu, X.Y. Qiao, T.-S. Chung, Dual-layer P84/polyethersulfone hollow fibers for pervaporation dehydration of isopropanol, *Journal of Membrane Science*, 294 (2007) 103–114.

150. S.A. Hashemifard, A.F. Ismail, T. Matsuura, Co-casting technique for fabricating dual-layer flat sheet membranes for gas separation, *Journal of Membrane Science*, 375 (2011) 258–267.

151. X. Ding, Y. Cao, H. Zhao, L. Wang, Q. Yuan, Fabrication of high performance Matrimid/polysulfone dual-layer hollow fiber membranes for O2/N2 separation, *Journal of Membrane Science*, 323 (2008) 352–361.

152. S.P. Sun, K.Y. Wang, N. Peng, T.A. Hatton, T.-S. Chung, Novel polyamide-imide/cellulose acetate dual-layer hollow fiber membranes for nanofiltration, *Journal of Membrane Science*, 363 (2010) 232–242.

153. S. Bonyadi, T.S. Chung, Flux enhancement in membrane distillation by fabrication of dual layer hydrophilic-hydrophobic hollow fiber membranes, *Journal of Membrane Science*, 306 (2007) 134–146.

154. L. Setiawan, L. Shi, W.B. Krantz, R. Wang, Explorations of delamination and irregular structure in poly(amide-imide)-polyethersulfone dual layer hollow fiber membranes, *Journal of Membrane Science*, 423–424 (2012) 73–84.

155. D. Li, T.-S. Chung, R. Wang, Morphological aspects and structure control of dual-layer asymmetric hollow fiber membranes formed by a simultaneous co-extrusion approach, *Journal of Membrane Science*, 243 (2004) 155–175.

156. N. Widjojo, T.S. Chung, W.B. Krantz, A morphological and structural study of Ultem/P84 copolyimide dual-layer hollow fiber membranes with delamination-free morphology, *Journal of Membrane Science*, 294 (2007) 132–146.

157. A.I. Schafer, A.G. Fane, T.D. Waite, *Nanofiltration—Principles and Applications*, Elsevier Advanced Technology, Oxford, 2005.

158. T. He, M. Frank, M.H.V. Mulder, M. Wessling, Preparation and characterization of nanofiltration membranes by coating polyethersulfone hollow fibers with sulfonated poly(ether ether ketone) (SPEEK), *Journal of Membrane Science*, 307 (2008) 62–72.

159. J.-H. Kim, P.-K. Park, C.-H. Lee, H.-H. Kwon, Surface modification of nanofiltration membranes to improve the removal of organic micro-pollutants (EDCs and PhACs) in drinking water treatment: Graft polymerization and cross-linking followed by functional group substitution, *Journal of Membrane Science*, 321 (2008) 190–198.

160. M.N. Abu Seman, M. Khayet, Z.I. Bin Ali, N. Hilal, Reduction of nanofiltration membrane fouling by UV-initiated graft polymerization technique, *Journal of Membrane Science*, 355 (2010) 133–141.

161. R. Bernstein, S. Belfer, V. Freger, Surface modification of dense membranes using radical graft polymerization enhanced by monomer filtration, *Langmuir*, 26 (2010) 12358–12365.

162. S. Béquet, J.-C. Remigy, J.-C. Rouch, J.-M. Espenan, M. Clifton, P. Aptel, From ultra-filtration to nanofiltration hollow fiber membranes: A continuous UV-photografting process, *Desalination*, 144 (2002) 9–14.

163. A. Akbari, S. Desclaux, J.C. Rouch, J.C. Remigy, Application of nanofiltration hollow fibre membranes, developed by photografting, to treatment of anionic dye solutions, *Journal of Membrane Science*, 297 (2007) 243–252.

164. A.S. Sarac, Redox polymerization, *Progress in Polymer Science*, 24 (1999) 1149–1204.

165. G. Zuo, R. Wang, Novel membrane surface modification to enhance anti-oil fouling property for membrane distillation application, *Journal of Membrane Science*, 447 (2013) 26–35.

166. L. Shi, R. Wang, Y. Cao, D.T. Liang, J.H. Tay, Effect of additives on the fabrication of poly(vinylidene fluoride-co-hexafluoropropylene) (PVDF-HFP) asymmetric microporous hollow fiber membranes, *Journal of Membrane Science*, 315 (2008) 195–204.

167. D.J. Lin, C.L. Chang, C.L. Chang, T.C. Chen, L.P. Cheng, Fine structure of poly(vinylidene fluoride) membranes prepared by phase inversion from a water/N-methyl-2-pyrollidone/poly(vinylidene fluoride) system, *Journal of Polymer Science, Part B: Polymer Physics*, 42 (2004) 830–842.

168. J.-J. Qin, R. Wang, T.-S. Chung, Investigation of shear stress effect within a spinneret on flux, separation and thermomechanical properties of hollow fiber ultrafiltration membranes, *Journal of Membrane Science*, 175 (2000) 197–213.

169. L. Shi, R. Wang, Y. Cao, Effect of the rheology of poly(vinylidene fluoride-co-hexafluropropylene) (PVDF-HFP) dope solutions on the formation of microporous hollow fibers used as membrane contactors, *Journal of Membrane Science*, 344 (2009) 112–122.

170. K.M. Persson, V. Gekas, G. Trägårdh, Study of membrane compaction and its influence on ultrafiltration water permeability, *Journal of Membrane Science*, 100 (1995) 155–162.

171. E. Jakobs, W.J. Koros, Ceramic membrane characterization via the bubble point technique, *Journal of Membrane Science*, 124 (1997) 149–159.

172. A. Jena, K. Gupta, Advances in pore structure evaluation by porometry, *Chemical Engineering and Technology*, 33 (2010) 1241–1250.

173. A.B. Abell, K.L. Willis, D.A. Lange, Mercury intrusion porosimetry and image analysis of cement-based materials, *Journal of Colloid and Interface Science*, 211 (1999) 39–44.

174. M. Wulff, Pore size determination by thermoporometry using acetonitrile, *Thermochimica Acta*, 419 (2004) 291–294.

175. E.P. Barrett, L.G. Joyner, P.P. Halenda, The determination of pore volume and area distributions in porous substances. I. Computations from nitrogen isotherms, *Journal of the American Chemical Society*, 73 (1951) 373–380.

176. W.B. Krantz, A.R. Greenberg, E. Kujundzic, A. Yeo, S.S. Hosseini, Evapoporometry: A novel technique for determining the pore-size distribution of membranes, *Journal of Membrane Science*, 438 (2013) 153–166.

177. T. Tsuru, T. Hino, T. Yoshioka, M. Asaeda, Permporometry characterization of microporous ceramic membranes, *Journal of Membrane Science*, 186 (2001) 257–265.

178. Y. Wyart, G. Georges, C. Deumié, C. Amra, P. Moulin, Membrane characterization by microscopic methods: Multiscale structure, *Journal of Membrane Science*, 315 (2008) 82–92.

179. J. Ren, Z. Li, F.S. Wong, A new method for the prediction of pore size distribution and MWCO of ultrafiltration membranes, *Journal of Membrane Science*, 279 (2006) 558–569.

180. D.B. Mosqueda-Jimenez, R.M. Narbaitz, T. Matsuura, G. Chowdhury, G. Pleizier, J.P. Santerre, Influence of processing conditions on the properties of ultrafiltration membranes, *Journal of Membrane Science*, 231 (2004) 209–224.

181. S. Derjani-Bayeh, V.G.J. Rodgers, Sieving variations due to the choice in pore size distribution model, *Journal of Membrane Science*, 209 (2002) 1–7.

182. J. Kassotis, J. Shmidt, L.T. Hodgins, H.P. Gregor, Modelling of the pore-size distribution of ultrafiltration membranes, *Journal of Membrane Science*, 22 (1985) 61–76.

16 Strategies to Use Nanoparticles in Polymeric Membranes

Bart Van der Bruggen, Ruixin Zhang, and Jeonghwan Kim

CONTENTS

16.1 Introduction 569
16.2 Self-Assembly of Nanoparticles on the Membrane Surface 571
16.3 Bulk Addition of Nanoparticles 574
16.4 Anchoring in/on the Membrane Surface 577
16.5 Layer-by-Layer Addition of Nanoparticles 582
16.6 Conclusions 584
References 584

16.1 INTRODUCTION

Membranes are effective tools for a wide range of advanced separations such as processing liquids, vapors, and gases. The majority of the applications make use of polymeric membranes, as they offer the potential of tailoring the separation for specific applications by adjusting the synthesis procedure. This often allows for inexpensive separations with a remarkably high performance. Nevertheless, polymeric membranes have intrinsic disadvantages, such as poor chemical and physical resilience, as well, which limit their potential.

For aqueous separations, polymeric membranes are used for the removal of particles, microorganisms, and organic matter from drinking water, wastewater, and industrial water. Compared with conventional treatment methods, membrane processes provide higher quality water, minimize disinfectant demand, are more compact, provide easier operational control, require less maintenance, and generate less sludge (AWWA Membrane Committee 2005). The wider use of membrane filtration, however, is impeded by colloidal deposition on membrane surface and/or adsorption into membrane pore matrix that produces *membrane fouling*. Fouling lowers the economic efficiency of membrane filtration by reducing the permeate flux, shortening membrane lifetime, and increasing the frequency of membrane cleaning, leading to the replacement of the membrane process.

Naturally occurring dissolved and colloidal organic matter (natural organic matter: NOM) is considered as a major contributor to membrane fouling in water treatment applications, including ultrafiltration (UF) and nanofiltration (NF) (Braghetta et al. 1997; Cho et al. 2000a). Organic molecules that constitute NOM are a special concern because they are abundant in surface water supplies and react with chlorine, a common disinfectant, to form carcinogenic chlorinated compound. The organics are also responsible for imparting color in natural water and forming complexes with heavy metals and hydrophobic organic pollutants such as pesticides (Cho et al. 2000a).

Modification of membrane surface chemistry has led to various claims of *low-fouling membranes*. NOM fouling has been observed to correlate with membrane properties, such as hydrophobicity, surface roughness, charge, and molecular weight cut-off (Jucker and Clark 1994). It has been reported that hydrophobic adsorption of NOM plays a key role in membrane fouling. Accordingly, the membrane surface property targeted for the modification has been its hydrophilicity (Kilduff et al. 2000; Wang et al. 2000; Taniguchi et al. 2003). UF membranes formed from polyethersulfone (PES) were modified by ultraviolet (UV)-assisted photochemical graft polymerization using hydrophilic monomers such as acrylic acid and *N*-vinyl-2-pyrrolidinone (Kilduff et al. 2000). The surface of porous PES membranes was modified by the polymerization of acrylamine using argon plasma and a low-temperature carbon dioxide plasma, both of which were shown to decrease fouling by increasing hydrophilicity of the membrane (Wavhal and Fisher 2002, 2003).

Such chemical modifications may be tedious, and few modifications were found feasible to apply on a commercial level. An alternative solution, potentially more simple or durable, is in the use of nanoparticles in the membrane structure. Studies have demonstrated that the assemblage of various nanoparticles into polymeric membranes can mitigate membrane fouling (Cortalezzi et al. 2002; Yan et al. 2006). Titanium dioxide (TiO_2) has attracted considerable attention in this regard because of its stability, commercial availability, and ease of preparation. Previous studies have shown that membrane fouling was significantly reduced by the introduction of TiO_2 nanoparticles into the polymeric membrane (Kwak et al. 2001). After they are deposited on the membrane surface, the hydrophilicity of membrane surface was improved and free water fraction was increased. Hydrophobic polyvinylidenefluoride (PVDF) membranes were assembled with different sizes of nanosized TiO_2 particles. It was found that smaller nanoparticles could significantly improve the antifouling property of PVDF membrane (Kim et al. 2003; Bae and Tak 2005). The hydrophilic modification of PVDF membranes resulted in the decrease of the adsorption and deposition of hydrophobic organics on the membrane surface.

Similar efforts have been made in other application areas. In gas separation, the addition of TiO_2 nanoparticles to polyvinyl acetate improved the thermal stability of the resulting membranes, which was demonstrated by an increase in the glass transition temperature (Ahmad and Hägg 2013). In this case, it was found that the addition of TiO_2 up to 10 wt% improved both the permeability and selectivity of the membranes for gas separation, including H_2, CO_2, O_2, and N_2. Similar observations were made on the effect of silica nanoparticles on the permeability of CO_2 and CH_4 for two types of nanocomposite membranes based on polyester urethane and polyether urethane (Hassanajili et al. 2013). Khan et al. (2013) studied mixed matrix membranes composed of acrylate-derivatized polysulfone and functionalized mesoporous MCM-41 and concluded that covalently linked MCM-41 fillers yielded a significantly higher CO_2/CH_4 and CO_2/N_2 selectivity, which was explained by the effect of the covalent link between the $-NH_2$ group of the filler and the acrylate of the polymer.

Another significant application area of mixed matrix membranes containing nanoparticles is pervaporation. For example, nanocomposite membranes made of sodium alginate/poly(vinyl pyrrolidone) blend polymers incorporated with phospho-tungstic acid ($H_3PW_{12}O_{40}$) nanoparticles were used in the ethanol dehydration by pervaporation (Magalad et al. 2013). Shirazi et al. (2012) used silica nanoparticles to enhance the performance of PDMS membranes in pervaporation. They concluded that incorporating silica nanoparticles promotes selectivity of the membranes in pervaporation, which was related to the polymer chains becoming more rigid and a decrease in the polymer free volume. It was also observed that permeation flux decreases as diffusion of the penetrants reduces in the presence of silica nanopar-ticles within the PDMS membranes. Metal organic frameworks have also been sug-gested to enhance the separation performance in pervaporation for the recovery of bioalcohols (Liu et al. 2011).

Many more examples can be given in these fields and in newer application fields, such as forward osmosis, but the general trend is that nanoparticles are used to enhance the permeability of liquids and gases, to maintain the permeability, and to improve the separation of the membranes, regardless of the membrane application. From the results of the above-mentioned studies and many other studies, it can be assumed that nanoparticles may indeed enhance membrane properties. The mecha-nisms leading to these effects, however, are less studied. It can be speculated that nanoparticles influence the morphology of the membranes, due to a shift in the ther-modynamics of the phase separation during the membrane synthesis. Unfortunately, this has not been elucidated or proven so far. Nevertheless, the many observations on the improved performance of membranes with added nanoparticles drive the need for further study of the effects that can be obtained by the addition of various nanopar-ticles to polymers used for synthesis of membranes, and the fundamental explanation of these effects. The aim of this chapter is to give an overview of various strategies for incorporating nanoparticles as part of a polymeric membrane, based on reported methods for the synthesis of mixed matrix membranes. The most prominent of these strategies are self-assembly, bulk addition, anchoring in/on the membrane surface, and layer-by-layer addition.

16.2   SELF-ASSEMBLY OF NANOPARTICLES ON THE MEMBRANE SURFACE

The most straightforward method for exploiting the functionalities of nanoparticles is to apply nanoparticles to the top layer of an existing membrane by self-assembly. A complete overview of this strategy and the potential for fouling mitigation is given by Kim and Van der Bruggen (2010). Self-assembly of nanoparticles is based on the immersion of a membrane or the top layer of a membrane into a dilute solution containing nanoparticles as a colloidal solution. After evaporation of the solvent, the nanoparticles remain on the membrane surface as a self-adhering thin layer, with a thickness that depends on the concentration of nanoparticles in the solution that has been used. No reaction other than spontaneous association of nanoparticles with the membrane polymer occurs. Of course, this is only possible for nanoparticles for which such interactions are possible, the most important being TiO_2. Such nanoparticles are of great interest for aqueous application, because of their hydrophilic and photocatalytic effect, and almost all applications in this method are intended for aqueous applications. Bae and Tak (2005) were the first to propose the self-assembly method for the addition of a thin layer of photocatalytic titania to a lab-made poly(ether)sulfone and a sulfonated poly(ether)sulfone membrane. They dipped the membranes in a solution of TiO_2 nanoparticles, synthesized *in situ* by a controlled hydrolysis of titanium tetraisopropoxide in 100 ml ethanol, to which they added drop-wise 900 ml of water. The *in situ* method has the advantage of allowing a precise control of the size of the nanoparticles, so that in general very small particle sizes can be obtained, 4–7 nm reported by Bae and Tak (2005). The attachment of the nanoparticles relies on hydrogen bonds and coordination of titanium, as shown in Figure 16.1. This mechanism was proposed by Luo et al. (2005), who followed a similar approach, although with a different method for synthesis of the nanoparticles, leading to a somewhat higher size range. A similar mechanism may take place for membranes containing -COOH groups (see Figure 16.2). Due to these

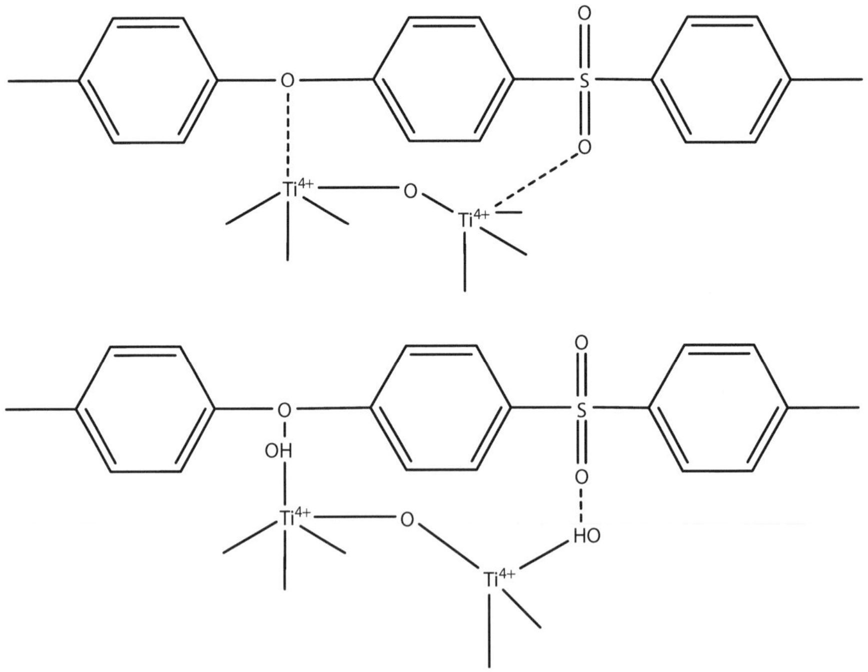

FIGURE 16.1 Mechanism of self-assembly of TiO_2 nanoparticles on a polyethersulfone membrane. Coordination of titanium with the ether bond is shown above; hydrogen bonding is shown below. (Adapted from Luo, M.-J. et al., *Appl. Surf. Sci.* 249, 76–84, 2005.)

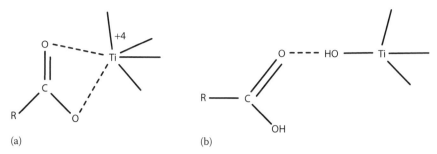

(a)   (b)

FIGURE 16.2   Self-assembly of TiO$_2$ nanoparticles on the membrane surface. (a) Bidentate coordination of carboxylate to Ti^{4+} and (b) H-bond between a carbonyl group and a surface hydroxyl group of TiO$_2$.

mechanisms, the procedure can be applied to a wide range of membranes, as shown by Bae et al. (2006), who applied the same procedure to polyacrylonitrile and PVDF membranes. Li et al. (2009b) made a blend of poly(styrene-alt-maleic anhydride) and poly(vinylidene) difluoride and also applied a similar procedure for self-assembly.

Bae and Tak applied a rather high concentration of TiO$_2$ in the solution (1 wt%); other authors later applied much lower concentrations. For example, Mansourpanah et al. (2009) used concentrations between 0.01 and 0.03 wt% TiO$_2$ for the functionalization of an unmodified membrane made of a blend of poly(ether)sulfone and polyimide, which was intended for NF, and an -OH functionalized membrane made of the same blend. Apart from the concentration, the procedure also depends on the dipping time; a longer dipping time results in more titania on the membrane surface. Because there is a large variety of dipping times in reported studies, it is very difficult to compare the results. Longer dipping times may even lead to a negative effect on the permeability because the nanoparticles may block the pores of the membrane. This was demonstrated by Rahimpour et al. (2008), who made a comparison of dipping times, ranging from 15 to 60 min.

The expected outcome of the self-assembly procedure is twofold. On the one hand, it is assumed that the nanoparticles make the membrane more hydrophilic, which would increase the permeability, and protect the membranes against flux decline due to fouling effects. On the other hand, the increase of the hydrophilic character is proven, but the improvement of the hydrophilicity seems to be compromised by the pore blocking effects, which were observed by Rahimpour et al. (2008). Therefore, the beneficial effects on permeability are not always observed. However, one may also consider another functionality of TiO$_2$, that is, its photocatalytic activity. This may be an underlying reason for the self-assembly approach, because it applies the nanoparticles on top of the surface of the membranes and, therefore, very accessible for light sources. This could open up a wide range of applications, in which photocatalytic degradation and filtration are combined, or even for organic synthesis, in a membrane reactor with simultaneous removal of the reaction product. Unfortunately, such applications cannot be found in the literature yet, which is mainly due to the long-term stability of the polymers when exposed to UV irradiation. Furthermore, the stability of the attachment of nanoparticles to the membrane may be problematic.

There are some indications that the self-assembly method yields sufficiently stable structures. Kwak et al. (2001) used a self-assembly method for TiO_2 nanoparticles with a size of 2 nm for aromatic polyamide reverse osmosis membranes, in view of flux enhancement; they applied X-ray photoelectron spectroscopy to observe the stability of the attachment of the nanoparticles and concluded that the adsorbed particles were sufficiently tightly bound to the surface to withstand various washing procedures and even harsh reverse osmosis-operating conditions.

In spite of the findings of Kwak et al. (2001), the long-term stability of the membrane association of the nanoparticles remains an issue. Other disadvantages of the self-assembly method are mainly related to the limitation in the use of nanoparticles that have sufficient interaction with the membrane polymer; in practice, there is a strong focus on TiO_2 as the only material that should be considered. Other metal oxides should at least partly have the same effect, although they have not been explored much, and it remains a question as to whether a method based on self-assembly would provide enough stability.

16.3   BULK ADDITION OF NANOPARTICLES

Bulk addition of nanoparticles refers to the use of nanoparticles as additives during the membrane synthesis process by phase inversion (Kim and Van der Bruggen 2010). Nanoparticles are dispersed in the polymer solution, which is then cast on a support layer and contacted with a nonsolvent, in the case of diffusion- or nonsolvent-induced phase separation (DIPS or NIPS). In the eventual membrane, the nanoparticles are present in the inner structure of the membrane, and not exclusively on the membrane surface. Therefore, the functionalities of the nanoparticles that were used can only be partly exploited. For example, catalytic activities would not be efficient, as the nanoparticles that should act as catalyst are shielded by the polymer material. This is a concern particularly for photocatalytic materials such as TiO_2, which are often employed for mixed matrix membranes.

On the other hand, the presence of hydrophilic nanoparticles in the bulk of the polymer might have a beneficial effect on water permeation, for aqueous applications. Hence, the entire pathway of water molecules permeating through the membrane is made more hydrophilic, so that one should expect an enhanced effect on the water flux. This, at least, is what would be the case if the changes in the morphology of the membrane are not considered, an aspect that is unfortunately not yet described in much detail. Tarabara (2009) has made the most complete analysis of the morphologic effects of additives in membrane synthesis. He outlines that the presence of nanoparticles affects the surface porosity of the membrane, the pore size of the skin layer, and the macrovoid morphology of the asymmetric support layer, among several other effects that may or may not influence transport. The thermodynamics of phase inversion would be significantly affected by the presence of nanoparticles. In DIPS/NIPS, the morphology of a membrane is defined by the shift in composition of the three-component system. This shift is visualized in the diagram of Figure 16.3. The left side of this figure shows the composition shift for delay demixing, in which there is a gradual change while remaining in the one-phase region, as opposed to the composition shift for instantaneous demixing (Figure 16.3b), where the two-phase region is reached

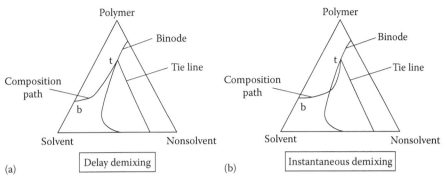

FIGURE 16.3 Schematic representation of delay demixing (a) and instantaneous demixing (b) in the synthesis of membranes by NIPS. The composition path as well as the position of the binode may change due to the presence of nanoparticles.

rapidly, followed by a further shift until the end point is reached. The influence of various parameters applied in the synthesis on the occurrence of macrovoids was already described for cellulose acetate membranes by Smolders et al. (1992). In general, membranes in which the phase transition occurs by delayed demixing have no macrovoids. These are formed when the phase transition occurs by instantaneous demixing. Local conditions of delayed demixing, due to processes such as the formation of nuclei by a change in the interfacial composition at the advancing coagulation front, may result in a shift in the demixing process. This may be induced by the presence of nanoparticles in the system, because they may hinder the diffusion pathway of solvents and nonsolvents. This should change the morphology of the membrane, probably in the sense that more macrovoids are formed, which may be smaller in size. Furthermore, the phase equilibrium itself may be influenced; the exact location of the two-phase region can shift due to the presence of nanoparticles, which can be assumed to have a stabilizing role in the solution. This would reduce the two-phase region, which promotes delayed demixing. The precise balance of these opposite effects, however, is still unclear.

Nasir et al. (2013) underline the mechanical, thermal, and chemical stability of mixed matrix membranes and conclude that they offer a high permeability and selectivity in gas separation, often better than pure polymeric materials. However, they also remark that the performance of such membranes is still below industrial expectations because of membrane defects and related processing problems, as well as the nonuniform dispersion of fillers in the membranes.

In addition to these thermodynamic and physico-chemical effects, the presence of nanoparticles may have a physical effect by reducing the compaction of the membranes (Tarabara 2009). This would be a secondary beneficial effect. The effect on the mechanical strength of the membranes, however, remains unclear to date, although some studies (Wu et al. 2008) observed a decrease of the mechanical strength.

Among the various types of nanoparticles applied in this way, TiO_2 is again dominant. For example, Wu et al. (2008) used this procedure for the addition of commercial rutile TiO_2 nanoparticles, with a reported size of 30 nm in the concentration range of 0–0.7 wt%. They reported problems with aggregation, which made a surface modification of TiO_2 nanoparticles necessary, which increased the dispersibility in the casting solution. Yang et al. (2007) estimated the upper concentration limit that can be applied before aggregation takes place and recommended values below 0.5 wt%. Several studies, however, reported the use of higher concentrations, such as Li et al. (2009a), who studied the concentration range between 0% and 15% of TiO_2 added to a solution of 15% PES in *N,N* dimethylacetamide and diethylene glycol as the solvent. Most studies report a beneficial effect on the water permeability and fouling mitigation, but the effect becomes negative about the threshold value where aggregation of nanoparticles caused pore blocking. Sotto et al. (2011) demonstrated that the largest beneficial effect was obtained at ultralow concentrations of TiO_2, below 0.1 wt% (Sotto et al. 2011). For a more complete overview of the potential in aqueous applications, the reader should refer to Kim and Van der Bruggen (2010). An in-depth review of the use of mixed matrix membranes for gas separation is given by Nasir et al. (2013).

Other nanoparticles used in mixed matrix membranes through a similar procedure of bulk addition include Al_2O_3, Ag, Au, carbon nanotubes (CNTs), SiO_2, and ZiO_2 and WS_2 for a variety of applications. Aqueous applications are dominant, and are typically related to flux enhancement, but nanoparticles may also be useful to improve the selectivity in processes including gas separation and pervaporation. The following list identifies some key applications of nanoparticles used in mixed matrix membranes formed by bulk addition:

- Al_2O_3 may be used to increase the hydrophilicity and mechanical strength of hydrophobic membranes such as those made of PVDF (Yan et al. 2006).
- Ag is a typical antifouling agent, which may be added by an *ex situ* synthesis method followed by addition to the casting solution, or by an *in situ* reduction of ionic silver by the polymer solvent (Taurozzi et al. 2008). The bactericidal effect depends on the release of ionic silver from the membrane, which is necessarily limited in time.
- Au can be used for local heating of the membrane using microwave energy, which may increase the permeability when a suitable setup is used (Li et al. 2013).
- CNTs are proposed for use in membranes by bulk addition for several functionalities, including conductivity and an improvement of the membrane morphology (Van der Bruggen 2012). In pervaporation, CNTs may yield a better selectivity by an improvement of the micromorphology (Choi et al. 2009). A review is given by Ismail et al. (2009) and by Van der Bruggen (2012).
- SiO_2 is applied for flux enhancement, as reported by Jadav and Singh (2009), and assumed to be caused by an increase in pore size, which is in line with the statements made above on membrane morphology.
- ZiO_2 has been used as a bulk material in mixed matrix polysulfone membranes, although in the size range of micrometers (Genné et al. 1996).
- Zeolites are known to enhance the fluxes and selectivities for membranes used in gas separation and pervaporation (Rezakazemi et al. 2012).
- WS_2 is a new material in membrane technology and may allow for enhanced permselectivity and fouling resistance; increased membrane lifetime is also speculated (Lin et al. 2013).

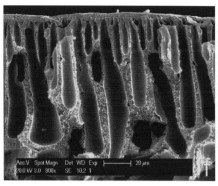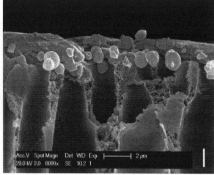

FIGURE 16.4 SEM images of the cross section of PES membranes enhanced with WS$_2$ nanoparticles (0.25% [WS$_2$/PES ratio]). (Reprinted from *J. Coll. Interf. Sci.*, 396, Lin, J.Y. et al., Novel PES membrane reinforced by nano-WS$_2$ with ultra-low concentration for enhanced permselectivity and improved fouling resistance, 120–128, Copyright 2013, with permission of Elsevier.)

Morphology of nanoparticles in the bulk of a polymer is shown in Figure 16.4 for WS$_2$ (Lin et al. 2013). It can be seen that the nanoparticles are mainly present in between finger-like macrovoids and particularly on top of these macrovoids, in the thin separating layer of the membrane.

16.4 ANCHORING IN/ON THE MEMBRANE SURFACE

Anchoring nanoparticles in/on the membrane surface refers to the use of specific anchoring groups or external forces to firmly fix nanoparticles in or on the membrane surfaces. One of the anchoring methods is the introduction of functional groups on the membrane surface in order to form chemical bonds with nanoparticles. For example, Mansourpanah et al. (2009) immersed PES/PI blend membranes in an aqueous solution containing DEA to induce –OH groups by nucleophilic and electrophilic addition in order to anchor TiO$_2$ nanoparticles on the membrane surface. This method was applied to improve the adhesion of TiO$_2$ nanoparticles and the hydrophilicity of the membrane surface. However, the main problems encountered with this anchoring method are that the modification method has to be carefully chosen for the specific membrane type, and there is a potential risk for the decomposition of the membrane surface.

Another anchoring method commonly used in practice is to incorporate nanoparticles into the top skin layer that is coated over a membrane surface. This method is usually employed in thin film composite (TFC) membrane synthesis. According to a general method for producing TFC membranes, a porous support membrane, such as a polysulfone UF membrane, is soaked in an aqueous solution containing amine monomers and then contacted with a water-immiscible solvent solution containing an acid reactant, such as a triacid chloride in hexane. The amine and acid chloride react at the interface of the low immiscible solutions to form an ultra-thin densely cross-linked PA layer on the membrane surface (Buonomenna 2013). The advantage of TFC membranes is that each layer can be optimized independently to achieve the maximum mechanical strength, chemical and thermal stability, and the desired separation performance. However, membrane fouling, demand of high water permeability for energy savings, and controlled selectivity in membrane design remain constraints of the TFC technology. In order to overcome these constraints, researchers attempted to incorporate functional nanomaterials into the TFC membranes.

Hoek et al. (2005) reported a new concept of TFN membranes (see Figure 16.5). In this new method, nanoparticles are predispersed either in the amine aqueous solution (Amini et al. 2013) or in the acid organic phase (Jeong et al. 2007) to form the nanocomposite PA top skin layer. By incorporating the nanoparticles inside the PA layer, the nanoparticles not only get firmly fixed by the PA skin layer but also influence the performance of the final membrane. Similarly, Jeong et al. (2007) used the same method to integrate NaA zeolite nanoparticles in the top skin layer. The final TFN membranes showed smoother and more hydrophilic, negatively charged surfaces. The pure water permeability was nearly twice that of the original TFC membrane with equivalent solute rejections at the highest nanoparticle loading. It was explained that the size of the zeolite molecular sieve particles matched the expected PA layer thickness and provided a flow path for water transport. Lind et al. (2009a) further compared the performance of zeolite

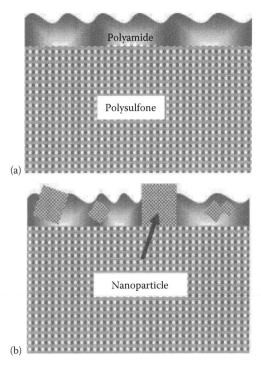

FIGURE 16.5 Conceptual illustration of (a) TFC and (b) TFN membrane structures. (Reprinted from *J. Memb. Sci.*, 294, Jeong, B.H. et al., Interfacial polymerization of thin film nanocomposites: A new concept for reverse osmosis membranes, 1–7, Copyright 2007, with permission of Elsevier.)

TFN membranes formed with sodium and silver cations exchanged within zeolite molecular sieve nanocrystals. The nanocomposite thin films containing zeolite nanocrystals in the silver form (AgA) exhibited more hydrophilic and smooth membrane surfaces and a better increase in water permeability than those in the sodium form (NaA). Although the AgA nanocrystals exhibited significant bactericidal activity, the shelter provided by the PA thin films greatly reduced the antimicrobial activity. Later on, the influence of the zeolite molecular sieve particle size on the TFN membrane performance was studied (Lind et al. 2009a). Smaller zeolites yielded better permeability enhancements, and larger zeolites gave more favorable surface properties. Recently, Pendergast et al. (2013) presented similar conclusions as those given by Jeong et al. (2007) and Lind et al. (2009b). In addition, it was reported that the incorporation of zeolite nanoparticles produced less cross-linked PA films than the pure PA films. The resistance to physical compaction was improved by embedment of nanoparticles inside the PA films. However, a clear fundamental insight in the mechanism of this behavior was not provided.

In addition to zeolite nanoparticles, other nanomaterials, such as TiO_2, silver, silica, and CNTs, incorporated in TFN membranes have been evaluated as well. As mentioned, the main purpose of incorporating TiO_2 into membranes is the hydrophilicity and the photocatalytic activity, which can significantly improve the antifouling performance and achieve an efficient membrane process. Although the self-assembly method can achieve the best performance of exposed TiO_2, the binding strength becomes the main drawback of this method. Incorporating TiO_2 nanoparticles into top skin layers is a reliable strategy to provide strong binding forces (Lee et al. 2008). However, the entirely embedded TiO_2 nanoparticles do not yield the same performance of high hydrophilicity and good photochemical reactivity as the exposed bare nanoparticles because of the shielding provided by the PA film. Additionally, above the critical concentration (5.0 wt%), the nanocomposite thin film was easily peeled off, which resulted in loss of the filtration performance (Lee et al. 2008).

Silver TFN membranes can maintain the bactericidal activity of silver nanoparticles. The silver composite thin film had a less compact structure and a rougher surface than the pure PA film (Mollahosseini and Rahimpour 2013). The silver TFN membranes were shown to have a greater antibiofouling performance, but the release of Ag is likely to influence the length of antibacterial and biofouling resistance performance (Lee et al. 2007b; Mollahosseini and Rahimpour 2013).

Silica nanoparticles incorporated in TFN membranes could significantly modify the polyamide network structure. The numbers of pores and the pore size increased with increasing silica content. The best membrane performance was obtained with a certain amount of silica loading. Higher concentration of silica resulted in a thicker PA film and exhibited superior thermal stability than the pure TFC membrane (Jadav and Singh 2009).

The CNTs in TFN membranes were reported to induce an enhanced permeate flux (Baroña et al. 2013). This is due to the atomic-scale smoothness of the nanotube walls enabling frictionless flow of fluids and molecular-ordering phenomena inside the nanopores, as exploited by Majumder et al. (2005) and Sholl and Johnson (2006). Due to the low hydrophilicity and the insolubility in solvent of CNTs, the use of functionalized nanotubes is suggested in most of the membrane modifications in order to improve CNT performance in membrane technology (Amini et al. 2013). After functionalization by hydrophilic groups, the CNTs enhanced the permeability of TFN membranes and maintained the high rejection of divalent ions (Amini et al. 2013; Baroña et al. 2013; Shen et al. 2013). Most of the explanations for the flux enhancement refer to the nanochannels provided by CNTs. However, considering the difficulties in longitudinally arranging the nanotubes, the nanochannel effect seems difficult to achieve. It is hard to distinguish whether the enhancement of permeability is due to the nanochannel effect or just because of the improved hydrophilicity of the membrane surfaces.

Another newly developed anchoring method to solve the fixation and the shielding problems arising in the self-assembly and TFN methods was proposed by Zhang et al. (2013). The robust binding performance is achieved by a mussel-inspired glue polydopamine. Mussels can attach on almost any surface type such as rocks, pilings, aquatic plants, and floating debris using strong adhesive byssus threads. This adhesive property is attributed to a special chemical dopamine (Lee et al. 2007a). Inspired by this function, dopamine was utilized as a versatile coating material and has been proven to self-polymerize on most surfaces of bulk materials in base conditions (Lee et al. 2007a; Waite 2008). Besides the simply coating applications, Zhang et al. (2013) found that dopamine is capable of self-polymerizing on the membranes as well as on the surfaces of nanoparticles. Mechanism of this behavior is explained in Figure 16.6. This robust link between the bulk membranes and the nanoparticles enables the nanomaterials to achieve the best surface properties.

The procedure of forming the nanoparticle–dopamine–membrane connection is simple. The dry membrane is contacted with fresh dopamine hydrochloride Tris buffer solution (pH around 8.5). After a certain deposition time, the solution is poured out. The soaked membrane is then contacted with the TiO_2 nanoparticles in a Tris buffer solution. The residual dopamine monomers on the membrane surface can further polymerize between the membrane surface and the nanoparticles, which forms a strong link between them. This new anchoring method can avoid the aggregation of nanoparticles and form a uniform coverage of nanoparticles throughout the membrane surfaces. The quantity of TiO_2 nanoparticles can be controlled by verifying the concentration of the nanoparticles suspension and the polymerization time of dopamine. In view of the filtration performance of membranes, a shorter polymerization time and a larger quantity of nanoparticles are more favorable. Because the shielding effect of the membrane matrix is greatly reduced in this anchoring method, the membrane modified by polydopamine-TiO_2 showed a remarkable antifouling performance with low static BSA surface adhesion, low additional membrane resistance, and less relative flux decline compared to the virgin membrane and the membranes modified by polydopamine or TiO_2.

Additionally, the dopamine modified TiO_2 is reported to absorb the visible light spectrum and can be excited by visible light energies below 3.2 eV (~387.5 nm) (Dimitrijevic et al. 2009). This property enables the polydopamine anchoring method to achieve visible light induced photocatalytic activity when it is used to anchor TiO_2. However, this visible light photocatalysis may degrade the surface structures of membranes. With these advantages and drawbacks, this polydopamine anchoring method for TiO_2 is suggested to be applied in dark conditions in order to achieve the best and

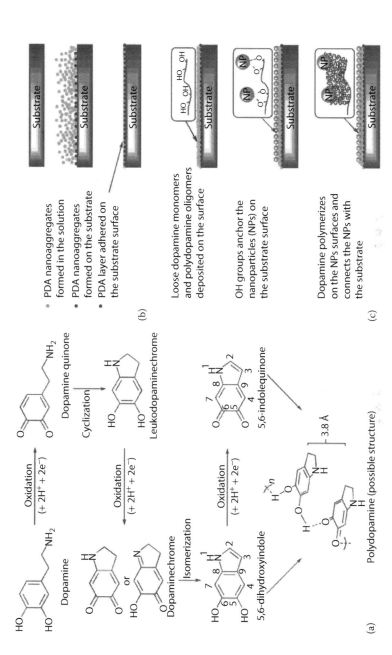

FIGURE 16.6 (a) First steps of self-polymerization of dopamine in aqueous solution; the resultant polydopamine is proposed to be comprised of intra and interchain noncovalent interactions, including charge transfer, π-stacking, and hydrogen bonding interactions; (b) possible deposition mechanism of polydopamine on the membrane surface; and (c) possible mechanism of binding TiO$_2$ nanoparticles onto the membrane surface. (Reprinted from *J. Memb. Sci.*, 437, Zhang, R.X. et al., Novel binding procedure of TiO$_2$ nanoparticles to thin film composite membranes via self-polymerized polydopamine, 179–188, Copyright 2013, with permission of Elsevier.)

long-term antifouling performance. If the function of photocatalysis is needed, an antioxidizing reagent should be added to the polydopamine solution or on the membrane surfaces. Because the anchoring method is newly developed, there are still many questions needed to be answered. For instance, when anchoring different nanomaterials or changing the conditions of self-polymerization, the binding, filtrating, and antifouling performance of the final modified membranes remain to be studied.

16.5 LAYER-BY-LAYER ADDITION OF NANOPARTICLES

Fabricating polymeric membranes with nanoparticles can be achieved by embedding the nanoparticles into a polymeric membrane or by coating of the membrane. As described in Section 16.3, nanoparticles as additives in the casting solution may result in difficulties to control their functionalities because they are embedded into the polymeric matrix. Embedding nanoparticles into polymeric membranes can shield the intrinsic effects of the nanoparticles, thereby limiting their functionality not only for membrane properties but also for water treatment or gas separation. The layer-by-layer deposition was developed to coat indium doped in oxide electrodes with Fe_2O_3 and TiO_2 particles (McKenzie et al. 2002). The layer-by-layer or directed assembly method was also proposed for the formation of nanofilm deposits based on a two-component dip coating technique and is considered as an established method for producing layered structures of nanoparticles on a membrane surface in a cost-effective way (Wang et al. 2010). The nanoparticles can be immobilized on the membrane surface in a multilayer fashion using covalent and electrostatic binding forces between a molecular binder and the nanoparticles, as shown in Figure 16.7. In general, the prepared membrane is immersed in the polyelectrolyte solution and rinsed with deionized water. Subsequently, the membrane is immersed in a stable colloidal solution consisting of nanoparticles with opposite charge to the polyelectrolyte and then rinsed with deionized water. This procedure can be repeated to obtain the desired number of coating layers of the nanoparticles on the membrane surface.

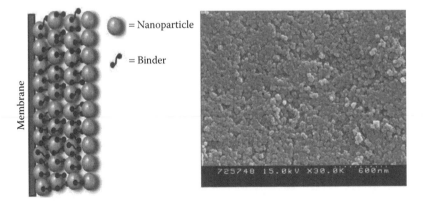

FIGURE 16.7 Schematic representation of nanoparticles-coated membrane (left) and SEM surface image of TiO_2-deposited membrane prepared by the layer-by-layer method (right). (Adapted from Kim, J.H. et al., *Micropor. Mesopor. Mater.*, 173, 121–128, 2013.)

By building up coating layers of nanoparticles on the membrane surface, the novel properties of nanoparticles such as catalytic activity or electrical conductivity can be used effectively for enhancing membrane surface properties. In addition, the hydrophilicity of the membrane can be augmented by increasing the number of coating layers composed particularly of hydrophilic nanoparticles.

For layer-by-layer assembly, binders are generally employed to link a stable coating layer of nanoparticles on the membrane. Because the deposition of nanoparticles by adsorption can be limited to a monolayer, the deposition process using nanoparticles and molecular binder needs to be continued to form a multilayer of nanoparticles on the membrane (McKenzie et al. 2002). Therefore, the selection of appropriate binders is important to form a stable coating layer. Polyelectrolyte solutions have been widely applied as a *linker molecule* to bind nanoparticles, but other organic binders can be employed depending on the type of nanoparticles and membranes. The formation of active membranes by employing layer-by-layer addition of nanoparticles was demonstrated for the case of Fe_2O_3 or TiO_2 nanoparticles with organic binders such as phytic acid (Karnik et al. 2006; Kim et al. 2013). Various polyanions such as poly(acrylic acid) and poly(styrene sulfonate) or polycations such as poly(ethyleneimine) (PEI) and poly(allylamine hydrochloride) have also been applied to bind nanoparticles (Wang et al. 2010). The layer-by-layer method was employed to coat nylon fibers with silver nanoparticles to investigate antimicrobial activity. For this study, cationic poly(diallyldimethyl ammonium chloride) was applied as molecular binder to deposit anionic poly(methacrylic acid) capped silver nanoparticles on fibrous materials (Dubas et al. 2006). A deposition procedure to coat polymeric membrane with nanoparticle can be applied, but the dipping method has been widely adopted in a layer-by-layer approach. Filtration was also applied to fabricate a coating layer of nanoparticles by controlling the volume of particle dispersion through membrane (Hirsjärvi et al. 2006). However, the desired pressure to be applied to the filtration cell is still unknown to avoid passage of the nanoparticles through the membrane and to avoid a disorder of the structured layer due to convective flows. The deposition method, on the other hand, can yield an unstable membrane in the sense that nanoparticles are gradually released to the solution in contact with membrane (Kim and Van der Bruggen 2010). Immobilization of nanoparticles on the membrane surface is a critical issue to be resolved for the layer-by-layer method, as the release of nanoparticles from the membrane surface may decrease the membrane performance and raise environmental safety concerns (Kim and Van der Bruggen 2010).

A stable permeate flux can be maintained through any number of coating layers, but the optimum number of layers should be determined. Both membrane permeability and functionalities such as antifouling properties can vary depending on the number of coating layers of nanoparticles. In the case of ceramic membranes, Karnik et al. (2006) reported that an iron oxide nanoparticle-based ceramic membrane was not damaged by the coating process. The number of coating layers of iron oxide nanoparticles on the TiO_2/Al_2O_3 composite membrane was found to have little effect on the membrane permeability. For the polymeric materials, the membrane functionalities, such as antibacterial effects, increased proportionally with the number of coating layers of silver nanoparticles (0, 10, and 20 layers). In addition, the permeability of a PES membrane coated by TiO_2 nanoparticles with the layer-by-layer method increased almost exponentially, as the number of coating layers increased from 1 to 20 (Kim et al. 2013). The layer-by-layer coating method has a great potential to fabricate functional membranes using nanoparticles as compared to self-assembly method, as indicated in Section 16.2, because this coating technique can form a deposited layer consisting of nanoparticles in a more organized manner, thereby improving their functionality. However, further study is required to determine the optimum number of coating layers along with the concentration of nanoparticles for the best performance of nanoparticle-coated membranes fabricated by the layer-by-layer technique. In addition, a more robust coating procedure is needed to immobilize nanoparticles on membranes in long-term operation, while maintaining their intrinsic properties.

16.6 CONCLUSIONS

Nanoparticles have a proven potential as constituents in membrane synthesis to improve the membrane performance. Several methods have been applied to produce mixed matrix membranes with nanoparticles with variable success. Part of the uncertainties might be related to the unknown thermodynamics of the systems for some methods, in particular the bulk addition method. The self-assembly method may be insufficiently stable in long-term operation, but this can be improved using the layer-by-layer approach. A promising alternative is the approach in which nanoparticles are anchored on the membrane surface, as found in thin-film nanocomposites or by adhering nanoparticles using polydopamine.

REFERENCES

Ahmad, J. and Hägg, M.B. 2013. Polyvinyl acetate/titanium dioxide nanocomposite membranes for gas separation. *J. Memb. Sci.* 445, 200–210.

Amini, M., Jahanshahi, M., and Rahimpour, A. 2013 Synthesis of novel thin film nanocomposite (TFN) forward osmosis membranes using functionalized multi-walled carbon nanotubes. *J. Memb. Sci.* 435, 233–241.

AWWA Membrane Technology Research Committee. 2005. Recent advances and research needs in membrane fouling. *J. AWWA* 97(8), 79–89.

Bae, T.-H., Kim, I.-C., and Tak, T.-M. 2006. Preparation and characterization of fouling resistant TiO_2 self-assembled nanocomposite membranes. *J. Memb. Sci.* 275, 1–5.

Bae, T.H. and Tak, T.M. 2005. Effect of TiO_2 nanoparticles on fouling mitigation of ultrafiltration membranes for activated sludge filtration. *J. Memb. Sci.* 249, 1–8.

Baroña, G.N.B., Lim, J., Choi, M., and Jung, B. 2013. Interfacial polymerization of polyamide-aluminosilicate SWNT nanocomposite membranes for reverse osmosis. *Desalination*, 325, 138–147.

Braghetta, A., DiGiano, F.A., and Ball, W.P. 1997. Nanofiltratioin of natural organic matter: pH and ionic strength effects. *J. Env. Eng.* 123(7), 628–641.

Buonomenna, M.G. 2013. Nano-enhanced reverse osmosis membranes. *Desalination*, 314, 73–88.

Cho, J., Amy, G., and Pellegrino, J. 2000a. Membrane filtration of natural organic matter: Factors and mechanisms affecting rejection and flux decline with charged ultrafiltration (UF) membrane. *J. Memb. Sci.* 164, 89.

Cho, J., Sohn, J., Choi, H., Kim, I.S., and Amy, G. 2000b. Effects of molecular weight cutoff, f/k ratio (a hydrodynamic condition) and hydrophobic interactions on natural organic matter rejection and fouling in membranes. *J. Water Supply Res. Technol.-AQUA* 51, 109–123.

Choi, J.H., Jegal, J., Kim, W.N., and Choi, H.S. 2009. Incorporation of multiwalled carbon nanotubes into poly(vinyl alcohol) membranes for use in the pervaporation of water/ethanol mixtures. *J. Appl. Polym. Sci.* 111(5), 2186–2193.

Cortalezzi, M.M., Roseb, J., Barronc, A.R., and Wiesnerd, M.R. 2002. Characterization of ceramic membranes derived from alumoxane nanoparticles. *J. Memb. Sci.* 205, 33.

Dimitrijevic, N.M., Rozhkova, E., and Rajh, T. 2009 Dynamics of localized charges in dopamine-modified TiO_2 and their effect on the formation of reactive oxygen species. *J. Am. Chem. Soc.* 131, 2893–2899.

Dubas, S.T., Kumlangdudsana, P., and Potiyaraj, P. 2006. Layer-by-layer deposition of antimicrobial silver nanoparticles on textile fibers. *Colloids Surf. A: Physicochem. Eng. Aspects* 289, 105–109.

Genné, I., Kuypers, S., and Leysen, R. 1996. Effect of the addition of ZrO_2 to polysulfone based UF membranes. *J. Memb. Sci.* 113, 343–350.

Hassanajili, S., Masoudi, E., Karimi, G., and Khademi, M. 2013. Mixed matrix membranes based on polyetherurethane and polyesterurethane containing silica nanoparticles for separation of CO_2/CH_4 gases. *Sep. Purif. Technol.* 116, 1–12.

Hirsjärvi, S., Peltonen, L., and Hirvonen, J. 2006. Layer-by-layer polyelectrolyte coating of low molecular weight poly(lactic acid) nanoparticles. *Colloids Surf. B Biointerfaces* 49, 93–99.

Hoek, E.M.V., Jeong, B.H., and Yan, Y. 2005. Nanocomposite membranes and methods of making and using same, U.S. Application 11/364,885, Priority: US 60/660,428.

Ismail, A.F., Goh, P.S., Sanip, S.M., and Aziz, M. 2009. Transport and separation properties of carbon nanotube-mixed matrix membrane. *Separ. Purif. Technol.* 70, 12–26.

Jadav, G.L. and Singh, P.S. 2009. Synthesis of novel silicaepolyamide nanocomposite membrane with enhanced properties. *J. Memb. Sci.* 328, 257–267.

Jeong, B.H., Hoek, E.M.V., Yan, Y., Subramani, A., Huang, X., Hurwitz, G., Ghosh, A.K., and Jawor, A. 2007. Interfacial polymerization of thin film nanocomposites: A new concept for reverse osmosis membranes. *J. Memb. Sci.* 294, 1–7.

Jucker, C. and Clark, M.M. 1994. Adsorption of aquatic humic substances on hydrophobic ultrafiltration membranes. *J. Memb. Sci.* 97, 37–52.

Karnik, B.S., Baumann, M.J., Masten, S.J., and Davies, S.H.R. 2006. AFM and SEM characterization of iron oxide coated ceramic membrane. *J. Mater. Sci.* 41, 6861–6870.

Khan, A.L., Klaysom, C., Gahlaut, A., and Vankelecom, I.F.J. 2013. Polysulfone acrylate membranes containing functionalized mesoporous MCM-41 for CO_2 separation. *J. Memb. Sci.* 436, 145–153.

Kilduff, J.E., Mattaraj, S., Pieracci, J.P., and Belfort, G. 2000. Photochemical modification of poly(ether sulfone) and sulfonated poly (sulfone) nanofiltration membranes for control of fouling by natural organic matter. *Desalination* 132, 133–142.

Kim, J. and Van der Bruggen, B. 2010. The use of nanoparticles in polymeric and ceramic membrane structures: Review of manufacturing procedures and performance improvement for water treatment. *Environ. Pollut.* 158(7), 2335–2349.

Kim, J.H., Sotto, A., Chang, J., Nam, D., Boromand, A., and Van der Bruggen, B. 2013. Embedding and surface coating by layer-by-layer deposition of TiO_2 nanoparticles on nanoporous polymeric films. *Micropor. Mesopor. Mater.* 173, 121–128.

Kim, S.H., Kwak, S.Y., Sohn, B.H., and Pa, T.H. 2003. Design of TiO_2 nanoparticle self-assembled aromatic polyamide thin-film-composite (TFC) membrane as an approach to solve biofouling problem. *J. Memb. Sci.* 211, 157–165.

Kwak, S.Y., Kim, S.H., and Kim, S.S. 2001. Hybrid organic/inorganic reverse osmosis (RO) membrane for bactericidal anti-fouling. 1. Preparation and characterization of TiO_2 nanoparticle self-assembled aromatic polyamide thin-film-composite (TFC) membrane. *Environ. Sci. Technol.* 35(11), 2388–2394.

Lee, H., Dellatore, S.M., Miller, W.M., and Messersmith, P.B. 2007a. Mussel-inspired surface chemistry for multifunctional coatings. *Science* 318, 426–430.

Lee, H.S., Im, S.J., Kim, J.H., Kim, H.J., Kim, J.P., and Min, B.R. 2008. Polyamide thin-film nanofiltration membranes containing TiO_2 nanoparticles. *Desalination* 219, 48–56.

Lee, S.Y., Kim, H.J., Patel, R., Im, S.J., Kim, J.H., and Min, B.R. 2007b Silver nanoparticles immobilized on thin film composite polyamide membrane: Characterization, nanofiltration, antifouling properties, *Polym. Adv. Technol.* 18, 562–568.

Li, J.-F., Xu, Z.-L., Yang, H., Yu, L.-Y., and Liu, M. 2009a. Effect of TiO_2 nanoparticles on the surface morphology and performance of microporous PES membrane. *Appl. Surf. Sci.* 255, 4725–4732.

Li, J.H., Xu, Y.Y., Zhu, L.P., Wang, J.H., and Du, C.H. 2009b. Fabrication and characterization of a novel TiO_2 nanoparticle self-assembly membrane with improved fouling resistance. *J. Memb. Sci.* 326, 659–666.

Li, Y., Verbiest, T., and Vankelecom, I. 2013. Improving the flux of PDMS membranes via localized heating through incorporation of gold nanoparticles. J. Memb. Sci. 428, 63–69.

Lin, J.Y., Zhang, R., Ye, W., Jullok, N., Sotto, A., and Van der Bruggen, B. 2013. Novel PES membrane reinforced by nano-WS_2 with ultra-low concentration for enhanced permselectivity and improved fouling resistance. *J. Coll. Interf. Sci.* 396, 120–128.

Lind, M.L., Ghosh, A.K., Jawor, A., Huang, X., Hou, W., Yang, Y., and Hoek, E.M.V. 2009a. Influence of zeolite crystal size on zeolite-polyamide thin film nanocomposite membranes. *Langmuir* 25(17), 10139–10145.

Lind, M.L., Jeong, B.H., Subramani, A., Huang, X.F., and Hoek, E.M.V. 2009b. Effect of mobile cation on zeolite-polyamide thin film nanocomposite membranes. *J. Mater. Res.* 24, 1624–1631.

Liu, X.L., Li, Y.S., Zhu, G.Q., Ban, Y.J., Xu, L.Y., and Yang, W.S. 2011. An organophilic pervaporation membrane derived from metal-organic framework nanoparticles for efficient ecovery of bio-alcohols. *Angew. Chem. Int. Ed.* 50(55), 10636–10639.

Luo, M.-J., Zhao, J.-Q., Tang, W., and Pu, C.-S. 2005. Hydrophilic modification of poly(ether sulfone) ultrafiltration membrane surface by self-assembly of TiO_2 nanoparticles. *Appl. Surf. Sci.* 249, 76–84.

Magalad, V.T., Gokavi, G.S., Ranganathaiah, C., Burshe, M.H., Han, C., Dionysiou, D.D., Nadagouda, M.N., and Aminabhavi, T.M. 2013. Polymeric blend nanocomposite membranes for ethanol dehydration-effect of morphology and membrane-solvent interactions. *J. Memb. Sci.* 430, 321–329.

Majumder, M., Chopra, N., Andrews, R., and Hinds, B.J. 2005. Nanoscale hydrodynamics: Enhanced flow in carbon nanotubes. *Nature* 438, 44–45.

Mansourpanah, Y., Madaeni, S.S., Rahimpour, A., Farhadian, A., and Taheri, A.H. 2009. Formation of appropriate sites on nanofiltration membrane surface for binding TiO_2 photo-catalyst: Performance, characterization, and fouling-resistant capability. *J. Memb. Sci.* 330, 297–306.

McKenzie, K.J., Marken, F., Hyde, M., and Compton, R.G. 2002. Nanoporous iron oxide membranes: Layer-by-layer deposition and electrochemical characterization of processes within nanopores. *New J. Chem.* 26, 625–629.

Mollahosseinia, A. and Rahimpour, A. 2013 A new concept in polymeric thin-film composite nanofiltration membranes with antibacterial properties. *Biofouling* 29(5), 537–548.

Nasir, R., Mukhtar, H., Man, Z., and Mohshim, D.F. 2013. Material advancements in fabrication of mixed-matrix membranes. *Chem. Eng. Technol.* 36(5), 717–727.

Pendergast, M.T.M., Ghosh, A.K., and Hoek, E.M.V. 2013. Separation performance and interfacial properties of nanocomposite reverse osmosis membranes. *Desalination* 308, 180–185.

Rahimpour, A., Madaeni, S.S., Taheri, A.H., and Mansourpanah, Y. 2008. Coupling TiO_2 nanoparticles with UV irriadiation for modification of polyethersulfone ultrafiltration membranes. *J. Memb. Sci.* 313, 158–169.

Rezakazemi, M., Shahidi, K., and Mohammadi, T. 2012. Hydrogen separation and purification using crosslinkable PDMS/zeolite A nanoparticles mixed matrix membranes. *Int. J. Hydrog. Energ.* 37(19), 14576–14589.

Shen, J.N., Yu, C.C., Ruan, H.M., Gao, C.J., and Van der Bruggen, B. 2013. Preparation and characterization of thin-film nanocomposite membranes embedded with poly(methyl methacrylate) hydrophobic modified multiwalled carbon nanotubes by interfacial polymerization. *J. Memb. Sci.* 442, 18–26.

Shirazi, Y., Ghadimi, A., and Mohammadi, T. 2012. Recovery of alcohols from water using polydimethylsiloxane-silica nanocomposite membranes: Characterization and pervaporation performance. *J. Appl. Polym. Sci.* 124(4), 2871–2882.

Sholl, D.S. and Johnson, J.K. 2006. Materials Science: Making high-flux membranes with carbon nanotubes. *Science* 312, 1003–1004.

Smolders, C.A., Reuvers, A.J., Boom, R.M., and Wienk, I.M. 1992. Microstructures in phase inversion membranes. Part 1. Formation of macrovoids. *J. Memb. Sci.* 73, 259–275.

Sotto, A., Boromand, A., Balta, S., Kim, J., and Van der Bruggen, B. 2011. Doping of polyethersulfone nanofiltration membranes: Antifouling effect observed at ultralow concentrations of TiO_2 nanoparticles. *J. Mater. Chem.* 21(28), 10311–10320.

Taniguchi, M., Kilduff, J.E., and Belfort, G. 2003. Low fouling synthetic membranes by UV-assisted graft polymerization: Monomer selection to mitigate fouling by natural organic matter, *J. Memb. Sci.* 222, 59–70.

Tarabara, V.V. 2009. Multifunctional nanomaterial-enabled membranes for water treatment. In: Savage et al. (Eds.), *Nanotechnology Applications for Clean Water*. Norwich: William Andrew Inc., pp. 59–75.

Taurozzi, J.S., Arul, H., Bosak, V.Z., Burban, A.F., Voice, T.C., Bruening, M.L., and Tarabara, V.V. 2008. Effect of filler incorporation route on the properties of polysulfone-silver nanocomposite membranes of different porosities. *J. Memb. Sci.* 325, 58–68.

Van der Bruggen, B. 2012. The separation power of nanotubes in membranes: A review. *ISRN Nanotechnology*, Article ID 693485, 17 pp.

Waite, J.H. 2008. Surface chemistry: Mussel power. *Nat. Mater.*, 7, 8–9.

Wang, F., Alazemi, M., Dutta, I., Blunk, R.H., and Angelopoulos, A.P. 2010. Layer by layer assembly of hybrid nanoparticle coatings for proton exchange membrane fuel cell bipolar plates. *J. Power Sources* 195, 7504–7560.

Wang, Y., Kim, J.H., Choo, K.H., Lee.Y.S., and Lee, C.H. 2000. Hydrophilic modification of polypropyrene microfiltration membranes by ozone-induced graft polymerization. *J. Memb. Sci.* 169, 269–276.

Wavhal, D. and Fisher, E. 2002. Modification of porous poly(ethersulfone) membranes by low-temperature CO_2 plasma treatment. *J. Polym. Sci. B Polym. Phys.*, 40, 2473.

Wavhal, D. and Fisher, E. 2003. Membrane surface modification by plasma-induced polymerization of acrylamide for improved surface properties and reduced protein fouling. *Langmuir*, 19, 79.

Wu, G., Gan, S., Cui, L., and Xu, Y. 2008. Preparation and characterization of PES/TiO_2 composite membranes. *Appl. Surf. Sci.* 254, 7080–7086.

Yan, L., Li, Y.S., Xiang, C.B., and Xianda, S. 2006. Effect of nano-sized Al_2O_3-particle addition on PVDF ultrafiltration membrane performance. *J. Memb. Sci.* 276, 162–167.

Yang, Y., Zhang, H., Wang, P., Zheng, Q., and Li, J. 2007. The influence of nano-sized TiO_2 fillers on the morphologies and properties of PSF UF membrane. *J. Memb. Sci.* 288, 231–238.

Zhang, R.X., Braeken, L., Luis, P., Wang, X.L., and Van der Bruggen, B. 2013. Novel binding procedure of TiO_2 nanoparticles to thin film composite membranes via self-polymerized polydopamine. *J. Memb. Sci.* 437, 179–188.

17 Surface Modification of Inorganic Materials for Membrane Preparation

Dipak Rana, Takeshi Matsuura, and Ahmad Fauzi Ismail

CONTENTS

17.1 Introduction 589
17.2 Surface Modification of Inorganic Membranes 591
17.3 Conclusions and Future Work 612
References 612

17.1 INTRODUCTION

Membrane separation processes have recently attracted considerable attention because they offer an effective separation alternative to traditional energy-intensive separation techniques, such as distillation and cryogenic separation. Of particular interest are inorganic membranes that have advantages over polymeric membranes because of their small pore sizes as well as excellent thermal and chemical stabilities. However, the rigid structure of inorganic membranes does not allow an easy adjustment of the pore sizes unlike polymeric membranes. Fabrication of organic/inorganic composite membranes by the surface modification of inorganic substrates has addressed this issue, enabling the development of novel membranes for improved performance.

Surface modification of inorganic substrates with organic compounds has long been investigated in the fields of microelectronics, biotechnology, and separation science. Properties such as adhesion, wettability, biocompatibility, and adsorption affinity of the surface can be altered and tuned to the specific requirements by the chemical or physical modification of the surfaces. In particular, the modification of inorganic oxide surfaces with organosilanes, studied for a long time for the purpose of liquid chromatography, has significantly influenced the methods developed to modify the surface of inorganic membranes. Hence many studies on the modification of inorganic membrane substrates with organosilanes have appeared in the literature.

Surface modification affects membrane performance by fine-tuning the membrane pores in the following two ways. First, it reduces the membrane pore size. Second, it alters membrane surface chemistry, resulting in the change in the interaction between the penetrant molecule and the pore wall. By reducing the pore size, the separation of feed components is enabled by the sieve mechanism, whereas by altering the surface chemistry of the pore, molecular transport by the surface flow mechanism is strongly affected. Some examples of the surface modification of inorganic substrates are summarized in Table 17.1 for membrane separation processes including microfiltration (MF), ultrafiltration (UF), pervaporation (PV), gas separation, and fuel cell applications.

Another approach to develop organic/inorganic membranes is to blend inorganic filler particles in organic polymeric material. Membranes fabricated are called mixed-matrix membranes (MMMs). Development of MMMs has been facilitated particularly as a method to combine the two unique features of organic and inorganic materials–high processibility of organic polymers and high performance, in terms of high selectivity and high permeability, of inorganic materials. Probably, the first attempt of MMM fabrication was made by te Hennepe et al. in 1987 for the separation of alcohol and water by PV. They incorporated silicate-1 in a membrane made of rubbery polydimethylsiloxane (PDMS) to enhance the sorption of alcohols in the membrane over water. The success of their attempt was due to the favorable interaction between the filler molecular sieves and polymers of relatively low glass transition temperature (T_g) such as the rubbery polymer PDMS. However, practical gas separation membranes are prepared mostly from rigid matrix polymers with high T_g, and the formation of MMMs is difficult for such polymers because of poor adhesion between the polymer and the filler particles. Surface modification of inorganic particles was hence attempted to improve the filler/polymer interaction and to remove the void space between the filler particles and polymer matrix, as

TABLE 17.1
Surface Modification of Inorganic Substrates

Base Inorganic Membranes	Surface-Modifying Agents	Membrane Types	Processes	Sources
Al_2O_3 tubular membrane	Nanosized ZrO_2 coating	Asymmetric	Microfiltration	Zhou et al. 2010a
TiO_2 or porous steel	Polycarbon	Asymmetric	Liquid filtration	Soldatov et al. 2006
Mesoporous silica	Hexamethyl disilazane	Asymmetric	Pervaporation	Jin et al. 2011
MFI-zeolite	Tetraethoxysilane	Asymmetric	Gas separation	Tang et al. 2009
Mesoporous Vycor glass	Dimethyloctadecyl chlorosilane	Asymmetric	Gas separation	Singh et al. 2005
Al_2O_3	Carbon molecular sieve	Asymmetric	Gas separation	Wang and Hong 2005
Mesoporous silica	3-Aminopropyl-triethyloxysilane	Asymmetric	Gas separation	Ostwal et al. 2011
MFI zeolite	Methyldiethoxysilane	Asymmetric	Gas separation	Zhu et al. 2010
Porous glass	Organosilane	Asymmetric	Fuel cell	Kikukawa et al. 2005

the presence of the void space lowers the selectivity of the membrane considerably due to gas leakage. The main stream of the particle surface functionalization was, again, silylation. Recently, as the use of nanoparticles has become more fashionable, surface modification has become more important to prevent the agglomeration of nanoparticles. Some examples of MMM fabrication are summarized in Table 17.2 for PV, gas separation, and membrane absorption.

Thus, the objective of this chapter is to review the surface modification of inorganic materials for membrane development based on the above two approaches, emphasizing the progresses made recently in this field.

17.2 SURFACE MODIFICATION OF INORGANIC MEMBRANES

Surface of an alumina MF membrane was modified by depositing a ZrO_2 layer (Zhou et al. 2010a). First, Al_2O_3 tubular MF membranes were dried at 110°C for 12 h in an oven, and then immersed in $ZrCl_4$ alcohol solution for 24 h at room temperature. After being washed by absolute alcohol, Al_2O_3 membranes were dried at 60°C for 4 h and heated in hot water for 6 h. The membranes were then calcined at 600°C for 2 h at heating rate of 1°C/min. Although the alumina membrane was soaked in $ZrCl_4$ alcohol solution, $ZrCl_4$ chemically adsorbed on the alumina surface and during the following *in situ* hydrolysis and calcination process, $ZrCl_4$ transformed into the nano-ZrO_2 coating without morphological changes. Transformation of $ZrCl_4$ to ZrO_2 can be denoted as follows:

$$ZrCl_4 + 9H_2O = ZrOCl_2 \cdot 8H_2O + 2HCl$$

$$ZrOCl_2 \cdot 8H_2O \rightarrow ZrO_2 + 2HCl + 7H_2O$$

It was observed by transmission electron microscopy (TEM) that a layer of aggregated nanosized ZrO_2 of 100 nm thickness was formed on top of the Al_2O_3 support membrane with a large number of hydroxyl (–OH) groups at the surface, rendering the membrane more hydrophilic, as confirmed by the contact angle measurement.

Operation of the virgin and modified Al_2O_3 membrane was compared by the separation of an oil–water emulsion, which consisted of 20[#] engine oil (1 g/L), Tween 80 (0.5 g/L), Span 80 (0.5 g/L), and distilled water. Stable oil–water emulsion had an average droplet size of 1.79 μm, and 90% of the oil droplets were between 0.67 and 7.4 μm. Filtration tests were conducted at the transmembrane pressure of 0.16 MPa, feed flow rate of 5 m/s, and at an operating temperature of 30°C. The membrane was back-flashed at the interval of 10 min. For the virgin Al_2O_3 membrane, the flux declined sharply in the first 60 min, from 446 to 159 L/m^2 h. The steady flux was only about 30% of the initial flux, implying that the Al_2O_3 membrane was seriously fouled. However, the flux of the modified membrane quickly reached a constant in 10 min. The flux declined from 506 to 441 L/m^2 h, which is 88% of the initial flux.

In another example of the surface modification of inorganic membrane, the deposition of pyrocarbon crystallites was performed by methane pyrolysis at 800°C–1000°C on top of TRUMEM composite membranes (TiO_2 on porous steel) (Soldatov et al. 2006). According to the X-ray data, the lattice of pyrocarbon crystallites of size

TABLE 17.2

Surface Modification of Inorganic Fillers in MMMs

Fillers	Binding Agents	Polymer Matrixes	Membrane Types	Processes	Sources
MMT	PVP or NVP	PVDF	Asymmetric	Ultrafiltration	Wang et al. 2012
Hydrophobic silicate		PDMS	Homogeneous	Pervaporation	te Hennepe et al. 1987
Nanofused silica	$NH_2-C_3H_6-Si(OC_2H_5)_3$	PDMS or CA	Asymmetric	Pervaporation	Peng et al. 2011
Silicate-1	Vinyltrimethoxy-silane (VTMS)	PDMS	Homogeneous	Pervaporation	Zhou et al. 2010b
ZSM-5	N-octyltriethoxy-silane (OTES)	PVDF	Homogeneous	Pervaporation	Liu et al. 2011a
Silicate-1		PDMS	Homogeneous	Pervaporation and gas separation	Jia et al. 1992
Silica	Mercaptosilane	PDMS	Asymmetric	Organic gas separation	Kim et al. 2009
Silica	$Si-C_8H_{17}, Si-C_{16}H_{33}$	PTMS		Organic gas separation	Gomes et al. 2005
Zeolite 4A	Plasticizer RDP Fyroflex® Di-butyl phthalate 4-Hydroxy benzophenone	Matrimid	Homogeneous	Gas separation	Mahajan et al. 2002
			Homogeneous and asymmetric	Gas separation	Chung et al. 2007
Zeolite-5	APTS	Polyimide	Homogeneous	No separation	Vankelecom et al. 1996
Modified zeolites (ZSM-2)	APTS	6FDA-6FpDA-DABA	Homogeneous	Gas separation	Pechaf et al. 2002
Benzylamine-modified C_{60}		Matrimid 5218	Homogeneous	Gas separation	Chung et al. 2003
BEA	Organosilane	Kraton G 1652	Homogeneous	Gas separation	Sirikittikul et al. 2008
Zeolite	APDEMS	Polyethersulfone	Homogeneous	Gas separation	Li et al. 2006

			Asymmetric		
Zeolite	3,5-DABA 3-APTMS	Polyimide	Asymmetric	Gas separation	Boroglu and Gurkaynak 2011
LTA, MFI	MgO_xH_y	Matrimid	Homogeneous	Gas separation	Lydon et al. 2012
Zeolite 4A	$Mg(OH)_2$ of whisker like morphology, thionyl chloride and Grignard reagent	–	Homogeneous	No separation	Shu et al. 2007a
Same as above	Same as above		Homogeneous		Shu et al. 2007b
Zeolite MFI	$Mg(OH)_2$ deposited solvothermally	Ultem Matrimid		Gas separation	Bae et al. 2009
Zeolite MFI and LTA	$Mg(OH)_2$ deposited either by Grignard treatment or solvothermally	Ultem	Homogeneous	No separation	Bae et al. 2011
Zeolite 5A	$Mg(OH)_2$ by sol-gel method	Polyimide 6FDA-DAM	Homogeneous	No separation	Liu et al. 2011b
MFI-zeolite	Grignard agent or salotherma approach	Polyvinyl acetate	Homogeneous	No separation	Lee et al. 2010
Silica-based sorbent	Polyamino acid	Cellulose		Membrane sorption	Ritchie et al. 1999
Clay (Laponite)	Sulfonic monomer, sodium salt of styrene sulfonic acid or 1,3-propanosulfonic acid	Nafion®	Homogeneous	Fuel cell	Fatyeyeva et al. 2011

$Lc = 40.0$ nm had hexagonal symmetry with $d_{002} = 3.368$Å. Deposition of pyrocarbon crystallites with Lc up to 1.5 nm was identified by X-ray photoelectron spectroscopy (XPS) and scanning electron microscopy (SEM). Coating the pore channels of membranes with pyrocarbon decreased the density of the electric charge on their surface by a factor of ~5.5. The temperature dependence of the hydrodynamic permeability coefficient was investigated in a dead end filtration scheme for the membranes with and without pyrocarbon-modification using ethanol, a polar, decane, and a nonpolar fluid as the feed. Electrostatic force and energy of the interaction of ethanol molecules with each other and the surface of pores were calculated; the results were compared with the O–H···O H-bonding energy. The research identified that the main reason for the formation of the alcohol adsorption layers was due to the formation of H-bonds between its molecules and O-H···O atoms on the surface of pores (Soldatov et al. 2006). It is noted that the hydrodynamic permeability coefficient increases as the temperature increases due to the H-bonding with oxygen atoms of TiO_2 membrane surface. However, the hydrodynamic permeability coefficient for the modified membrane is higher than that for the unmodified membrane and remains constant with temperature, as electrostatic interaction is the only interaction with pore walls after the deposition of pyrocarbon crystallites.

Silica membrane coated with a hydrophobic organic layer was used for PV (Jin et al. 2011). Commercial porous asymmetric alumina tubes (NGK Insulators, Ltd.; mean pore diameter, 0.1 μm; o.d., 10 mm; i.d., 7 mm; and length 50–100 mm) were used as the membrane supports. The support tubes were dip-coated with commercial silica-colloid sols, Snowtex OL (mean particle size of 40–50 nm; Nissan Chemical Industries, Ltd.) and Snowtex O (mean particle size of 10–20 nm; Nissan Chemical Industries, Ltd.), withdrawn at a rate of 1 mm/s, and then dried at room temperature. After dipping four times, the membranes were calcined at 873 K for 2 h in air with a heating and cooling rate of 2 K/min. The silica-coated alumina supports were soaked in 120 mL of dry toluene in a round three-neck flask immersed in an oil bath at 383 K. Then, 0.1 ml of 1,1,1,3,3,3- hexamethyldisilazane (HMDS, LS-7150; Shin-Etsu Chemical Co., Japan) per gram of support was added to the flask and the suspension was refluxed for approximately 24 h. The supports were removed carefully and washed with toluene for 4 h at 383 K to remove excess alkylsilanes, and then washed with ethanol to remove the excess toluene. Finally, the membrane was vacuum dried at 393 K. X-ray diffraction (XRD) analysis revealed that the silica layer formed by dip-coating had an amorphous structure. From the SEM image, the silica layer thickness was found to be 2 μm. Fourier transform infrared spectroscopy (FTIR) absorption spectra revealed the presence of –$Si(CH_3)_3$ groups at the membrane surface, rendering the membrane very hydrophobic. Brunauer–Emmett–Teller (BET) analysis detected the presence of pores with a mean pore radius of 1.16 nm with a narrow pore size distribution with a BET surface area of 225 m^2/g. PV experiments were conducted using a ternary feed solution of 5 wt% organics (ethanol or acetone)–94 wt% water–1 wt% acetic acid, which simulates industrial conditions of alcohol production by biomass fermentation or wastewater treatment in the pharmaceutical industry. Results from the PV experiments are summarized in Table 17.3. It is noted that with increasing temperature, the overall flux increases; however, the separation factors of both ethanol and acetone decrease.

The silica membranes possess a dual structure, namely, an organic–inorganic hybrid structure. Due to the surface hydrophobic structure and pore structure in the

TABLE 17.3
Summary of Pervaporation Results

Feed	Temperature, K	Flux, Kg/m² h	Separation Factor	
			Ethanol over Water	Acetone over Water
Ethanol soln[a]	303	0.92	8.4	
	313	1.24	8.2	
Acetone soln[b]	301	1.20		26
	313	1.85		24

Source: Jin, T. et al., *Desalination*, 280, 139–145, 2011. With permission.
[a] 5 wt% ethanol, 94 wt% water, 1 wt% acetic acid
[b] 5 wt% acetone, 94 wt% water, 1 wt% acetic acid

hydrophobic zeolite, the transport and separation mechanism was explained using the adsorption–diffusion model.

Mordenite Framework Inverted (MFI) zeolite membranes were synthesized on the inner surface of α-alumina tubes (Pall Corp.) from an aluminum-free precursor solution containing SiO_2, NaOH, H_2O, and template tetrapropylammonium hydroxide (TPAOH) by *in situ* hydrothermal crystallization at 453 K for 20 h (Tang et al. 2009). The resultant zeolite membrane had a thickness of 2–3 μm. The membrane surface was modified by the *in situ* catalytic cracking deposition of methyldiethoxysilane (MDES) molecules at the sites of [(tSi-O⁻)H⁺] whereas MDES vapor was carried by an equimolar H_2/CO_2 mixture flowing over the membrane surface at a pressure of 1.5 bar and a temperature of 723 K.

BET surface area, pore volume, and mean pore size were 454 ± 4.6 m²/g, 0.187 cm³/g, and 0.561 nm, respectively, before modification and were 428 ± 3.8 m²/g, 0.175 cm³/g, and 0.521 nm, respectively, after modification, indicating small changes in pore structure by the surface modification. From FTIR spectra, on the other hand, it was concluded that the final deposits in the MDES-modified membrane are likely to be $(OH)_3Si[O-Si\equiv]_{framework}$. Considering the MFI zeolitic channels of 0.36–0.47 nm, the effective pore size of the MDES-modified membrane is most likely <0.36 nm. As a result, the modification caused a drastic decrease in permeance for the gas molecules with $d_k > 0.3$ nm but caused only a small decrease in permeance for H_2 and He, which have $d_k < 0.3$ nm. The H_2/CO_2 permselectivity was 141 with a high H_2 permeance of 3.96×10^{-7} mol/m² s Pa at 723 K. The cutoff for permeance between H_2 and CO_2 suggests that the effective pore size of the modified MFI zeolite membrane was close to the CO_2 kinetic diameter of 0.33 nm, which agrees with the earlier estimate.

Modification of inorganic substrates (Vycor glass) by either octadecyltrichlorosilane (ODS) or octadecyldimethylchlorosilane (OCS) resulted in the enhancement of selectivity for the heavier hydrocarbon gases such as *n*-butane over smaller hydrocarbon gases such as methane and permanent gases such as nitrogen and helium, either due to the surface modification of the inorganic membrane surface or due to the formation of a polymer layer on the surface of the membrane pore (see Figure 17.1) (Singh et al. 2005).

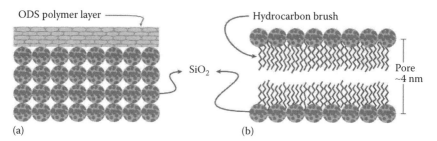

(a)   (b)

FIGURE 17.1 Hypothetical structures of the saline-modified Vycor porous glass: (a) octadecyltrichlorosilane and (b) octadecyldimethylchlorosilane. (Reprinted from Singh, R.P. et al., *J. Membr. Sci.*, 259, 34–46, 2005. With permission.)

The objective of the Singh et al. (2005) study was to gain insight into the structure of the saline layer of the ODS-modified membrane responsible for the high selectivity. A symmetric, mesoporous Vycor glass membrane (Corning, Inc., Corning, NY) was used as a support for the synthesis of organic–inorganic composite membranes. The membrane surface was salinized by either OCS or ODS. ODS reacts first with a trace amount of water on the glass surface and then a covalent bond is formed with the glass surface. OH groups of the hydrolyzed ODS react either with the surface silanols or with silanols present in other silane molecules to form cross-linking. The reaction scheme is as follows:

$$C_{18}H_{37}\text{-Si(Cl)}_3 + 3H_2O \dashrightarrow C_{18}H_{37}\text{-Si(OH)}_3 + 3HCl$$

$$C_{18}H_{37}\text{-Si(OH)}_3 + \text{-OH-Glass} \dashrightarrow C_{18}H_{37}\overset{\displaystyle |}{\underset{\displaystyle |}{\text{-Si}}}\text{-O-Glass}$$

On the other hand, OCS reacts directly with the membrane surface without forming cross-linking.

$$C_{18}H_{37}\text{-Si(CH}_3)_2\text{(OH)} + \text{-OH-Glass} \dashrightarrow C_{18}H_{37}\text{-Si(CH}_3)_2\text{-O-Glass}$$

A variety of reactions and pretreatment conditions were studied to optimize the loading of monochlorosilanes (OCS). For a typical OCS-modified membrane, the pure *n*-butane/nitrogen selectivity was 7.8 as compared to the mixed gas selectivity of 72.9 for a 50% mixture of *n*-butane in nitrogen. This difference was probably due to the pore-blocking effects of *n*-butane.

Carbon molecular sieve (CMS) membranes for separating gases were deposited on porous Al_2O_3 disks using $CH_3COCH_3 + CH_4$, by a remote inductively coupled-plasma (ICP) chemical vapor deposition (CVD) (Wang and Hong 2005). In particular, the newly proposed method consisted of surface treatment with high-energy ion bombardment and subsequent high-temperature pyrolysis. SEM imaging revealed that the top surface morphology did not change but the thickness of the top dense layer decreased considerably after pyrolysis. By optimizing the surface-treatment conditions, H_2/N_2 selectivity reached 50 with an extremely high H_2 permeance of approximately 1.6×10^{-6} mol/m² s Pa at 423 K. O_2/N_2 selectivity reached 2.5 and the O_2 permeance was approximately 2×10^{-7} mol/m² s Pa at 398 K. The authors concluded that the mechanism differs from the conventional pore-plugging mechanism of typical CVD, because bombardment with $CH_3COCH_3 + CH_4$ ions increases the H_2 permeances and the H_2/N_2 selectivity. Not only the modified surface structure of the CMS membrane but also the bulk structure must have contributed to the improvement in the separation efficiency.

High-temperature inorganic membranes have the potential to play an important role in the development of economical processes for precombustion and/or post-combustion CO_2 capture. For this purpose, Vycor membrane was surface-modified by 3-aminopropyltriethoxysilane (APTS) followed by heat treatment (Ostwal et al. 2011). Symmetric, mesoporous Vycor tubes (from Corning, Inc.) with a pore size of 4.7 nm were used as substrate/support. After thorough cleaning, the modification of the support by APTS was performed under an inert atmosphere (helium). The Vycor tube was filled with APTS solution and was then heated to about 383 K in an oven for 2–3 h or until all the silane was evaporated. Silane vapors reacted with the surface hydroxyl groups forming brush-like structures on the support pore walls (see Figure 17.2), as evidenced by ^{29}Si NMR. ^{13}C NMR, on the other hand, revealed the formation of carbamate species after the exposure of the APTS membrane to CO_2, indicating the following reaction at the membrane surface.

$$CO_2 + 2R-NH_2 \dashrightarrow R-NH_3^+ + R-NHCOO^-$$

BET analysis showed that the pore diameter decreased from 4.71 to 2.97 nm after the APTS modification. Unmodified silica membranes exhibited Knudsen diffusion

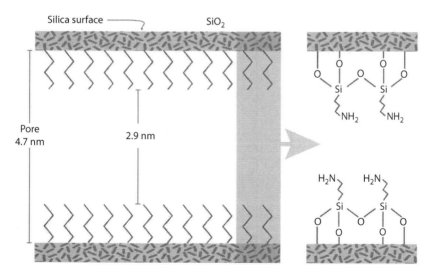

FIGURE 17.2   Reaction scheme of APTS modification. (Reprinted from Ostwal, M. et al., *J. Membr. Sci.*, 369, 139–147, 2011. With permission.)

behavior for the most gases, but there was also some contribution from the surface diffusion of heavier or interacting gases such as CO_2 and CH_4. Gas separation experiments were also performed on the modified membranes for both pure gas and mixed gases for a range of temperatures and feed gas compositions. CO_2/N_2 separation factor ranged from 1 to 10 depending on the temperature and feed composition. As the temperature increased, the CO_2/N_2 separation factor increased for a given feed composition or CO_2 feed partial pressure due to the decrease of binding energy of CO_2 to the amine groups as well as an increase in its diffusivity. For a given temperature, the CO_2/N_2 separation factor decreased with increased CO_2 content in the feed gas. This is due to the fact that at higher CO_2 concentrations in the feed, the amine reaction sites at the surface of the pore are saturated with CO_2 in the form of carbamate that hampers the movement of CO_2 gas.

CVD modification of MFI zeolite membrane was made for studying gas separation performance (Zhu et al. 2010). α-Alumina porous supports made of alumina powder (Alcoa, A-16), with a thickness of 2 mm and a diameter of 20 mm (average pore diameter, ~0.2 μm; porosity, ~45%), were used as the substrates, and MFI zeolite membranes were grown onto the surface of α-alumina porous supports by an *in situ* crystallization method. For the synthesis, a solution was prepared by dissolving 4.0 g of fumed silica (99.99%, Aldrich) into a solution containing 20 mL of 1 M TPAOH (Aldrich) solution and 0.28 g of NaOH (>97%, Aldrich) at 80°C under vigorous stirring. Zeolite membrane was grown on the surface of the α-alumina membrane was placed in an oven at 180°C for 4 h. CVD modification was further conducted with MDES vapor saturated in an equimolar H_2/CO_2 gas mixture flowing into the MFI zeolite membrane at 450°C. Crack free and continuous MFI zeolite membranes with a thickness of 2 μm were observed on the surface of the porous α-alumina surface. The authors concluded that the quality of the MFI zeolite membrane before the surface modification should be evaluated by H_2/CO_2 selectivity rather than H_2/SF_6 or N_2/SF_6 selectivity (Table 17.4). When the MFI-zeolite membrane is poorly prepared, there are defect pores through which gases flow by Knudsen mechanism at room temperature and higher H_2 selectivity is observed. When the membrane does not have defect pores, CO_2 transport is controlled by the surface flow at room temperature, resulting in lower H_2 selectivity. After CVD modification, the selectivity of poorly prepared MFI-zeolite membrane remains almost the same

TABLE 17.4

Change of H_2/CO_2 Separation Factor Before and After CVD Modification

Membrane Code	α_{H_2/CO_2} at Room Temperature Before Modification	α_{H_2/CO_2} at 450°C After Modification
M1	3.5	3.5
M2	0.56	8.2
M3	0.59	6.5
M4	0.67	6.1

Source: Zhou, H. et al., *Sep. Purif. Technol.*, 75, 286–297, 2010. With permission.

as before the CVD modification, whereas the H_2/CO_2 selectivity increases enormously because the pore size is now reduced and the sieve mechanism starts to control the selectivity. H_2/SF_6 or N_2/SF_6 selectivities, on the other hand, cannot serve as good criteria due to the swelling of the membrane in the presence of SF_6.

Nafion membranes that consist of perfluorosulfonate ionomers show degradation above 100°C. To overcome these limitations of the proton-conducting material for direct methanol fuel cell, it is desirable to develop thermally stable and nonswelling membranes with high proton conductivity over a wide temperature range. Two porous glasses PG-1 and PG-2 with pore diameters of 3.8 and 3.0 nm, respectively, were prepared from a phase-separated borosilicate glass. They were then treated with 3-mercaptopropyl trimethoxysilane to cover the surface with thiol groups, followed by treatment with concentrated nitric acid (Kikukawa et al. 2005). The reaction scheme is illustrated in Figure 17.3. After the surface treatment, the pore sizes

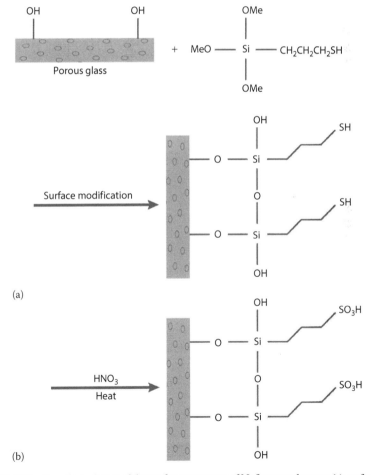

FIGURE 17.3 Reaction scheme of the surface treatment of Nafion membranes: (a) surface modification with 3-mercaptopropyl trimethoxysilane and (b) heat treatment with concentrated nitric acid. (Reprinted from Kikukawa, T. et al., *J. Membr. Sci.*, 259, 161–166, 2005. With permission.)

of PG-1 and PG-2 substrate were reduced to 3.4 and 2.2 nm, respectively. Proton conductivity of the PG-1-based hybrid membrane was 1.25×10^{-1} S/cm at 140°C under 100% relative humidity, whereas the proton conductivity was lower for the PG-2-based hybrid membrane under the same conditions. The membranes were thermally stable up to 140°C. They were also chemically stable in methanol because the framework structure of these membranes consisted of porous inorganic glass. Methanol diffusivity was not reported.

Hydrophilicity of hydrophobic PVDF is increased by incorporating surface-modified clay particles PVP-g-montmorillonite (MMT) (Wang et al. 2012). First, a reactive group (methacryloyl) was anchored on the surface of MMT through γ-methacryloxypropyltrim ethoxysilane hydrolysis in acidic solution. Then, N-vinylpyrrolidone (NVP) was polymerized to polyvinylpyrrolidone (PVP) using H_2O_2–$NH_3 \cdot H_2O$ as the initiator in aqueous solution. The reaction scheme is given in Figure 17.4, as confirmed by the FTIR analysis. All the membranes were prepared by a classical phase inversion method with PVDF as bulk material, N,N-dimethylacetamide (DMAC) as the solvent, PVP-g-MMT as the additive, and distilled water at room temperature as the nonsolvent coagulation bath. Water permeation flux of PVDF membrane was less than 10 L/m² h bar at the end of 1 h of UF, whereas the flux of PVDF/PVP-g-MMT membrane was ca 36 L/m² h bar when 6 wt% of PVP-g-MMT was added. In view of flux recovery after washing with deionized water, the PVDF membrane (ca 60%) was increased to 91.7% of PVDF/PVP-g-MMT membrane (with 6 wt% of PVP-g-MMT). Increased antifouling performance was due to increased hydrophilicity of the membrane surface as confirmed by contact angle measurement.

Inorganic–organic composite membrane and the MMM was developed in the initial article of te Hennepe et al. in the *Journal of Membrane Science* in 1987 for the separation of alcohol and water by PV (te Hennepe et al. 1987). They incorporated silicate-1 in the membrane made of rubbery PDMS to enhance the sorption of alcohols in the membrane over water, thus attempting to improve PV performance for the concentration of alcohols in the permeate from feed alcohol–water mixtures. The idea worked successfully, as shown by Table 17.5. From the table, it is obvious that both the separation factor and total volumetric flux increased with an increase in silicate content in the membrane.

In order to improve the separation performance of PDMS membrane, fumed silica, which consists of many spherical nonporous nanoscale particles, has been used. Fumed silica particles were incorporated because fumed silica is a cheap filler commonly used to reinforce silicone rubber (Peng et al. 2011). Silica surface was first modified by NH_2-C_3H_6-$Si(OC_2H_5)_3$ (KH-550) and added to PDMS solution in hexane

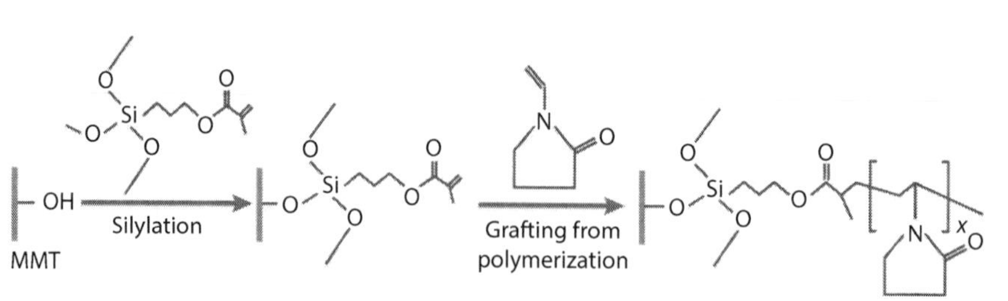

FIGURE 17.4 Reaction scheme of PVP-g-MMT formation. (Reprinted from Wang, P. et al., *Langmuir*, 28, 4776–4786, 2012. With permission.)

TABLE 17.5

Pervaporation Experimental Results of Alcohol–Water Pervaporation[a]

Alcohol	Silicate Content (wt%)	Separation Factor[b]	Total Volumetric Flux[c] (L/m² h)
Methanol	0	7.6	2.09
	30	9.5	3.51
	60	13.0	7.84
Ethanol	0	7.6	1.40
	30	14.9	3.62
	60	16.5	5.07
Propanol-2	0	9.5	2.65
	30	11.9	2.90
	60	23.0	3.86
Propanol-1	0	19.1	2.76
	30	21.2	2.91
	60	38.3	3.15

Source: te Hennepe, H.J.C., Bargeman, D., Mulder, M.H.V., and Smolders, C.A., *J. Membr. Sci.* 35, 39–55, 1987. With permission.

[a] Feed alcohol concentration, 5–5.5 wt % at 22.5°C

[b] Alcohol/water

[c] Normalized to *t* membrane thickness of 100 μm

together with a cross-linking agent and catalyst. The polymer solution was cast onto a cellulose acetate support membrane, which had been pretreated with water, before the solvent was evaporated in ambient air for 24 h and in an oven for 6 h at 80°C. The thickness of the top skin layer of the composite membrane was 20 μm. The reaction scheme on the silica surface modification is as follows:

$$Silica\text{-}OH + H_5C_2O\text{-}Si(OC_2H_5)_2\text{-}C_3H_6NH_2$$

$$\dashrightarrow Silica\text{-}O\text{-}Si(OC_2H_5)_2\text{-}C_3H_6NH_2 + HOSi(CH_3)_2\text{——}$$

$$\dashrightarrow Silica\text{-}O\text{-}Si(Si(CH_3)_2\text{——})_2\text{-}C_3H_6NH_2$$

where:

$H_5C_2O\text{-}Si(OC_2H_5)_2\text{-}C_3H_6NH_2$ and $HOSi(CH_3)_2$—are the coupling agent and cross-linking agent, respectively

Silica particles were well dispersed in the PDMS matrix, as confirmed by SEM imaging. Permeate side pressure was maintained at 10–30 kPa by a vacuum pump. In addition, nitrogen gas was used to bring out the permeation vapor. It was found that adding the modified silica particles significantly improved the PV performance. When the silica content was 5 wt%, the permeation flux was about 200 g/m² h, and the separation factor was 19 for the 5 wt% ethanol/water mixture at 40°C.

Another attempt to make silica nanoparticles compatible for a PDMS matrix was reported by Zhou et al. (2010b). In general, silane coupling agent has a formula $(OR)_3$–Si–X, where X represents a functional organic group, whereas OR corresponds to an easily hydrolyzable methoxy group. When silicate-1 and a saline coupling agent is brought into contact in a polar media hydrolysis and condensation, reactions occur simultaneously between the silane coupling agents and the silicalite-1 surface, rendering the surface hydrophobic. In this work, vinyltrimethoxysilane (VTMS) was used as the coupling agent and the reaction scheme was as shown below:

$$(CH_3O)_3Si\text{-}CH = CH_2 + 3H_2O \dashrightarrow (HO)_3Si\text{-}CH = CH_2 + 3CH_3OH$$

$$Silica\text{-}OH + (HO)_3Si\text{-}CH = CH_2 \dashrightarrow Silica\text{-}O\text{-}Si(OH)_2\text{-}CH = CH_2$$

PDMS and VTMS modified silicalite-1 particles were mixed in n-heptane and the solution was cast on a stainless steel plate, before being kept at room temperature overnight to evaporate the solvent. The plate was then placed in a vacuum oven to complete cross-linking. VTMS-modified silicalite-1 particles and hybrid membranes were characterized by FTIR, ^{29}Si Cross polarization (CP)-Magic angle spinning (MAS) nuclear magnetic resonance (NMR) (CP-MAS NMR), Differential scanning calorimetry (DSC), Thermogravimetric analysis (TGA), XRD, and SEM. Results showed that the coupling agent VTMS was readily grafted on the surface of silicalite-1 by a hydrolysis reaction and a condensation reaction. In addition, the chemical linking between the $-CH = CH_2$ group on the surface-modified silicalite-1 and $-Si-H$ on the PDMS substantially eliminated the nonselective voids inside the membrane. XRD patterns of the modified silicalite-1 and the unmodified silicate-1 were the same indicating that the crystalline structure of silica-1 was not changed by surface modification. SEM images showed that silicalite-1 particles were evenly dispersed in the PDMS matrix after the surface modification. Using a PDMS membrane with modified silicate-1 content of 68%, total flux of ca 90 g/m^2 h and a butanol separation factor over water of 145 was achieved from the ABE feed mixture of 5 g/L acetone + 1.2 g/L ethanol + 10 g/L butanol at the flow rate of 2 ml/min and 50°C.

Liu et al. (2011a) reported another attempt to improve inorganic filler particles and polymer matrix compatibility by surface coating of the particles. These researches used Zeolite-5 particles as inorganic fillers. Zeolite socony mobil (ZSM)-5 particles were modified in n-octyl chains using N-octyltriethoxysilane (OTES)/n-heptane solution, followed by the addition of a certain amount of PDMS polymer. Then, the suspension containing surface-modified ZSM-5 particles and PDMS was mixed with the cross-linked PDMS solution, which consisted of PDMS, n-heptane, tetraethoxysilane (TEOS), and dibutyltin dilaurate (PDMS:n-heptane:TEOS:dibutyltin dilaurate = 1:10:0.1:0.005, by weight), to prepare the casting dope. Subsequently, the ZSM-5 filled PDMS solution was dip-coated on the outer surface of the pre-treated tubular asymmetric ZrO_2/Al_2O_3 membrane. Removal of residual solvent and the cross-linking occurred at room temperature and at 120°C.

Silylation scheme is as follows:

$$(ZSM\text{-}5)\text{-}OH + (OC_2H_5)_3Si\text{-}(C_8H_{17}) \dashrightarrow (ZSM\text{-}5)\text{-}O\text{-}Si(OC_2H_5)_2\text{-}(C_8H_{17})$$

According to the molecular dynamics simulation, interaction energy between the (ZSM-5) and PDMS was −9.4 kcal/mol, whereas that between (ZSM-5)-O-$Si(OC_2H_5)_2$-(C_8H_{17}) and PDMS was −921.8 kcal/mol, indicating far greater compatibility between modified ZSM-5 and PDMS. SEM images revealed the uniform dispersion of surface-modified ZSM-5 particles in the PDMS matrix even at the highest zeolite loading of 40 wt%. FTIR analysis confirmed the decrease in the amount of hydroxyl groups at the ZSM-5 surface. The contact angle increased from 114.0° of PDMS without zeolite loading to 138.7° with 40 wt% loading of modified ZSM-5. Separation (ethanol/water) was increased considerably by the surface modification of the zeolite particles. Total flux of 408 g/m² h and a separation factor of 14 were obtained with the membrane with modified ZSM-5 loading of 40 wt% at 40°C.

Laboratory synthesized silicate-1 (typical molar composition: 15 NaOH.9 0 SiO_2.l0 TPABr.70 H_2O, crystallites lies 0.2 to 0.5 µm), two-component PDMS polymer, and iso-octane were mixed and PDMS was prepolymerized at 70°C to a viscosity appropriate for membrane casting (Jia et al. 1992). Suspension was cast on a Teflon plate and heat-cured at 80°C and dried before being subjected to PV of a water–ethanol mixture and O_2 and N_2 gas permeation experiments. Table 17.6 shows that the incorporation of silicate, commercial or laboratory made, improved both flux and selectivity (ethanol over water). Table 17.7 shows improvement in O_2, N_2 separation. By optimizing the membrane preparation conditions, a flux of 150 g/m² h and a separation factor of 34 were achieved.

Kim et al. (2009) fabricated membrane for propylene recovery from off-gas by incorporating silica nanoparticles of size 12–400 nm in a PDMS layer coated on a polysulfone (PSf) support. Hydrophilic fumed silica (Aerosil 200, particle size 12 nm) or silica nanoparticles prepared by a sol-gel method (particle size 300–400 nm) were incorporated in the PDMS layer to make asymmetric MMMs. To prevent particle agglomeration, silane coupling was attempted using three types of silane agents: 3-mercaptopropyltrimethoxysilane, 3-aminopropyl-trimethoxylsilane, and 3-methacryloxypropyl-trimethoxylsilane. Propylene/nitrogen mixture (volume ratio 15/85) was cooled to about 0°C before entering the membrane cell as feed to make propylene more condensable and a vacuum was applied on the permeate side.

TABLE 17.6

Comparison of PDMS Membranes Filled with Commercial or Laboratory Synthesized Silicate[a]

Silicate Type	Content, wt%	Flux[b] g/m² h	Separation Factor
–	0	25	7.3
UOP commercial	70	52	29
Laboratory synthesized	67	62	28
Laboratory synthesized	77	71	59

Source: Jia, M.-D. et al., *J. Membr. Sci.*, 73, 119–128, 1992. With permission.

[a] Feed ethanol content, 70 wt%; temperature, 22°C

[b] Normalized to membrane thickness of 10 µm

TABLE 17.7

Comparison of PDMS Membranes Filled with Laboratory Synthesized Silicate[a]

Silicate Type	Content, wt%	O_2 Permeability m³/m² h bar × 10^2	Ideal Selectivity
–	0	28.8	2.22
Laboratory synthesized	70	8.52	2.70
Laboratory synthesized	70	24.3	2.34

Source: Jia, M.-D. et al., *J. Membr. Sci.*, 73, 119–128, 1992. With permission.

[a] At 30°C

Successful coupling was confirmed by FTIR, and a better dispersion of coupled silica nanoparticles in PDMS matrix was attained. Although the propylene permeance of 42 gas permeance units (GPUs) with a separation factor of 8.5 is claimed in the abstract (separation factor of 7.5 is claimed in the conclusion), the effect of silica content on the performance of composite membranes performance plot (Figure 13 of Kim et al. 2009) shows a propylene permeance of 42 with a separation factor of ca 6, when sol-gel silica nanoparticles coupled with mercaptosilane content was 15 wt%.

Nanocomposite membranes based on poly(1-trimethylsilyl-1-propyne) (PTMSP) and silica were synthesized by the sol-gel copolymerization of TEOS with different organoalkoxysilanes in tetrahydrofuran solutions of PTMSP to study n-C_4H_{10}/CH_4 separation (Gomes et al. 2005). In comparison, nanocomposite membranes based on PTMSP were also prepared by dispersing silica particles with different functional groups (50% surface modification) into the PTMSP casting solution. In this chapter, only the results of the latter investigation are reported. Mixed gas permeation properties of membranes were determined with a feed containing 2 vol.% n-butane and 98% methane at 30°C. The feed and the permeate pressure were maintained at approximately 1.4 and 1 bar, respectively. Nitrogen was used as the sweep gas. The feed, retentate, and permeate compositions were determined with a gas chromatograph. When 30% of fumed silica (based on polymer content) was added to PTMSP (Lot # L-4D-4677-FN with smaller free volume), remarkable increases in both butane permeability and butane/methane selectivity were observed, as shown in Figure 17.5.

Interestingly, the performance enhancement was far greater when silica particles with hydrophilic surfaces were used. The authors attributed these results to unfavorable polymer/filler interaction, leading to an agglomeration of the long n-alkyl groups at the surface of the polymer. An increase of butane permeability up to sixfold of unfilled polymer was obtained. On the other hand, when PTMSP (Lot # 45-5641) of larger free volume was used, the addition of fumed silica increased the butane permeability but decreased the butane/methane selectivity.

It has been discussed above that successful MMMs can be formed for PV using polymers of relatively low T_g such as PDMS that have favorable interaction with the filler molecular sieves. However, practical gas separation membranes are prepared

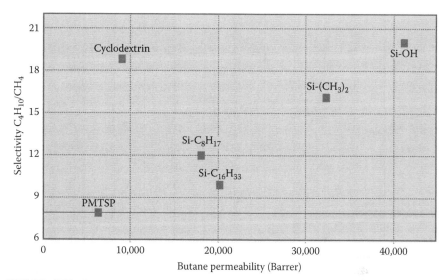

FIGURE 17.5 Butane permeation and selectivity data of PTMS-fumed silica MMMs for different silica-surface functional groups. (Reprinted from Gomes, D. et al., *J. Membr. Sci.*, 246, 13–25, 2005. With permission.)

from rigid matrix polymers with high T_g, and the formation of MMMs is difficult for such polymers because of poor adhesion between the polymer and the filler particles. To overcome these problems, plasticizers were used to adjust the T_g of the polymer matrix (Mahajan et al. 2002).

In this regard, there is an excellent review article on MMMs for gas separation, with a detailed discussion on the morphology of the interface between the inorganic particles and the polymer matrix (Chung et al. 2007). Unlike many other articles, this deals with asymmetric membranes for both flat sheets and hollow fibers aimed at the formation of an ultrathin defect-free mixed-matrix skin layer.

Silylation of borosilicate with APTS (γ-aminopropyl)triethoxysilane) was attempted as a tool to improve its incorporation in PI (polyimide) films (Vankelecom et al. 1996). This was achieved by a minimal coverage of zeolite crystals with APTS, as evidenced by the characterization of silylation through xylene sorptions, NMR spectroscopy, and measurements of the specific surface of the zeolites. Silylated zeolite was incorporated in PI films on which tensile strength, density, and xylene sorption were measured. Indeed, the density and tensile strength measurements on these composite PI membranes proved a better incorporation of borosilicates after silylation with APTS without changing xylene sorption.

MMMs of 4,4′-hexafluoroisopropylidenediphthalic anhydride (6FDA)- 4,4′-hexaflouroisopropyl-idene dianiline (6FpDA) -diaminobenzoic acid (DABA) (6FDA-6FpDA-DABA), a glassy PI, and modified zeolites (ZSM-2) were successfully fabricated using aminopropyltriethoxysilane (APTS) as the binder (Pechaf et al. 2002). Comparison of FTIR spectra revealed the presence of hydrogen bonding in the MMM solution. FESEM and TEM images did not reveal the presence of voids between the polymer and the zeolite. Changes in permeability for each gas correlated well with the change in the diffusion coefficient. Permeabilities of He, CO_2, and CH_4 all decreased, whereas those of O_2 and N_2 increased.

Membranes were prepared by physically blending a series of benzylamine-modified fullerene, C_{60} with pure Matrimid 5218 (Chung et al. 2003). The bulk Matrimid and the dispersed benzylamine-modified fullerene molecules showed strong interfacial interactions as demonstrated by a significant increase in the T_g of the film with C_{60} loading. Permeability decreased from the neat Matrimid membrane to 10 wt% loading of modified C_{60} loading for H_2 (25 to 17 Barrer, by 32%), O_2 (1.87 to 1.09 Barrer, by 45.2%), N_2 (0.275 to 0.161 Barrer, by 41.5%), CH_4 (0.199 to 0.1019 Barrer, by 45.2%), and CO_2 (from 7.15 to 3.79 Barrer, by 41.5%), respectively, due to the decrease in the gas diffusivity. Presence of the benzylamine-modified C_{60} seems to serve as impenetrable regions and rigidifying elements within the polymer matrix.

A systematic investigation on the modification of zeolite beta (BEA, SiO_2/Al_2O_3 ratio of 300, zeolyst; particle size 0.62 μm) by a series of organosilane compounds $R(CH_3)_nSiX_{(3-n)}$, where X is a chloro- or alkoxy-group with $n = 0$ or 2 and R is an alkyl chain varying from CH_3 to $C_{18}H_{37}$, was made (Sirikittikul el al. 2008). More specifically, the formula of the organic silanes is as follows: $(MeO)_3SiCH_3$, $(MeO)_3SiC_3H_7$, $(MeO)_3SiC_8H_{17}$, $(MeO)_3SiC_{16}H_{33}$, $(MeO)_3SiC_{18}H_{37}$, $(MeO)Si(CH3)_2C_{18}H_{37}$, $Cl_3SiC_{18}H_{37}$, $ClSi(CH_3)_2C_{18}H_{37}$. The results of FTIR and ^{29}Si CP/MAS NMR indicated that the alkylsilyl species were covalently bound onto the BEA surface. The modified BEA was dispersed in 20% solution of the elastomeric block copolymer, styrene-b-(ethylene-ran-butylene)-b-styrene (SEBS, Kraton G1652, Kraton Polymers) in toluene. The membranes were subjected to characterization and gas permeation experiments after removing the solvent thoroughly. Partition between the water and heptane phase showed that unmodified BEA particles were distributed in the water phase whereas all modified BEA particles were distributed in the heptane phase, indicating the higher hydrophobicity of modified BEA particles. SEM images revealed that unmodified BEA particles and BEA particles modified by CH_3-Si particles were poorly dispersed in the polymer matrix of the membrane. BEA particles modified by silane compounds with alkyl chains longer

TABLE 17.8

Gas Permeation Test Results of SEBS Membranes Containing 10 wt% Unmodified or Modified BEA Particles

Particles	Pure Gas Permeability, mL mm/m² day		
	Carbon dioxide	Oxygen	Ethylene
Without BEA	1560	800	4764
Unmodified BEA	1835	1216	4142
BEA modified with CH_3Si-	2089	1385	3529
BEA modified with C_3H_7Si-	4064	1162	3583
BEA modified with $C_8H_{17}Si$-	3604	1050	4527
BEA modified with $C_{16}H_{33}Si$-	5736	1070	4102
BEA modified with $C_{18}H_{37}Si$-	4667	915	3856

Source: Sirikittikul, D. et al., *Polym. Adv. Technol.*, 20, 802–810, 2008. With permission.

than CH_3 were well dispersed in the polymer matrix. Results of gas permeation tests are summarized in Table 17.8. Grafted chain length was found to have an effect on gas permeability. Carbon dioxide, oxygen, and ethylene permeabilities of the membranes containing the unmodified BEA were comparable to those of the CH_3Si-grafted BEA. Interestingly, the membranes containing the BEA grafted with C_3H_7 to $C_{18}H_{37}$ species showed the enhancement of the carbon dioxide permeability. Affinity of the long alkyl chain to carbon dioxide probably caused the increase of carbon dioxide permeability.

The molecular sieve, zeolite 4A, and zeolite 3A and 5A resulting from K^+ and Ca^{2+} exchange treatment of zeolite 4A, respectively, were surface-modified by a coupling agent (3-aminopropyl)-diethoxymethyl silane (APDEMS) (Li et al. 2006). The reaction scheme is as follows:

$$Zeolite(-OH)_2 + (C_2H_5O)_2Si(CH_3)-(CH_2)_3-NH_2$$
$$\cdots \rightarrow Zeolite\ (-O-)_2Si(CH_3)-(CH_2)_3-NH_2$$

Both elemental analysis and XPS spectra confirmed the chemical modification, whereas BET measurements showed no changes in zeolite surface area and total pore volume after the modification. Polyethersulfone (PES)–zeolite 3A, 4A, and 5A MMMs were fabricated by applying high processing temperatures to eliminate the void between polymer and zeolite phases. The thickness of the resultant dried MMMs was from 60 to 90 μm. SEM images confirmed good adhesion between the polymer and zeolite phases when modified zeolites were used. Pure gas permeability of He, H_2, O_2, N_2, CH_4, and CO_2 was obtained for each MMM. The test was performed for H_2 at 35°C and 3.55×10^5 Pa (3.5 atm) and at 35°C and 10.13×10^5 Pa (10 atm) for other gases. Interestingly, the permeability of PES-zeolite-4A-NH_2 MMM kept decreasing with an increase in zeolite loading whereas that of PES-zeolite-5A-NH_2 MMM showed a minimum, which was attributed to the change of the pore size of the zeolite particle. Typical data for CO_2/CH_4 gas pair was >1 Barrer CO_2 permeability and a CO_2/CH_4 selectivity of 47 with the PES-modified 4A membrane with a 50% filler loading. Prediction of MMM performance was attempted by the Maxwell model with good agreement.

Zeolite 4A (pore size 0.4 nm) and 13x (pore size 1.0 nm) were reacted with 3-aminopropyltrimethoxysilane (3-APTMS) (Boroglu and Gurkaynak 2011). Then, a zeolite containing monomer was synthesized by the amidization of 3,5-DABA and 3-APTMS modified zeolite in NMP. The reaction scheme is as follows:

$$Zeolite\ (-OH)_3 + (CH_3O)_3Si(CH_2)_3NH_2 \cdots \rightarrow Zeolite\ (-O-)_3Si(CH_2)_3NH_2$$

$$Zeolite\ (-O-)_3Si(CH_2)_3NH_2 + HOOC(C_6H_3)(NH_2)_2$$
$$\cdots \rightarrow Zeolite\ (-O-)_3Si(CH_2)_3NHCO(C_6H_3)(NH_2)_2$$

(New zeolite containing monomer for imidization)

The zeolite containing monomer called (DABA/N-propyltrimethoxysilane (PTMS) + zeolite) was mixed with pyromellitic dianhydride (PMDA) and 4,4-oxydianiline (ODA) in NMP to obtain a polyamic acid solution, which was cast onto a glass plate followed by solvent removal and the imidization of the cast film at high temperature. Thus,

TABLE 17.9

Gas Permeability of Poly(Imide Siloxane) and Poly(Imide Siloxane)-Zeolite MMMs

Membranes	Permeability, Barrer		
	O_2	N_2	Selectivity
PMDA-[ODA:(DABA/PTMS)][9:1]	0.96	0.17	5.64
PMDA-[ODA:(DABA/PTMS)] + 5% Zeolite 4A	0.93	0.12	7.81
PMDA-[ODA:(DABA/PTMS)] + 10% Zeolite 4A	0.92	0.12	7.80
PMDA-[ODA:(DABA/PTMS)] + 15% Zeolite 4A	0.85	0.10	8.05
PMDA-[ODA:(DABA/PTMS)] + 5% Zeolite 13×	2.06	1.74	1.18
PMDA-[ODA:(DABA/PTMS)] + 10% Zeolite 13×	2.05	1.80	1.14

Source: Boroglu, M.S. and Gurkaynak, M.A., *Polym. Bull.*, 66, 463–478, 2011. With permission.

zeolite particles were covalently bonded to the PI matrix. SEM images of the MMMs showed the interface between polymer and zeolite phases getting closer when surface-modified zeolite was used. Increase in glass transition temperature (T_g) confirmed that the polymer chain was becoming more rigid induced by the presence of zeolite. The results from the gas permeation experiments are summarized in Table 17.9. The permeation data shows that the permeability decreases and selectivity increases as the zeolite 4A loading increases whereas permeability increases significantly by adding zeolite 13× but selectivity is low due to the large pore size of zeolite 13×.

Many studies of zeolite polymer composite materials have focused primarily on interfacial tailoring to promote adhesion between the inorganic and polymer species in order to minimize interfacial defects. Intensive efforts to yield defect-free composite materials have been carried out using silane coupling agents, integral chain linkers, and polymer coatings on the molecular sieve surfaces. Unfortunately, they were not able to eliminate defects completely. In addition, the use of coupling agents is usually limited to a specific polymer-filler pair depending on the chemistry of the polymeric materials. A new approach was proposed to achieve a defect-free composite (Shu et al. 2007a; Bae et al. 2009; Lydon et al. 2012). Grignard reagent or solvothermal treatment was employed to create $Mg(OH)_2$ inorganic whisker or asperity nanostructures on the zeolite surface. The highly roughened zeolite surfaces promote adhesion at the polymer–particle interface through thermodynamically induced adsorption and physical entanglement of polymer chains in the whisker structures.

Lydon et al. (2012) used four such methods, including Grignard reagent, solvothermal, modified solvothermal, and ion exchange functionalization, to produce rough surfaces on zeolite MFI or Linde Type A (LTA). The modified particles were thoroughly characterized by many different methods, including TEM, High-resolution transmission electron microscopy (HRTEM), N_2 physisorption, multiscale compositional analysis (XPS, Energy-dispersive X-ray [EDX], and Inductively coupled plasma atomic emission spectroscopy [ICP-AES] elemental analysis), and diffraction (Energy-dispersive [ED] and X-ray diffraction [XRD]). The modified zeolite particles were incorporated into Matrimid, and permeation tests were conducted for the

CO_2–CH_4 gas pair. LTA/Matrimid MMMs containing ion exchange functionalized particles provided the best results with CO_2/CH_4 selectivity (~40), while maintaining a CO_2 permeability of ~10 Barrers.

Zeolite 4A (an aluminosilicate) was dealuminated through thionyl chloride and then treated with Grignard reagent (Shu et al. 2007a). This two-step reaction process created a very rough surface on the zeolite particles. Thionyl chloride partially removes aluminum from the zeolite 4A framework and yields NaCl and $AlCl_3$, which precipitate on the surfaces of zeolite particles, serving as nuclei for the $Mg(OH)_2$ crystal growth that occurs when methylmagnesium bromide (Grignard reagent) is reacted with 2-propanol in the following process. The crystals that look like whiskers and the rough surfaces of the modified particles provided enhanced interfacial adhesion in polymeric composites. In further research, the modified zeolite 4A of rough surface was dispersed (15 to 40 wt%) into poly(vinyl acetate) (PVAc) (Aldrich, Mw = 500 kD) and a polyetherimide, Ultem 1000 (GE Plastics, Mw = 56 kD) solutions to prepare casting dopes (Shu et al. 2007b). The polymer solution film was cast, dried, and annealed at temperatures above the T_g of the polymers. The film thickness was around 62.5 μm. SEM images showed that the particles were uniformly distributed in the films. PVAc and Ultem composites containing the modified particles exhibited defect-free interfaces. Dynamic mechanical analysis testing revealed that such composites have higher moduli as compared to those embedded with nontreated fillers with the same loadings. According to the gas permeation tests at the upstream pressure of 4.5 atm and 35°C, the performance improved from O_2 permeability of 0.4 Barrer and O_2/N_2 selectivity of 7.5 of neat Ultem to O_2 permeability of 0.5 Barrer to O_2/N_2 selectivity of 12 for the composite Ultem membrane with 40 wt% modified zeolite 4A loading. As for the test results for the CO_2/CH_4 gas pair, the performance improved from a permeability of 1.5 Barrer and a CO_2/CH_4 selectivity of 38 for neat Ultem to a permeability of 3 Barrer and a CO_2/CH_4 selectivity of 75 for 40 wt% modified zeolite 4A loading.

Another method of creating rough surfaces on zeolite crystals by solvothermal process is presented by Bae et al. (2009). Pure-silica zeolite MFI crystals of different sizes (5 μm, 2 μm, 300 nm, and 100 nm) were synthesized hydrothermally, and the solvothermal deposition of $Mg(OH)_2$ was performed at 160°C in a solvent mixture of ethylenediamine (EDA) and an $MgSO_4$ aqueous solution. Well-defined whiskers with high surface density formed on 2 and 5 μm MFI crystals. Morphology of the solids deposited on the 300 nm MFI particles had the appearance of *cotton balls*. The 100 nm MFI particles were coated with a much finer layer of $Mg(OH)_2$ and no whisker structure could be seen. Some examples for the improvement of the gas separation performance are given in Table 17.10.

The pure-silica MFI and aluminosilicate LTA were synthesized and then subjected to Grignard reagent treatment and solvothermal treatment (Bae et al. 2011). For the Grignard treatment, the zeolite particles were seeded with NaCl and Grignard reagent (CH_3MgBr) was added followed by the addition of 2-propanol to quench the Grignard reagent. For solvothermal treatment, zeolite was dispersed either in EDA or in diethylenetriamine before adding $MgSO_4$ solution drop-wise. Solvothermal treatment was then performed at high temperatures. Surface-treated zeolite particles were then incorporated into polyetherimide (Ultem) solution to prepare the casting of the matrix in order to compare with the untreated zeolites. Characterization revealed that surface treated LTA crystals are promising materials for MMM fabrication.

TABLE 17.10

Comparison of MMM Containing 300 nm Solvothermally Treated MFI Crystals at 35°C and 2 atm Upstream Pressure

Membranes	CO_2 Permeability, Barrer	CO_2/CH_4 Selectivity
Ultem	1.4	38
20% loading in Ultem	2.2	43
30% loading in Ultem	2.0	45
Matrimid	7.6	35
35% loading in Ultem	31	39

Source: Bae, T.-H. et al., *Am. Chem. Soc.*, 131, 14662–14663, 2009. With permission.

A similar result was obtained for the formation of $Mg(OH)_2$ nanowhiskers by a three-step sol-gel precipitation process on zeolite 5A (Liu et al. 2011b). Fluorinated PI FDA-DAM was used to prepare MMMs. Effect of nanoscale surface morphology of pure-silica MFI zeolite on the interfacial, mechanical, and thermal properties of pure-silica MFI zeolite/PVAc composites was investigated under different annealing conditions (Lee et al. 2010). Surface of zeolite particles was first seeded with NaCl and then treated with the Grignard reagent (CH_3MgBr). The particles were then dispersed in EDA solution followed by the drop-wise addition of $MgSO_4$ solution. The mixture was then subjected to further solvothermal treatment at 160°C in an oven for 12 h. $Mg(OH)_2$ was thus deposited on the particle surface.

Formation of $Mg(OH)_2$ inorganic whisker- or asperity-like nanostructures is obvious in Figure 17.6. Unmodified and surface-treated MFI zeolites were dispersed in DCM, and a desired amount of dried PVAc was added to prepare the casting dope. Composite films were prepared by casting the solution mixtures with a blade on ODS-treated glass substrate followed by the complete removal of solvent. Finally, composite films were annealed in a vacuum oven at various temperatures. PVAc composites containing surface-modified particles showed increased tensile strength and elongation at the break point as compared with composites containing unmodified zeolite.

(a)   (b)   (c)

FIGURE 17.6 SEM images of the surface of MFI zeolite particles (a) untreated, (b) Grignard-treated, and (c) solvothermally treated. (Reprinted from Lee, J.-H. et al., *Polymer*, 51, 5744–5755, 2010. With permission.)

Ritchie et al. (1999) prepared two novel membranes, one cellulose-based and the other silica-based MF membranes. Only the investigation with the silica-based membrane is reported here. The silica-based MF membranes used in this work as a substrate are manufactured by the extrusion of a mixture of oil, submicron-sized silica particles, and polyethylene resin, composed of about 70% silica particles in polyethylene. Gaps between the adjacent silica aggregate particles constitute the membrane pores and the convective flow path.

The substrate membrane was then derivatized by the permeation of a 5% solution (v/v) of silane 3-glycidoxypropyltrimethoxysilane (GOPS) in o-xylene, toluene, or hexane through the membrane under convective flow at 25°C–60°C for 2 h. Polyamino acid (poly-L-aspartic acid [PLAA] and poly-$[\alpha,\beta]$-DL-aspartic acid [PDAA], Mw: 6–36 kD) was then reacted to the expoxy end group with 100 mL of a 100 mg/L aqueous solution at pH 9.2–9.5. The reaction scheme is given as follows:

$$Silica\ (OH)_3 + (CH_3O)_3 \equiv Si\text{-}(CH_2)_3\text{-}OCH_2\text{-epoxy group}$$

$$\cdots\rightarrow Silica \equiv Si\text{-}(CH_2)_3\text{-}OCH_2\text{-epoxy group} + H_2N\text{-}R$$

$$\cdots\rightarrow Silica \equiv Si\text{-}(CH_2)_3\text{-}OCH_2\text{-}CH(OH)\text{-}CH_2\text{-}NH\text{-}R$$

Metal sorption experiments were conducted with feed solutions of 1000 mg/L of Pb^{2+}, Cu^{2+}, and Cd^{2+}. The feed pH was adjusted to 5.5 for Pb and Cd and 5.0 for Cu. The permeate was recycled several times to allow the sorbent to reach its sorption equilibrium capacity. The extent of metal sorption was calculated from the volume of permeate collected and the permeate analysis. Sorption capacity of the membranes is given in Table 17.11 for Pb and Cd. The water permeability of the membrane after silane attachment was 0.4×10^{-4} m^3/m^2 s bar.

High cost and sharp decline in its conductivity above 100°C are the drawbacks of Nafion® membranes, and many attempts have been made to improve these (Fatyeyeva et al. 2011). One such attempt is the fabrication of MMMs with surface-treated particles. Laponite clay was plasma-treated followed by the post-grafting of sulfonic monomer, sodium salt of p-styrene sulfonic acid, or 1,3-propane sultone. Plasma-processed clay particles were treated by PEG 1500 and freeze-dried

---

TABLE 17.11

Metal Sorption Capacity of the Membranes

Metals	Polyamino Acid	Sorption Capacity meq/g
Pb	PLGA	2.8
Pb	PLAA	1.2
Cd	PLAA	1.4
Pb	PLGA	2.5

Source: Ritchie, S.M.C. et al., *Langmuir*, 15, 6346–6357, 1999. With permission.

---

overnight in order to obtain the exfoliated clay. Nafion® was added to the exfoliated clay-DMF mixture to prepare the casting dope (4.5 wt% of exfoliated clay mixture and 95.5 wt% of Nafion®). The membrane was heated to remove the solvent and to enhance mechanical properties. Occurrence of the surface modification was confirmed, and the chemical structure of resulting products was characterized with FTIR, XPS, thermal analysis, and XRD. Nafion®/grafted Laponite (3 wt%) membranes had 20% improvement of power density at 0.7 V. Hybrid nanocomposite membranes showed better performance than pristine Nafion® membrane due to the incorporation of the inorganic hygroscopic charges to the Nafion® polymer matrix.

17.3 CONCLUSIONS AND FUTURE WORK

It is interesting to note that membranes fabricated by the surface modification of inorganic membranes are mostly asymmetric membranes, and these membranes have mostly been applied for PV and membrane gas separation. Despite excellent stability of inorganic membranes, especially in harsh organic environments, there are only a few examples for the treatment of organic liquids by reverse osmosis (RO)/nanofiltration (NF)/UF/MF processes using the membranes developed by this method. More work is needed to fine-tune the surface chemistry and the pore size of inorganic membranes for the separation of liquid mixtures. Pore-filling of MF inorganic membranes by surface functionalization is another interesting field to explore.

In contrast to surface modification of inorganic membranes, MMMs with surface-modified particles were fabricated as dense homogenous membranes to attempt at improving their intrinsic permeability as well as membrane separation performance. Chung's work of hollow-fiber MMM spinning is notably an exception. For practical applications, however, more asymmetric membranes should be fabricated and the permeance measured. Preparation of a thin defect-free membrane is necessary to obtain a large GPU value, which is not an easy task, and special membrane design is required. Similar to the surface-modified inorganic membranes, most of the MMMs development was made for PV (with rubbery polymers) and gas separation (with glassy polymers), aimed at enhancing filler/polymer compatibility and preventing filler agglomeration. Very few researches have been undertaken for the development of MMMs for water treatment by RO/NF/UF/MF. It should be noted that, to develop such MMMs, the functionalization of the filler surface is not to fill the gap between the filler and the polymer matrix. Rather, it is made for the fine-tuning of the size and the surface chemistry of the void space, which will eventually act as the membrane pore. It will be very interesting to apply MMMs for thin-film composite membranes, which has only recently been started.

REFERENCES

Bae, T.-H., Liu, J., Lee, J.S., Koros, W.J., Jones, C.W., and Nair, S. 2009. Facile high-yield solvothermal deposition of inorganic nanostructures on zeolite crystals for mixed matrix membrane fabrication. *J. Am. Chem. Soc.* 131:14662–14663.

Bae, T.-H., Liu, J., Thompson, J.A., Koros, W.J., Jones, C.W., and Nair, S. 2011. Solvothermal deposition and characterization of magnesium hydroxide nanostructures on zeolite crystals. *Microporous Mesoporous Maters.* 139:120–129.

Boroglu, M.S. and Gurkaynak, M.A. 2011. Fabrication and characterization of silica modified polyimide–zeolite mixed matrix membranes for gas separation properties. *Polym. Bull.* 66:463–478.

Chung, T.-S., Chan, S.S., Wang, R., Lu, Z., and He, C. 2003. Characterization of permeability and sorption in Matrimid/C_{60} mixed matrix membranes. *J. Membr. Sci.* 211:91–99.

Chung, T.-S. Jiang, L.Y., Li, Y., and Kulprathipanja, S. 2007. Mixed matrix membranes (MMMs) comprising organic polymers with dispersed inorganic fillers for gas separation. *Prog. Polym. Sci.* 32:483–507.

Fatyeyeva, K., Bigarré, J., Blondel, B., Galiano, H., Gaud, D., Lecardeur, M., and Poncin-Epaillard, F. 2011. Grafting of p-styrene sulfonate and 1,3-propane sultone onto Laponite for proton exchange membrane fuel cell application. *J. Membr. Sci.* 366:33–42.

Gomes, D., Nunes, S.P., and Peinemann, K.-V. 2005. Membranes for gas separation based on poly(1-trimethylsilyl-1-propyne)–silica nanocomposites. *J. Membr. Sci.* 246:13–25.

Jia, M.-D., Pememann, K.-V., and Behlmg, R.-D. 1992. Preparation and characterization of thin- film zeolite-PDMS composite membranes. *J. Membr. Sci.* 73:119–128.

Jin, T., Ma, Y., Matsuda, W., Masuda, Y., Nakajima, M., Ninomiya, K., Hiraoka, T., Fukunaga, J., Daiko, Y., and Yazawa, T. 2011. Preparation of surface-modified mesoporous silica membranes and separation mechanism of their pervaporation properties. *Desalination* 280:139–145.

Kikukawa, T., Kuraoka, K., Kawabe, K., Yamashita, M., Fukumia, K., Hirao, K., and Yazawa, T. 2005. Stabilities and pore size effect of proton-conducting organic–inorganic hybrid membranes prepared through surface modification of porous glasses. *J. Membr. Sci.* 259:161–166.

Kim, H., Kim, H.-G., Kim, S., and Kim, S.S. 2009. PDMS–silica composite membranes with silane coupling for propylene separation. *J. Membr. Sci.* 344:211–218.

Lee, J.-H., Zapata, P., Choi, S., and Meredith, J.C. 2010. Effect of nanowhisker-modified zeolites on mechanical and thermal properties of poly(vinyl acetate) composites with pure-silica MFI. *Polymer* 51:5744–5755.

Li, Y., Guana, H.-M., Chung, T.-S., and Kulprathipanja, S. 2006. Effects of novel silane modification of zeolite surface on polymer chain rigidification and partial pore blockage in polyethersulfone (PES)-zeolite A mixed matrix membranes. *J. Membr. Sci.* 275:17–28.

Liu, G., Xiangli, F., Wei, W., Liu, S., and Jin, W. 2011a. Improved performance of PDMS/ceramic composite pervaporation membranes by ZSM-5 homogeneously dispersed in PDMS via a surface graft/coating approach. *Chem. Eng. J.* 174:495–503.

Liu, J., Bae, T.-H., Esekhile, O., Nair, S., Jones, C.W., and Koros, W.J. 2011b. Formation of $Mg(OH)_2$ nanowhiskers on LTA zeolite surfaces using a sol–gel method. *J. Sol-Gel Sci. Technol.* 60:189–197.

Lydon, M.E., Unocic, K.A., Bae, T.-H., Jones, C.W., and Nair, S. 2012. Structure-property relationships of inorganically surface-modified zeolite molecular sieves for nanocomposite membrane fabrication. *J. Phys. Chem.* 116:9636–9645.

Mahajan, R., Burns, R., Schaeffer, M., and Koros, W.J. 2002. Challenges in forming successful mixed matrix membranes with rigid polymeric materials. *J. Appl. Polym. Sci.* 86:881–890.

Ostwal, M., Singh, R.P., Dec, S.F., Lusk, M.T., and Way, J.D. 2011. 3-Aminopropyltriethoxysilane functionalized inorganic membranes for high temperature CO_2/N_2 separation. *J. Membr. Sci.* 369:139–147.

Pechaf, T.W., Tsapatsis, M., Marand, E., and Davis, R. 2002. Preparation and characterization of a glassy fluorinated polyimide zeolite-mixed matrix membrane. *Desalination* 146:3–9.

Peng, P., Shi, B., and Lan, Y. 2011. Preparation of PDMS-Silica nanocomposite membranes with silane coupling for recovering ethanol by pervaporation. *Sep. Sci. Technol.* 46:420–427.

Ritchie, S.M.C., Bachas, L.G., Olin, T., Sikdar, S.K., and Bhattacharyya, D. 1999. Surface modification of silica- and cellulose-based microfiltration membranes with functional polyamino acids for heavy metal sorption. *Langmuir* 15:6346–6357.

Shu, S., Husain, S., and Koros, W.J. 2007a. Formation of nanostructured zeolite particle surfaces via a halide/Grignard route. *Chem. Mater.* 19:4000–4006.

Shu, S., Husain, S., and Koros, W.J. 2007b. A general strategy for adhesion enhancement in polymeric composites by formation of nanostructured particle surfaces. *J. Phys. Chem.* 111:652–657.

Singh, R.P., Way, J.D., and Dec, S.F. 2005. Silane modified inorganic membranes: Effects of silane surface structure. *J. Membr. Sci.* 259:34–46.

Sirikittikul, D., Fuongfuchata, A., and Booncharoen, W. 2008. Chemical modification of zeolite beta surface and its effect on gas permeation of mixed matrix membrane. *Polym. Adv. Technol.* 20:802–810.

Soldatov, A.P., Rodionova, I.A., and Parenago, O.P. 2006. The influence of pyrocarbon modification on the physicochemical characteristics of the surface of pores and transport properties of inorganic membranes. *Russian J. Phys. Chem.* 80:418–424.

Tang, Z., Dong, J., and Nenoff, T.M. 2009. Internal surface modification of MFI-type zeolite membranes for high selectivity and high flux for hydrogen. *Langmuir* 25:4848–4852.

te Hennepe, H.J.C., Bargeman, D., Mulder, M.H.V., and Smolders, C.A. 1987. Zeolite-filled silicone rubber membranes. Part 1. Membrane preparation and pervaporation results. *J. Membr. Sci.* 35:39–55.

Vankelecom, I.F.J., Van den Broeck, S., Merckx, E., Geerts, H., Grobet, P., and Uytterhoeven, J.B. 1996. Silylation to improve incorporation of zeolites in polyimide films. *J. Phys. Chem.* 100:3753–3758.

Wang, L.-J., and Hong, F.C.-N. 2005. Surface structure modification on the gas separation performance of carbon molecular sieve membranes. *Vacuum* 78:1–12.

Wang, P., Ma, J., Wang, Z., Shi, F., and Liu, Q. 2012. Enhanced separation performance of PVDF/PVP-g-MMT nanocomposite ultrafiltration membrane based on the NVP-grafted polymerization modification of Montmorillonite (MMT). *Langmuir* 28:4776–4786.

Zhou, J., Chang, Q., Wang, Y., Wang, J., and Meng, G. 2010a. Separation of stable oil–water emulsion by the hydrophilic nano-sized ZrO_2 modified Al_2O_3 microfiltration membrane. *Sep. Sci. Technol.* 75:243–248.

Zhou, H., Su, Y., Chen, X., Yi, S., and Wan, Y. 2010b. Modification of silicalite-1 by vinyltrimethoxysilane (VTMS) and preparation of silicalite-1 filled polydimethylsiloxane (PDMS) hybrid pervaporation membranes. *Sep. Purif. Technol.* 75:286–297.

Zhu, X., Wang, H., and Lin, Y.S. 2010. Effect of the membrane quality on gas permeation and chemical vapor deposition modification of MFI-type zeolite membranes. *Ind. Eng. Chem. Res.* 49:10026–10033.

18 Fabrication of Low-Fouling Composite Membranes for Water Treatment

Victor Kochkodan and Nidal Hilal

CONTENTS

18.1 Introduction 616
18.2 Surface Properties Affecting Membrane Fouling 617
    18.2.1 Hydrophilic–Hydrophobic Properties of Membrane Surfaces 617
    18.2.2 Surface Charge 619
    18.2.3 Surface Roughness 619
18.3 Development of Low-Fouling Composite Membranes through IP 621
18.4 Preparation of Low-Fouling Composite Polymer Membranes through
    Surface Grafting 622
    18.4.1 Photoinitiated Grafting on the Membrane Surface 623
        18.4.1.1 Membrane Modification by UV Graft
                Polymerization without the Use of a Photoinitiator 623
        18.4.1.2 Membrane Modification by
                UV Graft Polymerization with a Photoinitiator 625
    18.4.2 Redox- and Miscellaneously Initiated Grafting 627
        18.4.2.1 Redox-Initiated Grafting 627
        18.4.2.2 Enzymatic Grafting 628
    18.4.3 Plasma-Initiated Grafting and Initiated Chemical Vapor
        Deposition on the Membrane Surface 629
        18.4.3.1 Plasma-Induced Grafting of the Membrane Surface 629
        18.4.3.2 Initiated Chemical Vapor Deposition 630
18.5 Fabrication of Low-Fouling Composite Membranes through Physical
    Coating 632
    18.5.1 Coating through Casting 632
    18.5.2 Static/Dynamic Deposition of Protective Low-Fouling Layer
        on the Membrane Surface 635

18.6 Incorporation of Nanoparticles in Composite Membranes for Fouling
      Reduction 639
18.7 Conclusions 645
References 646

18.1 INTRODUCTION

A sharp growth in the world's population coupled with urbanization results in a rapidly increasing demand in water consumption. Water shortage has plagued many communities around the world [1,2]. It is reported that more than 1.2 billion people in the world lack access to safe drinking water [3]. In order to avoid possible global crises due to depleting available resources of clean and potable water, the need is inevitable to maximize the treatment of wastewater, minimize the discharge from water treatment plants, and tap into the virtually unlimited available saltwater in the oceans. During the last decade, pressure-driven membrane processes such as microfiltration (MF), ultrafiltration (UF), nanofiltration (NF), and reverse osmosis (RO) have been thoroughly investigated and used for water treatment and desalination [4]. However, one of the main problems arising upon the operation of the membrane units is the membrane fouling, which seriously hampers the application of membrane technologies [5].

Membrane fouling is an extremely complex phenomenon that has not been defined precisely yet. In general, the term is used to describe the undesirable deposition of retained particles, colloids, macromolecules, and salts at the membrane surface or inside the membrane pores. Depending on the membrane process and chemical nature of foulants, several types of fouling can occur in membrane systems, such as inorganic fouling or scaling, organic fouling, and colloidal/biocolloidal fouling [6,7].

Inorganic fouling or scale formation at the membrane surface results from the increased concentration of one or more inorganic salts such as $CaCO_3$, $CaSO_4$, $\cdot 2H_2O$, and $Ca_3(PO_4)_2$ beyond their solubility limits and their ultimate precipitation onto the membranes [8].

With organic fouling, dissolved organic compounds in water, such as proteins, humic substances, and polysaccharides, have been implicated as strong, irreversible membrane foulants in pressure-driven membrane processes [9]. Natural organic matter (NOM) is the main organic foulant for membrane treatment of surface water, brackish water, and seawater [10]. It has been demonstrated that the hydrophobic fraction of NOM was the major factor causing permeate flux decline due to the strong adsorption on the membrane surface, whereas the hydrophilic fraction of NOM had a relatively small effect on the membrane fouling [11].

Colloidal fouling refers to membrane fouling with colloidal and suspended particles in the size range of a few nanometers to a few micrometers. According to Buffle et al. [12,13], colloids in natural waters can be classified into the following general groups: (1) inorganic colloids such as silica, iron oxides/hydroxides, and hydroxides of heavy metals; (2) organic colloids such as aggregated NOM and proteins; (3) (bio)colloids such as bacteria, microorganisms, viruses, and other biological matter.

Membrane fouling with biocolloids is usually known as *biofouling* [14]. Biofouling is a dynamic process of microbial colonization and growth, which results in the formation of microbial biofilms on the membrane surface [15]. Membrane biofouling is caused mainly by *Corynebacterium, Pseudomonas, Bacillus, Arthrobacter, Flavobacterium,* and *Aeromonas* bacteria and to a lesser extent by fungi such as *Penicillium* and *Trichoderma* and other eukaryote microorganisms [16].

According to the osmotic-resistance filtration model, the formation of a foulant layer on the membrane surface may affect permeate flux in two different ways:

1. The foulant layer introduces an additional hydraulic resistance [17]. Consequently, this reduces the overall membrane permeability and decreases water flux at a fixed applied pressure. Alternatively, a higher pressure is required to maintain a constant permeate flux.
2. A porous cake layer may also affect membrane flux behavior by promoting a severe concentration polarization inside the unstirred cake layer [18]. The enhanced concentration polarization can significantly increase the solute concentration at the membrane surface, and thus drastically increase the osmotic pressure at the membrane surface.

Currently, there is a consensus in research that membrane fouling with organic compounds and biocolloids is mainly determined by the ability of the foulants to adhere to the membrane surface, influenced by hydrophobic interactions, hydrogen bonding, London–van der Waals attractions, and electrostatic interactions [9,19]. Therefore, one of the main strategies toward reducing membrane fouling is the prevention of the undesired adhesion interactions between a foulant and a membrane to inhibit or, at least, minimize the fouling process. This may be realized through the development of composite membranes containing a low-fouling top membrane layer, which plays a vital role in both fouling interactions and membrane separation. This chapter surveys the studies on the development of low-fouling composite membranes using interfacial polymerization, surface grafting, physical coating/adsorption, and surface modification of polymer membranes with nanoparticles.

18.2 SURFACE PROPERTIES AFFECTING MEMBRANE FOULING

Several of the surface characteristics of membranes such as hydrophobic/hydrophilic properties, membrane charge (zeta potential), and surface roughness are strongly related to fouling because they determine the interaction between the membrane and the foulants [20,21].

18.2.1 HYDROPHILIC–HYDROPHOBIC PROPERTIES OF MEMBRANE SURFACES

As known, the common evaluation approach for hydrophilic–hydrophobic properties of membrane is based on the evaluation of a contact angle formed between the liquid–gas tangent and membrane–liquid boundary [22]. Most commercial membranes for pressure-driven processes are made from hydrophobic polymers with high thermal, chemical, and mechanical stabilities such as polyvinylidenefluoride (PVDF), polyethersulfone (PES), polysulfone (PS), polypropylene (PP), polyacrylonitrile (PAN), polyamide (PA), and polyethylene (PE) [23]. Usually these materials are characterized by high contact angle values and are prone to the adsorption of the various solutes from feed streams. Recently, it was shown that membranes with larger contact angles adsorb more mass per unit area of a hydrophobic solute than a membrane with a smaller contact angle [24]. It has been well documented that membranes with hydrophilic surfaces are less susceptible to fouling with organic substances, microorganisms, and charged inorganic particles due to a decrease of the interaction between the foulant and the membrane surface [25,26].

Nabe et al. [27] found a direct correlation between the extent of membrane fouling and hydrophilic properties of the membrane surface. They showed that during the filtration of bovine serum albumin (BSA) solution through PS and different PS-modified membranes the relative flux decreases, which correlates with the membrane fouling, with an increase of contact angle of the membrane surface, and thus with an increase in membrane hydrophobicity (Figure 18.1).

The major reason for hydrophobic membrane fouling with organic compounds is that there are almost no hydrogen bonding interactions in the boundary layer between the membrane interface and water. Repulsion of water molecules away from the hydrophobic membrane surface is a spontaneous process with increasing entropy, and, therefore, foulant molecules have a tendency to adsorb onto the membrane surface and dominate the boundary layer. In contrast, the membrane with a hydrophilic layer possesses a high surface tension and is able to form hydrogen bonds with the surrounding water molecules to reconstruct a thin water boundary between the membrane and bulk solutions. It is difficult, therefore, for hydrophobic solutes to approach the water boundary and break its orderly structure [28]. For example, it was found that the improvement of membrane hydrophilicity can enhance the permeate flux of RO membrane [29]. This finding may be explained by the fact that a thin layer of bounded water exists on the surface of hydrophilic membranes due to the formation of hydrogen bonds. This layer can prevent or reduce undesirable adsorption or adhesion of hydrophobic foulants on the membrane surface (Figure 18.2a). Therefore,

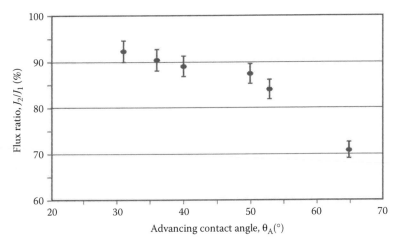

FIGURE 18.1 Correlation between contact angle and relative flux of PS-modified membranes. (Data from A. Nabe et al., *J Memb Sci*, 133, 57–72, 1997.)

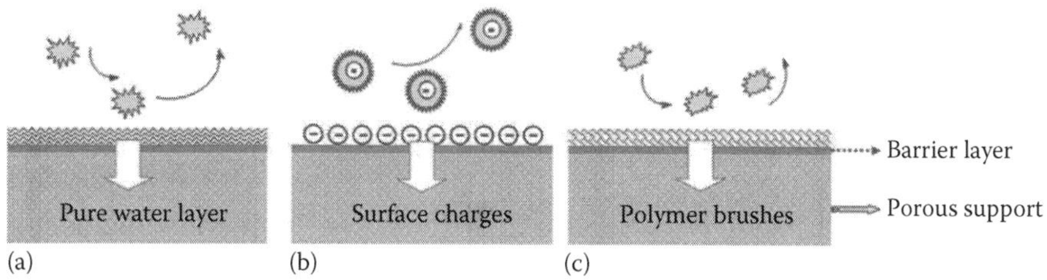

FIGURE 18.2 Schematic representation of various antifouling mechanisms with composite membranes: (a) thin layer of bounded water, (b) electrostatic repulsion, and (c) steric repulsion. (Data from G.D. Kang et al., *Water Res*, 46, 584–600, 2012.)

an increase in the hydrophilicity of the membrane surface is often a key goal for reducing membrane fouling with colloids, microorganisms, and organic pollutants.

It should be mentioned that the surface-bound hydrophilic polymer brushes may prevent adsorption of foulants onto membrane surface because of the steric repulsion mechanism [31]. When hydrophilic polymer chains are created on membrane surfaces, this diffuse hydrophilic layer will exert steric repulsion on the foulant that reaches the surface (Figure 18.2c). Extent of steric repulsion depends on the density, length, and regularity of grafted or deposited polymer brushes [32,33].

18.2.2 SURFACE CHARGE

The electrostatic charge of membranes is also very important for the reduction of membrane fouling when the foulants are charged. To reduce membrane fouling, it is appropriate to use a membrane carrying the same electrical charge as the foulants. When the surface and foulant have similar charge, electrostatic repulsion forces between the foulant and the membrane prevent the deposition of the foulants on the membrane, thereby reducing the fouling (Figure 18.2b) [34,35]. Therefore, there have been a number of attempts to reduce fouling by incorporating ionizable functional groups on the membrane surface. For example, a negative surface charge of the membrane may have a beneficial effect during proteins filtration at neutral pH, because most of proteins have negative charge at such conditions [23]. Because NOM are also negatively charged at pH values, which are typical for surface water and seawater, the negatively charged membranes are more resistant to NOM fouling [36]. Similarly, a positively charged membrane showed electrochemical repulsion against positively charged organic compounds such as proteins [37]. Thus, the development of low-fouling membranes should consider the charged properties of target foulants in feed streams.

18.2.3 SURFACE ROUGHNESS

There is a strong correlation between the membrane fouling and the surface roughness for RO and NF membranes. It was shown that the RO hydrophilic cellulose acetate (CA) membranes with smooth membrane surfaces are less prone to colloidal fouling compared to the relatively more hydrophobic and rougher PA membranes [38].

TABLE 18.1

Correlation between the Surface Roughness of Commercial RO/NF Membranes and Their Relative Fluxes for the Filtration of a 0.05 M NaCl Solution, Which Contained 200 mg/l Silica Particles (0.10 μm); pH = 6.8. Flux Decline Values Determined for 10 l of Permeate Volume Filtered

Membrane Type	Flux Decline, J/J_0, %	Average Roughness, nm	RMS Roughness, nm
Osmonics HL	13.9	10.1	12.8
Trisep X-20	38.3	33.4	41.6
Dow NF-70	46.9	43.3	56.5
Hydranautic LFC-1	49.3	52.0	67.4

Source: E.M. Vrijenhoek et al., *J Memb Sci*, 188, 115–128, 2001.

As seen in Table 18.1, fluxes for commercial RO (Hydranautics LFC-1, Trisep X-20) and NF (Dow-FilmTec NF-70, Osmonics HL) membranes during the filtration of NaCl solution with the addition of silica particles decrease when surface roughness of the membrane samples increase. A greater membrane roughness increases the total surface area to which foulants can be attached, and the ridge-valley structure favors the accumulation of foulants at the surface. Using atomic force microscopy (AFM), Vrijenhoek et al. [39] clearly showed that colloidal particles preferentially accumulate in the *valleys* of rough membranes, resulting in *valley clogging*, which causes more severe flux decline than with smooth membranes.

Abu Seman et al. [40] also reported that irreversible fouling of PES membranes strengthened with an increase of surface roughness during the filtration of humic acid (HA) solutions, whereas Hobbs et al. [41] found that the fouling became more severe when treating surface groundwater as the membrane surface roughness increased.

With an AFM adhesion force measurement technique, Bowen et al. [42] characterized an interaction force between a colloidal silica probe and a rough membrane surface. It was found that membrane surface roughness significantly reduced electrostatic repulsion between the colloid and the surface, and the valley regions experienced a greater adhesion force.

On the other hand, Boussu et al. [43] suggested that while colloidal fouling was affected by both membrane hydrophobicity and roughness, membrane hydrophobicity seems to play a more significant role for promoting fouling. Similarly, Park et al. [44] also observed low fouling potential for smooth and hydrophilic semiaromatic piperazine-based PA membranes compared to the more hydrophobic *m*-phenylenediamine-based fully aromatic membranes. These authors attributed the greater antifouling tendency to the large repulsive acid–base interaction for the more hydrophilic poly(piperazine) membranes.

In general, taking into account the importance of the properties of the top membrane layer on the fouling process, it might be concluded that the main goals in the development of low-fouling composite membrane consist of hydrophilization, smoothing, introduction of charged/bactericide groups, and polymer brushes on the membrane surface to minimize undesired interactions with the potential foulants to prevent or reduce the membrane fouling. So far the main techniques that are used for the development of composite membrane include interfacial polymerization (IP), photo/redox-initiated surface grafting, plasma treatment, coating of a thin protective layer onto the membrane, or surface modification of polymer membranes with nanoparticles. These methods of composite membrane fabrication will be discussed in detail in this chapter.

18.3 DEVELOPMENT OF LOW-FOULING COMPOSITE MEMBRANES THROUGH IP

IP is the most important method for the commercial fabrication of thin-film composite (TFC) RO and NF membranes. The first interracially polymerized TFC membranes were developed by Cadotte et al. [45] and represented a breakthrough in membrane performance for RO applications [46]. Original IP protocol involved soaking a microporous PS support in an aqueous solution of a polymeric amine and then immersing the amine-impregnated membrane into a solution of a di-isocyanate in hexane. The membrane was then cross-linked by heat treatment at 110°C [45]. The resulting TFC polyurea membrane had better salt rejection and higher water flux than that of an integrally skinned asymmetric CA membrane [46].

It should be noted that the IP reaction is self-inhibiting through the passage of a limited supply of reactants through the already-formed layer. This results in an extremely thin film of approximately 50 nm thickness [4]. A wide variety of TFC membranes have been successfully developed because of the significant advantages of IP in optimizing independently the properties of the skin layer and microporous substrate layer [47,48]. Various factors such as the concentration of monomers, solvent type, reaction time, and posttreatment conditions affect the structural morphology and composition of the barrier membrane layer [48–51].

Most of the NF and RO membranes produced by IP have a thin layer of PA on top of the membrane support. Among the active monomers used to form a functional PA layer on the RO/NF membranes, m-phenylenediamine (MPD) and trimesoyl chloride (TMC) are the most common (Figure 18.3). Other amine monomers for the production of TFC PA membranes include p-phenylenediamine [22], piperazine [52], triethylenetetramine [53], N-N'-diaminopiperazine and N-(2-aminoethyl)-piperazine [52], and poly(ethyleneimine) (PEI) [54].

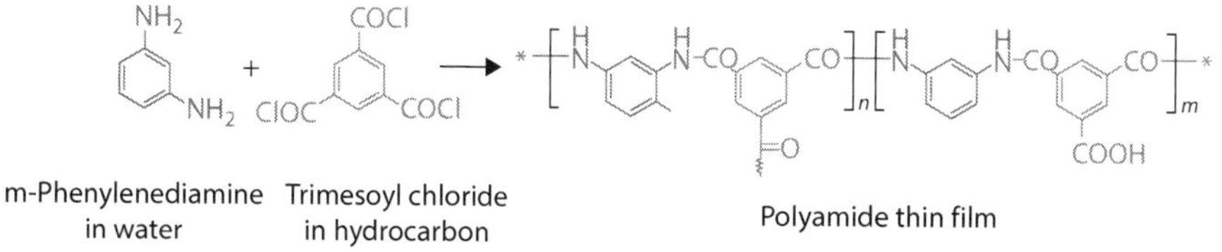

m-Phenylenediamine   Trimesoyl chloride
   in water   in hydrocarbon   Polyamide thin film

FIGURE 18.3 Commercial PA membrane derived from m-phenylenediamine (MPD) and trimesoyl chloride (TMC) through IP. (Data from W.J. Lau et al., *Desalination*, 287, 190–199, 2012.)

Recently, novel monomers have been suggested for the preparation of TFC membranes through the IP technique [52,54]. These monomers contain more functional or polar groups, so the prepared membrane exhibits a smoother surface or better hydrophilicity, which is advantageous to the improvement of antifouling properties of the membranes. For example, Li et al. [55] synthesized two novel tri and tetrafunctional biphenyl acid chlorides, 3,4′,5-biphenyl triacyl chloride and 3,3′,5,5′-biphenyl tetraacyl chloride, which were then used to prepare TFC RO membranes with MPD.

Liu et al. [56] developed a novel RO composite membrane prepared from 5-isocyanato-isophthaloyl chloride and MPD. Because the modified membrane had better hydrophilicity and a smoother surface compared to composite membranes prepared through IP of TMC/MPD and commercial energy-saving polyamide membrane, it showed better resistance to fouling during the filtration of surface water (the static contact angle was 28.5°, 44.3°, and 35.0° and the average roughness was 43.89, 54.36, and 160.2 nm, respectively). These results proved that the antifouling properties of RO membranes are closely correlated with their hydrophilicity and surface roughness.

In addition to the exploration of novel monomers for IP, efforts have been made for the improvement of IP through the addition of active organic modifiers into TMC or MPD solutions. Modifiers can participate in the reaction and are introduced into the functional barrier layer, thus improving the surface property and fouling resistance of resultant RO membranes. For example, Abu Tarboush et al. [22] added 4,4′-methylene bis(phenyl isocyanate) and PEGs of average molecular weight 200 and 1000 Da into the organic phase containing TMC to incorporate *in situ* hydrophilic surface-modifying macromolecules. Prepared membranes, which exhibited significantly higher surface hydrophilicity, were then subjected to long-term fouling studies using model foulants including sodium humate, silica particles, and chloroform spiked in the feeding NaCl solution. Results showed that the flux decline was significantly reduced after the incorporation of organic modifiers into the TFC membranes. A similar approach was also used by An et al. [57], who added polyvinyl alcohol (PVA) into piperazine solution during IP to prepare low-fouling NF membrane. It was also shown that the addition of an acid-acceptor, for example, salt of triethylamine with sulfonic acid in aqueous solution could speed-up the IP reaction by removing hydrogen halide by-products formed during amide bond formation [58,59].

18.4   PREPARATION OF LOW-FOULING COMPOSITE POLYMER MEMBRANES THROUGH SURFACE GRAFTING

Currently surface grafting is widely used by the research community to develop low-fouling composite membranes. In this method, grafted macromolecular chains are covalently bonded to the membrane surface. Surface grafting may be initiated by ultraviolet (UV)-irradiation, chemically, and by plasma or enzymatic treatment of the membrane surface. Choice of the specific graft polymerization technique depends on the chemical structure of the membrane and the desired characteristics after surface modification.

18.4.1 Photoinitiated Grafting on the Membrane Surface

Among the grafting techniques used for membrane modification, photoinitiated grafting is one of the most attractive techniques because of its simplicity, low cost of operation, mild reaction conditions, and a possibility of easy incorporation during the end stages of a membrane manufacturing process. Over the last decade, UV-initiated graft polymerization has been widely used for the surface modification of polymer membranes in attempts to develop composite membranes with enhanced resistance toward organic and biocolloidal fouling. Different hydrophilic monomers such as N-vinyl-2-pyrrolidinone (NVP), N-vinylformamide (NVF), N-vinyl-caprolactam (NVC), 2-hydroxylethyl methacrylate (HEMA), acrylic acid (AA), acrylamide (AAm), 2-acrylamidoglycolic acid (AAG), quaternized 2-(dimethylamino) ethyl metacrylate (qDMAEM), 2-acrylamido-methyl-propane sulfonic acid (AMPS), 3-sulfopropyl methacrylate (SPMA), poly(ethylene glycol) (PEG), poly(ethylene glycol) methacrylate (PEGMA), D-gluconamidoethyl methacrylate (GAMA) ([methacryloylamino]propyl)-dimethyl (3-sulfopropyl) ammonium hydroxide inner salt (MPDSAH inner salt), 2,4-phenylenediamine (PDA), ethylene diamine (EDA), and poly (dimethylsiloxane) have been used for the modification of MF, UF, NF, and RO neat membranes made of polymers, of various chemical properties, including PVDF, PES, PS, PP, PAN, and PE [20,23].

In general, the modification of the membrane surface by UV graft polymerization may be realized *with* or *without* using an appropriate photoinitiator. Formation of starter radicals for graft polymerization may be promoted by the addition of photoinitiators into a modification solution or on a membrane surface. On the other hand, when a chromophore on a polymer backbone absorbs UV light, it goes to an exited state that may dissociate into reactive free radicals, which initiate the grafting process.

18.4.1.1 Membrane Modification by UV Graft Polymerization without the Use of a Photoinitiator

This approach involves the direct generation of free radicals from the neat membrane polymers under UV irradiation. Yamagishi et al. [60,61] modified the surface of UF PS and PES membranes through UV grafting of vinyl monomers using the intrinsic photosensitivity of the PS/PES polymers and their ability to generate free radicals upon UV irradiation. It was shown that HEMA was the most effective monomer to decrease both static adsorption and membrane fouling during BSA filtration. UF fluxes for 0.1wt% BSA solutions were considerably higher (20%–27%) for the membranes grafted with HEMA than for initial (unmodified) membranes, whereas the BSA retentions were similar for both membranes. Adsorption isotherms showed that the amount of BSA adsorbed onto the HEMA-modified PES membrane was about 43, and 28% less than for the unmodified PES membranes for 1 and 2 mg/l of BSA in solution, respectively. Antifouling resistance of the modified membranes was improved due to an increase in their hydrophilicity. However, it was also shown that the pore size of the membrane increased on UV irradiation due to PES photodestruction.

Using the intrinsic photosensitivity of PES, 10 kDa UF membranes were modified through UV-induced surface grafting of NVP, NVF, and NVC [62]. The modified membranes have been shown to have increased hydrophilicity of 12%–25%, as measured by the static contact angle. This resulted in decreased BSA fouling compared to the unmodified PES membrane and a higher relative filtration flux compared with the commercial low-protein-adsorbing membranes cast from regenerated cellulose. The best-performing NVP-modified membrane showed a 25% increase in hydrophilicity, a 49% decrease in BSA fouling, and a 4% increase in BSA retention compared with the unmodified PES membrane.

As a rule, two methods are used for membrane modification by UV graft polymerization without the use of a photoinitiator: an immersion method, when the membranes are UV-irradiated while immersed in a monomer solution; and a dip method, when the membranes are dipped in a monomer solution, and then UV-irradiated in nitrogen or other inert gas [63]. In comparing the immersion and the dip methods, it should be noted that immersion modification usually requires a larger amount of monomer and may be less adaptable to a continuous process on an industrial scale. Moreover, it was shown that at the same conditions (time of irradiation, monomer concentration, etc.), the degree of grafting for the dip method is 2–3 times higher than that of the immersion method [63]. Abu Seman et al. [64] used the immersion method for UV-initiated graft polymerization of AA on the surface of NF PES membrane to reduce its organic fouling (Figure 18.4).

It was shown that at neutral pH, the UV-grafted membranes exhibited a lower irreversible fouling factor (7%–14% depending on modification conditions) than the ungrafted membrane (24%), due to the increase of hydrophilicity and negative surface charge of the modified membranes. It was also found that the proper selection of a monomer concentration and UV-irradiation time are very important for modification, because monomer cross-linking and chain scission of polymer backbone are two parallel competitive processes in UV-irradiation technique, and care should be taken as longer irradiation time may damage the membrane support [65].

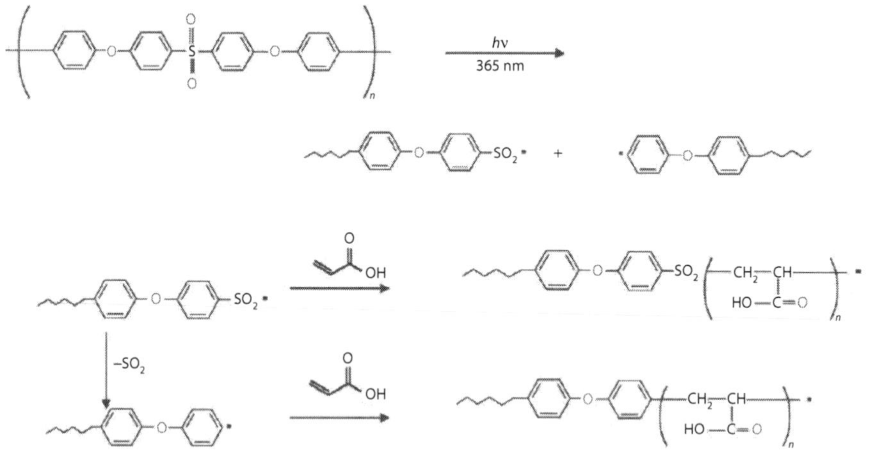

FIGURE 18.4 Photochemical modification of the PES membrane with AA. (Data from M.N.A. Seman et al., *J Memb Sci*, 355, 133–141, 2010.)

18.4.1.2 Membrane Modification by UV Graft Polymerization with a Photoinitiator

When a photoinitiator is used to initiate graft polymerization, the starting radicals are generated on the membrane surface by the reaction of a photoinitiator with a polymer membrane backbone under UV irradiation. Benzophenon (BP) or its derivatives are the most often used for the initiation of the UV-assisted graft polymerization of vinyl monomers on the surface of polymer membranes. In this case, the key step in the initiation of starting radicals for photoinitiated surface grafting is hydrogen abstraction from the polymer backbone.

The photoinitiator may be dissolved in the monomer solution or loaded on the membrane surface by adsorption. For example, Yamagishi et al. [60] modified PES membranes with HEMA using BP dissolved in the monomer solution followed by UV irradiation of the solution and the immersed membrane. This approach is relatively simple; however, its main drawback is a low local concentration of BP on the membrane surface because BP moves to the membrane surface only by diffusion. This results in a low grafting efficiency, whereas high bulk BP concentration usually causes a side reaction such as homopolymerization. In addition, the use of monomers that do not have a common solvent with BP is limited; for instance, BP is almost insoluble in water.

To improve the grafting process, the preliminary adsorption of a photoinitiator on the membrane surface was used [66]. In this way, the local BP concentration on the membrane surface was increased, whereas the BP concentration in the bulk of monomer solution was kept very low to reduce the homopolymerization process. Using this method, PAN membranes have been grafted with PEGMA to prepare a low-protein-adsorbing UF membrane with relatively high permeability [67].

Similarly, in an attempt to enhance the membrane resistance to biofouling, microfiltration PES and PVDF membranes were photochemically modified with a thin layer of a number of hydrophilic polymers: neutral poly(HEMA), negatively charged poly(AMPS), positively charged poly(AMPS), and bactericidal poly(qDMAEM) [68]. Higher grafting values were obtained for HEMA compared with qDMAEM and AMPS monomers. It was suggested that due to the electrostatic repulsion of similarly charged molecules, the possibility of growing grafted chains for positively charged qDMAEM and negatively charged AMPS monomers diminishes compared with HEMA. As a result, grafted poly(AMPS) and poly(qDMAEM) chains are shorter than those for poly(HEMA). In filtration tests with *Escherichia coli* suspensions, it was found that the chemical nature of the membrane surface significantly affected the ability of the modified membranes to recover their fluxes after washing. As seen in Table 18.2, the membranes, with either positively or negatively charged surfaces are more inclined to fouling compared with the membranes with neutral hydrophilic surfaces.

Obviously, the charged surface regions serve as convenient sites for the attachment of microorganisms due to the electrostatic attraction and the formation of ionic bonds [69]. The highest value of flux recovery (J_1/J_0) for the membrane modified by HEMA is explained by the absence of electrostatic attraction between the bacterial cell and the membrane surface (Table 18.2). As a result, the cells are more

TABLE 18.2

Performance of Surface-Modified PES Membranes in the Filtration of *E. coli* Suspensions

Membrane	$J_{c1}{}^a$	$J_{c2}{}^a$	J_{c2}/J_{c1}, %	$J_1/J_0{}^b$ %	Antibacterial Activity,[c] %
Initial PES	900	270	30	51	0
PES + HEMA	690	240	35	72	0
PES + AMPS	650	235	36	63	25
PES + qDMAEM	680	225	33	51	97

Source: V.M. Kochkodan et al., *Colloid J+*, 68, 267–273, 2006.

Degree of modification of the composite membranes is 500 ± 20 µg/cm². The concentration of suspension is 1.58×10^4 colonies forming units/cm³, operating pressure is 0.5 bars, and filtration time is 1 h.

[a] J_{c1} and J_{c2} are the membrane fluxes at the beginning and at the end of filtration cycle, respectively.

[b] J_0 and J_1 are the initial water flux and water flux after filtration of the bacterial suspension and washing with distilled water, respectively.

[c] Antibacterial activity is a ratio of the numbers of bacterial colonies that were grown on the initial and modified membranes at the identical conditions.

easily removed from the membrane surface during washing, as compared with the membranes with charged surfaces.

To reduce the formation of undesired homopolymer and cross-linked or branched polymer in the *grafting-from* approach with an initiator, a photo-induced living graft polymerization method for surface membrane modification was proposed by Ma et al. [70]. In the first step, BP abstracts hydrogen from the substrate to generate surface radicals and semipinacol radicals, which combine to form surface photoinitiators in the absence of monomer solutions (Figure 18.5). In the subsequent step,

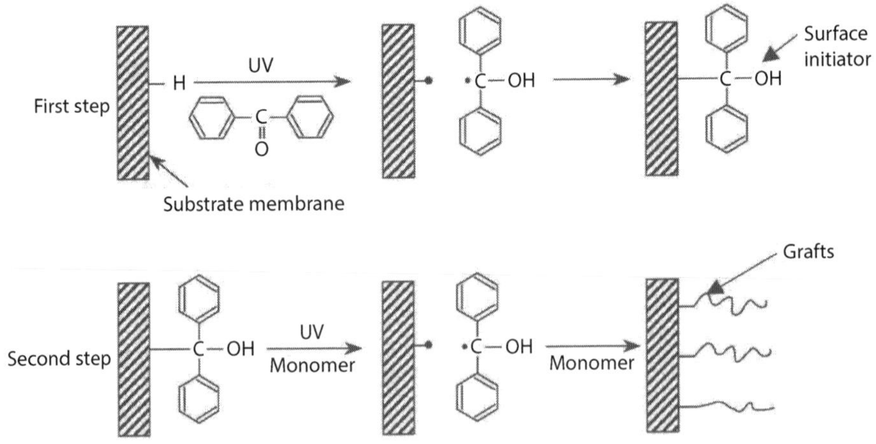

FIGURE 18.5 Schematic diagram of a photo-induced living graft polymerization on a membrane surface. (Data from H.M. Ma et al., *Macromolecules*, 33, 331–335, 2000.)

the monomer solutions are added onto the active substrate, and the surface initiators initiate the grafting process under UV irradiation.

In general, it should be noted that very detailed studies have been carried out on the surface membrane modification through UV-initiated graft polymerization, including investigations on grafting mechanisms, dependence of grafting efficiency on grafting time, monomer type, UV wavelength, and intensity. Attractive features of UV grafting are easy and controllable introduction of graft chains with a high density and their exact localization to the membrane surface. Furthermore, the covalent attachment of graft chains onto a polymer surface avoids their delaminating and assures the long-term chemical stability of introduced chains in contrast to physically coated polymer layers. Disadvantages of the modified membranes are usually their reduced fluxes compared with those of the unmodified membranes because the grafting layer adds extra hydraulic membrane resistance. Grafting may also increase manufacturing costs due to the additional use of organic solvents, monomers, and UV equipment.

18.4.2 REDOX- AND MISCELLANEOUSLY INITIATED GRAFTING

18.4.2.1 Redox-Initiated Grafting

In contrast to UV-initiated graft polymerization, the redox-initiated grafting allows the modification of the polymer membranes *in situ*, even inside commercial wound membrane elements [71]. Redox system most often used consists of potassium persulfate ($K_2S_2O_8$) and potassium metabisulfite ($K_2S_2O_3$) [71,72]. Redox reaction that occurs generates radicals on the membrane surface, upon which monomers can be attached:

$$S_2O_8^{2-} + S_2O_3^{2-} \rightarrow SO_4^{-*} + SO_4^{2-} + S_2O_3^{-*}$$

$$SO_4^{-*} + H_2O \rightarrow HSO_4^- + OH^*$$

In the initiation step, a monomer molecule is attached to the SO_4^{-*} and $S_2O_3^{-*}$ radicals. Subsequently, the propagation step occurs, through which the chain grows. Termination of the polymerization reaction can occur through the recombination of two growing chains, or a growing chain can subtract a radical from another growing chain, leading to the formation of a double bond [72].

The described redox system was used to generate starting radicals for graft polymerization of AA and other hydrophilic monomers on the surface of CA membranes [73]. It was shown that despite the gradual decrease of the flux, the surface-modified membranes had a lower protein sorption and a better and more reversible flux recovery after cleaning [74].

A similar approach was used for the *in situ* preparation of NF PES membranes containing hydrophilic functional groups such as SO_3H, COOH, or C(= O)NH_2 [75]. Testing the modified and unmodified membranes over a period of 30 days demonstrated that surface-modified composite membranes have better fouling-resistance characteristics. In the case of the unmodified membranes, the flux decreased from 41.65 to 19.21 l/m²h, whereas for the surface-modified membranes under similar conditions, the flux reduced from 46.75 to 31.62 l/m²h. However, in the case of NF

PA membranes, it was observed that polymerization could take place inside the pores of the base support membrane as a result of the penetration of the monomer through the active layer, particularly for high degrees of grafting [76].

Belfer et al. [77] followed a similar grafting procedure using a range of monomers, including AA, MAA, PEGM, and SPMA. It was also shown that when a high monomer concentration is used, the membrane permeability may significantly decrease (up to 30%) because the grafted chains can penetrate into the membrane and block the pores.

Van der Bruggen [72] used a redox system for methacrylic acid (MAA) grafting on the surface of NF PES membranes. It was shown that the contact angles of the modified membranes were in the range 59°–66° (compared to 70° for the unmodified membrane); however, the positive effect of hydrophilization is counterbalanced by the narrowing of pores, even to the extent that pores may be entirely blocked at high grafting degrees.

Wei et al. [78] used 2,2'-azobis(isobutyramidine) dihydrochloride, which can be thermally decomposed to generate starting radicals for graft polymerization of 3-allyl-5,5'-dimethylhydantoin on the surface of RO membrane. It was shown that the modified membranes had lower contact angles than those of the raw membranes, indicating the increase of surface hydrophilicity. After exposure to microbial cell suspension, the modified membranes showed slight decrease in pure water flux and less adsorption of microbial colonies on the surface, which verified the improvement of antibiofouling properties.

18.4.2.2 Enzymatic Grafting

This method uses enzymes to convert the substrate (monomer, oligomer, or polymer chains) into reactive free radical(s), which undergoes subsequent nonenzymatic reaction with the membrane [79–82]. For example, Nady et al. [83] used laccase from *Trametes versicolor* to create free radicals and graft phenolic acid monomers (e.g., gallic acid or 4-hydroxybenzoic acid) to the membrane.

Gullinkala and Escobar [84] used porcine pancreatic lipase to catalyze the polycondensation of PEG to the surface of the CA membranes (Figure 18.6). The virgin and modified membranes showed the comparable initial flux values and retentions

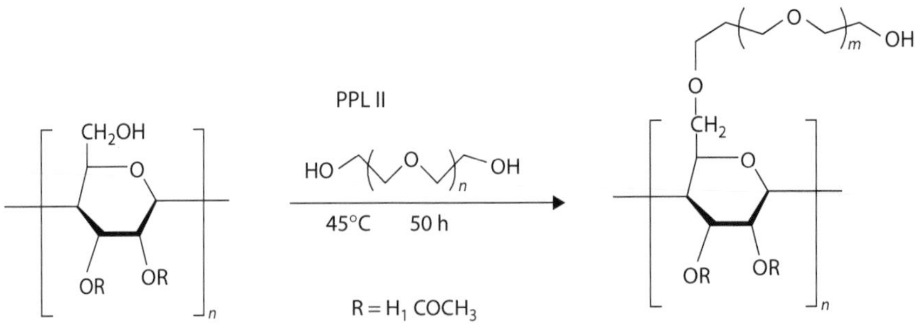

FIGURE 18.6 Polycondensation reaction of PEG to CA membrane in the presence of porcine pancreatic lipase (PPL II). (Data from T. Gullinkala and I. Escobar, *J Memb Sci*, 360, 2010, 155–164.)

during the filtration of dextran and BSA solutions. However, the flux recovery after UF of NOM followed by backwashing was quite different; the modified membrane regained nearly 97% of its initial flux value within 40 min of filtration, whereas the unmodified membrane recovered only by 85%. A lower HA accumulation on the membrane surface was also found for modified membranes. These improvements are believed to be due to the high flexibility of the high hydrophilic-grafted PEG chains that prevents the membrane fouling.

With respect to health and safety issues, enzymes offer the potential of eliminating the need for reactive reagents (and solvents) and their associated hazards. It should be noted, however, that the membrane modification through the described approach takes rather a long time, about 50 h [84].

18.4.3 PLASMA-INITIATED GRAFTING AND INITIATED CHEMICAL VAPOR DEPOSITION ON THE MEMBRANE SURFACE

18.4.3.1 Plasma-Induced Grafting of the Membrane Surface

Plasma modification of polymer membranes, including plasma-initiated grafting on the membrane surface, has been recently reviewed in several articles [85–87]. Plasma-generated radicals on the polymer surface are stable in vacuum but can react rapidly when exposed to gaseous monomers or monomer solutions. Starting from the surface radicals that are created during the plasma activation, the grafted macromolecular chains grow as shown in Figure 18.7.

Grafting density and length of grafted chains can be controlled by plasma parameters, including power, pressure, and sample disposition, and polymerization conditions such as monomer concentration and grafting time [89].

Gancarz et al. [90] compared the three different approaches to modify PS membranes with AA through plasma-initiated graft polymerization: (1) grafting in solution, the plasma-treated polymer membrane was exposed to air for 5 min and dipped into a deaerated aqueous solution of monomer; (2) grafting in vapor phase, when Ar plasma treatment on polymers was completed, a monomer vapor was introduced into the chamber; and (3) plasma polymerization of monomer vapors in a plasma reactor. It was shown that modified PS membranes prepared in a vapor phase possessed the highest flux.

A similar grafting process in an He or He/water vapor phase was used by Ulbricht and Belfort for the modification of UF PAN and PAN/PS membranes with AAm,

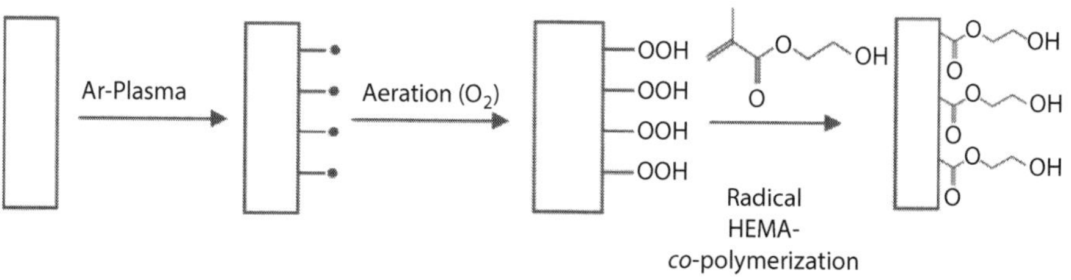

FIGURE 18.7 Schematic representation of plasma-induced HEMA grafting on the membrane surface. (Data from D. Tyszler et al., *Desalination*, 189, 119–129, 2006.)

MAA, and HEMA [91,92]. It was shown that a loss of hydraulic permeability due to the modification may be compensated by a higher permeate flux during the filtration of aqueous solutions due to reduced organic fouling.

Yu et al. [93] used air plasma-initiated grafting of PVP in attempts to reduce fouling of PP membrane in a submerged membrane bioreactor. It was shown that after continuous operation for about 50 h, flux recovery, reduction of flux, and relative flux ratio for modified membranes were 53% higher, 17.9% lower, and 79% higher, respectively, than those of the initial membranes. Water contact angle on the PVP-immobilized membrane showed a minimum value of 72.3°, approximately 57° lower than that of the unmodified membrane.

Kim and co-workers presented a study on surface nanostructuring of RO membranes through atmospheric pressure plasma-induced graft polymerization for the improvement of fouling resistance and flux performance [94]. The PA RO membranes were activated with impinging atmospheric plasma, followed by a solution free-radical graft polymerization of water-soluble monomers, including MAA and AAm, onto the surface of the membrane. The results showed that grafted polymer layers on the PA surface resulted in RO membranes of significantly lower mineral scaling propensity, compared to the commercial RO membrane with the same salt rejection.

Zou et al. [95] grafted a PEG-like hydrophilic polymer (trimethylene glycol dimethyl ether) onto PA RO membrane by plasma polymerization to reduce organic fouling. After modification, the surface contact angle was reduced from 32° to 7°, indicating an enhanced hydrophilicity of the modified membrane. After 210 min, the filtration of BSA or alginate solutions, no flux decline was found for the modified membrane, whereas a 27% reduction of the initial flux was observed for the untreated membrane. Moreover, the modified membrane was easily cleaned. The flux recovery after cleaning by water was only up to 99.5% for the modified membrane, whereas it was only 91.0% for the untreated one. The authors suggested that van der Waals forces were driving water molecules to create a protective layer around the introduced plasma polymer chains, and thus formed a protective layer over the modified membrane surface. A reasonable plasma polymerization time was recommended as 30 s to ensure the stability and long life span of the collated polymer film. It was shown that the longer plasma treatment time reduced the initial water flux by plugging the pores of the membrane layers, whereas at the shorter plasma treatment time, the deposited polymer layer was too thin to be stable and may have been lost during the filtration [95].

In general, it can be concluded that the advantageous features of membrane plasma treatment are very short treatment time and a possibility to precisely adjust the surface membrane properties without affecting the bulk of the membrane. However, it should be noted that the plasma-modification parameters are highly system-dependent, and there are some problems with the reproducibility of the properties of plasma-treated membranes. It is also not easy to scale up from an experimental setup to a larger plasma reactor.

18.4.3.2 Initiated Chemical Vapor Deposition

Initiated chemical vapor deposition (iCVD) is an all-dry free radical polymerization technique performed at low temperatures and low operating pressures [96]. Using this technique, an antifouling copolymer layer containing poly-(sulfobetaine) zwitterionic groups was covalently grafted on commercial RO membrane (Figure 18.8). For a 30 nm coating with a composition of 35% sulfobetaine, the permeation flux was reduced by 15% compared to unmodified RO membranes, but cell adhesion tests using *E. coli* showed that the modified RO membranes exhibited superior antifouling performance compared to the unmodified RO membranes.

Composition and the smoothness of the iCVD layer have been demonstrated to reduce the attachment of proteins and bacteria to the surface (Figure 18.9).

Because of the low substrate and filament temperatures, iCVD enables the synthesis of low roughness ultra thick and hydrophilic layers on virtually any porous/nonporous substrate with the systematic control of layer thickness through deposition time (typically ~ several minutes). Another advantage of this method is the avoidance of damage to the surface morphology and the membrane structure by the

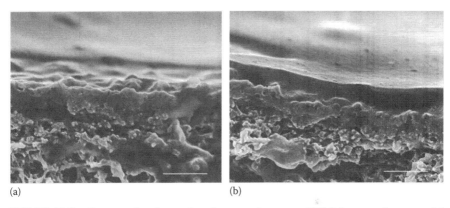

(a)   (b)

FIGURE 18.8 Cross-sectional scanning electron microscopy (SEM) images of commercial RO membrane (Koch TFC-HR) (a) before and (b) after iCVD surface modification. (Data from R. Yang et al., *Chem Mater*, 23, 1263–1272, 2011.)

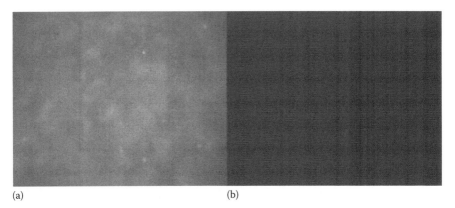

(a)   (b)

FIGURE 18.9 Fluorescence micrographs of commercial RO membrane surface incubated with fluorescently labeled *E. coli* (a) before and (b) after iCVD coating. Lack of fluorescence in (b) indicates the iCVD surface resists the attachment of the bacteria. (Data from R. Yang et al., *Chem Mater*, 23, 1263–1272, 2011.)

solvent. On the other hand, similar to plasma treatment, it seems difficult to scale up the iCVD coating from a small to a larger reactor.

18.5 FABRICATION OF LOW-FOULING COMPOSITE MEMBRANES THROUGH PHYSICAL COATING

Coating a thin layer of water-soluble polymers, surfactants, or polyelectrolytes from solution by physical adsorption is a simple and flexible technique to optimize hydrophilicity, smoothness, and surface charge of the membrane surface [97].

The first studies in this field dealt with the modification of UF membranes to reduce their fouling with proteins [98–100]. Kim et al. [98] showed that fouling of UF membranes with proteins may be reduced by surface adsorption with water-soluble polymers such as PVA, methylcellulose (MC), and PVP. The treatment provided an increase in initial UF flux and a slower flux decline. MC was the most effective of the polymers tested in enhancing UF flux, showing an average flux advantage of 30%–40% for the first usage. Nonionic, hydrophilic polymers were found to be most effective in minimizing lactoglobulin and ovalbumin adsorption as well as in decreasing membrane resistance during UF, while the application of surfactants and ionic polymers was less successful [99].

Thereafter, many hydrophilic polymers and polyelectrolytes, such as PVA, polyacrylic acid (PAA), PEG-based hydrogels, chitosan, poly(sodium 4-styrenesulfonate) (PSS), and PEI, have been coated on different MF, UF, NF, and RO membranes using casting [101–105], static adsorption (dipping) [106], or dynamic (flow-through) adsorption [107–109].

18.5.1 COATING THROUGH CASTING

Sagle et al. [104] used a drawdown coating of cross-linked PEG-based hydrogels modified with RO membranes to reduce their fouling. Cross-linked PEG-based hydrogels were synthesized through photoinitiated copolymerization of PEG diacrylate as the cross-linker and PEG acrylate, 2-hydroxyethyl acrylate, or AA as the co-monomers. It was evaluated that the coatings deposited on the membrane surface were approximately 2 μm thick. It was shown that water fluxes of coated membranes were smaller than those of the uncoated membranes, but the fouling of the modified membranes with cationic dodecyltrimethyl ammonium bromide (DTAB), anionic sodium dodecyl sulfate surfactants, and oil/water emulsions was reduced. In the filtration of oil/water emulsion made with DTAB, the flux of the base membrane after 24 h decreased to 26% of its initial value, whereas the water flux of a PEGDA-coated RO membrane was 73% of its initial value. It was shown that the membrane-surface charge correlates with fouling properties of the membranes; negatively charged membranes foul extensively in the presence of positively charged surfactants and experience minimal fouling in the presence of negatively charged surfactants.

Du et al. [105] improved hydrophilicity and surface smoothness of commercial UF PVDF membranes by surface coating with a PVA aqueous solution followed by solid-vapor interfacial cross-linking with glutaraldehyde. It was shown that during

UF of surface water of the Grand River with a total organic carbon content of 7 mg/l, the flux of the PVA/PVDF membrane was 14% higher than that of the unmodified PVDF membrane after 4 h of filtration and 95% higher after 18 h of filtration (Figure 18.10).

As seen in Figure 18.10, the flux of the modified membrane reached a plateau with increasing operating time, whereas the flux for the unmodified membrane kept decreasing, thus indicating that foulants continued to accumulate on the surface over time. Additionally, the cake-fouling layer could be more easily removed from the PVA-modified membrane by alkaline cleaning. Higher performance of the modified membrane was related to the increase in hydrophilicity and smoothness of the membrane surface after coating with the PVA layer.

Asatekin et al. [103] prepared novel composite NF membranes by casting the synthesized amphiphilic copolymer PAN-*graft*-poly(ethylene oxide) (PAN-*g*-PEO) on UF PAN membranes. The coated membranes were immersed in isopropanol for 30 min, and thereafter in a water bath. It was shown that during precipitation, the copolymer undergoes microphase separation, forming interpenetrating networks of PAN-rich and PEO-rich nanodomains. Transmission electron microscopy (TEM)

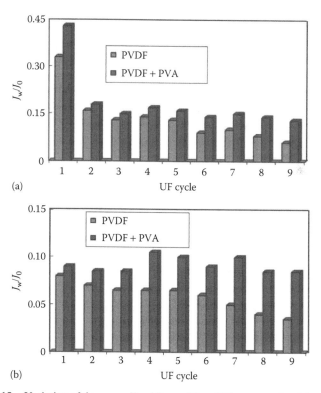

FIGURE 18.10 Variation of the normalized fluxes (Grand River water flux/clean water flux) between base PVDF and modified PVA/PVDF membranes: (a) at the beginning of UF (a) and (b) at the end of UF. Duration of UF cycle is 2 h, followed by the membranes cleaning with NaOH solution. (Data from J.R. Du et al., *Water Res*, 43, 4559–4568, 2009.)

reveals that PEO domains act as water-permeable nanochannels and provide the size-based separation capability of the membrane. A small decline in flux (15%) was observed in a 24 h dead-end filtration experiment with 1 g/l BSA solution for modified membrane, whereas the base UF membrane lost 81% of its flux irreversibly at the same conditions. It was concluded that a PEO *brush* layer, formed on the membrane surface, acts as a steric barrier to protein adsorption, endowing these membranes with exceptional fouling resistance.

Composite UF PVDF membranes modified with a self-assembling graft copolymer PVDF-*graft*-poly(oxyethylene) methacrylate showed good fouling resistance for BSA, HA, and sodium alginate at feed concentrations of 1000 mg/l and for activated sludge [104]. For example, dead-end filtration of activated sludge with 1750 mg/l volatile suspended solids resulted in constant flux throughout the 16 h filtration period. Interfacial force measurements with AFM showed the presence of steric foulant–membrane repulsive forces and a lack of adhesion forces between the foulant and the membrane. However, a possible ester bond linkage of the PEO side chain in acidic or basic media may restrict the application of the modified membranes.

Louie et al. [110] coated both high-flux (ESPA1 and ESPA3) and low-flux (SWC4) commercial RO PA membranes with a very hydrophilic block copolymer of polyether–PA (PEBAX® 1657). 1 wt% coating solutions were prepared by dissolving the copolymer in *n*-butanol under reflux conditions. The coating was applied to the membrane surface using a custom-made dip-coating apparatus, and then immediately dried in an oven at 60°C. The coating greatly reduced the surface roughness and during a long-term (106 days) fouling test with an oil/surfactant/water emulsion, the rate of flux decline was slower for coated than for uncoated SWC4 membranes, resulting in consistently higher flux values after 15 days (Figure 18.11).

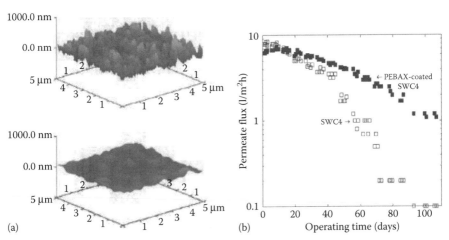

(a)   (b)

FIGURE 18.11 (a) AFM images of uncoated SWC4 (top) and SWC4 coated with 1 wt% PEBAX 1657 (bottom) and (b) flux through uncoated (open squares) and 1 wt% PEBAX 1657-coated (black squares) SWC4 membranes during the treatment of an oil/surfactant/ water emulsion. Data are plotted on a semilogarithmic scale. (Data from J.S. Louie et al., *J Memb Sci*, 280, 762–770, 2006.)

18.5.2 STATIC/DYNAMIC DEPOSITION OF PROTECTIVE LOW-FOULING LAYER ON THE MEMBRANE SURFACE

For coating of the membrane surface with hydrophilic polymers and polyelectrolytes, static and dynamic adsorption techniques may be used. Static adsorption implies simple immersion (dipping) of porous support in appropriate polymer solution, whereas the modified polymer/polyelectrolyte solution is forced to move on/through the support membrane for deposition of a protective low-fouling polymer layer when dynamic adsorption is used.

A nearly neutrally charged NF membrane has been developed by adsorption of a layer of negatively charged sulfonated poly(ether ether ketone) onto the surface of a positively charged NF membrane [97]. When using BSA, HA, and sodium alginate as the model foulants, the modified membrane exhibited much better fouling resistance than both the positively and the negatively charged membranes. Foulants would be less likely to deposit onto the membrane due to the elimination of the charge interaction between the membrane and the foulants.

Charged membranes were prepared by coating UF PES membranes with sulfonated poly(2,6-dimethyl-1,4-phenylene oxide) [111]. Membranes were dipped in methyl methacrylate-based comb polymers with short oligoethylene glycol side chains that provided the membrane with long-term, biorepellant surfaces; cell-lysate flux recovery increased from 47% for unmodified membrane to 94% for the coated membrane after a five-cycle filtration–washing process [112]. It is claimed that this is caused by the hydrophilic PEO groups on the surface.

N′carboxymethyl chitosan/poly(ethersulfone) (CM-CS/PES) composite membranes were prepared by immersing PES MF membranes into CM-CS solutions and cross-linking with glutaraldehyde [113]. Streaming potential measurements indicate that CM-CS/PES composite membranes possess a weak positive charge at low pH and a rather strong negative charge at high pH. Therefore, the negative electrostatic repulsion interactions between membrane and protein molecules at pH 6–8 (i.e., above BSA isoelectric point) resist protein fouling at these conditions.

Ba et al. [97] used the dynamic deposition of water-soluble polymers such as PVA, PAA, and polyvinyl sulfate-potassium salt (PVS) on the surface of positively charged P84-PEI membrane to form a protective coating layer to improve membrane fouling resistance. PVA, PAA, and PVS as the coating materials represented neutral, partially charged, and highly charged polyelectrolytes, respectively. Surface coating experiments were carried out in a cross-flow filtration cell through circulation of a 50 mg/l PVA, PAA, or PVS polyelectrolyte aqueous solution over the base membrane for 8–12 h. It was shown that by applying these coatings, the hydrophilicity, smoothness, and surface charge may be modified and optimized. This reduced the membrane fouling with BSA, HA, and sodium alginate. Membrane surface charge was observed to play the most important role in foulant adsorption. The uncoated membrane had a strong positive charge, so that foulants such as BSA, HA, and sodium alginate were adsorbed quickly and firmly. The PVA-coated membrane also had a positive charge, and fouling by negatively charged materials such as HA and sodium alginate was still high. PVS-coated and PAA-coated membranes had a low surface charge, and, as a result, the fouling with BSA and HA was diminished due to a reduction in the charge interactions.

UF PES membranes with negatively charged sulfonic acid groups on the surface were obtained by the filtration of an aqueous solution of PSS for about 100 min using a dead-end filtration cell [108]. It was shown that thin porous membranes are primarily modified only on the top surface because the PSS macromolecules are not able to enter the pores. However, for membranes with wider pores, PSS permeation resulted in the formation of charged groups on both the surface and pore walls of the membrane. The major difference between the modified and unmodified membranes was found in their flux recovery after UF of the PEG and dextrans solutions. Flux recovery ratios of >90% were obtained for the modified membranes compared with 55% for unmodified membranes. Thus, the surface-modified membranes have better *cleanability* and *antifouling* characteristics than the base membranes.

Boributh et al. [109] compared three different methods for modifying MF PVDF membranes with chitosan to reduce BSA fouling. These were as follows: (1) an immersion method, when the membrane was immersed in a chitosan solution for a fixed time; (2) a flow-through method, when the chitosan solution was filtered through the membrane; and (3) the combined flow-through and the surface-flow method. It was shown that the membranes modified by a combined flow-through and a surface-flow method showed improved antifouling properties compared with others. This is due to the deposition of chitosan both on the surface and in the pores, resulting in the prevention of BSA adsorption. For a membrane modified by immersion, the chitosan was deposited only on the membrane surface. Therefore, BSA could be adsorbed easily on the pore walls, which led to a high flux decline and irreversible fouling.

In the case of polyelectrolytes, layer-by-layer (LbL) assembly may be used for the preparation of a low-fouling protective layer on the membrane surface [114]. In this method, a polycation and a polyanion are alternately deposited on a porous substrate and are adsorbed by electrostatic interaction (Figure 18.12).

Advantages of the LbL method are that film thickness can be controlled at the nanometer scale [115] to allow for high water flux. Further, properties of the deposited layer can be optimized by varying the types of polyelectrolytes [116] and deposition conditions [117]. Surface charge of the film can be either positive or negative depending on whether the outer film is specified with a polycation or polyanion. It should be noted that the membrane surface with rough morphology can be significantly smoothed by LbL polyelectrolyte deposition (Figure 18.13). LbL assembly is, therefore, expected to reduce membrane fouling.

To reduce membrane fouling of RO membrane, Ishigami et al. [118] used LbL assembly between PSS (M_w = 70 kDa) and poly(allylamine hydrochloride) (M_w = 70 kDa). Polyelectrolytes solutions of 10 g/l were alternately supplied to the membrane surface for 30 min in a cross flow cell at operating pressure of 0.75 MPa.

As seen in Figure 18.14, the antifouling capability of modified membranes is improved with the increasing number of deposited polyelectrolyte layers. This occurs because the membrane surface becomes hydrophilic and the surface roughness becomes smooth with increasing number of layers. On the other hand, the true water permeability increases up to four layers and then declines, as the increased

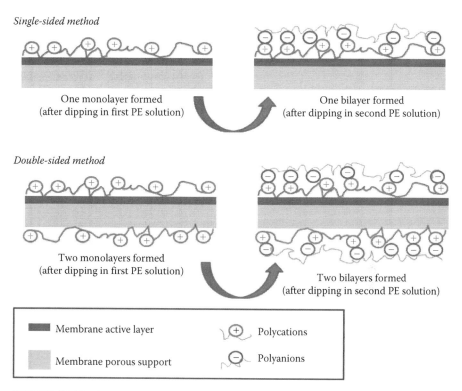

Single-sided method

One monolayer formed
(after dipping in first PE solution)

One bilayer formed
(after dipping in second PE solution)

Double-sided method

Two monolayers formed
(after dipping in first PE solution)

Two bilayers formed
(after dipping in second PE solution)

■ Membrane active layer   ⊕ Polycations

  Membrane porous support   ⊖ Polyanions

FIGURE 18.12 Schematic representation of membrane modification by the LbL method when only the active side of the membrane (single-sided deposition) or both sides of the support membrane are exposed and deposited with polyelectrolyte macmolecules (double-sided deposition). (Data from L.Y. Ng et al., *Adv Colloid Interfac*, 197–198, 85–107, 2013.)

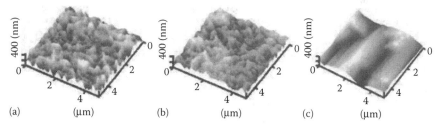

(a) (μm) (b) (μm) (c) (μm)

FIGURE 18.13 AFM images of the outer surface of polyelectrolyte multilayered RO membrane: (a) unlayered (original RO membrane), (b) six-layered, and (c) 12-layered. (Data from T. Ishigami et al., *Sep Purif Technol*, 99, 1–7, 2012.)

hydrodynamic resistance for the thicker assembled layers negates the antifouling improvement. In addition, it is seen that the water permeability of an even numbered layer is higher than those of its preceding and successive odd number layers. BSA, with isoelectric point 4.8, is negatively charged in the feed solution at $pH = 7$. Adsorption of negatively charged BSA onto the negative-charged surface is suppressed by electrostatic repulsion in addition to the enhanced hydrophilicity on the

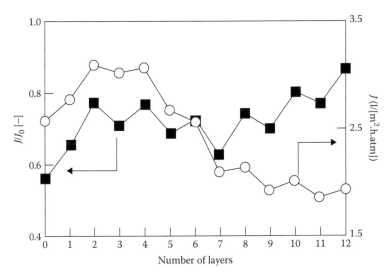

FIGURE 18.14 Effect of the number of layers on water permeability (unfilled circles) and relative permeability (black squares) of the polyelectrolyte multilayered RO membrane after 120 min filtration of 950 mg/l BSA solution. Relative permeability is defined as the permeability at 120 min (J) divided by the initial water permeability (J_0). (Data from T. Ishigami et al., *Sep Purif Technol*, 99, 1–7, 2012.)

membrane surface in the case of an even numbered layer. On the other hand, for odd-numbered layers, the contributions to the antifouling capability are hydrophilicity enhancement on the membrane surface and surface smoothing. Thus, the antifouling capability is more pronounced when even-numbered layers are applied to the original membrane.

Malaisamy et al. [119] showed that the magnitude of the zeta potential increased with an increase in the number of bilayers on the surface of composite PA membranes modified by the LbL method with PSS/poly(diallyldimethylammonium chloride) (PDADMAC). It was also found that the positive or negative sign of the zeta potential of the membranes was dependent on the last polyelectrolyte layer on top of the membrane surface. Thus, by controlling the type of final polyelectrolyte layer and the number of bilayers, the membrane zeta potential, and thus separation performance and antifouling suitability of the membranes, can be controlled.

Su et al. [120] used both static and dynamic adsorption for LbL deposition of PDADMAC and PSS on the surface of UF PS membranes. It was shown that the method used to deposit the polyelectrolyte layers did not significantly alter the membrane surface hydrophilicity, and it was found that the membrane hydrophilicity was only greatly improved when more than five bilayers were adsorbed onto the membrane surface.

In general, because of its efficiency, the coating method is used by manufacturers to produce the so-called low fouling or fouling-resistant TFC PA membranes [121–123]. Tang et al. [124] characterized several widely used commercial RO and NF PA membranes by AFM, TEM, contact angle measurement, and streaming potential analysis and found that some commercial RO membranes were coated with aliphatic polymeric alcohol (which seemed to be PVA). They found that the PVA-coated membranes, in general, have neutral, more hydrophilic, and less rough surface compared with the membranes without PVA coating. In this way, the rough and relatively hydrophobic PA rejection layer is masked by a neutral, smooth, and hydrophilic layer, with the carboxylic group no longer available (or less accessible) for specific interaction with ions in the feedwater.

At this point, it should be mentioned that depending on the adsorption affinity with the membrane surface, the adsorbed coating layer can be stable or removable. The thin-coated films prepared through the deposition of positively and negatively charged polyelectrolyte show good stability due to the electrostatic attraction between the membrane surface and the deposited layers [125]. On the other hand, for hydrogen-bonded modified layers, the strength of hydrogen bonding between the membrane surface and the deposited layer can be altered by making changes in solution pH; thus, these layers can be removed and replaced [126,127]. For example, the cleaning procedure for PVA-coated membranes includes membrane treatment with HCl at pH 2 and stirring for 15–20 h [97]. Thus, if membrane fouling occurs, the PVA layer and the attached foulants can be removed by acid cleaning to refresh the membrane. It may be much easier and more cost-effective to remove and replace the coating layer instead of replacing the membrane.

In general, the adsorbed coatings are relatively simple to apply, and the process can be performed in commercial membrane elements. In addition, the type of coating can be tailored to the specific application of interest. It should be noted, however, that the deposition of coating layers may reduce water flux after modification. For example [119], a 50% loss in the pure water flux was recorded in comparison to the membrane before LbL modification with polyelectrolytes. Therefore, for practical purposes, the coating layer should have an inherently high water permeability and be made sufficiently thin to maintain the water flux as much as possible.

However, despite the flexibility of the coating methods to change the hydrophilicity, smoothness, and charge of the membrane surface, their main drawback is the limited stability of the modified layer with time, because of the possible desorption of the coated/adsorbed polymers from the membrane surface into the bulk of the feed solutions. As a result, the antifouling properties of modified membranes may be gradually deteriorated during long-term operation. Additionally, the solubility of polymers/polyelectrolytes in water might contribute to the contamination of the feed solution in some cases.

18.6 INCORPORATION OF NANOPARTICLES IN COMPOSITE MEMBRANES FOR FOULING REDUCTION

Use of nanoparticles in modifying polymeric membranes has received much attention during the last few years in attempts to enhance flux and reduce fouling [55,128,129]. Hybrid low-fouling inorganic/organic composite membranes can be prepared by directly coating or depositing inorganic particles onto the membrane surface. Kwak et al. [130] performed one of the first studies in this field. TiO_2

nanoparticles of approximate size 2 nm were immobilized through self-assembly with terminal functional groups on the surface of PA RO membranes. X-ray photoelectron spectroscopy demonstrated quantitatively that TiO_2 nanoparticles were tightly self-assembled with a sufficient bonding strength to the membrane, which meant that particles could withstand various washing procedures and RO operating conditions. Self-assembly mechanism of fixing TiO_2 nanoparticles on the membrane surface with COOH functional groups may include bonding with the two oxygen atoms of the carboxylate group through a bidentate coordination to Ti^{4+} cations or through the formation of a hydrogen bond between a carbonyl group and the surface hydroxyl group of TiO_2 [130].

Self-assembly procedure was also used by Bae et al. [132] for modifying sulfonated PES membranes and by Mansourpanah et al. [133] for coating PES/PI blend membrane and OH functionalized PES/PI membrane with TiO_2 nanoparticles. Luo et al. [131] also applied a similar approach for the deposition of TiO_2 nanoparticles onto the PES membranes. For the PES membranes, self-assembly can be due to the coordination of the sulfone group and the ether bond to Ti^{4+}, or by a hydrogen bond between the sulfone group and the ether bond and a surface hydroxyl group of TiO_2, a result of the strong electronegativity of oxygen in the ether bond and the sulfone group of the PES (Figure 18.15).

Self-assembly of the TiO_2 nanoparticles on a membrane surface is usually realized by dipping the porous membrane support in a colloidal suspension of TiO_2. The concentration of aqueous colloidal suspension of TiO_2 may vary from 0.01 through 0.03 wt% [133] to 1 wt% [55], whereas the time of immersion of the porous supports in the suspension was suggested to be 1 h [131] or 1 week [128]. Rahimpour et al. [134] studied the effect of dipping time in a 0.03 wt% TiO_2 colloidal suspension by comparing 15, 30, and 60 min of dipping. They concluded that a 15-min immersion

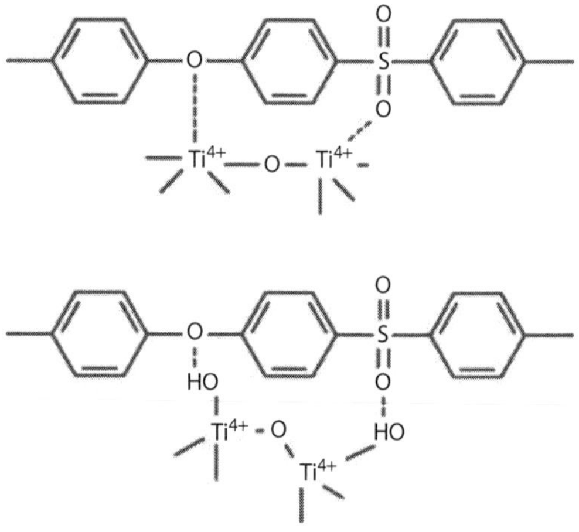

FIGURE 18.15 Mechanism of self-assembly of TiO_2 nanoparticles by (i) a coordination of a sulfone group and an ether bond to Ti^{4+} and (ii) an H-bond between the sulfone group and an ether bond and the surface hydroxyl group of TiO_2. (Data from M.L. Luo et al., *Appl Surf Sci*, 249, 76–84, 2005.)

yielded the best performance in terms of permeability and hypothesized that longer dipping times led to more pore plugging.

Another method to prepare hybrid organic/inorganic composite membranes is the incorporation of inorganic particles in a top membrane layer through the IP process. Lee et al. [135] applied an *in situ* IP procedure on PES support for preparing composite nanoparticles-based membranes. In this procedure, commercial TiO$_2$ nanoparticles of 30 nm were dispersed in TMC solution. The PES support was first immersed into aqueous m-phenyl diamine with 0.05 wt% NaOH; the excess reagent was removed from the surface, so that a controlled reaction was obtained on subsequent immersion in the solution of TMC in 1,1-dichloro-1-fluoroethane. As a result, a thin modified layer with immobilized nanoparticles was obtained on the surface of the PES support.

Using the same approach, Jeong et al. [136] prepared a thin-film nanocomposite (TFN) PA RO membrane (Figure 18.16) by dispersing 0.004%–0.4% (w/v) of zeolite nanoparticles with a size of 50 nm to 150 nm in TMC solution (Figure 18.16). The prepared zeolite–PA membrane surface showed enhanced hydrophilicity, more negative charge, and lower roughness, implying a strong potential use as an antifouling membrane.

Rana et al. [19] added 0.25 wt% of silver salt (silver nitrate, silver citrate hydrate, or silver actate) into aqueous MPD phase instead of organic TMC phase to prepare hybrid organic/inorganic RO membrane. Results showed that silver salts incorporated in the TFC membranes indeed improve the antibiofouling property.

Lee et al. [137] prepared PA/Ag nanocomposite membranes by an *in situ* IP between aqueous MPD and organic TMC solutions containing 10 wt% of silver nanoparticles. Hybrid membranes were shown to sharply reduce fouling with *Pseudomonas* bacteria. Scanning electron microscopy (SEM) measurements confirmed that all *Pseudomonas* cells were made inactive on the modified membrane surface, whereas water fluxes and salt rejections remained unchanged. It was shown that despite the fact that most of the Ag particles remained on the membrane surface, some nanoparticles were also encapsulated in the bulk of the active layer, reducing the antifouling activities.

Bae et al. [138] prepared two types of TiO$_2$-immobilized UF membranes, TiO$_2$ entrapped in the polymer body and TiO$_2$ deposited on the membrane surface, and

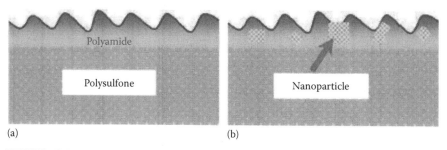

(a)   (b)

FIGURE 18.16   Schematic representation of (a) TFC and (b) TFN membrane structures. (Data from B.H. Jeong, E.M.V. Hoek, Y.S. Yan, A. Subramani, X.F. Huang, G. Hurwitz, A.K. Ghosh, A. Jawor, Interfacial polymerization of thin film nanocomposites: A new concept for reverse osmosis membranes, *J Memb Sci*, 294, 1–7, 2007.)

applied them to an activated sludge filtration to evaluate their fouling mitigation effect. It was shown that the entrapment of the TiO_2 nanoparticles in the membranes increased the hydrophilicity of their surfaces. Water contact angles were changed from 87.6° for neat PS to 73.1° for PS-TiO_2, from 86.7° for PVDF to 81.1° for PVDF-TiO_2, and from 45.5° for PAN to 43.1° for PAN-TiO_2 membranes. Such hydrophilization leads to a reduction in membrane fouling during the filtration of activated sludge, which contains a great number of various organic and microbiological foulants. TiO_2-entrapped membrane showed a lower flux decline than the neat polymeric membrane (Figure 18.17).

On the other hand, as shown in Figure 18.17, the TiO_2-deposited membrane showed a greater fouling mitigation effect compared with that of the TiO_2-entrapped membrane. Obviously, the degree of fouling mitigation is mainly affected by the surface area of the TiO_2 nanoparticles, which are located on the membrane surface and are exposed to the feed solution. In the case of the TiO_2-deposited membrane, the degree of surface modification was higher than that for the TiO_2-entrapped membrane, and the fouling mitigation effect significantly improved.

It was also shown [132] that the cake layer resistance of the modified membrane, which is a major influence on membrane fouling during the filtration of the activated sludge, dramatically decreased compared with that of the initial PES membrane. As the introduction of nanoparticles increases the hydrophilicity of the polymeric membrane surfaces, the adsorbed foulants on the modified membranes can be more

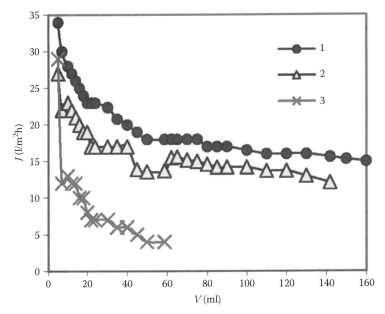

FIGURE 18.17 Flux decline behavior of neat and TiO_2 modified membranes during the activated sludge filtration: 1—TiO_2-deposited PS membrane, 2—TiO_2-entrapped PS membrane, and 3—unmodified PS membrane. (Data from T.H. Bae, T.M. Tak, Preparation of TiO_2 self-assembled polymeric nanocomposite membranes and examination of their fouling mitigation effects in a membrane bioreactor system, *J Memb Sci*, 266, 1–5, 2005.)

readily dislodged by the shear force than those on the unmodified PES membranes. As a result, the hydrophilic modification of the membrane surface by the introduction of the TiO_2 nanoparticles inhibits the hydrophobic interactions between the organic foulants and the membrane surface.

TiO_2 is a photocatalytic material and has been widely used for the disinfection and decomposition of organic compounds. These properties make it interesting as a self-cleaning coating. It has been demonstrated that the antifouling potential of the TiO_2-modified membranes is much better realized with the application of UV irradiation. Rahimpour et al. [134] compared TiO_2-entrapped PES membranes and self-assembled TiO_2-coated membranes with and without UV irradiation during filtration of nonskimmed milk. The initial pure water flux and the milk–water permeation of TiO_2-entrapped membranes were low compared with the unmodified PES membrane. However, the antifouling property and long-term flux stability were significantly enhanced. UV illumination further improved the membrane performance and antifouling properties, and the UV-irradiated TiO_2-deposited membranes had increased flux and higher antifouling properties compared with the TiO_2-entrapped membranes. The authors suggested that the membranes with TiO_2 nanoparticles on their surface and radiated by UV light obtained two main characteristics, namely, photocatalytic properties to decompose the organic compounds adsorbed on the membrane surface and superhydrophilicity, which result in a decrease in the contact angle between water and the membrane surface [134]. Therefore, foulants such as fats and proteins may be decomposed by photocatalysis and then removed from the surface by feed flow. Furthermore, with the increase in the membrane hydrophilicity, there is a competition between the adsorption of water and the foulants molecules, which leads to the improved removal of the pollutants from the membrane surface.

Photocatalytic bactericidal effect of the TFC PA membrane with the deposited TiO_2 (hybrid membrane) was examined by the evaluation of water flux at exposure to microbial suspensions *under* and *without* black UV illumination [129,139]. Figure 18.18 shows the plot of the water flux of the hybrid and the neat TFC PA membranes with and without UV light illumination after exposure to *E. coli* microbial cells. As shown in the figure, water flux decreases for all the membranes due to microbial fouling. Flux reduction should be attributed to membrane biofouling. Although the flux for the neat membrane is initially more than that for hybrid membrane, it drops more rapidly. Flux of the hybrid membrane under UV is greater than that of neat membrane in the same condition after a day of fouling. Within three days, the flux of neat membrane in the dark condition is decreased to 55.4 gfd, and that for hybrid membrane with UV illumination, however, is 66.1 gfd.

Mechanism of the bactericidal action of TiO_2 under black UV light is based on the formation of $OH^·$, $O_2^{-·}$, $HO_2^·$ radicals in water [140]. Adhesion of the bacterial cells to the TiO_2 particles controlled by the hydrophobic and charge interactions allows the active oxygen-containing species to reach and damage the bacterial cell wall. Due to the strong photobactericidal properties under UV treatment, the modified membranes are capable of inhibiting the growth of microorganism on the membrane surface, and thus membrane biofouling is reduced. This reduction in biofouling was demonstrated when TiO_2-modified membranes were used for surface water filtration,

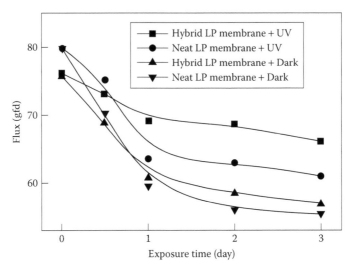

FIGURE 18.18 Water flux of the hybrid and the neat low pressure (LP) membranes with and without UV light after exposure to microbial cells. (Data from S.H. Kim, S.Y. Kwak, B.H. Sohn, T.H. Park, Design of TiO$_2$ nanoparticle self-assembled aromatic polyamide thin-film-composite [TFC] membrane as an approach to solve biofouling problem, *J Memb Sci*, 211 157–165, 2003.)

as the fluxes of the modified membranes were 1.7–2.3 times higher compared with those for the control samples [139].

Membrane biofouling may be also reduced through surface membrane modification with Ag nanoparticles. Silver-loaded PAN hollow fibers were prepared through the dry jet–wet spinning technique from a dope containing 0.5 wt% silver nitrate [141]. It was shown that at Ag-loading of 0.1 wt%, bacteria growth for both *E. coli* and *Staphylococcus aureus* was not observed on the membrane surface. Antibacterial activity of the modified membrane is attributed to trace amounts of silver ion released from the fiber [142]. Antibacterial effect of silver is related to its interactions with sulfur and phosphorous of amino acids in the bacterial cell wall, most notably through the thiol groups (S-H) present in the amino acid cysteine [143]. Such interactions can change the metabolic activity of bacteria and prohibit their growth [144]. However, after flushing with water for 60 days, the silver content in the hollow fibers decreased from 0.19% to 0.1%, while still keeping the antibacterial activities of *E. coli* and *S. aureus*. During the membrane process operation, the permeating water reduces the silver content of the hollow fibers, thus requiring periodic replenishment.

High antibacterial activity toward *E. coli* and *S. aureus* was also found with CA membranes modified with Ag nanoparticles [145]. However, a significant loss of silver was found; as a result when permeated with water, the antibacterial activity of the membranes disappeared after five days [146]. Loss of the entrapped silver nanoparticles was also reported for modified PS membranes, which have a high antimicrobial activity toward *E. coli*, *Pseudomonas mendocina*, and the MS2 bacteriophage [147].

In general, it may be concluded that despite the endeavor described earlier to develop low-fouling membranes using various nanoparticles, further research is still needed to investigate the combined effects of water chemistry, the nature of nanoparticles, and the coating conditions on the modified-membrane performance and fouling mitigation. Similar to surface coating with organic modifiers, the coating or deposition of inorganic particles onto the RO membrane surfaces also faces the problem of loss or leaching. Therefore, further studies should be performed to strengthen the combination between the membrane surface and inorganic particles to maintain the long-term improvement of antifouling properties. Also, careful control and monitoring of the nanoparticles released from the modified membranes are necessary to minimize potential (eco)toxicity effects [148].

18.7 CONCLUSIONS

As can be seen from the presented discussion, remarkable progress has been made in the fabrication of low-fouling composite membranes for water treatment applications in pressure-driven membrane processes. Numerous studies have shown that increasing the membrane hydrophilicity, reducing the roughness, or imposing charged groups and bactericidal agents on the membrane surface may reduce membrane fouling with organic compounds, colloids, and microorganisms. However, despite the extensive knowledge that exists on how to *tailor* the surface properties of the membrane and cross-sectional morphology by the selection of the appropriate fabrication methods, there is still a challenge to produce reliable composite membranes with antifouling properties, high mechanical strength, and minimal thickness of the membrane barrier layer to provide a high flux.

IP is probably one of the most versatile and cost-effective membrane fabrication techniques for the preparation of low-fouling composite membranes with a low surface roughness and various hydrophilic/hydrophobic or charged properties. As to other fabrication methods, many factors need to be considered in the overall process of membrane preparation, such as uniformity and reproducibility, together with precise control over functional groups. UV/redox-initiated graft polymerization and the coating of the membranes with hydrophilic polymer layers have the advantages of simplicity, low cost of operation, and mild reaction conditions. On the other hand, the modified membranes usually have reduced fluxes compared with the unmodified membranes because the grafting/deposited layer adds extra hydraulic membrane resistance. It should be pointed out that most of the studies on the development of low-fouling composite membranes have been performed on a small-lab scale. Thus, currently there is a need for the scaling-up of modification procedures to thoroughly investigate the possibility of their incorporation into the end stages of a membrane manufacturing process. Special efforts should be also focused on reducing the cost of membrane fabrication, especially for plasma and iCVD treatment methods.

Antifouling properties of the fabricated composite membrane were evaluated mainly with model colloidal particles, organic foulants such as BSA and HA, or bacterial suspensions, such as *E. coli* or *S. aureus* in short-term laboratory tests. There is a lack of studies on the estimation of antifouling properties of low-fouling composite membranes with real polycomponent feed streams, especially for their long-term applications. Therefore, further research is still needed in all these fields.

It should also be mentioned that totally nonfouling membranes are practically impossible to develop, because apart from the surface membrane properties, membrane fouling also depends on the hydrodynamics inside the membrane elements/ cells as well as on the feed stream parameters, such as solute concentration and pH [20]. Nevertheless, the development of low-fouling composite membranes using targeted tailoring of membrane surface properties in combination with the optimization of operation conditions in the membrane elements and feed pretreatment, undoubtedly, will reduce membrane fouling, extending the effective and long-term membrane performance.

REFERENCES

1. M.A. Shannon, P.W. Bohn, M. Elimelech, J.G. Georgiadis, B.J. Marinas, A.M. Mayes, Science and technology for water purification in the coming decades, *Nature*, 452 (2008) 301–310.
2. M. Elimelech, W.A. Phillip, The future of seawater desalination: Energy, technology, and the environment, *Science*, 333 (2011) 712–717.
3. M.A. Montgomery, M. Elimelech, Water and sanitation in developing countries: Including health in the equation, *Environ Sci Technol*, 41 (2007) 17–24.
4. R.W. Baker, *Membrane Technology and Applications*, Wiley, New York, 2004.
5. K. Scot, R. Hughes, *Industrial Membrane Separation Technology*, Blackie Academics & Professional, London, 1996.
6. K. Kimura, Y. Hane, Y. Watanabe, G. Amy, N. Ohkuma, Irreversible membrane fouling during ultrafiltration of surface water, *Water Res*, 38 (2004) 3431–3441.
7. H.C. Flemming, Reverse osmosis membrane biofouling, *Exp Therm Fluid Sci*, 14 (1997) 382–391.
8. C.A.C. van de Lisdonk, J.A.M. van Paassen, J.C. Schippers, Monitoring scaling in nanofiltration and reverse osmosis membrane systems, *Desalination*, 132 (2000) 101–108.
9. I.C. Escobar, E.M. Hoek, C.J. Gabelich, F.A. DiGiano, Y.A. Le Gouellec, P. Berube, K.J. Howe et al., Committee Report: Recent advances and research needs in membrane fouling, *J Am Water Works Ass*, 97 (2005) 79–89.
10. A.S. Al-Amoudi, A.M. Farooque, Performance restoration and autopsy of NF membranes used in seawater pretreatment, *Desalination*, 178 (2005) 261–271.
11. J.A. Nilson, F.A. DiGiano, Influence of NOM composition on nanofiltration, *J Am Water Works Ass*, 88 (1996) 53–66.
12. J. Buffle, G.G. Leppard, Characterization of aquatic colloids and macromolecules.1. Structure and behavior of colloidal material, *Environ Sci Technol*, 29 (1995) 2169–2175.
13. J. Buffle, K.J. Wilkinson, S. Stoll, M. Filella, J.W. Zhang, A generalized description of aquatic colloidal interactions: The three-colloidal component approach, *Environ Sci Technol*, 32 (1998) 2887–2899.
14. H.C. Flemming, G. Schaule, T. Griebe, J. Schmitt, A. Tamachkiarowa, Biofouling—The Achilles heel of membrane processes, *Desalination*, 113 (1997) 215–225.
15. H.C. Flemming, G. Schaule, Biofouling on membranes—A microbiological approach, *Desalination*, 70 (1988) 95–119.
16. J.S. Baker, L.Y. Dudley, Biofouling in membrane systems—A review, *Desalination*, 118 (1998) 81–89.

17. W. Yuan, A.L. Zydney, Humic acid fouling during ultrafiltration, *Environ Sci Technol*, 34 (2000) 5043–5050.
18. E.M.V. Hoek, M. Elimelech, Cake-enhanced concentration polarization: A new fouling mechanism for salt-rejecting membranes, *Environ Sci Technol*, 37 (2003) 5581–5588.
19. D. Rana, Y. Kim, T. Matsuura, H.A. Arafat, Development of antifouling thin-film-composite membranes for seawater desalination, *J Memb Sci*, 367 (2011) 110–118.
20. D. Rana, T. Matsuura, Surface modifications for antifouling membranes, *Chem Rev*, 110 (2010) 2448–2471.
21. C. Bellona, J.E. Drewes, P. Xu, G. Amy, Factors affecting the rejection of organic solutes during NF/RO treatment—A literature review, *Water Res*, 38 (2004) 2795–2809.
22. B.J. Abu Tarboush, D. Rana, T. Matsuura, H.A. Arafat, R.M. Narbaitz, Preparation of thin-film-composite polyamide membranes for desalination using novel hydrophilic surface modifying macromolecules, *J Memb Sci*, 325 (2008) 166–175.
23. M. Ulbricht, Advanced functional polymer membranes, *Polymer*, 47 (2006) 2217–2262.
24. K. Kimura, G. Amy, J. Drewes, Y. Watanabe, Adsorption of hydrophobic compounds onto NF/RO membranes: An artifact leading to overestimation of rejection, *J Memb Sci*, 221 (2003) 89–101.
25. N. Hilal, O.O. Ogunbiyi, N.J. Miles, R. Nigmatullin, Methods employed for control of fouling in MF and UF membranes: A comprehensive review, *Sep Sci Technol*, 40 (2005) 1957–2005.
26. A.G. Fane, C.J.D. Fell, A review of fouling and fouling control in ultrafiltration, *Desalination*, 62 (1987) 117–136.
27. A. Nabe, E. Staude, G. Belfort, Surface modification of polysulfone ultrafiltration membranes and fouling by BSA solutions, *J Memb Sci*, 133 (1997) 57–72.
28. F. Liu, N.A. Hashim, Y.T. Liu, M.R.M. Abed, K. Li, Progress in the production and modification of PVDF membranes, *J Memb Sci*, 375 (2011) 1–27.
29. Q. Li, X.H. Pan, C.Y. Hou, Y. Jin, H.J. Dai, H.Z. Wang, X.L. Zhao, X.Q. Liu, Exploring the dependence of bulk properties on surface chemistries and microstructures of commercially composite RO membranes by novel characterization approaches, *Desalination*, 292 (2012) 9–18.
30. G.D. Kang, Y.M. Cao, Development of antifouling reverse osmosis membranes for water treatment: A review, *Water Res*, 46 (2012) 584–600.
31. F.Q. Nie, Z.K. Xu, P. Ye, J. Wu, P. Seta, Acrylonitrile-based copolymer membranes containing reactive groups: Effects of surface-immobilized poly(ethylene glycol)s on anti-fouling properties and blood compatibility, *Polymer*, 45 (2004) 399–407.
32. T. McPherson, A. Kidane, I. Szleifer, K. Park, Prevention of protein adsorption by tethered poly(ethylene oxide) layers: Experiments and single-chain mean-field analysis, *Langmuir*, 14 (1998) 176–186.
33. P. Wang, K.L. Tan, E.T. Kang, K.G. Neoh, Plasma-induced immobilization of poly(ethylene glycol) onto poly(vinylidene fluoride) microporous membrane, *J Memb Sci*, 195 (2002) 103–114.
34. B. Van der Bruggen, M. Manttari, M. Nystrom, Drawbacks of applying nanofiltration and how to avoid them: A review, *Sep Purif Technol*, 63 (2008) 251–263.
35. A. Al-Amoudi, R.W. Lovitt, Fouling strategies and the cleaning system of NF membranes and factors affecting cleaning efficiency, *J Memb Sci*, 303 (2007) 6–28.
36. S.K. Hong, M. Elimelech, Chemical and physical aspects of natural organic matter (NOM) fouling of nanofiltration membranes, *J Memb Sci*, 132 (1997) 159–181.
37. K. Kato, E. Uchida, E.T. Kang, Y. Uyama, Y. Ikada, Polymer surface with graft chains, *Prog Polym Sci*, 28 (2003) 209–259.
38. M. Elimelech, X.H. Zhu, A.E. Childress, S.K. Hong, Role of membrane surface morphology in colloidal fouling of cellulose acetate and composite aromatic polyamide reverse osmosis membranes, *J Memb Sci*, 127 (1997) 101–109.

39. E.M. Vrijenhoek, S. Hong, M. Elimelech, Influence of membrane surface properties on initial rate of colloidal fouling of reverse osmosis and nanofiltration membranes, *J Memb Sci*, 188 (2001) 115–128.

40. M.N.A. Seman, M. Khayet, N. Hilal, Development of antifouling properties and performance of nanofiltration membranes modified by interfacial polymerisation, *Desalination*, 273 (2011) 36–47.

41. C. Hobbs, S.K. Hong, J. Taylor, Effect of surface roughness on fouling of RO and NF membranes during filtration of a high organic surficial groundwater, *J Water Supply Res T*, 55 (2006) 559–570.

42. W.R. Bowen, T.A. Doneva, Atomic force microscopy studies of membranes: Effect of surface roughness on double-layer interactions and particle adhesion, *J Colloid Interf Sci*, 229 (2000) 544–549.

43. K. Boussu, A. Belpaire, A. Volodin, C. Van Haesendonck, P. Van der Meeren, C. Vandecasteele, B. Van der Bruggen, Influence of membrane and colloid characteristics on fouling of nanofiltration membranes, *J Memb Sci*, 289 (2007) 220–230.

44. N. Park, B. Kwon, I.S. Kim, J.W. Cho, Biofouling potential of various NF membranes with respect to bacteria and their soluble microbial products (SMP): Characterizations, flux decline, and transport parameters, *J Memb Sci*, 258 (2005) 43–54.

45. J.E. Cadotte, R.J. Petersen, R.E. Larson, E.E. Erickson, New thin-film composite seawater reverse-osmosis membrane, *Desalination*, 32 (1980) 25–31.

46. J. Cadotte, R. Forester, M. Kim, R. Petersen, T. Stocker, Nanofiltration membranes broaden the use of membrane separation technology, *Desalination*, 70 (1988) 77–88.

47. W.J. Lau, A.F. Ismail, N. Misdan, M.A. Kassim, A recent progress in thin film composite membrane: A review, *Desalination*, 287 (2012) 190–199.

48. R.J. Petersen, Composite reverse-osmosis and nanofiltration membranes, *J Memb Sci*, 83 (1993) 81–150.

49. A.P. Rao, N.V. Desai, R. Rangarajan, Interfacially synthesized thin film composite RO membranes for seawater desalination, *J Memb Sci*, 124 (1997) 263–272.

50. I.J. Roh, A.R. Greenberg, V.P. Khare, Synthesis and characterization of interfacially polymerized polyamide thin films, *Desalination*, 191 (2006) 279–290.

51. A.K. Ghosh, B.H. Jeong, X.F. Huang, E.M.V. Hoek, Impacts of reaction and curing conditions on polyamide composite reverse osmosis membrane properties, *J Memb Sci*, 311 (2008) 34–45.

52. S. Verissimo, K.V. Peinemann, J. Bordado, Influence of the diamine structure on the nanofiltration performance, surface morphology and surface charge of the composite polyamide membranes, *J Memb Sci*, 279 (2006) 266–275.

53. S.H. Huang, C.L. Li, C.C. Hu, H.A. Tsai, K.R. Lee, J.Y. Lai, Polyamide thin-film composite membranes prepared by interfacial polymerization for pervaporation separation, *Desalination*, 200 (2006) 387–389.

54. A.R. Korikov, R. Kosaraju, K.K. Sirkar, Interfacially polymerized hydrophilic microporous thin film composite membranes on porous polypropylene hollow fibers and flat films, *J Memb Sci*, 279 (2006) 588–600.

55. L. Li, S.B. Zhang, X.S. Zhang, G.D. Zheng, Polyamide thin film composite membranes prepared from 3,4',5-biphenyl triacyl chloride, 3,3',5,5'-biphenyl tetraacyl chloride and m-phenylenediamine, *J Memb Sci*, 289 (2007) 258–267.

56. L.F. Liu, S.C. Yu, L.G. Wu, C.J. Gao, Study on a novel antifouling polyamide-urea reverse osmosis composite membrane (ICIC-MPD)—III. Analysis of membrane electrical properties, *J Memb Sci*, 310 (2008) 119–128.

57. Q.F. An, F. Li, Y.L. Ji, H.L. Chen, Influence of polyvinyl alcohol on the surface morphology, separation and anti-fouling performance of the composite polyamide nanofiltration membranes, *J Memb Sci*, 367 (2011) 158–165.

58. M.H. Liu, S.C. Yu, J. Tao, C.J. Gao, Preparation, structure characteristics and separation properties of thin-film composite polyamide-urethane seawater reverse osmosis membrane, *J Memb Sci*, 325 (2008) 947–956.

59. M.A. Kuehne, R.Q. Song, N.N. Li, R.J. Petersen, Flux enhancement in TFC RO membranes, *Environ Prog*, 20 (2001) 23–26.

60. H. Yamagishi, J.V. Crivello, G. Belfort, Development of a novel photochemical technique for modifying poly(arylsulfone) ultrafiltration membranes, *J Memb Sci*, 105 (1995) 237–247.

61. H. Yamagishi, J.V. Crivello, G. Belfort, Evaluation of photochemically modified poly(arylsulfone) ultrafiltration membranes, *J Memb Sci*, 105 (1995) 249–259.

62. J. Pieracci, J.V. Crivello, G. Belfort, Photochemical modification of 10 kDa polyethersulfone ultrafiltration membranes for reduction of biofouling, *J Memb Sci*, 156 (1999) 223–240.

63. J. Pieracci, D.W. Wood, J.V. Crivello, G. Belfort, UV-assisted graft polymerization of N-vinyl-2-pyrrolidinone onto poly(ether sulfone) ultrafiltration membranes: Comparison of dip versus immersion modification techniques, *Chem Mater*, 12 (2000) 2123–2133.

64. M.N.A. Seman, M. Khayet, Z.I. Bin Ali, N. Hilal, Reduction of nanofiltration membrane fouling by UV-initiated graft polymerization technique, *J Memb Sci*, 355 (2010) 133–141.

65. M.N.A. Seman, M. Khayet, N. Hilal, Comparison of two different UV-grafted nanofiltration membranes prepared for reduction of humic acid fouling using acrylic acid and N-vinylpyrrolidone, *Desalination*, 287 (2012) 19–29.

66. M. Ulbricht, Photograft-polymer-modified microporous membranes with environment-sensitive permeabilities, *React Funct Polym*, 31 (1996) 165–177.

67. M. Ulbricht, H. Matuschewski, A. Oechel, H.G. Hicke, Photo-induced graft polymerization surface modifications for the preparation of hydrophilic and low-protein-adsorbing ultrafiltration membranes, *J Memb Sci*, 115 (1996) 31–47.

68. V.M. Kochkodan, N. Hilal, V.V. Goncharuk, L. Al-Khatib, T.I. Levadna, Effect of the surface modification of polymer membranes on their microbiological fouling, *Colloid J+*, 68 (2006) 267–273.

69. M. Pasmore, P. Todd, S. Smith, D. Baker, J. Silverstein, D. Coons, C.N. Bowman, Effects of ultrafiltration membrane surface properties on Pseudomonas aeruginosa biofilm initiation for the purpose of reducing biofouling, *J Memb Sci*, 194 (2001) 15–32.

70. H.M. Ma, R.H. Davis, C.N. Bowman, A novel sequential photoinduced living graft polymerization, *Macromolecules*, 33 (2000) 331–335.

71. S. Belfer, Y. Purinson, R. Fainshtein, Y. Radchenko, O. Kedem, Surface modification of commercial composite polyamide reverse osmosis membranes, *J Memb Sci*, 139 (1998) 175–181.

72. B. Van der Bruggen, Comparison of redox initiated graft polymerisation and sulfonation for hydrophilisation of polyethersulfone nanofiltration membranes, *Eur Polym J*, 45 (2009) 1873–1882.

73. S. Belfer, Y. Purinson, O. Kedem, Surface modification of commercial polyamide reverse osmosis membranes by radical grafting: An ATR-FTIR study, *Acta Polym*, 49 (1998) 574–582.

74. S. Belfer, J. Gilron, Y. Purinson, R. Fainshtain, N. Daltrophe, M. Priel, B. Tenzer, A. Toma, Effect of surface modification in preventing fouling of commercial SWRO membranes at the Eilat seawater desalination pilot plant, *Desalination*, 139 (2001) 169–176.

75. A.V.R. Reddy, J.J. Trivedi, C.V. Devmurari, D.J. Mohan, P. Singh, A.P. Rao, S.V. Joshi, P.K. Ghosh, Fouling resistant membranes in desalination and water recovery, *Desalination*, 183 (2005) 301–306.

76. V. Freger, J. Gilron, S. Belfer, TFC polyamide membranes modified by grafting of hydrophilic polymers: An FT-IR/AFM/TEM study, *J Memb Sci*, 209 (2002) 283–292.

77. S. Belfer, R. Fainchtain, Y. Purinson, O. Kedem, Surface characterization by FTIR-ATR spectroscopy of polyethersulfone membranes-unmodified, modified and protein fouled, *J Memb Sci*, 172 (2000) 113–124.

78. X.Y. Wei, Z. Wang, Z. Zhang, J.X. Wang, S.C. Wang, Surface modification of commercial aromatic polyamide reverse osmosis membranes by graft polymerization of 3-allyl-5,5-dimethylhydantoin, *J Memb Sci*, 351 (2010) 222–233.

79. Y. Tsujii, K. Ohno, S. Yamamoto, A. Goto, T. Fukuda, Structure and properties of high-density polymer brushes prepared by surface-initiated living radical polymerization, *Adv Polym Sci*, 197 (2006) 1–45.

80. G.F. Payne, M.V. Chaubal, T.A. Barbari, Enzyme-catalysed polymer modification: Reaction of phenolic compounds with chitosan films, *Polymer*, 37 (1996) 4643–4648.

81. L.H. Shao, G. Kumar, J.L. Lenhart, P.J. Smith, G.F. Payne, Enzymatic modification of the synthetic polymer polyhydroxystyrene, *Enzyme Microb Tech*, 25 (1999) 660–668.

82. G. Kumar, P.J. Smith, G.F. Payne, Enzymatic grafting of a natural product onto chitosan to confer water solubility under basic conditions, *Biotechnol Bioeng*, 63 (1999) 154–165.

83. N. Nady, M.C.R. Franssen, H. Zuilhof, M.S.M. Eldin, R. Boom, K. Schroen, Modification methods for poly(arylsulfone) membranes: A mini-review focusing on surface modification, *Desalination*, 275 (2011) 1–9.

84. T. Gullinkala, I. Escobar, A green membrane functionalization method to decrease natural organic matter fouling, *J Memb Sci*, 360 (2010) 155–164.

85. M. Bryak, I. Gancarz, Plasma modification of polymer membranes, in: N. Hilal, M. Khayet, C.J. Wright (Eds.) *Membrane Modification: Technology and Applications*, CRC Press, 2012, pp. 179–214.

86. M. Bryjak, I. Gancarz, K. Smolinska, Plasma nanostructuring of porous polymer membranes, *Adv Colloid Interfac*, 161 (2010) 2–9.

87. V.M. Kochkodan, V.K. Sharma, Graft polymerization and plasma treatment of polymer membranes for fouling reduction: A review, *J Environ Sci Heal A*, 47 (2012) 1713–1727.

88. D. Tyszler, R.G. Zytner, A. Batsch, A. Brugger, S. Geissler, H.D. Zhou, D. Klee, T. Melin, Reduced fouling tendencies of ultrafiltration membranes in wastewater treatment by plasma modification, *Desalination*, 189 (2006) 119–129.

89. Z.P. Zhao, J.D. Li, D. Wang, C.X. Chen, Nanofiltration membrane prepared from polyacrylonitrile ultrafiltration membrane by low-temperature plasma: 4. Grafting of N-vinylpyrrolidone in aqueous solution, *Desalination*, 184 (2005) 37–44.

90. I. Gancarz, G. Pozniak, M. Bryjak, A. Frankiewicz, Modification of polysulfone membranes. 2. Plasma grafting and plasma polymerization of acrylic acid, *Acta Polym*, 50 (1999) 317–326.

91. M. Ulbricht, G. Belfort, Surface modification of ultrafiltration membranes by low-temperature plasma. 1. Treatment of polyacrylonitrile, *J Appl Polym Sci*, 56 (1995) 325–343.

92. M. Ulbricht, G. Belfort, Surface modification of ultrafiltration membranes by low temperature plasma.2. Graft polymerization onto polyacrylonitrile and polysulfone, *J Memb Sci*, 111 (1996) 193–215.

93. H.Y. Yu, Z.K. Xu, Y.J. Xie, Z.M. Liu, S.Y. Wang, Flux enhancement for polypropylene microporous membrane in a SMBR by the immobilization of poly(N-vinyl-2-pyrrolidone) on the membrane surface, *J Memb Sci*, 279 (2006) 148–155.

94. M.M. Kim, N.H. Lin, G.T. Lewis, Y. Cohen, Surface nano-structuring of reverse osmosis membranes via atmospheric pressure plasma-induced graft polymerization for reduction of mineral scaling propensity, *J Memb Sci*, 354 (2010) 142–149.

95. L. Zou, I. Vidalis, D. Steele, A. Michelmore, S.P. Low, J.Q.J.C. Verberk, Surface hydrophilic modification of RO membranes by plasma polymerization for low organic fouling, *J Memb Sci*, 369 (2011) 420–428.

96. R. Yang, J.J. Xu, G. Ozaydin-Ince, S.Y. Wong, K.K. Gleason, Surface-tethered zwitterionic ultrathin antifouling coatings on reverse osmosis membranes by initiated chemical vapor deposition, *Chem Mater*, 23 (2011) 1263–1272.

97. C.Y. Ba, D.A. Ladner, J. Economy, Using polyelectrolyte coatings to improve fouling resistance of a positively charged nanofiltration membrane, *J Memb Sci*, 347 (2010) 250–259.

98. K.J. Kim, A.G. Fane, C.J.D. Fell, The performance of ultrafiltration membranes pretreated by polymers, *Desalination*, 70 (1988) 229–249.

99. L.E.S. Brink, D.J. Romijn, Reducing the protein fouling of polysulfone surfaces and polysulfone ultrafiltration membranes—Optimization of the type of presorbed layer, *Desalination*, 78 (1990) 209–233.

100. M. Nystrom, Fouling of unmodified and modified polysulfone ultrafiltration membranes by ovalbumin, *J Memb Sci*, 44 (1989) 183–196.

101. M.C. Wilbert, J. Pellegrino, A. Zydney, Bench-scale testing of surfactant-modified reverse osmosis/nanofiltration membranes, *Desalination*, 115 (1998) 15–32.

102. A. Asatekin, E.A. Olivetti, A.M. Mayes, Fouling resistant, high flux nanofiltration membranes from polyacrylonitrile-graft-poly(ethylene oxide), *J Memb Sci*, 332 (2009) 6–12.

103. A. Asatekin, A. Menniti, S.T. Kang, M. Elimelech, E. Morgenroth, A.M. Mayes, Antifouling nanofiltration membranes for membrane bioreactors from self-assembling graft copolymers, *J Memb Sci*, 285 (2006) 81–89.

104. A.C. Sagle, E.M. Van Wagner, H. Ju, B.D. McCloskey, B.D. Freeman, M.M. Sharma, PEG-coated reverse osmosis membranes: Desalination properties and fouling resistance, *J Memb Sci*, 340 (2009) 92–108.

105. J.R. Du, S. Peldszus, P.M. Huck, X.S. Feng, Modification of poly(vinylidene fluoride) ultrafiltration membranes with poly(vinyl alcohol) for fouling control in drinking water treatment, *Water Res*, 43 (2009) 4559–4568.

106. A. Maartens, E.P. Jacobs, P. Swart, UF of pulp and paper effluent: Membrane fouling-prevention and cleaning, *J Memb Sci*, 209 (2002) 81–92.

107. N. Li, Z.Z. Liu, S.G. Xu, Dynamically formed poly (vinyl alcohol) ultrafiltration membranes with good anti-fouling characteristics, *J Memb Sci*, 169 (2000) 17–28.

108. A.V.R. Reddy, D.J. Mohan, A. Bhattacharya, V.J. Shah, P.K. Ghosh, Surface modification of ultrafiltration membranes by preadsorption of a negatively charged polymer I. Permeation of water soluble polymers and inorganic salt solutions and fouling resistance properties, *J Memb Sci*, 214 (2003) 211–221.

109. S. Boributh, A. Chanachai, R. Jiraratananon, Modification of PVDF membrane by chitosan solution for reducing protein fouling, *J Memb Sci*, 342 (2009) 97–104.

110. J.S. Louie, I. Pinnau, I. Ciobanu, K.P. Ishida, A. Ng, M. Reinhard, Effects of polyether-polyamide block copolymer coating on performance and fouling of reverse osmosis membranes, *J Memb Sci*, 280 (2006) 762–770.

111. A. Hamza, G. Chowdhury, T. Matsuura, S. Sourirajan, Sulphonated poly(2,6-dimethyl-1,4-phenylene oxide)-polyethersulphone composite membranes—Effects of composition of solvent system, used for preparing casting solution, on membrane-surface structure and reverse-osmosis performance, *J Memb Sci*, 129 (1997) 55–64.

112. J. Hyun, H. Jang, K. Kim, K. Na, T. Tak, Restriction of biofouling in membrane filtration using a brush-like polymer containing oligoethylene glycol side chains, *J Memb Sci*, 282 (2006) 52–59.

113. Z.P. Zhao, Z. Wang, S.C. Wang, Formation, charged characteristic and BSA adsorption behavior of carboxymethyl chitosan/PES composite MF membrane, *J Memb Sci*, 217 (2003) 151–158.

114. L.Y. Ng, A.W. Mohammad, C.Y. Ng, A review on nanofiltration membrane fabrication and modification using polyelectrolytes: Effective ways to develop membrane selective barriers and rejection capability, *Adv Colloid Interfac*, 197–198 (2013) 85–107.

115. G. Decher, Fuzzy nanoassemblies: Toward layered polymeric multicomposites, *Science*, 277 (1997) 1232–1237.

116. O.Y. Lu, R. Malaisamy, M.L. Bruening, Multilayer polyelectrolyte films as nanofiltration membranes for separating monovalent and divalent cations, *J Memb Sci*, 310 (2008) 76–84.

117. J.D. Mendelsohn, C.J. Barrett, V.V. Chan, A.J. Pal, A.M. Mayes, M.F. Rubner, Fabrication of microporous thin films from polyelectrolyte multilayers, *Langmuir*, 16 (2000) 5017–5023.

118. T. Ishigami, K. Amano, A. Fujii, Y. Ohmukai, E. Kamio, T. Maruyama, H. Matsuyama, Fouling reduction of reverse osmosis membrane by surface modification via layer-by-layer assembly, *Sep Purif Technol*, 99 (2012) 1–7.

119. R. Malaisamy, A. Talla-Nwafo, K.L. Jones, Polyelectrolyte modification of nanofiltration membrane for selective removal of monovalent anions, *Sep Purif Technol*, 77 (2011) 367–374.

120. B.W. Su, T.T. Wang, Z.W. Wang, X.L. Gao, C.J. Gao, Preparation and performance of dynamic layer-by-layer PDADMAC/PSS nanofiltration membrane, *J Memb Sci*, 423 (2012) 324–331.

121. C.R. Bartels, M. Wilf, K. Andes, J. Iong, Design considerations for wastewater treatment by reverse osmosis, *Water Sci Technol*, 51 (2005) 473–482.

122. M. Wilf, S. Alt, Application of low fouling RO membrane elements for reclamation of municipal wastewater, *Desalination*, 132 (2000) 11–19.

123. C.Y.Y. Tang, Y.N. Kwon, J.O. Leckie, Probing the nano- and micro-scales of reverse osmosis membranes—A comprehensive characterization of physiochemical properties of uncoated and coated membranes by XPS, TEM, ATR-FTIR, and streaming potential measurements, *J Memb Sci*, 287 (2007) 146–156.

124. C.Y.Y. Tang, Y.N. Kwon, J.O. Leckie, Effect of membrane chemistry and coating layer on physiochemical properties of thin film composite polyamide RO and NF membranes II. Membrane physiochemical properties and their dependence on polyamide and coating layers, *Desalination*, 242 (2009) 168–182.

125. M.L. Bruening, D.M. Dotzauer, P. Jain, L. Ouyang, G.L. Baker, Creation of functional membranes using polyelectrolyte multilayers and polymer brushes, *Langmuir*, 24 (2008) 7663–7673.

126. S.A. Sukhishvili, S. Granick, Layered, erasable polymer multilayers formed by hydrogen-bonded sequential self-assembly, *Macromolecules*, 35 (2002) 301–310.

127. E. Kharlampieva, S.A. Sukhishvili, Ionization and pH stability of multilayers formed by self-assembly of weak polyelectrolytes, *Langmuir*, 19 (2003) 1235–1243.

128. J.F. Li, Z.L. Xu, H. Yang, L.Y. Yu, M. Liu, Effect of TiO_2 nanoparticles on the surface morphology and performance of microporous PES membrane, *Appl Surf Sci*, 255 (2009) 4725–4732.

129. S.H. Kim, S.Y. Kwak, B.H. Sohn, T.H. Park, Design of TiO_2 nanoparticle self-assembled aromatic polyamide thin-film-composite (TFC) membrane as an approach to solve biofouling problem, *J Memb Sci*, 211 (2003) 157–165.

130. S.Y. Kwak, S.H. Kim, S.S. Kim, Hybrid organic/inorganic reverse osmosis (RO) membrane for bactericidal anti-fouling. 1. Preparation and characterization of TiO_2 nanoparticle self-assembled aromatic polyamide thin-film-composite (TFC) membrane, *Environ Sci Technol*, 35 (2001) 2388–2394.

131. M.L. Luo, J.Q. Zhao, W. Tang, C.S. Pu, Hydrophilic modification of poly(ether sulfone) ultrafiltration membrane surface by self-assembly of TiO(2) nanoparticles, *Appl Surf Sci*, 249 (2005) 76–84.

132. T.H. Bae, I.C. Kim, T.M. Tak, Preparation and characterization of fouling-resistant TiO_2 self-assembled nanocomposite membranes, *J Memb Sci*, 275 (2006) 1–5.

133. Y. Mansourpanah, S.S. Madaeni, A. Rahimpour, A. Farhadian, A.H. Taheri, Formation of appropriate sites on nanofiltration membrane surface for binding TiO_2 photo-catalyst: Performance, characterization and fouling-resistant capability, *J Memb Sci*, 330 (2009) 297–306.

134. A. Rahimpour, S.S. Madaeni, A.H. Taheri, Y. Mansourpanah, Coupling TiO_2 nanoparticles with UV irradiation for modification of polyethersulfone ultrafiltration membranes, *J Memb Sci*, 313 (2008) 158–169.

135. H.S. Lee, S.J. Im, J.H. Kim, H.J. Kim, J.P. Kim, B.R. Min, Polyamide thin-film nanofiltration membranes containing TiO_2 nanoparticles, *Desalination*, 219 (2008) 48–56.

136. B.H. Jeong, E.M.V. Hoek, Y.S. Yan, A. Subramani, X.F. Huang, G. Hurwitz, A.K. Ghosh, A. Jawor, Interfacial polymerization of thin film nanocomposites: A new concept for reverse osmosis membranes, *J Memb Sci*, 294 (2007) 1–7.

137. S.Y. Lee, H.J. Kim, R. Patel, S.J. Im, J.H. Kim, B.R. Min, Silver nanoparticles immobilized on thin film composite polyamide membrane: Characterization, nanofiltration, antifouling properties, *Polym Advan Technol*, 18 (2007) 562–568.

138. T.H. Bae, T.M. Tak, Preparation of TiO_2 self-assembled polymeric nanocomposite membranes and examination of their fouling mitigation effects in a membrane bioreactor system, *J Memb Sci*, 266 (2005) 1–5.

139. V. Kochkodan, S. Tsarenko, N. Potapchenko, V. Kosinova, V. Goncharuk, Adhesion of microorganisms to polymer membranes: A photobactericidal effect of surface treatment with TiO_2, *Desalination*, 220 (2008) 380–385.

140. F.M. Salih, Enhancement of solar inactivation of *Escherichia coli* by titanium dioxide photocatalytic oxidation, *J Appl Microbiol*, 92 (2002) 920–926.

141. D.G. Yu, M.Y. Teng, W.L. Chou, M.C. Yang, Characterization and inhibitory effect of antibacterial PAN-based hollow fiber loaded with silver nitrate, *J Memb Sci*, 225 (2003) 115–123.

142. G.J. Zhao, S.E. Stevens, Multiple parameters for the comprehensive evaluation of the susceptibility of Escherichia coli to the silver ion, *Biometals*, 11 (1998) 27–32.

143. J.T. Trevors, Silver resistance and accumulation in bacteria, *Enzyme Microb Tech*, 9 (1987) 331–333.

144. Q.L. Feng, J. Wu, G.Q. Chen, F.Z. Cui, T.N. Kim, J.O. Kim, A mechanistic study of the antibacterial effect of silver ions on *Escherichia coli* and *Staphylococcus aureus*, *J Biomed Mater Res*, 52 (2000) 662–668.

145. W.L. Chou, D.G. Yu, M.C. Yang, The preparation and characterization of silver-loading cellulose acetate hollow fiber membrane for water treatment, *Polym Advan Technol*, 16 (2005) 600–607.

146. W.K. Son, J.H. Youk, T.S. Lee, W.H. Park, Preparation of antimicrobial ultrafine cellulose acetate fibers with silver nanoparticles, *Macromol Rapid Comm*, 25 (2004) 1632–1637.

147. K. Zodrow, L. Brunet, S. Mahendra, D. Li, A. Zhang, Q.L. Li, P.J.J. Alvarez, Polysulfone ultrafiltration membranes impregnated with silver nanoparticles show improved biofouling resistance and virus removal, *Water Res*, 43 (2009) 715–723.

148. K. Tiede, M. Hassellov, E. Breitbarth, Q. Chaudhry, A.B.A. Boxall, Considerations for environmental fate and ecotoxicity testing to support environmental risk assessments for engineered nanoparticles, *J Chromatogr A*, 1216 (2009) 503–509.

19 Fabrication of Polymer Nanocomposite Membrane by Intercalating Nanoparticles for Direct Methanol Fuel Cell

Juhana Jaafar, Ahmad Fauzi Ismail, Mohd Hafiz Dzarfan Othman, and Mukhlis A. Rahman

CONTENTS

19.1 Introduction 656
19.2 Types of Nanocomposite Membrane by Structure 657
19.3 Preparation of Exfoliated Nanocomposite Membranes............................ 658
    19.3.1 Incorporation of Advanced Additives... 658
19.4 Membrane Solution Preparation Method.. 664
    19.4.1 Melt Blending 664
    19.4.2 Solution Blending 665
    19.4.3 *In Situ* Polymerization 666
    19.4.4 Sol-Gel.. 666
19.5 Preparation of SPEEK/Cloisite15A/TAP Nanocomposite Membrane 667
    19.5.1 Effect of Nanoclay Intercalation on Membrane Morphological Structure... 668
    19.5.2 Effect of Nanoclay Intercalation on Transport Mechanism of Water, Proton, and Methanol of the Membrane 669
    19.5.3 Effect of Nanoclay Intercalation on Physicochemical Properties of the Membrane 673
19.6 Future Improvements in the Fabrication of Polymer Nanocomposite Membrane.. 675
19.7 Conclusion 675
References.. 676

19.1   INTRODUCTION

Polymer electrolyte membrane (PEM) is the heart of the direct methanol fuel cell (DMFC) system, which acts as an electrolyte for proton transfer from anode to cathode, as well as providing a barrier to the pathway of electrons between the electrodes [1]. Perfluorinated (PFI) proton exchange membranes (PEMs) from DuPont, Dow Chemical, Asahi Glass, and Asahi Chemical companies, which have been used as electrolyte membranes in PEM fuel cell (PEMFC), possess outstanding chemical and mechanical stabilities. They perform excellently not only in PEMFC but also in DMFC [2,3].

However, PFI membranes have the following two shortcomings: (1) they are very expensive due to the complicated production process, and (2) they have high methanol permeability in DMFCs [3–5]. Methanol permeation across the PFI membranes in DMFC limits their performances. Methanol crossover is particularly troublesome because it leads to (1) cathode depolarization, with a loss of fuel cell power, when methanol contacts the cathode catalyst and oxidation occurs; (2) excessive water production and possible electrode flooding when methanol is chemically oxidized in the air at cathode; and (3) a reduction in the fuel efficiency of the entire system [6].

Due to these problems, research on the development of alternative PEM has been carried out to minimize the shortcomings. Developments in preparing new membranes can be classified into three different branches such as (1) synthesizing new polymers based on nonfluorinated backbones [7,8]; (2) incorporating inorganic fillers such as montmorillonite (MMT) [9], palladium alloy [10], silicon [11], titanium oxide [12], and zeolite [13] into parent polymer matrices; and (3) sulfonated polymers [14,15].

Nonfluorinated polymers such as sulfonated polyimides, polystyrene sulfonic acid, sulfonated poly(arylene ether sulfones), polyphosphazene, polybenzimidazole, sulfonatedpolysulfone, sulfonated poly(phthala-zinone ether ketone), and sulfonated poly(ether ether ketone) have been developed as a potential electrolyte membrane for DMFC [4,6,16–23].

Recently, great interest has been paid to the preparation of polymer-inorganic nano-composite materials in order to overcome the high methanol permeability problem in sulfonated poly(ether ether ketone) (SPEEK) membranes comprising of high surface area nanostructured particles [24]. Exclusive characteristics of polymer-inorganic nanocomposites rely on the dispersion of nanoscale clay layers in the polymer matrix, which strongly depends on the interfacial structural properties [25,26]. From the morphological viewpoint, exfoliated/delaminated polymer-inorganic nanocomposite rather than intercalated or ordinary nanocomposites were considered as promising structures with great potential to form high performance nanocomposites [26,27].

Many studies are now focused on the interfacial voids issue, which is due to the poor adhesion of the hydrophobic polymer and the hydrophilic inorganic filler surface. There are numerous studies that have introduced new polymer-inorganic electrolyte membrane by modifying the external surface of the inorganic filler itself to produce homogeneous polymer-inorganic membrane [25,28–32]. However, studies focusing on the particular issue of filling the interface spacing between the polymer and inorganic with a compatibilizer are rarely reported [27,33].

Therefore, the development of polymer-clay nanocomposite based on the dispersion of nanoscaled clays into SPEEK matrices assisted by a compatibilizer was found to be an interesting approach. Due to its high length to width ratio, that is, 70–150, which is crucial to significantly reduce the methanol crossover and provide a good conductivity value, that is, 10^{-4} at room temperature, the commercially available organically modified MMT clay, namely, Cloisite15A®, and 2,4,6-triaminopyrimidine (TAP) as a compatibilizer were selected to develop new PEM, with the enhancement of overall characteristics in DMFC.

19.2 TYPES OF NANOCOMPOSITE MEMBRANE BY STRUCTURE

Generally, there are two terms used to describe the structure of polymer/inorganic nanocomposites, intercalated and delaminated/exfoliated nanocomposite, as illustrated in Figure 19.1b and c, respectively. As described in Figure 19.1a, a conventional composite or no-chain penetration composite structure is obtained when a layered silicate is dispersed in polymer matrix without any modification to the filler. In the intercalated structure case, the inorganic fillers sustain the self-assembled, well-ordered multilayer structures. Polymer chains are inserted into the gallery space between parallel individual clay layers. In the case of delaminated/exfoliated structures, the interlayer spacing can be in the order of the radius of hygration of the polymer. Individual clay layers are no longer close enough to interact with the gallery cations of the adjacent layers. Therefore, the clay layers may be considered to disperse well in the organic polymer. Thus, the nanocomposite has a monolithic

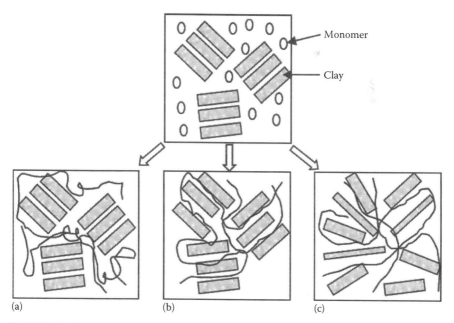

FIGURE 19.1 Typical dispersion states of a nanoclay within a polymer: (a) no-chain penetration, (b) ordered-chain intercalation, and (c) disordered clay exfoliation. (Data from Tsapatsis, M. et al., U.S. Patent 7,087,288, 2006.)

structure at the microscale. This behavior may lead to dramatic changes in the mechanical and physical properties of the new materials [34,35].

Membrane structure can be controlled by either the degree of cross-linking of the polymer matrix or the types of connection bonds between the polymer and inorganic phases in the nanocomposite material, covalent bond, van der Waals force, or hydrogen bonds [36]. It was reported that polymer-inorganic nanocomposite membrane prepared using solution blending formed van der Waals force or hydrogen bonds, and this bonding resulted in an aggregation of inorganic ingredients in the membranes. On the other hand, in the *in situ* polymerization method, inorganic nanoparticles with functional groups can be connected with polymer chains by covalent bonds. However, it is still difficult to avoid the aggregation of inorganic nanoparticles in the fabricated membranes [30,36].

Wang and Dong [37] have successfully developed an exfoliated organophilic MMT clay dispersed into polymer PVDF-HFP using sonication. It was believed that sonication successfully enhanced the clays delamination and reduced or avoided the aggregation of inorganic fillers in the membranes. In the other approach, in order to exfoliate clay sheets in polymeric membrane, first, the polymer solution was obtained by replacing alcoholic co-solvent with an organic solvent such as dimethyl acetamide and dimethyl formamide (DMF) and then the compatible polymer-inorganic nanocomposite solution could be obtained [38]. These new preparation techniques for polymer/inorganic-based PEM membranes have shown good results in reducing the methanol crossover through the membrane. However, they either sacrifice the mechanical strength of the native commercial PEM or cause an increase of the membrane resistance.

19.3 PREPARATION OF EXFOLIATED NANOCOMPOSITE MEMBRANES

19.3.1 INCORPORATION OF ADVANCED ADDITIVES

Although there have been improvements in nanocomposite membrane performance by advances in the external surface modification of inorganic filler techniques and methods employed to produce homogeneous nanocomposite solution, there are still many studies that have reported the problems regarding failing in the production of homogenous nanocomposite membranes [39]. This shows that a novel technique for the fabrication of homogeneous nanocomposite membranes with high membrane performance in DMFC systems is still required to be developed.

Intial reasons for producing homogeneous nanocomposite membranes developed due to the poor compatibility of the inorganic surface and the polymer where the polymer chains could not tightly contact the inorganic nanoparticles, thus forming a narrow gap surrounding the inorganic nanoparticles, as shown in Figure 19.2. This caused the methanol diffusion path to shorten, and thus the apparent methanol diffusivity and permeability were increased. This also explained why the addition of nanoparticles enhanced methanol permeability. Once the nanoparticle surface was compatible with the polymer, nanogaps could not form any more due to the tight contact between the polymer and filler particles [25,36].

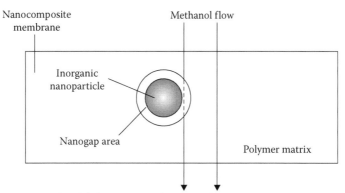

Methanol diffusivity is greatly increased in the nanogap area

FIGURE 19.2   Illustration of nanogap formation in nanocomposite membrane. (Data from Cong, H. et al., *Separation and Purification Technology*, 55, 281–291, 2007.)

Addition of a functional inorganic, which includes the inorganic filler with the presence of functional groups such as sulfonics; phosphonic acids; and quarternary ammonium salts, or nonfunctional inorganic such as weak ionic groups (e.g., carboxylic acids; hydroxyls; and primary, secondary, and tertiary amine groups) into the organic material is usually practiced in PEMs to improve their mechanical and thermal stabilities and performance. For decades, in the development process, perfluorosulfonated membranes and their composites with inorganic materials have dominated and exhibited better thermal and mechanical stabilities and proton conductivities for DMFC. Concurrently, composites of hydrocarbon polymers, such as poly(ether ether ketone), poly(benzimidazole), poly(sulfone), and poly(vinyl alcohol), have been developed as alternatives to the high-cost PFI membranes [40]. Development of new membranes that can conduct protons with little or no water and prevent methanol permeability is the greatest challenge in the fuel cell community.

One of the promising strategies to enhance the performance of PEMs is through improving water management and incorporating nanometer-sized particles such as layered silicates, titania, and zirconia, which can act as a water reservoir. Layered silicates are a class of inorganic materials that are naturally layered in structure. Layered silicates refer to the following categories of clays: (1) natural clays, such as smectites, and (2) synthetic-layered silicates, such as magadiite and mica. Among the available natural clays, MMT and hectorite belong to the smectite family and are the most commonly used layered silicates in nanocomposites due to their high cation-exchange capacities, surface areas, surface reactivities, and adsorptive properties [34,41]. However, the addition of highly crystalline layered silicates does not contribute to the formation of a uniform polymer solution because its chemical or physical interaction is limited to the external surface (active surface). Therefore, the selection of amorphous silica clay is more favorable [35].

Modification of well-ordered silicate materials such as MMT into more amorphous clay, for instance, the Cloisite clay, is crucial in order to improve the processability and performance of the polymer-inorganic materials. Cloisite15A possesses the largest basal spacing (*d* spacing), that is, 31.5Å as compared to the other commercial

Cloisite clay available in the market. Therefore, Cloisite15A is more flexible and, thus, can be fabricated according to the application commonly used in modifying the properties of the ionic polymers [42]. Table 19.1 shows the characteristics of Cloisite clay as stated by Southern Clays Co.

There are some studies that have been conducted on the implementation of Cloisite15A filler in nanocomposites. Some reported Cloisite15A addition provides the best compatibility with the polymer base. For instance, Kim et al. [43] reported that it exhibited the best homogeneity with poly(ethylene oxide) as compared to other types of Cloisite clay tested in their study. On the other hand, Hasani-Sadrabadi et al. [39] reported that by loading Cloisite15A into sulfonated poly(2,6-dimethyl-1,4-phenylene oxide) as well as into sulfonated poly(ether ether ketone), the membranes exhibited lower performance than that of parent polymer membranes, which was attributed to poor contact ability [39]. Based on these findings, it is beneficial to incorporate a compatibilizer such as TAP into the dope, instead of just Cloisite15A. This is a promising idea to enhance the compatibility between SPEEK and Cloisite15A particles, thus increasing the degree of dispersion of Cloisite15A particles.

A new approach based on filling the interface space between the polymer and the inorganic filler with compatibilizer, to improve the methanol barrier properties of polymer-inorganic membranes, suggested to enhance homogeneity for DMFC application, is rarely reported. There are required properties for a compatibilizer to be incorporated with polymer and filler. A compatibilizer should be from a low molecular weight material that could interact simultaneously with fillers and polymers. It should also have the ability to induce strong bonding such as hydrogen bonding with the chemical structure of fillers and polymers. In addition to making chemical bonds, they must be soluble in the solvent used to make the polymer dope solution. In order to prevent the compatibilizer from evaporating during dope formulation, the compatibilizer must be in a solid form at room temperature. This is one of the requirement characteristics of a compatibilizer should have as an additive in composite solution [33]. TAP is a kind of compatibilizer that possesses all the characteristics that a compatibilizer should have. Figure 19.3 illustrates the chemical structure of TAP.

It is expected that TAP will form hydrogen bonds with both the polymer (SPEEK) and the inorganic filler (Cloisite15A) due to the presence of NH_2 functional groups in its chemical structure. Figure 19.4 illustrates the expected bonding formed by TAP between the surfactant Cloisite15A and SPEEK.

Interaction mechanism illustrated in the model of Figure 19.5 could also further reduce the methanol crossover problem related to DMFC. Figure 19.6 illustrates the water, methanol, and proton transport model in SPEEK/Cloisite15A nanocomposite membrane that is attached by TAP. From the transport model, illustrated in the

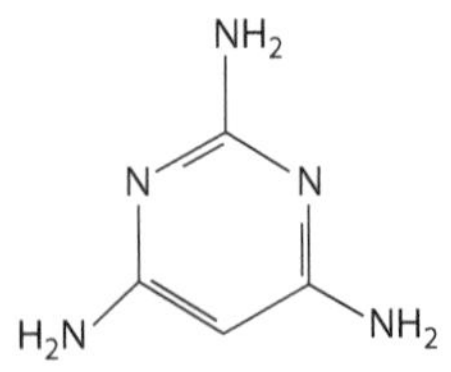

FIGURE 19.3   Chemical structure of 2,4,6-triaminopyrimidine (TAP).

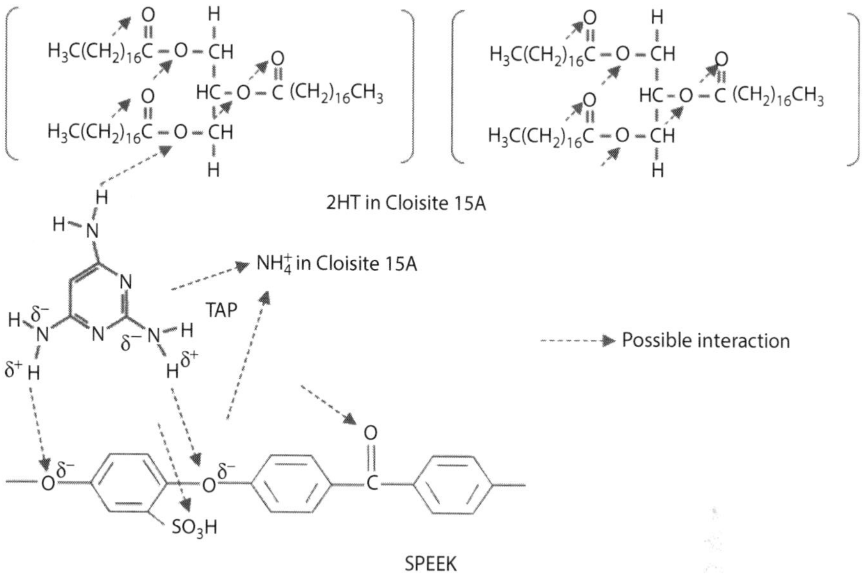

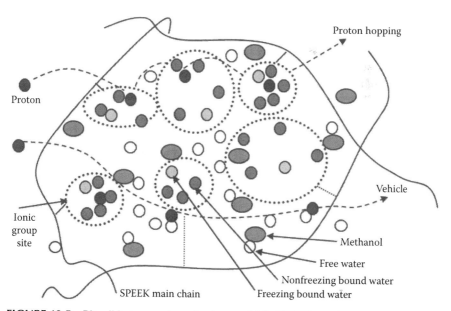

FIGURE 19.4 Plausible interaction model between Cloisite15A clay, 2,4,6-triaminopyrimidine and SPEEK.

FIGURE 19.5 Plausible transport mechanism model in SPEEK membrane.

TABLE 19.1
Characteristics of Various Cloisite Clays

Sample Code	Organic Modifiers (Surfactants)	Chemical Structure of Surfactant	Anion	Interlayer d-Spacing (Å)
CloisiteNA$^+$	None	None	None	11.7
Cloisite10A®	Dimethyl, benzyl, hydrogenated tallow, and quaternary ammonium		Chloride	19.2
Cloisite15A®	Dimethyl, dihydrogenated tallow, and quaternary ammonium		Chloride	31.5
Cloisite20A®	Dimethyl, dihydrogenated tallow, and quaternary ammonium		Chloride	24.2

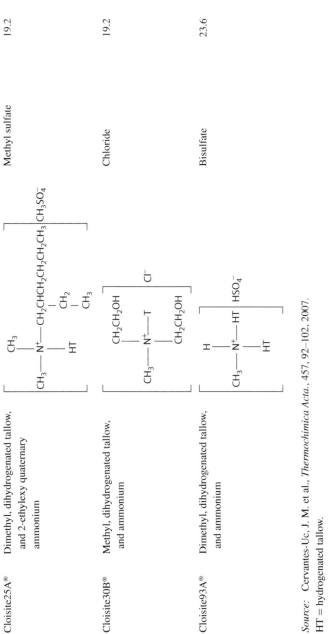

Cloisite25A®	Dimethyl, dihydrogenated tallow, and 2-ethylexy quaternary ammonium		Methyl sulfate	19.2
Cloisite30B®	Methyl, dihydrogenated tallow, and ammonium		Chloride	19.2
Cloisite93A®	Dimethyl, dihydrogenated tallow, and ammonium		Bisulfate	23.6

Source: Cervantes-Uc, J. M. et al., *Thermochimica Acta.*, 457, 92–102, 2007.

HT = hydrogenated tallow.

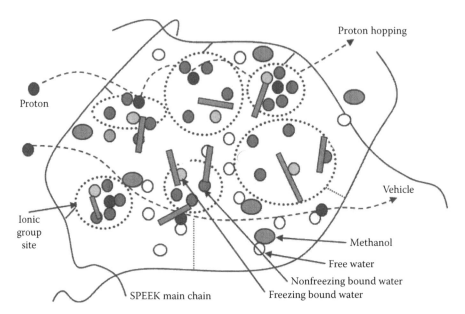

FIGURE 19.6 Plausible transport mechanism model of SPEEK/Cloisite15A/TAP nanocomposite membrane.

figures, it is expected that there is more free water presence in the SPEEK model as compared to SPEEK/Cloisite15A/TAP model. From the model proposed by this study [44], it is expected that the presence of Cloisite15A significantly reduced the amount of free water content in the ionic cluster channel, thus reducing the amount of methanol transport across the nanocomposite membrane.

19.4   MEMBRANE SOLUTION PREPARATION METHOD

When the appropriate approach has been identified, the preparation of the nanocomposite membrane solution then has to be very carefully selected. It is crucial in order to minimize the strong aggregation of the nanofillers because there is huge difference between the polymer and inorganic materials in their properties.

Generally, there are four preparation methods that are most commonly used: melt blending, solution blending, *in situ* polymerization, and sol-gel.

19.4.1   MELT BLENDING

In this method, a molten thermoplastic polymer is blended with an organophilic layered silicate. The mixture is then annealed at a temperature above the glass transition temperature of the polymer, thereby forming a nanocomposite. Figure 19.7 illustrates the melt blending method. Vaia et al. [45] applied the melt blending method based on a predominantly enthalpic mechanism. By maximizing the number of polymer–host interactions, the unfavorable loss of conformational entropy associated with intercalation of the polymer can be overcome leading to exfoliated nanostructure.

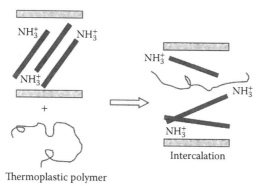

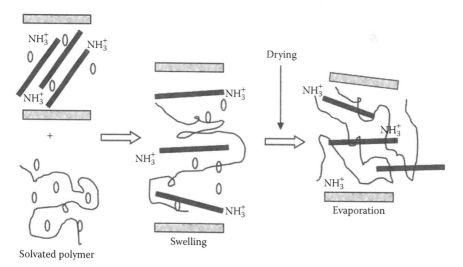

FIGURE 19.7 Schematic representation of the melt blending technique. Polymer and inorganic phases connected by van der Waals force or hydrogen bonds. (Data from Tsapatsis, M. et al., U.S. Patent 7,087,288, 2006.)

19.4.2 SOLUTION BLENDING

Solution blending is a simple way to fabricate polymer-inorganic nanocomposite membranes. First, a polymer is dissolved in a solvent to form a solution, and then inorganic nanoparticles are added into the solution and dispersed by stirring. The nanocomposite membrane is cast by removing the solvent [36]. Figure 19.8 illustrates the solution blending technique. Despite their advantages, the batch processes are not suitable for mass production due to time-consuming procedures for the removal of solvents and fluctuation of mechanical and electrochemical properties of each batch [46]. Because there are still some drawbacks of using this method, a modification on this

FIGURE 19.8 Schematic representation of the solution blending technique. Polymer and inorganic phases connected by van der Waals force and/or hydrogen bonds. (Data from Tsapatsis, M. et al., U.S. Patent 7,087,288, 2006.)

method is necessary. Solution intercalation method is one of the techniques that has been developed from the modification of the solution blending method.

In this method, the polymer and inorganic fillers were dissolved in separate containers containing similar types of solvent. Once both solutions were dissolved, these mixtures were mixed together and were continuously stirred to form a homogeneous polymer-inorganic solution. This simple method was implemented by some researchers and modified to meet their application needs [25,30,32,46]. For instance, the use of the solution intercalation method along with sonication curing is also extensively used to fabricate polymer-inorganic nanocomposite membranes. Sonication is a good method to aid the delamination process and preparation of the polymer-clay nanocomposite. It was reported that delamination/exfoliation polymer-inorganic nanocomposite can be enhanced by controlling the sonication conditions, such as temperature and operating time [37,47,48]. Yoonessi et al. [47] carried out the sonication of dicyclopentadiene (DCPD)/organically modified MMT nanocomposite for 1–3 h under nitrogen at 20 kHz and wave amplitude of 30. Results showed that increasing sonication time increased the d-spacing recorded by the X-ray diffraction pattern, which indicated that further delaminated DCPD nanocomposites were obtained by the sonication curing process.

19.4.3 *In Situ* Polymerization

In this method, the nanoparticles are mixed well with organic monomers, and then the monomers are polymerized. There are often some functional groups such as hydroxyl and carboxyl on the surface of inorganic particles, which can generate initiating radicals, cations, or anions under high-energy radiation, plasma, or other circumstances, to initiate the polymerization of the monomers on their surface. For example, Nunes et al. [49] used this technique to fabricate poly(ether imide)/SiO$_2$ nanocomposite membranes.

In this nanocomposite membrane preparation method, inorganic nanoparticles with functional groups can be connected with polymer chains by covalent bonds. However, the problem of the aggregation of inorganic fillers in the fabricated membranes still remains. Figure 19.9 shows the *in situ* polymerization method.

19.4.4 Sol-Gel

Sol-gel method is the most widely used preparation technology for nanocomposite membranes. In this method, organic monomers, oligomers or polymers, and inorganic nanoparticles precursors are hydrolyzed and condensed into well-dispersed nanoparticles in the polymer matrix. The advantage of this method is obvious; the reaction conditions are moderate, usually room temperature and ambient pressure, and the concentrations of organic and inorganic components are easy to control in the solution. Additionally, the organic and inorganic ingredients are dispersed at the molecular or nanometer level in the membranes, and thus the membranes are homogenous [36]. Figure 19.10 illustrates the sol-gel method.

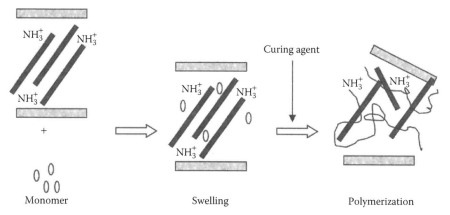

FIGURE 19.9 Schematic representation of the *in situ* polymerization method. Polymer and inorganic phase connected by covalent bonds. (Data from Tsapatsis, M. et al., U.S. Patent 7,087,288, 2006.)

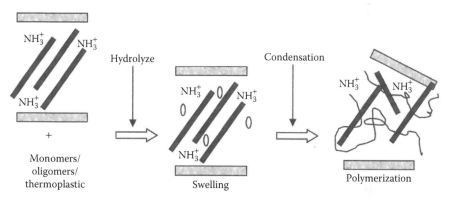

FIGURE 19.10 Schematic representation of the sol-gel method. Polymer and inorganic phases connected by covalent bonds. (Data from Tsapatsis, M. et al., U.S. Patent 7,087,288, 2006.)

19.5 PREPARATION OF SPEEK/CLOISITE15A/ TAP NANOCOMPOSITE MEMBRANE

Based on the above-mentioned methods, solution blending has been found to be the simplest method to be employed. Consequently, the authors have employed the technique with some modifications. This modified technique was known as the *solution intercalation method*.

In order to prepare the homogenous SPEEK nanocomposite membrane, 10 wt% of SPEEK solution was first prepared by dissolving SPEEK in DMSO. Desired amounts of Cloisite15A and TAP were added to a small amount of DMSO, separately, in another container and the mixture was vigorously stirred for 24 h at room temperature. Consequently, each Cloisite15A and TAP mixture was combined in another container, and the mixture was again vigorously stirred for 24 h at room temperature. The latter mixture was then added to the SPEEK solution, so that the total amount of DMSO was 90 ml. The SPEEK-containing mixture was again vigorously stirred for 24 h at room temperature to produce a homogeneous solution. For instance, in order to prepare SP63/2.5CL/5.0TAP, 10 wt% of SPEEK at DS of 63% was dissolved in 60 ml of DMSO. In another container, 0.25 g of Cloisite15A was added to 15 ml of DMSO. 0.5 g of TAP was added to 15 ml of DMSO in another container. They were then stirred thoroughly in a separate container and mixed together and stirred again to obtain a homogeneous dope solution. Before proceeding to the casting process, the mixture was heated to 100°C to evaporate the remaining DMSO solvent.

The polymer dope was then cast on a glass plate with a casting knife to form a thin film. The thickness of the membranes was controlled by adjusting the height of casting knife from the glass plate. The resultant film was then dried in a vacuum oven for 24 h at 80°C. The membrane was further dried for 6 h at 100°C to remove the residual solvent. After being detached from the glass plate, by immersing the membrane together with the glass plate into water bath, the membrane was dried for three days in a vacuum oven at 80°C. Finally, the membrane was treated with 1M sulfuric acid solution for a day at room temperature and subsequently rinsed with water several times to remove the remaining acid and to assure that the sulfonate was in acid form. Similar treatment was applied to other tested membranes.

19.5.1 EFFECT OF NANOCLAY INTERCALATION ON MEMBRANE MORPHOLOGICAL STRUCTURE

Figure 19.11a–e depicts the FESEM membrane surface micrographs of SP63/2.5CL, SP63/2.5CL/1.0TAP, SP63/2.5CL/2.5TAP, SP63/2.5CL/5.0TAP, and SP63/2.5CL/7.5TAP membranes, respectively. Designation of the membrane was based on the loading of the additives. For instance, SPEEK with a degree of sulfonation of 63% incorporated with 2.5 wt% of Cloisite15A and 1.0 wt% of TAP was designated as SP63/2.5CL/1.0TAP. From the figure, it is obvious that the SP63/2.5CL/5.0TAP nanocomposite membrane exhibited the most homogenous structure with smaller sizes of distributed Cloisite15A particles as compared to the other tested membranes. Addition of 1.0 wt% of TAP into SPEEK/Cloisite15A membrane did not show any obvious improvement from SP63/2.5CL sample. The largest particle size observed was about 280 nm for both SP63/2.5CL and SP63/2.5CL/1.0TAP membranes. When 2.5 wt% of TAP was loaded, the relative percentage of smaller Cloisite15A particles was increased. It seemed that both SP63/2.5CL/2.5TAP and SP63/2.5CL/5.0TAP membranes have a similar morphological structure. The range of Cloisite15A particle size observed from SP63/2.5CL/5.0TAP membranes was bigger (~20–160 nm) as compared to that of SP63/2.5CL/2.5TAP (~20–120 nm). However, the larger particles that can be seen in SP63/2.5CL/2.5TAP membrane are more than double in number as in the SP63/2.5CL/5.0TAP membrane. Apparently, the addition of 5.0 wt% of TAP has significantly decreased the free volume unfilled with the nanoclay particles throughout the polymer matrix as compared to the SP63/2.5CL/2.5TAP membrane. This suggests that the addition of 5.0 wt% of TAP is the optimum loading to compatibilize 2.5 wt% of Cloisite15A particles in the SPEEK matrices. Interestingly,

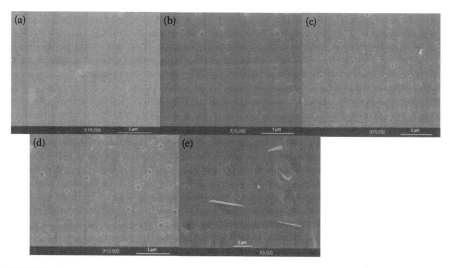

FIGURE 19.11 FESEM images of membrane surface of (a) SP63/2.5CL, (b) SP63/2.5CL/1.0TAP, (c) SP63/2.5CL/2.5TAP, (d) SP63/2.5CL/5.0TAP, and (e) SP63/2.5CL/7.5TAP.

it was observed that the Cloisite15A particle distribution occurred just under the membrane skin layer. There is no single Cloisite15A particle observed on top of the membrane surface. This observation confirms the assumption made in this study that all Cloisite15A particles inserted into SPEEK were not lost during the membrane preparation process. In addition, at 7.5 wt% of TAP incorporation, the TAP particles appear in the SPEEK matrices at lower SEM magnification. Apparently, some of the TAP particles did not dissolve during the dope formulation process due to the excessive amount of TAP used.

19.5.2 Effect of Nanoclay Intercalation on Transport Mechanism of Water, Proton, and Methanol of the Membrane

Typically, the transport of protons in a cation exchange membrane is greatly affected by the presence of water molecules. Therefore, water provides a better medium for protons to be extremely mobile, more than in the common ions environment. This is the result of the fact that the movement of protons in water does not take place through normal diffusion, but through a hopping process where hydrogen bonds between water molecules are converted into covalent bonds [50].

Migration of proton in water occurs by two different processes. The first is a free-solution diffusion process, as shown in Figure 19.12a, in which the proton and the associated water of hydration diffuse through the water phase. The second process is a *proton hopping*, in which the proton moves by a sequence of steps involving the formation and breakage of hydrogen bonding of a series of water molecules. This second mechanism, involving proton hopping, is known as *Grotthuss mechanism* (see Figure 19.12b). This mechanism involves a proton hopping from H_3O^+ to a neighboring H_2O molecule which, in turn, ejects one of the

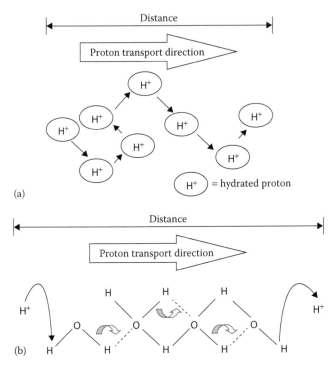

FIGURE 19.12 Schematic representation of proton transport through (a) free-solution diffusion and (b) proton-hop mechanism (Grotthuss mechanism). (Data from Jiang, W., Preparation and characterization of pore-filled cation exchange membranes, Ph.D. Thesis, McMaster University, Hamilton, Ontario, Canada, 1999.)

protons, which is to form a hydrogen bonding with a neighboring proton. Thus, a proton *hops* (is transported) from one water molecule to another. As a result, a succession of proton hopping steps leads effectively to proton transport through the water. In Grotthuss mechanism, the original proton entering the membrane is not the proton coming out of the membrane because it is not the proton itself that is transported but only the charge of the proton (not its mass); charge is passed on from one water molecule to another [51].

The proton transport through a hydrated ion-exchange membrane can, in principle, occur through both the above-mentioned processes. The approach to explain transport in an ion-exchange membrane is that the water in an ion-exchange material can be understood as bound (or *nonfreezable, associated*) water and free (*freezable, bulk-like*) water. Free water resembles bulk water with the major interactions being hydrogen bonds. Bound water molecules, however, are in a vicinity of the materials, whereas free water is further away from them. Strength of the bonding between the bound water molecule and the material is greater than that of hydrogen bonding in free water. Thus, free water is bulk-like but bound water is assumed to be part of the materials. Free-solution diffusion of a proton through a hydrated membrane can, in principle, involve both free and bound water. However, free diffusion is expected

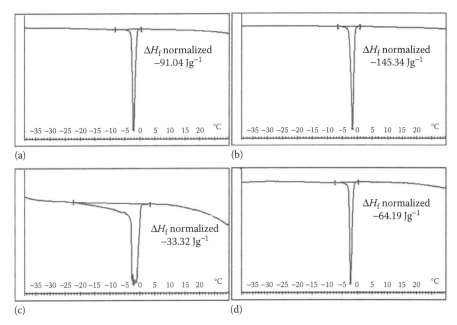

FIGURE 19.13 DSC melting curves of hydrated membranes of (a) SP63, (b) SP63/2.5CL, (c) SP63/2.5CL/5.0TAP, and (d) Nafion®112.

to be much slower in bound water. On the other hand, the proton hopping or the Grotthuss mechanism can occur in both free and bound water. This transport phenomenon in an ion-exchange membrane explains the ion transport in polyelectrolyte membranes [52].

Figure 19.13a–d presents differential scanning calorimetric (DSC) measurements to determine the type of water content of SP63, SP63/2.5CL, SP63/2.5CL/5.0TAP, and Nafion®112 membranes. This measurement is crucial to further understand the water-retention capability of the prepared PEMs.

There are three states of water found in membrane: (1) freezable bound water, weakly bound to clusters and embedded particles in the polymer; (2) free water corresponding to water occupied in the free volume of a membrane, crystallizing at a higher temperature than freezable bound water; and (3) nonfreezable water, strongly bound to clusters and embedded particles in the polymer and shows no thermal transition by DSC ice-melting diagrams [53]. Because the endothermic peaks in DSC ice-melting diagrams are attributed to the freezable water, that is, freezable bound water and free water, the amount of the freezable water in the nanocomposite membranes can be estimated from the DSC profile. Figure 19.13 shows the endothermic peaks corresponding to the freezing free and freezing bound water. The melting enthalpy is obtained through the integration and normalization in units of Jg^{-1} of the swollen membrane. Latent heat of water (333 Jg^{-1}) was taken for the calculation. Based on the calculation, free water and nonfreezing bound water in water uptake is summarized in Table 19.2.

TABLE 19.2

Composition of Water in SPEEK, SPEEK Nanocomposite, and Nafion®112 Membranes

Sample	Total Water (%)	ΔH_f Normalized (Jg^{-1} sample)[a]	ΔH_f per Mass Water (Jg^{-1} Water)[b]	Freezing Water/Total Water (%)[c]	Nonfreezing Water/Total Water (%)	Freezing Water/Sample (%)	Non-freezing Water/Sample (%)
SP63	29.70	91.04	306.53	92.05	7.95	27.34	2.36
SP63/2.5CL	54.87	145.34	264.88	79.54	20.46	43.64	11.23
SP63/2.5CL/5.0TAP	26.19	33.32	127.22	38.21	61.79	10.00	16.19
Nafion®112	21.43	64.19	299.53	89.95	10.05	19.28	2.15

[a] Obtain from the DSC measurement.

[b] ΔH_f per mass water (Jg^{-1} water) = ΔH_f normalized (Jg^{-1} sample)/total water (%).

[c] Freezing water/total water (%) = ΔH_f per mass water (Jg^{-1} water)/melting enthalpy of pure water, that is, 333 Jg^{-1}.

19.5.3 Effect of Nanoclay Intercalation on Physicochemical Properties of the Membrane

It was suggested that there are several factors that may contribute to good overall membrane characteristics based on proton conductivity and methanol permeability. Some of the factors include the following: (1) appropriate water uptake, (2) homogeneous dispersion of inorganic filler (exfoliated nanocomposite structure), and (3) high polar cationic surfactant in the inorganic clay layers. However, it is important to further discuss other possible factors that may also contribute to the improvement of both characteristics. It is natural for a membrane with low water uptake to have low methanol permeability. It widely accepted that methanol permeability generally increases with the amount of soaked water, but it dominantly occurs through free water (freezing water) inside the interconnected membrane structure channels and insignificantly through nonfreezing bound water associated with the ionic sites. As shown in Table 19.2, SP63/2.5CL/5.0TAP nanocomposite membrane exhibited the lowest amount of free water content, and hence the presence of low methanol permeability was not surprising. This was because methanol molecules were transported through the central space filled with free water [54].

On the other hand, an increase in the proton conductivity of SP63/2.5CL/5.0TAP nanocomposite membrane can be interpreted in the following way. The sorbed water molecules are present mostly in the ionic cluster domains and ionic cluster channels. In particular, in the ionic cluster channels, the water molecules exist in two different forms. One is the protonated water (mostly nonfreezing bound water) that is bound strongly to the ionic site and the other is free water that occupies the central space free from the influence of the ionic sites. Proton transfer through the ionic cluster channel occurs by two different mechanisms: (1) near the channel wall through the bound water, in which proton is transported by the Grotthuss mechanism, hopping from one ionic site to the other, and (2) through free water and a vehicle mechanism, in which proton diffusion is facilitated by the water molecules moving through the interconnected central channel space. But the contribution from the Grotthuss mechanism is more essential [53]. Results presented in Table 19.2 indicate an increase in the bound water (nonfreezing water/sample) content of the SPEEK membrane with the inorganic filler (Cloisite15A). Thus, the water-retention capacity of the nanocomposite membranes was enhanced due to the presence of inorganic fillers, which was expected to reduce the dehydration of membrane. Therefore, it can be said that the enhanced proton conductivity of the SP63/2.5CL/5.0TAP nanocomposite membrane is attributed to the water-retention capability [53–56]. From these results, it is clear that the incorporation of appropriate amounts of Cloisite15A and TAP into SPEEK has synergistic effects to lower the methanol permeability and to increase the proton conductivity.

In general, the incorporation of silicate layers into polyelectrolytes restricts the accessible nanometric channels for the migration of the polar elements such as hydrogen ions, water, and methanol molecules [26]. However, at appropriate Cloisite15A (2.5 wt%) loading and the formation of a homogenous nanocomposite membrane with TAP, a sufficient water uptake, acceptable proton conductivity, and low methanol permeability were obtained. Table 19.3 summarizes the membrane physicochemical

TABLE 19.3

Improvement on Physicochemical Properties of SPEEK Nanocomposite Membranes with Various TAP Loadings

Sample	Water Uptake (%)	Reduction (%)	Membrane Thickness[a] (μm)	Membrane Resistance[a] (Ω cm²)	Proton Conductivity (mScm⁻¹)	Difference (%)	Methanol Permeability × 10⁸ (cm²s⁻¹)	Reduction (%)
SP63	29.70 ± 0.10	NA	60 ± 2.00	0.73 ± 0.031	6.23 ± 0.21	NA	57.6 ± 3.8	NA
SP63/2.5CL	54.87 ± 0.07	NA	67 ± 1.73	0.82 ± 0.015	7.84 ± 0.36	NA	8.97 ± 0.05	NA
SP63/2.5CL/1.0TAP	21.86 ± 0.86	60.16	71 ± 1.00	0.65 ± 0.024	8.18 ± 0.20	+4.16	5.85 ± 0.22	34.78
SP63/2.5CL/2.5TAP	25.62 ± 0.44	53.31	74 ± 0.58	0.52 ± 0.022	10.6 ± 0.40	+26.04	2.19 ± 0.15	75.59
SP63/2.5CL/5.0TAP	26.19 ± 0.27	52.27	71 ± 1.15	0.35 ± 0.026	16.3 ± 0.11	+51.90	1.30 ± 0.21	85.51
SP63/2.5CL/7.5TAP	18.02 ± 0.49	67.16	75 ± 0.58	0.80 ± 0.04	7.04 ± 0.40	−11.36	0.903 ± 0.04	89.93
Nafion®112	21.43 ± 0.74	NA	60 ± 0.10	0.39 ± 0.011	11.6 ± 0.38	NA	156.00 ± 3.6	NA

[a] Membrane thickness/ohmic resistance for proton conductivity measurement.

NA, Not measured; +, Increment; −, Reduction.

properties in terms of water uptake, proton conductivity, and methanol permeability as well as percentage of the improvement in comparison with SP63/2.5CL membrane.

19.6 FUTURE IMPROVEMENTS IN THE FABRICATION OF POLYMER NANOCOMPOSITE MEMBRANE

Contribution of electrospinning in nanofiber fabrication has already burgeoned since the sixteenth century, when the first electrostatic attraction of a liquid was recorded by William Gilbert [57]. From that time onward, many researchers have produced nanofibers for various applications. Compared to the conventional methods, such as wet spinning, dry spinning, and melt spinning, electrospinning can produce a much larger specific surface area and smaller pore size with fiber diameters in the range of 10 ~ 1000 nm, whereas conventional methods only produce fibers with diameter in the range of 5 ~ 500 μm [58]. Other than electrospinning processes, drawing, template synthesis, phase separation, and self-assembly are the other methods used in developing nanofibers [59]. However, due to the versatility possessed by electrospinning, it is more useful in developing highly porous, patterned, nanofibrous polymeric materials [60]. Other than producing a nanofiber, the advantages possessed by electrospinning are due to its low cost, capability, and high speed, which gives it a high potential in producing nanocomposite fiber [61]. Unique properties, such as extremely long, large surface area, complex pore size, and complex alignment, on either woven or nonwoven electrospun nanofibers make it practicable in various applications [62–65], especially in PEMs. Lately, SPEEK has been electrospun and incorporated with SiO_2 as a supported PEM for fuel cell application by Lee et al. [66]. They impregnated SiO_2/SPEEK nanofiber mat into Nafion solution in order to get a dense membrane for PEMFC applications. Nafion impregnated SiO_2/SPEEK nanofiber mat membrane had excellent proton conductivity compared with cast Nafion and SPEEK. Neppalli et al. [67] has reported in their work that clay can behave as a nucleating agent as well as an obstacle to the polymer mobility, and this depended on the quantity and the dispersion state of the clay within the polymer matrix. Fortunately, the electrospinning process favors the elongation of the chains and the ordering of the polymer [68]. Li et al. [69] conducted an experiment regarding the incorporation of silver ions within the polymer matrix through electrospinning. However, the use of silver ion within the SPEEK polymer was not effective for DMFC applications. It can be concluded that electrospinning is an advanced process for the large-scale production of long continuous nanocomposite fibers from a variety of polymers [59] and inorganic fillers for sophisticated PEMs.

19.7 CONCLUSION

In conclusion, in order to overcome dispersion problems commonly found with the use of clay, the introduction of TAP as a compatibilizer was found to be desirable. The success in the selection of inorganic material additives was more pronounced by employing an appropriate modified membrane solution preparation method, namely, the intercalation method. In addition to the selection of the best solution preparation method, the application of advanced technology, such as electrospinning, in producing nanostructures of clay particles is also interesting and is worthy of future studies.

REFERENCES

1. Xing, P., Robertson, G. P., Guiver, M. D., Mikhailenko, S. D., Wang, K., and Kaliaguine, S. (2004). Synthesis and Characterization of Sulfonated Poly (ether ether ketone) for Proton Exchange Membranes. *Journal of Membrane Science.* 229: 95–106.
2. Zaidi, S. M. J., Mikhailenko, S. D., Robertson, G. P., Guiver, M. D., and Kaliaguine, S. (2000). Proton Conducting Composite Membranes from Polyether Ether Ketone and Heteropolyacids for Fuel Cell Applications. *Journal of Membrane Science.* 173: 17–34.
3. Chang, J. H., Park, J. H., Kim, C. S., and Park, O. O. (2003). Proton-Conducting Composite Membranes Derived from Sulfonated Hydrocarbon and Inorganic Materials. *Journal of Power Sources.* 124: 18–25.
4. Li, L., Zhang, J., and Wang, Y. (2003). Sulfonated Poly (ether ether ketone) Membranes for Direct Methanol Fuel Cell. *Journal of Membrane Science.* 226: 159–167.
5. Dillon, R., Srinivasan, S., Arico, A. S., and Antonucci V. (2004). International Activities in DMFC R&D: Status of Technologies and Potential Applications. *Journal of Power Sources.* 127: 112–126.
6. Carter, R. (2003). Sulfonated polyphosphazene based membranes for use in direct methanol fuel cells. Ph.D. Dissertation. Tulane University, New Orleans, LA.
7. Antonucci, P. L., Arico, A. S., Creti, P., Ramunni, E., and Antonucci, V. (1999). Investigation of a Direct Methanol Fuel Cell Based on a Composite Nafion-Silica Electrolyte for High Temperature Operation. *Solid State Ionics.* 125: 431–437.
8. Jones, D. J. and Roziere, J. (2008). Advances in the Development of Inorganic-Organic Membranes for Fuel Cell Applications. *Advances in Polymer Science.* 215: 219–264.
9. Jung, D. H., Cho, S. Y., Peck, D. H., Shin, D. R., and Kim, J. S. (2003). Preparation and Performance of a Nafion/Montmorillonite Nanocomposite Membrane for Direct Methanol Fuel Cell. *Journal of Power Sources.* 118: 205–211.
10. Ma, Z. Q., Cheng, P., and Zhao, T. S. (2003). A Palladium-Alloy Deposited Nafion Membrane for Direct Methanol Fuel Cells. *Journal of Membrane Science.* 215: 327–336.
11. Jung, D. H., Cho, S. Y., Peck, D. H., Shin, D. R., and Kim, J. S. (2002). Performance Evaluation of a Nafion/Silicon Oxide Hybrid Membrane for Direct Methanol Fuel Cell. *Journal of Power Sources.* 106: 173–177.
12. Yoon, S. R., Hwang, G. H., Cho, W. I., Oh, I. H., Hong, S. A., and Ha, H. Y. (2002). Modification of Polymer Electrolyte Membranes for DMFCs using Pd Films Formed by Sputtering. *Journal of Power Sources.* 106: 215–223.
13. Libby, B., Smyrl, W. H., and Cussler, E. L. (2003). Polymer-Zeolite Composite Membranes for Direct Methanol Fuel Cells. *AIChE Journal.* 49: 991–1001.
14. Roelofs, K. S., Hirth, T., and Schiestel, T. (2010). Sulfonated Poly(ether ether ketone)-based Silica Nanocomposite Membranes for Direct Ethanol Fuel Cells. *Journal of Membrane Science.* 346: 215–226.
15. Tian, S. H., Shu, D., Wang, S. Xiao, J. M., and Meng, Y. Z. (2010). Sulfonated Poly(Fluorenyl Ether Ketone Nitrile) Electrolyte Membrane with High Proton Conductivity and Low Water Uptake. *Journal of Power Sources.* 195: 97–103.
16. Glipa, X., El Haddad, M., Jones, D. J., and Roziere, J. (1997). Synthesis and Characterization of Sulfonated Polybenzimidazole: A Highly Conducting Proton Exchange Polymer. *Solid State Ionics.* 97: 323–331.
17. Hasiotis, C., Deimede, V., and Kontoyannis, C. (2001). New Polymer Electrolytes Based on Blends of Sulfonated Polysulfones with Polybenzimidazole. *Electochimica Acta.* 46: 2401–2406.

18. Lufrano, F., Gatto, I., Staiti, P., Antonucci, V., and Passalacqua, E. (2001). Sulfonated Polysulfone Ionomer Membranes for Fuel Cells. *Solid State Ionics.* 145: 47–51.

19. Woo, Y., Oh, S. Y., Kang, Y. S., and Jung, B. (2003). Synthesis and Characterization of Polimide Membranes for Direct Methanol Fuel Cell. *Journal of Membrane Science.* 220: 31–45.

20. Smitha, B., Sridhar, S., and Khan, A. A. (2003). Synthesis and Characterization of Proton Conducting Polymer Membranes for Fuel Cells. *Journal of Membrane Science.* 225: 63–76.

21. Kaliaguine, S., Mikhailenko, S. D., Wang, K. P., Xing, P., Robertson, G., and Guiver, M. (2003). Properties of SPEEK Based PEMs for Fuel Cell Application. *Catalysis Today.* 82: 213–222.

22. Lee, H. C., Hong, H. S., Kim, Y. M., Choi, S. H., Hong, M. Z., Lee, H. S., and Kim, K. (2004). Preparation and Evaluation of Sulfonated-Fluorinated Poly(arylene ether)s Membranes for Proton Exchange Membrane Fuel Cell (PEMFC). *Electrochimica Acta.* 49: 2315–2323.

23. Gil, M., Ji, X., Li, X., Na, H., Hampsey, J. E., and Lu, Y. (2004). Direct Synthesis of Sulfonated Aromatic Poly(Ether Ether Ketone) Proton Exchange Membranes for Fuel Cell Applications. *Journal of Membrane Science.* 234: 75–81.

24. Staiti, P., Arico, A. S., Baglio, V., Lufrano, F., Passalacqua, E., and Antonicci, V. (2001). Hybrid Nafion-Silica Membranes Doped with Heteropolyacids for Application in Direct Methanol Fuel Cells. *Solid State Ionics.* 145: 101–107.

25. Thomassin, J. M., Pagnoulle, C., Caldarella, G., Germain, A., and Jerome, R. (2006). Contribution of Nanoclays to the Barrier Properties of a Model Proton Exchange Membrane for Fuel Cell Application. *Journal of Membrane Science.* 270: 50–56.

26. Hasani-Sadrabadi, M. M., Dashtimoghadam, E., Sarikhani, K., Majedi, F. S., and Khanbabaei, G. (2010a). Electrochemical Investigation of Sulfonated Poly(Ether Ether Ketone)/Clay Nanocomposite Membranes for Moderate Temperature Fuel Cell Applications. *Journal of Power Sources.* 195: 2450–2456.

27. Villaluenga, J. P. G., Khayet, M., Valentin, J. L., Seoane, B., and Mengual, J. I. (2007). Gas Transport Properties of Polypropylene/Clay Composite Membranes. *European Polymer Journal.* 43: 1132–1143.

28. Thomassin, J. M., Pagnoulle, C., Caldarella, G., Germain, A., and Jerome, R. (2005). Impact of Acid Containing Montmorillonite on the Properties of Nafion Membranes. *Polymer.* 46: 11389–11395.

29. Kim, Y., Lee, J. S., Rhee, C. H., Kim, H. K., and Chang, H. (2006). Montmorillonite Functionalized with Perfluorinated Sulfonic Acid for Proton Conducting Organic-Inorganic Composite Membranes. *Journal of Power Sources.* 162: 180–185.

30. Lin, Y. F., Yen, C. Y., Ma, C. C. M., Liao, S. H., Hung, C. H., and Hsiao, Y. H. (2007a). Preparation and Properties of High Performance Nanocomposite Proton Exchange Membrane for Fuel Cell. *Journal of Power Sources.* 165: 692–700.

31. Lin, Y. F., Yen, C. Y., Hung, C. H., Hsiao, Y. H., and Ma, C. C. M. (2007b). A Novel Composite Membranes Based on Sulfonated Montmorillonite Modified Nafion for DMFCs. *Journal of Power Sources.* 168: 162–166.

32. Chuang, S. W., Chung, S. L., and Hsu, H. C. L. (2007). Synthesis and Properties of Fluorine-Containing Polybenzimidazole/Montmorillonite Nanocomposite Membranes for Direct Methanol Fuel Cell Applications. *Journal Power Sources.* 168: 172–177.

33. Yong, H. H., Park, H. C., Kang, Y. S., Won, J., and Kim, W. N. (2001). Zeolite-Filled Polyimide Membrane Containing 2, 4, 6-Triaminopyrimide. *Journal of Membrane Science.* 188: 151–163.

34. Tsapatsis, M., Jeong, H. K., and Nair, S. (2006). Layered Silicate Material and Applications of Layered Materials with Porous Layers. (U.S. Patent 7,087,288.)

35. Carrado, K. A. (2000). Synthetic Organo- and Polymer–Clays: Preparation, Characterization, and Materials Applications. *Applied Clay Science.* 17: 1–23.
36. Cong, H., Radosz, M., Towler, B. F., and Shen, Y. (2007). Polymer-Inorganic Nanocomposite Membranes for Gas Separation. *Separation and Purification Technology.* 55: 281–291.
37. Wang, M. and Dong, S. (2007). Enhanced Electrochemical Properties of Nanocomposite Polymer Electrolyte Based on Copolymer with Exfoliated Clays. *Journal of Power Sources.* 170: 425–432.
38. Song, M. K., Park, S. B., Kim, Y. T., Kim, K. H., Min, S. K., and Rhee, H. W. (2004). Characterization of Polymer-Layered Silicate Nanocomposite Membranes for Direct Methanol Fuel Cells. *Electrochimica Acta.* 50: 639–643.
39. Hasani-Sadrabadi, M. M., Ghaffarian, S. R., Mokarram-Dorri, N., Dashtimoghadam, E., and Majedi, F. S. (2009b). Characterization of Nanohybrid Membranes for Direct Methanol Fuel Cell Applications. *Solid State Ionics.* 180: 1497–1504.
40. Lobato, J., Cañizares, P., Rodrigo, M. A., Linares, J. J., Fernández-Fragua, A. (2006). Application of Sterion® Membrane as a Polymer Electrolyte for DMFCs. *Chemical Engineering Science.* 61: 4773–4782.
41. Yang, D., Li, J., Jiang, Z. Y., Lu, L., and Chen, X. (2009). Chitosan/TiO_2 Nanocomposite Pervaporation Membranes for Ethanol Dehydration. *Chemical Engineering Science.* 64: 3130–3137.
42. Cervantes-Uc, J. M., Cauich-Rodríguez, J. V., Vázquez-Torres, H., Garfias-Mesías, L. F., and Paul, D. R. (2007). Thermal Degradation of Commercially Available Organoclays Studied by TGA–FTIR. *Thermochimica Acta.* 457: 92–102.
43. Kim, S., Hwang, E. J., Jung, Y., Han, M., and Park, S. J. (2008). Ionic Conductivity of Polymeric Nanocomposite Electrolytes Based on Poly(Ethylene Oxide) and Organo-Clay Materials. *Colloids and Surfaces A: Physicochemical Engineering Aspects.* 313–314: 216–219.
44. Juhana Jaafar, Ismail A. F., and Matsuura, T. (2009). Preparation and Barrier Properties of SPEEK/Cloisite/TAP Nanocomposite Membrane for DMFC Application. *Journal of Membrane Science.* 345: 119–127.
45. Vaia, R. A., Ishii, H., and Giannelis, E. P. (1993). Synthesis and Properties of Two-Dimensional Nanostructures by Direct Intercalation of Polymer Melts in Layered Silicates. *Chemistry Material.* 5: 1694–1696.
46. Kim, T. K., Kang, M., Choi, Y. S., Kim, H. K., Lee, W., Chang, H., and Seung, D. (2007). Preparation of Nafion-Sulfonated Clay Nanocomposite Membrane for Direct Methanol Fuel Cells via a Film Coating Process. *Journal of Power Sources.* 165: 1–8.
47. Yoonessi, M., Toghiani, H., Kingery, W. L., Pittman Jr, C. U. (2004). Preparation, Characterization, and Properties of Exfoliated/Delaminated Organically Modified Clay/ Dicyclopentadiene Resin Nanocomposites. *Macromolecules.* 37: 2511–2518.
48. Zhou, Y., Pervin, F., Rangari, V. K., and Jeelani, S. (2007). Influence of Montmorillonite Clay on the Thermal and Mechanical Properties of Conventional Carbon Fiber Reinforced Composites. *Journal of Materials Processing Technology.* 191: 347–351.
49. Nunes, S. P., Peinenmann, K. V., Ohlorogge, K., Alpers, A., Keller, M., and Pires, A. T. N. (1999). Membranes of Poly(Ether Imide) and Nanodispersed Silica. *Journal of Membrane Science.* 157: 219–226.
50. Moilanen, D. E., Spry, D. B., and Fayer, M. D. (2008). Water Dynamics and Proton Transfer in Nafion Fuel Cell Membranes. *Journal of the American Chemical Society.* 24(8): 3690–3698.
51. Park, S. H., Park, J. S., Yim, S. D., Park, S. H., Lee, Y. M., and Kim, C. S. (2008). Preparation of Organic/Inorganic Composite Membranes Using Two Types of Polymer Matrix via a Sol–Gel Process. *Journal of Power Sources.* 181: 259–266.

52. Jiang, W. (1999). Preparation and characterization of pore-filled cation exchange membranes. Ph.D. Thesis. McMaster University, Hamilton, Ontario.
53. Norddin, M. N. A., Ismail, A. F., Rana, D., Matsuura, T., Mustafa, A., and Tabe-Mohammadi, A. (2008). Characterization and Performance of Proton Exchange Membranes for Direct Methanol Fuel Cell: Blending of Sulfonated Poly(Ether Ether Ketone) with Charged Surface Modifying Macromolecule. *Journal of Membrane Science.* 323: 404–413.
54. Karthikeyan, C. S., Nunes, S. P., Prado, L. A. S. A., Ponce, M. L., Silva, H., Ruffmann, B., and Schulte, K. (2005). Polymer Nanocomposite Membranes for DMFC Application. *Journal of Membrane Science.* 254: 139–146.
55. Sambandam, S. and Ramani, V. (2007). SPEEK/Functionalized Silica Composite Membranes for Polymer Electrolyte Fuel Cells. *Journal of Power Sources.* 170: 259–267.
56. Tripathi, B. P., Kumar, M., and Shahi, V. K. (2009). Highly Stable Proton Conducting Nanocomposite Polymer Electrolyte Membrane (PEM) Prepared by Pore Modifications: An Extremely Low Methanol Permeable PEM. *Journal of Membrane Science.* 327: 145–154.
57. Tucker, N., Stanger, J. J., Staiger, M. P., Razzaq, H., and Hofman, K. (2012). The History of the Science and Technology of Electrospinning from 1600 to 1995. *Journal of Engineered Fibers and Fabrics (SPECIAL ISSUE-July-FIBERS).* pp: 63–73.
58. Frenot, A. and Chronakis, I. S. (2003). Polymer Nanofibers Assembled by Electrospinning. *Science and Technology*, 8(1): 64–75.
59. Huang, Z.-M., Zhang, Y.-Z., Kotaki, M., and Ramakrishna, S. (2003). A Review on Polymer Nanofibers by Electrospinning and Their Applications in Nanocomposites. *Composite Science and Technology*, 63: 2223–2253.
60. Zucchelli, A., Fabiani, D., and Gualandi, C. (2009). An Innovative and Versatile Approach to Design Highly Porous, Patterned, Nanofibrous Polymeric Materials. *Journal of Material Science.* 44: 4969–4975.
61. Zhang, D., Karki, A. B., Rutman, D., Young, D. P., Wang, A., Cocke, D., Ho, T. H., and Guo, Z. (2009). Electrospun Polyacrylonitrile Nanocomposite Fibers Reinforced with Fe_3O_4 Nanoparticles: Fabrication and Property Analysis. *Polymer*, 50(17): 4189–4198.
62. Fang, J., Wang, X., and Lin, T. (2011). Functional Applications of Electrospun Nanofibers. In Lin, T. (Ed.) *Nanofibers-Production, Properties and Functional Applications* (pp. 287–326). InTech.
63. Cavaliere, S., Subianto, S., Savych, I., Jones, D. J., and Rozière, J. (2011). Electrospinning: Designed Architectures for Energy Conversion and Storage Devices. *Energy Environmental Science*, 4(12): 4761–4785.
    Cebe, P. and Hong, S.-D. (1986). Crystallization Behaviour of Poly (ether ether ketone). *Polymer*, 27.
64. Sautther, B. P. (2005). *Continuous Polymer Nanofibers Using Electrospinning.* NSF-REU Summer 2005 Program, August 5. Universiti of Illinois, Chicago, IL.
65. Thavasi, V., Singh, G., and Ramakrishna, S. (2008). Electrospun Nanofibers in Energy and Environmental Applications. *Energy Environmental Science.* 1: 205–221.
66. Lee, C., Jo, S. M., Choi, J., Baek, K.-Y., Truong, Y. B., Kyratzis, I. L., and Shul, Y. G. (2013). SiO_2/Sulfonated Poly Ether Ether Ketone (SPEEK) Composite Nanofiber Mat Supported Proton Exchange Membranes for Fuel Cells. *Journal of Material Science.* 48: 3665–3671.
67. Neppalli, R., Wanjale, S., Birajdar, M., and Causin, V. (2013). The Effect of Clay and of Electrospinning on the Polymorphism, Structure and Morphology of Poly (Vinylidene fluoride). *European Polymer Journal.* 49: 90–99.

68. Wang, Y., Li, M., Rong, J., Nie, G., Qiao, J., Wang, H., Wu, D., Su, Z., Niu, Z., and Huang, Y. (2013). Enhanced Orientation of PEO Polymer Chains Induced by Nanoclays in Electrospun PEO/Clay Composite Nanofibers. *Colloid and Polymer Science*. 291: 1541–1546.

69. Li, X., Hao, X., and Na, H. (2007). Preparation of Nanosilver Particles into Sulfonated Poly (Ether Ether Ketone) (S-PEEK) Nanostructures by Electrospinning. *Materials Letters*. 61: 421–426.

20 Effects of Solvent and Blending on the Physical Properties of Sulfonated Poly(Ether Ether Ketone)

A Promising Membrane Material for PEMFC

Amir-Al-Ahmed, Abdullah S. Sultan, and S.M. Javaid Zaidi

CONTENTS

20.1 Introduction 681
20.2 Membrane Preparation and Physical Properties Evaluation 685
    20.2.1 Effect of Solvent 685
        20.2.1.1 On Morphology... 685
        20.2.1.2 On Thermal Properties.. 686
        20.2.1.3 On Solution Viscosity 687
        20.2.1.4 On Rheology 691
    20.2.2 Effect of Blending Component.. 692
        20.2.2.1 On Morphology... 692
        20.2.2.2 On Thermal Properties.. 695
        20.2.2.3 On Solution Flow 698
20.3 Conclusion 703
Acknowledgment 705
References.. 706

20.1 INTRODUCTION

Polymer electrolyte membrane (PEM) is one of the key components of a PEM fuel cell. It serves both as an electrolyte and as a separator to prevent direct physical mixing of the hydrogen from the anode and oxygen supplied to the cathode. PEM contains ionizable groups, in which protons (mobile cations) are electrostatically associated with the fixed anionic charges on the polymer. The protons migrate through the membrane by hopping sequentially from one fixed charge to another and are stabilized during this transfer by several water molecules. The essential property requirements of a PEM are chemical and electrochemical stability, adequate mechanical stability, and strength to ensure dimensional stability when swollen and under tension, surface properties compatible with bonding catalytic electrodes to the membrane, low permeability for reactants and products, and high ionic conductivity for large current densities and low internal resistances [1–3]. Present state-of-the-art fuel cell technology depends on perfluoronatedionomer membrane Nafion®. Nafion is costly (~$800 per m²); though its chemical and oxidative resistance is excellent, its durability and reactant permeability become problematic when thinner membranes are used [1,4]. As a result, a major thrust in PEM fuel cell research is engaged in developing new membrane materials. PEMs based on aromatic hydrocarbon polymers are viewed as an alternative option. These polymers are attractive for several reasons, such as low cost, easier processing parameters, and high temperature use. In recent years, a variety of new ionomers have been prepared and characterized as membranes for use in PEM fuel cells. These include sulfonated polyimides [5–9], sulfonated poly(arylene ether sulfones) [10–17], sulfonated poly(arylene ether phosphine oxide)s [18,19], and sulfonated polyketones [20–24]. These materials showed some promises with respect to conduction, stability, methanol crossover, and water transport, though none has as yet been commercialized as a viable replacement for Nafion.

The family of poly(arylene ether ketone)s (PAEKs) consists of polymers with different ratios of arylene ether and arylene ketone groups. Among the various structures of PAEKs, the most common variant is poly(ether ether ketone), (PEEK), though other PAEKs are also studied. Two procedures are normally used to obtain sulfonated PEEK (SPEEK) [25]: (1) postsulfonation of PEEK and (2) direct copolymerization of sulfonated and unsulfonated segments. Generally, PEEK is sulfonated by electrophilic aromatic substitution. This substitution takes place on the phenyl ring containing the highest electron density, typically the orthoposition between the ether segments. The first systematic sulfonation procedure of PEEK was reported by Jin et al. in 1985 [26]. Common sulfonating agents are concentrated sulfuric acid, fuming sulfuric acid, and chlorosulfonic acid [27]. Fuming sulfuric acid and chlorosulfonic acid are much stronger sulfonating agents than concentrated sulfuric acid but have less control over the sulfonation reactions accompanied by side reactions [28]. Degree of sulfonation (DS) of PEEK in concentrated sulfuric acid mainly depends on the concentration of sulfuric acid, the sulfonation temperature, and time. When PEEK is sulfonated with excess amount of concentrated sulfuric acid, the kinetics of the reaction is first order with the concentration of the PEEK chain repeat unit, and inversely proportional to the concentration of the sulfonated PEEK units as reported by Shibuya and Porter [29,30]. The latter effect was attributed to a competing desulfonation reaction that occurs for arylsulfonic acids in an acidic medium [31]. Huang et al. [32], however, reported that the sulfonation of PEEK followed first-order kinetics and depends only on the concentration of the PEEK repeat unit. This postsulfonation process can be subdivided into heterogeneous and homogeneous sulfonation processes [33]. In the case of heterogeneous sulfonation, the dissolution and sulfonation of

PEEK occur simultaneously, whereas in the homogeneous sulfonation technique, first PEEK is dissolved in a solvent (e.g., methylsulfonic acid), and subsequently sulfonated by the addition of sulfuric acid. The DS of SPEEK prepared by heterogeneous sulfonation at room temperature was strongly dependent on sulfonation time. Homogeneous sulfonation was time consuming, because of the lower concentration of sulfuric acid in the reaction mixture. SPEEK obtained with both sulfonation types exhibited similar characteristics. The DS strongly influences the processability and stability of the polymer [34,35]. If the sulfonation degree is too high (>90%), SPEEK is water soluble, and if the DS is too low (<40%), it is not soluble in standard solvents used for membrane casting. Lack of control over the degree and location of functionalization, degradation of the polymer, and unwanted side reaction are major drawbacks of the postsulfonation process. On the other hand, the copolymerization of sulfonated and unsulfonated segments gives control over the chain length regulation, the use of monomers containing two sulfonic acid groups, and control of the DS and polymer structure without crosslinking reaction [25,36]. The first reported sulfonated monomer used in the direct copolymerization of SPEEK was 2-fluorobenzenesulfonate [37]. SPEEK prepared by copolymerization showed improved membrane properties in comparison to postsulfonated SPEEK [36,38].

Similar to Nafion, the structure of sulfonated PAEK nanoseparates when material is humidified or wetted. Kreuer [39] made a comparison between the morphology of Nafion and sulfonated poly(ether-ether-ketone-ketone) (SPEEKK) based on a network model. He observed that the transport properties and morphological stability between both systems were clearly different. These differences were ascribed to the more hydrophobic and flexible backbone and the more acidic sulfonic acid groups of Nafion in comparison to SPEEKK. Therefore, Nafion exhibits wide water channels, a good connected percolated hydration structure, and no dead-end channels. For SPEEKK, on the other hand, the hydrophilic/hydrophobic regions are less separated. The highly branched hydration structure consists of small channels with many dead ends. This is represented schematically in Figure 20.1.

When these structures are compared, it can be observed that in wet systems, proton transport as well as fuel crossover in Nafion are high, and in the case of PAEK, these incidents are low. A comparative study was reported by Xue and Yin [2] on methanol permeability and proton conductivity of SPEEK with DS between 59% and 93% with Nafion. Methanol permeability strongly depends on the DS and increased in that range from 27×10^{-8} to 154×10^{-8} cm^2 s^{-1}. The lowest permeability was 6.5 times lower than for Nafion, whereas this factor for proton conductivity was just 2.4. These results were obtained at 22°C. The fuel crossover of PAEKs membranes is relatively low, but membrane modifications should lead to further decrease in fuel crossover while maintaining or improving the proton conductivity. Obviously, the stability features (e.g., swelling and mechanical) should be optimal to obtain good performance in fuel cell tests [1,2].

Many nonfluorinated specialty polymers have been functionalized, characterized, and tested for the application in direct alcohol fuel cells but the most promising polymer type is PEEK. Sulfonation of these polymers leads to structures that can nanoseparate similar to Nafion. Sulfonation degree has a significant impact on

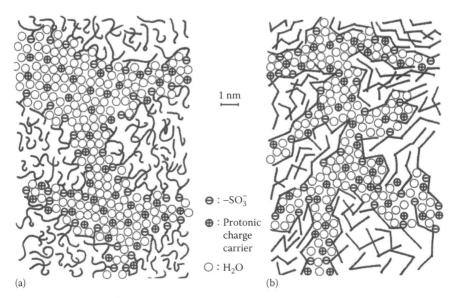

FIGURE 20.1 Schematic representation of the microstructures of (a) Nafion and (b) a SPEEKK. (Data from Kreuer, K. D., *J. Membr. Sci.*, 185, 29–39, 2001.)

swelling behavior, fuel crossover, and proton conductivity. SPEEK is applicable with sulfonation degrees ranging from 40% to 90%. With higher DS, the proton conductivity is high but generally lacks proper stability. That is why researchers have tried several methods to improve its properties such as cross-linking, blending with other polymers, and even blending with inorganic particles. Presence of an inorganic phase resulted in a reduction in alcohol crossover in nearly all cases. Similar to perfluorosulfonic acid modifications, the inorganic materials used by various groups are metal oxides, prepared by the sol-gel method, or commercially obtained layered silicates, zeolites, and zirconium phosphates. Heteropolyacids were also used but in combination with an inorganic phase to prevent leaching out. However, there are reports that after a certain time, there is a leaching of inorganic particles, and this affects the performance of the membrane [40–43].

SPEEK 1.6 (DS 53%, Mw ~110,000) was selected for this study because of its moderate DS and good proton conductivity; it does not dissolve in water or methanol, whereas at the same time retains good mechanical properties [40,44–47]. To get a better understanding of the effects of the solvents and temperature on the physical properties of SPEEK 1.6, it was dissolved in three different solvents (dimethylacetamide [DMAc], dimethylsulfoxide [DMSO], and *N,N*-dimethylformamide [DMF]), and its solution viscosity and rheology were studied; surface morphology and thermal properties of the casted membrane were also investigated. A detailed comparative study was reported in our previous study [44,46]. In another investigation, to study the effect of blending components on SPEEK 1.6, it was blended with polyetherimide (PEI), polysulfone (PS), and boron orthophosphate (BPO$_4$). Surface morphology, thermal, and rheological properties were investigated.

20.2   MEMBRANE PREPARATION AND PHYSICAL PROPERTIES EVALUATION

As described in Section 20.1, pure SPEEK 1.6 membrane was prepared by solution casting method [44]. 1.5 g of SPEEK 1.6 was dissolved in 50 mL of N,N-DMAc at room temperature and then stirred at 60°C for 12 hours to get a concentrated solution. The resulting viscous solution was cast on a glass plate using a casting knife. This membrane was dried in low vacuum at 60°C for 6 hours and then at 80°C for another 6 hours and again at 100°C for 12 hours to remove any trace amount of solvent. Finally this membrane was immersed into de-ionized water for 15 min and again dried under low vacuum at 60°C for 6 hours. The same procedure was used for membrane preparation with DMSO and DMF. Physical properties of these membranes were investigated by scanning electron microscopy (SEM) (JEOL-JSM-6460A); Differential scanning calorimetry (DSC2910 [METTLER TOLEDO]) [44]. For solution viscosity measurement (Cannon-Fenske [USA] viscometer), five different solutions of pure SPEEK 1.6 were prepared by dissolving 0.1, 0.05, 0.025, 0.0125, and 0.00625 g of polymer in 100 mL of the different solvents: DMAc, DMSO, and DMF. For rheology, pure SPEEK 1.6 was dissolved in DMAc, DMSO, and DMF, and these solutions were stirred for about 12 hours at 100°C to get a thick gel; precautions were taken to prevent any air bubble formation. Both dynamic strain sweep and dynamic frequency sweep tests were carried out (range 2–2000 g for normal force; 2–2000 g-cm for torque). These measurements were performed in a descending frequency, ω, from 10^2 to 10^{-1} rad s^{-1}. Both the steady-shear and oscillatory experiments were carried out at room temperature (25°C). After careful evaluation of the results, we found that DMAc is comparatively a better solvent for SPEEK, so all the blend samples were prepared in DMAc [44,45]. SPEEK 1.6 was blended with PEI (SPEEK:PEI-1 [98:2], SPEEK:PEI-2 [95:5], SPEEK:PEI-3 [90:10], SPEEK:PEI-4 [85:15], and SPEEK:PEI-5 [75:25]); PS (SPEEK:PS-1 [98:5], sPEEK:PS-2 [95:10], SPEEK:PS-3 [90:20], SPEEK:PS-4 [85:30], and SPEEK:PS-5 [75:40]); and BPO$_4$ (SPEEK:BPO$_4$-1 [90:5], SPEEK:BPO$_4$-2 [80:10], SPEEK:BPO$_4$-3 [70:20], SPEEK:BPO$_4$-4 [60:30], and SPEEK:BPO$_4$-5 [50:40]) to study the effect of the different blending constituents on the physical properties of the membrane and membrane materials. SEM, DSC, and rheological properties of these blends were investigated (at room temperature) [45]. Solution viscosity of the selected blend samples were studied at different temperatures.

20.2.1   Effect of Solvent

20.2.1.1   On Morphology

SPEEK 1.6 membranes were prepared from its solution of DMAc, DMF, and DMSO by a casting method. SEM images of the cross-sections of these membranes were taken and presented in Figure 20.2. During the membrane preparation steps, the polymer–solvent mixture and even the casted membranes are often exposed to heating, sometimes well above 100°C, to eliminate the high boiling solvents. All these processes facilitate chemical interactions between polymer–solvent, polymer–polymer, and even solvent–solvent. The processing stage has an influence on the morphology of the membrane and eventually the performance. Influence of solvents

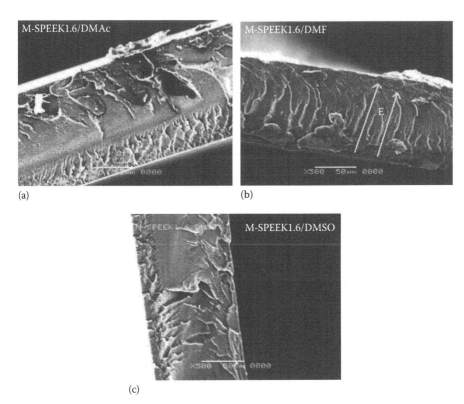

(a)

(b)

(c)

FIGURE 20.2 SEM images for the fracture surface SPEEK 1.6 membrane into (a) DMAc, (b) DMF, and (c) DMSO. (Data from Sultan, A. S. et al., *Macromol. Symp.*, 313–314, 182–193.)

on membrane properties and its effect on proton conductivity was reported by K. Nishida et al. [48]. Here we observed that all these three solvents are quite good for membrane preparation and all the prepared membranes showed good uniform morphology. However, we observed several differences in other properties (thermal, viscosities, and so on), which are discussed in the Sections 20.2.1.2 and 20.2.1.3.

20.2.1.2 On Thermal Properties

Robertson et al. [49] studied the effect of solvents on the properties of polyelectrolyte membranes. They observed that these solvents readily interact with SPEEK. Among them, DMF and DMSO forms hydrogen bonding even at low temperature, but DMAc exhibits such a bonding at high temperatures only, around 140°C [49]. DSC analysis of SPEEK 1.6 membranes prepared from its solution of DMAc, DMF, and DMSO is presented in Table 20.1. It was observed that the glass transition temperature increased from DMAc, DMF to DMSO, respectively, and this is mainly due to the different stages of interaction between these solvents with the acid group of SPEEK. Comparing values of T_g for solid SPEEK 1.6 without solvent with that of SPEEK 1.6 membrane dissolved in DMAc, a reduction of about 6°C was observed. Most likely this was attributed to the partial decomposition of DMAc. It is known that DMAc as well as DMF decompose at their normal boiling points to give dimethylamide (DMA) and other by-products [49]. Two decomposition

TABLE 20.1

DSC Analysis of SPEEK 1.6 Powder and Membranes in Argon Atmosphere

Polymer	Solvent	T_g (°C)	T_{d1} (°C)	T_{d2} (°C)
PEEK		146.43		
SPEEK 1.6 (solid)		200.87	314.20	n.r.
SPEEK 1.6 (membrane)	DMAc	196.35	320.45	361.5
SPEEK 1.6 (membrane)	DMF	203.77	306.90	366.2
SPEEK 1.6 (membrane)	DMSO	210.19	327.30	–

n.r., not recorded.

temperatures (T_d) were obtained, and the values of T_{d1} were in the 305°C–320°C range, whereas T_{d2} was in the range of 355°C–360°C. The first one is mainly due to the splitting of the sulfonic acid groups, and the second one corresponds to the sulfonic acid decomposition. These new results are also similar to our previously reported ones [44].

20.2.1.3 On Solution Viscosity

Several correlation equations were used to model the viscometric data obtained for pure SPEEK 1.6 solutions in different solvents at different temperatures.

Huggins equation [50–52]:

$$\frac{\eta_{sp}}{c} = [\eta] + K_H [\eta]^2 c$$

Fouss–Strauss equation [53,54]:

$$\frac{\eta_{sp}}{c} = \frac{A}{1 + B\sqrt{c}} + D$$

Fedors equation [55]:

$$\frac{1}{2(\eta_r^{0.5} - 1)} = \frac{1}{[\eta]} \frac{1}{c} - \frac{1}{c_m [\eta]}$$

The linear regression coefficients (R^2) for equations were determined. Parameters obtained by fitting the data in Fedors model are presented in Table 20.3 [44]. In our previous study we observed that the higher the DS, then the higher the intrinsic viscosity ($[\eta]$) at the same temperature, and $[\eta]$ was also dependent on the properties of the solvent [44]. The polymer concentration parameter (C_m) in the Fedors correlation depends on solvent quality and molecular interaction, which shows that DMAc is a strong solvent for SPEEK (Table 20.2). The values of parameter B in the Fouss–Strauss correlation also depends on the polyelectrolyte–solvent interaction; this again indicates strong interaction between SPEEK and DMAc.

TABLE 20.2

Intrinsic Viscosity Calculated Using Fedors Correlation Parameters, C_m, B, and K, for SPEEK Solution at Different Temperatures

Polymer	Solvent	21°C	40°C	50°C	60°C	C_m[a]	B[a]	K[a]
		$[\eta]_{Fedors}$						
SPEEK 1.6	DMAc	1.4086	1.4569	1.5011	1.5952	0.6467	6.4679	0.7933
SPEEK 1.6	DMSO	2.0036	1.8867	1.8965	2.0238	0.5343	1.2347	0.5895
SPEEK 1.6	DMF	1.3697	1.4275	1.5220	1.5340	0.5605	1.9113	0.6265

Source: Sultan, A. S. et al., *Macromol. Symp.*, 313–314, 182–193.
[a] Values are reported at room temperature.

The power parameter k in Korecz–Csákvári–Tü dos model [56]

$$\frac{\eta_{sp}}{c} = \left(a_0 + a_1 c\right)c^{k-1}$$

which can be treated as a deviation factor from natural behavior of polymeric solution ($k = 1$), and the solution of DMSO showed more polyelectrolyte behavior compared to DMAc and DMF.

Effect of temperature on viscosity was studied by Arrhenius equation

$$\eta = \eta e^{\beta/T}, \text{ with } \beta = \frac{E}{r}$$

where
  E is the Helmholtz activation energy
  r is the gas constant (8.314 J mol^{-1} K^{-1})

Table 20.3 shows values of η_o and β with the corresponding correlation coefficient (R^2) for all the solutions used in this study. The model of Yang et al. [57]

$$\Omega = \frac{\eta}{\eta_0 e^{\beta/T}} = A'X + B' + 1$$

was used to generalize the combined effects of concentration and temperature.

The plot of parameter Ω versus X should be linear, and A′ and B′ (two constant parameters) can be calculated from the slope and intercept, respectively. Figure 20.3 shows the relation between resistance (Ω) and reactance (X) for SPEEK 1.6 dissolved in DMAc, DMSO, and DMF, respectively, which gives the linear values. The two constant parameters A′ and B′ that were calculated from the slope and intercept are presented in Table 20.3. The constant A′ is related to the polyion conformation. Low Huggins coefficients (η_0), less than 0.5, indicates good interaction between the polymer and solvent. The –SO$_3$H group in the polymer backbone facilitates the interaction process. Thus, the intermolecular repulsive interactions between charged groups along

TABLE 20.3

Arrhenius Parameters and A′, B′ Used in Yang et al. Model for Predicting the Effect of Temperature and Concentration on Apparent Viscosity

Polymer	Solvent	β	η_0	R^2	ρ^a(g cm^{-3})	RI[a,b]	A′	B′	R^2
SPEEK 1.6	DMAc	1336.2	0.0201	0.9998	0.9350	1.4362	−0.635	0.9021	0.9999
SPEEK 1.6	DMSO	1576.8	0.0193	0.9992	0.9420	1.4280	−0.836	1.1851	0.9961
SPEEK 1.6	DMF	1229.4	0.0230	1.000	1.089	1.4760	−0.552	0.7932	0.9991

Source: Sultan, A. S. et al., *Macromol. Symp.*, 313–314, 182–193.

[a] Values are reported at room temperature.

[b] RI = Refractive index.

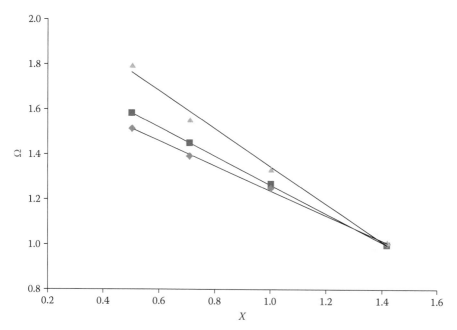

FIGURE 20.3 Parameter Ω versus the scaling variable X (reactance) for SPEEK 1.6 solution: (▲) DMSO, (■) DMAc, and (♦) DMF. (Data from Sultan, A. S. et al., *Macromol. Symp.*, 313–314, 182–193.)

the chain increase and the expansion of the polymer chain consequently take place. SPEEK solution in DMSO exhibited greater reduction in viscosity values than DMF and DMAc, indicating more expanded chain in the case of this polyelectrolyte [44].

Concentration dependence of reduced viscosity of the solution at different temperatures is shown in Figure 20.4. A sharp increase in the reduced viscosity (η_{sp}/C) at low concentration has been observed. This is a characteristic behavior of polyelectrolyte solution [58–60]. This occurs in the polyelectrolyte solutions because of the expansion of the long polyionic chain, which is caused by the progressively

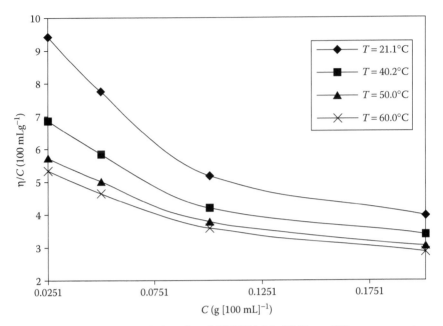

FIGURE 20.4 Plot of reduced viscosity of SPEEK 1.6 in DMAc at different temperatures. (Data from Sultan, A. S. et al., *Macromol. Symp.*, 313–314, 182–193.)

enhanced dissociation of ionizable groups as the concentration decreases, and therefore the intensification of intermolecular repulsive interaction between the ionic groups on the chain [58,59,61]. Abdullah et al. [44] also reported that the SPEEK 1.3 samples showed an increase in the viscosity gradient between the concentrations 0.05–0.0125 g/100 ml. This was due to the overlapping and/or interpenetrating of molecular chains with each other, which increases the viscosity. This overlapping occurs at a certain concentration, and this sudden transition corresponds to the onset of coil overlapping between the polymer chains in the solution. However, rearranging the above Fouss–Strauss equation, and plotting (C/η_{sp}) versus $\sqrt{C}$, a straight line was observed with intrinsic viscosity as the reciprocal of the intercept, $([\eta] = 1/\text{intercept})$, as presented in Figure 20.5.

Results correlated by the Fouss–Strauss empirical model have been verified to be accurate in the semidilute condition, in which η_{sp} increases monotonically to the value of $(A + D)$ with decreasing concentration. However, in the literature [62,63], it was found that the specific viscosity (η_{sp}) of the diluted polyelectrolyte solution, especially with low salt concentration, shows an intriguing anomalous concentration dependence, in which η_{sp} has a maximum value at certain concentration of C_{max}. In general, the Fouss–Strauss empirical equation describes the behavior of polyelectrolyte successfully, but it has not involved the relationship for uncharged polymers as a limiting case, which is the advantage of the Tüdos model over the Fouss–Strauss empirical model [56]. An alternative explanation for this characteristic behavior was given by Eisenberg and Pouyet and rephrased by Nashida et al. [48], where the intermolecular electrostatic interaction between polyions play a role. Another possible

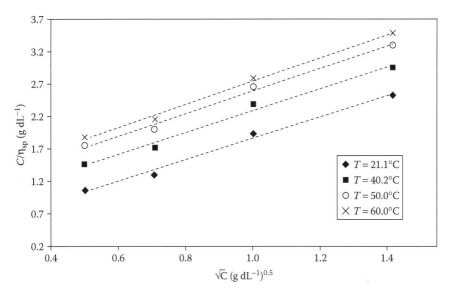

FIGURE 20.5 Variation of reduced viscosity reciprocal (η_{sp}/C) as a function of square root of concentration ($\sqrt{C}$) of SPEEK 1.6 in DMAc at different temperatures based on Fouss–Strauss equation. (Data from Sultan, A. S. et al., *Macromol. Symp.*, 313–314, 182–193.)

explanation is the electroviscous effect due to the interaction between a polyion and counter-ions. Correlating the data with Huggins model, where the reduced viscosity (η_{sp}/C) was plotted against concentration (C), and the observed nonlinear behavior is a typical characteristic of polyelectrolytes; nonionic polymers give a linear relationship for this model [50–52].

20.2.1.4 On Rheology

Dynamic (η') and steady-shear (G') measurements of these three SPEEK gels were carried out at room temperature (21°C ± 0.5) at different frequencies (ω) during step rate tests. Measurements were performed in the frequency (ω) range of 100–0.1 rad s^{-1}. A shear-thinning behavior was observed for all three samples. Dynamic viscosity decreased with increasing shear rate, and all these solutions exhibited a limiting dynamic viscosity value at high frequency. Abdullah et al. [44] reported that at lower frequency ($\omega = 1$), values of G' are also sensitive to DS. Rheological properties of the polymer solution are sensitive to the concentration [61,64,65]. As the concentration increases, coils begin to overlap and finally entanglements are expected to form, and hence dynamic viscosity increases. This was also explained by polymer–solvent and polymer–polymer interaction. While preparing a dense solution, if the polymer–solvent interaction is of sufficient strength, the polymer chains will tend to extend themselves, and the evaporation of solvent from such solution may result in collapsed coils. On the other hand, if the polymer–polymer interaction is significantly stronger than polymer–solvent interaction, helices are formed and macromolecular aggregates occur during desolvation. Formation of helices and folds depends on the reactivity of the polymer

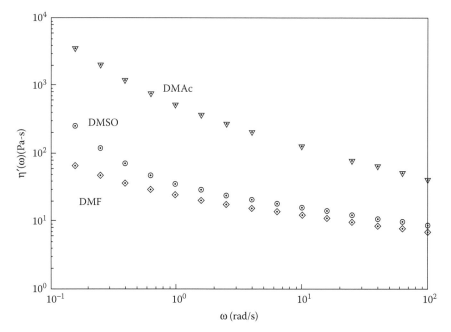

FIGURE 20.6 Effects of solvents on $\eta'(\omega)$ for pure SPEEK 1.6 samples at room temperature ($\gamma° = 15\%$). (Data from Sultan, A. S. et al., *Macromol. Symp.*, 313–314, 182–193.)

with itself, especially the hydrogen atoms in the chain. Configuration and type of the side group also have a role in the folding of the molecules [66]. For SPEEK, the sulfonic group ($-SO_3H$) is expected to form hydrogen bonds with the solvent and with other SPEEK molecules, which results in molecular folding at higher concentration. Effects of solvents on dynamic viscosity of SPEEK 1.6 are shown in Figure 20.6. The solution in DMF shows lower values of $\eta'(\omega)$ as compared to DMSO and DMAc.

20.2.2 Effect of Blending Component

20.2.2.1 On Morphology

SEM images of SPEEK 1.6:PEI blend membranes are presented in Figure 20.7. The first four samples (SPEEK 1.6:PEI-1 to SPEEK 1.6:PEI-4) showed homogeneous surface morphology, but at 25%, PEI concentration phase separation was observed. Morphological properties of PEEK:PEI blends has also been studied by several researchers. Jin et al. [67] observed a phase separation during PEEK crystallization, which leads PEI segregation into an amorphous phase between the PEEK texture units. Frigione et al. [68] found that blends with 80/20 weight ratios of PEEK/PEI had homogeneous membrane texture. Again Zaidi et al. [40] investigated the morphological behavior of SPEEK:PEI blends with 5%, 15%, and 25% of PEI contents and reported phase segregation occurs in a similar way, as reported by the earlier researchers. They also observed that small (<1 μm)

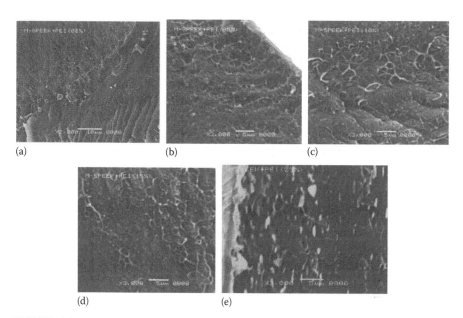

FIGURE 20.7 SEM images for SPEEK 1.6:PEI blend membranes prepared in DMAc: (a) 98:2, (b) 95:5, (c) 90:10, (d) 85:15, and (e) 75:25. (Data from Sultan, A. S. et al., *Eur. Polym. J.*, 47, 2295–2302.)

spherical particles of PEI were formed at 5% PEI content, and it increased up to 1 μm length at 15% PEI and ~1–2 μm length at 25% of PEI concentration. A similar behavior was also observed in this study (Figure 20.5). In this case the phase separation occurred at 25% PEI concentration, which is similar to the findings of previous research [4,40]. Because of the opposite (acidic and basic) nature of SPEEK 1.6 and PEI, it is obvious that there will be a polymer–polymer interaction, and this leads to a semiporous surface morphology. Again, with the increased PEI concentration, there will be more unbounded PEI units, which also facilities the phase separation.

SEM images of SPEEK 1.6:PS composite membranes are presented in Figure 20.8 (90/10, 80/20, 70/30, 60/40, and 50/50 SPEEK/PS). A trend of average pore size can be detected in the images of low PS content; the higher the PS weight percent, the bigger is the average pore size. In fact, the pore size of the membrane needs to be enlarged to facilitate a better conductivity, whereas the thickness must be reduced to give a higher permeability [69]. Influence on morphology of SPEEK as an additive to PS was studied by Bowen et al. (2002) [70]. They observed that at low percentage of PS (10%), the blend exhibits moderate uniform surface morphology with small finger-like particles of PS distributed over the surface. A similar trend was reported by Arozak et al. [71,72], who also observed small finger-like particles coming out of the surface of the membranes. Figure 20.8c shows a special characteristic of SPEEK:PS blends, where large (2–7 μm) finger-like structure appears to be dominant at this composition. The effect of PS in enhancing phase separation tends to be more important in the case of high percentages of PS (40% and 50%), as shown

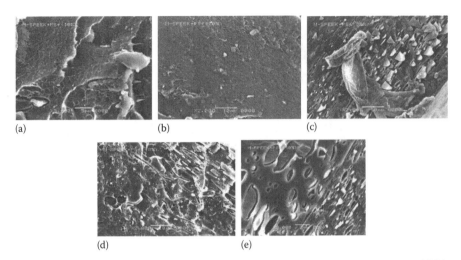

FIGURE 20.8 SEM images for SPEEK 1.6:PS blends prepared in DMAc: (a) 90/10, (b) 80/20, (c) 70/30, (d) 60/40, and (e) 50/50.

in Figure 20.8, respectively. This phase separation also indicates the low miscibility of PS into SPEEK [73]. Arzak et al. [71,72] also reported that at 30% PS composition, both PS and SPEEK phases are strongly deformed. This effective contribution to the strength of the membrane makes this composition undeniably good for membrane mechanical properties.

Phase separation for the higher second component composition was observed for all the blends. Several expressions have been used to predict phase separation in polymer blends, and the most often used one depends on composition and viscosity, where continuous phases exist by the semi-imperial expression

$$\left[\frac{\eta_1 \varphi_2}{\eta_2 \varphi_1} \right] \approx 1$$

where:

η and φ are the viscosity and the composition, respectively, of each component

Using values of viscosity obtained in this work, the calculated fraction was found to be 1.037 at 25% PEI and 1.048 at 30% PS composition. This is in agreement with morphologies observed by SEM, where the phase separation seems to take place at PS content of around 30%. Introducing solid electrolyte, such as BPO_4, into the polymer matrix had a profound impact on the polymer structure. SEM images of the SPEEK:BPO_4 are shown in Figure 20.9. These inorganic moieties incorporated microporosity in the surface of the membrane. It was observed that with the increasing BPO_4 concentration, the surface became less homogenous and at 40 and 50 wt% of BPO_4, a phase separation is clearly visible. This is similar to the findings of Zaidi et al. [40] for 40 and 60 wt% BPO_4 embedded in the SPEEK matrix.

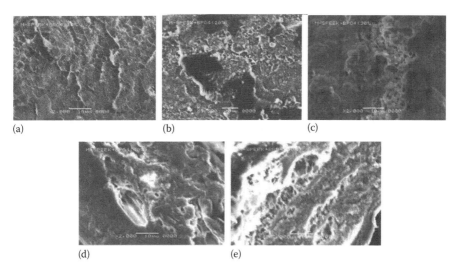

(a) (b) (c)

(d) (e)

FIGURE 20.9 SEM images of SPEEK 1.6:BPO$_4$ blend membranes prepared in DMAc:
(a) 90:10, (b) 80:20, (c) 70:30, (d) 60:40, and (e) 50:50.

20.2.2.2 On Thermal Properties

To investigate the compatibility of SPEEK 1.6:PEI, SPEEK 1.6:PS, and SPEEK
1.6:BPO$_4$ blends, DSC analysis was performed to measure T_g as a function of composition. Glass transition behavior of solid polymer blends has been used as a measure
of component miscibility, where each glass transition observed in the blend reflected
a distinct segmental relaxation phase over the size scale inherent to the segmental
motions [74]. SPEEK 1.6:PEI blends showed a single glass transition temperature
ranging between 195°C and 215°C (Table 20.4). Appearance of a single glass transition temperature at a temperature intermediate between the pure components of T_g is
indicative of molar homogeneity. In Figure 20.10, the experimental T_g values were
lower than the values expected from the Gorden–Taylar model ranging from 0.10
to 1.0 wt% and higher than the model's values ranging from 0.0 to 0.10 wt%. The
former is called *negative deviation* and the latter is called *positive deviation*. This
positive deviation in T_g from the free-volume model is often cited as an indication
of strong intermolecular interaction in the blends [75]. Although the observation of
a single glass transition does not provide a thermodynamic justification, establishing
miscibility between components, complementary methods such as neutron scattering and fluorescence spectroscopy have confirmed the correlation between single T_g
behavior and intimate mixing at the molecular level [74]. However, the value of specific heat capacity at T_g of the SPEEK 1.6:PEI blends was studied by Chun and others
[75], and they observed that the difference in the ΔC_p values between the experimental and additivity rule values were negative. Two decomposition temperatures (T_d)
were recorded for all blends. The first one is believed to be due to the splitting of the
sulfonic acid group, and the second one corresponds to the sulfonic acid decomposition. A maximum decomposition temperature ranging from 525°C to 570°C was
observed, which corresponds to the decomposition of the main chain of PEEK.

TABLE 20.4

DSC Analysis of SPEEK 1.6:PEI Blends and Pure Polymer Membranes Prepared in DMAc

Polymer Blend	T_g (°C)	T_m (°C)	T_{d1} (°C)	T_{d2} (°C)
SPEEK 1.6	195.72		318.5	361.3
PEI	212.24	264.33		
SPEEK 1.6:PEI-1	211.53		317.2	361.5
SPEEK 1.6:PEI-2	215.38		319.5	363.2
SPEEK 1.6:PEI-3	199.05		313.3	354.0
SPEEK 1.6:PEI-4	197.83		319.7	362.3
SPEEK 1.6:PEI-5	184.29		316.3	359.5

Source: Sultan, A. S. et al., *Eur. Polym. J.*, 47, 2295–2302.

Data obtained for the DSC analysis of SPEEK 1.6:PS blends are presented in Table 20.5. Here also, one glass transition temperature was observed ranging from 181°C to 211°C. At higher PS content, a negative deviation of T_g was observed, and this negative deviation indicated weak intermolecular interaction in the blends. SPEEK 1.6:PS blends with 10, 30, 40 wt% PS containing samples are in the positive side of the free-volume model, and with 20 and 50 wt% PS are in the negative side of the free-volume model (Figure 20.10b). This weak interaction also supports the findings of SEM analysis, where the phase separation starts around 30 wt% PS concentration. In this case, we also observed two decomposition temperatures (T_d) (Table 20.5). Values of T_{d1} were varying between 331°C and 323°C, whereas the values of T_{d2} were ranging from 352°C to 365°C. The former is attributed to the splitting-off sulfonic acid groups and the latter to the sulfonic acid decomposition.

Blending inorganic acid (BPO$_4$) with partially sulfonated PEEK increased the proton conductivity value exceeding that of the pure SPEEK [40]. DSC analysis for SPEEK 1.6:BPO$_4$ were also carried out and tabulated in Table 20.6. One glass transition temperature was observed ranging from 194°C to 219°C, and all the values are in the positive side of the free-volume model (Figure 20.10). This positive deviation is cited as an indication of strong intermolecular interaction in the blends. The amount of water absorbed is high in the SPEEK 1.6:BPO$_4$ composite compared to either SPEEK 1.6:PEI or SPEEK 1.6:PS blends due to the emerging porosity of the composite membranes, and this effect was confirmed by a water-evaporation peak in DSC. However, SPEEK 1.6:BPO$_4$ showed one decomposition temperature (T_d), and the values of T_d were varying between 318°C and 333°C. This temperature is believed to be due to the splitting-off sulfonic acid group as a combined effect with the porosity. The absence of the second decomposition temperature is attributed to the reaction between the sulfonic acid group and BPO$_4$, which also supports the higher average T_{max} (580°C) for SPEEK 1.6:BPO$_4$ blends, whereas for SPEEK 1.6:PEI blends it was 545°C, and for SPEEK 1.6:PS it was 525°C. The free-volume

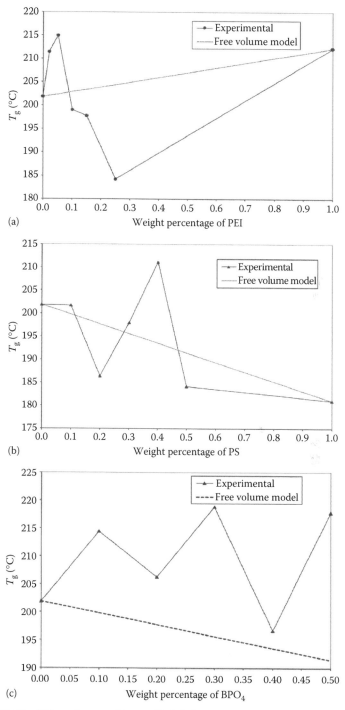

FIGURE 20.10   Effect of composition on T_g (°C) for (a) SPEEK 1.6:PEI, (b) SPEEK:PS, and (c) SPEEK 1.6:BPO$_4$ blends.

TABLE 20.5
DSC Analysis of SPEEK 1.6:PS Blends and Pure Polymer
Membranes Prepared in DMAc

Polymer Blend	T_g (°C)	T_m (°C)	T_{d1} (°C)	T_{d2} (°C)
SPEEK 1.6	195.72	–	318.5	361.3
PSU	181.03	264.33	–	–
SPEEK 1.6:PS-1	201.78	–	313.3	352.5
SPEEK 1.6:PS-2	186.42	–	323.5	365.3
SPEEK 1.6:PS-3	198.0	–	312.7	360.0
SPEEK 1.6:PS-4	211.23	–	315.5	362.5
SPEEK 1.6:PS-5	184.24	–	318.7	364.2

TABLE 20.6
DSC Analysis of SPEEK 1.6:BPO$_4$ Blends and Pure
Polymer Membranes Prepared in DMAc

Polymer Blend	T_g (°C)	T_{d1} (°C)	T_{d2} (°C)
SPEEK 1.6	195.72	318.5	361.3
BPO$_4$			
SPEEK 1.6:BPO$_4$-1	214.45	308.5	343.5
SPEEK 1.6:BPO$_4$-2	206.27	325.7	–
SPEEK 1.6:BPO$_4$-3	218.92	337.3	–
SPEEK 1.6:BPO$_4$-4	196.68	327.0	–
SPEEK 1.6:BPO$_4$-5	217.81	333.3	–

model indicates a strong interaction between the polymer and inorganic acid, and this is maybe why the phase separation starts at a higher percentage of BPO$_4$ as compared to PEI and PS.

20.2.2.3   On Solution Flow

20.2.2.3.1   On Rheology

Dynamic and steady-shear measurements were carried out at room temperature (25°C) for all the blends. We observed a shear-thinning behavior, as the dynamic viscosity decreases with increasing shear rate. Polymer solutions exhibit a limiting dynamic viscosity value at high frequency, and the magnitude of this value is mainly a function of the type of polymer despite the effect of concentration. Rheological properties of the polyelectrolyte blends are sensitive to the concentration of the polymers, interaction between the polymers, and also interaction between polymers and solvents [76]. As explained earlier, with increasing concentration, the chains or coils begin to overlap and finally entanglements are formed, and it increases the viscosity. While preparing thicker polyelectrolyte solution, if the polymer–solvent interaction is of sufficient strength, the polymer chains will tend to extend themselves and the evaporation of solvent may cause coils to collapse. Again, if the polymer–polymer interaction is stronger than polymer–solvent interaction, helices are formed and macromolecular aggregates may also occur. The configuration and type of the side group play important roles in the folding process [77,78]. Sulfonic acid groups ($-SO_3H$) attached to the backbone of SPEEK are expected to form hydrogen bonds with the solvent, though the number of interacting groups can be very small in this case, and also with O = molecules of other constituents in the blends, which will result in molecules folding and/or coiling. Linear viscoelastic properties of blends have been investigated using frequency-sweep measurements with low amplitude strain. Frequency dependence of the storage (elastic) moduli G' for the blends and pure polymer are plotted in Figure 20.11, and it was observed that for most of the blends there was a terminal regime, which covers almost the entire frequency window, indicating that these samples were viscous. Pure polymers showed higher G' values compared to the blends; this is mainly due to the polymer–polymer interactions within the blends. But in the case of SPEEK:PEI blends (Figure 20.11a), G' value increases as the PEI concentration is raised to 15%; above this, the G' decreases, which is believed to be due to the formation of a cross-linked network between the two polymers. An unusual behavior for the samples with 10% PEI contents was observed; this was attributed to a lack of homogeneity and also may be some air bubbles trapped inside the thick gel-type blend. In the case of SPEEK:PS blends (Figure 20.11b),

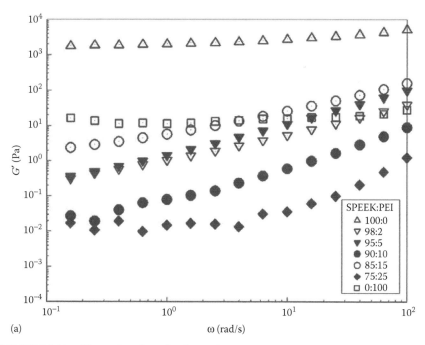

(a)

FIGURE 20.11 G' as a function of ω for (a) SPEEK:PEI blends and pure polymers. (Data from Sultan, A. S. et al., *Eur. Polym. J.*, 47, 2295–2302.)

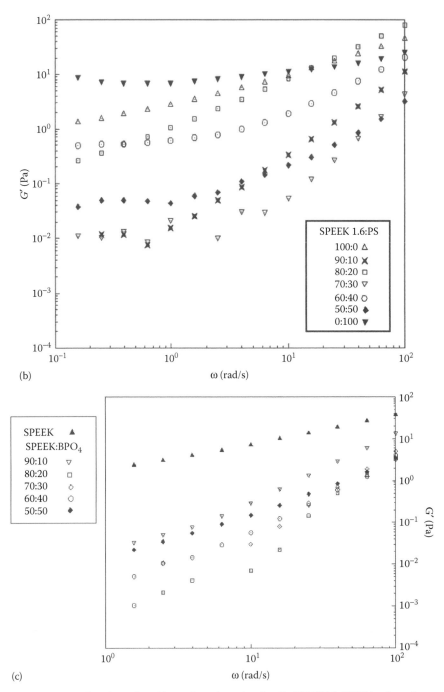

FIGURE 20.11 (Continued) G′ as a function of ω for (b) SPEEK 1.6:PS blends and pure polymers and (c) SPEEK:BPO$_4$ and pure SPEEK 1.6 (T_{test} = 25°C, γ = 15%).

20%, 30%, and 40% PS containing blends showed higher G'. SPEEK:PBO$_4$ samples gave lower G' compared to pure SPEEK 1.6 and a sharp decrease of G' compared to polymer–polymer blends. A lower G' value for SPEEK:PBO$_4$ blends was attributed to the slightly greater interaction between the SPEEK and BPO$_4$ and the lower size of the inorganic acid. G' value increases for blends from 20 to 50 wt% of BPO$_4$, with an exception in the case of 10 wt% BPO$_4$. For the 10 wt% of BPO$_4$ containing sample (SPEEK:BPO$_4$-1), the concentration of BPO$_4$ is low, and it predominately showed similar characteristic properties to pure SPEEK, which was also observed in the DSC study, where SPEEK:BPO$_4$–1 showed two decomposition temperatures like pure SPEEK.

In Figure 20.12, the plots between dynamic viscosity, η', and angular frequency, ω, for all the blends are presented. The value of dynamic viscosity decreases with the angular frequency, and for the pure polymers this decrease is sharp compared to that of the blends. The value of dynamic viscosity is lower for the blends than the pure polymers, and here also the reasoning is the same, that the polymer–polymer interaction is much lower in pure polymer solution. The η' values increased for the SPEEK:PEI blends up to 15% of PEI contents and then decreased for 25% of PEI, and again an unusual behavior for the 10% PEI was observed. This was also attributed to the lack of homogeneity and may be some air bubbles trapped inside the thick gel of the blends. A similar trend was also observed in the case of SPEEK:PS samples. For SPEEK:PBO$_4$ samples, dynamic viscosity also decreased with the increasing

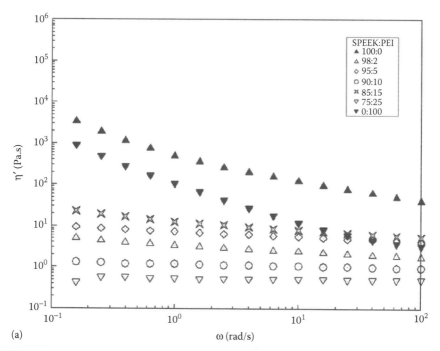

(a)

ω (rad/s)

FIGURE 20.12 η' as a function of X for (a) SPEEK:PEI blends and pure polymers. (Data from Sultan, A. S. et al., *Eur. Polym. J.*, 47, 2295–2302.)

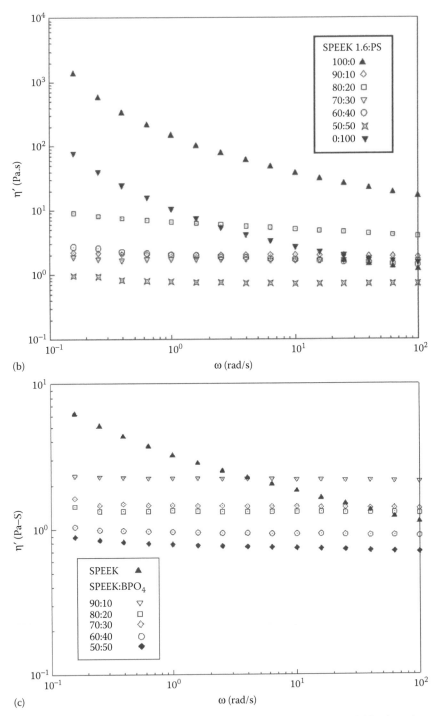

FIGURE 20.12 (*Continued*) η' as a function of X for (b) SPEEK 1.6:PS blends and pure polymers and (c) SPEEK:BPO$_4$ blends and of pure SPEEK 1.6 ($T_{test} = 25°C$, $\gamma° = 15\%$).

angular frequency for pure SPEEK, but for the blends η' values remained stable in this frequency range. η' decreased for the blends with 10% to 50 wt% BPO_4 content, with an exception for 20 wt% BPO_4, which is mainly because of the lack of homogeneity and may be some air bubbles trapped inside the thick blend gel.

20.2.2.3.2   On Viscosity

After analyzing the results obtained for the SEM, DSC, and rheology study, we found SPEEK 1.6:PEI-4 (85:15) and SPEEK 1.6:PS-2 (80:20) have better prospects. Therefore, we further studied the effect of temperature and concentration on the reduction of viscosity of these two blends. Concentration dependence of the reduced viscosity of the blends at different temperatures is presented in Figure 20.13. A sharp increase in the reduced viscosity (η_{sp}/c (dLg^{-1})) with the increasing temperature was observed for the low concentration range [79,80,81], which is the characteristic behavior of diluted polyelectrolyte solution. This occurs in the polyelectrolyte solution because of the expansion of the long polyionic chain, and also because of the enhanced dissociation of ionizable groups as the concentration decreases. Therefore, there is an intensification of the intermolecular repulsive interaction between the ionic groups on the chain [80,81]. SPEEK:PEI-4 showed better linearity at all temperatures and concentrations. But in the case of SPEEK 1.6:PS-2, at 25°C and 60°C, a very small abrupt increase in the viscosity gradient between the concentration 0.05–0.0125 g/100 ml was observed (Figure 20.13b). This mainly occurs due to the overlapping and/or interpenetrating of the molecular chains with each other, which can suddenly increase the viscosity. This overlapping occurs at a certain concentration, and this sudden transition corresponds to the onset of coil overlapping between the polymer chains in the solution. Data obtained from the viscometric experiments were linearized using several well-known equations such as Fouss–Strauss [82,83], Fedors [84], and Korecz [85], but Fouss–Strauss equation gave the best linearity (Figure 20.14).

20.3   CONCLUSION

We have selected SPEEK 1.6 as a promising membrane material for proton exchange membrane fuel cells (PEMFCs) because of its moderate proton conductivity and at the same time good work stability. Morphological, thermal, and flow patterns of PEMFCs have been studied in different solvents successfully, and we found DMAc comparatively a better solvent. Blends were prepared using SPEEK 1.6 and two other organic polymers PEI and PS and also with an inorganic acid using DMAc as a solvent. All the blends showed homogeneous surface morphology at lower second component composition, but phase separation was observed at higher concentration, that is, 25% weight ratios and above, which also supports the findings by other researchers. In DSC analysis, a single glass transition temperature (T_g) was found for all the blends. Both dynamic and steady-shear measurements were carried out at room temperature (25°C) for pure SPEEK 1.6, PEI, and PS polymers, as well as the blends. Here also, 15% PEI- and 20% PS-loaded blends showed better dynamic and steady-shear modulus, which is close to pure polymers. Similarly, in the temperature dependence viscosity study, we observed that these two blends have better

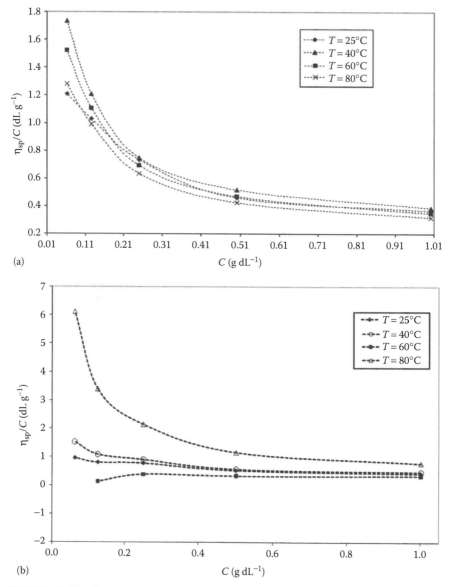

FIGURE 20.13 Concentration dependence of the reduced viscosity of (a) SPEEK 1.6:PEI (85:15) (Data from Sultan, A. S. et al., *Eur. Polym. J.*, 47, 2295–2302.) and (b) SPEEK 1.6:PS (80:20) blends at different temperatures.

performance. For SPEEK 1.6:BPO$_4$, the dynamic and steady-shear modulus is high for the first blend and slowly decreased with the increasing BPO$_4$ concentration. It has been observed from these investigations that PEI is a better blending polymer for SPEEK, and 15% PEI contents is the optimum concentration for composite membrane preparation. Work is in progress to study the electrochemical properties of these composite membranes to be used in fuel cells.

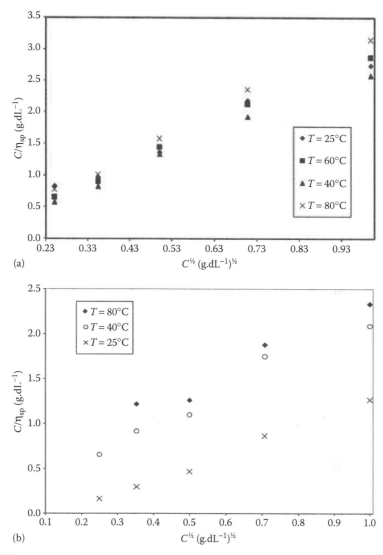

FIGURE 20.14   Representation of Fouss–Strauss equation for (a) SPEEK 1.6:PEI (85:15) (Data from Sultan, A. S. et al., *Eur. Polym. J.*, 47, 2295–2302.) and (b) SPEEK 1.6:PS (80:20) blends at different temperatures.

ACKNOWLEDGMENT

One of the authors thankfully acknowledges the Center of Research Excellence in Renewable Energy (CoRE-RE), King Fahd University of Petroleum and Minerals (KFUPM), Saudi Arabia, for providing excellent research facilities. Authors also thank the King Abdul Aziz City for Science and Technology (KACST) through National Science Technology and Innovation Plan (NSTIP) office at KFUPM for funding this work under the Project (10-ENE1374-04).

REFERENCES

1. Peighambardoust, S. J., Rowshanzamir, S., and Amjadi, M. 2010. Review of the proton exchange membranes for fuel cell applications. *Int. J. Hydrogen Energy* 35:9349–9384.
2. Xue, S. and Yin, G. 2006. Methanol permeability in sulfonated poly(etheretherketone) membranes: A comparison with Nafion membranes. *Eur. Polym. J.* 42(4):776–785.
3. Li, X., Zhao, C., Lu, H., Wang, Z., and Na, H. 2005. Direct synthesis of sulfonated poly(ether ether ketone ketone)s (SPEEKKs) proton exchange membranes for fuel cell application. *Polymer* 46:5820–5827.
4. Swier, S., Chun, Y. S., Gasa, J., Shaw, M. T., and Weiss, R. A. 2005. Sulfonated poly(ether ketone ketone) ionomers as proton exchange membranes. *Polym. Eng. Sci.* 45(8):1081–1091.
5. Gebel, G., Aldebert, P., and Pineri, M. 1993. Swelling study of perfluorosulphonate-dionomer membranes. *Polymer* 34(2):333–339.
6. Kim, H. and Litt, M. H. 2003. Synthesis and characterization of sulfonated polyimide polymer electrolyte membranes. *Macromol. Res.* 11(6):458–466.
   Gunduz, N. and McGrath, J. E. 2000. Synthesis and characterization of sulfonated polyimides. *Polym. Preprints* 41(1):182–183.
7. Bai, H. and Ho, W. S. W. 2008. Synthesis and characterization of new sulfonated polyimide copolymers and blends as proton-exchange membranes for fuel cells. *J. Environ. Eng. Manage.* 18(5):289–300.
   Gunduz, N. and McGrath, J. E. 2001. Wholly aromatic five- and six-membered ring polyimides containing pendant sulfonic acid functional groups. *Polym. Preprints* 41(2):1565–1566.
8. Zhang, Y., Litt, M., Savinell, R. F., Wainright, J. S., and Vendramini, J. 2000. Molecular design of polyimides toward high proton conducting materials. *Polym. Preprints* 41(2):1561–1562.
9. Hong, Y. T., Einsla, B., Kim, Y., and McGrath, J. E. 2002. Synthesis and characterization of sulfonated polyimides based on six-member ring as proton exchange membrane. *Polym. Preprints* 43(1):666–667.
10. Nolte, R., Lefjeff, K., Bauer, M., and Mulhaupt, R. 1993. Partially sulfonated poly(arylene ether sulfone). *J. Membr. Sci.* 83:211–220.
11. Johnson, B. C., Yilgör, İ., Tran, C., Iqbal, M., Wightman, J. P., Lloyd, D. R., and McGrath, J. E. 1984. Synthesis and characterization of sulfonated poly(acrylene ether sulfones). *J. Polymer. Sci. Polymer Chem. Ed.* 22:721–737.
12. Chao, H. S. and Kelsey, D. R. 1986. Process for preparing sulfonated poly(aryl ether) resins, U.S. Patent 4,625,000.
13. Coplan, M. J. and Götz, G. 1983. Heterogeneous sulfonation process for difficulty sulfonatablepoly(ether sulfone), U.S. Patent 4,413,106.
14. Wang, F., Hickner, M., Ji, Q., Harrison, W., Mecham, J., Zawodzinski, T. A., and McGrath, J. E. 2001. Synthesis of highly sulfonated poly(arylene ether sulfone) random (statistical) copolymers via direct polymerization. *Macromol. Symp.*175:387–396.
   Wang, F., Ji, Q., Harrison, W., Mecham, J., Formato, R., Kovar, R., Osenar, P., and McGrath, J. E. 2000. Synthesis of sulfonated poly(arylene ether sulfone)s via direct polymerization. *Polym. Preprints* 41(1):237–238.
15. Harrison, W., Shobha, H., Wang, F., Mecham, J., Glass, T., O'Conor, K., and McGrath, J. E. 2000. Influence of bisphenol structure on the direct synthesis of sulfonated poly(arylene ether)s. *Polym. Preprints* 41(2):1239.
16. Harrison, W. L., Wang, F., Mecham, J. B., Bhanu, V. A., Hill, M., Kim, Y. S., and McGrath, J. E. 2003. Influence of the bisphenol structure on the direct synthesis of sulfonated poly(arylene ether) copolymers. *I. J. Polym. Sci. A Polym. Chem.* 41:2264–2276.

17. Lufrano, F., Gatto, I., Staiti, P., Antonucci,V., and Passalacqua, E. 2001. Sulfonated polysulfone ionomer membranes for fuel cells. *Solid State Ionics* 145:47–51.

18. Shobha, H. K., Smalley, G. R., Sankarpandian, M., and McGrath, J. E. 2000. Synthesis and characterization of sulfonated poly(arylene ether)s based on functionalized triphenyl phosphine oxide for proton exchange membranes. *Polym. Preprints* 41(1):180–181.

19. Kim, Y. S., Wang, F., Hickner, M., Zawodzinski, T. A., and McGrath, J. E. 2003. Fabrication and characterization of heteropolyacid (H3PW12O40)/directly polymerized sulfonated poly(arylene ether sulfone) copolymer composite membranes for higher temperature fuel cell applications. *J. Membr. Sci.* 121(1/2):263–282.

20. Schmeller, A., Ritter, H., Ledjeff, K., Nolte, R., and Thowirth, R. 1993. Polymer electrolyte membrane and process for its manufacture, Eur. Patent 0574791 A2.

21. Helmer-Metzmann, F., Osan, F., Schneller, A., Ritter, H., Ledjeff, K., Nolte, R., and Thorwirth, R. 1995. Polymer electrolyte membrane and process for the production thereof, U.S. Patent 5,438,082.

22. Kobayashi, T., Rikukawa, M., Sanui, K., and Ogata, N. 1998. Proton-conducting polymers derived from poly(ether-etherketone) and poly(4-phenoxybenzoyl-1,4-phenylene). *Solid State Ionics* 106:219–225.

23. Bauer, B., Jones, D. J., Rozière, L., Tchicaya, L., Alberti, G., Casciola, M., Massinelli, L., Peraio, A., Besse, S., and Ramunni, E. 2000. Electrochemical characterisation of sulfonated polyetherketone membranes. *J. New Mat. Electroch. Syst.* 3:93–98.

24. Chun, Y. S. and Weiss, R. A. 2004. Thermal behavior of poly(ether ketone ketone)/thermoplastic polyimide blends. *J. Appl. Polym. Sci.* 94:1227–1235.
Swier, S., Chun, Y.-S., Gasa, J., Shaw, M. T., and Weiss, R. A. 2005. Sulfonated poly(ether ketone ketone) ionomers as proton exchange membranes. *Polym. Eng. Sci.* 45:1081–1091.

25. Hickner, M. A., Ghassemi, H., Kim, Y. S., Einsla, B. R., and McGrath, J. E. 2004. Alternative polymer systems for proton exchange membranes (PEMs), *Chem. Rev.* 104(10):4587–4612.

26. Jin, X., Bishop, M. T., Ellis, T. S., and Karasz, F. E. 1985. A sulphonated poly(aryl ether ketone). *Br. Polym. J.* 17(1):4–10.

27. Roziere, J. and Jones, D. J. 2003. Non-fluorinated polymer materials for proton exchange membrane fuel cells. *Annu. Rev. Mater. Sci.* 33(1):503–555.

28. Bishop, M. T., Karasz, F. E., Russo, P. S., and Langkey, K. H. 1985. Solubility and properties of a poly(aryl ether ketone) in strong acids. *Macromolecules* 18(1):86–93.

29. Shibuya, N. and Porter, R. S. 1992. Kinetics of PEEK sulfonation in concentrated sulfuric acid. *Macromolecules* 25(24):6495–6499.

30. Shibuya, N. and Porter, R. S.1994. A reconsideration of the kinetics of aromatic sulfonation by sulfuric acid. *Macromolecules* 27:6267–6271.
Jagur-Grodzinski, J. 2007. Polymeric materials for fuel cells: Concise review of recent studies. *Polym. Adv. Technol.* 18(10):785–799.

31. Cerfontain, H. 1968. *Mechanistic Aspects in Aromatic Sulfonation and Desulfonation.* Wiley, New York.

32. Huang, R. Y. M., Shao, P., Burns, C. M., and Feng, X. 2001. Sulfonation of poly(ether ether ketone)(PEEK): Kinetic study and characterization. *J. Appl. Polym. Sci.* 82:2651–2660.

33. Do, K. N. T. and Kim, D. 2008. Comparison of homogeneously and heterogeneously sulfonated polyetheretherketone membranes in preparation, properties and cell performance. *J. Power Sources* 185(1):63–69.

34. Xing, P., Robertson, G. P., Guiver, M. D.,Mikhailenko, S. D.,Wang, K., and Kaliaguineet, S. 2004. Synthesis and characterization of sulfonated poly(ether ether ketone) for proton exchange membranes. *J. Membr. Sci.* 229(1/2):95–106.

35. Wilhelm, F. G., Pünt, I. G. M., van der Vegt, N. F. A., Strathmann, H., and Wessling, M. 2002. Cation permeable membranes from blends of sulfonated poly(ether ether ketone) and poly(ether sulfone). *J. Membr. Sci.* 199:167–176.

36. Gil, M., Ji, X., Li, X., Na, H., Hampsey, J. E., and Lu, Y. 2004. Direct synthesis of sulfonated aromatic poly(ether ether ketone) proton exchange membranes for fuel cell applications. *J. Membr. Sci.* 234(1/2):75–81.

37. Wang, F., Chen, T., and Xu, J. 1998. Sodium sulfonate-functionalized poly(ether ether ketone)s. *Macromol. Chem. Phys.* 199(7):1421–1426.

38. Lakshmi, R. T. S. M., Meier-Haack, J., Schlenstedt, K., Vogel, C., Choudhary, V., and Varma, I. K. 2005. Sulphonatedpoly(ether ether ketone) copolymers: Synthesis, characterisation and membrane properties, *J. Membr. Sci.* 261(1/2):27–35.

39. Kreuer, K. D. 2001. On the development of proton conducting membranes for hydrogen and methanol fuel cell. *J. Membr. Sci.* 185:29–39.

40. Zaidi, S. M. J., Mikhailenko, S. D., Robertson, G. P., Guiver, M. D., and Kaliaguine, S. 2000. Proton conducting composite membrane from polyether ether ketone and heteropolyacids for fuel cell. *J. Membr. Sci.* 173:17–34.
    Zaidi, S. M. J. 2005. Preparation and characterization of composite membranes using blends of SPEEK/PBI with boron phosphate. *Electrochim. Acta* 50:4771–4777.

41. Rani, G. S., Beera, M. K., and Pugazhenthi, G. 2012. Development of sulfonated poly(ether ether ketone)/zirconium titanium phosphate composite membranes for direct methanol fuel cell. *J. Appl. Polym. Sci.* 124:E45–E56.

42. Ismail, A. F., Othman, N. H., and Mustafa, A. 2009. Sulfonated polyether ether ketone composite membrane using tungstosilicic acid supported on silica–aluminium oxide for direct methanol fuel cell (DMFC). *J. Membr. Sci.* 329:18–29.

43. Handayani, S. and Dewi, E. L. 2011. Influence of silica/sulfonated polyetherether ketone as polymer electrolyte membrane for hydrogen fueled proton exchange membrane fuel cells. *Internat. J. Sci. Eng.* 2(2):26–30.

44. Sultan, A. S., Al-Ahmed, A., and Zaidi, S. M. J. 2012. Viscosity, rheological and morphological properties of sulfonated poly(ether ether ketone): PEM fuel cell membrane material. *Macromol. Symp.* 313–314:182–193.

45. Sultan, A. S., Al-Ahmed, A., and Zaidi, S. M. J. 2011. Reduced viscosity, rheology and morphological properties of sulfonated poly (ether ether ketone): Polyetherimide blends. *Eur. Polym. J.* 47:2295–2302.

46. Bowen, W. R., Doneva, T. A., and Yin, H. 2002. The effect of sulfonated poly(ether ether ketone) additives on membrane formation and performance. *Desalination* 145:39–45.

47. Yee, R. S. L., Zhang, K., Bradley, P., and Ladewig, B. P. 2013. The effects of sulfonated poly(ether ether ketone) ion exchange preparation conditions on membrane properties. *Membranes* 3:182–195.

48. Nishida, K., Kaji, K., Kanaya, T., and Fanjat, N. 2002. Determination of intrinsic viscosity of polyelectrolyte solutions. *Polymer* 43:1295–1300.

49. Robertson, G., Serguei, P., Mikhailenko, D., Wang, K., Xing, P., Guiver, M. D., and Kaliaguine, S. 2003. Casting solvent interactions with sulfonated poly(ether ether ketone) during proton exchange membrane fabrication. *J. Membr. Sci.* 219:113–121.

50. Zhang, Y. X., Da, A. H., Butler, G. B., and Hogen-Esch, T. E.1992. A fluorine-containing hydrophobically associating polymer. I. Synthesis and solution properties of copolymers of acrylamide and fluorine-containing acrylates or methacrylates. *J. Polym. Sci., Part A: Polym. Chem.* 30:1383–1391.

51. Ng, W. K., Tam, K. C., and Jenkins, R. D. 1999. Evaluation of intrinsic viscosity measurements of hydrophobically modified polyelectrolyte solutions. *Eur. Polym. J.* 35:1245–1252.

52. Ng, W. K., Tam, K. C., and Jenkins, R. D. 2001. Rheological properties of methacrylic acid/ethyl acrylate co-polymer: Comparison between an unmodified and hydrophobically modified system. *Polymer* 42:249–259.

53. Dragan, S. and Ghimici, L. 2001. Viscometric behaviour of some hydrophobically modified cationic polyelectrolytes. *Polymer* 42(7):2887–2891.
    Terayama, H. 1955. Viscosities of aqueous solutions of potassium cellulose sulfate in the presence of other electrolytes. *J. Polym. Sci.* 15:575–590.

54. He, M. J., Chen, W. X., and Dong, X. X. 1990. *Physics of High Polymers*. Revised ed., Fudan University Press: Shanghai, China, Chapter 4.

55. Fedors, R. F. 1979. An equation suitable for describing the viscosity of dilute to moderately concentrated polymer solutions. *Polymer* 20(2):225–228.

56. Korecz, L., Csákvári, E., and Tüdos, F. 1988. Physical chemistry of polyelectrolytes. *Polym. Bull.* 19:493–500.

57. Yang, J., Liu, N., Yu, D., Peng, C., Liu, H., Hu, Y., and Jiang, J. 2005. Viscosity of polyelectrolyte solutions: Experiment and a new model. *Ind. Eng. Chem. Res.* 44:8120–8126.

58. Pavlov, G. M., Gubarev, A. S., Zaitseva, I. I., and Sibileva, M. A. 2006. Determination of intrinsic viscosity of polyelectrolytes in salt-free solutions. *Russ. J. Appl. Chem.* 79:1407–1412.

59. Kitano, T., Hashmi, S. A. R., and Chand, N. 2004. Influence of steady shear flow on dynamic viscoelastic properties of un-reinforced and Kevlar, glass fiber reinforced LLDPE. *Bull. Mater. Sci.* 27(5):409–415.

60. Ghimici, L. and Popescu, F. 1998. Determination of intrinsic viscosity for some cationic polyelectrolytes by Fedors method. *Eur. Polym. J.* 34:13–16.

61. Ydens, I., Moins, S., Degée, P., and Dubois, P. 2005. Solution properties of well-defined 2-(dimethylamino)ethyl methacrylate-based (co)polymers: A viscometric approach. *Eur. Polym. J.* 41:1502–1509.

62. Shibayama, M., Uesaka, M., Inamoto, S., Mihara, H., and Nomura, S. 1996. Analogy between swelling of gels and intrinsic viscosity of polymer solutions for ion-complexed poly(vinyl alcohol) in aqueous medium. *Macromolecules* 29:885–891.

63. Cohen, J., Priel, Z., and Rabin, Y. 1988. Viscosity of dilute polyelectrolyte solutions. *J. Chem. Phys.* 88:7111–7116.
    Cohen, J. and Priel, Z. 1990. Viscosity of dilute polyelectrolyte solutions: Temperature dependence. *J. Chem. Phys.* 93:9062–9068.

64. Wu, Q., Shangguan, Y., Zhou, J., Song, Y., and Zheng, Q. 2009. Steady and dynamic rheological behaviors of sodium carboxymethyl cellulose entangled semi-dilute solution with opposite charged surfactant dodecyl-trimethylammonium bromide. *J. Colloid Interface Sci.* 339: 236–242.

65. Kang, R., Ahn, S. W., Lee, S. J., Lee, B., and Lee, S. S. 2011. Medium viscoelastic effect on particle segregation in concentrated suspensions under rectangular microchannel flows. *Korea-Australia Rheo. J.* 23(4):247–254.

66. Fried, J. R. 1995. *Polymer Science and Technology*, Prentice-Hall, Upper Saddle River, NJ.

67. Jin, X., Bishop, M. T., Ellis, T. S., and Karasz, F. E. 1985. A sulphonatedpoly(aryl ether ether ketone). *Br. Polym. J.* 17:4–9.

68. Frigione, M., Carlo Naddeo, C., and Acierno, D. 1996. Crystallization behavior and mechanical properties of poly(aryl ether ether ketone)/poly(ether imide) blends. *Polym. Eng. Sci.* 36(16):2119–2128.

69. Ismail, A. F. and Hafiz, W. A. 2002. Effect of polysulfone concentration on the performance of membrane-assisted lead acid battery, *Songklanakarin J. Sci. Technol.*, 24(Suppl.):815–821.

70. Bowen, W. R., Doneva, T. A., and Yin, H. 2002. The effect of sulfonated poly(ether ether ketone) additives on membrane formation and performance. *Desalination* 45:39–45.

71. Arzak, A., Eguiazábal, J. I., and Nazábal, J. 1997. Biphasic compatible blends from injection-molded poly(ether ether ketone)/polysulfone. *J. Appl. Polym. Sci.* 65:1503–1510.
72. Arzak, A., Eguiazábal, J. I., and Nazábal, J. 1997. Solid state behaviour and properties of compression mouldedpoly(ether ether ketone)/polysulfone blends. *Macromol. Chem. Phys.* 198: 1829–1842.
73. Hwang, J. W., Cho, K., Yoon, T. H., and Park, C. E. 2000. Effects of molecular weight of polysulfone on phase separation behavior for cyanate ester/polysulfone blends. *J. Appl. Polym. Sci.* 77:921–927.
74. Kalogeras, I. M. and Brostow, W. 2009. Glass transition temperatures in binary polymer blends. *J. Polym. Sci., Part B: Polym. Phys.* 47:80–95.
    Menczel, J. D. and Prime, R. B., eds. 2009. *Thermal Analysis of Polymers, Fundamentals and Applications.* Wiley, New York, 2009.
75. Chun, Y. S., Kwon, H. S., Kim, W. N., and Yoon, H. G. 2000. Compatibility studies of sufonated poly (ether-ether ketone)- poly(ether-imide)- polycarbonate ternary blends. *J. Appl. Polym. Sci.* 78: 2488–2494.
76. Erwin, B. M., Cloitre, M., Gauthier, M., and Vlassopoulos, D. 2010. Dynamics and rheology of colloidal star polymers. *Soft Matter* 6:2825–2833.
77. De Kee, D. and Wissbrun, K. F. 1998. Polymer rheology. *Phys. Today* 51(6):24–29.
78. Meng, Y. Z., Tjog, S. C., and Hay, A. S. 1998. Morphology, rheological and thermal properties of the melt blends of poly (phthalazinone ether ketone sulfone) with liquid crystalline copolymer. *Polymer* 39(10):1845–1850.
79. Ghimici, L. and Popescu, F. 1998. Determination of intrinsic viscosity for some cationic polyelectrolytes by Fedors method. *Eur. Polym. J.* 34(1):13–16.
80. Pavlov, G. M., Gubarev, A. S., Zaitseva, I. I., and Sibileva, M. A. 2006. Determination of intrinsic viscosity of polyelectrolytes in salt-free solutions. *Russ. J. Appl. Chem.* 79:1407–1412.
81. Mitchell, J. A. and Umberger, J. C. 1958. Viscosity-concentration relations of cellulose acetate. *Ind. Eng. Chem. Chem. Eng. Data Series* 3(1):124–128.
82. Chun, Y. S., Kwon, H. S., Kim, W. N., and Yoon, H. G. 2000. Compatibility studies of sufonated poly (ether-ether ketone)-poly(ether-imide)- polycarbonate ternary blends. *J. Appl. Polym. Sci.* 78:2488–2494.
83. Dannenberg, K., Ekdunge, P., and Lindbergh, G. 2000. Mathematical model of the PEMFC. *J. Appl. Electrochem.* 30:1377–1387.
84. Dutta, S., Shaimpalee, S., and Van Zee, J. W. 2000. Three dimensional numerical simulation of straight channel PEM fuel cells. *J. Appl. Electrochem.* 30:133–146.
85. Chiu, H. T. and Wang, J. H. 1999. A study of rheological behavior of a polypyrrole modified UHMWPE gel using a parallel rheometer. *Polym. Eng. Sci.* 39:1769–1775.

Index

Note: Locators followed by "*f*" and "*t*" denote figures and tables in the text

1/Ka, 447
2,4,6-triaminopyrimidine (TAP), 660, 660*f*, 667–675
    characteristics of, 662*t*–663*t*
    FESEM images of, 669*f*
    interaction model, 661*f*
    transport mechanism model, 661*f*, 664*f*
2-dimensional crystal structure, 412
2-hydroxylethyl methacrylate (HEMA), 623, 629*f*
3-aminopropyltriethoxysilane (APTS), 597, 597*f*

A

Acetone, 24, 47, 200–201
Acid hydrolysis, 195–196, 196*f*
Acidic medium, 170, 682
Acidic solutions, 109, 175, 202, 600
Acrylic acid, 117–119, 570, 583
Activation polarization resistance, 340
Additives
    affecting PVDF membrane fabrication, 274–276
    classification, 274
    in dope formulation effect of membrane formation, 527–529
    incorporation in nanocomposite membranes, 658–664
    nonsolvent additive in spinning dopes, 223–224
    polymeric, 223
Adsorbent, 165, 176, 408
Adsorption membranes, 180
Adsorption property, 181
Agglomeration, 206, 208, 497, 591
Air bubble, 685, 699, 701, 703
Air-gap, 17, 18*f*, 218, 228–229, 530–531
Air-gap MD (AGMD), 403
Alcohol solutions, 132, 168
Alginate membranes, 163–164
Alginic acid, 162–165
    applications in membrane technology
        pervaporation, 162–164, 163*f*, 164*f*
        removal of metal ions, 164–165
    structure of, 162*f*
Aliphatic polymeric alcohol, 639
Alkanolamine, 286

Alumina hollow-fiber membranes
    controlled sintering process for permeable, 330–335
        mechanical evaluation of, 335*f*
        parameters, 331*t*
        pore size distribution, 333*f*
        porosity of sponge-like structure, 334*f*
        SEM images, 331*f*
        thermal-treatment temperature, 332*f*, 333
        water permeation flux, 334*f*
    for wastewater treatment, asymmetric, 326–330, 326*f*
        mercury intrusion, 329*f*
        peak loading and mechanical strength, 329*f*
        pore size distribution, 328, 328*f*
        pure water permeation, 330*f*
        SEM images, 327*f*
Amine aqueous solution, 294, 578
Amine groups, 165, 180, 302–303
Amino acids, 138, 644
Ammonia stripping, 292
Amphipathic structure, 298
Amphiphilicity, 195
Amphiphilic outer layer, 388
Amphoteric composite, 170
Anchoring nanoparticles in/on membrane surface, 577–582
Annealing process, 360–362
Anode inner layer, 368, 370
Anode structure, 369
Anodic functional layer (AFL)
    extrusion rate, 338*f*
    mechanical strength, 339*f*
Antibacterial activity, 495, 626*t*, 644
Anti-biofouling property, 496, 500, 628, 641
Antifouling mechanisms, 22, 109, 619*f*
    property, 108, 178, 180, 200, 570, 643
Antifouling resistance, 623
Antimicrobial activity, 496, 579, 583, 644
Antioxidizing reagent, 582
Anti-solvent, 198
AQP membranes, 25, 390–395
    preparation, 391–392, 392*f*, 393*t*
    properties and applications, 392–394
    water-specific transport, 394*f*
Aquaglyceroporins, 392

Aquaporins (AQPs) membrane. *See* AQP
    membranes
Aqueous media, 162, 164, 196, 551
Aqueous separations, 569
Aromatic hydrocarbons, 81, 82*t*, 85*t*, 86*f*, 422
Arrhenius equation, 688, 689*f*
Arrhenius law, 447–448
Art progress, state of
    multilayer ceramic hollow-fiber membranes
        co-sintering, 324–325
        overview, 321
        parameters for morphology control, 323
    single-layer ceramic hollow-fiber membranes
        overview, 320–321
        parameters for morphology control,
            321–323
        sintering, 324
As-spun fibers, 218–219, 235, 240
As-spun multibore hollow fibers, 236
Atomic force microscopy (AFM) images, 127
    ICD_{theory} value, 128*f*
    membrane surface morphologies, 111*f*, 130*f*
    NCC crystals of Kimwipe tissue paper,
        196, 197*f*
    polyelectrolyte multilayered RO
        membrane, 637*f*
    SWC4 membranes, 634, 634*f*
Atom transfer radical polymerization
    (ATRP), 406
Attenuated total reflection-FTIR (ATR-FTIR),
    95, 96*f*, 117, 118*f*
Au-substrate, 418
Autothermal membrane reactor, 479*f*
Average diameter, 404, 541
Axial force, 444
Azeotrope, 27

B

Back-scattered electron (BSE) mode, 337, 338*f*
Bactericidal action, 643
Bead-on-a-string morphology, 53
Beads, 53, 165, 175, 537, 540–541
Benzophenon, 21, 625
Berghmans point, 9
Binodal curves, 8–10, 16–17, 518–519
Biocolloidal fouling, 616, 623
Biocompatibility, 56–57, 109, 177, 388, 401,
    406, 589
Biodegradability, 162, 176
Biodegradable polymer, 75, 204, 206, 210
Biofouling, 25, 65, 108, 496, 616, 616–617
Biomedical membranes, 165, 204
Biomimetic membranes, 25, 391, 393*t*, 394–395
Biopolymers, 35, 161–162, 182–183, 208
Bisphenol A ethoxylate dimethacrylate (BEMA),
    209, 209*f*

Blending component effect in membrane
        preparation
    on morphology, 692–694
    on solution flow, 698–703
    on thermal properties, 695–698
Boltzmann constant, 144
Bone mineral, 237
Bore fluid (BF), 217, 224–226, 225*f*, 524, 529
    injection, 15
Boron orthophosphate (BPO_4), 684, 695*f*
Bovine serum albumin (BSA), 618
Brackish water, 32, 490, 494
Brackish water RO (BWRO) membranes, 20
Brittleness, 292
Brunauer–Emmett–Teller (BET) analysis, 594
Brush-painting technique, 373*f*
Bucky papers (BPs), 386–387
Bulk addition of nanoparticles, 574–577
    applications in mixed matrix
        membranes, 576
    morphology, 577*f*

C

Cake-fouling layer, 633
Capacitive deionization, 387, 491
Capillary membranes, 15, 319
Carbodiimide coupling reaction, 388
Carbonaceous materials, 385
Carbon molecular sieve (CMS) membranes, 26,
    491, 596
Carbon nanotubes (CNTs) membrane. *See* CNT
    membranes
Carboxymethyl cellulose (CMC), 199
    desalination performance of, 206*t*
Cast film, 254–257, 263
Casting flat-sheet membranes, 524, 524*f*
Casting solution, 254
Catalyst
    application of nanofibers for, 406–407
    impregnation in microreactor pore structures,
        362, 363*f*, 364–365
    solution, 362, 364
Catalytic convertor substrate, generation,
        341–343
    channel wall surface, 342, 342*f*
    fabrication process, 342
    reduced catalyst loading, 343
    reduced pressure drop, 343
    reduced substrate volume, 342–343
Catalytic hollow-fiber membrane microreactor
    (CHFMMR), 357, 358*f*
    SEM image, 361*f*
    YSZ hollow fibers to develop, 359*f*
Cation, 182, 422, 498, 669
Cationic dyes, 176
Cauchy's inequality, 534

Cellulose, 176–181
  applications in membrane technology
    pervaporation, 179–180
    removal of metal ions and dye from
        aqueous solutions, 180–181
    UF and NF membrane, 177–178, 179*f*
  desalination performance of PVA, 206*t*
  fillers, 203–210
  forms of
    derivatives, 198–199, 199*t*
    MFC, 197–198, 198*f*
    NCC, 195–197, 196*f*
  hydrogen bonding existing, 177*f*
  membranes, 200–203
    in different applications, 204*t*
    SEM images of CA, 201*f*
  structure of, 176*f*, 177*f*
  TEM images
    microfibrils, 197, 198*f*
    whiskers from banana waste, 206*f*
Cellulose acetate (CA) membranes, 11, 170, 177*f*,
        195, 395, 619
  membrane SEM images, 201*f*, 202*f*
  polycondensation reaction of PEG to,
        628, 628*f*
Cellulose macro fibrils, 197
Cellulose nanocrystals (CNs), 195, 208–209
Cellulosic fillers, 203–210
Ceramic hollow-fiber membranes
  applications in high-temperature
        energy conversion systems,
        356–378
    microreactors, 356–365
    microtubular SOFC, 366–378
  classification, 349*f*
  fabrication of, 350–356
    precursors extrusion, 351–354, 354*f*
    sintering process, 354–356
    spinning suspension preparation,
        350–351
  multilayer
    co-sintering, 324–325
    overview, 321
    parameters for morphology control, 323
  progression of, 348*f*
  single-layer
    overview, 320–321
    parameters for morphology control,
        321–323
    sintering, 324
Ceramic layers, 457, 459
Ceramic membranes, 318–319, 326*f*, 348
Ceramic particles, 324–325, 330, 350–351,
        354–355
Ceramic substrate, 341–343, 342*f*
Cerium gadolinium oxide (CGO), 368
Chain entanglement, 16, 230–231, 526

Chemical and temperature resistant, 292
Chemical cross-linking, 119, 547, 548*t*
Chemical dopamine, 580
Chemical linking, 602
Chemical modification, 165, 175–176, 402,
        570, 624*f*
Chemical potential difference, 517, 518*f*
Chemical resistance, 21, 24, 108, 216, 251, 348
Chemical vapor deposition (CVD), 370,
        389, 596
  separation factor, 598*t*
Chitosan, 165–176
  applications in membrane technology
    pervaporation, 168–170, 169*t*, 170*f*
    removal of metal ion and dyes, 174–176
    UF and NF membranes, 170–174
  cross-linking of sulfated, 171*f*, 172*f*
  derivatives, 166*t*–167*t*
  formation of *N*-propylphosphonic, 173*f*
  interaction with metal ions, 174*f*
  structure of CH, 165*f*
Chromatography, 136, 285
Chromophore, 117, 623
Clay particles, 31, 237, 611, 676
Cloisite15A, 659–660, 667–675
  characteristics of, 662*t*–663*t*
  FESEM images of, 669*f*
  interaction model, 661*f*
  transport mechanism model, 661*f*, 664*f*
Cloud points, 552
CNT membranes, 164, 385–390, 385–390,
        576, 576
  dense-array outer-wall, 385
  preparation, 385
  properties and applications, 386–390, 390*t*
  template-synthesized, 385
  transport mechanisms, 386
CO$_2$
  absorption and stripping performance,
        membrane contactor for, 288*f*,
        299–301
    conditions highlighted, 301
    using gas–liquid, 300*t*
  removal, 240, 290, 295, 299, 302, 308
  scrubbing, 299
  stripping, 285–309
CO$_2$/CH$_4$ selectivities, 423
CO$_2$/N$_2$ separation factor, 388, 422, 598
Coagulant temperature, 530
Coated membranes, 278, 632–633, 635
Coaxial electrospinning, 55–57, 55*f*, 56*f*,
        398, 398*f*
  multichannel, 398, 399*f*
Coaxial spinneret, 56, 400
Co-condensation, 457
Co-extrusion, 337, 367–370, 547–548
Coherent film, 543

Cohesive energy density, 515
Collector modifications of electrospinning
    system
  controlled deposition, 60–62
    near-field electrospinning, 61–62
    rotating disk, 61
    rotating drum, 60–61
  electrostatic deflection, 62, 62*f*
Colloidal fouling, 616
Colloidal suspension, 640
Combined dry-jet wet extrusion and sintering
    technique, 354
Compact membrane reactor, 475, 477*f*,
    478*f*, 479*f*
Completely wetted modes, membrane, 293
Complexing agent, 175, 180
Composite films, 204, 208, 610
Composite membrane, preparation, 542–552
  advantages, 542
  chemical cross-linking, 547, 548*t*
  chemical grafting, 551–552
  dip-coating, 550–551, 550*f*
  dual layer co-extrusion/co-casting,
    547–550
  interfacial polymerization, 542–543
    factors affecting, 543
    mixed-matrix, 543
    polyamide reactions, 542, 543*f*
  multilayer polyelectrolyte deposition,
    544–547, 544*f*
    factors affecting, 544–547, 547*f*
    polyanions, 544, 545*f*
    polycations, 544, 546*f*
Composite Pd membranes, 442–445
Conductive solutions, 540
Conductivity, 52–53, 63, 540
Conical micro-channels, 357, 364–365
Contact angle (CA), 113, 125, 127, 268, 271*f*,
    295, 308
Controlled reaction, 641
Controlled sintering process, 330–335, 331*t*
Conventional multi-step fabricating
    processes, 336
Coordination polymers, 415
Co-sintering profile, 319, 324–325, 336
Co-solvent, 527, 529, 658
Covalent attachment, 627
Critical point, 10
Cross-flow filtration, 635
Cross-linking agent, 163, 170, 179, 407, 601
Cross-linking reaction, 547
Crystalline
  lattices, 419
  material, 195
  polymers, 10, 253
Crystallinity, 88, 252
Crystallization dominant, 219

Crystallization of PLA membrane
  photo-induced crystallization, 93–99
    Norrish II mechanism for
      photo-oxidation, 95*f*
    thermal properties of, 98*t*
  solvent-induced crystallization, 81–93
    density and degree of swelling, 85*t*
    HSP, 81, 82*t*–83*t*, 84*f*
    physical properties of, 83–93
    solubility tests, 81, 83
    thermal properties of, 89*t*
    X-ray analysis of, 92*t*
  thermally induced crystallization, 76–81, 77*t*
Crystallization temperature, 8, 78, 267
$Cu_3(btc)_2$ seeding method, 422
Cyclodextrins (CDs), 181–182
  toroid structure of, 181*f*

D

Dead-end Pd-membrane reactor, 468*f*
Debye length, 144
Decomposition temperature, 695–696
Degree of polymerization (DP), 198
Degree of substitution (DS), 199
Deionized water, 119, 147, 293, 359, 417, 501, 582
Delamination, 18, 362, 370, 550, 666
Delayed demixings, 261, 261*f*, 575*f*
Delayed solidification, 13
Delay time, 10
Demixing process, 7, 221, 226, 261, 268, 521, 575
Dense-array outer-wall CNT membrane, 385
Dense gas extraction, 286, 292
Dense metal layer, 446, 448, 451
Densification, 110, 324–325, 328, 333, 358
Deposition of polyelectrolytes, 113, 117, 119, 121,
    123, 136, 150
Desalination
  global market share, 490*f*
  MMMs for
    performance evaluation, 498–503
    technological needs, 494–496
    types, 491–494
Desorption rate, 294
Desulfonation reaction, 682
Diethanolamine solution, 293
Differential scanning calorimetry (DSC)
    thermograms, 78
  analysis of SPEEK 1.6 membranes
    in argon atmosphere, 687*t*
    PEI blends, 696*t*
    PS blends, 698*t*
  curves, 88*f*
  first heating scan, 78*f*
  hydrated membranes, 671, 671*f*
  membrane density as crystallinity, 90*f*
  VUV-irradiated, 97*f*

Diffusion
  coefficient, 448, 451
    of silver, 456
  length of, 455, 456*f*
  welding, 455, 457*f*, 458*f*
Diffusion-induced PS (DIPS), 7
  phase diagrams for, 7–12, 8*f*
Diffusivity, 455–456
Dimethyl formamide (DMF), 263, 270*f*
Dimethyl sulfoxide (DMSO) systems, 353, 353*f*,
    652, 684
Dip-coating process, 115, 550–551
Dipropylene glycol, 200
Direct methanol fuel cell (DMFC) system, 656
Disk/flat sheet, 348–349
Distilled water, 119, 291, 302
Donnan effect, 112, 138, 151
Donnan exclusion mechanism, 171, 178
Dopamine, self-polymerization, 580, 581*f*
Dope, 254
  rheology, 229–232, 552
  viscoelastics properties, 16
Double-sided method, 121, 121*f*
Double-walled CNTs (DWCNTs), 385
Drug delivery, 57, 402, 412
Dry-jet wet extrusion technique, 350–351
Dual-layer hollow fiber membrane, 238
  co-extrusion
    conditions with electrolyte thicknesses,
        369*t*
    process of precursor, 368*f*
  morphology and physical properties of,
      369–371
  reduced
    bending strength, 372*f*
    gas permeability of, 373*f*
    SEM images of, 369*f*, 371*f*
  single-step fabrication of electrolyte/anode,
      367–369, 367*f*
  spinning suspension for, 368*t*
Dual-layer spinneret, 232–233, 233*f*
Dual-opposite-spinneret electrospinning (DOSE),
    397, 397*f*
Dual structure, 594
Dyes
  cationic, 176
  removal in membrane technology
    cellulose, 180–181
    chitosan, 174–176
Dynamic viscosity, 691–692, 698, 701

E

Elasticity, 176, 240, 407, 538
Electrical conductivity, 540, 583
Electrical potential difference, 533
Electric field density, 58

Electrochemical diazonium grafting, 388
Electrochemical measurements, 372
Electrochemical membrane reactor, 470
Electrochemical synthesis of MOF
    membranes, 417
Electrode, 542
Electrodialysis, 490–491, 490*f*
Electroless plating (ELP) technique, 360, 445,
    457, 459
Electrolyte layer, 340, 367, 370, 372, 375, 378
Electro-osmosis, 388
Electrophilic aromatic substitution, 682
Electroplating, 457
Electrospinning polymer modifications, 63
Electrospinning process, 47–49, 395, 396*f*
  bicomponent fiber, 397*f*
  coaxial, 398, 398*f*
  electrospun polymer, 49*f*, 65*f*, 66*f*
  fluid drop, 536*f*
  industrial and commercial applications,
      63–66
  jet, 535*f*
  materials, 49–50
  multichannel coaxial, 399*f*
  nanofibrous membranes through, 531–541,
      534*f*
    complex processes governing, 532–538
    intrinsic properties of polymer solution,
        effects, 538–541
    operating conditions effects, 541
  parameters, 50–54
    applied voltage, 50–52
    control parameters, 51*f*
    electrode distance, 52
    solution concentration and conductivity,
        52–53
    solution mass flow rate, 53–54
  Science Direct Scholar, 533*f*
  setup for, 532*f*
  single-emitter, 48*f*
  stages, 533
  target electrodes, 538, 539*f*
Electrospinning system, modification, 54–63
  collector modifications
    controlled deposition, 60–62
    electrostatic deflection, 62, 62*f*
  electrospinning polymer modifications, 63
  emitter modifications, 54–60
    coaxial electrospinning, 55–57, 55*f*, 56*f*
    multiple spinnerets, 59–60
    needleless electrospinning, 57–59, 59*f*
Electrospraying, 47
Electrospun nanofibrous membranes
    (ENMs), 402
  applications, 409, 409*t*–410*t*
  with single component, 395–396, 396*f*
Electrostatic attraction, 52, 625, 639, 675

Electrostatic charge, 619
Electrostatic interactions, 112, 148, 174, 544,
        594, 636
Electrostatic layer-by-layer, 399
Electrostatic repulsion, 47, 56, 106, 545, 620, 635
Elongation tension, 218, 229
Emitter modifications, electrospinning system,
        54–60
    coaxial electrospinning, 55–57, 55f, 56f
    multiple spinnerets, 59–60
    needleless electrospinning, 57–59, 59f
Endocrine disrupting chemicals, 551
Endothermic peaks, 88, 96, 671
Energy consumption, 108, 193, 286, 318, 492,
        494, 503
Energy conversion systems, hollow fiber
        membrane applications in, 356–378
    microreactors, 356–365
        background, 356–357
        deposition of membrane onto, 360–362
        fabrication of hollow-fiber support for,
            357–359
        impregnation of catalyst into pore
            structures, 362–365
        morphology of hollow-fiber support for,
            359–360
        performance, 365
    microtubular SOFC, 366–378
        background, 366–367
        development of SOFC, 372–374
        fabrication of electrolyte/anode dual-layer
            hollow fiber, 367–369, 367f
        fuel cell performances, 374–378,
            376f, 377f
        morphology and physical properties
            of dual-layer hollow fibers,
            369–372
Energy dispersive X-ray spectroscopy (EDS)
        analysis, 361
Enthalpy of mixing, 514–515
Entropy, 515, 618
    of mixing, 514
Enzymatic grafting, 628–629
Enzymatic transformation, 286
Equilibrium contact angle, 403
Ethanol steam reforming (ESR), 357
    CHFMMR for, 358f
Ethanol/water mixtures, 28, 226, 386, 601
Ethylene chlorotrifluoroethylene (ECTFE),
        27–28
Ethylene-*co*-vinyl acetate (EVA), 205
    permeation behavior, 207f
Ethylenediamne, 180
Evaporation-induced crystallization of MOF
        membranes, 418
Evaporation induced phase separation, 218
Evaporation-induced PS (EIPS), 7

Evaporation-precipitation method, 182, 201
External coagulant, 224–228
Extrusion process, 342, 349–351
Extrusion speed, 323

F

Fabrication process/techniques, 342
    core–shell nanofibers, 398f, 400f
    and modification of nanomaterials for
        MMMs, 496–498
    of MOF membranes, 418–419
    for water-treatment process, 194t
Fabrication technology, R&D on PVDF
        hollow-fiber
    mixed-matrix hollow fibers, 236–239
        multilayer, 238–239, 238f
        single-layer, 236–237
    spinneret design in, 232–236
        dual-layer spinneret, 232–233, 233f
        multibore spinneret, 235–236,
            235f, 236f
        multichannel rectangular spinneret,
            233–235, 234f
    TIPS in spinning, 239–240
Fedors equation, 687
Feed solution, 106, 142, 150, 250–251, 545,
        639, 642
Fermentation, 286, 594
Fiber diameter, 50, 52–53, 62, 229, 405
Fick's law, 289–290
    first law, 448–449
    second law, 447–448
Field emission SEM (FESEM) images, 129, 557
    membrane surface, 669f
    porous alumina substrate, 133f
Final sintering, 5, 324, 354–355
Finger-like single-tube membrane module, 469f
Fixed-bed reactor, 365, 365f
Flake form of graphite, 411
Flat membrane reactor, 475, 477f, 478f
Flory-Huggins theory, 515
Flow configurations, 285
Fluid displacement, 556
Fluorinated silica (fSiO$_2$), 292
Fluorocopolymers, 253
Flux recovery, 173, 600, 625, 627, 630,
        635–636
Forward osmosis (FO) membranes, 491, 544,
        547, 571
    application, 493
    in MMMs types for desalination, 492–493
Fossil fuels, 356, 437
Fouling, 494
Fouling mitigation effect, 642
Fourier transform infrared (FTIR) spectroscopy,
        84, 594

Fouss–Strauss empirical model, 690
Fouss–Strauss equation, 687
Free energy, 278, 499, 516–517
Free-falling speed, 524, 531
Free radicals, 117, 623, 628, 630
Free-solution diffusion process, 669–670, 670*f*
Free volume, 169, 604, 696–698
Fuel cells, application of nanofibers in, 405
Fully dense structure, 355
Functional groups, 182, 419, 627, 645, 659, 666
Functionality of TiO$_2$, 573
Functionalized mesoporous MCM-41, 571
Functionalized MWCNTs (f-MWCNTs), 389
Functionalized nanotubes, 579–580
Functional nanomaterials, 495, 578

G

Gamma-butyrolactone (GBL), 30
Gas adsorption–desorption method, 556
Gas channeling, 296
Gas–liquid operation, 286, 300*t*
Gas membranes, 294
Gas permeability, 224, 296, 413, 607
    poly(imide siloxane) and poly(imide
      siloxane)-zeolite MMMs, 608*t*
    of SEBS membranes, 606*t*
Gas separation (GS), 26–27, 262, 348, 356, 390,
    415–416, 424
Gas stripping MD (GSMD) technique, 404
Gas-tightness, 339–340, 371–372
Gelation point, 220, 529
Gelation time, 9, 11
Gel-layer synthesis of MOF membranes, 418
Geometric surface area (GSA), 318, 341–343
Gibbs free energy, 514
    binary and ternary systems, 515–516
    of mixing, 7
Gibbs–Thomson equation, 556
Gilbert, William, 46
Glass transition temperature, 520, 570, 608,
    686, 695
Glassy polymers, 11, 217, 219
Gold nanoparticles, 389
Graphene membrane, 385, 411*f*
    application and properties, 412–415
    preparation, 411–412
Graphene oxide (GO), 175, 412
Grotthuss mechanism, 669–670

H

Hansen solubility parameter (HSP), 81
    of organic solvents, 82*t*–83*t*, 84*f*
    relationship between swelling and, 86*f*
Heating rate, 336, 339, 591
Heat treatment, 63, 210, 597

Heavy metal removal, 405
Hemofiltration, 203
Hen Egg White Lysozyme (HEWL)
    crystallization, 32
Henry's law, 286, 288
Heterogeneous sulfonation, 682–683
High electron mobility, 412
High-energy radiation, 180, 666
Higher grafting values, 625
High pressure liquid chromatography (HPLC)
    analysis, 407
High-temperature heat treatment, 367
Hildebrand solubility parameter, 12
Hollow-fiber membranes, 319
    ceramic. *See* Ceramic hollow-fiber
      membranes
    effect of bore fluid on, 530*f*
    multilayer polyelectrolyte deposition, 544*f*
    spinning, 524, 525*f*
      dual-layer, 549*f*
    triple-layer. *See* Triple-layer membranes
    UV-photo-grafting, 551, 551*f*
    for wastewater treatment, asymmetric
      alumina, 326–330, 326*f*
      mercury intrusion, 329*f*
      peak loading and mechanical
        strength, 329*f*
      pore size distribution, 328, 328*f*
      pure water permeation, 330*f*
      SEM images, 327*f*
Hollow-fiber spinning process, 218–219, 218*f*
    fabrication technology, R&D on PVDF
      mixed-matrix hollow fibers, 236–239
      spinneret design in, 232–236
      TIPS in spinning, 239–240
    polymer dope stresses, 218–219
    spinning parameters on PVDF
      air-gap distance and take-up speed,
        228–229
      dope rheology, 229–232, 230*f*,
        231*f*, 232*f*
      internal (bore fluid) and external
        coagulant chemistry, 224–226,
        225*f*, 227*f*
      nonsolvent additives in spinning dopes,
        223–224
      polymer concentration, 221–223, 222*f*
      solubility parameters, 228*t*
Homogeneous polymer solution, 239, 319,
    513, 524
Homogeneous sulfonation, 682–683
Homopolymerization, 625
Homopolymers, 252, 626
Huggins equation, 687
Humic acid rejection test, 130
Hybrid organic-inorganic material, 415
Hydrated membrane, 670, 671*f*

Hydrogen
  bonds, 168, 618, 658, 669, 692, 699
  interaction with metals, 438–445
    properties of hydrogenated Pd alloys, 438–442, 439f, 440f, 441f, 442f
    stability of composite Pd membranes, 442–445, 443f
  mass transfer through metals
    hydrogen diffusion, 447–448
    hydrogen permeation, 448–452, 449f, 451f, 463f
    hydrogen solubility, 446–447, 446f
Hydrogenation, 438, 445, 464, 480
Hydrogen flux, 446–447
Hydrophilic layer, 547, 618
Hydrophobic foulants, 618
Hydrophobicity, 15, 30, 224, 238, 296, 298
Hydrophobic membrane, 27, 29, 31, 253, 308, 618
Hydrothermal methods, 278, 416
Hydroxyl group, 161, 176, 179, 198–199, 640
HYFLON 60XAD, 278
Hyper-branched aromatic PA-grafted silica (HBP-g-silica), 502

I

Ideal carbon dioxide flux, 388
Immersion modification, 624
Immersion precipitation process, 13, 20, 256f
  parameters affecting PVDF membrane fabrication, 276
  rationalization of, 255–262
    growth of polymer-rich nuclei, 258, 258f
    kinetic aspects, 260–262
    thermodynamics aspects, 256–259
    three-component phase diagram, 257f, 259f, 260f
Immiscibility-induced phase separation (I²PS) process, 233
Immiscible viscous liquids, 398–399
Inductively coupled plasma (ICP), 596, 608
Inertia effects, 176
Infinite selectivity, 357, 365
Initiated chemical vapor deposition (iCVD), 630–632
Inorganic filler, 168, 236, 656, 659
Inorganic materials, 104, 591, 659, 684
Inorganic membranes, 26, 104–105, 589, 612
Inorganic salts, 178, 527, 616
Inorganic substrates, surface modification, 589, 590t, 591–612
  fillers in MMMs, 592t–593t
  metal sorption capacity of membranes, 611t
  reaction on silica, 601
  Vycor glass, 595, 596f
In situ polymerization method, 426, 497, 666, 667f

In situ synthesis approach, 497, 498f
Instantaneous demixing, 11, 261, 261, 261f, 296, 522, 527, 529, 575f
Integral asymmetric structure, 513
Interchain noncovalent interactions, 581f
Interconnected-cellular structure, 229
Interfacial hydrodynamic instabilities, 13
Interfacial polymerization (IP), 109–110, 494, 501, 542–543, 621
Interfacial voids, 656
Intermetallic diffusion, 659
Internal coagulant, 217, 322
Internal combustion engines, 356
Interstitial pores, 389
Intramolecular bonds, 95
Intrinsic photosensitivity, 623
Inward radial force, 530
Ionic cluster, 664, 673
Ionic complexation degree (ICD_{theory}), 127, 128f
Ionic conductivity, 375, 682
Ionizable functional groups, 619
Ionizable groups, 681, 689–690, 703
Irreversible solute adsorption, 494
Isomers, 182
Isoreticular imidazolate frameworks, 420

J

Joint European Torus (JET) housekeeping waste, 472, 475f

K

Kelvin equation, 557
Kinetic separation, 419
Knudsen diffusion behavior, 388, 597–598
Korecz–Csákvári–Tü dos model, 688

L

Layer-by-layer (LbL) method, 115, 399, 636
  addition of nanoparticles, 582–584, 582f
  membrane modification, 637f
Layered silicates, 657f, 659, 684
Length of diffusion, 455, 456f
Less volatile nonsolvent, 255
Leveque's correlation, 289
Lignins, 182–183, 183f
Linker molecule, 583
Liqui-Cel membrane, 304–305
Liquid absorbent for membrane contactor, 301–304, 304t
Liquid entry pressure (LEP), 224, 250
Liquid–liquid demixing, 10, 217, 219, 229, 262, 266, 521, 521
Liquid–liquid interfacial coordination mechanism, 418

Liquid–liquid operation, 286
Loose RO membranes, 493
Low critical surface tension, 302
Low energy consumption, 193, 318, 492
Low-fouling composite membranes
    development through IP, 621–622
    fabrication through physical coating, 632–639
        casting, 632–634
        static/dynamic deposition, 635–639
Low-fouling composite polymer membranes
        through surface grafting, 622–632
    enzymatic grafting, 628–629
    iCVD, 630–632
    photoinitiated grafting, 623
        UV graft polymerization with, 625–627
        UV graft polymerization without, 623–624
    plasma-induced grafting, 629–630, 629f
    redox-initiated grafting, 627–628
Low-fouling membranes, 570, 645
Low water flux, 493

M

Macromolecular chains, 622, 629
Macromolecules, 95, 252, 526, 616
Macroscopic expansion, 440
Macroscopic transport model, 386
Macrostructure, 353, 354f, 355
Macrovoid, 13–17, 221–232, 522–524
Mark–Houwink equation, 53
Mass transfer
    coefficient, 289
    flux, 182
    resistances, 288–290, 289f, 451
    through metals, hydrogen
        hydrogen diffusion, 447–448
        hydrogen permeation, 448–452, 449f,
            451f, 463f
        hydrogen solubility, 446–447, 446f
Matched sintering behavior, 325, 339–340
Materials, electrospinning, 49–50
Mechanical disintegration, 197
Mechanical strength, 46, 63, 113, 339–340
Melt blending method, 664, 665f
Melting temperatures, 78, 96, 239, 252
Membrane
    porosity, 6, 106, 294, 324, 330–331
    properties, 4, 32, 113, 263, 308, 531
    selective layer, 544
Membrane characteristics
    membrane wetting, 293–295
        modes, 293
        potential of, 293–294
        prevention, 294–295
    polymer, 290–293
    structure and surface modification, 295–298
Membrane charge, 617

Membrane contactors (MCs), 4
    categories, 28f
    CO_2 absorption and stripping performance,
        288f, 299–301
        conditions highlighted, 301
        using gas–liquid, 300t
    field test/commercial pilot plant operation of,
        306t–307t
    future direction, 305, 308
    hydrophobic and hydrophilic, 29f
    liquid absorbent for, 301–304, 304t
    mass transfer resistance, 288–290, 289f
    membrane material for, 291t
    membrane preparation for, 28–34
        gas/liquid contactors, 29–30
        liquid/liquid contactors, 30
        membrane distillation, 30–31
        OD and membrane crystallizers, 31–32
    principles, 286, 288
    recent development, 304–305
    in summary, 287t
Membrane crystallization (MCr), 32
Membrane distillation (MD), 28, 30–31, 250, 387
    application of nanofibers, 403–404
    membranes in, 250–251
Membrane fabrication, factors affecting PVDF
    additives, effects of, 274–276
    dissolving temperature, effect of, 276–277, 277f
    nonwoven support, effect of, 278
    optimization of immersion precipitation
        parameters, 276
    phase inversion, 262–274
        coagulation bath composition and
            temperature, effect of, 268–273, 270f,
            271f, 272f
        exposure time between casting and
            precipitation, effect of, 262–263, 263f
        PVDF blending with polymers, effect of, 274
        PVDF concentration, effect of, 268, 269f
        PVDF grade, effect of, 267
        solvent choice, effect of, 263–267, 264f,
            265f, 266f, 267f
    posttreatment, 278
    rasonic irradiation, 277–278
    TiO_2 coating, 278
Membrane formation, effects on
    casting/spinning conditions, 524f, 525f,
        529–531
        air-gap, 530–531, 531f
        coagulant temperature, 530
        composition of coagulant, 529
        exposure to air, 529
        take-up speed, 531
    dope formulation, 525–529
        additives, 527–529
        polymer, 525–526
        solvent, 526–527

Membrane fouling, 108, 296, 495, 569, 569–570
  incorporation of nanoparticles for fouling
    reduction, 639–645
  surface properties affecting, 617–621
    hydrophilic 1/N hydrophobic properties,
     617–619
    surface charge, 619
    surface roughness, 619–621, 620*t*
Membrane–liquid boundary, 617
Membrane materials, 4, 4*t*, 5*t*, 104–114, 152, 206,
  216, 253
  for MCs, 291*t*
  NF application
    drawbacks, 107–109, 109*t*
    overview, 105–107
  NF membrane modification or fabrication
    polyelectrolytes as alternative for,
     112–114
    polyelectrolyte solution parameters for, 114
    to produce, 109–112
Membrane preparation, 685–703
  effect of blending component
    on morphology, 692–694
    on solution flow, 698–703
    on thermal properties, 695–698
  effect of solvent
    on morphology, 685–686
    on solution viscosity, 687–691
    on SPEEK 1.6 membranes, 692, 692*f*
    on thermal properties, 686–687
  laser interference lithography, 6*f*
  for MC, 28–34
    gas/liquid contactors, 29–30
    liquid/liquid contactors, 30
    membrane distillation, 30–31
    OD and membrane crystallizers, 31–32
  PI, peculiarities of hollow-fiber, 15–18, 18*f*
  for pressure-driven separation processes, 19–28
    gas separation, 26–27
    microfiltration, 20–21
    NF and solvent-resistant NF, 22–24
    pervaporation, 27–28
    properties of, 19*t*
    reverse osmosis, 24–25
    ultrafiltration, 21–22
  techniques, 4–6
    materials and applications, 4, 4*t*
    sintering, 5, 5*t*, 6*f*
    stretching, 5, 6*f*
    template leaching, 6
    track-etching, 5–6, 6*f*
    via PI, 7
Membrane preparation principles, phase
  inversion, 514–524
  macrovoid formation, 522–524, 523*f*, 528*f*
  phase diagram and membrane formation,
   519–522

  polymer solutions
    demixing of, 516–519
    solidification mechanisms, 520
    thermodynamics of, 514–516
  thermodynamics and kinetics influence on
    membrane morphology, 522
Membranes, 3
  adsorption, 402, 406–407, 426
  -based separation
    advantages, 161–162
    removal of metal ions in alginic acid
     applications, 164–165
  biofouling, 108, 617
  cellulose, 200–203
    in different applications, 204*t*
    SEM images of CA, 201*f*, 202*f*
  classification, 193
  formation, kinetics of, 522
  fouling, 108
  lifespan, 109
  metal sorption capacity, 611*t*
  microreactor, 356, 365
  modules, 32, 408, 415, 438–440, 464–465
  nanocomposite membrane, solution
    preparation method
    melt blending, 664, 665*f*
    in situ polymerization, 666, 667*f*
    sol-gel, 666–667, 667*f*
    solution blending, 665–666, 665*f*
  permeability, 14, 108, 131, 151–152, 251
  polymeric, 33*t*, 34*f*
  pore wetting, 30, 250
  reactor, 390, 437–480
  repulsion model for multibipolar, 137*f*
  roughness, 620
  surface
    anchoring in/on, 577–582
    self-assembly of nanoparticles,
     571–574
  technology, 193. *See also* Membrane
    technology, applications in
  types of structures, 26
Membranes/solutions, characterizations
  membrane separation, 553–558
    fluid displacement, 556
    fluid transition, 556–557
    microscopic techniques, 557
    solute retention, 557–558
    techniques for membrane pore,
     554*t*–555*t*
  polymer dope, 552–553
    cloud point, 552
    dope viscosity and rheology, 552–553
Membrane technology, applications in
  alginic acid
    pervaporation, 162–164, 163*f*, 164*f*
    removal of metal ions, 164–165

cellulose
  pervaporation, 179–180
  removal of metal ions and dye from
      aqueous solutions, 180–181
  UF and NF membrane, 177–178, 179*f*
chitosan
  pervaporation, 168–170, 169*t*, 170*f*
  removal of metal ion and dyes, 174–176
  UF and NF membranes, 170–174
Mercury intrusion, 327, 332
Metallic membrane, 5, 362
Metal membranes, fabrication of, 452–464
  other, 460–464, 462*f*, 463*f*
  Pd-based composite membranes, 457–459,
      461*f*, 462*f*, 464*f*
  Pd-based self-supported membranes,
      453–457
    diffusion bonding joint, 456, 458*f*
    rolling mills, 454*f*
    rupture pressure, 453, 453*f*
Metal-organic frameworks (MOFs) membrane.
      See MOF membranes
Metal oxides, 22, 405, 412–415, 491–492, 684
Metals
  hydrogen mass transfer through
    hydrogen diffusion, 447–448
    hydrogen permeation, 448–452, 449*f*,
        451*f*, 463*f*
    hydrogen solubility, 446–447, 446*f*
  removal of ions in membrane technology
    alginic acid, 164–165
    cellulose, 180–181
    chitosan, 174–176
  sorption capacity of membranes, 611*t*
Metastable state, solution, 8–11, 257, 257*f*, 517
Methacrylic acid (MAA) grafting, 628
Methanol crossover, 656–658, 660, 682
Methanol diffusion path, 658
Methanol permeability, 656, 659, 673, 675, 683
Methyldiethoxysilane (MDES), 595
Methyl methacrylate (MMA), 501
Micellar-enhanced filtration (MEF), 405–406
Microbial cell suspension, 628
Microchannels, 318, 320, 327, 342, 364–365
Microcracks, 339–340
Microcrystalline cellulose (MCC), 196
Microdroplets, 17
Microfibrillated cellulose (MFC), 195, 197–198
  energy-intensive processes, 197–198
  hydrophilic nature, 198
  TEM image of, 197, 198*f*
Microfibrils, 195
Microfiltration (MF), 4, 20–21, 318, 590, 616
  application of nanofibers in, 402–403
Microreactor, hollow-fiber membrane
  background, 356–357
  deposition of membrane onto, 360–362

fabrication of hollow-fiber support for,
    357–359
impregnation of catalyst into pore structures,
    362–365
morphology of hollow-fiber support for,
    359–360
performance, 365
Microsieves, 6
Microstructures, 337–339
Microtubular SOFC
  apparatus of single, 374*f*
  applications in hollow fiber membrane,
      366–378
    background, 366–367
    development of SOFC, 372–374
    fabrication of electrolyte/anode dual-layer
        hollow fiber, 367–369, 367*f*
    fuel cell performances, 374–378,
        376*f*, 377*f*
    morphology and physical properties of
        dual-layer hollow fibers, 369–372
  single-step formation of triple-layer
      membranes, 335–341
    cell performance, 340
    compositions of layer, 336, 337*f*
    dimensions of dual- and, 340*t*
    mechanical strength and gas-tightness of
        electrolyte, 339–340, 339*f*
    morphology and microstructure, 337–339
Microwave radiation, 417
Miller index, 90
Miscibility, 7–8
Miscible phases, 28, 30
Mixed-matrix hollow fibers
  multilayer, 238–239, 238*f*
  single-layer, 236–237
Mixed-matrix membranes (MMMs), 385, 590
  comparison of, 610*t*
  for desalination. *See* MMMs for desalination
  modification of nanomaterials and
      fabrication, 496–498
  silica-surface functional groups, 605*f*
MMMs for desalination
  performance evaluation, 498–503
  technological needs, 494–496
  types, 491–494
    FO membranes, 492–493
    NF, 493–494
    RO membranes, 491–492
MOF-5, 416–418, 420–422, 425
MOF membranes, 415–425
  application, 419–424
    dicarboxylic acid linkers, 422*f*
    for gas separation, 425*t*
  challenges in engineering, 424
  characterization and properties, 419
  MMM with, 423

MOF membranes (*Continued*)
    preparation
        colloidal deposition method for, 417
        direct growth/deposition, 416–417
        electrochemical synthesis, 417
        evaporation-induced crystallization, 418
        gel-layer synthesis, 418
        LBL or liquid phase epitaxy, 417
        methods for fabrication, 418–419
        synthesis of, 416
Molar homogeneity, 695
Molecularly imprinted membranes (MIMs), 407
Molecular sieves, 413
Molecular weight (Mw), 16, 114
Molecular weight cut-off (MWCO), 19, 106, 122,
        151, 557
Molecular weight distribution, 252
Molten polymer solution, 239
Monochlorosilanes, 596
Monoethanolamine, 292
Monovalent electrolyte, 114
Monovalent ions, 19, 142, 170
Montmorillonite, 656
Mordenite Framework Inverted (MFI) zeolite
        membranes, 595
    comparison of MMM containing, 610*t*
    SEM images of, 609, 610*f*
m-phenylenediamine (MPD), 22, 621–622, 621*f*
    aromatic polyamide based on, 23*f*
Multibore spinneret, 235–236, 235*f*, 236*f*
Multichannel coaxial electrospinning system, 399*f*
Multichannel rectangular spinneret, 233–235, 234*f*
Multichannel tubes, 326
Multifluidic compound-jet electrospinning, 398
Multilayer ceramic hollow-fiber membranes
    co-sintering, 324–325
    overview, 321
    parameters for morphology control, 323
Multilayer composite membranes, 175
Multilayer mixed-matrix hollow fibers,
        238–239, 238*f*
Multilayer polyelectrolyte deposition, 544–547, 544*f*
Multiple spinnerets, 59–60
Multistage flash (MSF), 491
Multitube membrane module, 476*f*
Multiwalled CNTs (MWCNTs), 170, 385, 501
Multiwalled nanotubes (MWNTs), 385

N

Nafion membranes, 405, 599, 599*f*, 611–612,
        682–683
    microstructures of, 684*f*
Nanoclay intercalation effect of SPEEK/
        Cloisite15A/TAP membrane
    on membrane morphological structure,
        668–669

on physicochemical properties, 673,
        674*t*, 675
    on transport mechanism of water, proton, and
        methanol, 669–671
Nanocomposite membranes, 604
    improvements in fabrication of
        polymer, 675
    membrane solution preparation method
        melt blending, 664, 665*f*
        in situ polymerization, 666, 667*f*
        sol-gel, 666–667, 667*f*
        solution blending, 665–666, 665*f*
    preparation of exfoliated, 658–664
        incorporation of additives, 658–664
        nanogap formation, 659*f*
    preparation of SPEEK/Cloisite15A/TAP,
        667–675
        characteristics of, 662*t*–663*t*
        FESEM images of, 669*f*
        interaction model, 661*f*
        transport mechanism model, 661*f*, 664*f*
    types by structure, 657–658
Nanocrystalline cellulose (NCC), 195–197
    AFM images, 196, 197*f*
    using acid hydrolysis, 196*f*
Nanoenhanced membranes (NEMs), 25
Nanofibers, 46, 395–410
    applications, 46
    fabrication methods, 47*f*
    preparation
        ENMs with single component, 395–396,
            396*f*
        with side-by-side components, 396–401
    properties and applications, 401–408
        for adsorption, 405–406
        for catalysts, 406–407
        in fuel cells, 405
        MD, 403–404
        for membrane separation process, 408
        other, 407
        in RO, UF, NF, and MF, 402–403
Nanofibrous membranes through electrospinning,
        531–541, 534*f*
    complex processes governing, 532–538
        bending deformation with looping and
            spiraling trajectories, 536–537, 537*f*
        onset of jetting and rectilinear jet
            development, 533–536
        solidification and deposition on counter
            electrodes/substrates, 537–538
    intrinsic properties of polymer solution,
        effects
        electrical conductivity, 540
        polymer concentration and dope viscosity,
            538–540
        solvent, 540–541
        surface tension, 540

operating conditions effects
  ambient environment, 541
  applied voltage, 541
  other parameters, 541
Nanofiltration (NF) membrane. *See* NF
      membranes
Nanoparticles (NPs), 20, 165, 495–497, 546
  bulk addition, 574–577
  in composite membranes for fouling
      reduction, 639–645
  in/on membrane surface, anchoring, 577–582
  layer-by-layer addition, 582–584, 582*f*
  self-assembly
    on membrane surface, 571–574, 572*f*, 573*f*
    of TiO$_2$, 640*f*
Nanosilica, 238
NanoSpider™ range, 65
Nanostructured flat membranes, 403
Nanotechnology, 490, 493, 504
Nanowhiskers, 196, 610
Nascent fiber, 217, 228, 524, 531
Nascent membrane, 521, 523, 529
Natural organic matter (NOM), 616
Natural polysaccharides, 161
Near-field electrospinning, 61–62
Needleless electrospinning method, 57–59, 59*f*
Negative deviation, 695
NF membranes, 4, 493
  AFM images, 111*f*
  application of nanofibers in, 402–403
  categories, 493–494
  characteristics and performances using
      polyelectrolytes
    changes in membrane separation layer
        thickness, 143–145, 144*f*
    membrane pure water permeability or
        flux, 130–134, 135*f*, 135*t*
    membrane surface and cross-sectional
        morphology, 127–129
    membrane surface charge property, 129–130
    membrane surface hydrophilicity,
        125–127, 126*f*
    performance stability of, 145–147
    solute rejection and selectivity
        capability, 134–143, 136*f*, 137*f*, 139*f*,
        140*f*, 140*t*
  drawbacks in application, 109*t*
    low solute rejection and selectivity in NF,
        107–108
    membrane fouling, 108
    membrane life-time limitation, 108–109
  in MMMs types for desalination, 493–494
  modification/fabrication for polyelectrolytes
    commercially available and used, 116*t*
    drawbacks on applications of, 147–149
    proposed future studies on, 150–151
    zeta potential of unmodified, 134*f*

modification/fabrication to produce, 109–112
salt rejection tests, 111*f*
and solvent-resistant NF, 22–24
surface modification using polyelectrolytes,
    115–125
  chemical cross-linking of PECs, 119
  dynamic self-adsorption of, 122–123,
      124*t*, 125
  static self-adsorption of, 119, 121–122,
      124*t*
  UV-grafting, 117–119
Nickel (Ni) membrane
  device for preparing tube, 463*f*
  external surface, 462*f*
  laminated, 465*f*, 466*f*
N-methyl-2-pyrrolidone (NMP), 353, 353*f*, 526
N,N-dimethylacetamide (DMAc), 526, 600
N,N-dimethylformamide (DMF), 526
Nonfluorinated backbones, 656
Nonfluorinated polymers, 656
Nonporous membranes, 512–513
Nonsolvent, 7, 15–16, 200, 230–231, 255, 261,
    264, 269, 321, 352, 522, 528, 552
Nonsolvent coagulation bath, 271, 352
Nonsolvent-induced phase separation (NIPS), 7,
    216, 513–514
  cloud points, 552
  isothermal phase diagram, 553*f*
  membrane formation, 521*f*
  nascent hollow fiber through, 219*f*
  polymers used in, 526*t*
  synthesis of membranes, 575*f*
  ternary phase diagram for, 520*f*
Nonsolvent internal coagulant, 352
Nonwetted modes, membrane, 293
Nonwoven support effect, PVDF membrane
    fabrication, 278
Novel surface modifying macromolecules
    (nSMMs), 298
Nuclear magnetic resonance (NMR), 84
Nucleating agent, 203–204, 675
Nucleation and growth (NG), 9, 521
Nucleus, 9, 521, 523–524

O

O$_2$ permeability, 604*t*
Octadecyldimethylchlorosilane (OCS), 595–596
Octadecyltrichlorosilane (ODS), 595–596
Ohmic area specific resistance (ASR), 375, 378
Ohmic resistance, 375, 378
On-board hydrogen generation, 356–357
Open-circuit voltage (OCV), 340, 341*f*, 375
Organically modified montmorillonite, 657, 666
Organic ligands, 416, 424
Organic membranes, 193
Organic modifiers, 622, 645

Organic monomers, 666
Organic solute, 493
Organic solvent, 81, 93, 165, 529
Organic solvent permeation, 86–88
Organosilane compounds, 589, 666
Organosilanes, 589
Orifice size, 541
Osmotic distillation (OD), 28
    and membrane crystallizers, 31–32
Osmotic pressure, 106, 523
Osmotic-resistance filtration model, 617
Overall mass transfer coefficient, 289, 293, 404
Oxidant reduction reaction, 373–374
Oxygen ion, 375
    conducting material, 366
Oxygen separation, 349–350

P

Palladium (Pd)-alloy membrane, 175, 357, 438.
        See also Pd–Ag membrane
    composite
        in fabrication of metal membranes,
            457–459, 461*f*, 462*f*, 464*f*
        stability of, 442–445, 443*f*
    properties of hydrogenated, 438–442, 439*f*,
        440*f*, 441*f*, 442*f*
    reactor, 464–479
        autothermal membrane, 479*f*
        compact flat-and-frame Pd permeator,
            477*f*, 478*f*
        dead-end, 468*f*
        finger-like single-tube, 469*f*
        module, 468*f*
        Pd–Ag membrane, 474*f*, 475*f*
        PERMCAT, 471–472, 472*f*, 473*f*
        single-tube, 467*f*
        WGS, 476*f*
PAN membranes, 122, 138, 145, 625
Partially wetted modes, membrane, 293, 295
Particulate removal, 402
Pd–Ag membrane, 357, 360, 362, 365
    alkaline electrolyser using, 470*f*, 471*t*
    ceramic strains, 444
    diffusion bonding joint, 456, 458*f*
    diffusion welding, 458*f*
    dilatometric behavior, 442*f*
    electrical resistivity, 441*f*
    foils to flat metal supports, 461*f*
    hydrogen solubility, 439*f*
    laminated nickel membrane, 465*f*, 466*f*
    ohmic heating, 468, 469*f*, 475*f*
    peeling of, 445*f*
    reactor, 474*f*, 475*f*
    shear stress, 444, 445*f*
    tensile strength, 441*f*
    thin-wall permeator, 459*f*

Pd-composite membranes, 438, 457
Perfluorosulfonated membranes, 659
PERMCAT reactor, 471–472, 472*f*, 473*f*
Permeability, 20, 151, 204, 208, 251, 392, 419,
        439, 460, 492, 498, 553, 573, 606
Permeability coefficient, 450
Permeation
    driving force of Sieverts' expression, 451*f*
    flux, 228, 335, 402, 571
    hydrogen, 438–439, 440*f*, 448–452, 449*f*, 463*f*
Perturbations, 59, 540
Pervaporation (PV), 27–28, 162, 590
    alcohol–water, 601*t*
    applications in membrane technology
        alginic acid, 162–164, 163*f*, 164*f*
        cellulose, 179–180
        chitosan, 168–170, 169*t*, 170*f*
    results, 595*t*
PES photodestruction, 623
PES support, 641
Phase inversion (PI), 4, 352
    extrusion/sintering technique for ceramic
        hollow fiber fabrication, 350–356
        precursors, 351–354, 354*f*
        sintering process, 354–356
        spinning suspension preparation, 350–351
        steps, factors and influences, 351*t*
    mechanisms of PVDF, 219–221, 220*f*
    membrane preparation, 7, 254–255, 513–531
        membrane fabrication through immersion
            precipitation, 524–525
        membrane formation, effects on. *See*
            Membrane formation, effects on
        principles, 514–524
    parameters affecting PVDF membrane
        fabrication, 262–274
        coagulation bath composition and
            temperature, effect of, 268–273, 270*f*,
            271*f*, 272*f*
        exposure time between casting and
            precipitation, effect of, 262–263, 263*f*
        PVDF blending with polymers, effect of,
            274
        PVDF concentration, effect of, 268, 269*f*
        PVDF grade, effect of, 267
        solvent choice, effect of, 263–267, 264*f*,
            265*f*, 266*f*, 267*f*
    peculiarities of hollow-fiber membrane
        preparation, 15–18, 18*f*
    thermodynamic principles of, 7–15
        phase diagrams for TIPS and DIPS, 7–12,
            8*f*, 9*f*, 11*f*
        solubility parameters, 12–13
        trade-off between thermodynamic,
            kinetic, and membrane morphology,
            13–15
    viscous fingering-induced, 319–320

Phase separation (PS), 4, 226, 239, 257–258, 319, 319, 513, 522, 694
  phenomenological description, 7
Phonon theory, 386
Photo-activation stage, 117
Photocatalytic degradation, 573
Photocatalytic effect, 573
Photographic films, 199
Photo-induced crystallization of PLA, 93–99, 94
  Norrish II mechanism for photo-oxidation, 95f
  thermal properties of, 98t
Photoinitated copolymerization, 632
Photoinitiated grafting, 623
  UV graft polymerization with, 625–627
  UV graft polymerization without, 623–624
Photo-initiator, 117, 551
Physical absorbent, 293
Physical vapor deposition, 457
Pin-holes formation, 362
Piperazine (PIP), 22
  polyamide based on, 23f
  reaction betweem TMc and, 23f
PLA membrane, crystallization of
  photo-induced crystallization, 93–99
    Norrish II mechanism for photo-oxidation, 95f
    thermal properties of, 98t
  solvent-induced crystallization, 81–93
    density and degree of swelling, 85t
    HSP, 81, 82t–83t, 84f
    physical properties of, 83–93
    solubility tests, 81, 83
    thermal properties of, 89t
    X-ray analysis of, 92t
  thermally induced crystallization, 76–81, 77t
Plasma-induced grafting, 629–630, 629f
Plasma membranes, 392
Plasma parameters, 629
Plasma reactor, 629–630
Plasma treatment, 111, 630
Plasticizers, 605
Plateau, 633
Pluronic copolymers, 527
Poiseuille viscous flow models, 389
Polarity, 538, 543
Polarization microscopy (POM) images of PLA, 79, 80f, 87f
Poly(1-trimethylsilyl-1-propyne) (PTMSP), 604
Polyacrylonitrile (PAN), 21, 64, 122, 170
Polyamide (PA) membrane, 65, 210, 543, 617, 621, 621f
  rejection layer, 492
Polyanions, 150, 583
Poly(arylene ether ketone)s (PAEKs), 682
Polycarbonate membranes, 387
Polycations, 112, 114, 121–123, 150, 152, 544, 546, 583, 636–637

Poly(diallyldimethylammonium) chloride (PDADMAC), 122
  bilayers of, 134f
  flux of, 143f
  thickness measurements of, 146f
  variation in contact angle, 126f
Polydimethylsiloxane (PDMS), 590
  laboratory synthesized silicate, comparison of, 600, 603t, 604t
Polydopamine, 580, 582, 584
Polyelectrolyte complexes (PECs), 24, 112, 180
  chemical cross-linking, 119
  helical structures of, 113f
  multilayered, 131f
  PEC-modified membrane, 129f
  proposed mechanism, 120f
Polyelectrolytes, 544
  deposition, multilayer, 544–547, 544f
  deposition of, 121
  dipping methods, 121f
  layer, 123, 125, 130, 142, 147, 638
  NF membrane modification/fabrication
    as alternative for, 112–114
    commercially available and used, 116t
    drawbacks on applications of, 147–149
    proposed future studies on, 150–151
    solution parameters for, 114
    zeta potential of unmodified, 134f
  NF membranes characteristics and performances using
    changes in membrane separation layer thickness, 143–145, 144f
    membrane pure water permeability or flux, 130–134, 135f, 135t
    membrane surface and cross-sectional morphology, 127–129
    membrane surface charge property, 129–130
    membrane surface hydrophilicity, 125–127, 126f
    performance stability of, 145–147
    solute rejection and selectivity capability, 134–143, 136f, 137f, 139f, 140f, 140t
  NF membranes surface modification using, 115–125
    chemical cross-linking of PECs, 119
    dynamic self-adsorption of, 122–123, 124t, 125
    static self-adsorption of, 119, 121–122, 124t
    UV-grafting, 117–119
  solution, 114, 121, 138, 143, 544, 583, 703
  UV-grafting of, 117
    acrylic acid grafted on PES NF membranes, 117–119, 118f, 135f
    sodium p-styrenesulfonate grafted on PSF UF membranes, 119
  zeta potential of, 547f

Poly(ether ether ketone) (PEEK), 682
Polyetherimide (PEI) blend membranes, 684
   SPEEK 1.6 membranes
      DSC analysis of, 687*t*
      SEM images for, 693*f*
Polyethersulfone (PES) membranes
   SEM images, 577*f*
Polyethersulfone (PESf) membranes, 330,
      332*f*, 353, 353*f*
Polyethylene (PE), 5, 617
Poly(ethylene glycol) (PEG), 200, 274, 389
   polycondensation reaction, 628, 628*f*
Poly(ethylene glycol) methyl ether methacrylate
      (PEGMA), 209, 209*f*
Poly(ethylene oxide) (PEO), 527, 528*f*
Polyhydroxybutyrate-*co*-valerate (PHBV),
      203–204
   SEM images of, 205*f*
Polyimide, 24
Poly(lactic acid) (PLA), 75
   chemical structure of, 76*f*
   crystallization of. *See* PLA membrane,
      crystallization of
   number of references, 76*f*
   VUV irradiation, 76
      ATR-FTIR, 95, 96*f*
      DSC thermograms, 97*f*
      properties, 94
      SEM images of, 99*f*
      UV–vis spectra of, 94*f*
      WAXD patterns, 97*f*
   WAXD patterns of, 76, 77*f*
   types, 90, 91*f*
   VUV-irradiated, 97*f*
   waveform separation of Gaussian
      functions, 93*f*
Polymer electrolyte membrane (PEM), 357,
      656, 681
Polymer electrolyte membrane fuel cell
      (PEMFC), 357
Polymeric additives, 223, 274, 384, 527, 569
Polymeric membranes, 3–35, 105, 117, 146, 151
Polymeric nanofiber membranes, 401, 532
Polymer–inorganic nanocomposite
      materials, 656
Polymerization
   in composite membrane preparation,
      interfacial, 542–543
      factors affecting, 543
      mixed-matrix, 543
      polyamide reactions, 542, 543*f*
   conditions, 629
   reaction, 419–420, 542–543, 627
Polymer–polymer interaction, 253, 691, 699, 701
Polymer precipitation, 10, 14, 254, 319, 322
Polymer-rich phase, 7, 9, 11, 263, 276, 352, 522

Polymer(s)
   binder precipitation, 322
   chain entanglement, 221
   chemistry, 268, 495
   concentration, 10, 15–16, 221–223,
      538–540, 687
   degradation, 117, 497
   dispersion states of nanoclay, 657*f*
   dope characteristics, 552–553
   in dope formulation effects, 525–526
   effect of PVDF blending with, 274
   improvements in fabrication of
      nanocomposite membrane, 675
   matrix, 28, 389, 423, 496, 606, 658
   membrane, 290–293
   mobility, 675
   nonfluorinated, 656
   PVDF applications
      homo and copolymers, 252*f*
      organic solvents for, 253, 254*t*
      properties of, 252–253
      structure and crystallinity, 252
   reactivity, 419–420, 542–543
   surface energy of, 295, 295*t*
   ternary phase diagram, 352*f*
   used in NIPS, 526*t*
Polymer solutions
   effects in nanofibrous membranes preparation
      electrical conductivity, 540
      polymer concentration and dope viscosity,
         538–540
      solvent, 540–541
      surface tension, 540
   membrane preparation principles
      demixing of, 516–519
      solidification mechanisms, 520
      thermodynamics of, 514–516
Polymer–solvent interaction, 691, 698–699
Poly(methyl methacrylate) (PMMA), 501
Polyphenylsulfone (PPSU), 24, 25*f*
Polypropylene (PP), 5, 180
Polysaccharides, 114, 161–162, 181, 183–184
Polystyrene sulfonate (PSS), 122
   adsorption of bilayers, 133*f*
   bilayers, 134*f*
   flux of, 143*f*
   porous alumina with, 133*f*
   variation in contact angle, 126*f*
Polysulfone (PS), 617, 618*f*
Polytetrafluoroethylene (PTFE), 5, 253,
      290, 362
Polyvinyl acetate, 63
Poly(vinyl alcohol) (PVA) membrane, 205
   desalination performance of, 206*t*
Polyvinylidenefluoride hollow-fiber membrane
      contactor (PVDF HFMC), 292

Poly(vinylidene fluoride) (PVDF) membrane, 216, 250, 526
  at different extrusion, 297*f*
  factors affecting membrane fabrication
    additives, effects of, 274–276
    dissolving temperature, effect of, 276–277, 277*f*
    nonwoven support, effect of, 278
    optimization of immersion precipitation parameters, 276
    phase inversion parameters, 262–274
    posttreatment, 278
    rasonic irradiation, 277–278
    TiO₂ coating, 278
  hollow-fiber. *See* PVDF hollow-fiber
  homo and copolymers, 252*f*
  isothermal phase diagram, 553*f*
  methods for modifying MF, 636
  nanofibrous membranes, 403
  normalized fluxes, 633*f*
  organic solvents for, 253, 254*t*
  phase inversion mechanisms of, 219–221, 220*f*
  properties of, 252–253
  structure and crystallinity, 252
Polyvinylpyrrolidone (PVP), 200
  SEM images of, 201*f*
Poor adhesion, 590, 604–605, 656
Pore blocking, 573, 576
Pore diameter, 141, 237
Pore-forming additive, 16, 30, 527
Pore size distribution (PSD), 250, 318, 553, 556
Porocritical extraction, 287*t*
Porosity, 5, 46, 60, 251, 273, 328, 404, 696
Porous alumina support, 128–129, 142, 420
Porous cake layer, 617
Porous fillers, 164
Porous stainless steel tubes, 423
Positive deviation, 695
Power density, 336, 340, 366, 375
Pre-sintering process, 354
Pressure-driven membrane, 19, 492–493, 616, 645
Pressure drop, 292, 343
Protonated water, 673
Proton conductivity, 405, 600, 673, 683, 703
Proton hopping process, 669–670, 670*f*
Proton transfer
  free-solution diffusion, 669–670, 670*f*
  hydrated ion-exchange, 670
  proton-hop, 669–670, 670*f*
  through ionic cluster, 673
PS blends, 684
  SPEEK 1.6 membranes
    DSC analysis of, 698*t*
    SEM images for, 694*f*

Pure liquid substance, 12
Pure water permeability (PWP), 18, 130–131, 328–330, 543, 578
PVA-modified membrane, 633
PVDF hollow-fiber
  fabrication technology, R&D on
    mixed-matrix hollow fibers, 236–239
    spinneret design in, 232–236
    TIPS in spinning, 239–240
  membranes classification, 223
  spinning parameters on
    air-gap distance and take-up speed, 228–229
    dope rheology, 229–232, 230*f*, 231*f*, 232*f*
    internal (bore fluid) and external coagulant chemistry, 224–226, 225*f*, 227*f*
    nonsolvent additives in spinning dopes, 223–224
    polymer concentration, 221–223, 222*f*
    solubility parameters, 228*t*
PVP-*g*-montmorillonite (MMT), 600, 600*f*

Q

Quantum effects, 178
Quaternized chitosan, 166*t*

R

Ram extrusion-based process, 318, 324, 375, 378
Rasonic irradiation, PVDF membrane fabrication, 277–278
Rayleigh instability, 53, 536
Rectilinear jet, 533–536
Redox-initiated grafting, 627–628
Rejection rates, 64, 107, 200, 391, 402, 404
Relative centrifugation force, 196–197
Relative permeability, 638*f*
Renewable source, 356, 437
Repulsion of water, 618
Research and development (R&D), 216
  on PVDF hollow-fiber fabrication technology
    mixed-matrix hollow fibers, 236–239
    spinneret design in, 232–236
    TIPS in spinning, 239–240
Reverse osmosis (RO) membranes, 4, 24–25, 489–490
  application of nanofibers in, 402–403
  fluorescence micrographs of, 631*f*
  loose, 493
  in MMMs types for desalination, 491–492
  performances, 503*f*
  polyelectrolyte multilayered
    AFM images of, 637*f*
    relative permeability, 638*f*
  SEM images of, 631*f*

Rheological hindrance, 14
Root mean square (RMS), 127
Rotating disk, 61
Rotating drum, 60–61
Rubbery polymer, 590
Rupture pressure, 453, 453*t*

S

Salt rejection, 106, 143, 171, 404, 499
Salts additives, 144*f*
Scale formation, 616
Scaling, 108, 151, 503, 616
Scanning electron microscopy (SEM) images,
    557, 594
  alumina porous membrane, 134*f*
  CA membranes, 201*f*, 202*f*
  cathode layer using brush-painting
      technique, 373*f*
  ceramic hollow-fiber precursor, 354*f*
  CNT BP, 387*f*
  MFI zeolite particles, 610*f*
  PES membranes, 577*f*
  of PLA, 79, 80*f*, 87*f*, 99*f*
  PVDF membranes
    and contact angle measurements, 271*f*
    at different extrusion, 297*f*
    using DMF, 270*f*
  reduced dual-layer hollow fibers, 369*f*, 371*f*
  RO membrane, 631*f*
  for SPEEK 1.6 membranes, 686*f*
    BPO₄ blend membranes, 695*f*
    PEI blend membranes, 693*f*
    PS blends, 694*f*
Seawater desalination, 200, 403–404, 492–493
Seawater RO (SWRO) membranes, 20
Secondary bonds, 95
SEM–EDS technique, 362, 363*f*
Semicrystalline, 10, 273
Semicrystalline polymer, 11, 217, 219, 239, 273
Semipermeable membrane, 414, 492
Separation factor, 107, 134, 136, 168, 414
Separation performance, 105, 122, 151, 217, 426,
    571, 600
Shear force, 444
Shear rate, 229, 691
Shear stress, 444, 445*f*
Shift effect, 466
Sieve mechanism, 590, 598–599
Sieverts' law of permeation, 447, 450, 451*f*
Sieving effect, 105
Silane agents, 500, 597, 603
Silica-based MF membrane, 611
Silica membrane, 594
Silica nanoparticles, 496, 502, 570–571, 579, 603
Silicon micro machining technology, 6
Single-emitter electrospinning apparatus, 48*f*

Single-layer ceramic hollow-fiber membranes
  overview, 320–321
    parameters for morphology control, 320*t*,
        321–323
    sintering, 324
Single-layer mixed-matrix hollow fibers,
    236–237
Single-walled CNTs (SWCNTs), 385
Single-walled nanotubes (SWNTs), 385
Sintering process, 5, 5*t*, 323–325, 354
  ceramic hollow fiber fabrication, 354–356
  controlled, 330–335, 331*t*
  co-sintering of multilayer hollow-fiber
      membranes, 324–325
    adhesion between layers, 325
    material composition, 325
    profile, 325
    sintering behavior, 325
  development of ceramic microstructure, 355*f*
  final, 355
    parameters associated with stages of, 324, 325*t*
    single-layer hollow-fiber membranes, 324
Size exclusion mechanism, 138–139, 142
Small molecular weight additives, 223
Sodium alginate, 162–165, 635
Sodium montmorillonite, 169
Sodium *p*-styrenesulfonate (SSS), 119
Soft coagulants, 529
Sol-gel method, 237, 497, 666–667, 667*f*
Sol-gel precursor, 398, 400f
Solid electrolyte, 366, 694
Solidified polymer filaments, 396
Solid–liquid demixing, 11, 219, 223, 226, 239
Solid oxide fuel cells (SOFC), 319
  microtubular. *See* Microtubular SOFC
  operation principles of, 366, 366*f*
  single-step formation of triple-layer
      membranes for microtubular, 335–341
    cell performance, 340
    compositions of layer, 336, 337*f*
    dimensions of dual- and, 340*t*
    mechanical strength and gas-tightness of
        electrolyte, 339–340, 339*f*
    morphology and microstructure, 337–339
Solubility coefficient, 447
Solubility parameter, 13–14, 16, 81, 226, 228*t*,
    231–232, 514–515
Solute rejection, 107–108, 553
Solution blending method, 665–666, 665*f*
Solution intercalation method, 667
Solutions/membranes, characterizations
  membrane separation, 553–558
    fluid displacement, 556
    fluid transition, 556–557
    microscopic techniques, 557
    solute retention, 557–558
    techniques for membrane pore, 554*t*–555*t*

polymer dope, 552–553
    cloud point, 552
    dope viscosity and rheology, 552–553
Solvent effect in membrane preparation
    on morphology, 685–686
    on solution viscosity, 687–691
    on SPEEK 1.6 membranes, 692, 692*f*
    on thermal properties, 686–687
Solvent evaporation, 61, 255, 424, 522, 540
Solvent-induced crystallization of PLA, 81–93
    density and degree of swelling, 85*t*
    HSP, 81, 82*t*–83*t*, 84*f*
    physical properties of, 83–93
    solubility tests, 81, 83
    thermal properties of, 89*t*
    X-ray analysis of, 92*t*
Solvent/nonsolvent exchange, 11, 268, 319, 353, 522
Solvent-resistance nanofiltration (SRNF)
        membranes, 20
Solvothermal methods, 416, 608
Sonication, 658, 666
Specific viscosity, 690
SPEEK 1.6 membranes, 685
    DSC analysis of
        in argon atmosphere, 687*t*
        PEI blends, 696*t*
        PS blends, 698*t*
    dynamic viscosity, 701, 701*f*–702*f*
    effects of solvents, 692, 692*f*
    effects of temperature, 697*f*
    reduced viscosity of, 690*f*, 691*f*
    resistance *vs.* reactance, 688, 689*f*
    SEM images for, 686*f*
        BPO$_4$ blend membranes, 695*f*
        PEI blend membranes, 693*f*
        PS blends, 694*f*
    storage (elastic) moduli, 699, 699*f*–700*f*
    viscometric data, 687, 688*t*
SPEEK nanocomposite membrane, 667, 674*t*
Spinneret, 18, 54, 217, 229, 232, 396–397, 531
Spinneret design in hollow-fiber fabrication
    dual-layer spinneret, 232–233, 233*f*
    multibore spinneret, 235–236, 235*f*, 236*f*
    multichannel rectangular spinneret, 233–235,
        234*f*
Spinneret of electrospinning system, 54
Spinning sites, 58, 58*f*
Spinning suspensions, 349, 368
Spinodal boundary, 257
Spinodal curves, 8, 10, 269, 519
Spinodal decomposition (SD), 9, 521
Sponge-like regions, 354, 356, 364
Spray pyrolysis, 457
Sputtering, 457, 459
Stability test, 110*f*, 443
Stable state, solution, 516
Static deposition, 119, 123, 131, 136, 138

Steady-shear measurement, 698, 703
Steep concentration gradient, 353
Strong coagulant, 224, 529
Strong grain boundary, 324
Styrene-*b*-(ethylene-ranbutylene)-*b*-styrene
        (SEBS), 606, 606*t*
Sulfated chitosan (SCS), 170
Sulfonated PEEK (SPEEK) membranes, 656,
        667–675, 682
    composition of water, 672*t*
    FESEM images of, 669*f*
    interaction model, 661*f*
    physicochemical properties of, 674*t*
    transport mechanism model, 661*f*, 664*f*
Sulfonated PES (SPES), 550
Sulfonated poly(ether-ether-ketone-ketone)
        (SPEEKK), 683, 684*f*
Sulfonating agents, 682
Sulfonation degree, 683–684
Sulfonation time, 683
Sulfonic acid groups, 168–169, 636, 696, 699
Supercritical fluid–liquid operation, 286
Supercritical fluids, 419
Supplementary elongational stress, 17
Surface grafting, 31, 622
Surface modification
    process, 298
    using polyelectrolytes, NF membranes,
        115–125
        chemical cross-linking of PECs, 119
        dynamic self-adsorption of, 122–123,
            124*t*, 125
        static self-adsorption of, 119, 121–122, 124*t*
        UV-grafting, 117–119
Surface modifying macromolecules (SMMs), 31,
        274, 298
Surface photo-oxidation, 94, 98
Surface reactions, 447, 450
Surface roughness, 127, 278, 619–621, 634
Surface tension, 13, 301–302, 540, 618
Suspension viscosity, 319, 322
Sweep gas, 27, 288, 299
Swelling, 85*t*, 168
Symmetric membranes, 5, 322
Symmetric microporous membrane, 239
Symmetric structure, 262, 269–271, 349
Synthetic membranes, 193
Synthetic polymers, 195, 210
Syringe pump, 396–397

T

Take-up speed, 15, 218, 228–229, 524, 531
Taylor cone, 48, 48*f*, 58*f*, 533
Temperature-induced PS (TIPS), 7, 267
    phase diagrams for, 7–12, 8*f*, 9*f*
Template-synthesized CNT membranes, 385

Terminal functional groups, 639–640
Ternary phase diagram, 10–11, 519, 526–527
Tetrahydrofuran, 53, 205
Tetramethylurea, 263–264
Thermal gelation, 254
Thermally induced crystallization of PLA, 76–81, 77t
Thermally induced phase separation (TIPS), 217, 513
  binary phase diagram for, 519f
  membrane formation, 521f
  in PDVF hollow-fiber spinning, 239–240
Thermal stability, 63, 169, 205, 253, 277, 348
Thermal treatment temperature, 79, 333, 335
Thermodynamic equilibria, 365
Thermodynamics
  and kinetics influence on membrane morphology, 522
  polymer solutions, 514–516
  principles of PI, 7–15
    phase diagrams for TIPS and DIPS, 7–12, 8f, 9f, 11f
    solubility parameters, 12–13
    trade-off between thermodynamic, kinetic, and membrane morphology, 13–15
Thermogravimetric analysis (TGA), 331, 332f
Thermolysis, 354–355
Thermoporometry, principle, 556
Thin-film composite (TFC) membranes, 19, 22, 170, 577, 578f, 621, 641f
  advantages, 22
  mechanism for preparation of, 23f
  PA membrane
    photocatalytic bactericidal effect, 643
    water flux, 643, 644f
  structure of, 22f
Thin-film nanocomposite (TFN) membrane, 492, 498, 578f, 579
  formation of, 500f
  PA RO membrane, 641, 641f
  synthesis, 502f
Thioglycolic acid, 180
Three-dimensional network, 520
Three-dimensional structure, 81, 393
Three-point bending test, 328, 370
Three-zone model, 139, 141f
Threshold voltage, 59
Titanium dioxide (TiO$_2$), 236, 496, 570
  flux decline behavior, 642f
  nanoparticles self-assembly, 640f
  self-assembly, 572f, 573f
Titanium tetraisopropoxide, 572
Transmission electron microscopy (TEM) images, 591
  cellulose
    microfibrils, 197, 198f
    whiskers from banana waste, 206f

Transport mechanisms, 384, 386, 448
Triacid chloride, 577
Triethylene glycol (TEG), 20, 529
Triethyl phosphate (TEP), 13
Trimesoyl chloride (TMC), 22, 621–622, 621f
  polyamide based on, 23f
  reaction between PIP and, 23f
Trimethyl phosphate (TMP), 263–264
Triple-layer membranes
  microtubular SOFC, single-step formation, 335–341
    cell performance, 340
    compositions of layer, 336, 337f
    dimensions of dual- and, 340t
    mechanical strength and gas-tightness of electrolyte, 339–340, 339f
    morphology and microstructure, 337–339
Triple-orifice spinneret, 549f
Tungsten inert gas (TIG) welding, 455, 455f

U

Ultrafiltration (UF) membrane, 4, 21–22
  application of nanofibers in, 402–403
  and NF membranes applications in membrane technology
    cellulose, 177–178, 179f
    chitosan, 170–174
Ultrapure hydrogen, 438, 446, 452, 472
Ultraviolet (UV) irradiation, 75, 93, 95, 131, 622, 623
Uncoated membranes, 632
Undesired homopolymer, 626
Unstable state, solution, 517
UV-assisted grafting, 298
UV-grafting of polyelectrolytes, 117
  acrylic acid grafted on PES NF membranes, 117–119, 118f, 135f
  sodium p-styrenesulfonate grafted on PSF UF membranes, 119
UV-initiated grafting, 117, 623, 627
UV-photo-grafting setup, 551, 551f
UV–visible (UV–vis) spectra of PLA, 94

V

Vacuum MD (VMD), 404
Vacuum UV (VUV) irradiation, 76
  ATR-FTIR, 95, 96f
  DSC thermograms, 97f
  properties, 94
  SEM images of, 99f
  UV–vis spectra of, 94f
  WAXD patterns, 97f

Valley clogging, 620
Van der Waals force, 385, 630, 658
Vapor-induced PS (VIPS), 7
Vapor–liquid operation, 286
Vapor pressure, 556–557
Vertically aligned CNTs (VA-CNTs),
    385, 389
V group metals, 460
Vinyl monomers, 623, 625
Vinyltrimethoxysilane (VTMS), 602
Viscosity gradient, 690, 703
Viscous fingering, 319, 321–322, 353
Viscous fingering-induced phase inversion
    process, 319–320
Vitrification, 9, 520
Volatile organic compounds (VOCs), 404
Volatile solvent, 254–255, 263, 529

W

Wall effects, 450
Water chemistry, 645
Water gas shift (WGS) reactor, 438, 476f
Water in membrane
    DSC measurements, 671, 671f
    SPEEK nanocomposite and
        nafion®112, 672t
    states of, 671
Water insoluble additives, 178
Water intrusion, 221, 225–226
Water permeability, 16, 492, 498, 501,
    526, 639
Water-permeable nanochannels, 633–634
Water softening, 20, 494
Weak coagulant, 529
Well-order multilayer structures, 657

Wide-angle X-ray diffraction (WAXD) patterns
    of PLA, 76, 77f
    types, 90, 91f
    VUV-irradiated, 97f
    waveform separation of Gaussian functions, 93f

X

Xerogel, 362, 417
X-ray diffraction (XRD), 362, 404, 594
X-ray photoelectron spectroscopy (XPS), 574,
    594, 640
XRD spectrum, 362, 363f

Y

Yang–Cussler correlation, 290
Young–Laplace equation, 294, 556
Young modulus, 64–65, 197, 210
Yttria-stabilized zirconia (YSZ) hollow fiber,
    357–359
    to develop CHFMMR, 359f
    fluxes of N_2, 364, 364f
    nickel on, 363f
    XRD spectra for, 363f
    zirconium represents, 361f

Z

Zeolite membrane, 26, 595, 598
Zeolite molecular sieve particles, 578
Zeolite socony mobil (ZSM)-5 particles,
    602–603
Zero air-gap, 322
Zeta potential, 130, 151, 638
Zwitterionic groups, 630–631

Printed and bound by CPI Group (UK) Ltd, Croydon, CR0 4YY

22/10/2024

01777615-0020